Calculus

Modeling and Application

Calculus

Modeling and Application

David A. Smith Lawrence C. Moore
Duke University

D. C. HEATH AND COMPANY
Lexington, Massachusetts Toronto

Address editorial correspondence to:
D. C. Heath and Company
125 Spring Street
Lexington, MA 02173

Acquisitions: Charles Hartford
Development: Kathleen Sessa-Federico
Editorial Production: Melissa Ray and Craig Mertens
Design: Cornelia Boynton
Art Editing: Gary Crespo
Production Coordination: Charles Dutton
Permissions: Margaret Roll
Cover: Linda Manly Wade

Published simultaneously in Canada.

Printed in the United States of America.

International Standard Book Number: 0-669-32787-5

Library of Congress Catalog Number: 95-78713

10 9 8 7 6 5 4 3 2 1

To our patient and supportive wives,

Dorothy and Dutton,

who helped us accomplish
a seemingly infinite task
in finite time.

Preface

Calculus is the study of change. The concepts of calculus enable us to model processes that change and to describe properties of these processes that remain constant in the midst of change. Now change has come to the **learning** of calculus—change driven by the need to respond to the revolution in technology and by our increased awareness of how students learn. This text is an outgrowth and agent of that change. We wrote this book initially to support the reformed calculus course developed at Duke University with the support of the National Science Foundation. Since 1990, various preliminary versions have been used at many other colleges across the country. The experiences of dozens of teachers and many thousands of students have helped us refine our work to the form of this first edition.

In our development of the course and the text we were guided by the following goals:

- Students should be able to use mathematics to structure their understanding of and investigate questions in the world around them.
- Students should be able to use calculus to formulate problems, to solve problems, and to communicate their solutions of problems to others.
- Students should be able to use technology as an integral part of this process of formulation, solution, and communication.
- Students should work and learn cooperatively.

The course we developed to serve these goals emphasizes

- real-world problems,
- hands-on activities,
- discovery learning,
- writing and revision of writing,
- teamwork,
- intelligent use of available tools, and
- high expectations of students.

Our *Instructor's Guide* explains how the various components of the course fit together and suggests a variety of ways it can be taught—most of them based on actual experience at one or more campuses. In this Preface we concentrate on how the book itself serves our goals and emphases.

Real-world contexts We provide a real-world setting for each concept and calculational rule. For example, both differentiation and the exponential function appear early in Chapter 2 in connection with natural growth of populations. The Chain Rule is introduced as part of the modeling process for reflection and refraction of light. Both improper integrals and polynomial approximation of functions result from an investigation of models for the distribution of data. Throughout, we emphasize differential equations and initial value problems — the main connection between calculus and applications in the sciences and engineering.

Discovery learning There are three features in the book that enhance discovery learning: Exploration Activities, Checkpoints, and Examples. We encourage students to construct their own knowledge by attempting Exploration Activities embedded in the text. These activities invite students to explore new concepts and problems, and to see what the issues are before they are discussed in detail in the text. Each of these activities is followed by a discussion that leads into the new ideas. In addition, we encourage students to check their understanding by attempting Checkpoint calculations, which appear after the appropriate new ideas have been introduced; their answers are given at the ends of the sections. The combination of Exploration Activities, Checkpoints, and Examples provides a wealth of engaging illustrations of the central concepts and techniques.

Problem solving Throughout the text we emphasize general principles and practice in problem solving. At the end of each section and of each chapter we have provided a range of exercises, from practice calculations to additional development of concepts. Answers are provided at the back of the book for approximately one-fourth of the calculational exercises. (If an answer is provided, the exercise number is underlined in the text.) The exercises are intended primarily for individual students, although the more challenging ones also work well for group activities in or out of class. In addition, each chapter has a number of projects, which are more open-ended explorations designed for investigation by small groups of three or four students. They may be used in a variety of ways — for in-class activities that are written up in homework style or as a basis for more formal reports.

Technology We assume each student has access to a graphing calculator, but no particular make or model is assumed. We encourage students to supplement our figures with their own graphs and plots of data, and we help them find ways to use their calculators for checking other work, such as symbol manipulation.

We assume that the solution of equations and definite integration are either buttons on the calculator or available programs. We study Newton's Method, Simpson's Rule, and other numerical techniques to understand how calculators and computers work — in particular, how they use ideas from calculus. In our study of the Fundamental Theorem, we emphasize construction of antiderivatives by definite integration, a process easily carried out by a calculator.

Laboratories We have written this text so that no particular laboratory activity is required — the text stands on its own. However, we expect that many students using this text will participate in more extensive investigations using programmable calculators or computer software. We have developed companion laboratory manuals that motivate, clarify, and deepen student understanding. These manuals, listed at the end of this Preface, are available for a wide range of technologies. Lab activities are an excellent source of writing projects, ranging from fill-in-a-paragraph answers to full-scale reports. At the ends of some chapters in this book there are Optional Lab Readings designed to provide background for particular lab projects. (They can also serve as supplementary readings.)

Chapter Content

In Chapter 1 we study the concept of function in a context of mathematics as a tool for modeling. Because of our emphasis on technology, we also discuss numerical calculation and significant digits in a section that may be unique for a book at this level.

Chapter 2 begins the study of rates of change in the context of exponential growth. Here we introduce the derivative and the natural exponential function. In addition, we take our first look at slope fields and initial value problems. We conclude the chapter with semilog and log-log plots and their uses for discovering growth patterns in data.

In Chapter 3 we take a more detailed look at initial value problems in the context of Newton's Law of Cooling—to solve a murder mystery. We also study falling bodies—first without air resistance, but one of the end-of-chapter projects explores linear resistance and links the falling-body problem to a number of other exponential decay situations, including Newton's Law of Cooling.

In Chapter 4 we obtain the remainder of the basic calculational tools for differential calculus and examine the interplay between function graphs and values of the derivatives. Throughout we raise issues of antidifferentiation along with differentiation. Our emphasis on simple differential equations allows us to address topics needed for physics and engineering courses well in advance of the more difficult topic of the definite integral.

Chapter 5 introduces Euler's Method for numerical solution of initial value problems to examine the SIR-model of epidemics. We also look at both continuous and discrete models for the evolution of prices in a simple economy. Here we see instability in the discrete model, a harbinger of the discussion of chaos in Chapter 7.

In Chapter 6 we study trigonometric functions and their derivatives in the context of periodic motion—modeling springs and pendulums.

Chapter 7 investigates symbolic solutions of separable differential equations, continuing our thread of modeling population growth. Here we obtain symbolic solutions to the logistic equation and investigate a superexponential model for world population growth. The discrete form of the logistic differential equation—equivalent to an Euler's Method approximation to the continuous model—leads to a unit on chaos.

In Chapter 8 we introduce the definite integral and obtain the Fundamental Theorem of Calculus from our understanding of Euler's Method for approximation of the solution of an initial value problem. We stress the role of the Fundamental Theorem in solving problems of antidifferentiation, because calculators can construct functions by definite integration to a variable upper limit. The traditional role for finding definite integrals remains important for problems that depend on parameters, as is often the case with mathematical models.

Chapter 9, on integral calculus, parallels Chapter 4, on differential calculus. Here we investigate some of the uses of the definite integral with emphasis on how one decides that the calculation of a particular integral is what is needed in a given situation. We look at numerical methods for evaluating integrals from the point of view, "What might your calculator or computer be doing when you press that key?" Because not everyone has constant access to a symbolic computer system, we include an extensive section on the use of the Integral Table. Finally, to motivate integrals of trigonometric functions, we include a brief introduction to Fourier approximations for analyzing complex periodic phenomena.

In Chapter 10 we explore continuous distributions of data, in particular exponential and normal distributions, as a context for introduction of improper integrals and a more detailed investigation of the concept of limit. In particular, the distribution function for the standard normal distribution leads to the error function. In Chapter 11 we look at how a computer or calculator might be generating values of this and other more common transcendental functions. This leads to polynomial approximation and then to infinite series as convenient and very useful representations of functions.

Early versions of this text were distributed by the authors under the title of *The Calculus Reader*. Revised preliminary versions were published by D. C. Heath under the titles *The Calculus Reader* and *Project CALC: Chapters 1–6*. A companion volume for multivariable calculus is already being distributed by D. C. Heath in preliminary form. The multivariable material will be published in a separate volume.

Supplements

Instructor's Guide This manual explains how to use this text with a variety of approaches to calculus. We discuss:

- working with groups,
- assigning, responding to, and grading written work,
- getting students to read mathematics,
- integration of classroom and laboratory activities, and
- the use of gateway tests to check competency in symbol manipulation.

In addition, we give detailed suggestions for using the text on a week-by-week basis.

Complete Solutions Manual This manual contains complete solutions for all of the exercises in the text.

Laboratory Manuals The following manuals contain laboratory activities designed for a once-a-week lab that is closely correlated with the text material. These activities usually focus on interactive explorations and modeling of real-world problems with real data. They include problems from biology, physics, economics, epidemiology, and statistics.

- **TI-82/85 Laboratory Manual**
- **Mathematica Laboratory Manual**
- **Derive Laboratory Manual**
- **Maple Laboratory Manual**
- **HP-48 Laboratory Manual**
- **Mathcad for Windows Laboratory Manual**

Acknowledgments

We are grateful for the support of the National Science Foundation through grants USE88-140832, DMS-8951909, DUE-8953961, and DUE9153272. In particular, we wish to thank Louise Raphael, John Bradley, John Kenelly, and James Lightbourne for their support. We also thank the members of our Advisory Board, Wade Ellis, Harley Flanders, Morton Lowengrub, Alan Schoenfeld, and Paul Zorn, for their advice and encouragement.

We began our project in collaboration with faculty at the North Carolina School of Science and Mathematics, who have subsequently developed their own materials directed toward the secondary school market. These materials have been published by Janson Publications under the title *Contemporary Calculus Through Applications*. In particular, we acknowledge our debt to Kevin Bartkovich, John Goebel, Lawrence Gould (deceased), and Jo Ann Lutz.

Angelika Langen deserves special thanks for keeping us organized and helping with whatever needed doing. We also thank Ann Tunstall and the rest of the staff in the Mathematics Department at Duke University. Four department chairs have supported this project: Mike Reed, Bill Pardon, David Schaeffer, and John Harer. We wish to acknowledge the help of our other colleagues at Duke who taught the course, offered suggestions, and contributed ideas. They include Lewis Blake, Jack Bookman, Robert Bryant, Elizabeth Dempster, Patty Dunn, David Kraines, Blaik Mathews, Sam Morris, David Morrison, Chris Odden, Emily Puckette, and Henry Suters. In addition, Mathews, Odden, and Suters worked on the text and other materials.

We want to thank the Duke undergraduates who took the first Project CALC classes. We learned as much from them as they did from us. Many of these students stayed on to help with the course; in particular, we thank Lee Miller, Mark Nugent, and David Vanderweide. We also thank Orin Day for his technical help, especially in the hectic first year.

The administration at Duke has supported us solidly in this effort. We especially want to thank Richard White, Dean of Trinity College at Duke. It has been our good fortune to have the help and support of the Duke University Writing Program, especially George Gopen, director, and Alec Motten, consultant to our project.

We would like to thank the following colleagues who reviewed the manuscript and made many helpful comments: Steve Benson, Santa Clara University; Marc Frantz, Indiana University-Purdue University; Ronald Freiwald, Washington University; Richard Hill, Michigan State University; Ronald Jeppson, Moorhead State University; Steven Leth, University of Northern Colorado; Len Lipkin, University of North Florida; Michael Meck, Southern Connecticut State University; David Meredith, San Francisco State University; Ronald Miech, University of California, Los Angeles; Gertrude Okhuysen, Mississippi State University; Mary Platt, Salem State College; Thomas Ralley, Ohio State University; Joseph Stephen, Northern Illinois State University; and Sandra Taylor, College of the Redwoods.

We gratefully acknowledge the help of schools that site-tested the materials: Albertson College; Alverno College; Big Bend Community College; Boston University; Bowdoin College; Brunel University (UK); Bryan College; California State University, Chico; Central Oregon Community College; Colby-Sawyer College; College of St. Francis; Dakota Wesleyan University; Duke University; East Carolina University; Evergreen State College; Frostburg State University; Hood College; Kenyon College; Lynchburg College; Marietta College; McPherson College; Medgar Evers College; Mercer University; Mid Michigan Community College; Middle Tennessee State University; Nazarene College; Northern Michigan University; Pennsylvania State University; Principia College; Randolph-Macon College; Raritan Valley Community College; Rockhurst College; Saint Andrew's Presbyterian College; St. Gregory's College; Saint Mary's University; San Diego State University; Seattle Central Community College; Texas A & I University; University of California, Irvine; University of California, Santa Cruz; University of

Mississippi; University of North Florida; University of Northern Colorado; Vermont Technical College; Virginia Wesleyan College; Weber State University; West Virginia University; Westmont College; Widener University; and Yale University.

We also thank our many colleagues who generously offered their advice at various stages of the project, especially Charles Alexander, Steve Amgott, Bill Barker, Marcelle Bessman, David Bressoud, Rob Cole, George Dimitroff, Lee Gerber, Franz Helfenstein, Roger Higdem, Kendell Hyde, King Jamison, Sam Thompson, Alvin Kay, Len Lipkin, Robert Mayes, Betty Mayfield, Bill Mueller, Sharon Pedersen, Susan Pustejovsky, Anita Salem, Richard Shores, Keith Stroyan, Catherine Tackman, William Trott, Steve Unruhe, and Jim White.

This book was created and typeset in EXP, a product of Brooks/Cole Publishing Co. We are thankful for assistance provided by Simon Smith and Bob Evans.

Finally, we wish to thank the people at D. C. Heath who guided us through the multiple revisions and production: Charlie Hartford, Acquisitions Editor; Kathy Sessa-Federico, Development Editor; Melissa Ray and Craig Mertens, Production Editors; Cia Boynton, Designer; Gary Crespo, Art Editor; Carolyn Johnson, Editorial Associate; and Chuck Dutton, Production Coordinator.

D. A. S.
L. C. M.

Contents

Introduction

What is this book about?

Most mathematics books are about answers — and how to get them. This book is about *questions* — and what to do about them. Like the world around us, this book has more questions than answers. Indeed, our questions *are* the questions of the world around us. Almost everything of importance in our world is moving or changing, and **calculus is the mathematical language of motion and change.**

To give you some idea of the importance of our subject, we will pose some questions that calculus might help us answer. You won't find the answers in the back of the book — indeed, answers that fit neatly in books are seldom real solutions to real problems. Here we go.

- Are we in the midst of a global population explosion? If so, what resources will we exhaust first: food, fuel, or terrestrial space? As a response to such a crisis, should we colonize outer space? If so, what would it take to do that, and how do we go about it? Can people survive in large numbers on the moon or Mars? Can we move enough of them there to make any difference? If so, what are the scientific, engineering, economic, political, sociological, theological, and biomedical problems we would have to solve? How do we solve them?

- Suppose we find there *is* a population crisis, but there is no viable solution to the problem of space colonization. What problems would we have to solve to continue our existence in relative peace on Earth? Population control? Waste management? Pollution control? Technological advances in computers, consumer goods, weapons, communications? Arms control or reduction? Management of international relations? Peace through strength or strength through peace? Economic growth or economic stability?

- Suppose there is *no* impending population explosion — population may be self-limiting. What then? Will we see world population level off at some stable number? If so, how big can we expect that number to be? Would its sheer size lead us to grapple with a host of other problems, such as extreme scarcity of resources and drastically lowered standards of living?

- If there is no leveling-off point, will there be oscillations in the population level? If so, will these be wild swings between very high and very low levels, or will they be modest variations at manageable levels? If the latter — which would suggest that population problems need not be high on our priority scale — what *are* the important problems of a society and a world that appear to be changing ever more rapidly?

These are challenging questions about *change*, more precisely about the *rates* at which dynamic quantities change and about the consequences we can determine from those rates. Calculus provides us with the conceptual framework and many of the computational tools for the quantitative and qualitative study of rates of change, and that's what this book is about.

Why study calculus?

We often ask our beginning students why they are taking calculus. Here are a few of the most common answers to this question.

- It's required for my major.
- I have always had a mathematics course, and this was the next one in line.
- My parents said I had to take it.
- I like mathematics.
- Everyone says mathematics is important — I just felt that I ought to do it.
- Calculus is central to understanding the development of philosophy and science in the last three centuries. Without a thorough grasp of this fundamental branch of mathematics, one cannot be considered an educated person.

Well, honesty compels us to admit that no one has actually given the last response, but our hope springs eternal.

Communication and cooperation

Often a problem comes to us in the form of data: An object falls through the air, and we observe data consisting of distances fallen at, say, ten different times. Can we tell how far the object had fallen at some time other than those at which we made the observations? Can we predict how a similar object will fall in the future? In this case, the theory comes to us from physics. The language of the theory is a mixture of English and calculus, and the calculations necessary to answer the questions require the same mix.

We concentrate on the use of calculus to solve problems. What has English — reading, writing, speaking — to do with solving problems? Problem solving requires deciding what should be done, executing the calculations, and interpreting the results. The environment for this intellectual activity is language, English in our case. Until you can describe what you have done, why you did it, and what it means, you have not solved the problem. For this reason we expect you to write up the projects on which you will work — in class or lab, or on your own time.

This textbook, your calculator or computer, and your instructor are all important resources for learning about calculus and the art of problem solving. There is another resource just as important as those already mentioned: your fellow students. We expect you to work on projects in teams, to talk about what you are doing, to explain your ideas and insights to each other. In the course of this work, you will find your fellow students an excellent source of help for understanding the course in general. Whatever your question, it is likely that somebody else in the class has considered it already and has some ideas for an answer. The key here is to talk to one another. When you do not understand why one thing follows from another, say so. When you do not see the evidence to support a conclusion, say so.

Learning is a cooperative — not a competitive — activity. We are about to embark on a great cooperative adventure: learning calculus. *Bon voyage!*

CHAPTER

1

Relationships

As students of calculus, nothing is more important to us than **functions.** *Other things may be more important to you in other contexts—for example, social justice, Mozart, or baseball. But this course is about functions—their various representations, their rates of growth and decay, and their uses in solving problems in many different disciplines. Thus, in this introductory chapter, we take up one big question:*

What is a function?

It's very important that we have a common understanding about the answer to this question. Otherwise, it will be difficult to make much sense of the rest of the course. Furthermore, our answer to this big question may or may not agree with your previous use of the word "function" in other mathematics courses. We will start with your prior understanding and work toward a new understanding that will serve us better for your study of calculus and its uses.

1.1 Related Variables

No doubt you already know an answer to the big question — What's a function? — from your previous study of mathematics. In fact, we will use your prior knowledge as our starting point for discussion and refinement of this central concept. Of course, only *you* can provide that prior knowledge.

Exploration Activity 1

Somewhere that you can keep track of it, write your present answer to the big question. You may use definitions, examples, symbols, or whatever you think appropriate for explaining to someone else what a function is.

The word "function," like all the other words we shall use in this course, belongs to the English language. As such, it has a definition — or several:

> **function** (fungk′shən) *n.* **1.** The action for which a person or thing is particularly fitted or employed. **2. a.** Assigned duty or activity. **b.** Specific occupation or role: *in his function as attorney.* **3.** An official ceremony or formal social occasion. **4.** Something closely related to another thing and dependent on it for its existence, value, or significance: *Growth is a function of nutrition.*[1]

Do you see any point of contact between that dictionary definition and what you wrote in response to Exploration Activity 1? Perhaps not — and therein lies a fundamental difficulty in the study of mathematics. Math books and math teachers seem to use words from the English language, but often with meanings that seem arbitrary and unrelated to common usage. Actually, as we shall see, definition **4.** is rather close to the meaning we shall establish in this chapter. If you wrote something like that, give yourself a pat on the back. If nothing you wrote looks like that, don't despair — by the end of this chapter we will have connected your prior experience with the concept of "function" as the word will be used in this course.

You may have noticed that the title of this chapter is not "Functions" but "Relationships." The key word in definition **4.** is "related," and we will establish our meaning for the word "function" within the more general context of relationships.

Variables and Data

We start with examples of quantitative relationships. Think for a moment about each of the following questions. You are not expected to know answers to these questions, just to be willing to think about them.

- The United States has a serious dropout problem. What is the relationship between state expenditures on teacher salaries and high school graduation rates?

1. Copyright © 1991 by Houghton Mifflin Company. Reprinted by permission from THE AMERICAN HERITAGE DICTIONARY, SECOND COLLEGE EDITION.

- For adults, high blood pressure is linked to weight. Is there a similar relationship for children or adolescents?
- Will the world be seriously overpopulated in 20 years?

These questions are different in many respects, but answering each requires collection, organization, and interpretation of data. Each requires analysis of the relationship between *two variables*. In the first question, the two variables are the state expenditure on teacher salaries and the high school graduation rate. In the second question, the variables are blood pressure and weight. And in the third, the variables are time and population.

We can classify possible relationships between pairs of variables in four categories:

- One variable may have a causative effect on the other. For example, we expect that blood pressure in adults of the same height depends in some way on weight—higher weight causes higher blood pressure—but probably not the other way around.
- Other times there is a relationship between the variables, but it is not one of cause and effect. For example, at any given time, some number is the actual population of the world—but we would not consider either time or population to be a cause of the other.
- We might find that there is no relationship at all between the two variables. For example, we do not expect a relationship between the distance from a student's home to college and her height.
- Finally, we may not know if there is a relationship between two variables. For example, there might be some relationship between the annual dollar amount of imports from Mexico to the United States and the annual dollar amount of exports from the United States to Mexico—but, without any data, we have no way of knowing.

Checkpoint 1

For each of the following descriptions, state whether you think there is a relationship between the two variables and if either of them might have a causative effect on the other.

(a) Annual crop yield and annual rainfall in a given area.

(b) Temperature of an object measured in degrees Fahrenheit and temperature measured in degrees Celsius.

(c) The number of students entering Michigan State University in a given year and the year of entry.

(d) State expenditures on teachers' salaries and high school graduation rates.

To determine whether a relationship exists between two variables, we must analyze pairs of data—each pair consisting of a value of the first variable and a corresponding value of the second variable. Sometimes these data are gathered from a well-designed, carefully controlled scientific experiment, as might be the case for a study of blood pressure or crop yields. Other times we want to analyze data that already exist in the world around us, such as census data on populations.

Exploration Activity 2

The September 7, 1987, *U.S. News and World Report* claims that "spending heavily on teachers doesn't always yield a bumper crop of graduates." The comment is followed by the data in Table 1.1. Study this list of paired data to determine whether you agree with the statement made by the magazine.

Table 1.1 Public school spending per pupil and high school graduation rates

State	*Spending*	*Graduation Rate*	*State*	*Spending*	*Graduation Rate*
Alaska	$8,842	67.1%	California	$3,751	65.8%
New York	$6,299	62.7%	Iowa	$3,740	86.5%
Wyoming	$6,229	74.3%	Maine	$3,650	78.6%
New Jersey	$6,120	77.3%	Texas	$3,584	63.2%
Connecticut	$5,532	80.4%	New Mexico	$3,537	71.9%
Dist. of Columbia	$5,349	54.8%	North Carolina	$3,473	70.3%
Massachusetts	$4,856	76.3%	Nebraska	$3,437	86.9%
Delaware	$4,776	69.9%	New Hampshire	$3,386	75.2%
Pennsylvania	$4,752	77.2%	Indiana	$3,379	76.4%
Wisconsin	$4,701	84.0%	Missouri	$3,345	76.1%
Maryland	$4,659	77.7%	Louisiana	$3,237	54.7%
Rhode Island	$4,574	67.6%	North Dakota	$3,209	86.1%
Vermont	$4,459	83.4%	South Dakota	$3,190	85.1%
Hawaii	$4,372	73.8%	Georgia	$3,167	62.6%
Minnesota	$4,241	90.6%	Kentucky	$3,107	68.2%
Oregon	$4,236	72.7%	South Carolina	$3,005	62.4%
Kansas	$4,137	81.4%	West Virginia	$2,959	72.8%
Colorado	$4,129	72.2%	Tennessee	$2,842	64.1%
Montana	$4,070	82.9%	Arkansas	$2,795	75.7%
Florida	$4,056	61.2%	Arizona	$2,784	64.5%
Illinois	$3,980	74.0%	Oklahoma	$2,701	71.1%
Michigan	$3,954	71.9%	Alabama	$2,610	63.0%
Virginia	$3,809	73.7%	Idaho	$2,555	76.7%
Washington	$3,808	74.9%	Mississippi	$2,534	61.8%
Ohio	$3,769	76.1%	Utah	$2,455	75.9%
Nevada	$3,768	63.9%			

It is hard to tell from the table how the two variables are related. Minnesota has a 90.6% graduation rate with an expenditure of $4241 per pupil, while North Dakota has an 86.1% graduation rate with only $3209 per pupil. Then there is Alaska with only a 67.1% graduation rate and an expenditure of $8842 per pupil. To find information hidden in the numbers, we need some way to organize the data so that we can see meaning without getting lost in a horde of numbers. Then maybe we can decide whether or not the graduation rate is related to educational spending.

Scatter Plots

One way to display the general relationship between two variables is to make a **scatter plot**, that is, a graph in a rectangular coordinate system of all the pairs of data. In Figure 1.1 we show a computer-generated scatter plot of the fifty-one data points in Table 1.1. Making such a graph is easy if there are not too many points. For this many points, it would be tedious to make the plot by hand. You should learn how to make scatter plots with your graphing calculator or a computer.

Figure 1.1 Graduation rate versus public school spending

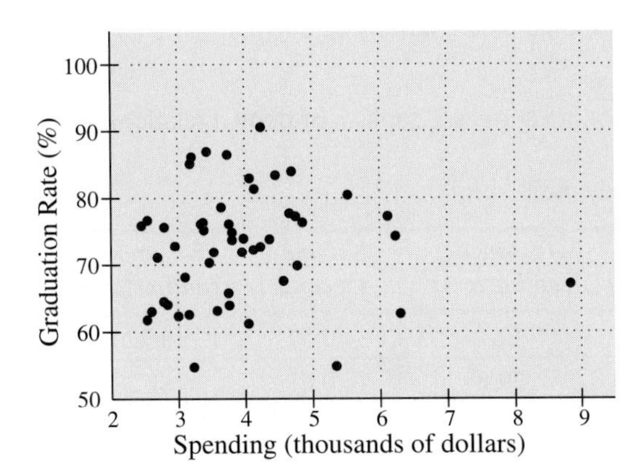

For each state, the pair of numbers consisting of a spending amount and a graduation rate is represented by one point in the plane. State spending is plotted on the horizontal axis, and graduation rate is plotted on the vertical axis. For example, the pair representing Montana is (\$4070, 82.9%). When making a scatter plot, it really does not matter which variable is plotted on which axis. If we suspect one variable to be dependent on the other, however, we usually plot the dependent variable on the vertical axis.

Exploration Activity 3

(a) Find the plotted point you think represents Montana in Figure 1.1.

(b) Are you surprised by the point that lies far to the right of the others? Which state does that point represent?

(c) Study the scatter plot carefully. Do you agree with the *U.S. News and World Report* claim that "spending heavily on teachers doesn't always yield a bumper crop of graduates"? Does there appear to be any relationship between state spending and the graduation rate?

(d) Which display do you find easier to interpret, the table or the graph?

To find Montana, look first for the intersection of the \$4000 line and the 80% line. Just to the right and above are two points. The higher one is Montana, the other is Kansas. The point far to the right of all the others is Alaska, a state with very high spending per student and a relatively low graduation rate.

The scatter plot indicates little or no relationship. Most people find the graphical representation easier to interpret, since one may identify a possible relationship — or lack of one — with a single glance.

Checkpoint 2

The data in Table 1.2 were gathered by a team of two students in a physics lab. Each used a motion detector to measure, at a number of closely spaced times, the distance a falling object had traveled. Times that appear twice in the table were selected by both students, and each selected some times not used by the other. The scatter plot for these data is shown in Figure 1.2.

(a) Do you think there should be a relationship between time and distance for an object falling in a gravitational field?

(b) Does the scatter plot in Figure 1.2 confirm or deny a relationship?

Table 1.2 Falling body data

Time (seconds)	*Distance (centimeters)*	*Time (seconds)*	*Distance (centimeters)*
0.16	12.1	0.57	150.2
0.24	29.8	0.61	182.2
0.25	32.7	0.61	189.4
0.30	42.8	0.68	220.4
0.30	44.2	0.72	254.0
0.32	55.8	0.72	261.0
0.36	63.5	0.83	334.6
0.36	65.1	0.88	375.5
0.50	124.6	0.89	399.1
0.50	129.7		

Figure 1.2 Scatter plot of falling body data

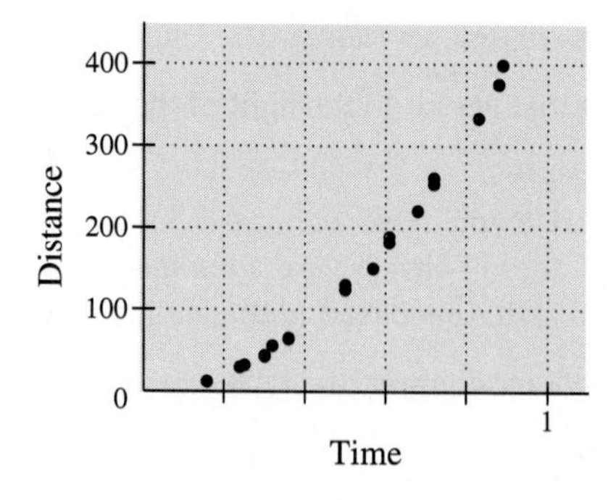

Exploration Activity 3 and Checkpoint 2 are extreme cases of scatter plots in opposite directions — *lots of scatter* and *not very much scatter.* The next example is an "in between" case.

Exploration Activity 4

Tables 1.3, 1.4, and 1.0 show the total numbers of rebounds, assists, personal fouls committed, and points scored by individual basketball players on three professional basketball teams during the 1992–93 season.[2] Study the "Personal Fouls" and "Points Scored" columns in Tables 1.3, 1.4, and 1.0, and think about whether there is a relationship between the number of personal fouls that a basketball player commits in a season of play and the number of points he scores. In the scatter plot shown in Figure 1.2, we have plotted (for the three teams combined) personal fouls on the horizontal axis and points scored on the vertical axis.

(a) What is the general shape of the plot in Figure 1.2?

(b) How strong is the relationship between these variables? How well could you predict the points scored by a player committing 75 personal fouls?

(c) In general, if you were given a value for one variable, how confident would you be in predicting a value for the other variable?

(d) Does a change in one variable *cause* the other to change, or is this a situation where they simply change together as the result of some underlying factor?

(e) Would you expect the coach to encourage his players to commit lots of fouls in order to be sure of scoring many points? If not, what underlying factors could be influencing both of our variables?

Table 1.3 Chicago Bulls statistics, 1992–93

Player	*Rebounds*	*Assists*	*Personal Fouls*	*Points Scored*
Jordan	522	428	188	2541
Pippen	621	507	219	1510
Grant	729	201	218	1017
Armstrong	149	330	169	1009
S. Williams	451	68	230	422
Cartwright	233	83	154	354
King	207	71	128	408
Armstrong	149	330	169	1009
Tucker	71	82	65	356
Perdue	287	74	139	341
Paxson	48	136	99	246
McCray	158	81	99	222
Blanton	3	1	1	6
C. Williams	31	23	24	81
Courtney	2	1	9	11
Walker	58	53	63	80
Nealy	64	15	41	69
English	6	1	5	6

Table 1.4 Boston Celtics statistics, 1992–93

Player	*Rebounds*	*Assists*	*Personal Fouls*	*Points Scored*
Gamble	246	226	185	1093
Parish	740	61	201	994
Brown	246	461	203	874
McHale	358	73	126	762
Douglas	162	508	166	618
Adelnaby	337	27	189	578
Fox	159	113	133	453
McHale	358	73	126	762
Battle	11	2	2	14
Pinckney	43	1	13	32
Webb	10	2	11	39
Kleine	346	39	123	257
Kofoed	1	10	1	17
Bagley	7	20	11	23
Williams	55	5	29	36
Wolf	3	0	2	1

2. Source: *The Sporting News 1993–94 NBA Guide.*

Table 1.5 Cleveland Cavaliers statistics, 1992–93

Player	*Rebounds*	*Assists*	*Personal Fouls*	*Points Scored*
Daugherty	726	312	174	1432
Price	201	602	105	1365
Nance	668	223	223	1268
Ehlo	403	254	170	949
Wilkins	214	183	154	890
Williams	415	152	171	738
Brandon	179	302	122	725
Sanders	170	75	150	454
Ferry	279	137	171	573
Battle	29	54	39	223
Phills	17	10	19	93
Lane	53	17	32	59
Kerr	7	11	2	12
Guidinger	64	17	48	51

Figure 1.3 Points scored versus personal fouls

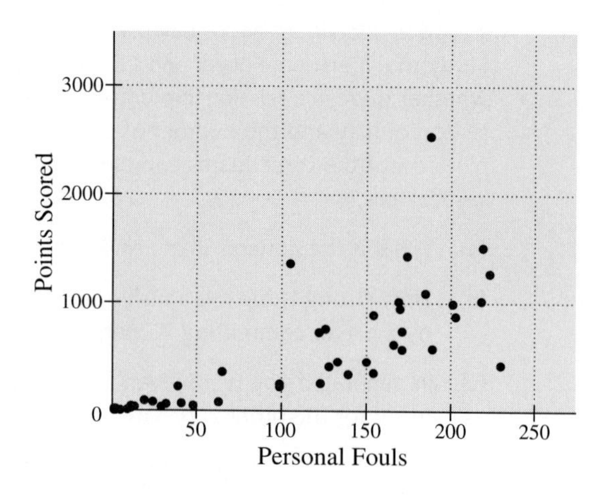

There seems to be a positive relationship between the number of fouls committed and the number of points scored. That is, "more points" and "more fouls" seem to go together—an increasing trend as you move from left to right. In fact, if we ignore the two outlying points (Jordan and Price), the relationship is roughly linear. It looks like a player with 75 fouls might have around 350 points. However, we would not expect one variable to be a reliable predictor of the other. We certainly do not expect either points or fouls to cause the other. It is possible that both are related to the number of minutes played—or perhaps they are both related to a player's aggressiveness. In particular, it would not be a sensible strategy to order a player to foul more in order to score more.

Answers to Checkpoints

1. (a) There is likely to be a relationship, and it is likely that rainfall is a cause of yield—but not the other way around.

   (b) There is a definite relationship represented by the formula $F = \frac{9}{5}C + 32$. Neither measurement causes the other.

   (c) There is a definite relationship: In any given year, there is some number of students who enter MSU. Neither the year nor the number of students causes the other.

   (d) In the absence of data, it is difficult to tell whether there is a relationship. (See Exploration Activity 2.)

2. (a) Most people think there should be a relationship.

   (b) The plot supports a definite relationship.

Exercises 1.1

The first eight exercises review some basic graphing skills that will be used in this chapter and throughout the course. Each of these requires a 4×4 grid like that shown in Figure 1.4. Start each exercise by making such a grid on your paper.

Figure 1.4 Practice grid

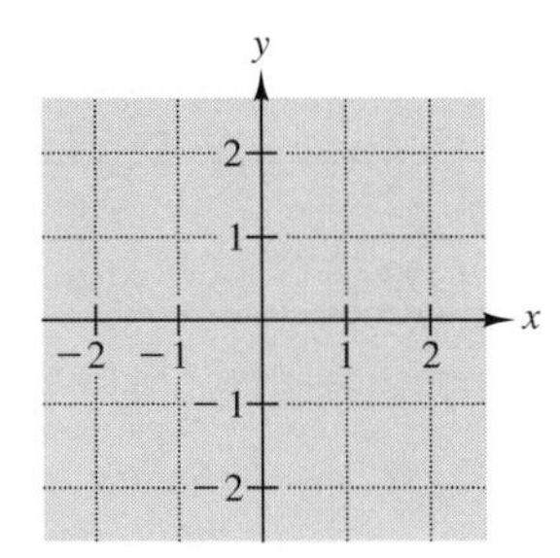

1. Plot the points $(1, 1)$ and $(-1, -2)$. Draw a line through these points.
2. Plot the points $(1.5, 2)$, $(0.5, 0)$, and $(-0.5, -2)$. How many lines can you draw through pairs of these points?
3. Draw a line through the origin and the point $(-2, 1)$.
4. Plot the points $(-2, 1)$, $(1, -0.5)$, and one other so that the three points are vertices of a right triangle. Draw the triangle.
5. Draw a line through the origin with slope $\frac{1}{2}$. Find an equation for your line.
6. Draw a line with slope $\frac{3}{2}$ through the point $(-1, -0.5)$. Find an equation for your line.
7. Draw a line with slope $-\frac{1}{2}$. Find an equation for your line.
8. Draw a line through the points $(-1, -0.5)$ and $(1.5, 2)$. Find an equation for your line.
9. Do the falling body data in Table 1.2 agree or disagree with what you think should happen? If there are discrepancies with what you think should happen, how do you think those discrepancies might arise in the process of data gathering?

The next twelve exercises require graph paper. Start each exercise by labeling and scaling your axes in a way that is reasonable for that exercise.

For Exercises 10–19, do the following:

(a) Make a scatter plot of the data in the given table.

(b) Describe the shape of the graph.

(c) Explain what the shape of the graph indicates about the relationship between the variables.

10. **Table 1.6** Related variables

x	0.8	1.5	3.2	2.6	1.9	2.4	3.5	0.6	2.1
y	0.7	2.1	10.5	6.8	3.5	5.9	12.4	0.3	4.3

11. **Table 1.7** Distance traveled by a falling object

Time (seconds)	*Distance (meters)*
1	5
2	19
3	44
4	78
5	123
6	176
7	240
8	313
9	396
10	489

12. **Table 1.8** Third class postage

Weight (ounces)	*Postage ($)*
0 to 1	0.25
1 to 2	0.45
2 to 3	0.65
3 to 4	0.85
4 to 6	1.00
6 to 8	1.10
8 to 10	1.20
10 to 12	1.30
12 to 14	1.40
14 to 16	1.50

13. **Table 1.9** Temperature and pressure of a gas in a closed container

Temperature (°K)	*Pressure (mm Hg)*
263	752
268	755
278	777
293	811
298	834
303	840
308	854
318	892
323	906

14. **Table 1.10** Firearm homicides per 100,000 population ages 0–19[3]

Year	*Homicides per 100,000*
1985	1.5
1986	1.7
1987	2.1
1988	1.5
1989	2.5
1990	3.9
1991	4.7
1992	4.3

15. **Table 1.11** Imports from Mexico to the United States ($ billion)[4]

Year	*Imports from Mexico ($B)*
1987	20.2
1988	23.2
1989	27.2
1990	30.2
1991	31.2
1992	35.2
1993	39.3

16. **Table 1.12** Exports to Mexico from the United States ($ billion)[5]

Year	*Exports to Mexico ($B)*
1987	14.6
1988	20.6
1989	24.9
1990	28.4
1991	33.3
1992	40.6
1993	41.6

17. **Table 1.13** Average salaries of major league players[6]

Year	*Salary ($ thousands)*
1973	37
1975	45
1977	76
1979	114
1981	187
1983	289
1985	371
1987	412
1989	497
1991	851
1993	1076

18. **Table 1.14** Automatic teller machines[7]

Year	*ATMs (thousands)*
1980	20
1981	27
1982	34
1983	40
1984	58
1985	60
1986	65
1987	70
1988	73
1989	78
1990	81
1991	84
1992	87

19. **Table 1.15** ATM transactions

Year	*Transactions (millions)*
1980	100
1981	140
1982	180
1983	200
1984	270
1985	300
1986	310
1987	340
1988	380
1989	430
1990	490
1991	530
1992	610

20. Use the data in Tables 1.14 and 1.15 to construct a table of numbers of transactions per ATM over the years 1980 to 1992. Make a scatter plot of the information in your new table. What is the shape of the graph? What conclusions (if any) can you draw about numbers of transactions per ATM?

21. Make a scatter plot to analyze the relationship between rebounds and assists for the basketball data provided in Tables 1.3, 1.4, and 1.0. Use the graph to comment on important characteristics of the relationship.

22. Figure 1.5 shows scatter plots of revenues and expenses for Major League Baseball from 1983 to 1992, both in millions of dollars.
    (a) In which years did baseball make a profit?
    (b) In which years did baseball lose money?
    (c) What was baseball's best financial year during the period shown?

23. (a) Use the information in Figure 1.5 to make a table of profits for Major League Baseball from 1983 to 1992. (Treat a loss as a negative profit.)
    (b) Use your table to make a scatter plot of profits for Major League Baseball.
    (c) Find the total profit (or loss) for Major League Baseball over the entire period shown.

Figure 1.5 Major League Baseball revenues and expenses in millions of dollars[8]

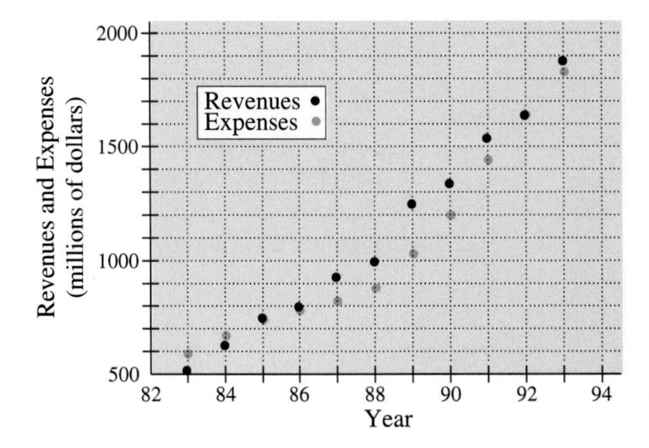

---

3. Source: Office of the Chief Medical Examiner, as reported in *The News & Observer*, June 13, 1993.
4. Source: U.S. Department of Commerce, as reported by *USA Today*, September 30, 1994.
5. Source: U.S. Department of Commerce, as reported by *USA Today*, September 30, 1994.
6. Source: Major League Baseball Players Association.
7. Source for Tables 1.14 and 1.15: *Bank Network News*, as reported by Associated Press, June 6, 1993.
8. Source: Associated Press, July 11, 1993.

1.2 Mathematical Models

When a relationship between varying quantities is suggested by a scatter plot, we may want to describe it mathematically by an *equation* that summarizes the way the two variables are related. Such an equation is called a **mathematical model**.

A good model simplifies the phenomenon it represents and gives us the ability to predict. If we can find an equation of a line or curve that closely fits a scatter plot, we can focus on the important characteristics of the relationship between the variables without the clutter of a scatter plot. We also can use this equation to predict the values of one variable for specific values of the other variable. Sometimes we use the model to **interpolate**, or estimate values between observed values. Sometimes we use the model to **extrapolate**, or predict values outside the region of observations.

Exploration Activity 1

Figure 1.6 is adapted from a newspaper article about advertising expenditures by major drug manufacturers. In this figure, we plot the sales figures in billions of dollars versus the year.

Figure 1.6 Sales by the pharmaceutical industry in North Carolina[9]

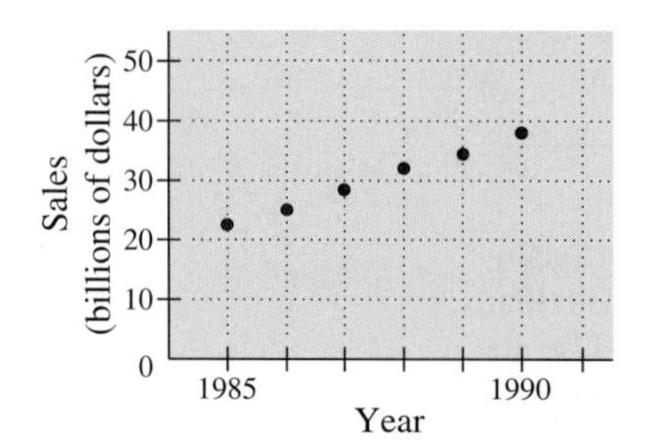

(a) On Figure 1.6, draw a straight line that seems to fit the data reasonably well.

(b) Find an equation for your line of the form $y = mx + b$, where x stands for the year and y stands for sales (in billions of dollars).

(c) Use your equation to estimate the sales by North Carolina drug companies in each of the years 1986, 1987, 1988, and 1989, and compare your model estimates with the graphical data in Figure 1.6.

(d) Now use your equation to estimate the sales in 1991, i.e., to predict the next data point beyond the time at which the article was written.

Our estimated equation for the line is $y = 3.2x - 6330$, but there is no unique answer. Once we estimated the first and last points, we calculated a slope from those two points. Then we used that slope and one of the points to determine the equation for the line. This equation leads to an estimate of \$41.2 billion for the sales by North Carolina drug companies in 1991. Whether you agree with that estimate or not depends not only on the correctness of your (and our) calculations but also on how accurately you (and we) could

9. Source: *The News & Observer*, April 14, 1991.

read sales numbers from the graph. You also have to accept our implicit assumption that the linear pattern would continue for at least one more year.

Important If you had difficulty with any of the steps in the preceding paragraph, stop right now and straighten out whatever went wrong. What are the coordinates of the two points determining your line? Do the y-coordinates make sense in terms of the graph? How do you calculate slope from the coordinates of the two points? Once you know the slope m, how do you find b? You don't have the luxury of substituting $x = 0$, because that is not a year in which we know anything about drug sales. However, you can substitute either of your known pairs of x and y to get an equation in which b is the only unknown. Once you have sensible values for m and b, you can do the rest of the arithmetic with your calculator.

Have you noticed that very few of our activities and exercises have clearly determined "right" answers? That's often the way mathematics works in the real world. There are answers that are more sensible and answers that are less sensible and answers that are clearly *wrong*—but not always answers that are clearly right. Your job is to find answers that are *sensible*—and to be prepared to defend those answers.

Exploration Activity 2

Figure 1.7 shows data on the percentage of U.S. citizens among those earning doctorates in the mathematical sciences at universities in the United States. The graph shows a clear decreasing trend, starting about 1977 and extending another dozen years. Find a straight line that approximates this portion of the data. How can you check your equation?

Figure 1.7 Percentage of U.S. citizens among those earning doctorates in the mathematical sciences[10]

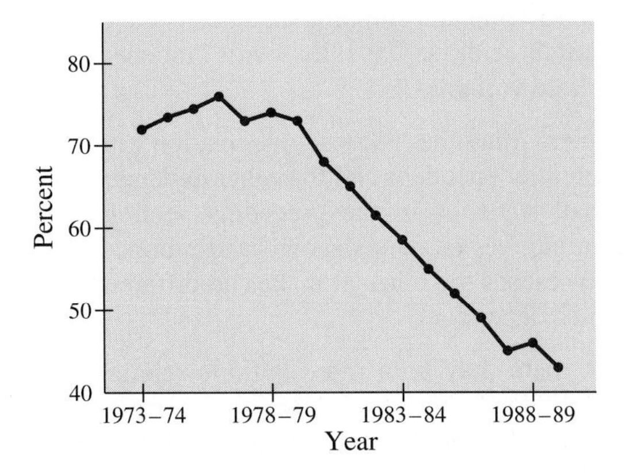

A straight line through the points for 1976–77 and 1989–90 comes reasonably close to all the intermediate points. For simplicity, let's identify the x-coordinates of those points as 1977 and 1990. We estimate from the graph that the corresponding y-coordinates are 79 and 43 (percent), respectively. The slope m of the line through the points $(1977, 79)$ and $(1990, 43)$ is

$$\frac{43 - 79}{1990 - 1977} \approx -2.77.$$

10. Source: *Focus*, Mathematical Association of America, January–February 1991.

Using a point-slope form, we find that an equation of the line through our two points is $y - 43 = -2.77(x - 1990)$, or, in simplified form,

$$y = -2.77x + 5555.$$

Check If we substitute $x = 1977$ in this formula, we get $y \approx 79$, as we should. For $x = 1984$, we get $y \approx 59$. How does that compare with the graph?

Checkpoint 1

(a) From the graph in Figure 1.7, make a table of estimated percentages in each year from 1978–79 on.

(b) Add another column to your table with percentages estimated from the formula $y = -2.77x + 5555$.

(c) If the trend in the 1980s were to continue for another decade, what would the percentage be in 1999–2000?

(d) If the trend continued until there were no U.S. citizens receiving mathematics doctorates in U.S. universities, in what year would you expect that to occur?

Independent and Dependent Variables

In some relationships between varying quantities, it may be that one of the quantities determines the values of the other, in which case we say the second quantity "depends on" the first. When this is true, the quantity that depends on the other is called the **dependent** variable; the other is called the **independent** variable. For example, because the stopping distance for your car depends on the speed you are traveling, speed is the independent variable and stopping distance is the dependent variable.

There are three other ways the words "independent" and "dependent" may be used to describe related variables:

- First, there may be a loose association, in which case either variable may be considered independent and the other dependent. We saw an example of this with the basketball statistics in the preceding section. There appears to be an association between number of points scored and number of fouls committed, but neither points nor fouls causes the other. It makes no difference which variable is called independent and which dependent.

- Second, there may be a more definite relationship in which one of the variables is naturally independent. In the drug sales example (Exploration Activity 1), there is a definite number representing the total sales each year. The date on the calendar does not cause drug sales to happen, nor does a certain level of sales cause a year to happen. Nevertheless, we speak of time as the independent variable and sales as the dependent variable. We could label the variables the other way around, but that would not seem natural.

- Third, it is possible that a value of each of the variables determines—even causes—a corresponding value of the other. For example, the amount people are willing to pay for bonds is related to the effective rate of return on the bonds. At times the rate might determine the price, but sometimes the price also determines the rate. We might declare either variable to be independent and the other dependent as we set up a description of this relationship.

Representing Relationships

We can display or describe the relationship (if any) between two quantities or variables in many different ways. Here are some of the ways that are already familiar:

- List all the pairs of values that occur together.
- Make a table or chart that displays some or all of the pairs in the relationship.
- Give a formula or equation that relates the two variables.
- Draw a graph in a rectangular coordinate system with the dependent variable on the vertical axis and the independent variable on the horizontal axis.

You will spend a fair amount of time in this course studying graphs of relationships between two variables. So far you have seen scatter plots and line graphs of data. We will often use graphs of formulas, whether derived from known relationships or from fitting formulas that approximate data. And a graph also can be a *conceptual picture* of a relationship for which we know neither data nor formulas.

For example, Figure 1.8 shows a typical *learning curve* for a concept such as "function." The independent variable is time, and the dependent variable is the amount learned. (Note that time alone is not sufficient to cause learning, but learning certainly depends on time spent working at it.) Initially, the learner may know nothing about the concept. With any new learning challenge, the pace of learning is likely to be slow at first, then pick up (the steep part of the curve), then slow down again as there is little more to be learned in order to master the concept. The shape of the curve in Figure 1.8 captures these features of the learning process. Of course, you can imagine variations of the process that would lead to curves with somewhat different shapes.

Figure 1.8 Learning curve for a concept

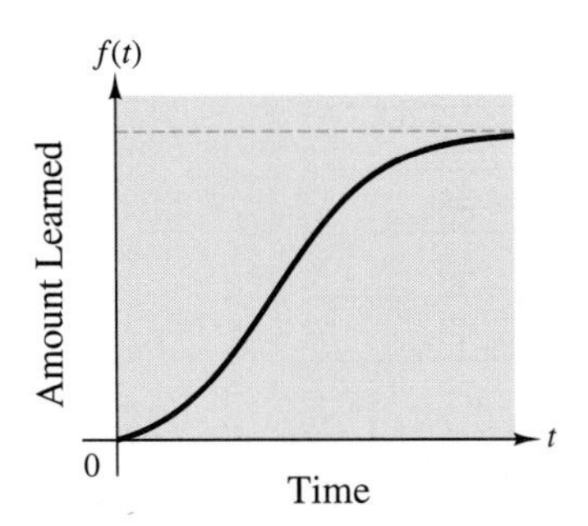

Exploration Activity 3

(a) Sketch a graph to represent the relationship between the number of customers checking out at a grocery store and the average amount of time each person must wait in a checkout line.

(b) Does the number of customers determine the waiting time, or is it the other way around? Which is the independent variable in this relationship?

(c) What characteristics of the relationship can you determine even without specific information about the time required for a customer to complete the checkout process?

(d) As the number of people waiting to check out increases, what can you say about their average waiting time?

Your graph should show an increasing relationship. It also might show other features, such as jumps if certain numbers of customers trigger opening another register, thereby shortening waiting time. If we assume that this does not happen, then the number of customers determines the waiting time, and *number of customers* is likely to be the independent variable. As the number of customers increases, the waiting time also increases. We do not know the coordinates of any particular points on the graph, nor do we know the precise steepness of the graph. Nevertheless, we can represent an important characteristic of the relationship by the fact that the graph is increasing.

In this section we have seen that we can use formulas and graphs to model real phenomena represented by data. Our models thus far have relied on your prior knowledge of linear functions and straight lines. As we develop more powerful mathematical tools, we will be able to model more complex phenomena.

Answers to Checkpoint

1. (a), (b)

Year	*Estimated from graph*	*Calculated from formula*
1978–79	74	73
1979–80	73	70
1980–81	67	68
1981–82	64	65
1982–83	61	62
1983–84	58	59
1984–85	54	57
1985–86	50	54
1986–87	48	51
1987–88	45	48
1988–89	47	45
1989–90	43	43

(c) 15 (d) 2005–06

Exercises 1.2

For each of the relationships in Exercises 1–16:

(a) Identify the two quantities that vary, and decide which should be represented by an independent variable and which by a dependent variable.

(b) Sketch a reasonable graph to show a possible relationship between the two variables. (You should be concerned only with the basic shape of the graph, not with particular points.)

(c) Write a sentence or two to justify the shape and behavior of your graph.

1. The height of a baseball after being hit into the air.
2. The speed of an egg dropped from the tenth floor of a building.
3. The height of an individual as he or she ages.
4. The amount of time it takes a person to run a mile as he or she ages.
5. The amount of money in a savings account over an extended period of time.
6. The size of a person's vocabulary from birth onwards.
7. The number of bacteria in a laboratory culture over a period of time.
8. The daily high temperature at your local airport over a calendar year.
9. The number of people absent from class and the day of the academic year.
10. The temperature of an ice-cold drink left in a warm room over a period of time.
11. The number of hours of daylight and the day of the year.
12. The cost of a pizza and the diameter of the pizza.
13. A safe following distance for your car at a given speed.
14. Ideal weight at a given height, as represented by a chart in your doctor's office.
15. The income tax on a given taxable income.
16. The sales tax on a given purchase.
17. (a) Sketch a graph of the relationship in Table 1.16. (You may have done this already.)
    (b) Identify dependent and independent variables.
    (c) Discuss any conclusions you can draw from the shape of your graph.

Table 1.16 Third class postage

Weight (ounces)	*Postage ($)*
0 to 1	0.25
1 to 2	0.45
2 to 3	0.65
3 to 4	0.85
4 to 6	1.00
6 to 8	1.10
8 to 10	1.20
10 to 12	1.30
12 to 14	1.40
14 to 16	1.50

18. (a) Sketch a graph of the relationship presented in Table 1.17.

Table 1.17 Federal income tax

Taxable Income	*Income Tax*
Less than $10,000	$0
$10,000 but less than $50,000	15% of income over $10,000
$50,000 or more	$6000 + 27% of income over $50,000

    (b) Identify dependent and independent variables.
    (c) Discuss any conclusions you can draw from the shape of your graph.

1.3 Relations and Functions

The technical term for the relationships we have considered so far is **relation**. We use this term whenever values of one variable are paired with values of another variable, whether or not the two variables appear to be related.

Definition A **relation** is a pairing of the values of one varying quantity with values of another varying quantity.

Let's consider the relation that pairs the price you pay for an airplane ticket and the distance that you fly. Although one might think that the cost of a ticket should depend on how far you plan to fly, it often happens that two tickets for flights of the same length (even on the same plane!) have different prices. The price of a ticket is not *uniquely* determined by the length of the flight. Given a value for the independent variable (distance), there is not just one value for the dependent variable (price).

We contrast this example with the drug sales relation from Exploration Activity 1 in the preceding section. As we have seen already, in each year there is a definite number representing the total sales by drug companies in a given state. Thus, for each value of the independent variable time (a calendar year), there is *just one* value of the dependent variable, sales. There are many important relations like this, in which each value of the independent variable is paired with a *unique* value for the dependent variable. Whenever this is true, the relation is called a **function**.

Definition A **function** is a pairing of the values of one varying quantity with the values of another varying quantity in such a way that each value of the first variable is paired with *exactly one* value of the second variable.

In the airplane ticket relation, we are *not* guaranteed a unique value for the dependent variable, so this relation is *not* a function.

Your fitted line for the drug sales relation (Exploration Activity 1 in the preceding section) provides a second example of a function. That line has a formula of the form $y = mx + b$, where m and b are whatever numbers you calculated to make the line fit the data. For each value of the independent variable x (a year), there is a unique value for the dependent variable y, the predicted sales figure. For integer values of x between 1985 and 1990, these y-values should be very similar to the actual sales figures, but this formula-based function is *not the same function* as the number pairing represented by the actual sales figures.

We hasten to add that it is *not* having a formula such as $y = mx + b$ that makes a relation a function; rather, *it is the pairing of values of the independent variable with* ***unique*** *values of the dependent variable*. We have to be aware of the distinction between functions that are defined by data (or graphs of data) and functions that are defined by formulas. The latter are often useful *models* of reality, but they are not in fact that reality.

We will often find it useful to think of a function as a *process* that associates each permissible first coordinate with a unique second coordinate—a pairing process, with the special condition that the second value in each pair is uniquely determined by the first. In this course, it is important to think about functions as being dynamic, as "doing something" to an x-value to get a y-value. What a function does to each value of the independent variable is called a **rule** for the function. For example:

- If a function is defined by a two-column table of data, the dynamic interpretation (or rule) of the function is, "For each number in the left column, look for the paired number in the right column."
- If a function is defined by a graph, the rule is, "For each first coordinate on the horizontal axis, go up to the corresponding point on the graph, and then locate the second coordinate on the vertical axis."
- For the model function defined by the formula $y = -2.77x + 5555$, the dynamic interpretation is, "For each number x, multiply x by -2.77, and then add 5555 to get the corresponding y."

The "for each number" parts of the preceding examples bring us to another important characteristic of a function: the set of numbers that are allowed to be values of the independent variable. This set is called the **domain** of the function. Thus, if a function is defined by a table, its domain is the set of numbers in the left-hand column. If it is defined by a graph, its domain is the set of first coordinates of points on the graph. If it is defined by a formula, the domain is the set of numbers that can legitimately be substituted for the independent variable in the formula.

By combining the ideas of domain and rule, we have another way—in addition to the pairing definition—to characterize the concept of function:

1. We start with a collection of numbers that constitute allowable *inputs* to the function. This set is the **domain** of the function.
2. We have a **rule** for associating with each input a unique *output* or *result* or *value* of the function.

The connection between the domain/rule description and the pairing definition is this: The domain identifies which numbers can be first elements of a pair, and the rule tells us what number is the second element that goes with each first element.

Checkpoint 1

State as explicitly as you can the domain and rule for each of the following functions.

(a) For each member of your class, weight (in pounds) as a function of Social Security number.

(b) Drug sales as a function of time, as defined by Figure 1.9.

(c) $y = x^2$.

(d) $y = \sqrt{x}$.

Figure 1.9 Sales by the pharmaceutical industry in North Carolina

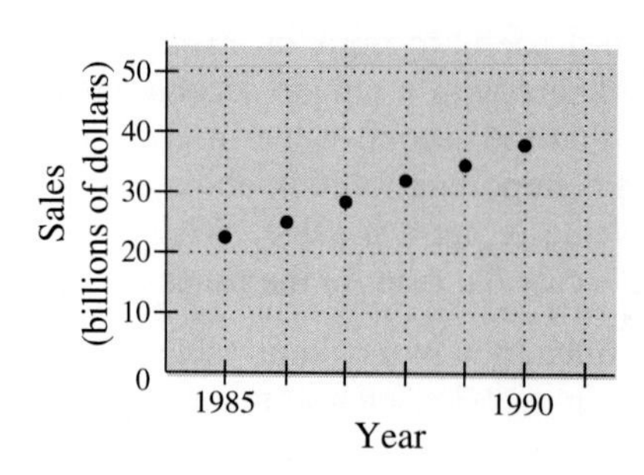

In this section we have formalized the distinction between *relation*—any pairing of numbers from two sets of numbers—and *function*—a pairing in which each first number is associated with exactly one second number. We have seen that a function can be described dynamically by specifying its *domain* (set of first numbers) and a *rule* for finding the second number associated with each number in the domain. For functions and relations arising in the real world, the pairing is likely to be represented by a data table or a graph. For model functions that approximate real phenomena, the association of variables is likely to be represented by a formula or a graph.

Answers to Checkpoint

1. (a) The domain is the set of Social Security numbers of members of your class. The rule is to assign to each Social Security number the weight of the person who owns that number.

   (b) The domain is the set of dates 1985, 1986, 1987, 1988, 1989, 1990. The rule is to assign to each date the y-coordinate of the corresponding point on the graph, that is, the drug sales in that year.

   (c) The domain is the set of all real numbers, and the rule is to square each number.

   (d) The domain is the set of nonnegative numbers, and the rule is to take the square root of each number in the domain.

Exercises 1.3

For each of the relationships in Exercises 1–16:

(a) Determine whether the relation is or is not a function, and explain why you think it is or is not.

(b) If the relation is not a function, can you think of additional conditions you could impose on the description that would make it a function?

(c) If the relation is a function, describe its domain and a rule for determining its values.

1. The height of a baseball after being hit into the air.
2. The speed of an egg dropped from the tenth floor of a building.
3. The height of an individual as he or she ages.
4. The amount of time it takes a person to run a mile as he or she ages.
5. The amount of money in a savings account over an extended period of time.
6. The size of a person's vocabulary from birth onwards.
7. The number of bacteria in a laboratory culture over a period of time.
8. The daily high temperature at your local airport over a calendar year.

9. The number of people absent from class and the day of the academic year.
10. The temperature of an ice-cold drink left in a warm room over a period of time.
11. The number of hours of daylight and the day of the year.
12. The cost of a pizza and the diameter of the pizza.
13. A safe following distance for your car at a given speed.
14. Ideal weight at a given height, as represented by a chart in your doctor's office.
15. The income tax on a given taxable income.
16. The sales tax on a given purchase.

Your calculator has a lot of keys called function keys. In the following exercises we examine whether these keys satisfy our description of a function. Your conclusions may depend on what calculator you have. Don't try to look up the answers—experiment with the calculator itself. For each of the following function keys, (a) graph the function, (b) decide whether the graph is indeed the graph of a function in the sense defined here, and (c) if so, describe the domain.

17. [sin] 18. [cos] 19. [tan]
20. [$\sin^{-1}$] 21. [$\cos^{-1}$] 22. [$\tan^{-1}$]
23. [x^2] 24. [log] 25. [ln]
26. [$\sqrt{\ }$] 27. [x^{-1}] 28. [e^x]

29. (a) Make a scatter plot of the data in Table 1.18. (You may have done this already.)

Table 1.18 Related variables

x	0.8	1.5	3.2	2.6	1.9	2.4	3.5	0.6	2.1
y	0.7	2.1	10.5	6.8	3.5	5.9	12.4	0.3	4.3

(b) Explain why the table defines a function.
(c) What is the domain of this function? What is a rule for the function?

30. (a) Make a scatter plot of the data in Table 1.19. (You may have done this already.)
(b) Explain why the data in Table 1.19 defines a function. What is a function of what?
(c) What is the domain of this function? What is a rule for the function?

Table 1.19 Distance traveled by a falling object

Time (seconds)	*Distance (meters)*
1	5
2	19
3	44
4	78
5	123
6	176
7	240
8	313
9	396
10	489

31. Some thermometers measure temperatures on both Fahrenheit and Celsius scales. These side-by-side scales suggest a functional relationship.
(a) Define Fahrenheit temperature as a function of Celsius temperature by a simple formula.
(b) Now express Celsius temperature as a function of Fahrenheit temperature.
(c) Graph both these functions. Would you describe either of these functional relationships as "linear"? Why or why not?
(d) Is there a temperature that is assigned the same number on both the Fahrenheit and Celsius scales? If so, what is it, and how do you find it? Is it a high temperature, a low temperature, or an in-between temperature?

32. Consider the relationship between the amount of money earned on a part-time job and the number of hours worked.
(a) Identify the two quantities that vary, and decide which should be represented by an independent variable and which by a dependent variable.
(b) Sketch a reasonable graph to show a possible relationship between the two variables. (You should be concerned only with the basic shape of the graph, not with particular points.)
(c) Write a sentence or two to justify the shape and behavior of your graph.

(d) Determine whether the relation is or is not a function, and explain why you think it is or is not.
(e) If the relation is not a function, can you think of additional conditions you could impose on the description that would make it a function?
(f) If the relation is a function, describe its domain and a rule for determining its values.

33. Consider the relationship between the postage on a first class letter and the weight of the letter.
(a) Identify the two quantities that vary, and decide which should be represented by an independent variable and which by a dependent variable.
(b) Sketch a reasonable graph to show a possible relationship between the two variables. (You should be concerned only with the basic shape of the graph, not with particular points.)
(c) Write a sentence or two to justify the shape and behavior of your graph.
(d) Determine whether the relation is or is not a function, and explain why you think it is or is not.
(e) If the relation is not a function, can you think of additional conditions you could impose on the description that would make it a function?
(f) If the relation is a function, describe its domain and a rule for determining its values.

34. Does Table 1.20 define a function? Why or why not?

Table 1.20 Falling body data

Time (seconds)	*Distance (centimeters)*	*Time (seconds)*	*Distance (centimeters)*
0.16	12.1	0.57	150.2
0.24	29.8	0.61	182.2
0.25	32.7	0.61	189.4
0.30	42.8	0.68	220.4
0.30	44.2	0.72	254.0
0.32	55.8	0.72	261.0
0.36	63.5	0.83	334.6
0.36	65.1	0.88	375.5
0.50	124.6	0.89	399.1
0.50	129.7		

1.4 Dichotomies

You're driving along the highway, impatient to reach your destination. You look at the dashboard clock, and you look at the speedometer. You have 28 minutes to go, your destination is still 25 miles away, and you have to park the car. Are you going fast enough?

For now, we ignore the risk of a speeding ticket, and we focus on the two familiar devices just mentioned: clock and speedometer. Together, they represent a function: Given a time on the clock, there is a simultaneous speedometer reading that gives the speed of the car at that instant of time—and there is *just one* such reading. In mathematical language, *speed is a function of time*.

Many of the functions we will study in this course will be functions of time. In the speedometer-clock example, the time variable is **discrete**; i.e., given a time value, there is a *next* value (except at the end of the domain) and there is a *previous* value (except at the beginning of the domain). For the digital clock on your dashboard, time moves in steps of one minute—or perhaps one second, if your clock is that precise.

You may prefer to think of time as moving **continuously**, i.e., with no identifiable next or previous time after or before any given instant. Some automobile dashboard clocks are *analog* clocks; that is, they have hands rather than digits. Furthermore, the speedometer *always* has a reading, whether or not the digital clock has just recorded a new time, so we may think of speed as a function of **continuous** time. Since the speedometer is not likely to be a perfectly accurate measuring device, there is a slightly different, highly theoretical function that represents the *real* speed of the car as a function of continuous time.

Consider the function determined by the U.S. Census data every ten years since 1790 (see Exercise 24). On one hand, we know that the census is not a perfectly accurate measurement. On the other hand, we can imagine that there is some *exact* U.S. population at any instant of continuous time. Unlike a real speed, which may or may not change continuously, the exact, real, highly theoretical, and unobservable population *cannot* change continuously. Why not? (Also see Exercise 25.)

These dichotomies:

- ***discrete*** *versus* ***continuous***
- ***real*** *versus* ***observable*** *or* ***measurable***
- ***exact*** *versus* ***approximate***

will be frequently recurring themes throughout our study of the interaction between mathematics and the world around us.

All three of these dichotomies interact with and overlap a fourth:

- ***calculational*** *versus* ***conceptual*** *functions*

The functions we *observe*, whether as tabulated data, graphical relationships, or measurements, can never be *known* at more than a handful of inputs, and the calculations we do with these functions can never be *exact*.

On the other hand, we can often **model** these functions by mathematical formulas, about which we know (or at least can hope to know) everything. (You did this yourself in Exploration Activity 1 of Section 1.2 when you fit a linear formula to drug sales data.) You can easily be seduced into believing that these formulas are the most important objects of study, that they represent exactness or truth, and that everything else is, at best, approximate. In fact, the true relationships of science, social science, and engineering are very seldom described exactly by formulas. Nevertheless, formulas that approximate these realities are often convenient models because we can study and manipulate the formulas to gain insights that might remain obscured by messy reality.

Exercises 1.4

In the first sixteen exercises in Sections 1.2 and 1.3 you constructed functional relationships from the pairs of variables given here in Exercises 1–16. Which of your constructed functions are functions of time, and which are not? Explain.

1. Height of a baseball, time after being hit.
2. Speed of a falling egg, time after being dropped.
3. Height of an individual, age of the individual.
4. Time to run a mile, age of the individual.
5. Amount in a savings account, time on deposit.
6. Size of vocabulary, age of the individual.
7. Number of bacteria in a culture, time the culture has been growing.
8. High temperature, day of the year.
9. Number absent, day of the academic year.

10. Temperature, time the fluid has been warming.

11. Number of hours of daylight, day of the year.

12. Cost of a pizza, diameter of the pizza.

13. Safe following distance, speed of your car.

14. Ideal weight, height.

15. Income tax, taxable income.

16. Sales tax, purchase price.

17. In the example of speed of a car (measured by the speedometer) as a function of time (measured by the dashboard clock), what is the domain of the function? What is a rule for the function?

18. Besides the speedometer, what other device on the dashboard of a car expresses some quantity as a function of time? What quantity is so expressed?

19. When you are driving a car, is time a function of speed? Why or why not?

20. Table 1.21 shows annual rainfall recorded in central North Carolina over an 11-year period.
    (a) Plot the rainfall data as a function of time.
    (b) What was the average annual rainfall for this 11-year period?
    (c) The 30-year average annual rainfall for central North Carolina is about 42 inches. Was this recent 11-year period unusually dry, unusually wet, or about average?

Table 1.21 Rainfall recorded at Raleigh-Durham International Airport[11]

Year	*Rainfall (inches)*
1982	44.35
1983	47.23
1984	38.17
1985	36.95
1986	42.09
1987	37.66
1988	54.15
1989	37.55
1990	54.15
1991	35.46
1992	43.18

21. (a) Sketch a graph of speed as a function of time for a typical trip to get a pizza—one way or round trip, but be sure to explain your graph.
    (b) Sketch a graph of distance traveled for the same run, also as a function of time.

22. What function of time is measured by the odometer in a car? Is the measurement discrete or continuous? Is the function itself discrete or continuous? Explain your answers in each case.

23. Consider the situation described in this sentence: *The child's temperature has been rising for the last two hours, but not as rapidly since we gave her the antibiotic an hour ago.* Sketch a graph of temperature as a function of time, consistent with the situation described.[12]

24. Every ten years, the U.S. Census Bureau attempts to count the population of the United States. The results of these efforts are shown in Table 1.22.

Table 1.22 U.S. Census data, 1790–1980

Census Date	*Population (millions)*	*Census Date*	*Population (millions)*
1790	3.929	1890	62.948
1800	5.308	1900	75.996
1810	7.240	1910	91.972
1820	9.638	1920	105.711
1830	12.866	1930	122.775
1840	17.069	1940	131.669
1850	23.192	1950	150.697
1860	31.443	1960	179.323
1870	38.558	1970	203.185
1880	50.156	1980	226.546

(a) On graph paper, make a scatter plot of the data in Table 1.22.
(b) What are the independent and dependent variables in the table and in your graph? Do the census data represent a function? If so, what is a function of what? What is the domain? What is a rule for the function?
(c) Use the graphical information in your scatter plot to sketch your own idea of the *real* U.S. population function over the continuous time interval from 1789 to the present.
(d) Do you see your graph as being continuous or discrete? Why?

(e) Now imagine a view through a high-powered microscope of a very small portion of your graph, say, between 2:30 P.M. and 3:17 P.M. last Wednesday. Sketch what you think this microscopic view should look like. Is this graph continuous or discrete? Why?

25. In one of our examples of a function of continuous time, we imagined an *exact* population of the United States at each instant of time. While it may seem obvious that there should be such an exact population (whether it can be measured or not), one can argue that no such thing exists. Is there an exact instant that an individual is born or dies? Is there an exact instant that an immigrant enters the country or an emigrant leaves? Discuss these questions with one or two classmates, and write a paragraph summarizing the discussion.

26. Ziggy, Tom Wilson's lovable cartoon character, tells us: "Time is just Nature's way of keeping everything from happening all at once." Explain how this statement is related to the concept of function of time.

---

11. Source: National Weather Service, cited in *The News & Observer*, January 1, 1993.

12. Adapted from *Calculus Problems for a New Century*, edited by Robert Fraga, MAA Notes, Vol. 28, 1993.

1.5 Words

Words, words, words!
I get words all day through,
First from him, now from you!
Is that all you blackguards can do?
Eliza Doolittle in *My Fair Lady*

By now you may share some of Eliza's anger and frustration. The outburst was directed at her would-be suitor, Freddie, but she was really angry at her chief tormentor, Professor Henry Higgins (the "him" in the lines quoted above), who was trying to mold her in his image of an English gentlewoman by freeing her from the straitjacket of a Cockney dialect. We intend to achieve the unintended result of Professor Higgins' efforts: by freeing you from your linguistic straitjackets, to enable you to become independent thinkers.

Lerner and Loewe's musical *My Fair Lady* was based on Shaw's play *Pygmalion*, which in turn was based on a Greek myth whose central character was a sculptor who fell in love with one of his statues that was brought to life by Aphrodite, the goddess of love and beauty. The theme is a common one that you may encounter in many other courses. Indeed, you are part of an ancient tradition that runs through thousands of years of literature, art, and education.

Thus we consider it vitally important that you learn and use correctly the language of mathematics, that is, of quantitative reasoning. We have devoted these many pages of words—with very few mathematical symbols—to examination of the word "function" precisely because this word represents the concept that will be our principal object of study.

We have a small confession to make. When we quoted *The American Heritage Dictionary* on the definition of "function," we left out something: its so-called math definition. Here is the missing part of the quotation:

> **function** (fungk′shən) *n.* **5.** *Math.* **a.** A variable so related to another that for each value assumed by one there is a value determined for the other. **b.** A rule of correspondence between two sets such that there is a unique element in one set assigned to each element in the other.[13]

Perhaps you see why we left this out earlier. First, in stark contrast to the other four definitions, these two statements are rather difficult to read and interpret. Why would a lexicographer write four straightforward definitions for different meanings of a word and then apparently obfuscate the technical meaning? Second, this appears to be *two* quite different definitions for only one concept—or do mathematicians use the one word "function" for two rather different concepts?

The dictionary definition does not tell us that **5.a.** and **5.b.** actually define the *same* concept or that the concept is merely an abstraction of definition **4.** We trust that you have figured all this out from our examples, exercises, labs, and classroom activities. We summarize our definitions here.

Definitions A **function** is a pairing of the values of one varying quantity with the values of another varying quantity in such a way that each value of the first variable is paired with *exactly one* value of the second variable. The first variable is called **independent**, and the second variable is called **dependent**. The set of all values of the first variable is called the **domain** of the function.

Checkpoint 1

(a) In definition **5.a.**, two *variables* are mentioned, and each is referred to twice. Find these four references in the definition, and label each as **independent** or **dependent**.

(b) Definition **5.b.** refers to "one set" and "the other." Which is the domain? Find the appropriate words in the definition, and label them **domain**.

In Exploration Activity 1 of Section 1.1, you wrote your initial understanding of the word "function." The time has come to review your description. Have we made a connection yet? If not, our failure to connect may result from your use of formulas to define, describe, or exemplify functions. You may have noticed that hardly any formulas have turned up yet. We will see lots of formulas later—when they serve some useful purpose—and you will have many opportunities to use all those wonderful things you have learned about algebra and trigonometry. For now, the point we want to make is this: "Function" and "formula" are not synonyms! And "equation" is not a synonym for either "function" or "formula."

Checkpoint 2

You don't need a math book to sort out the meanings of these words — a dictionary will do. We have given you the dictionary definition of "function." Look up and write down the definitions of

(a) **formula**

(b) **equation**

(c) **expression**

Each will have several definitions — select the mathematical definitions.

The word "expression" may not be part of the vocabulary you learned in previous mathematics courses, but its definition may be one that you routinely confuse with "function" or "formula" or "equation." In fact, "formula" and "expression" are closely related — little harm results from using them interchangeably. "Equation" is a very special kind of formula or expression, one that asserts the *equality* of two quantities — each of which also may be described by a formula or expression.

None of this — equation, formula, expression — has any direct connection with *function*. However, the *rule* associating values of a dependent variable with values of an independent variable *might* be expressed by an equation or a formula or an expression. This almost never happens in the real world. However, as we have seen already, reality often can be modeled by formulas, and the functional rules embodied in those formulas may turn out to be convenient and/or useful approximations to reality.

Thus we will move freely among

- the **real** — but not completely knowable — functions that relate varying quantities in physics, biology, chemistry, engineering, or economics,
- the **observed** functions represented by data sampled from measurements of the real variables, and
- mathematical **model** functions that approximate one or both of the other types.

Only in this last category will we see any formulas. Indeed, this is the *purpose* of formulas: to give us approximate models of reality that are subject to manipulations that can, in turn, help us to understand reality.

Answers to Checkpoints

1. (a) "A *variable* [independent] so related to *another* [dependent] that for each value of *one* [independent] there is a value determined for the *other* [dependent]. [*Note*: This seems to define the function as its independent variable rather than as the relation between variables.]

   (b) The domain is "the other" set.

2. From *The American Heritage Dictionary*:

   (a) The seventh definition of *formula* is "A mathematical statement, esp. an equation, of a rule, principle, answer, or other logical relation."

   (b) The third definition of *equation* is "A linear array of mathematical symbols separated into left and right sides that are designated at least conditionally equal by an equal sign."

   (c) The third definition of *expression* is "A designation of any symbolic mathematical form, such as an equation."

1.6 Historical Background (Optional Reading)

Mental constructs, such as the concept of function, change through history as scholars get a better grip on important ideas. The meanings of the words used to represent these concepts also change, and for the same reason. The following text is quoted from *An Introduction to the History of Mathematics* by Howard Eves[14]; the footnotes are added.

> The concept of function, like the notions of space and geometry, has undergone a marked evolution, and every student of mathematics encounters various refinements of this evolution as his studies progress from the elementary courses of high school into the more advanced and sophisticated courses of the college postgraduate level.
>
> The history of the term *function* furnishes another interesting example of the tendency of mathematicians to generalize and extend their concepts. The word *function*, in its Latin equivalent, seems to have been introduced by Leibniz[15] in 1694, at first as a term to denote any quantity connected with a curve, the radius of curvature, and so on. Johann Bernoulli,[16] by 1718, had come to regard a function as any expression made up of a variable and some constants, and Euler,[17] somewhat later, regarded a function as any equation or formula involving variables and constants. This latter idea is the notion of a function formed by most students of elementary mathematics courses. The Euler concept remained unchanged until Joseph Fourier (1768–1830) was led, in his investigations of heat flow, to consider so-called trigonometric series.[18] These series involve a more general type of relationship between variables than had previously been studied, and, in an attempt to furnish a definition of function broad enough to encompass such relationships, Lejeune Dirichlet (1805–1859)[19] arrived at the following formulation: A **variable** is a symbol that represents any one of a set of numbers; if two variables x and y are so related that whenever a value is assigned to x there is automatically assigned, by some rule or correspondence, a value to y, then we say y is a (single-valued) **function** of x. The variable x, to which values are assigned at will, is called the **independent variable**, and the variable y, whose values depend upon those of x, is called the **dependent variable**. The permissible values that x may assume constitute the **domain of definition** of the function, and the values taken on by y constitute the **range of values** of the function.
>
> The student of mathematics usually meets the Dirichlet definition of function in his introductory course in calculus. The definition is a very broad one and does not

14. H. Eves, *An Introduction to the History of Mathematics,* 5th Edition, Saunders College Publishing, 1983, p. 462.
15. Gottfried Wilhelm Leibniz (1646–1716), one of the two principal inventors of the Calculus. The other was Isaac Newton (1642–1727). Many of Leibniz's and Newton's important contributions will appear later in this course.
16. 1667–1748; a follower of Leibniz who greatly extended the power and applicability of calculus.
17. Leonhard Euler (1707–1783), a student of Johann Bernoulli, described by Eves as "far and away the most prolific writer in the history of [mathematics]" (Eves, op. cit., p. 328). The work of Euler will play a very important role throughout this course.
18. We will touch on these ideas of Fourier in the second semester of this course. He wrote: "The deep study of nature is the most fruitful source of mathematical discovery."
19. A student of and eventually successor at Göttingen University to the great genius Carl Friedrich Gauss (1777–1855). Some of the lesser accomplishments of Gauss (mostly from his youth) also will appear in this course.

imply anything regarding the possibility of expressing the relationship between x and y by some kind of analytic expression; it stresses the basic idea of a relationship between two sets of numbers.

So, you see, our purpose so far in this chapter has been to induce you to move from an eighteenth-century understanding of function (that of Euler) to a nineteenth-century understanding (that of Dirichlet) — but with some sense of purpose for doing so.

1.7 Functions as Objects

Doing Versus Being

So far we have stressed the ***process*** interpretation of function: A function is a rule for *doing something* to each number in a certain set of numbers (the domain) to get some particular result:

- Put in an x, get out a corresponding y.
- For each time t, find a distance s.
- Process each allowable input to produce the corresponding output.
- For each entry in column 1, find the corresponding entry in column 2.
- For each x-coordinate, go up the graph, then over to the y-axis to find the corresponding height.

Each such function rule contains one or more active verbs that collectively specify an action to be carried out on each number in the domain.

Now we turn our attention to functions as ***objects***. You should think of the grammarian's meaning of the word "object": the noun that receives the action of the verb. Functions often will be the objects to which we *do things*. That is, the operations of calculus will be ones we carry out on functions to produce other functions. Thus it is important to understand a function as a single object that can itself be the input to some process.

The object interpretation of function is important because the answers to our questions, the solutions to our problems, will often turn out to be functions or collections of functions, not just numbers or particular values of variables.

Whatever a function is as an object, it is obviously a more complicated object than a single number or a single variable or even a small collection of these things — the sorts of things you are used to seeing as answers. On the other hand, it is not a more complicated object than a graph. Indeed, many graphs are just pictorial representations of the very objects we have called functions. Thus, if you start from your mental image of *graph* as an object, you won't be too far away from having a mental image of *function* as an object. The hard part will be to connect the geometric image with algebraic properties of functions.

Keep in mind the characteristic that separates functions from relations in general: the fact that each first coordinate is paired with exactly one second coordinate. We can use this characteristic property to decide whether a particular graph is or is not a representation of a function.

Exploration Activity 1

(a) For each of the graphs in Figure 1.10, decide whether the graph is or is not the graph of a function. If it is not, state explicitly what keeps it from being the graph of a function.

(b) In one complete sentence, state in your own words a simple geometric way to decide whether a graph is or is not the graph of a function.

Figure 1.10 Which graphs represent functions?

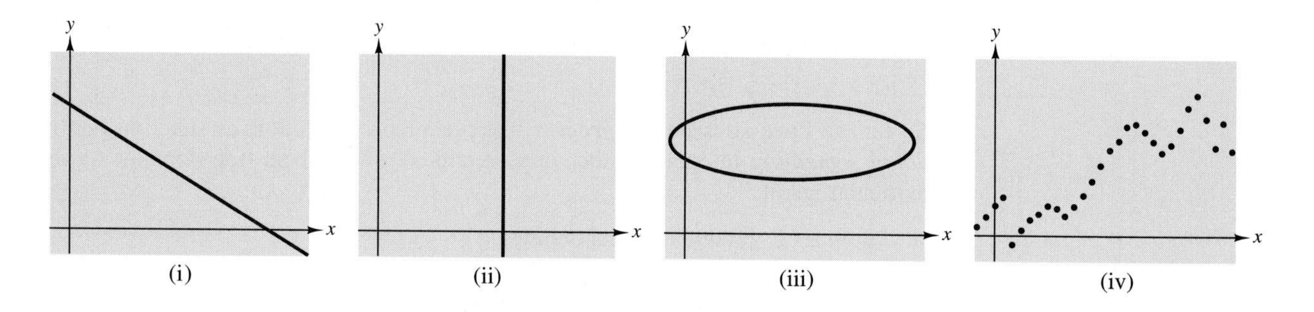

In these examples, (i) and (iv) represent functions, while (ii) and (iii) do not. In (ii), the graph is a vertical line. Thus every y-value is associated with the same x-value. In order to have a function, every x-value must be associated with a unique y-value. In (iii), we have a similar problem. Any vertical line drawn through the main part of the ellipse will intersect the curve twice. Again, we have x-values associated with more than one y-value. A simple geometric test to decide whether a graph represents a function is

Every vertical line may intersect the graph at most once.

In order to discuss algebraic properties of the objects called functions, we need some notation. It is easy to talk about processes in words, but it gets very awkward to do algebraic operations with words alone. Thus we abbreviate words and phrases to single symbols. In a given discussion, each function is usually abbreviated to a single symbol, such as f (for function) or g (next letter after f) or ϕ (phi, the closest thing to f in the Greek alphabet). The independent (input) and dependent (output) variables for a given function are also abbreviated to single characters, such as t for time and s for position.

Using such a shorthand, we abbreviate the entire sentence

The position of a falling body is a function of the time it has been falling

to

$$s = f(t).$$

The input to a function is inserted in parentheses following the function name, and the combined symbol is read "f of t."

Additive Functions

Here's an important algebraic question about functions for which the answer has to be a whole class of functions: What functions are additive?

Definition A function is called **additive** if what it does (as a process) to the sum of any two inputs is the sum of the corresponding outputs. In symbols, a function f is additive if

$$f(a+b) = f(a) + f(b)$$

for every pair of numbers a and b in its domain.

Example 1 Consider the function whose rule is "multiply by 5." Is this function additive?

Solution In symbols, this function is defined by the formula

$$f(x) = 5x.$$

We calculate that

$$f(a+b) = 5(a+b) = 5a + 5b = f(a) + f(b),$$

so the "multiply by 5" function is indeed additive. ■

Here is a procedure for deciding whether a function is additive or not:

1. If you suspect the function is not additive, choose particular numbers a and b in the domain of the function, and calculate $f(a+b)$ and $f(a)+f(b)$. If these two numbers turn out to be different, you have shown that the function is not additive. If they turn out to be the same, try two other numbers a and b.
2. On the other hand, if you suspect the function is additive, try to show that by an algebraic calculation like the one in Example 1.

Exploration Activity 2

(a) For each of the following functions, decide whether the function is additive or not. Write a one-sentence reason for your conclusion.

(i) $f(x) = -2x$ (ii) $f(x) = -2x + 7$ (iii) $f(x) = x^2$

(iv) $f(x) = \dfrac{2}{x}$ (v) $f(x) = \sqrt{x}$ (vi) $f(x) = \log x$

(b) Describe the largest class of functions you can think of that you know for sure are all additive. How do you know for sure?

(c) Would you describe additive functions as "relatively common" or "relatively rare"?

There is only one additive function in this group. Functions (ii) through (vi) are not additive. For example, in (iii) we have

$$f(1+1) = f(2) = 2^2 = 4,$$

but

$$f(1) + f(1) = 1 + 1 = 2.$$

Similarly, in (v) we have

$$f(9+16) = f(25) = 5,$$

but

$$f(9) + f(16) = 3 + 4 = 7.$$

On the other hand, the function in (i) is additive. This function is a member of the class of functions f with descriptions of the form

$$f(x) = cx.$$

Any function in this class is additive. To see this, we calculate

$$f(a+b) = c(a+b) = ca + cb = f(a) + f(b).$$

As you have seen, additive functions are rather rare.

The additivity condition, $f(a+b) = f(a) + f(b)$, is an example of a **functional equation**, that is, an equation involving an unknown function that might or might not be satisfied by any particular function. Your work on Exploration Activity 2 was an attempt to separate functions that satisfy this equation from functions that don't.

Multiplicative Functions

Definition A function is called **multiplicative** if it satisfies the functional equation

$$f(ab) = f(a)f(b),$$

that is, if what it does to any product is to produce the product of its results on the individual factors.

Example 2 Is the "multiply by 5" function multiplicative?

Solution Let's try two numbers in its domain, say $a = 2$ and $b = 3$. Then $f(a) = 5 \cdot 2 = 10$, and $f(b) = 5 \cdot 3 = 15$, so $f(a)f(b) = 150$. On the other hand, $f(ab) = f(6) = 30$, so this function is definitely not multiplicative. ■

Exploration Activity 3

(a) For each of the following functions, decide whether the function is multiplicative or not. Write a one-sentence reason for your conclusion.

(i) $f(x) = -2x$ (ii) $f(x) = -2x + 7$ (iii) $f(x) = x^2$

(iv) $f(x) = \dfrac{2}{x}$ (v) $f(x) = \sqrt{x}$ (vi) $f(x) = \log x$

(b) Describe the largest class of functions you can think of that you know for sure are all multiplicative. How do you know for sure?

(c) Would you describe multiplicative functions as "relatively common" or "relatively rare"?

In this list, the functions described in (i), (ii), (iv), and (vi) are not multiplicative. For example, for (i) we have

$$f(1 \cdot 1) = f(1) = -2,$$

but

$$f(1) \cdot f(2) = 0 \cdot \log 2 = 0.$$

For (vi),

$$f(1 \cdot 2) = f(2) = \log 2,$$

but

$$f(1) \cdot f(1) = (-2) \cdot (-2) = 4.$$

On the other hand, the square and square root functions are both multiplicative. In fact, any power function f with a description of the form $f(x) = x^k$ is multiplicative. To see this, we calculate

$$f(a \cdot b) = (a \cdot b)^k = a^k \cdot b^k = f(a) \cdot f(b).$$

Again, multiplicative functions would seem to be rather rare.

Symmetries: Odd and Even Functions

Other important functional equations describe symmetries—of functions that happen to have symmetries.

Definitions A function has **even symmetry** (or is an **even function**) if it satisfies the functional equation

$$f(-a) = f(a).$$

Similarly, a function has **odd symmetry** (or is an **odd function**) if it satisfies

$$f(-a) = -f(a).$$

Example 3 What is the symmetry of the squaring function (see Figure 1.11),

$$f(x) = x^2?$$

Solution This function has even symmetry, because

$$f(-a) = (-a)^2 = a^2 = f(a).$$

Figure 1.11 The graph of $y = x^2$

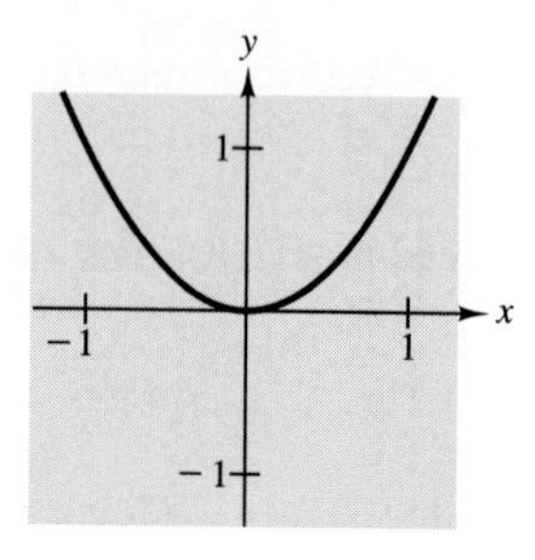

Example 4 What is the symmetry of the "multiply by 5" function (see Figure 1.12),

$$f(x) = 5x\,?$$

Solution This function has odd symmetry, because

$$f(-a) = 5(-a) = -5a = -f(a).$$

Power functions with even and odd powers provide additional simple examples — as well as a reason for the names "even" and "odd."

Figure 1.12 The graph of $y = 5x$

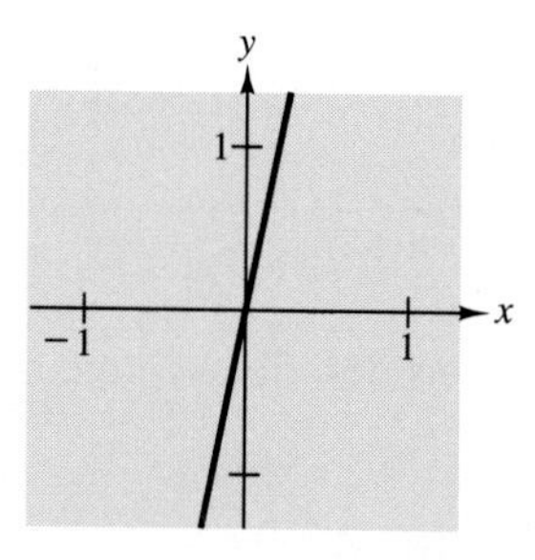

Example 5 We have just seen in Example 3 that the squaring function is even. In a similar manner, if

$$f(x) = x^3,$$

then

$$f(-a) = (-a)^3 = -a^3 = -f(a),$$

so the cubing function is odd (see Figure 1.13).

Figure 1.13 The graph of $y = x^3$

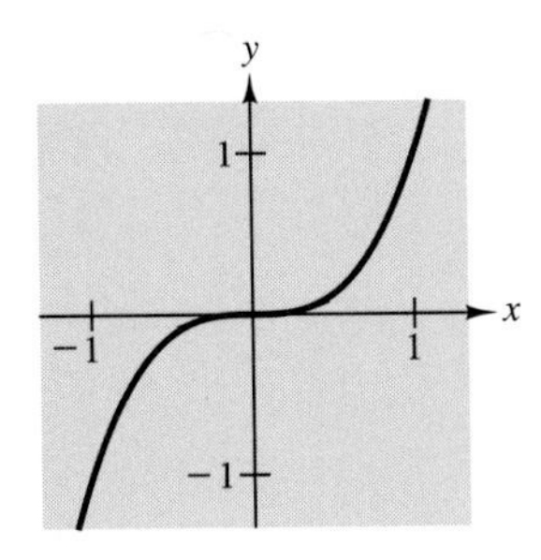

Exploration Activity 4

In Figure 1.14 we have (twice) sketched half the graph of a function, the half for positive values of the independent variable.

(a) In the left-hand graph, assume the function is *even*, and sketch the rest of the graph. (You may copy the page first if you don't want to draw in the book.) For a typical point (a, b) on the graph, we have indicated the locations of the points $(-a, b)$, $(a, -b)$, and $(-a, -b)$. Think about which of these points must lie on the graph of an even function. Apply the same reasoning to all the other points on the graph.

(b) In the right-hand graph, assume the function is *odd*, and sketch the rest of the graph.

Figure 1.14 Make the left graph even and the right graph odd

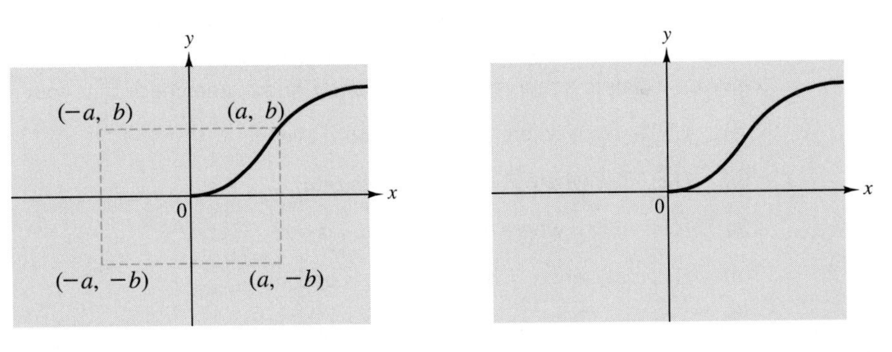

If the function graphed in Figure 1.14 is even, then the value assigned to $-a$ is the same as the value assigned to a; i.e., $(-a, b)$ is on the graph. This means that the portion of the graph on the left side of the y-axis may be obtained from the portion on the right by rotating around the y-axis. On the other hand, if the function is odd, then the value assigned to $-a$ is the negative of the value assigned to a; i.e., $(-a, -b)$ is on the graph. This means that the portion of the graph on the left side of the y-axis may be obtained from the portion on the right by rotating around the y-axis and then rotating about the x-axis (or vice versa). See Figures 1.15 and 1.16 for the even and odd cases, respectively.

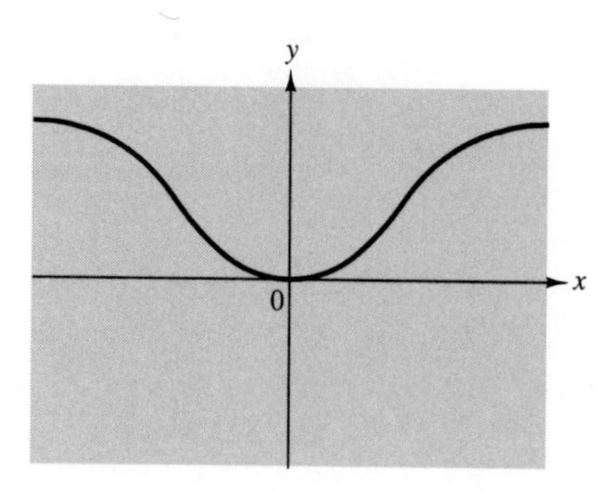

Figure 1.15 Even extension of the graph

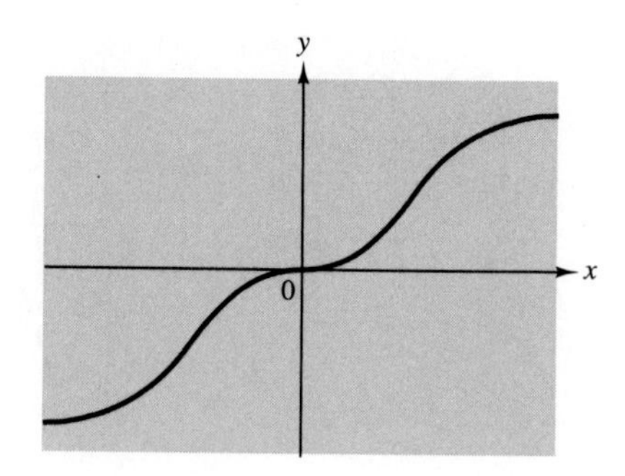

Figure 1.16 Odd extension of the graph

Checkpoint 1

For each of the following functions, decide whether the function is even, odd, or neither. Graph each of the functions to confirm your answers.

(a) $f(x) = x^3 + 4x$

(b) $f(x) = x^3 + 4x^2$

(c) $f(x) = |x|$

Checkpoint 2

For each of the following functions, decide whether the function is even, odd, or neither. For each case, give an example of appropriate functions f and g, and check that your answer is right for your example.

(a) $f(x) + g(x)$, where f and g are both even

(b) $f(x)\, g(x)$, where f and g are both odd

(c) $f(x) + g(x)$, where f is even and g is odd

(d) $f(x)\, g(x)$, where f is even and g is odd

The Algebra of Functions

The function descriptions in Checkpoints 1 and 2 suggest some possibilities for building functions from other functions. Indeed, there is a natural way to apply the operations of algebra (addition, subtraction, multiplication, division, extraction of roots, absolute value, and so on) to function-objects to produce new functions.

Example 6 We can describe the function $f(x) = x^3 + 4x$ [see part (a) of Checkpoint 1] as the **sum** of the cubing function $c(x) = x^3$ and the "multiply by 4" function $m(x) = 4x$. In Figure 1.17 we show the graphs of c, m, and f. Note that y-coordinates on the graph of f are sums of the corresponding y-coordinates on the graphs of c and m. ■

Figure 1.17 The function $f(x) = x^3 + 4x$ as a sum of simpler functions

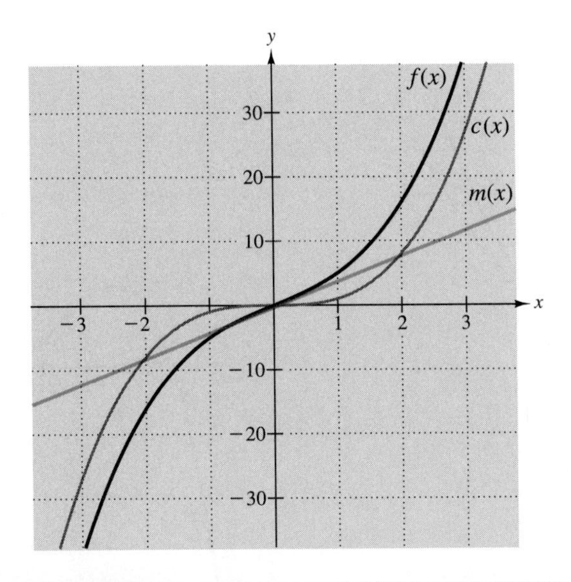

Definitions The function s defined by $s(x) = f(x) + g(x)$ is the **sum** of the functions $f(x)$ and $g(x)$. The function p defined by $p(x) = f(x)g(x)$ is the **product** of the functions $f(x)$ and $g(x)$. The function d defined by $d(x) = f(x) - g(x)$ is the **difference** of the functions $f(x)$ and $g(x)$. The function q defined by $q(x) = f(x)/g(x)$ is the **quotient** of the functions $f(x)$ and $g(x)$. Similar definitions apply for other operations on functions, such as **negative, square root**, and **absolute value**.

For example, the functions described in Checkpoint 2 are sums and products of the (unspecified) functions f and g.

Exploration Activity 5

Express the function $f(x) = x^3 + 4x^2$ [see Checkpoint 1(b)] in terms of operations on simpler functions. Is there more than one way to do this? Use your calculator to graph your simpler functions and f as a visual check of your answer.

One way to express f in terms of operations on simpler functions is to write it as the sum of the cubing function $c(x) = x^3$ and the quadratic function $h(x) = 4x^2$. (See Figure 1.18.)

Another possibility is to write f as a product:

$$f(x) = s(x) \cdot g(x),$$

where $s(x) = x^2$ and $g(x) = x + 4$. (See Figure 1.19.)

Figure 1.18 The function $f(x) = x^3 + 4x^2$ as a sum of simpler functions

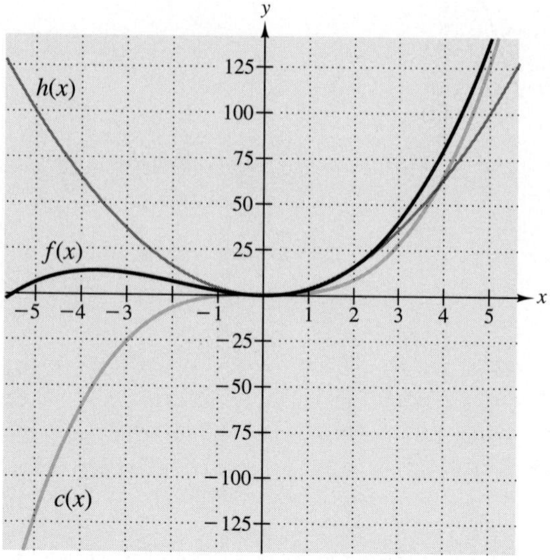

Figure 1.19 The function $f(x) = x^3 + 4x^2$ as a product of simpler functions

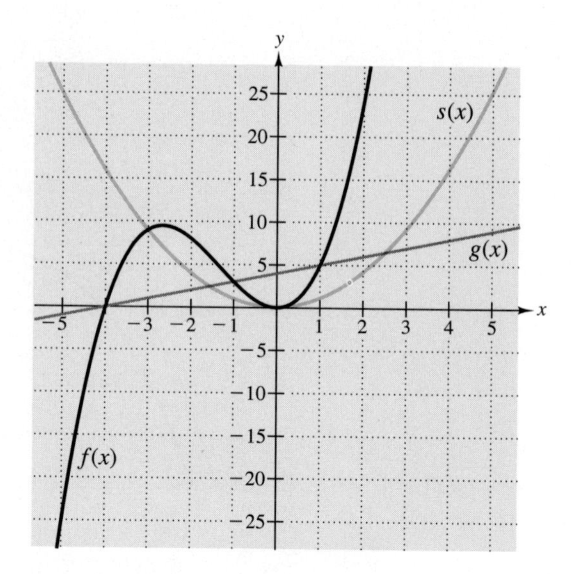

Combining functions with algebraic operations is not just an abstract game—it's a powerful conceptual tool for expressing relationships.

Example 7 Three functions are represented graphically in Figure 1.20: exports, imports, and trade balance, each as a function of time. The figure is adapted from a newspaper article reporting a monthly trade deficit that was the smallest in 5 years. What algebraic operation between functions is represented by this picture?

Figure 1.20 The merchandise trade balance over a decade[20]

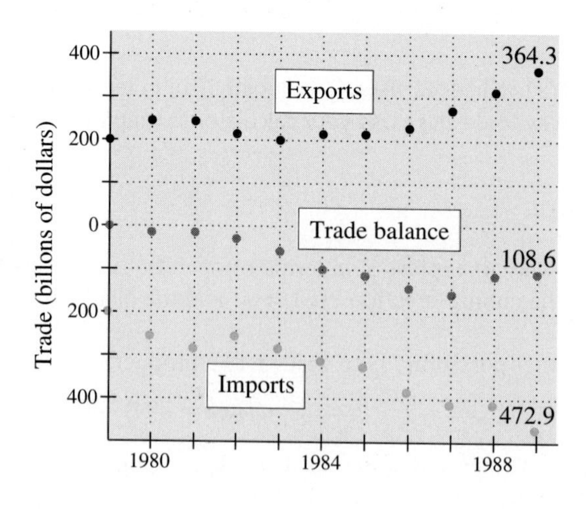

20. Source: Associated Press, reported in *The News & Observer*, February 17, 1990.

Solution Notice the artist's device of using a positive vertical scale in both directions. If we attach negative signs to the numbers below zero, the bottom graph represents the negative of the imports function. With or without that modification, Figure 1.20 represents graphically the relationship

$$\text{Exports} - \text{Imports} = \text{Trade balance.}$$

In any given year, this equation states something obvious about subtracting two numbers to get a third. For example,

$$\begin{aligned}\text{Exports}\,(1989) - \text{Imports}\,(1989) &= 364.3 - 472.9\\ &= -108.6 = \text{Trade balance}\,(1989).\end{aligned}$$

But by applying the subtraction to *functions* rather than numbers, we can express the entire relationship in Figure 1.20 by a single, simple formula. ■

What is a function? As we have seen, it's a pairing of the values of one variable with values of another variable in such a way that each value of the first variable is paired with exactly one value of the second variable. We express that idea in terms of inputs and outputs, in terms of first column and second column, in terms of x-coordinates and y-coordinates, in terms of domain values and corresponding function values, and in many other ways.

The act of pairing is a *process*, and it is often important to view functions dynamically, that is, as processes doing something to the input values to produce the output values. On the other hand, a function (that is, the collection of all the paired values) is an *object*, and thus we can do things to functions as well. In particular, functions can be treated as algebraic objects, albeit more complicated ones than the familiar numbers, literal constants, and variables. We can combine functions by the algebraic operations of addition, subtraction, multiplication, and division. We can form new functions from old ones by taking negatives, reciprocals, and square roots.

Answers to Checkpoints

1. (a) Odd. (b) Neither odd nor even. (c) Even.

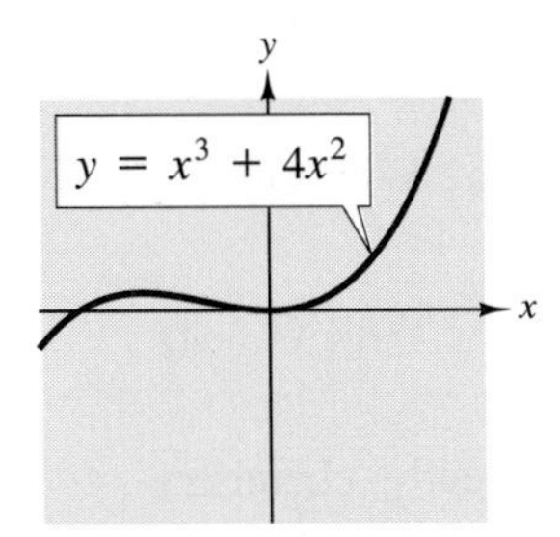

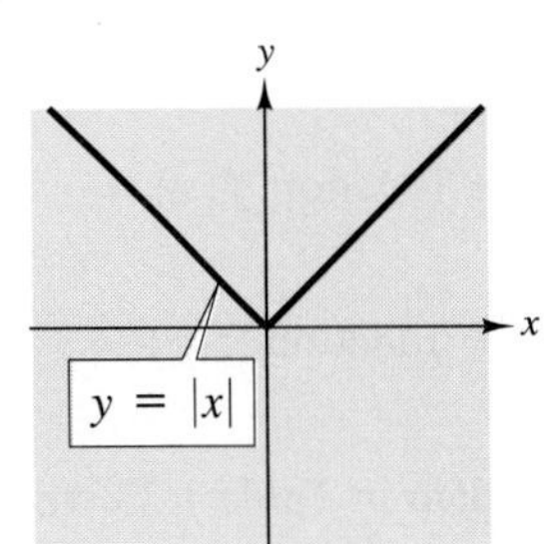

2. (a) Even. (b) Even.

   (c) Neither, e.g., $f(x) = x,\ g(x) = x^2$. (d) Odd.

Exercises 1.7

For each of the following functions, express the function in terms of operations on simpler functions.

1. $f(x) = x^3 - 4x^2$
2. $g(x) = -x^3 + 4x^2$
3. $h(x) = |x^3 - 4x^2|$
4. $j(x) = \sqrt{x^3 + 4x^2}$
5. $\phi(t) = t^{3/2}$
6. $\psi(t) = |t^{3/2}|$
7. $\lambda(t) = |t|^{3/2}$
8. $\mu(t) = |t|^{-3/2}$

For each of the following functions, graph the function along with the simpler functions that were your answer to the corresponding Exercise 1–8.

9. $f(x) = x^3 - 4x^2$
10. $g(x) = -x^3 + 4x^2$
11. $h(x) = |x^3 - 4x^2|$
12. $j(x) = \sqrt{x^3 + 4x^2}$
13. $\phi(t) = t^{3/2}$
14. $\psi(t) = |t^{3/2}|$
15. $\lambda(t) = |t|^{3/2}$
16. $\mu(t) = |t|^{-3/2}$
17. Is the function $f(x) = x^2 + 7$ additive? Why or why not?
18. Is the function $f(x) = x^2 + 7$ multiplicative? Why or why not?
19. Is the function $f(x) = x^2 + 7$ even? Why or why not?
20. Is the function $f(x) = x^2 + 7$ odd? Why or why not?
21. The data in Table 1.23 were collected by chemistry students who varied the temperature of a gas in a closed container and recorded the pressure exerted by the gas.
    (a) Make a scatter plot of the data in Table 1.23. (You may have done this already.)
    (b) When volume is held constant, is pressure related linearly to temperature? If so, how do you interpret the slope and y-intercept of the line?
    (c) What change in pressure is brought about by a one-degree change in temperature?
    (d) What change in temperature would cause a one-millimeter change in pressure?

Table 1.23 Temperature and pressure of a gas in a closed container

Temperature (°K)	*Pressure (mm Hg)*
263	752
268	755
278	777
293	811
298	834
303	840
308	854
318	892
323	906

22. Figure 1.21 shows the market shares of U.S., Japanese, and European companies for wafer steppers, devices that are important for manufacturing semiconductors.
    (a) The figure shows the graphs of three different functions. What is the relationship among these three functions?
    (b) What conclusions can you draw from the figure?

Figure 1.21 Market shares for wafer steppers: X marks the spot[21]

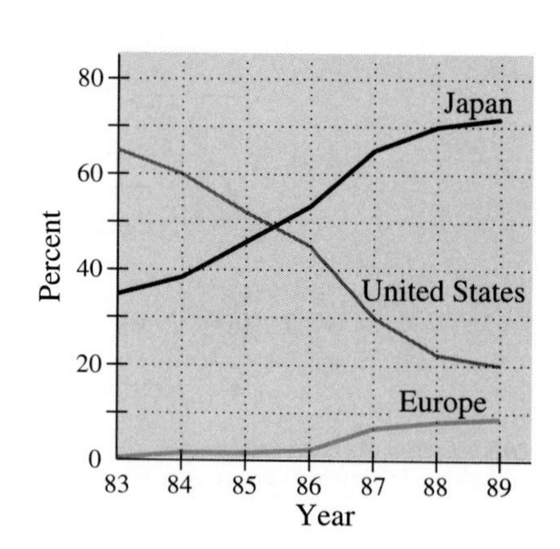

23. Table 1.24 shows the number of Democratic governors in office each year from 1980 through 1993. All the other governors were Republicans until 1991, when two independent governors were inaugurated.
   (a) Let $D(t)$ be the function of time represented by Table 1.24, and let $R(t)$ be the function whose value at time t is the number of Republican governors. What is the relationship between D and R, at least up to 1991?
   (b) What is the domain of each of the functions D and R?
   (c) Make a table of values of $R(t)$.
   (d) Plot both functions, $D(t)$ and $R(t)$, on the same coordinate axes. What do you observe about these graphs, at least up to 1991?

Table 1.24 Numbers of Democratic governors[22]

Year	*Number*
1980	31
1981	27
1982	27
1983	34
1984	35
1985	34
1986	34
1987	26
1988	26
1989	28
1990	29
1991	27
1992	28
1993	30

24. Table 1.25 shows the revenues and net income of AMR Corporation, parent company of American Airlines, over the period 1985–1992.
   (a) Plot each of these functions of time on separate coordinate axes.
   (b) What would you predict about AMR's revenues in 1993?
   (c) What were AMR's total revenues from 1985 through 1992?
   (d) What was AMR's net income over the same period of time?
   (e) Make up another question that can be answered from these data, and answer it.

Table 1.25 Revenues and net income of AMR Corp[23]

Year	*Revenues ($ billion)*	*Net Income ($ million)*
1985	6.1	346
1986	5.9	275
1987	7.2	201
1988	8.7	440
1989	10.2	415
1990	11.6	−50
1991	13.0	−220
1992	14.4	−935

25. Table 1.26 shows numbers of highway fatalities and numbers of miles driven in North Carolina from 1986 to 1992.

Table 1.26 Highway deaths and miles driven in North Carolina[24]

Year	*Deaths*	*Miles Driven (millions)*
1986	1645	52.9
1987	1595	54.6
1988	1580	57.7
1989	1460	60.8
1990	1375	62.7
1991	1351	64.9
1992	1231	NA

   (a) Let $D(t)$ and $M(t)$ be the functions whose values are given in the second and third columns of Table 1.26, each as a function of time represented by the first column. Plot the functions D and M on separate coordinate systems.
   (b) Estimate the missing entry in the "Miles Driven" column, and add that point to your plot.
   (c) Let $P(t)$ stand for the number of deaths per million miles. How is P related to D and M? Add a column to the table for $P(t)$.
   (d) Plot the function $P(t)$. What conclusions do you draw?
   (e) Make up another question that can be answered from these data, and answer it.
   (f) What do think might be the cause(s) of declining highway deaths in North Carolina?

26. Table 1.27 shows the total number of people in the armed forces and the percentage of women, over the period 1984–1992.
    (a) Let $T(t)$ be the total number function and $P(t)$ the percentage of women function. Plot each of these functions of time on separate coordinate axes.
    (b) Let $W(t)$ be the number of women in the armed forces at time t, and let $M(t)$ be the number of men in the armed forces. How are the functions W and M related to the functions T and P?
    (c) Add columns to Table 1.27 for the functions W and M, and fill in the numbers.
    (d) Plot the functions W and M on separate coordinate axes.
    (e) In what year between 1984 and 1992 were the largest number of women in the armed forces? In what year were the largest number of men?
    (f) Make up another question that can be answered from these data, and answer it.

Table 1.27 Total number in the armed forces and percentage of women[25]

Year	*Total Number (millions)*	*Women (%)*
1984	2.14	9.5
1985	2.15	9.8
1986	2.16	10.1
1987	2.17	10.2
1988	2.13	10.4
1989	2.12	10.7
1990	2.04	11.0
1991	2.03	10.9
1992	1.76	11.6

27. Table 1.28 shows the dollar amounts of trade between the United States and Mexico over a recent seven-year period.
    (a) Make a scatter plot of the two functions of time represented in Table 1.28.
    (b) Add to your graph a plot of the sum of these two functions. Interpret the sum function in terms of trade.
    (c) Carefully draw a model linear function that approximately fits each of the three data functions. Find equations for each of your three lines.

Table 1.28 Trade between the United States and Mexico ($ billion)[26]

Year	*Imports from Mexico ($B)*	*Exports to Mexico ($B)*
1987	20.2	14.6
1988	23.2	20.6
1989	27.2	24.9
1990	30.2	28.4
1991	31.2	33.3
1992	35.2	40.6
1993	39.3	41.6

    (d) Is your linear function for the sum equal to the sum of the other two linear functions? Explain.
    (e) Each of your lines has a slope. What do those slopes mean in terms of trade?
    (f) Two of your three lines cross within the time frame shown. What is the significance of the crossing point in terms of trade?
    (g) If the trends represented by your lines continued for another seven years, what would you predict the trade figures to be in the year 2000?
    (h) What happened in 1993 that might alter these trade trends? In what direction would you expect each of the trends to change? Explain.

<u>28</u>. Here's another important functional equation:

$$f(ab) = f(a) + f(b).$$

Functions that satisfy this equation **turn products into sums**, a useful thing to do, because sums are usually easier to work with than products. From your previous experience with algebra, you should know at least one such function. What is it? [*Hint*: Look again at Exploration Activity 3.] Do you know more than one such function?

29. What about functions that **turn sums into products**:

$$f(a+b) = f(a)f(b)?$$

Do you know any of those? [*Hint*: The operation of turning sums into products undoes the operation of turning products into sums. You know a function that turns products into sums. What undoes the effect of that function?]

30. (a) Define what it would mean for a function to be **subtractive**. Can you find any functions that satisfy your definition?

    (b) Define what it would mean for a function to be **divisive**. Can you find any functions that satisfy your definition?

31. Can you think of a function that is both additive and multiplicative? Can you think of a second one?

---

21. Source: Associated Press, reported in *The News & Observer*, December 17, 1989.
22. Source: National Governors' Association, cited in *USA Today*, November 16, 1992.
23. Source: AMR Corp., cited in *The News & Observer*, January 21, 1993.
24. Source: N.C. Highway Patrol, cited in *The News & Observer*, January 1, 1993.
25. Source: Department of Defense, cited in *The News & Observer*, April 11, 1993.
26. Source: U.S. Department of Commerce, as reported by *USA Today*, September 30, 1994.

1.8 Inverse Functions

In the preceding section we stressed the importance of function-as-object because we will often do operations to functions. Here we single out one of those operations for special attention, namely the operation of inverting or undoing a function. As with the other concepts in this chapter, this will be a frequently recurring theme throughout the course. We begin with an example that you may find of personal interest, namely, car payments.

The monthly payment on a new car depends on four factors: (1) the price of the car, (2) the size of the down payment, (3) the annual interest rate, and (4) the term of the loan. These quantities are related (for reasons we will see later) by the formula

$$p = \frac{(P - D)\, r\, (1 + r)^n}{(1 + r)^n - 1},$$

where

p is the monthly payment,

P is the price of the car,

D is the down payment,

r is the monthly interest rate (as a decimal fraction), and

n is the number of months required to pay back the loan.

The interest rate and the duration of the loan are usually determined by the seller, and the down payment is agreed on by the buyer and seller. Once these three quantities are set, the monthly payment p and the price P of the car are the only true variables in the formula. The price of the car determines the monthly payment, so price is usually thought of as the independent variable and the monthly payment is a function of the price.

Exploration Activity 1

(a) Suppose you want to buy a $15,000 car, and you have $2400 for a down payment. The annual interest rate, because the dealer is trying to move cars off the lot, is a very favorable 3.9% (equivalent to 0.325% per month), and the term of the loan will be 48 months. What does the dealer tell you your monthly payment will be?

(b) Your income and your other expenses limit you to a monthly payment of $230. Can you afford the $15,000 car? How expensive a car can you afford?

Using the payment formula and her handy calculator, the dealer determines that your monthly payment will be $283.93. Thus the pair of numbers (15000, 283.93) is one of the pairs

(price of car, monthly payment)

that make up the function defined by the payment formula when $D = 2400$, $r = 0.00325$, and $n = 48$.

Unfortunately, you cannot afford a monthly payment of $283.93, so you will have to shop around to find a car you can afford. In this situation, the monthly payment is no longer the independent variable. Now the price you can afford depends on the monthly payment you can afford, and the roles of independent and dependent variables have been reversed. In effect, there is a new function involved in this problem, namely, the pairing relation

(monthly payment, price of car).

This second function has a special—and obvious—relationship to the first function. The two functions are said to be **inverses** of each other.

When you substitute $p = 230$ in the payment formula, along with $D = 2400$, $r = 0.00325$, and $n = 48$, you can then solve for P to find out what price car you can afford. You should have found that the most expensive car you can afford costs about $12,600.

Exploration Activity 2

(a) *Don't* substitute $p = 230$ in the formula, or the other known values. Solve for P to find a *formula* for the inverse function, i.e., for P as a function of p (with D, r, and n still assumed to be constants).

(b) When you substitute the known values for D, r, and n in the original formula, the resulting formula expresses p as a *linear* function of P. What is the slope of that linear function?

(c) In part (a), you found that P is also a linear function of p. When you substitute the known values for D, r, and n in part (b), what slope do you get?

(d) What is the relationship between the two slopes in parts (b) and (c)?

The result of solving the payment formula

$$p = \frac{(P - D)\, r\, (1 + r)^n}{(1 + r)^n - 1},$$

for price P is

$$P = D + \frac{[(1 + r)^n - 1]p}{r(1 + r)^n}.$$

In particular, when we substitute $D = 2400$, $r = 0.00325$, and $n = 48$ in both formulas, we get

$$p = 0.0225P - 54.1$$

for payment as a function of price and

$$P = 2400 + 44.4p$$

for price as a function of payment. Except for our rounding, the slopes 0.0225 and 44.4 are reciprocals of each other.

The fact that these inverse linear functions have reciprocal slopes is not an accident. In fact, for any monthly interest rate r, the slope of the payment function is

$$\frac{r\,(1 + r)^n}{(1 + r)^n - 1},$$

and the slope of the price function is

$$\frac{(1 + r)^n - 1}{r(1 + r)^n},$$

clearly reciprocals of each other.

Tabular Representation of Inverse Functions

A calculator is a useful tool for exploring the concept of inverse functions. Suppose you are doing some calculation, and the number 37.958127 is on your calculator display. If you were to accidentally press the square root key, the new display would be 6.16101671804. How could you retrieve your original display? It should be clear that pressing the x^2 key will achieve this result, since squaring undoes the effect of taking a square root. This sequence of events can be understood in terms of inverses. The square root key acts as a function that pairs the input 37.958127 with the output 6.16101671804. The number 6.16101671804 is then the input to a second function, which is the inverse of the first. This second function squares its input, and so 6.16101671804 is paired with 37.958127. Check the calculations in this paragraph on your own calculator—the keystrokes on your calculator may be slightly different.

If the pairs of numbers that constitute a function are known explicitly—for example, by being listed in a table—it is very easy to display explicitly the inverse function: Just write all the pairs in reverse order—or, equivalently, interchange the columns of the table.

Exploration Activity 3

We explore the tabular concept of the previous paragraph with the drug-sales function that you studied in Section 1.2. In Figure 1.22 we repeat Figure 1.6.

Figure 1.22 Sales by the pharmaceutical industry in North Carolina

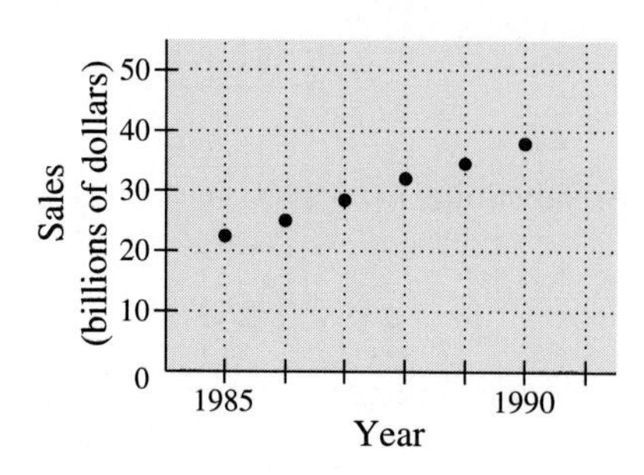

(a) Make a table of the data given graphically in Figure 1.22, and find the inverse function from your table.

(b) In what year did the sales pass $30 billion?

If you use nearest-integer values for sales, your table might look like Table 1.29. To answer a question about the year for a given sales figure, you could read the table backwards, or you could actually construct the inverse function (Table 1.30). Either way, it is clear that sales passed $30 billion in 1987. Of course, this was just as clear from the graph in Figure 1.22.

Table 1.29 Pharmaceutical sales in given years

x *(Year)*	y *(Sales)*
1985	22
1986	25
1987	28
1988	32
1989	35
1990	38

Table 1.30 Years for given pharmaceutical sales

y *(Sales)*	x *(Year)*
22	1985
25	1986
28	1987
32	1988
35	1989
38	1990

Graphical Representations of Inverse Functions

The graphs of a function and its inverse have a special relationship: For each point (a, b) on the graph of f, there is a point (b, a) on the graph of the inverse of f. The points (a, b) and (b, a) are mirror images in the line $y = x$ (see Figure 1.23). Thus the graph of f can be reflected about the line $y = x$ to obtain the graph of the inverse of f. Notice that this reflection takes the y-axis to the x-axis and vice versa; that is, it interchanges x- and y-coordinates.

Figure 1.23 Mirroring in the line $y = x$

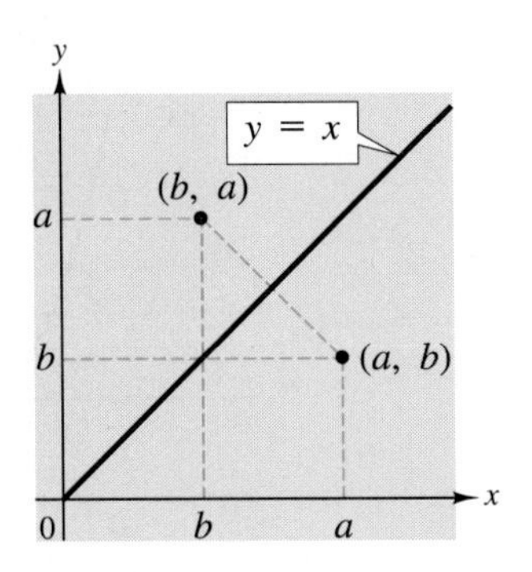

We illustrate this mirroring with an entire figure (not just a point) by repeating Figure 1.22 along with its reflection in Figures 1.24 and 1.25. Check that plotted points in the reflected graph match up with the entries in Table 1.30.

Figure 1.24 Pharmaceutical sales

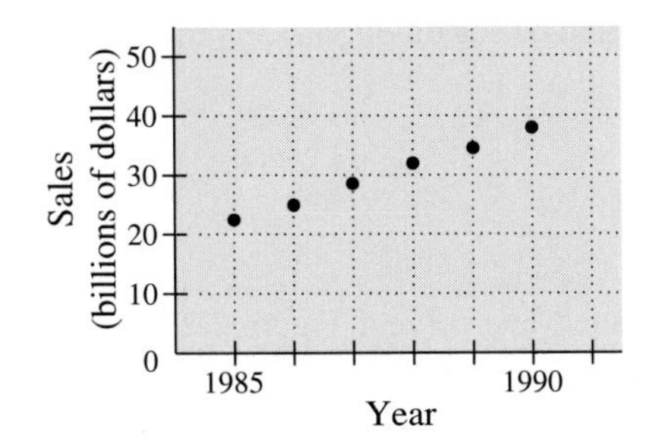

Figure 1.25 Reflection of Figure 1.24 about the line $y = x$

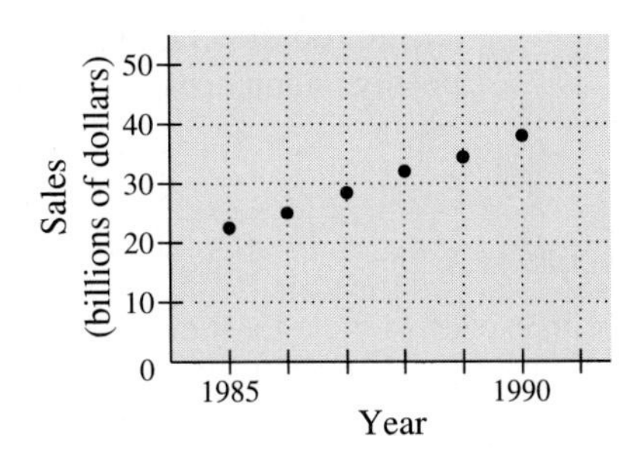

Checkpoint 1

In Exploration Activity 1 of Section 1.2 you calculated the slope of the line representing the drug-sales function in Figure 1.24. (If you don't have a record of that, do it again.) What is the slope of the inverse function? How are these two slopes related?

Following the same geometric idea for inverting functions, we show in Figures 1.26 and 1.27 the graphs of $y^2 = x$ and $y = x^2$. Each of these equations is obtained from the other by interchanging x and y. Each of the graphs is obtained from the other by reflecting in the line $y = x$, which amounts to the same thing.

Figure 1.26 Positive and negative square roots

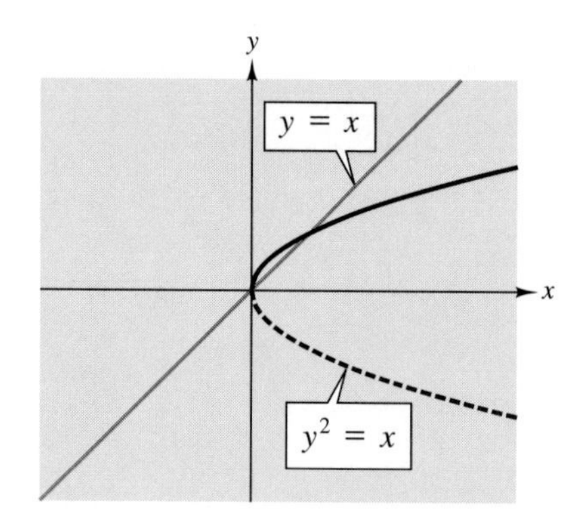

Figure 1.27 Squaring positive and negative numbers (mirroring Figure 1.26 in the line $y = x$)

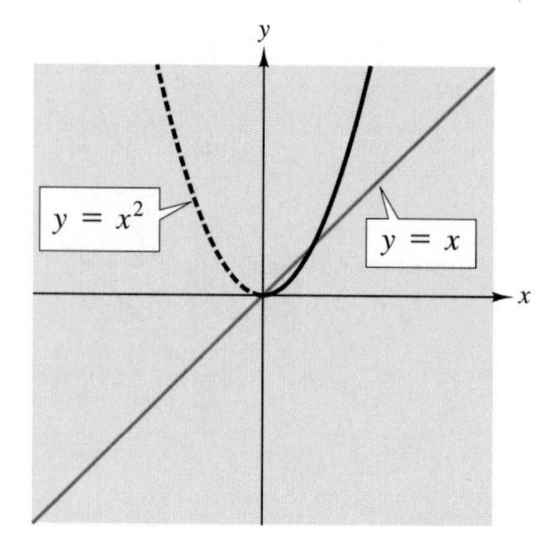

However, these figures reveal a new complication, suggested by the dashed and solid parts of the two graphs. The upper (solid) part of the graph of $y^2 = x$ is also the graph of the (positive) square root function, $y = \sqrt{x}$. The inverse of this function is *not* the entire squaring function but only the part shown with a solid curve in Figure 1.27. That is, the inverse of the square root function is the function defined by $y = x^2,\ x \geq 0$. Since the square root function has only positive numbers (and 0) for its *outputs*, its inverse has only positive numbers (and 0) for its *inputs*.

Checkpoint 2

Write an explicit formula for the function represented by the dashed-line curve in Figure 1.26. What is the inverse of this function? Where does it appear in Figure 1.27?

Suppose we focus now on the squaring function in Figure 1.27. The domain for this function consists of all real numbers x—positive, negative, and zero. That is, the entire graph in Figure 1.27, solid and dashed, is the graph of the squaring function. What's the inverse of this function? Well, if we invert all the pairs of numbers that represent points on the graph, we get all the points of the graph in Figure 1.26—solid and dashed. That's a perfectly good relation, but it's not a function. (Why? Refer back to your criterion in Exploration Activity 1 of Section 1.6.) That's the new complication: *The inverse of a function may not be a function!*

However, we can usually produce a function whose inverse is also a function if we restrict the domain in some appropriate way—such as considering the squaring function only for positive numbers. Your calculator already knows this, as we will see when we study its various inverse functions.

Symbolic Representation of Inverse Functions

Inversion of functions has a standard notation, one that you have probably already used in connection with inverse trigonometric and inverse logarithmic functions: The symbol f^{-1} represents the inverse of the function f. We hasten to add that you must interpret this notation with care: It looks like it ought to mean $1/f$, but it clearly does not. After all, the squaring and the square root functions are not reciprocals of each other.

Exploration Activity 4

(a) If $f(x) = x^2$, what function is $1/f$?

(b) If $f(x) = \sqrt{x}$, what function is $1/f$?

(c) If $f(x) = x^3$, what function is $1/f$?

(d) If $f(x) = x^3$, what function is f^{-1}?

(e) Let $f(x) = x^3$. Draw the graphs of f, $1/f$, and f^{-1}. Explain the features of the three graphs that tell you which goes with each function.

The reciprocals in the first three parts of Exploration Activity 4 are, respectively, x^{-2}, $x^{-1/2}$, and x^{-3}. In contrast to the last of these, the inverse of the cubing function is the cube root function, $x^{1/3}$. We show the graphs of the cube, reciprocal cube, and cube root functions in Figure 1.28. Observe that the cube and cube root functions are mirror images of each other in the line $y = x$. On the other hand, the reciprocal cube function does not have the shape of either of the others. Its values are big where x^3 is small, and vice versa.

Figure 1.28 Graphs of $y = x^3$ (black), $y = \sqrt[3]{x}$ (grey), and $y = x^{-3}$ (colored)

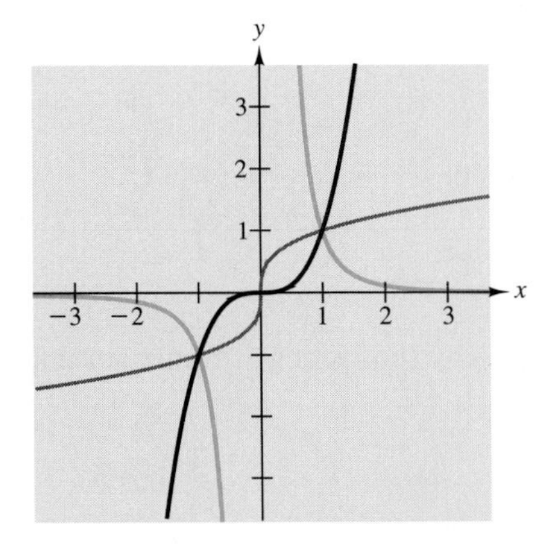

Example 8 Figure 1.29 shows the graph of $f(x) = 1/(2 - x)$. What is the inverse relation? Is it a function? What is a formula for this function?

Figure 1.29 Graph of $f(x) = \dfrac{1}{2-x}$

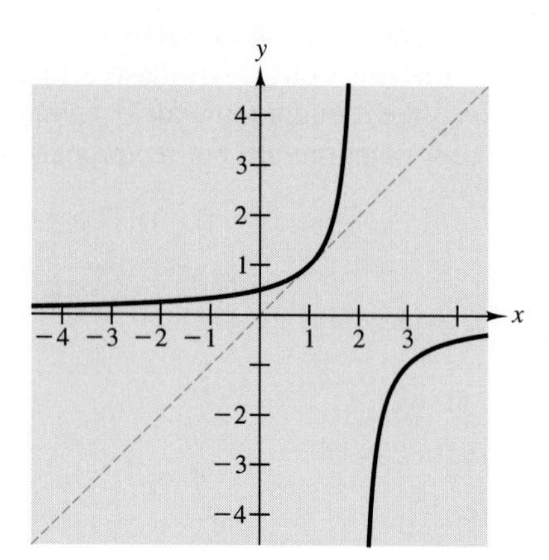

Solution In principle, all we have to do is flip the graph of $f(x)$ over the graph of $y = x$ (dotted line in Figure 1.29), and we will have a graph of the inverse relation. This will help you answer the first two questions, but it won't help with the third. For that, we need to do some algebra.

First, the graph of f is also the graph of the equation $y = 1/(2 - x)$. If we interchange the variables x and y in this equation, we get

$$x = \frac{1}{2-y},$$

which is an equation that defines the inverse relation—but it does not tell us how to express y as a function of x. For that, we need to solve the equation for y. We begin by taking reciprocals on both sides:

$$\frac{1}{x} = 2 - y.$$

Then we add y to both sides:

$$y + \frac{1}{x} = 2.$$

Now we may solve for y by subtracting $\frac{1}{x}$ from both sides:

$$y = 2 - \frac{1}{x}.$$

This last equation shows us that the inverse relation is indeed a function—each x determines a unique y—and it also gives us a formula for f^{-1}. ■

We graph this inverse function in Figure 1.30. In case there is any doubt about whether this is the reflection of the graph of f around the line $y = x$, in Figure 1.31 we show both graphs, f with a black line and f^{-1} with a grey line.

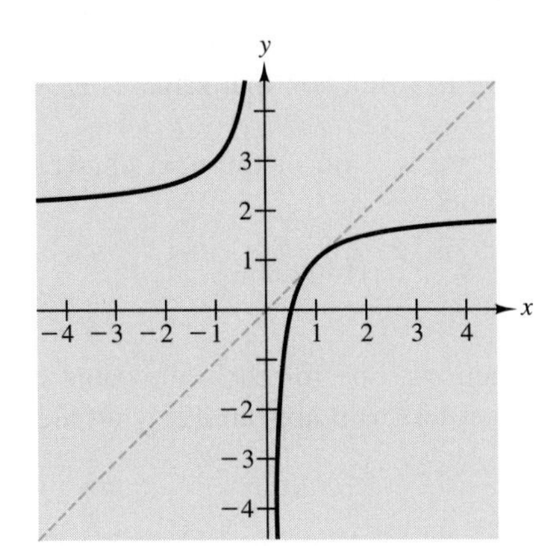

Figure 1.30 Graph of $f^{-1}(x) = 2 - \dfrac{1}{x}$

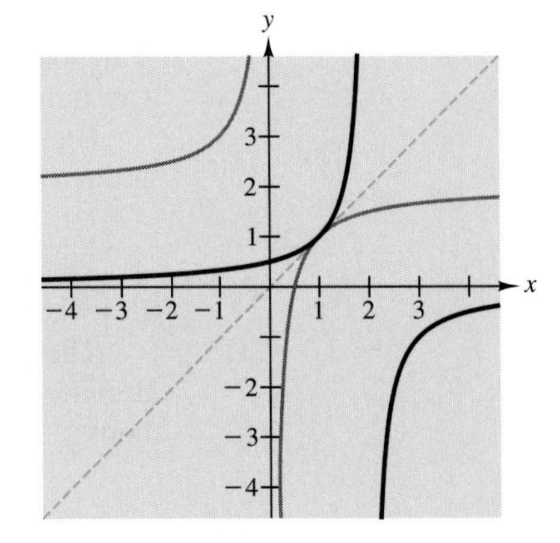

Figure 1.31 Combined graphs of f and f^{-1}

Checkpoint 3

(a) What is the domain of the function f in Example 8?

(b) What is the domain of f^{-1}?

(c) How is the domain of f^{-1} related to the set of values of f?

(d) How is the domain of f related to the set of values of f^{-1}?

Checkpoint 4

Find the inverse relation of the function

$$g(x) = \frac{x}{2 - x}.$$

Is the inverse a function? Graph both g and its inverse to confirm your analysis.

Logarithms

In Exercises 28 and 29 of Section 1.7—which you may or may not have done—we asked what functions turn sums into products and what functions turn products into sums. Whatever the answers are, you should recognize now that the verbal descriptions of these two classes of functions are inverses of each other. More specifically, if a given function is in the first of these classes, its inverse (if it's a function) will be in the second class—and vice versa.

From your previous study of algebra, you should know at least one function that turns products into sums — that's the whole point of **logarithms**, and you know from the function keys on your calculator that **log** is a function. You also know properties of logarithms, one of which is

$$\log AB = \log A + \log B,$$

where A and B are any numbers that have logarithms. (What numbers have logarithms?) This logarithmic property says that **log** is a function that satisfies the functional equation for turning products into sums.

Perhaps more important for what we will do in the next chapter is the logarithmic property of turning exponents into factors:

$$\log A^B = B \log A.$$

But what's a logarithm?

There are many different **log** functions, one for each allowable **base**. We recall the definition of **logarithm base** b, more or less as it appeared in your algebra or precalculus book:

Definition $y = \log_b x$ if and only if $x = b^y$.

Observe that "taking the logarithm (base b)" and "exponentiation (base b)" are inverse processes: Each undoes what the other does. Thus, if x is calculated as b raised to the y power (exponentiation base b applied to y), then y is recovered from x by taking the logarithm with the same base. Similarly, if y is calculated from x by taking a logarithm, then x is recovered from y by raising the base to the y power. The meaning of "take a logarithm" is "undo exponentiation."

Checkpoint 5

Check the assertions of the preceding paragraph with your calculator using base 10.

(a) Select a number y of your choice, and raise 10 to the y power. Apply [log] or [$\log_{10}$] to the result, and observe whether you get y back.

(b) Now select a positive number x, and calculate $\log_{10}$ of it. Use the resulting number as an exponent, with 10 as base, and see whether you get x back.

(c) Repeat parts (a) and (b) with several other values of y and x.

Checkpoint 6

In Figure 1.32, we show graphs of $y = 10^x$ and $y = \log_{10} x$. What do you observe about these graphs? How do your observations relate to our earlier discussion of graphical representations of inverse functions?

Figure 1.32 Exponentiation and logarithm base 10

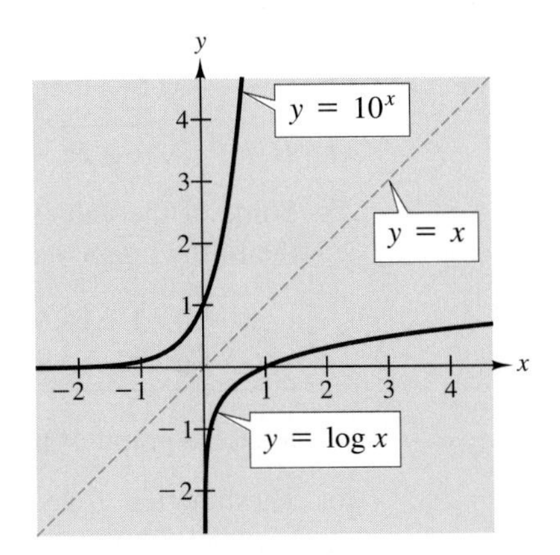

In this section we have studied the important operation of inverting a function, that is, of acting on the function *as object* to reverse the function *as process*. As we will do with many topics in this course, we have viewed inversion of functions three ways: symbolically, numerically, and graphically. Each of the three ways is important, and we review them here.

Symbolic inversion If a function is defined by a formula $y = f(x)$, then inversion means interchanging the symbols x and y to get a new equation, $x = f(y)$, which defines the inverse relation. If we can solve this equation for y, and if each x determines a unique y, then the resulting equation, $y = g(x)$, defines the inverse function. That is, $g = f^{-1}$. However, even if we can't solve the new equation for y, there still may be an inverse function—we just won't know a formula for it.

Numerical inversion This means at least two different—but closely related—things. In the car-loan and calculator-square-root examples, we did numerical calculations to find single values of x when we knew the values of y. That's often easy—with the help of a calculator. And when it's not easy, it's almost always possible—often with the help of a computer. On the other hand, a function may be known only as a table of data. In this case, inversion is trivial—just interchange the columns in the table.

Graphical inversion This is almost as easy as inverting a data table: Just flip the graph over the line $y = x$. When you draw graphs with a computer or calculator, you will learn just how easy this is.

Answers to Checkpoints

1. 3.2 and $5/16$. These are reciprocals of each other.
2. $y = -\sqrt{x}$. The inverse is $y = x^2$ for $x \leq 0$. This is the dashed curve on the left in Figure 1.27.
3. (a) All numbers x except 2. (b) All numbers x except 0.
   (c) They are the same. (d) They are the same.
4. $y = 2x/(x + 1)$, a function.
5. $\log_{10} 10^y = y$ for all numbers y. $10^{\log x} = x$ for all positive numbers x.
6. The graphs are mirror images in $y = x$. The functions are inverses of each other.

Exercises 1.8

For each of the following functions, (a) graph the function, (b) graph the inverse relation, and (c) decide whether the inverse is a function.

1. $f(x) = x^3$
2. $g(x) = \dfrac{1}{x^2}$
3. $h(x) = -\frac{1}{2}x + 3$
4. $f(t) = \dfrac{3}{t+5}$
5. $g(t) = \dfrac{3}{t^2+5}$
6. $h(t) = \dfrac{3t}{t+5}$
7. $F(x) = 2^x$
8. $G(x) = \sqrt{1+x}$
9. $H(x) = \sqrt{1+x^2}$
10. $F(t) = \log_{10} t$
11. $G(t) = \sqrt{t-1}$
12. $H(t) = \sqrt{1-t^2}$

For each of the following functions, (a) state the domain of the function, and (b) if the inverse relation is a function, state its domain as well.

13. $f(x) = x^3$
14. $g(x) = \dfrac{1}{x^2}$
15. $h(x) = -\frac{1}{2}x + 3$
16. $f(t) = \dfrac{3}{t+5}$
17. $g(t) = \dfrac{3}{t^2+5}$
18. $h(t) = \dfrac{3t}{t+5}$
19. $F(x) = 2^x$
20. $G(x) = \sqrt{1+x}$
21. $H(x) = \sqrt{1+x^2}$
22. $F(t) = \log_{10} t$
23. $G(t) = \sqrt{t-1}$
24. $H(t) = \sqrt{1-t^2}$
25. Some of the values of a function $f(x)$ are given in the following table:

x	0	1	3	6	8
$f(x)$	0	2	3	4	5

Sketch a graph of the inverse of the function f.

26. The function $f(x) = x^{2/3} + 3$, $x \geq 0$, is graphed in Figure 1.33. Copy the graph, and on the same axes, sketch a graph of $f^{-1}(x)$.

Figure 1.33 Graph of $f(x) = x^{2/3} + 3$, $x \geq 0$

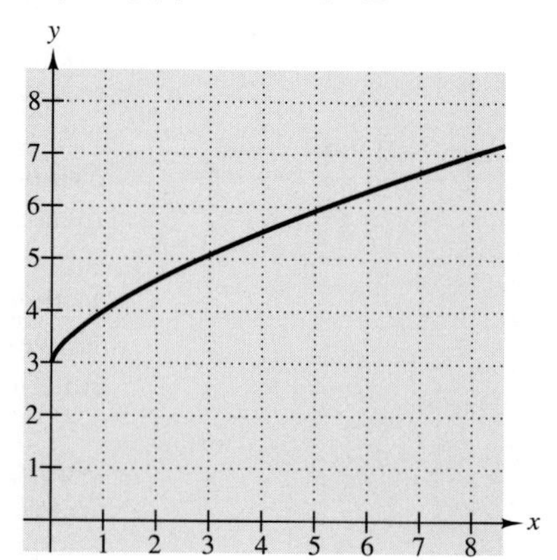

27. If $f(x) = x^{2/3} + 3$ for $x \geq 0$, find a formula for f^{-1}.
28. If $f(x) = x^{2/3} + 3$ for $x \geq 0$, make a table of values for f^{-1} with at least five entries.
29. Show that the inverse of any linear function of the form $y = mx + b$ (with $m \neq 0$) is a linear function whose slope is $1/m$.
30. Does a linear function with slope 0 have an inverse? Why or why not?
31. The function $f(x) = \frac{1}{8}x^3 + x$ is graphed in Figure 1.34. Copy the graph, and on the same axes, sketch a graph of $f^{-1}(x)$.

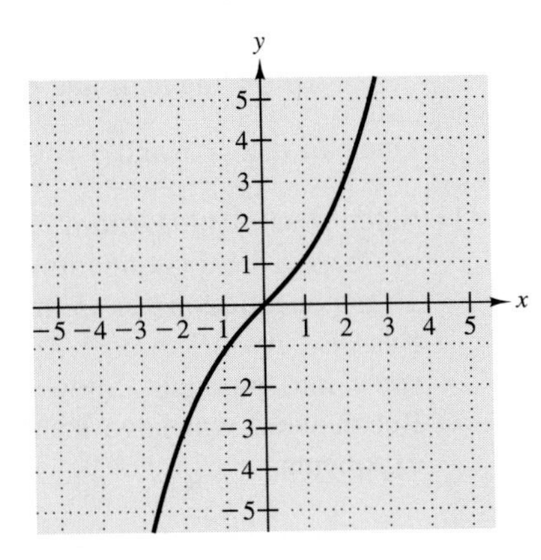

Figure 1.34 Graph of $f(x) = \frac{1}{8}x^3 + x$

32. (a) If $f(x) = \frac{1}{8}x^3 + x$, make a table of values for f^{-1} with at least five entries.
    (b) What's the problem with finding a formula for f^{-1} in this case?

33. A function $f(x)$ is graphed in Figure 1.35. Copy the graph, and on the same axes, sketch a graph of $f^{-1}(x)$.

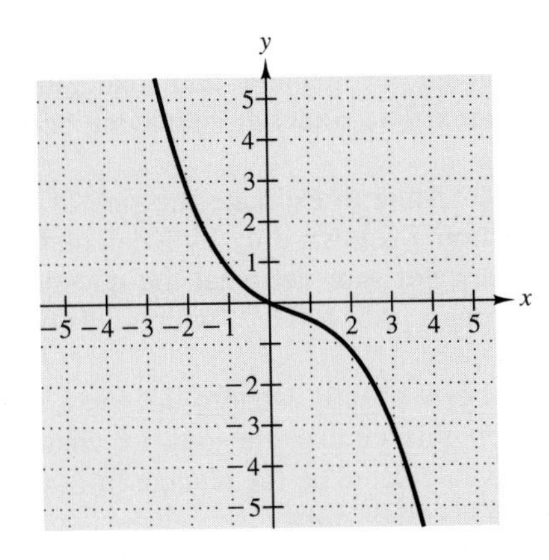

Figure 1.35 Graph of a function $f(x)$

34. (a) If $f(x)$ is the function graphed in Figure 1.35, make a table of values for f^{-1} with at least five entries.
    (b) What's the problem with finding a formula for f^{-1} in this case?

35. A function $f(x)$ is graphed in Figure 1.36. Decide whether or not this function has an inverse, and give a reason for your answer.

36. The function f graphed in Figure 1.36 has what kind of symmetry—even, odd, or neither? Give a reason for your answer.

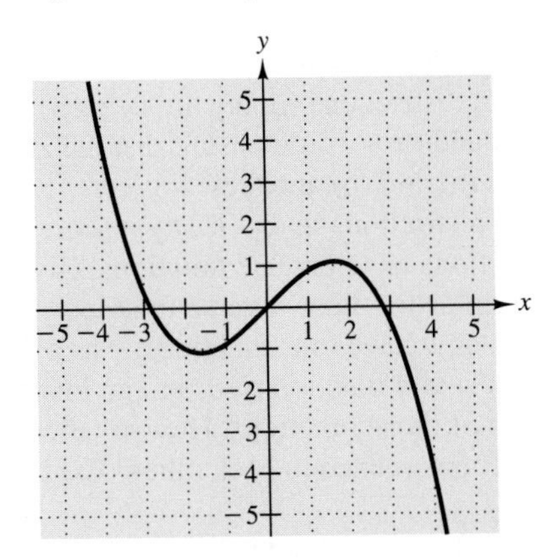

Figure 1.36 Graph of a function $f(x)$

37. (a) Make a scatter plot of the data in Table 1.31. (You may have done this already.)

Table 1.31 Related variables

x	0.8	1.5	3.2	2.6	1.9	2.4	3.5	0.6	2.1
y	0.7	2.1	10.5	6.8	3.5	5.9	12.4	0.3	4.3

    (b) Construct a table that defines the inverse function. Explain why the inverse is actually a function. What is the domain of the inverse function?
    (c) Calculate the square roots of the y-values, and graph points of the form $(x, \sqrt{y})$. Describe the shape of this graph.
    (d) Now calculate the squares of the x-values, and graph points of the form (x^2, y). Describe the graph.
    (e) What can you say about the relationship between the last two graphs? What can you say about the relationship between x and y?

38. (a) Explain why an even function (if its graph has more than one point) cannot have an inverse function.
    (b) Explain why the inverse of an odd function (if it has an inverse function) is odd.

39. (a) Does the equation $x^2 + y^2 = 4$ define y as a function of x? Graph the equation, and explain your conclusion.
    (b) Does the equation $x^2 + y^2 = 4$ define x as a function of y? Explain.
    (c) What restrictions (if any are needed) on values of x and/or y would ensure that the answer to (a) is "yes"? Explain.

(d) What restrictions (if any are needed) on values of x and/or y would ensure that the answer to (b) is "yes"? Explain.

(e) What restrictions (if any are needed) on values of x and/or y would ensure that the answers to both (a) and (b) are "yes"?

(f) The restrictions in (e) (if any) ensure that both functions have inverse functions. Give explicit formulas for those inverse functions.

(g) You now have four functions: y as a function of x, x as a function of y, and an inverse of each of those functions. How are these four functions related to each other? Explain.

40. For any given base b, there are two pairs of numbers x and y that we can easily see will satisfy

$$y = \log_b x \text{ if and only if } x = b^y.$$

One pair is $x = 1$ and $y = 0$. What's the other?

41. Since "base 10 logarithm" is defined as the inverse of "base 10 exponentiation," it may be that "base 10 exponentiation" is an answer to the question in Exercise 29 of Section 1.7 about functions that turn sums into products. Show that this is true. [*Hint*: Recall what you have learned about properties of exponents.]

1.9 What's Significant About a Digit?[27]

You need to know the answer to this question when your calculus or chemistry instructor asks you to report a numerical answer to, say, four significant digits. This question is not specifically about functions, the central theme of this chapter, but it is related to our numeric representations of functions—and to how we make sense out of those representations.

Here's a true story about a student, a calculator, and a calculus exam. We will call the student "Anna"—not her real name. To protect the innocent, both the calculator and the exam shall remain nameless.

A question on the exam asked Anna to estimate the rate of change, on a very short interval starting at 0.5, of a function f known only by a calculator button that had never been discussed in class. If you are not sure yet what the question means, think of it as asking for the slope of a line through $(0.5, f(0.5))$, which will be calculated as *rise over run*.

Being a good student, Anna knew exactly what to do. First, she chose her very short interval to be $[0.5, 0.501]$. Given this choice, the rate of change (think *slope*) of the function over the interval can be calculated as the change in function values (think *rise*) divided by the length of the interval (think *run*):

$$\frac{f(0.501) - f(0.5)}{0.501 - 0.5}.$$

Anna's calculator told her that $f(0.5)$ was 0.523598775598, which she wrote down as 0.524—and that $f(0.501)$ was 0.524753861551, which she recorded as 0.525. Next, she calculated the rate of change:

$$\frac{f(0.501) - f(0.5)}{0.501 - 0.5} = \frac{0.001}{0.001} = 1.$$

27. Based on an article with the same title by D. A. Smith in *The College Mathematics Journal* 20 (1989), pp. 136–139.

Actually, Anna wrote the final step as "≈ 1"—she knew that this was only an estimate, and that was all the problem asked for. However, by her calculator procedure, she forced the answer to have less accuracy than was available. If the problem had specified an accuracy, her answer might have been wrong.

What was wrong with Anna's procedure? There was no need for her to write down, round off, or re-enter either of the function values. The calculator's value for $f(0.501) - f(0.5)$ (with two presses of the function key and one of $\boxed{-}$) is 0.001155085952, so the end of the calculation might have been

$$\frac{0.001155}{0.001} = 1.155,$$

an answer that differs from Anna's by about 15%. There was no way Anna could have known *exactly* what she was trying to estimate—that's why we sometimes have to estimate! But she should have known—more important, you should know—that sloppy use of meaningful digits can lead to wrong answers.

In this section we will explore some examples and exercises on significance and then describe a test by which you can decide whether you are responding appropriately to assignments that call for numerical answers—which means most assignments.

There are at least these three sources of concern about significance of digits:

- The accuracy of available data
- The finite precision of your calculating device
- Loss of significance due to the way we manipulate the numbers

(Anna's calculation is in the third category.) We comment briefly on each of these and provide examples of each.

Accuracy of Measurements

In Section 1.4 we described several dichotomies by which we can sort our functions into various conceptual categories. For example, the position of a moving object is a *real* function of time that has an *exact* value at every instant of time—but we may *know* only an *observable approximation* to this exact value and only at selected instants of time. That is, we may have only a finite list of numbers, such as the distance numbers in Table 1.2. These numbers are necessarily limited in significance by the precision of the device that made the measurements, and it would be silly to expect that we can ever gain additional significance by doing calculations on the numbers.

Exploration Activity 1

Suppose a population of 240,000,000 people grows by 2% per year for three years. What will the population be at the end of the three years?

If P_0 stands for the starting population (that is, $P_0 =$ 240,000,000), then the population at the end of the first year will be $P_0 + 0.02P_0$ or $P_0 \times 1.02$. Each year the population will get multiplied again by 1.02, so at the end of three years the population will be $P_0 \times (1.02)^3$. When we substitute 240,000,000 for P_0 and multiply by $(1.02)^3$, our calculator says the new population is 254,689,920. (Check this with your own calculator.) But the original data had only two significant digits (2 and 4), or maybe three —we can't be sure about the first 0—and we can't possibly know more precise information about population at a later date simply by doing calculations with an abstract model. Thus the best answer to our question is "about 250,000,000"—an answer with the same number of significant digits as we started with. It would not be wrong to say "about 255,000,000," as long as we understand that there is no solid reason to believe the second 5. It definitely is wrong to report 254,689,920 as the answer, because this implies that you know the population to a high degree of accuracy, and you certainly don't.

Moral Calculation steps *never* add significant digits, so don't always believe all the digits you see on your calculator.

Checkpoint 1

Suppose the population of 240,000,000 continues to grow at 2% per year for 10 years. What would the population be at the end of that time?

Accuracy of Calculating Devices

Some of our important functions are mathematical models of reality that are defined by formulas—usually formulas that give an exact value for every input number but which are at best approximations to the reality being modeled. Reality aside, it may be important to know something about exact values—and our ability to know these values may be limited by the accuracy of our calculator or computer.

Example 1 If you ask your calculator for a value of $\log_{10}(3)$, it might respond with 0.4771212547 or .47712125472. (Try your own calculator—you may get some other response.) How many of these digits are significant?

Solution The leading zero in the first answer is not significant. In fact, leading zeros are *never* significant. Whether you do or do not put a zero in front of a decimal fraction is a matter of style and clarity. The first answer has 10 significant digits (count them), and the second has 11. Actually, the second answer confirms for us that all 10 digits of the first answer are correct, but we have to trust the manufacturer of the second calculator that the eleventh digit is correct — or check it against some more accurate source. Neither of these answers is the exact value of $\log_{10}(3)$, because exact representation of that number as a decimal expansion would require infinitely many digits. ■

Moral Finite machines (calculators and computers) cannot produce more significant digits than they have been programmed to produce. In particular, they cannot produce an exact numerical answer if that answer requires an infinite decimal expansion.

Without some clever programming, none of our calculators can give us more than ten or twelve or perhaps fifteen significant digits of a number such as $\log_{10}(3)$. There are computer programs capable of arbitrarily high-precision arithmetic. We asked one of those programs for a twenty-significant-digit value for $\log_{10}(3)$, and it responded 0.47712125471966243729. This shows us that (1) the eleventh decimal place is not 2, but (2) the eleven-place calculator responded with a value correctly rounded to eleven significant digits.

Checkpoint 2

(a) Suppose you are working on a problem whose answer is $\log_{10}(3)$, and you have been asked to report your answer to seven significant digits. What do you write down?

(b) Give a twelve-significant-digit approximation to $\log_{10}(3)$.

Loss of Significance

The information we get from our calculator or computer may be corrupted—in many different ways—by the sequence of operations we choose to perform. Sometimes this loss of significance is preventable, sometimes not. We have already given an example of preventable loss of significance in our tale of Anna and the exam problem.

Moral *(from Anna's tale)* Don't discard digits in an intermediate result. The only time you should round off is at the *end* of your calculation.

Exploration Activity 2

(a) Find an approximation to $\pi/\sqrt{2}$ that is as accurate as your calculator can provide. Write down all the digits. How many digits in your answer do you believe are correct?

(b) Write down a five-significant-digit approximation to $\pi/\sqrt{2}$.

(c) Use the approximations $\pi \approx 3.14$ and $\sqrt{2} \approx 1.414$ to calculate a decimal approximation to $\pi/\sqrt{2}$. How many digits of the answer are significant?

A twelve-place calculator reports that $\pi/\sqrt{2} = 2.22144146908$. (Your calculator may be different.) From this we see that a five-significant-digit approximation would be 2.2214. The result from $3.14/1.414$ is 2.22065.... The error is about 0.0008, which means the third decimal place cannot be assumed accurate, so the answer (as an approximation to $\pi/\sqrt{2}$) has only three significant digits.

Moral You should not expect more significant digits in any answer than are in the *least* accurate input to the calculation.

Example 2 Now a loss of significance that is not preventable. If you ask a ten-significant-digit calculator for a value of π, it will respond 3.141592654, which is correctly rounded to nine decimal places. If you ask the same calculator for a value of $355 \div 113$, you will see 3.14159292—almost the same number. What happens if we subtract the smaller number (π) from the larger ($355 \div 113$)? If we were doing it by hand, the calculation would look like this:

$$\begin{array}{r} 3.141592920 \\ -\,3.141592654 \\ \hline 0.000000266 \end{array}$$

It looks like our answer has only three significant digits! And our calculator confirms this by reporting 2.66×10^{-7} as the answer from the subtraction. Subtraction of nearly equal numbers can be a *significance killer*. In particular, subtraction of two ten-significant-digit numbers that agree in the first seven digits produces an answer that has only three significant digits.

Moral Watch out for *disastrous cancellations*. If you can't arrange your work to avoid them, at least be aware that your numbers have fewer significant digits as a result.

There is another significance issue in Example 2—not the point of that example, but nevertheless important. The calculator answer for $355 \div 113$ had only eight decimal places—nine significant digits. But we know the calculator carries ten significant digits. A reasonable—and correct—deduction is that the ninth decimal place must be 0. Calculators often drop trailing zeros *even when those digits are significant*. Notice that we provided the missing 0 in our hand calculation.

Anna's subtraction of the nearly equal numbers 0.525 and 0.524 resulted in a cancellation that was disastrous for her: From numbers with only three significant digits, she ended up with an answer that had only one significant digit. However, if she had prevented the *preventable* part of the problem—loss from twelve down to three significant digits—it wouldn't have mattered much that she had an *unpreventable* loss of two digits.

Checkpoint 3

(a) Use your calculator to evaluate the expression

$$\sqrt{x^2+1}-x$$

at $x = 42,709$. The correct answer, to fourteen significant digits, is $1.1707134326055 \times 10^{-5}$. How many digits of your answer are significant?

(b) Evaluate the same expression at $x = 1,342,709$. The correct answer, to fourteen significant digits, is 3.7238150634272. How many digits of your answer are significant? (To see how to get better approximations, do Exercise 5 at the end of the section.)

How to Tell Whether a Digit Is Significant

The significance of digits in an approximation has to be measured relative to an exact value that is being approximated. Regardless of where the decimal point is located, we ignore 0's to the left of the first nonzero digit. They function only as place holders and do not tell us anything about digits of the number being approximated. Starting from the first nonzero digit on the left, we ask of each digit in turn how well it matches the corresponding digit in the exact number. We will say it **matches** if the error in the approximation is less than 5 in the next decimal place. For example, if we are checking for a match in the thousands place, the next place, reading from left to right, is the hundreds place. To have a match in the thousands place, the error has to be less than 500.

Example 3 Suppose the exact number being approximated is 1,342,709, and the following numbers are calculated as proposed approximations:

1,340,000 1,341,624 1,342,000 1,343,000

How many digits of each approximation are significant?

Solution The number 1,340,000 is a three-significant-digit approximation, because the error, 2709, is less than 5000. Notice that trailing zeros are sometimes necessary, even when they are not significant.

Similarly, 1,341,624 is also a three-significant-digit approximation to 1,342,709, for exactly the same reason. The fact that it has other nonzero digits is irrelevant—its error is less than 5000 but not less than 500.

The number 1,342,000 is yet another three-significant-digit approximation to 1,342,709, but *not* a four-significant-digit approximation, because its error, 709, is not less than 500.

On the other hand, 1,343,000 *is* a four-significant-digit approximation, because its error,

$$|1{,}343{,}000 - 1{,}342{,}709| = 291,$$

is less than 500. Thus, from the exact number 1,342,709, we can report a four-significant-digit approximation (1,343,000) by correct rounding in the fourth place from the left. ■

Definitions Leading zeros are *never* significant. The decimal digit in the 10^k place of an approximation y to a number x is **significant** if it is not a leading zero, and if

$$|x - y| < 0.5 \times 10^k.$$

The number y is an n **significant digit approximation** to x if, after discarding leading zeros, the first n of the digits of y (reading from left to right) are significant in the sense just defined.

We will abbreviate the phrase "four significant digits" by **4SD**, and similarly for other numbers of SDs. We provide two more examples to illustrate the meaning of SD, and then we ask you to check your understanding in the exercises.

Example 4 Find a 4SD approximation to $\pi/40$.

Solution A calculator approximation to $\pi/40$ is 0.078539816.... The fourth significant digit of this approximation is in the 10^{-5} position, so we need an error of less than 5 in the sixth decimal place. We get this by rounding to five decimal places: 0.07854. Note that significance is not directly connected with numbers of decimal places. ■

Example 5 Find a 4SD approximation to $\dfrac{20013}{10006}$.

Solution A calculator approximation is 2.00009994... . Thus a 4SD approximation is 2.000. It does not make sense to report this answer as 2 or even as 2.0. The first suggests either an exact answer or a 1SD answer—only the context could make it clear which you intended. The second definitely suggests 2SD. Thus sometimes trailing zeros are not only significant but necessary to convey that significance. ■

In this section we have offered some cautions about the use of calculators and computers as tools for representing numerical values. Most such devices present every answer with the same number of digits, whether or not those digits are meaningful. It is *your* responsibility—not your calculator's—to keep track of significant digits and to avoid the errors that arise from using either too few or too many digits at various points in your problem-solving processes.

Answers to Checkpoints

1. About 290,000,000.
2. (a) 0.4771213 (b) 0.477121254720
3. (a) You should get 1.1707×10^{-5}; all five digits are significant.

   (b) You should get 3×10^{-7}; this has no significant digits.

Exercises 1.9

1. To four decimal places (5SD), π is 3.1416, and $\sqrt{2}$ is 1.4142. If you divide these approximations, how many significant digits of $\pi/\sqrt{2}$ will you get? Don't guess—do the calculation to be sure.
2. (a) Write down all the digits of your calculator's approximation to $9\pi/5$.
   (b) Write down a 4SD approximation to $9\pi/5$.
   (c) Explain why 5.66 is not a correct 3SD approximation to $9\pi/5$. That is, explain why it is not legitimate to do repeated rounding. What is the correct 3SD approximation?
3. Suppose your bank pays quarterly interest on savings accounts at an annual rate of 7%. That is, the interest added at the end of each quarter is calculated at a rate of $0.07/4 = 0.0175$.
   (a) If you deposit $160 at the start of a year, how much money will be in your account at the end of the year?
   (b) How many significant digits are there in your answer?
4. (a) Experiment with your calculator to determine the smallest positive number it can recognize and display.
   (b) Similarly, determine the largest number it can recognize and display.
   (c) Estimate the total number of different numbers your calculator can recognize and display. Is this number finite and moderately large, finite and enormously large, or infinite? Why?
5. The disastrous cancellations in Checkpoint 3 actually are avoidable, but only by transforming the expression to another form. Rewrite the expression $\sqrt{x^2+1} - x$ by rationalizing the numerator. That is, multiply and divide by $\sqrt{x^2+1} + x$, and simplify. Evaluate the resulting expression at each of the numbers in parts (a) and (b) of Checkpoint 3.

Chapter 1 Summary

We started this chapter with the big question:

What's a Function?

We have now answered that question in many different ways. In particular, we have seen that "function" is linguistically distinct from "formula." Formulas often define functions, but not always. In fact, functions may be defined by

- formulas
- graphs
- data tables
- verbal descriptions
- conceptual relationships
- physical, biological, chemical, and other relationships

Functions are everywhere, and the *concept* of function is a powerful tool for dealing with the complexity we see all around us.

We learned in this chapter that all the following may be modeled or represented by formulas:

- relations between real-world variables that definitely are not functions
- relations that may not be functions
- relations that definitely are functions but have no apparent formulas

We can classify our functions by a trichotomy: *real, observed, model.* These functions have the characteristics summarized in Table 1.32.

We can classify our variables by a dichotomy: *discrete* or *continuous*. Some varying quantities jump from one value to the next, like a digital clock. These variables are **discrete**. They may have only a finite number of distinct values, or they may have infinitely many. Other varying quantities change smoothly, in such a way that there is never a *next* or a *previous* value. Such variables are called **continuous**. For example, we usually think of time proceeding in this way, and we model this change by the second hand on an analog clock.

The two clock types highlight the fact that we often model a single varying quantity (time, in this case) by both discrete and continuous variables. Our ability to move easily between discrete and continuous models will give us a broad range of conceptual and computational tools for representing and attacking the problems posed in this course.

Perhaps the most important — but still unstated — message of this chapter is that all of this effort is *for something*. We are going to deal with — and solve — problems that you will recognize as being important in a variety of different ways. In the process, we will see the power of mathematical abstraction to isolate essential features of a problem, to clear away irrelevant clutter, to lead to recognition of or reduction to a problem whose solution is already known. Our first step in that direction has been to come to grips with the simple, but extremely powerful, concept of FUNCTION.

Table 1.32 Characteristics of functions by type

Type	*Relation to Reality*	*Knowable*	*Expressible by Formulas*	*Exact Calculations*
Real	Exact	Not completely	No	No
Observed	Approximate	In finite terms	No	No
Model	Approximate	Exactly	Usually	Often

Chapter 1 Exercises

Some of these exercises require graph paper. Start each such exercise by drawing horizontal and vertical axes and labeling scales clearly.

1. Draw the line through the points $(2.8, 2)$ and $(-3.7, 5)$. Find an equation for this line.
2. Plot four points on the line whose equation is $4x - 3y = 9$. Draw in the line.
3. The line in the preceding exercise is the graph of a function—what function? (Find a way to describe the function in words and/or symbols.)
4. The function in the preceding exercise has an inverse function. Describe it in words and/or symbols. Draw its graph.

Each of Exercises 5–8 describes a function. In each case,
(a) decide whether there is an inverse function,
(b) if so, describe the inverse function as best you can,
(c) if not, explain why the function has no inverse.

5. $y = \dfrac{x}{1 - x}$

6. **Table 1.33** Distance traveled by a falling object

Time (seconds)	1	2	3	4	5	6	7	8	9	10
Distance (meters)	5	19	44	78	123	176	240	313	396	489

7. **Table 1.34** Federal income tax

Taxable Income	*Income Tax*
Less than \$10,000	\$0
\$10,000 but less than \$50,000	15% of income over \$10,000
\$50,000 or more	\$6000 + 27% of income over \$50,000

8. **Figure 1.37** Graph of a function

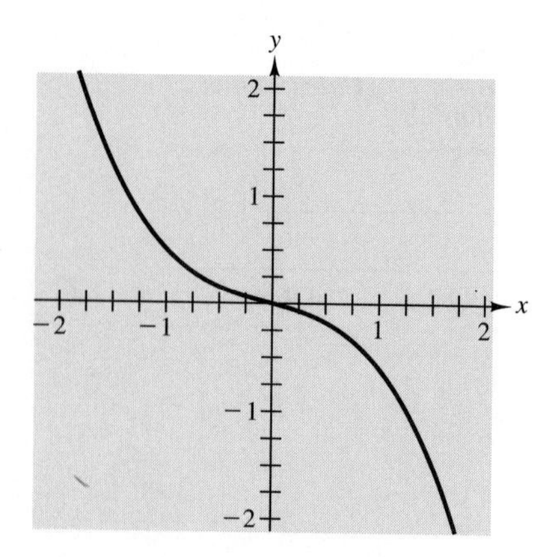

9. Figure 1.38 shows the number of deer in a forest at time t years after the beginning of a conservation study.
   (a) During which of the following time periods did the deer population decline at the rate of 50 deer per year?
   (i) 1 to 2 (ii) 1 to 3 (iii) 1 to 4 (iv) 2 to 3 (v) 5 to 6
   (b) When was the deer population increasing most rapidly?
   (c) Approximately how fast was the deer population increasing or decreasing at time $t = 1.5$? Your answer should include appropriate units.[28]

Figure 1.38 Data from a conservation study

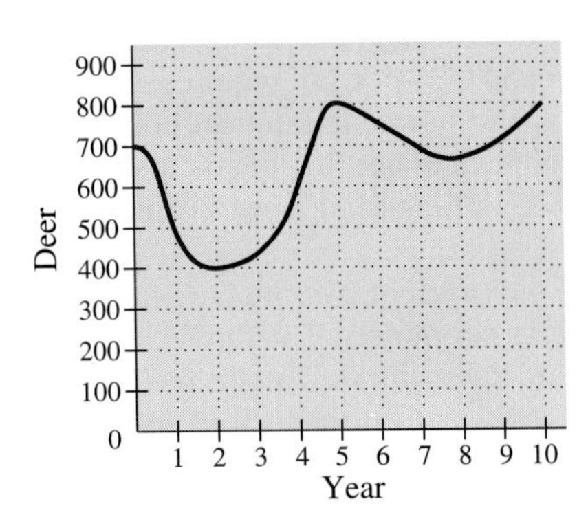

10. Table 1.35 shows data on total applications and applications by African-Americans to the Duke University Graduate School in the years 1985 and 1992.

Table 1.35 Applications and enrollments, Duke University Graduate School[29]

	African-American Applications	*Total Applications*	*African-Americans Enrolled*	*Total Enrolled*
1985	67	3190	26	1715
1992	205	6654	71	2297

   (a) How did the percentage of African-American applicants admitted compare with the percentage of all applicants admitted in 1985? In 1992?
   (b) In 1985, what percentage of applicants were African-American? In 1992?
   (c) In 1985, what percentage of students enrolled were African-American? In 1992?
   (d) Suppose the numbers of applicants and of African-American applicants grew linearly from 1985 to 1992. How many applications were there in 1990, and how many of these came from African-American students?
   (e) Make up another question that can be answered from these data. Then answer your question.

28. Exercise adapted from *Calculus Problems for a New Century*, edited by R. Fraga, MAA Notes, Vol. 28, 1993.
29. Source: *The News & Observer*, December 20, 1992.

11. A 20-watt compact fluorescent light bulb is sold for \$15.99. The package containing this bulb promises \$55 in energy savings over the life of the bulb, which is estimated to be $10,000$ hours. (That's the average life for such bulbs.) This bulb is as bright as a 75-watt incandescent bulb, for which the average life is 750 hours. Incandescents are sold in packs of four bulbs at prices ranging from \$1.99 a pack to \$3.99. Assume that electricity costs 10 cents per kilowatt-hour, and analyze the promise of \$55 in savings for the fluorescent. Explain your reasoning. If you disagree with that number, explain as best you can how the manufacturer can legally put such a claim on the package.

12. A flow restrictor is a low-cost device (a beveled washer) that can be inserted in a shower head to slow the rate of water flow. In this problem you will use some facts and some assumptions to determine the significance of using such a device.
Facts: The flow restrictor reduces the water flow rate through the shower head by 42%. A gallon of water weighs 8.36 pounds. Heating a pound of water 1°F requires one British thermal unit (Btu) of heat energy. Electric energy is measured in kilowatt-hours (kwh); one kwh is equivalent to 3412 Btus. An electric power plant emits 1.5 pounds of carbon dioxide for every kilowatt-hour of energy produced. At 33% efficiency, a power plant produces twice as much waste heat as useful energy.
Assumptions: A family averages four showers a day, each of five minutes' duration. The shower flow rate without the restrictor is six gallons per minute. The family's electric water heater heats water from 60 to 120°F. Their cost for electricity is 10 cents per kilowatt-hour.

(a) How much water can this family save in a year by using the flow restrictor?

(b) How much energy will they save in a year?

(c) How much money will they save in a year?

(d) What will be the yearly reduction in carbon dioxide emissions?

(e) What will be the yearly reduction in thermal pollution from the power plant?

(f) If this family is your family, how many years of flow-restricted showers would it take to pay for one year of your college tuition?[30]

13. A student who was helping at a yard sale watched as a customer looked through the pile of jeans for sale. The man would bend his hand back at the wrist, bend his arm at the elbow, and then wrap the waist of the pants around his forearm. The man explained that his waist was the same size as his forearm, so he never needed to try slacks on for size. The student decided to gather some data and model the relationship between forearm circumference (along the arm) and waist size. The scatter plot in Figure 1.39, with forearm circumference (measured in inches) on the horizontal axis, is a graph of his data set.

(a) Find an equation of the form $y = mx + b$ to fit the points. (You might want to experiment by moving a clear ruler or dark thread through the scatter plot until you find the line that seems to fit the points best.) After you find an equation of the line you feel is best, comment on the criteria you used to determine this line. Compare your line with those found by other students.

(b) What information is provided by the slope of the line you found in part (a)? What information is provided by the y-intercept? Use the model to predict your waist size. How accurate is the prediction? What message or advice would you give to a person whose

30. Exercise adapted from "Quantitative Aspects of an Energy Conservation Device," by J. R. Shanebrook, *NLA News*, May 1991.

data point lies above the line? What message or advice would you give to a person whose data point lies below the line?

(c) In parts (a) and (b), you considered two different relations: that represented by the data plotted in Figure 1.39 and that represented by your fitted linear equation. For one of these relations, but not for the other, each value of the independent variable determines a unique value of the dependent variable. Which of the two relations has this property? Why? Which of the two relations (if either) is a function?

Figure 1.39 Waist measurement (inches) versus forearm circumference (inches)

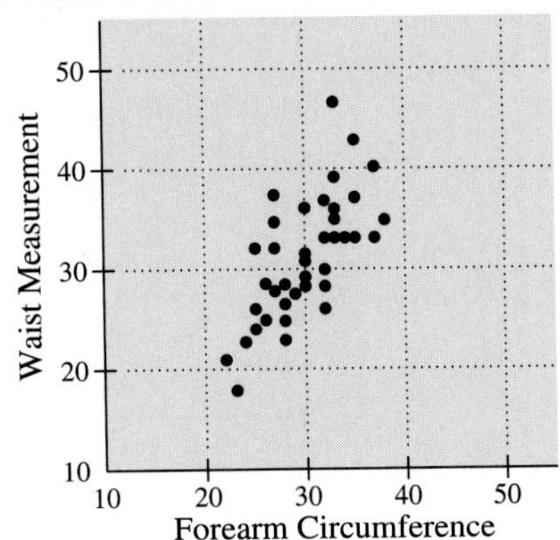

14. (a) It appears from the data in Table 1.36 that average income in the United States depends on the level of education a person has achieved. In particular, the table suggests three distinct functional relationships with average income as a function of years of education, one each for males, females, and the combined average. State the domain of each of these functions. Make a separate table to represent each of the three functions. (Assume that a high school dropout has completed 10 years of education.)

Table 1.36 Average income by education level[31]

Education	*Men*	*Women*	*Combined*
High school dropout	$13,655	$7,004	$10,326
High school graduate	$21,583	$11,143	$15,886
College degree	$37,002	$19,215	$28,406
Graduate degree	NA	NA	$38,604

(b) Make a scatter plot for each of your three functions. If you put all three plots on a single coordinate grid, use different point symbols to distinguish them. You may do the plotting on graph paper or use a graphing calculator or a computer.

(c) For each of the three average income functions, find a linear function that approximates the data reasonably well. What do the slopes of these linear functions represent?

(d) What do you think a person with two years of college could expect to earn on average? An eleventh-grade dropout? A person with five years of postsecondary education? How do your answers depend on gender?

(e) Estimate the missing entries in the table.[32]

31. Source: Census Bureau, Department of Commerce.
32. Problem based on an article in *TI-81 Newsletter*, February 1992.

15. The Hanford, Washington, Atomic Energy Plant has been a plutonium production facility since World War II, and some of the wastes have been stored in pits in the same area. Radioactive waste has been seeping into the Columbia River since that time, and eight Oregon counties and the city of Portland have been exposed to radioactive contamination. Table 1.37 lists the number of cancer deaths per 100,000 residents for Portland and these counties. It also lists an index of exposure that measures the proximity of the residents to the contamination. The index assumes that exposure is directly proportional to river frontage and inversely proportional both to the distance from Hanford and to the square of the county's (or city's) average depth away from the river. Figure 1.40 is a scatter plot of the data in Table 1.37.

Table 1.37 Cancer deaths in Oregon[33]

County/City	*Index*	*Deaths*
Umatilla	2.5	147
Morrow	2.6	130
Gilliam	3.4	130
Sherman	1.3	114
Wasco	1.6	138
Hood River	3.8	162
Portland	11.6	208
Columbia	6.4	178
Clatsop	8.3	210

Figure 1.40 Scatter plot of data in Table 1.37

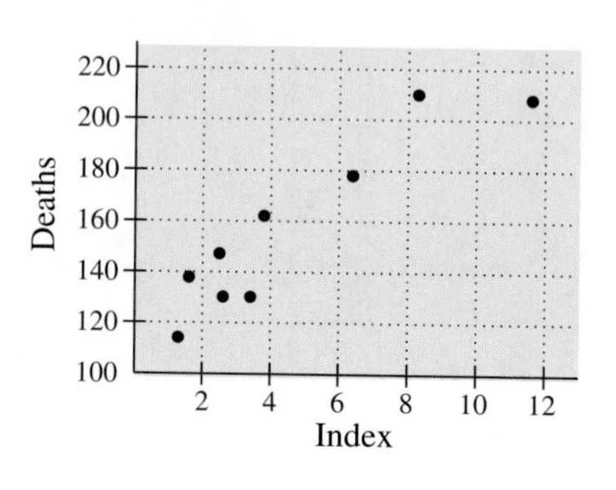

(a) Find an equation of a line that seems to best fit the data in the scatter plot.

(b) What information does the algebraic equation provide? In particular, what is the significance of the y-intercept in the equation? Explain why it should give the cancer death rate when the index of exposure is zero, that is, when there is no radioactive contamination. Does it seem to give a reasonable value for the number of cancer deaths per 100,000 residents in an area without exposure to radioactive contamination?

(c) What values can each of the variables x and y assume? What real numbers are not reasonable values for each of the variables?

(d) What is the slope of your fitted line? What is its significance in terms of the situation being modeled?

16. The graph in Figure 1.41 represents the "math pipeline." It shows the number of ninth-grade students studying mathematics in 1972 and the numbers of the same students at various stages of preparation for and achievement of a Ph.D. in the mathematical sciences. The graph appears to show a nearly linear relationship. What quantity do you think is linear as a function of time? Explain your answer.

33. Source: R. Fadeley, *Journal of Environmental Health* 27 (1965), pp. 883–897.

Figure 1.41 The math pipeline[34]

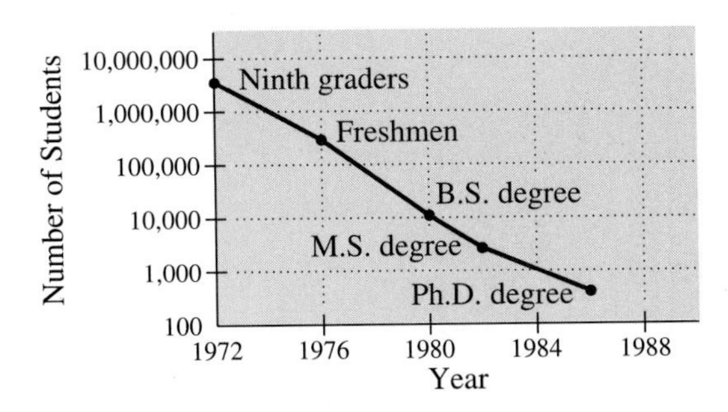

17. (a) A sociological study examined the process by which doctors decide to adopt a new drug. The doctors were divided into two groups. The doctors in group A had little interaction with other doctors and so received most of their information via mass media. The doctors in group B had extensive interaction with other doctors and so received most of their information via word of mouth. For each group, let $f(t)$ be the number who have learned about a new drug after t months. Match the graph of $f(t)$ for each of the groups to one of the graphs in Figure 1.42. Explain your choice.

Figure 1.42 Spread of new information

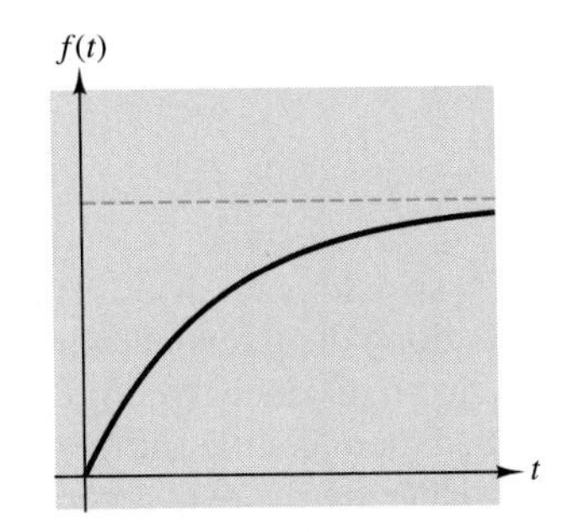

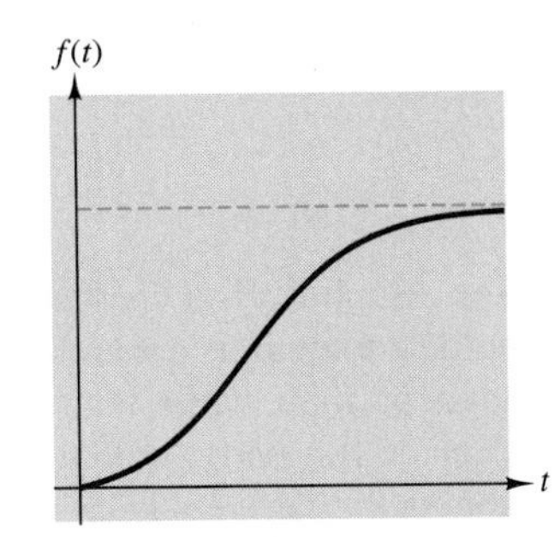

(b) The graphs in Figure 1.42 also describe the learning curves of two different types of jobs. If t is the time on the job, $f(t)$ describes how much of the required job skills the individual possesses. One type of job is skilled or semiskilled, such as an operator of a word processor. The other type is primarily unskilled, such as a french fry chef at a fast-food restaurant. Which of the graphs describes which learning curve, and why?[35]

18. According to Hooke's Law, the force (weight) required to stretch a spring beyond its natural length is proportional to the distance stretched (see Figure 1.43).

(a) Express by a simple formula the force as a function of distance stretched.

(b) Sketch the graph of this formula.

(c) What happens to the spring when its elastic limit is exceeded? Hooke's Law approximates the force required for stretching only up to the spring's elastic limit—add to your sketch what you think the force function looks like beyond that elastic limit.

---

34. Source: *Everybody Counts: A Report to the Nation on the Future of Mathematics Education*, National Academy Press, 1989.
35. Problem adapted from *Calculus Problems for a New Century*, edited by R. Fraga, MAA Notes, Vol. 28, 1993.

Figure 1.43 Stretching a spring

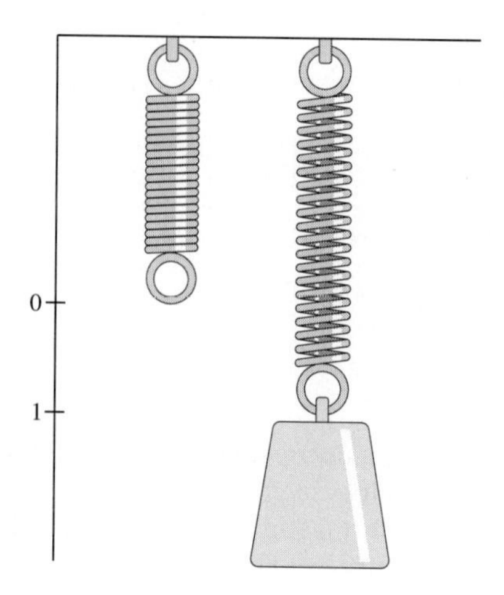

Chapter 1 Projects

1. **Speedometer-Odometer** Suppose you are testing your car's speedometer against the odometer. (You assume the odometer is accurate.) You drive along the interstate and try to hold your speed constant at 50 mph, while your friend takes down distance readings at regular one-minute intervals. (You do this early Sunday morning when there is little traffic so that you do not get run off the road by drivers wanting to go 65!) Here are the data recorded by your friend:

Time (minutes)	1	2	3	4	5	6	7	8	9	10
Distance Traveled (miles)	0.8	1.6	2.4	3.2	4.0	4.8	5.7	6.4	7.3	8.1

(a) Were you able to hold the speed relatively constant? Explain briefly.

(b) What is your best estimate of the speed at which the car was traveling?

(c) How far had the car traveled in the first 7.3 minutes?

(d) If the car continued moving in this manner, how far would it have gone in 12 minutes?

(e) Find an algebraic expression for a function $s(t)$ that gives the distance the car has traveled as a function of time t for values of t under consideration.

As in the previous scenario, we have data on the distance a car has traveled. In this case, the measurements were made every half-minute.

Time (minutes)	0.5	1.0	1.5	2.0	2.5	3.0	3.5	4.0	4.5	5.0
Distance Traveled (miles)	0.06	0.25	0.56	1.00	1.56	2.25	3.06	4.00	5.06	6.25

(f) What can you say about the average speeds?

(g) Find an algebraic expression for a function that approximates the average speeds.

(h) Using this function for speed, find a function that approximates the distance traveled at each time t.

(i) Compare your approximation to the actual data given. How accurate is your formula? Where is it most accurate? Where is it least accurate?

(j) Use your approximation to estimate the distance the car had traveled at 3.3 minutes.

(k) Assuming that the car continued traveling in the same manner, estimate how far it would have gone in 8 minutes.

(l) How realistic is this data for cars you know about?

2. **Growth of a Fruit Fly Population** We consider the growth of fruit flies in a favorable laboratory environment: unlimited food, unlimited space, and no predators. Our objective is to find an approximating function that we can describe by a formula and that we can use to estimate the population at any (reasonable) time, without actually counting the flies. The real population function has only integer values—we do not count pieces of flies!—but we allow our approximating function to assume fractional values. When we interpret our estimates from our approximating function, we have to remember not to be too impressed by a prediction of, say, 788.025 flies on day 15. Table 1.38 shows the data obtained by counting the flies at the same time each day for ten days.

Table 1.38 Fruit fly data

Day Number	*Number of Flies*
0	111
1	122
2	134
3	147
4	161
5	177
6	195
7	214
8	235
9	258
10	283

(a) Calculate the growth in the population for each of the 10 days. (Add a column to the table to show the daily growth.) Determine the *rate* of growth for each of the 10 days. What can we say about the rate of growth of the population? In particular, is it constant? What else can you say about the rate of growth?

(b) Biologists argue that, for populations of this type, the rate of growth should be proportional to the population. How can we test whether the data in hand support this theory?

(c) Carry out your test. If the data support the theory, how can you estimate the proportionality constant? Assuming there is one, what is your best estimate of the proportionality constant?

(d) We need to know what sort of a function has a rate of change proportional to the function itself. Decide which of the following functions have rates of change proportional to their own values at times $t = 0, 1, 2\ldots$. For those which do, find the constant of proportionality. (Show your calculations and conclusions for each of the six cases.)

(i) $f(t) = t^2$ (ii) $f(t) = 2^t$ (iii) $f(t) = 7 \cdot 2^t$

(iv) $f(t) = 10^t$ (v) $f(t) = t^3$ (vi) $f(t) = 7 \cdot t^3$

(e) Can you find a formula for a function $f(t)$ whose rate of change (for one-unit time steps) is k times $f(t)$, where k is the constant of proportionality you obtained in part (c)? Can you find one that *also* has the value 111 at $t = 0$? (Explain how to find such a function, and state your conclusion about such a function.)

(f) Check the values of your formula-defined function from part (e) against the fruit fly data in Table 1.38. (Add another column to the table.) Does the formula approximate the data reasonably well? If not, can you adjust the formula to fit better — without violating the conditions of part (e)?

(g) What does your formula predict for the fruit fly population on day 20? How confident are you of this prediction?

CHAPTER 2

Models of Growth: Rates of Change

Writing in 1798, the British economist Thomas Malthus made some dire predictions about the human population of the Earth. His concept of the problem is shown in Figure 2.1. He observed that the food supply was growing only linearly as a function of time and that the population was growing at a much faster rate. Thus, even if the food supply were more than adequate at the moment, population growth would soon outstrip the ability of people to feed themselves. Unless some disaster wiped out large portions of the population first, widespread famine would be the inevitable result. Malthus may have been wrong about both the food supply and the population. Nevertheless, we have been hearing, at least since the 1960s, even more dire predictions about overpopulation early in the twenty-first century, so it is clearly important to know what growth rates tell us about population size and overuse of essential resources.

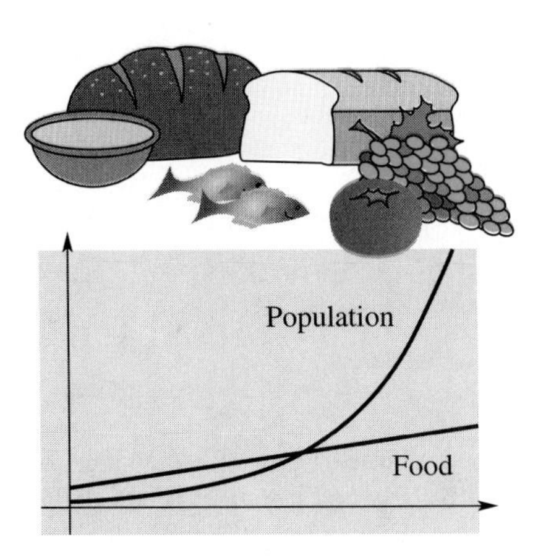

Figure 2.1

To analyze the population problem as Malthus saw it, we must understand and compare the growth rate of the world's human population and the growth rate of the food supply. Growth rates are examples of rates of change of functions, and the study of rates of change is the subject of the first course in calculus.

2.1 Rates of Change

Average Rates of Change: Slopes of Linear Equations

In the introduction to this chapter we mentioned Malthus's belief in linear growth of the world's food supply. Probably, linear functions are familiar from your previous study of mathematics. You associate them with linear equations, such as

$$y = mx + b,$$

where x is the input (also called the **independent variable**), y is the output (also called the **dependent variable**), m is the slope, and b is the y-intercept (i.e., the value of the function when $x = 0$). For linear equations, we know that

$$\text{slope} = \frac{\text{rise}}{\text{run}}.$$

That is, the slope of the line connecting two points (x_1, y_1) and (x_2, y_2) is the ratio described by *each* of the following expressions—all of which say the same thing (see Figure 2.2):

- rise over run
- change in y over change in x
- $y_2 - y_1$ over $x_2 - x_1$

Figure 2.2 Slope equals rise over run

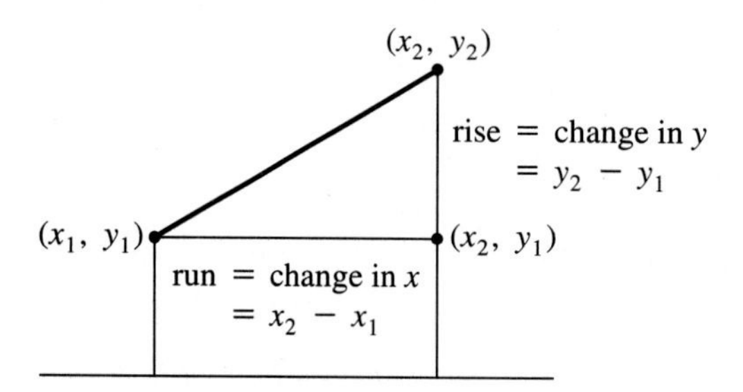

Lines are very special curves. The essence of linearity is that the slope of a line does not depend on which two points are selected for the computation. The ratio of rise to run is the same no matter where you start and no matter how long the run is or in which direction.

Exploration Activity 1

(a) Devise a method for drawing a line with slope 2. Pick a point you want the line to go through. Then decide how you will find a second point on the line. Draw your line, and use two other points on it to check its slope.

(b) Repeat for slope $-\frac{1}{2}$. Do you notice anything special about the relationship between this line and the one in part (a)?

Suppose we want to draw a line with slope 2. Then we need only mark a point P, move one unit to the right, move up two units, and mark a second point Q. The line connecting these two points (see Figure 2.3) has the desired slope. In Figure 2.4 we draw the line with slope $-\frac{1}{2}$. The two lines drawn in Exploration Activity 1 are perpendicular.

Figure 2.3 A line with slope 2

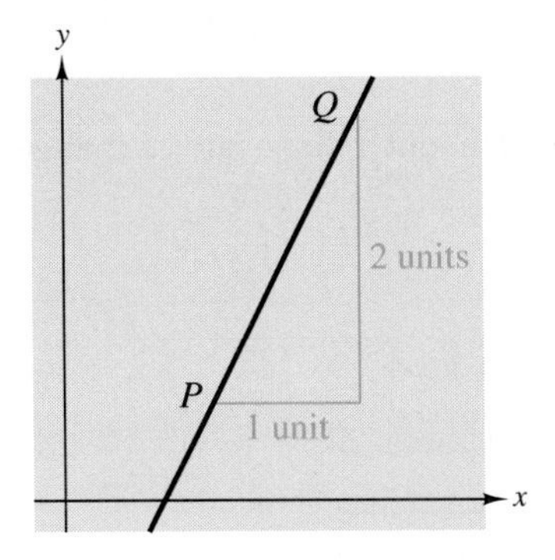

Figure 2.4 A line with slope $-\frac{1}{2}$

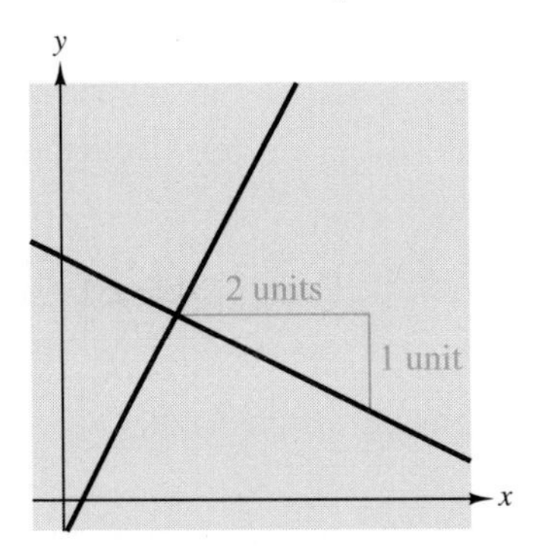

Checkpoint 1

(a) Draw a coordinate grid whose x- and y-scales are the same.

(b) Draw a line with slope $\frac{1}{4}$.

(c) Draw a line perpendicular to the line in part (b). What is the slope of this second line?

Average Rates of Change for More General Functions

In calculus we study rates of change in general, not just for linear functions. Our first step is to rewrite the equation

$$\text{slope} = \frac{\text{rise}}{\text{run}}$$

in the equivalent form

$$\text{average rate of change} = \frac{\text{the change in the dependent variable}}{\text{the change in the independent variable}}.$$

A linear function has only one rate of change: the slope of its graph. In this case, we may talk about *the* rate of change. However, the average rate of change of a more general function varies from one interval to another. In the next activity we ask you to compute some average rates of change in a familiar setting.

Exploration Activity 2

An object is dropped from a height. Table 2.1 records at each second the distance it has fallen.

(a) What is the average speed of the object for the first five seconds? For the first three seconds? For the first second?

(b) Estimate the average speed for the first 4.5 seconds.

(c) Suppose the height from which the object is dropped is 489 meters; what is the average speed for the last two seconds of its fall?

(d) Use your calculator to graph the data in Table 2.1. Change your window to plot just the portion of the graph that represents the last two seconds. How does your answer to (c) relate to what you see in the graph?

(e) Does it make sense to talk about a single rate of change for this function? Why or why not?

Table 2.1 Falling body data

Time (seconds)	*Distance (meters)*
1.0	5.0
2.0	19.4
3.0	44.1
4.0	78.0
5.0	122.8
6.0	175.8
7.0	240.0
8.0	312.8
9.0	396.1
10.0	489.0

You should have found that the average speeds are different for different intervals of time. For example, the average speed over the first 3 seconds is 14.7 meters per second, while in the first second it is only 5.0 meters per second.

You also can estimate average speeds over intervals whose endpoints are not in the table. The way to do that is to first estimate distances at the endpoint times. For example, at 4.5 seconds we might estimate the distance fallen to be 99 meters—a little closer to 78.0 than to 122.8 because the object falls farther in the second half of the $[4, 5]$ time interval than in the first half. That would make the average speed for the first 4.5 seconds 22 meters per second.

The average speed over the last two seconds is 88.1 meters per second, which is also the slope of the line segment connecting the data points corresponding to 8 and 10 seconds. When you zoom in on this part of the data graph, you find that the last three points lie almost on a straight line, so the average speed should closely approximate the actual speed throughout this time interval.

Checkpoint 2

Compute the average rate of change of the function defined by $y = x^2$ over each of the following intervals.

(a) $[0, 2]$

(b) $[3, 4]$

(c) $[4.5, 5]$

Mathematical Notation: The Difference Quotient

The phrases you see over and over—"rate of change," for example—are clearly important. In fact, we are repeating them frequently to stress that importance. However, once we have made our point, it becomes equally important that we reveal the standard notation for the important concepts.

The next concept we shorten to a symbol is "change," specifically, the change in a variable from one value to another.

Definition If x_1 and x_2 are values of the variable x, the **change** from x_1 to x_2 is the difference, $x_2 - x_1$. We write

$$\Delta x = x_2 - x_1$$

to abbreviate "The change in the variable x is the difference between the second value and the first."

In a similar manner, we may abbreviate $P_2 - P_1$ by ΔP, $s_2 - s_1$ by Δs, $t_2 - t_1$ by Δt, and so on. We combine these symbols to obtain a notation for the average rate of change of one variable with respect to another.

Definition If $s = f(t)$, then the **average rate of change** of s as t changes from t_1 to t_2 is the ratio of the change in s to the change in t, i.e.,

$$\frac{\Delta s}{\Delta t}.$$

This quotient of differences is called, naturally enough, a **difference quotient**.

Answers to Checkpoints

1. (a) and (b) There are many possible answers. Your lines do not have to pass through $(0, 0)$, but they should be parallel to the ones shown here.

   (c) The slope of the second line is -4.

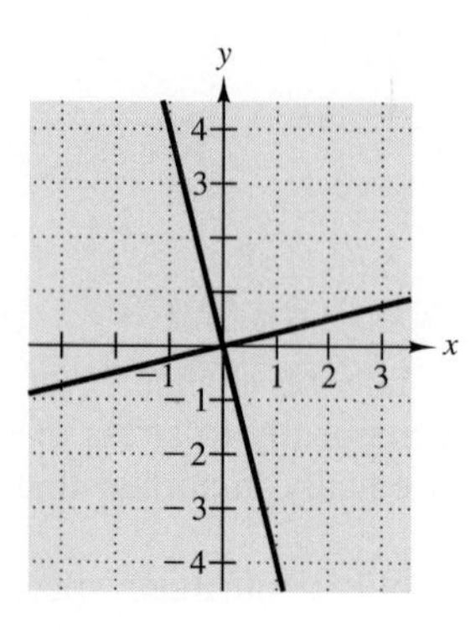

2. (a) 2 (b) 7 (c) 9.5

Exercises 2.1

For Exercises 1–4, draw a line with the indicated slope.

1. slope $-\frac{2}{3}$
2. slope $\frac{3}{5}$
3. slope $-\frac{3}{4}$
4. slope -2

Find the slope of each of the following lines.

5.

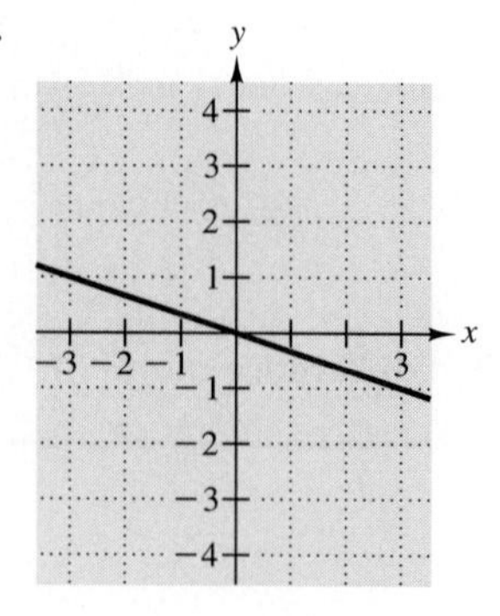

6.

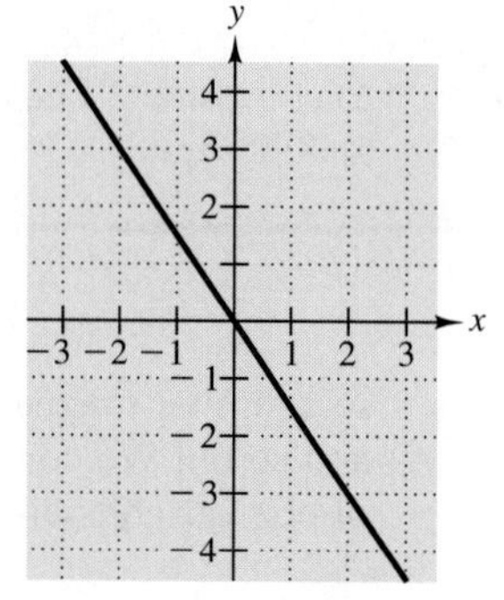

7.

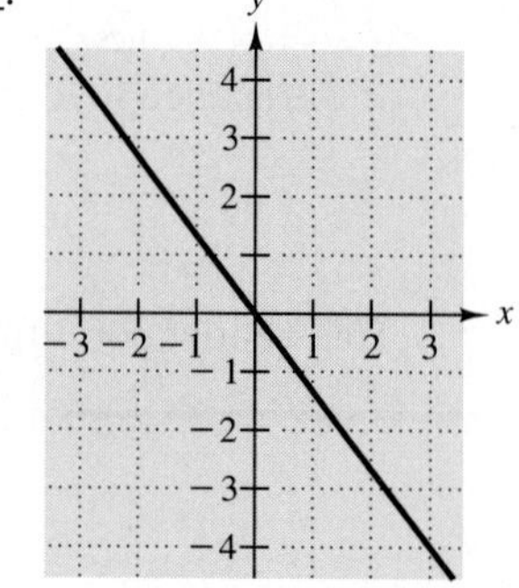

8.

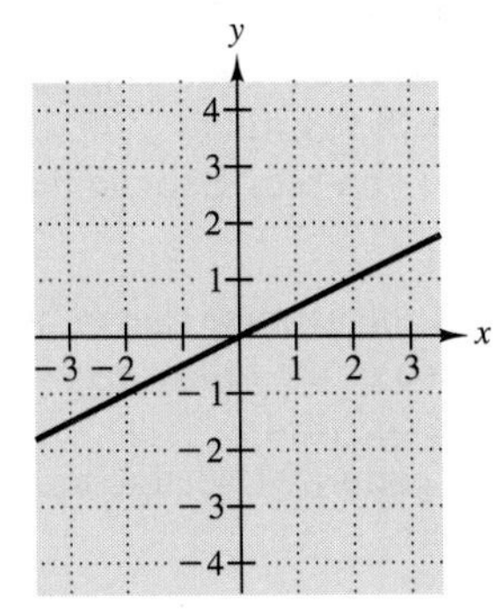

Find an equation for each of the following lines in the xy-plane.

9. The line with slope 1.5 through the point $(-1, 2.3)$.

10. The line through the points $(2.1, 1.7)$ and $(-1.5, 4.2)$.

11. The line with slope $\frac{1}{2}$ through the point $(5, -2)$.

12. The line through the points $(3, -7)$ and $(1, -3)$.

13. The line with slope 1.7 through the point $(1.5, -2.3)$.

14. The line through the points $(1.3, -0.7)$ and $(1.5, -3.2)$.

Sketch each of the following lines. (See Exercises 9–14.)

15. The line with slope 1.5 through the point $(-1, 2.3)$.

16. The line through the points $(2.1, 1.7)$ and $(-1.5, 4.2)$.

17. The line with slope $\frac{1}{2}$ through the point $(5, -2)$.

18. The line through the points $(3, -7)$ and $(1, -3)$.

19. The line with slope 1.7 through the point $(1.5, -2.3)$.

20. The line through the points $(1.3, -0.7)$ and $(1.5, -3.2)$.

For each of the following lines, choose two x values, x_1 and x_2, and calculate $\Delta y/\Delta x$. Then choose another pair of x values, and calculate $\Delta y/\Delta x$. (See Exercises 9–14.)

21. The line with slope 1.5 through the point $(-1, 2.3)$.

22. The line through the points $(2.1, 1.7)$ and $(-1.5, 4.2)$.

23. The line with slope $\frac{1}{2}$ through the point $(5, -2)$.

24. The line through the points $(3, -7)$ and $(1, -3)$.

25. The line with slope 1.7 through the point $(1.5, -2.3)$.

26. The line through the points $(1.3, -0.7)$ and $(1.5, -3.2)$.

Put your calculator in radian mode. Find the average rate of change of the sine function (use the SIN button) over each of the following intervals.

27. $[0, \pi/4]$

28. $[0, \pi/2]$

29. $[0, \pi]$

30. $[\pi/2, \pi]$

Put your calculator in degree mode. Find the average rate of change of the sine function (use the SIN button) over each of the following intervals.

31. $[0°, 45°]$

32. $[0°, 90°]$

33. $[0°, 180°]$

34. $[90°, 180°]$

Put your calculator in radian mode. Find the average rate of change of the cosine function (use the COS button) over each of the following intervals.

35. $[0, \pi/4]$

36. $[0, \pi/2]$

37. $[0, \pi]$

38. $[\pi/2, \pi]$

Put your calculator in degree mode. Find the average rate of change of the cosine function (use the COS button) over each of the following intervals.

39. $[0°, 45°]$

40. $[0°, 90°]$

41. $[0°, 180°]$

42. $[90°, 180°]$

43. A common way to check the accuracy of your speedometer is to drive one or more measured miles, holding the speedometer at a constant 60 mph and checking your watch at the start and end of the measured distance.

    (a) Explain the method. How does your watch tell you whether the speedometer is accurate?

    (b) Why shouldn't you use your odometer to measure the mile(s)?

44. (a) Make a table of the average rate of change of $y = x^2$ over each of the following intervals: $[0,1]$, $[1,2]$, $[2,3]$, $[3,4]$.
    (b) What pattern do you see? Explain this pattern by algebra.

45. (a) Make a table of the average rate of change of $y = x^3$ over each of the following intervals: $[0,1]$, $[1,2]$, $[2,3]$, $[3,4]$.
    (b) Can you find a pattern? Use algebra to confirm (or find) a pattern.

2.2 The Derivative: Instantaneous Rate of Change

In this section we move from average rates of change to instantaneous rates of change. To begin, we look at the instantaneous rates of change of an object falling near the surface of the earth without significant resistance by the air. This investigation will lead to the first major calculus concept: the derivative.

Zooming In: Local Linearity

We resume our study of the falling body problem, for which elementary physics provides a theoretical model (i.e., a formula) for distance fallen s as a function of time t :

$$s = ct^2,$$

where c is a constant that depends on the gravitational force and on the units of measurement. In this investigation we will measure time in seconds and distance in meters, so c is approximately 4.90 meters per second.

The question we address is this: How can we use the formula for distance as a function of time to determine the *instantaneous speed* of the falling object at any instant of its fall?

Exploration Activity 1

(a) Use your graphing tool to graph $s = 4.90\,t^2$. Zoom in several times in the vicinity of $t = 2$. What does the graph look like after you zoom in? What value seems reasonable for the instantaneous speed at $t = 2$?

(b) Repeat for $t = 5$.

Figure 2.5 shows the graph of our assumed relationship between time t and distance s. (For reasons that will be clear later, we have kept c in our vertical scale rather than multiplying the numbers out.) Notice that s (distance fallen, not height above the ground) increases as the object falls. At $t_1 = 5$ seconds, the object has fallen $s_1 = c(5)^2 = 25c$ meters. The point $(5, 25c)$ is marked on the graph.

In Figure 2.6 we repeat Figure 2.5 with an added zoom box, and we show the result of zooming in by magnifying the portion of the graph in the zoom box. We continue the zooming process in Figure 2.7 by magnifying the portion of the graph in a smaller zoom box, still centered at $(5, 25c)$.

Figure 2.5 Graph of $s = ct^2$

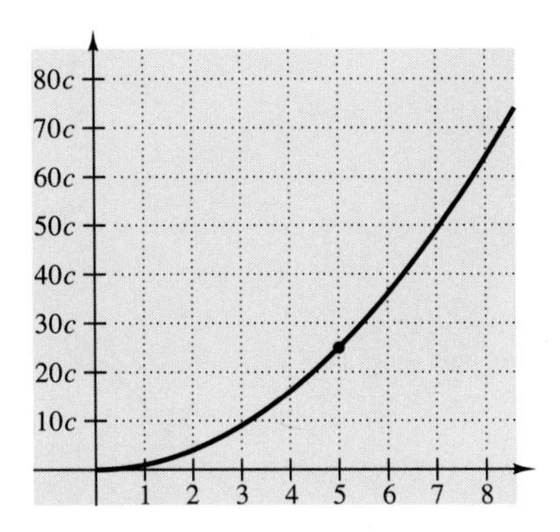

Figure 2.6 Zooming in near $t = 5$

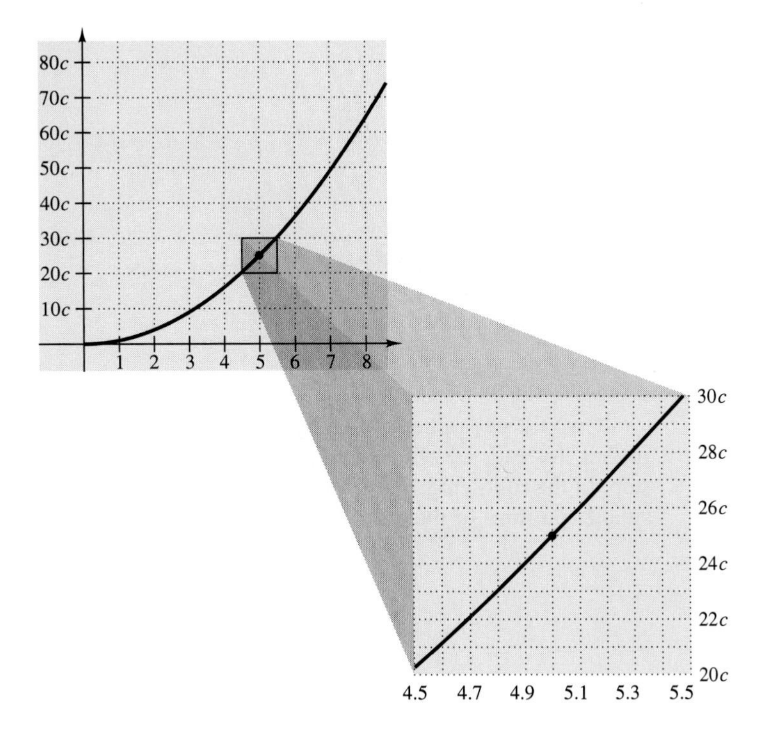

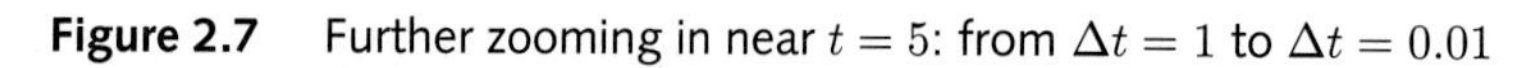

Figure 2.7 Further zooming in near $t = 5$: from $\Delta t = 1$ to $\Delta t = 0.01$

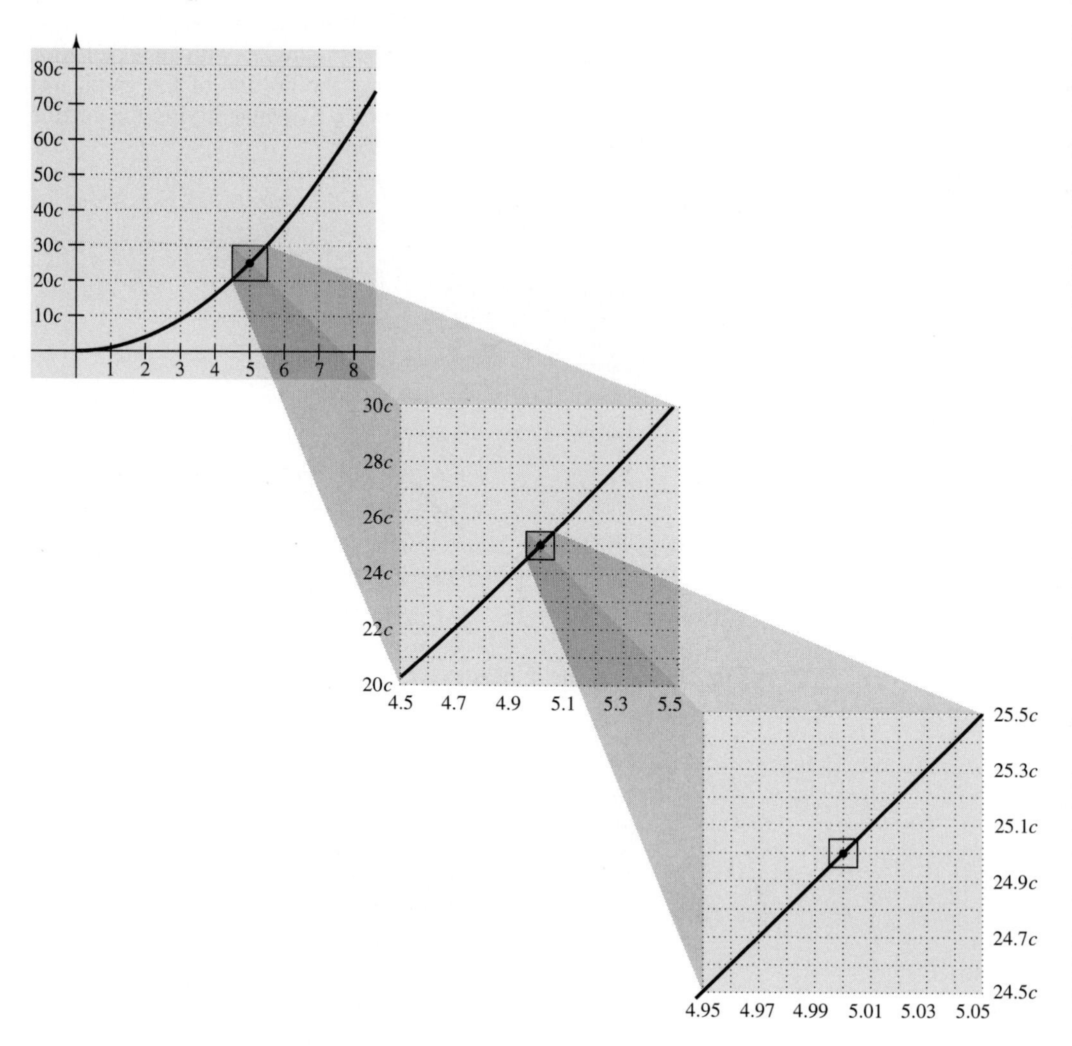

It would get rather unwieldy to repeat the whole sequence each time we zoom in further. Instead, we will identify the zoom box and show its magnification for each zoom step. In the following figures we start from the second step in Figure 2.7 and examine still smaller portions of the graph near $t = 5$ at still higher magnifications.

Figure 2.8 From $\Delta t = 0.1$ to $\Delta t = 0.01$

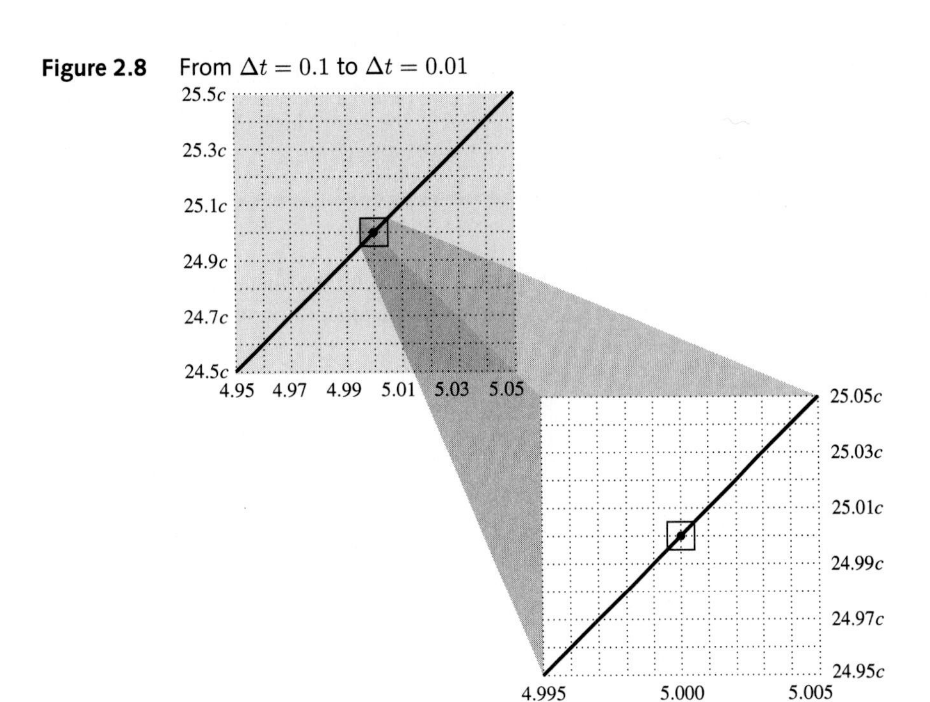

Figure 2.9 From $\Delta t = 0.01$ to $\Delta t = 0.001$

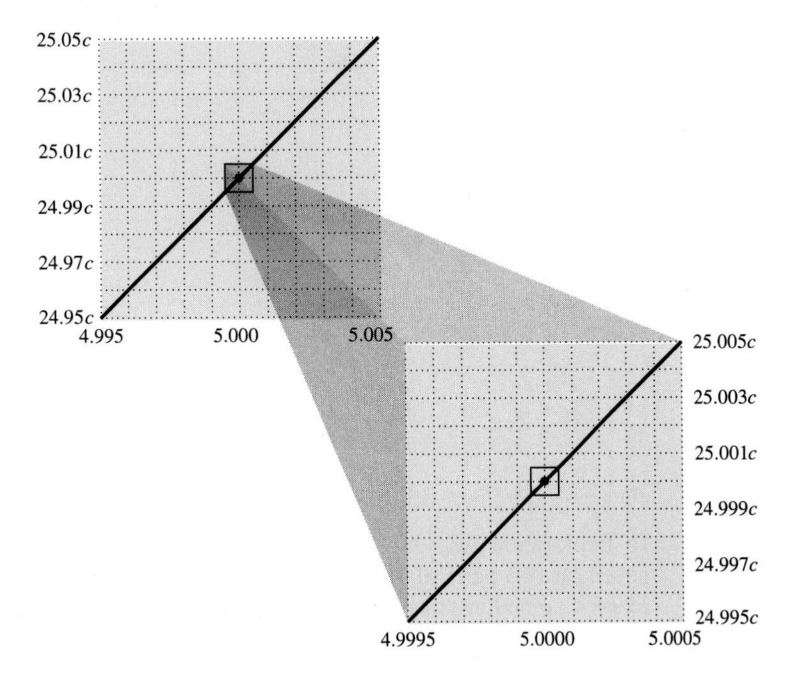

What you saw in Exploration Activity 1—and what leaps out from the sequence of right-hand pictures in Figures 2.6, 2.7, 2.8, and 2.9—is the emerging *straightness of the curves*. Lay a straightedge along each curve to check. This straightness you see as you look at shorter and shorter segments of the curve is a property of most well-behaved curves.

Definition The graph of a function $y = f(t)$ is said to be **locally linear** at a point (t_0, y_0) on the graph if, in the locality of the point, the curve looks like a straight line. In this case we also say that the function f is **locally linear** at t_0.

Now there is no problem calculating the instantaneous speed at $t = 5$, the rate at which s is increasing relative to t, because, on a straight line, that rate is just the slope, i.e., rise over run.

Checkpoint 1

(a) We repeat in Figure 2.10 the final zoom box in Figure 2.9. Use this figure to calculate the instantaneous speed at the instant $t = 5$. (Your answer should contain the constant c.)

(b) Recall that c is approximately 4.90. Find the numerical value of the instantaneous speed (in meters per second) of an object that has been falling for 5 seconds.

(c) Does this agree with your estimate in Exploration Activity 1?

Figure 2.10 A zoom-in view of the graph of $f(t) = ct^2$ with $\Delta t = 0.001$

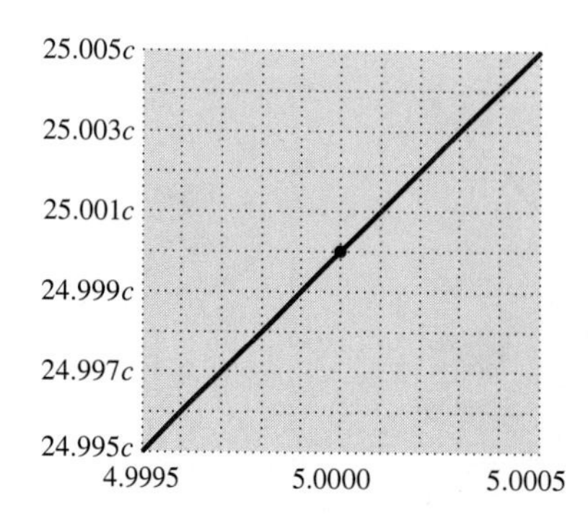

Graphical Calculation of Rates of Change

In Exploration Activity 1 we used the graphical viewpoint to determine instantaneous speed (at two particular instants) from the position function for a falling object. We also learned something that will enable us to use our graphing tool more effectively. We saw that we could get very good approximations to the rates of change at the instants $t = 2$ and $t = 5$ by calculating difference quotients of the form

$$\frac{\Delta s}{\Delta t}$$

(that is, a slope, a rise-over-run) with a *small* value of Δt.

Next, we will do that all at once for *every* value of t. That should give us a good approximation to a function whose value at each t is the instantaneous rate of change at t. We write

$$f(t) = ct^2$$

in order to have a name for the position function for the falling object, i.e.,

$$s = f(t),$$

where s is position at time t. We write $g(t)$ for the difference quotient with a suitably small Δt, say, $\Delta t = 0.001$:

$$g(t) = \frac{\Delta s}{\Delta t} = \frac{f(t + 0.001) - f(t)}{0.001}.$$

This new function $g(t)$ is *not* the same function as the instantaneous-rate-of-change function at t, but its value at each t should be very close to the instantaneous rate of change. Thus a graph of $g(t)$ should closely approximate a graph of the unknown instantaneous speed function.

We'll continue to ignore the fact that we actually know a value for the constant c. Instead, we'll study the whole family of functions of the form $f(t) = kt^2$.

Exploration Activity 2

Graph the approximate speed function g for $k = 1$. The graph has a familiar shape; what is it? Repeat for $k = \frac{1}{2}$ and $k = -\frac{1}{2}$. What seems to be true for the approximate speed functions for any k?

In Figures 2.11–2.18 we show the graphs of $f(t)$ and $g(t)$ for four different values of k. In particular, Figure 2.12 shows the graph of $g(t)$ for $k = 1$, which looks a lot like a straight line with slope 2. Since this line goes through the origin, an approximate formula for $g(t)$ is $2t$.

Figure 2.11 $f(t) = kt^2$ for $k = 1$

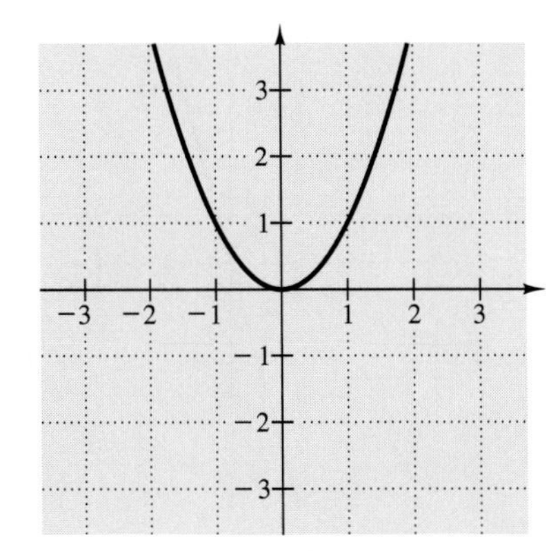

Figure 2.12 $g(t) = \dfrac{f(t + 0.001) - f(t)}{0.001}$ for $k = 1$

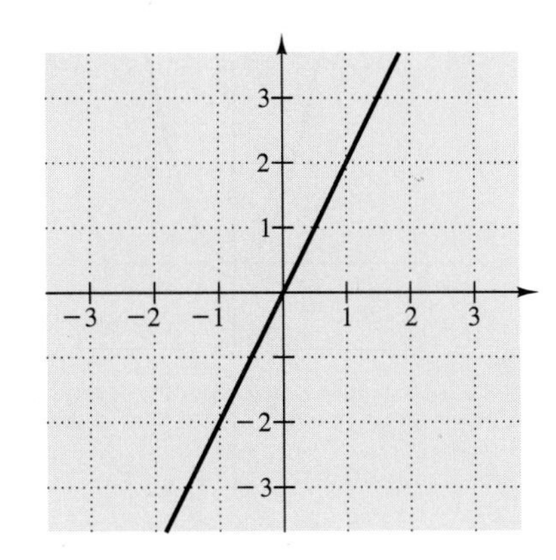

Figure 2.13 $f(t) = kt^2$ for $k = \frac{1}{2}$

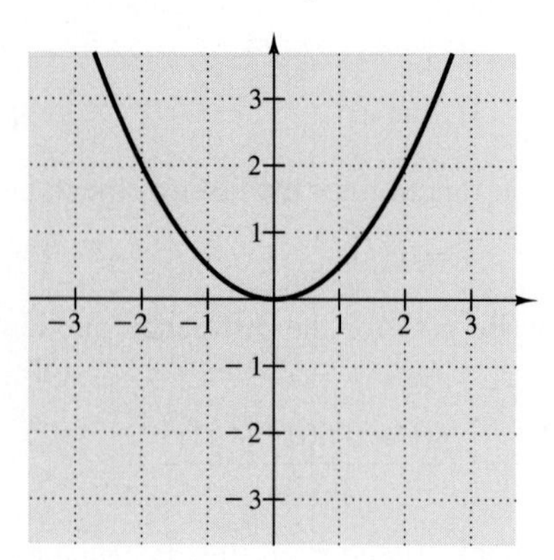

Figure 2.14 $g(t) = \dfrac{f(t + 0.001) - f(t)}{0.001}$ for $k = \frac{1}{2}$

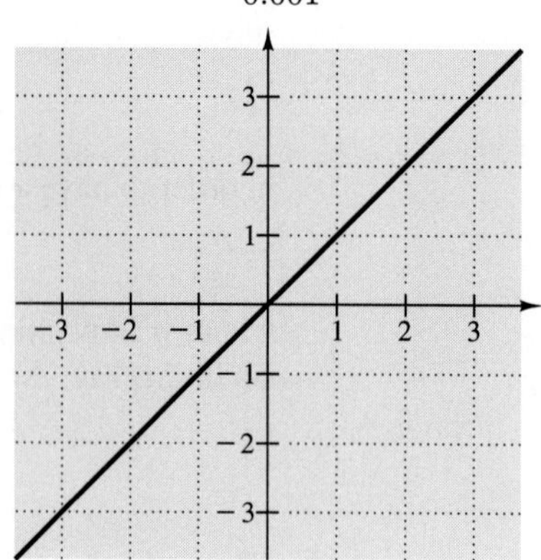

Figure 2.15 $f(t) = kt^2$ for $k = -\frac{1}{2}$

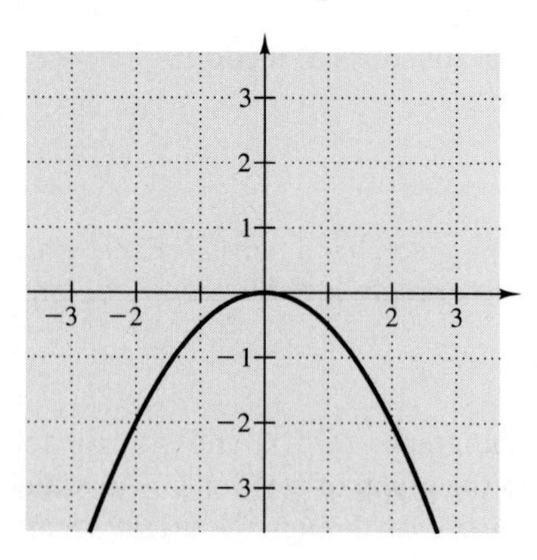

Figure 2.16 $g(t) = \dfrac{f(t + 0.001) - f(t)}{0.001}$ for $k = -\frac{1}{2}$

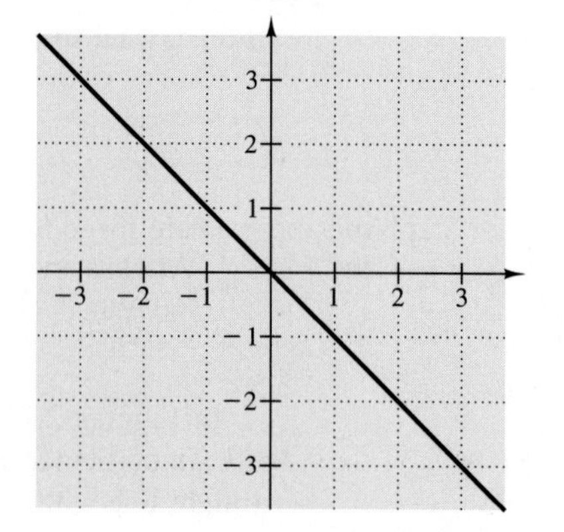

Figure 2.17 $f(t) = kt^2$ for $k = 1.8$

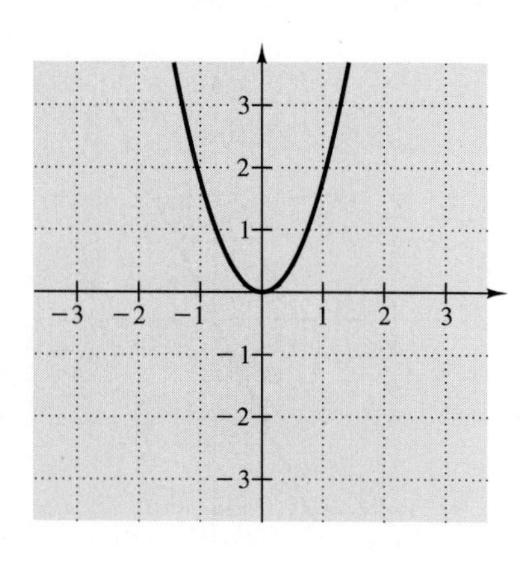

Figure 2.18 $g(t) = \dfrac{f(t + 0.001) - f(t)}{0.001}$ for $k = 1.8$

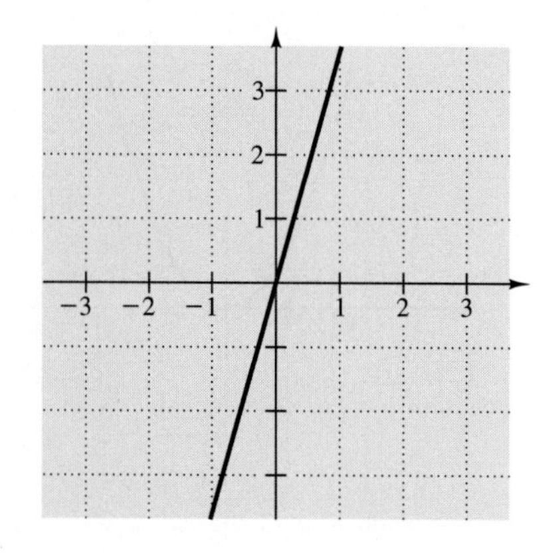

Checkpoint 2

(a) What is the slope of the graph in Figure 2.14? Write a simple linear formula for a function that approximates $g(t)$ when $k = \frac{1}{2}$.

(b) What is the slope of the graph in Figure 2.16? Write a simple linear formula for a function that approximates $g(t)$ when $k = -\frac{1}{2}$.

Checkpoint 3

Our fourth case is a little harder. Use your calculator to graph g for $k = 1.8$, and confirm Figure 2.18. Then use the trace feature of your calculator to find two points on the line, and use them to find the slope.

Exploration Activity 3

(a) Fill in the table below with an appropriate formula for the instantaneous speed function in each case.

$f(t)$	*Instantaneous speed at time* t
t^2	
$\frac{1}{2}t^2$	
$-\frac{1}{2}t^2$	
$1.8\,t^2$	

(b) If $f(t) = kt^2$ gives position at time t for an object moving in a straight line, what is your best guess at a formula for the instantaneous speed of the object at time t?

Your table should suggest that the instantaneous rate of change of the function $f(t) = kt^2$ is $2kt$.

Checkpoint 4

Check this formula against the result of Checkpoint 1; does it give the right answer when $t = 5$ and $k = c$?

Algebraic Calculation of Rates of Change

Using a high-tech graphing tool, supplemented by guessing, we have conjectured a formula for the instantaneous speed of a falling object whose position at time t is $s = ct^2$:

$$\text{speed at time } t = 2ct.$$

Our next approach is to use a low-tech tool: algebra. That's harder than looking at computer-drawn pictures but also, as we shall see, more satisfying, because there will not be any guessing at formulas.

In Figures 2.6 through 2.10 we used time intervals centered on $t = 5$, the time of interest. For purposes of our algebraic calculation, we will use intervals that have $t_1 = 5$ and $t_2 =$ some other time. This has the advantage that only one of the two time values changes when we change Δt. For the time being, think of using only the right-hand half

In Figures 2.6 through 2.10 we used time intervals centered on $t = 5$, the time of interest. For purposes of our algebraic calculation, we will use intervals that have $t_1 = 5$ and $t_2 =$ some other time. This has the advantage that only one of the two time values changes when we change Δt. For the time being, think of using only the right-hand half of each of the final zoom boxes in Figures 2.6 through 2.9. We repeat those boxes here as Figures 2.19 to 2.22.

Figure 2.19 The graph of $f(t) = ct^2$ from $t = 4.5$ to $t = 5.5$

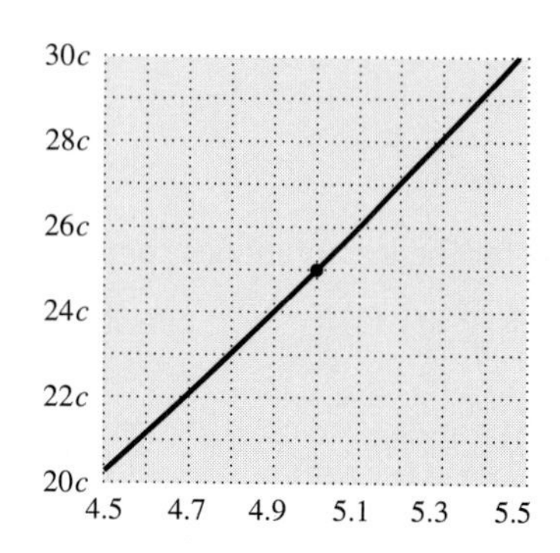

Figure 2.20 The graph of $f(t) = ct^2$ from $t = 4.95$ to $t = 5.05$

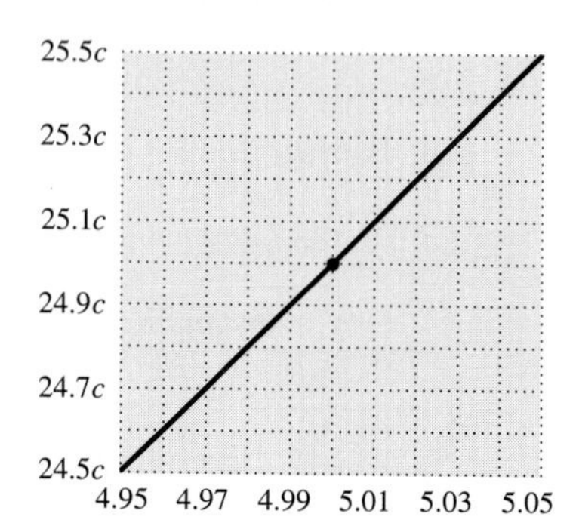

With $t_1 = 5$, we consider first the small change in time, $\Delta t = 0.5$, that leads to $t_2 = 5.5$ seconds, the right-hand edge of Figure 2.19. The distance fallen at time t_2 is $s_2 = (5.5)^2 c = 30.25c$ meters. Thus the average speed (rate of change of distance) from t_1 to t_2 is

$$\frac{\Delta s}{\Delta t} = \frac{s_2 - s_1}{t_2 - t_1} = \frac{30.25c - 25c}{0.5} = 10.5c \text{ meters/second.}$$

This only approximates the speed *at* t_1, but we get a closer approximation by choosing a smaller Δt, as in Figure 2.20, namely, $\Delta t = 0.05$. Then

$$\frac{\Delta s}{\Delta t} = \frac{s_2 - s_1}{t_2 - t_1} = \frac{25.5025c - 25c}{0.05} = 10.05c \text{ meters/second.}$$

Checkpoint 5

We still have only an approximation to the instantaneous speed at $t = 5$, but the approximation is better, because the time interval is smaller. What do you think the average speed will be if we choose $\Delta t = 0.005$, as in Figure 2.21? Would it surprise you if the answer is $10.005c$ meters per second? Check it! Then do the calculation one more time, as in Figure 2.22, with $\Delta t = 0.0005$.

Figure 2.21 The graph of $f(t) = ct^2$ from $t = 4.995$ to $t = 5.005$

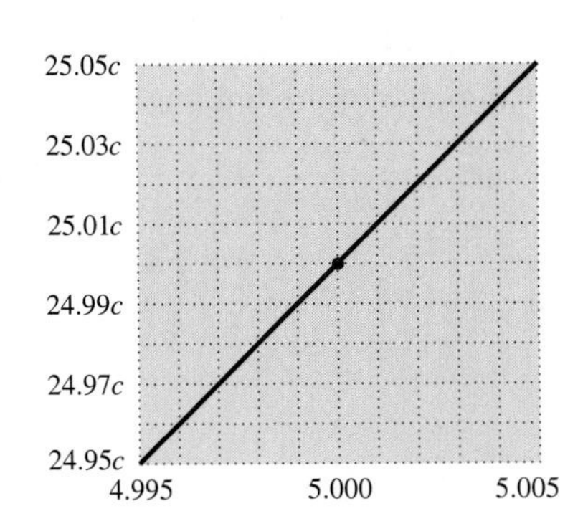

Figure 2.22 The graph of $f(t) = ct^2$ from $t = 4.9995$ to $t = 5.0005$

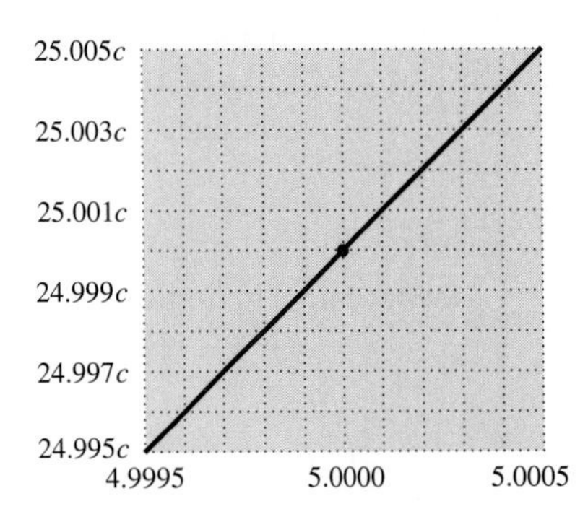

Now we observe that our speed calculation for the falling object does not depend on the specific instant $t_1 = 5$. Indeed, the whole process often turns out to be *easier* if we do it algebraically (for an arbitrary but unspecified t_1) rather than arithmetically, as we show in the next example.

Example 1 (a) Calculate algebraically $\Delta s/\Delta t$ for $s = ct^2$.

(b) What happens to the result as Δt becomes smaller and smaller?

Solution

(a) *Algebraic Step* — *Reason*

$$\frac{\Delta s}{\Delta t} = \frac{s_2 - s_1}{t_2 - t_1}$$
This is the definition of $\Delta s/\Delta t$.

$$= \frac{ct_2^2 - ct_1^2}{t_2 - t_1}$$
We substituted from our formula for s.

$$= \frac{c(t_2 - t_1)(t_2 + t_1)}{t_2 - t_1}$$
We factored the numerator.

$$= c(t_2 + t_1)$$
We canceled the factor $t_2 - t_1$.

$$= c(t_1 + \Delta t + t_1)$$
We substituted for t_2.

$$= c(2t_1 + \Delta t)$$
We collected terms.

(b) Now we can see clearly what happens as Δt becomes smaller and smaller. The Δt term in our computed average speed disappears, and the average speed approaches $2ct_1$. This agrees with our result when t_1 is 5 (Checkpoint 1), and it also should confirm your conjectured formula in Exploration Activity 3, since it tells us the instantaneous speed at *every* instant t_1. ■

Checkpoint 6

If the position s of a falling body at time t is given by $s = ct^2$, where $c = 4.90$, find the instantaneous speeds at the following times: $t = 1$, $t = 2.5$, $t = 17$, and $t = \sqrt{34}$.

The Derivative

To calculate instantaneous rate of change from average rate of change, we have to let the change in the independent variable (Δt in the example above) approach zero without actually letting it equal zero. We abbreviate the process of letting Δt approach zero by "$\Delta t \to 0$." Now we give a mathematical name to the limiting value of average rates of change.

Definition The limiting value of the difference quotient

$$\frac{\Delta s}{\Delta t}$$

as $\Delta t \to 0$ is called the **derivative of s with respect to t** (at the particular value of t in question) and is denoted by

$$\frac{ds}{dt}.$$

Of course, when the change in the independent variable approaches 0, the change in the dependent variable usually does also. Thus ds/dt is what $\Delta s/\Delta t$ approaches when Δt and Δs both approach zero. Think back to the zooming process in Figures 2.6 through 2.10. When we look at a segment of the graph of s (as a function of t) that is so small it appears straight, then ds/dt is the slope of that straight line.

We now have a notation,

$$\frac{ds}{dt},$$

to represent the instantaneous rate of change. In the transition from difference quotient to derivative, each upper-case delta, standing for *difference*, is replaced by a lower-case d, which stands for *differential.*

The calculation in Example 1 shows that if

$$s = ct^2,$$

then

$$\frac{ds}{dt} = 2ct.$$

There is still more notation and terminology, even without a single additional concept. The instantaneous rate of change (think *derivative*) of position or distance is what we have been calling "instantaneous speed"—at least for an object that moves in only one direction along a straight line. For motion in general, this rate is called **velocity**. We abbreviate the sentence

"Velocity is the instantaneous rate of change of position"

to

$$v = \frac{ds}{dt}.$$

Finally, when we want to focus on functional notation, we use a notation for "the derived function": If $s = f(t)$, then $v = f'(t)$ (which is read, "f prime of t"). The notations f' and ds/dt are interchangeable. Each will be used when it is convenient for the task at hand. Thus, for Example 1,

$$s = f(t) = ct^2$$

and

$$f'(t) = \frac{ds}{dt} = 2ct.$$

Exploration Activity 4

Suppose $s = 2t^2 + 3$.

(a) Use the zooming feature on your calculator to determine the values of ds/dt at $t = 1$, $t = 2$, and $t = 3$.

(b) Conjecture a formula for ds/dt that agrees with your calculation in part (a).

(c) Use algebra to calculate $\Delta s/\Delta t$, where $\Delta t = t_2 - t_1$.

(d) From your calculation in part (c), write down a formula for ds/dt at t_1.

(e) Write down a formula for ds/dt as a function of t. [This requires only a simple modification of your formula in part (d).] Does your formula agree with your conjecture in part (b)?

(f) Graph $y = 2t^2$ and $y = 2t^2 + 3$ together. Explain geometrically why the two functions should have the same derivative at every t.

The derivative of $s = 2t^2 + 3$ is the same as the derivative of $y = 2t^2$, namely, $4t$. Geometrically, this is clear, since the graph of $s = 2t^2 + 3$ is just an upward translation of the graph of $y = 2t^2$. In particular, at $t = 1,\ 2,$ and $3,$ the derivative assumes the values $4,\ 8,$ and 12, respectively.

In this section we have seen that, for most functions f, if you zoom in on the graph of $y = f(t)$ near a point (t_0, y_0), the graph appears to be a straight line. This property of the function is called "local linearity"; the slope of the apparent straight line is the value of the derivative of f at t_0. From an algebraic point of view, the derivative of f at t_0 is the limiting value of the difference quotients $\Delta y/\Delta t$ as Δt approaches 0. Here Δt is the difference $t_1 - t_0$ between a nearby point t_1 and t_0, and Δy is the corresponding difference of y-values.

We can think of the derivative as the instantaneous rate of change of f at t_0. In particular, if y represents distance and t represents time, the derivative is called "speed" or "velocity."

We denote the derivative by dy/dt if we want to emphasize the variable notation or by $f'(t)$ if we want to display the function name f.

Answers to Checkpoints

1. (a) $10c$ meters per second (b) 49 meters per second
2. (a) $1,\ t$ (b) $-1,\ -t$
3. 3.6
5. $10.0005c$
6. $9.8,\ 24.5,\ 166.6,$ about 57.14

Exercises 2.2

In Exercises 1–3, $f(t)$ is the reciprocal function:

$$f(t) = \frac{1}{t}.$$

1. Sketch the graph of $f(t)$. Check your graph with your graphing calculator.
2. Let $g(t)$ be the difference quotient function for a difference of $\Delta t = 0.001$:

$$g(t) = \frac{f(t+0.001) - f(t)}{0.001}.$$

   (a) Carefully select 8 values of t, and calculate the corresponding values of $g(t)$. (Your careful selection should enable you to sketch the graph of g.)
   (b) Sketch the graph of $g(t)$. Check your graph with your graphing calculator.
   (c) Describe the graph of $g(t)$ in words. For example, you might say how it compares with the graph of $f(t)$.
3. (a) Calculate the difference quotient $\Delta f/\Delta t$ algebraically for an arbitrary Δt.
   (b) Simplify your algebraic expression for $\Delta f/\Delta t$ until the Δt in the denominator cancels. What does the difference quotient approach as $\Delta t \to 0$?
   (c) What is the derivative of the reciprocal function?
   (d) Graph the derivative function you just obtained in part (c). How does it compare with the function g in the previous Exercise?

In Exercises 4–6, $f(t)$ is the cubing function:

$$f(t) = t^3.$$

4. Sketch the graph of $f(t)$. Check your graph with your graphing calculator.
5. Let $g(t)$ be the difference quotient function for a difference of $\Delta t = 0.001$:

$$g(t) = \frac{f(t+0.001) - f(t)}{0.001}.$$

   (a) Carefully select 8 values of t, and calculate the corresponding values of $g(t)$. (Your careful selection should enable you to sketch the graph of g.)
   (b) Sketch the graph of $g(t)$. Check your graph with your graphing calculator.
   (c) Describe the graph of $g(t)$ in words.
6. (a) Calculate the difference quotient $\Delta f/\Delta t$ algebraically for an arbitrary Δt.
   (b) Simplify your algebraic expression for $\Delta f/\Delta t$ until the Δt in the denominator cancels. What does the difference quotient approach as $\Delta t \to 0$?
   (c) What is the derivative of the function $f(t) = t^3$?
   (d) Graph the derivative function you just obtained in part (c). How does it compare with the function g in the preceding exercise?

If the position s of a falling body at time t is given by $s = ct^2$, where $c = 4.90$, find the instantaneous speeds at the following times.

7. $t = 3$
8. $t = 2.34$
9. $t = \sqrt{7}$

Suppose $s = 3t^2 + 5$. Use the zooming feature on your calculator or computer program to determine the values of ds/dt at each of the following times.

10. $t = 0.5$

11. $t = 4$

12. $t = 7$

13. Explain why the derivative of $s = 3t^2 + 5$ should be $6t$. (See Exploration Activity 4.) Use this to check your answers to Exercises 10–12.

14. If a and b are constants, find a formula for the derivative of $y = at + b$.

15. If a, b, and c are constants, find a formula for the derivative of $y = at^2 + bt + c$. Check to see that your formula agrees with your answer to Exercise 13.

16. Suppose gasoline sells for \$1.19 a gallon in June and for \$1.25 a gallon in July. What can you say about the average price of a gallon of gasoline during those 2 months?[1]

17. (See Exercise 17 at the end of Section 1.2.)
    (a) What is the rate of change of temperature in degrees Fahrenheit with respect to temperature in degrees Celsius?
    (b) What is the rate of change of temperature in degrees Celsius with respect to temperature in degrees Fahrenheit?
    (c) What relationship do you observe between these two rates?

18. (See Exercise 18 at the end of Chapter 1.) Suppose a weight of 10 ounces stretches a spring 1 inch.
    (a) For arbitrary weights hung from the same spring (within the elastic limit), what is the rate of change of weight with respect to displacement?
    (b) What is the rate of change of displacement with respect to weight?
    (c) What relationship do you observe between these two rates?

In Exercises 19–24, use a graphical approach to approximate the derivative of the function at $t = 1$ and $t = 2$.

19. $\log t$

20. $t \log t$

21. $\sin t$

22. $\tan t$

23. 2^t

24. 10^t

25. Figures 2.23, 2.24, and 2.25 show successive zoom-ins on the graph of $f(t) = |t|$ in the vicinity of the point $(0, 0)$.
    (a) Is this function locally linear at that point? Why or why not?
    (b) What is the average slope of the graph from $-\Delta t$ to Δt for a small value of Δt?
    (c) Can you use such an average slope to estimate the rate of change at 0? Why or why not?
    (d) What is the average slope of the graph from 0 to Δt for a small value of Δt? From $-\Delta t$ to 0?
    (e) What do you conclude about instantaneous rate of change at $t = 0$?

Figure 2.23 The absolute value function

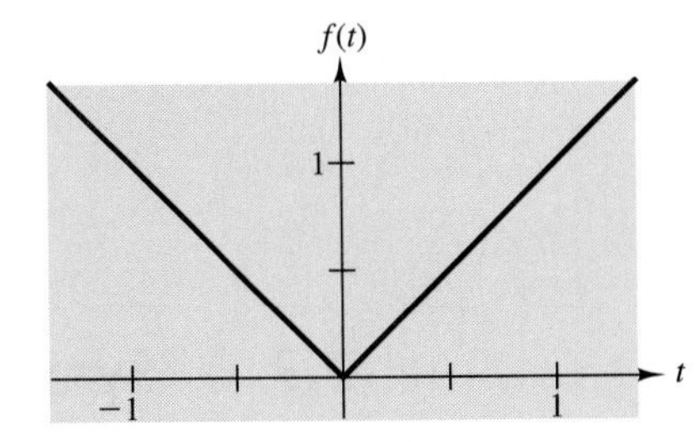

Figure 2.24 The absolute value function

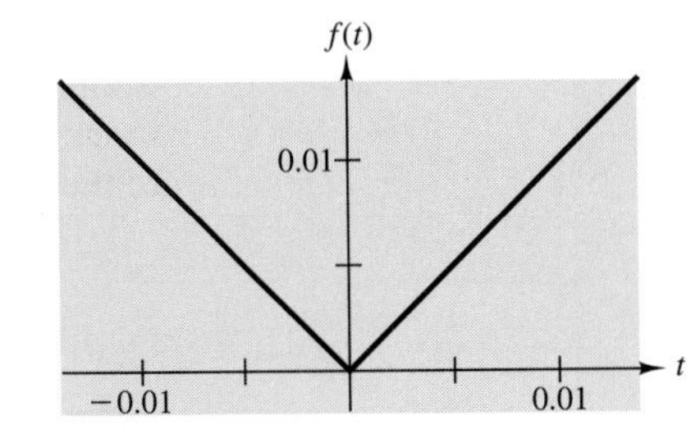

Figure 2.25 The absolute value function

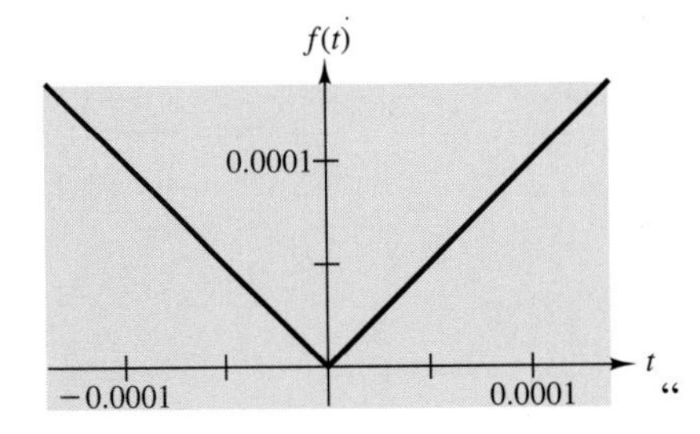

26. Use your calculator to zoom in on the graph of $f(t) = |t^3|$ in the vicinity of $(0, 0)$.
    (a) Is this function locally linear at that point? Why or why not?
    (b) What is the average slope of the graph from $-\Delta t$ to Δt for a small value of Δt?
    (c) Can you use such an average slope to estimate the rate of change at 0? Why or why not?
    (d) What is the average slope of the graph from 0 to Δt for a small value of Δt? From $-\Delta t$ to 0?
    (e) What do you conclude about instantaneous rate of change at $t = 0$?

27. Use your calculator to zoom in on the graph of $f(t) = |t|^{3/2}$ in the vicinity of $(0, 0)$.
    (a) Is this function locally linear at that point? Why or why not?
    (b) What is the average slope of the graph from $-\Delta t$ to Δt for a small value of Δt?
    (c) Can you use such an average slope to estimate the rate of change at 0? Why or why not?
    (d) What is the average slope of the graph from 0 to Δt for a small value of Δt? From $-\Delta t$ to 0?
    (e) What do you conclude about instantaneous rate of change at $t = 0$?

28. Use your calculator to zoom in on the graph of $f(t) = |t|^{2/3}$ in the vicinity of $(0, 0)$.
    (a) Is this function locally linear at that point? Why or why not?
    (b) What is the average slope of the graph from $-\Delta t$ to Δt for a small value of Δt?
    (c) Can you use such an average slope to estimate the rate of change at 0? Why or why not?
    (d) What is the average slope of the graph from 0 to Δt for a small value of Δt? From $-\Delta t$ to 0?
    (e) What do you conclude about instantaneous rate of change at $t = 0$?

---

1. From *Calculus Problems for a New Century*, edited by Robert Fraga, MAA Notes Number 28, 1993.

2.3 Symbolic Calculation of Derivatives: Polynomial Functions

In the last section we saw that the derivative was a measure of the instantaneous rate of change of a variable. We can calculate the derivative by zooming in on the graph of the function in question. We also saw how to use algebra to find the derivative of functions of the form kt^2. In this section we will consider more general polynomial functions.

Derivatives of Power Functions

Before we tackle the problem of finding the derivative of a polynomial function, we start with the simpler question of finding the derivative of a power function. We know some special cases:

$$\frac{d}{dt}k = 0$$ (If there is no change, the rate of change is zero.)

$$\frac{d}{dt}kt = k$$ (The rate of change of a linear function is the *slope* of its graph.)

$$\frac{d}{dt}kt^2 = 2kt$$ (Look back at Example 3 in Section 2.2.)

What about the general power function t^n, where n may be any positive integer? We repeat the cases we know (with k set equal to 1). Note that $t^0 = 1$ and $t^1 = t$.

$$\frac{d}{dt} 1 = 0$$

$$\frac{d}{dt} t = 1$$

$$\frac{d}{dt} t^2 = 2t$$

Let's do one more case by direct calculation of $\Delta s/\Delta t$, where $s = t^3$. We will do this one by writing

$$\Delta s = s(t + \Delta t) - s(t).$$

Here are the steps with justifications:

Algebraic Step	*Reason*
$\dfrac{\Delta s}{\Delta t} = \dfrac{(t+\Delta t)^3 - t^3}{\Delta t}$	We substituted for s.
$= \dfrac{\left[t^3 + 3t^2(\Delta t) + 3t(\Delta t)^2 + (\Delta t)^3\right] - t^3}{\Delta t}$	We expanded $(t + \Delta t)^3$.
$= \dfrac{3t^2(\Delta t) + 3t(\Delta t)^2 + (\Delta t)^3}{\Delta t}$	We removed the brackets: $t^3 - t^3 = 0$.
$= 3t^2 + 3t(\Delta t) + (\Delta t)^2$	We divided through by Δt.

Now, as $\Delta t \to 0$, the second and third terms on the right also approach zero, so

$$\frac{d}{dt} t^3 = 3t^2.$$

Exploration Activity 1

(a) Carry out a similar calculation to find the derivative of t^4. You may use either the factoring technique (Example 1 in Section 2.2) or the binomial expansion technique just illustrated.

(b) Make a conjecture about the derivative

$$\frac{d}{dt} t^n$$

for a general positive integer n.

You should have found that

$$\frac{d}{dt} t^4 = 4t^3.$$

Based on the examples so far, we may formulate a general rule for differentiating power functions of the form t^n.

Power Rule If n is any positive integer, then

$$\frac{d}{dt} t^n = nt^{n-1}.$$

Next, we consider derivatives of functions such as $f(t) = 6t^3$. Recall that we have already shown that

$$\frac{d}{dt} ct^2 = 2ct = c \frac{d}{dt} t^2.$$

For general derivatives of this form, we need the Constant Multiple Rule.

Constant Multiple Rule If $f(t)$ is any function that has a derivative, and c is any constant, then

$$\frac{d}{dt} cf(t) = c \frac{df}{dt}.$$

That is, constant factors can be factored out of derivative calculations.

Example 1 Use the Constant Multiple Rule to calculate the derivative of $6t^3$.

Solution

$$\frac{d}{dt} 6t^3 = 6 \frac{d}{dt} t^3 = 6\left(3t^2\right) = 18t^2$$

The Constant Multiple Rule follows from some simple algebra with difference quotients. If $s = cf(t)$, then

$$\begin{aligned}\frac{\Delta s}{\Delta t} &= \frac{cf(t+\Delta t) - cf(t)}{\Delta t} \\ &= c\,\frac{f(t+\Delta t) - f(t)}{\Delta t}.\end{aligned}$$

As $\Delta t \to 0$, we obtain

$$\frac{d}{dt} cf(t) = c \frac{df}{dt}.$$

We need one more differentiation rule to be able to differentiate general polynomials: the Sum Rule. In words, the Sum Rule says that the derivative of the sum of two functions is the sum of the derivatives of the two functions. This can be established in much the same way as the Constant Multiple Rule.

Sum Rule If $f(t)$ and $g(t)$ are any functions that have derivatives, then

$$\frac{d}{dt}[f(t)+g(t)] = \frac{df}{dt} + \frac{dg}{dt}.$$

Polynomials are just sums of terms of the form kt^n for various values of k and n. We now know how to differentiate each such term, and we also know how to differentiate sums.

Example 2 Apply the Sum, Constant Multiple, and Power Rules to differentiate the polynomial $3t^5 - 2t^3 + 7t^2 + 3$.

Solution

$$\begin{aligned}\frac{d}{dt}\left(3t^5 - 2t^3 + 7t^2 + 3\right) &= \frac{d}{dt}\left(3t^5\right) + \frac{d}{dt}\left(-2t^3\right) + \frac{d}{dt}\left(7t^2\right) + \frac{d}{dt}(3)\\ &= 3\frac{d}{dt}t^5 - 2\frac{d}{dt}t^3 + 7\frac{d}{dt}t^2 + 0\\ &= 3\cdot 5t^4 - 2\cdot 3t^2 + 7\cdot 2t\\ &= 15t^4 - 6t^2 + 14t\end{aligned}$$

■

Checkpoint 1

Find the derivative of each polynomial.

(a) $5t^3 + 7t^2 - 3t + 6$

(b) $t^4 - 18t^2 - 10t + 39$

In this section we derived the Power Rule and the Constant Multiple Rule for differentiation, and we stated the Sum Rule. With these three rules in hand, we can easily find the symbolic derivative of any polynomial.

Answers to Checkpoint

1. (a) $15t^2 + 14t - 3$ (b) $4t^3 - 36t - 10$

Exercises 2.3

Find the derivative of each of the following functions.

1. $2t^5 - 7t^3 + 4t^2 - 6t + 1$
2. $t^6 + 2t^4 - 4t^2 - 6$
3. $7t + 6$
4. $3t^2 + 7t + 6$
5. $3t^4 - t^3 + 7t^2 - t + 17$

6. $-t^3 + 5t^2 - 27t + 13$
7. 25.6
8. $(1 + 2t)^3$
9. $t^5 - 7t^4 + 2t^2 - 6t + 1$
10. $t^5 + 2t^4 - 3t^2 - 8$
11. $4t^4 - t^3 + 2t^2 - t + 7$
12. $13 - 27t + 5t^2 + t^3$
13. Use either the factoring technique or the expansion technique to explain how the Power Rule follows for arbitrary n.
14. Use difference quotients to derive the Sum Rule.
15. It follows from your calculation of derivatives of polynomials that

$$\frac{d}{dx}(mx + b) = m.$$

That is, the instantaneous rate of change of a linear function (at every x) is the same thing as its average rate of change, i.e., its slope. Explain why this is true.

16. Suppose $f(t)$ is any function that has a derivative, and C is any constant. Show that the function $g(t) = f(t) + C$ has the same derivative as $f(t)$.
17. (a) Graph the curve $y = 1/t^2$.
    (b) Zoom in on a representative selection of points, and estimate the derivative of $1/t^2$ at those points.
    (c) Use these estimations to formulate a rule for the derivative of $1/t^2$.
    (d) Check your rule by making graphical estimations of the derivative at two additional points.
18. Suppose the radius r of an inflating spherical balloon is given by the formula

$$r(t) = 2 + \frac{t}{2} \text{ centimeters.}$$

    (a) What is the rate of change of the radius?
    (b) Use your calculator to obtain graphical estimates of the rate of change of the surface area at $t = 2,\ 3,$ and 4 seconds. Formulate a general rule for the rate of change of the surface area, and check it at two additional points.
    (c) Use your calculator to obtain graphical estimates of the rate of change of the volume at $t = 2,\ 3,$ and 4 seconds. Formulate a general rule for the rate of change of the volume, and check it at two additional points.
    (d) Is this a realistic model for an inflating balloon? Why or why not?

2.4 Exponential Functions

When we take up population growth in Section 2.5, we will study functions P of time t whose rate of change is proportional to P itself, that is, functions P such that

$$\frac{dP}{dt} = kP$$

for some P. In this section we investigate what sort of functions have this property. We begin by considering another context in which the rate of growth is proportional to the amount present.

Interest Rates and Exponential Growth

Your savings account grows—because interest is added—at a rate proportional to the amount present in your account.

Exploration Activity 1

Suppose you start with \$300 in an account that pays interest once a year (i.e., interest is "compounded annually") at a rate of 5%, and you do not make any additional deposits or withdraw any of the money.

(a) Find the balance at 1, 2, 3, 4, and 5 years.

(b) Find the rate of growth for each of those years.

(c) For each of those years, find the ratio of the rate of growth to the balance at the start of the year.

At the end of one year, you will have the original \$300 plus $\$300 \times 0.05$ (or \$15). That is, your balance will grow by \$15 to become $300 + 15 = \$315$. At the end of two years, your interest will be $\$315 \times 0.05$ (or \$15.75), so your new balance will be $315 + 15.75 = \$330.75$. At the end of three years, your interest will be $\$330.75 \times 0.05 = \$16.54$, and your new balance will be $330.75 + 16.54 = \$347.29$. Notice that the *growth* (interest) each year is the same number as the *rate of growth*—because the time step Δt is 1 year. For the second year the ratio of the rate of growth to the starting balance is $15.75/315 = 0.05$; that is, the rate of growth is proportional to the beginning balance, with proportionality constant 0.05. The same is true for each of the other years.

We repeat the calculation from Exploration Activity 1 in another way. At the end of the first year, your new balance will be the starting amount plus interest. That is, your balance will be

$$\$300 + \$300 \times 0.05 = \$300 \times 1.05 = \$315.$$

At the end of the second year, you will have

$$\$315 + \$315 \times 0.05 = \$315 \times 1.05 = \$300 \times (1.05)^2.$$

(Use your calculator to check that this last expression gives \$330.75, as you got in Exploration Activity 1 by adding.) At the end of the third year, you will have $\$300 \times (1.05)^3$, or \$347.29. In general, after n years you will have $\$300 \times (1.05)^n$. Notice how the repeated *addition* of interest also can be calculated by repeated *multiplication* by (1 + the interest rate).

If you start with A dollars instead of \$300, then the amount you would have after n years is $A(1.05)^n$ dollars. If the percentage interest rate is $r\%$ rather than 5%, then the decimal interest rate is $k = r/100$, and the amount you have after n years is $A(1+k)^n$ dollars. Notice that time (represented by n) appears in the exponent in the expression $A(1+k)^n$.

Checkpoint 1

Suppose you invest \$300 at an annual interest rate of 7.35% compounded annually. How much would you have in 15 years?

Derivatives of Exponential Functions

Our discussion of interest rates and bank balances indicates that the functions we seek, functions with growth rate proportional to the amount present, might be *exponential* functions with constant base and variable exponent. To check this, we want to find the derivative of an exponential function of the form b^t. The base b might be 2 (if we want to talk about things doubling) or 10 (to invert the common logarithm function discussed in

Chapter 1) or 1.05 (if we want to talk about bank balances) or any other constant that makes sense as a base for exponentiation. As we did in finding the derivative of kt^2, we will approach this task first from a graphical point of view (with the aid of computer-drawn graphics) and then from a numerical point of view (with the aid of your calculator).

In our study of functions f of the form $f(t) = kt^2$, we saw that we could get a very good approximation to the derivative function at every instant by calculating difference quotients $\Delta f/\Delta t$ with a small value of Δt. We use the same idea now to see what we can learn about the derivative of an exponential function.

We write $f(t) = b^t$ in order to have a name for the exponential function with base b, and we write $g(t)$ for the difference quotient of f with a suitably small Δt, say, $\Delta t = 0.001$:

$$g(t) = \frac{\Delta f}{\Delta t} = \frac{f(t + 0.001) - f(t)}{0.001}.$$

This new function $g(t)$ is not the same function as $f'(t)$, but its value at each t should be very close to $f'(t)$. Thus a graph of $g(t)$ should closely approximate a graph of $f'(t)$.

Exploration Activity 2

We can determine whether $f'(t) = cf(t)$ for some constant c by asking whether the ratio $f'(t)/f(t)$ is a constant function. And since $g(t)$ approximates $f'(t)$, we can examine that question graphically by determining whether the graph of $g(t)/f(t)$ looks approximately constant. Graph this ratio for $b = 2$, 3, and 4. What do your graphs tell you?

In Figures 2.26 through 2.28 we graph $f(t)$, $g(t)$, and their ratio $g(t)/f(t)$ for the three different choices of the base b. In each case, the graph of $f(t) = b^t$ is the solid black curve, the graph of $g(t)$ is the dashed curve, and the graph of the ratio is the colored curve. The graphs of the ratios provide strong evidence that the derivative of an exponential function, no matter what the base, is indeed a constant multiple of the function itself.

Figures 2.26–2.28 Graphs of $f(t) = b^t$ (*solid*), difference quotients $g(t)$ (*dashed*), and their ratios (*colored*) for $b = 2,\ 3,$ and 4

Figure 2.26

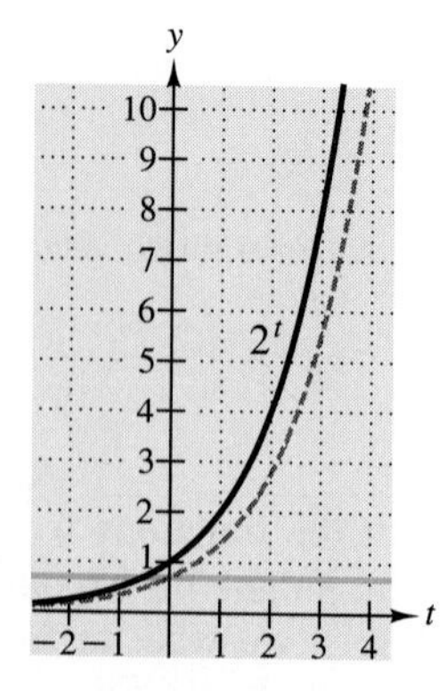

Figure 2.27

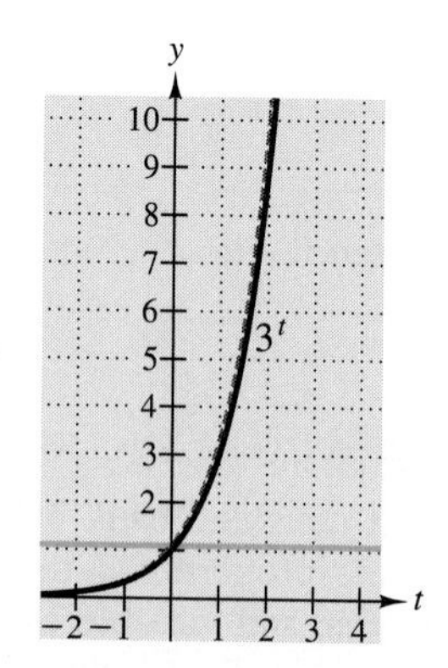

Figure 2.28

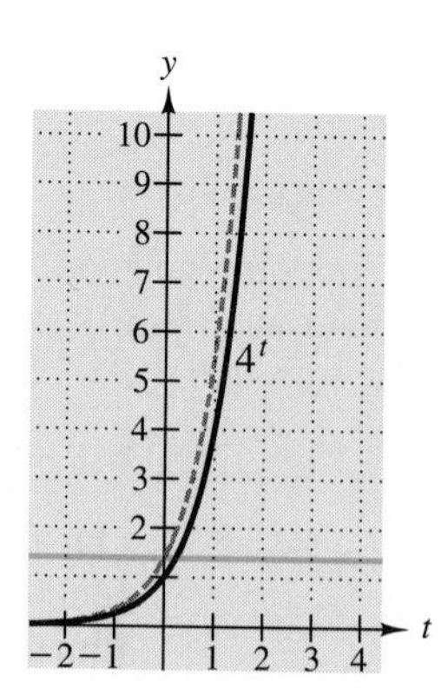

Figures 2.26 through 2.28 reveal more: The constant c in the formula $f'(t) = cf(t)$ is not the same for every choice of b. For $b = 2,\ 3,$ and $4,$ c is approximately $0.7,\ 1.1,$ and $1.4,$ respectively.

Figures 2.29-2.31 Graphs of $f(t) = b^t$ (*solid*), difference quotients $g(t)$ (*dashed*), and their ratios (*colored*) for $b = 5,$ $6,$ and 7

Figure 2.29

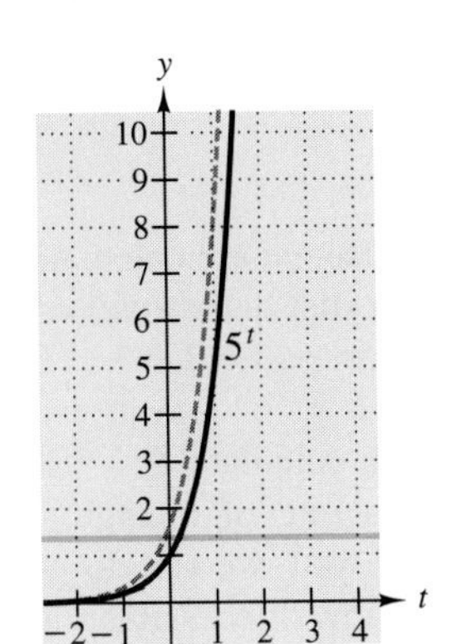

Figure 2.30

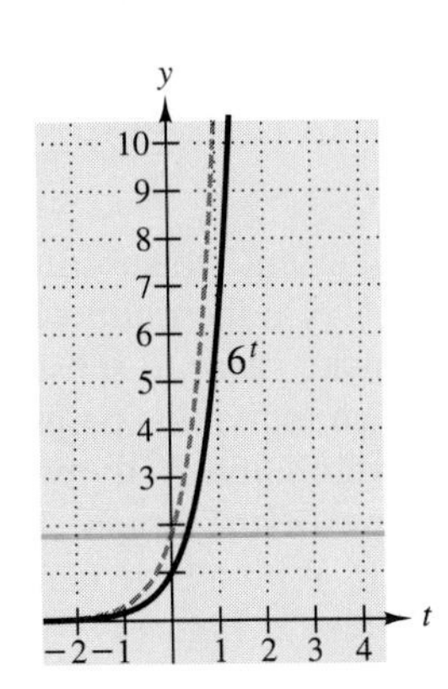

Figure 2.31

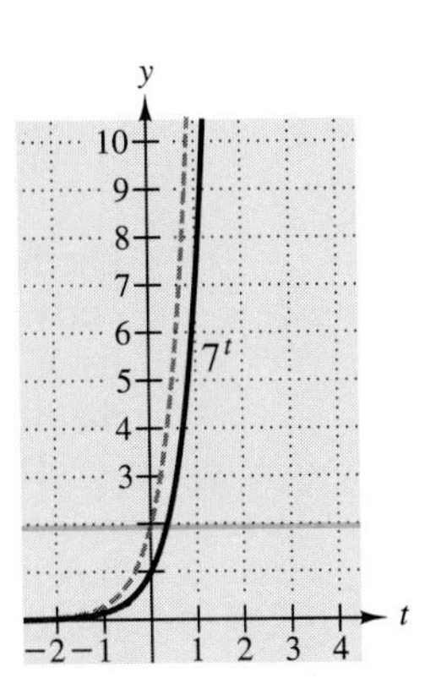

For more information about possible values of c, we include the corresponding graphs for $b = 5,\ 6,$ and 7 (Figures 2.29 to 2.31). Larger values of b produce larger values of c. There appears to be some (nonconstant) functional relationship between b and c. To indicate the dependence of c on b, we write $c = L(b)$, and we rewrite the formula $f'(t) = cf(t)$ in the form

$$\frac{d}{dt}\, b^t = L(b)\, b^t.$$

Let's be clear about what this formula says. For each fixed choice of base b, there is a constant, which we have just named $L(b)$, that is the proportionality constant relating the derivative to the exponential function.

Observation In each of the six cases graphed in Figures 2.26 through 2.31, the horizontal line $y = L(b)$ has the same y-intercept as the graph of the (approximate) derivative $g(t)$. This can't be an accident!

Exploration Activity 3

(a) Zoom to find

$$\frac{d}{dt}\, b^t \quad \text{at} \quad t = 0$$

for $b = 2,\ 3,$ and $4.$ Compare these numbers with $L(2),\ L(3),$ and $L(4).$

(b) Use the equation

$$\frac{d}{dt}\, b^t = L(b)\, b^t$$

to explain carefully why $L(b)$ must be the value of

$$\frac{d}{dt}\, b^t \quad \text{at} \quad t = 0,$$

no matter what the value of b is.

The key to the calculation is the observation that $b^0 = 1$. Notice that the expression b^t stands for two different things in the equation

$$\frac{d}{dt}\, b^t = L(b)\, b^t.$$

On the left, it does not make sense to substitute values of t because we are calculating the derivative of the *function* b^t. On the right, we have a formula for that derivative, which is itself a function. It does make sense to substitute a value of t—in particular, $t = 0$—on the right to find a slope at an individual point.

The observation in Exploration Activity 3 gives us a way to calculate $L(b)$ for each value of b: Find the slope of the graph (or the instantaneous rate of change) of the function $f(t) = b^t$ at $t = 0$. With a little help from our graphing tool, we can do just that for the particular case of $b = 2$. Figures 2.32, 2.33, and 2.34 show successive zooms in the vicinity of the point $(0, 1)$ on the graph of $f(t) = 2^t$.

Figure 2.32 Graph of $f(t) = 2^t$

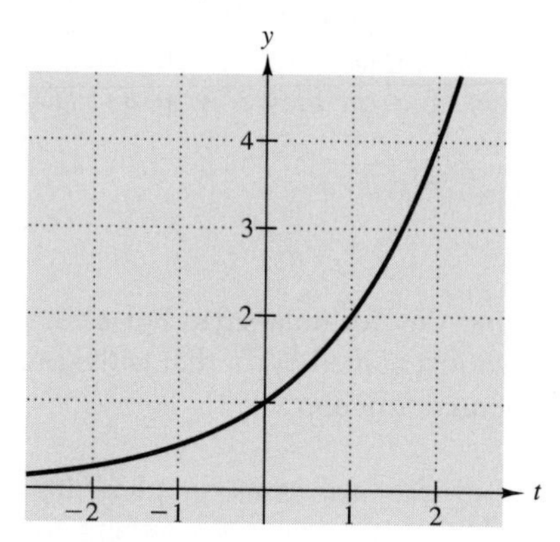

Figure 2.33 Zooming in by a factor of 0.01

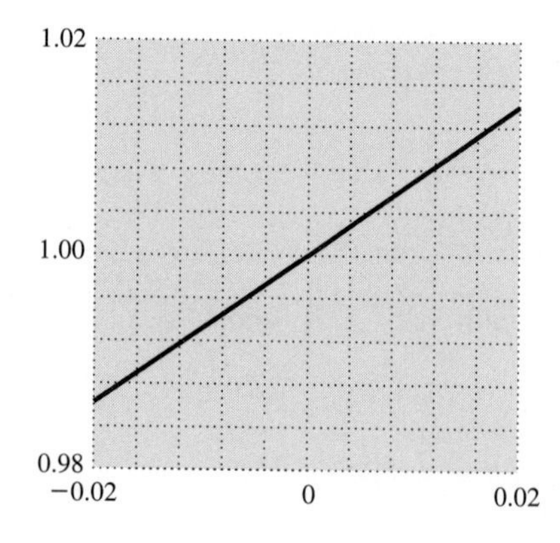

Figure 2.34 Zooming by another factor of 0.01

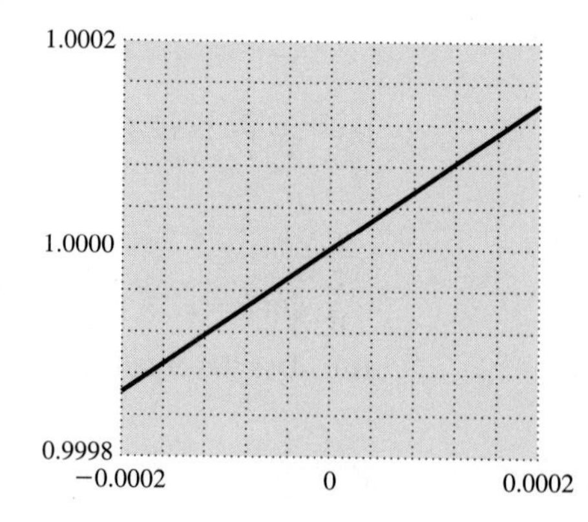

Checkpoint 2

Use Figures 2.32, 2.33, and 2.34 to calculate an approximate value for $L(2)$.

The Natural Base

We summarize what we know so far about derivatives of exponential functions. For a given b, we have

$$\frac{d}{dt}\, b^t = L(b)\, b^t$$

where $L(b)$ is the value of the derivative at $t = 0$. This number also may be described as

$$\text{the limiting value as } \Delta t \to 0 \text{ of } \frac{b^{\Delta t} - 1}{\Delta t}.$$

This description of $L(b)$ defines it as a function of b but does not tell us how to find values of that function. In Exploration Activities 2 and 3 we have determined three such values approximately, those for $b = 2$, 3, and 4. Table 2.2 shows those values together with three additional ones:

Table 2.2 Values of $L(b)$

b	2	3	4	5	6	7
$L(b)$	0.6931...	1.0986...	1.3862...	1.6094...	1.7917...	1.9459...

In Figure 2.35 we plot this scanty information about $L(b)$ as a function of b.

Figure 2.35 Known values of $L(b)$ as a function of b

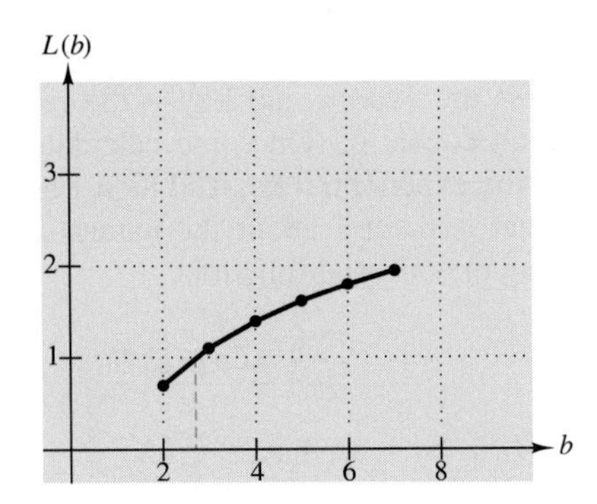

Exploration Activity 4

Figure 2.35 suggests the possibility of a smooth curve that passes through the points $(b, L(b))$ for $b = 2,\ 3,\ 4,\ 5,\ 6,$ and 7. To enhance this suggestion, we have "connected the dots" with straight-line segments that should approximate such a curve.

(a) Pencil in the curve being approximated. (Think back—or look back—to our work with logarithms in Chapter 1.)

(b) The figure now suggests a way to finish our derivative calculation, at least for one base b, without having to calculate $L(b)$. The curve suggested in Figure 2.35 crosses the horizontal line $L(b) = 1$ for some b. Estimate from your graph about how big b has to be to make $L(b) = 1$.

One way to estimate b is to use linear interpolation — you have the coordinates of the points in the Table 2.2. In class you may do a numerical experiment to determine the first few digits of this natural base, the number b such that $L(b) = 1$. The entire world of science and technology knows the natural base for exponentials as e. The symbol e for the natural base was first used by the eighteenth-century Swiss genius Léonhard Euler (pronounced "oiler"), who modestly named this number for himself.

Definition The number e is the value of b such that $L(b) = 1$.

The exponential function with e as its base has a special name.

Definition The **natural exponential function, exp**, is the function defined by

$$\exp(t) = e^t.$$

Your calculator knows the natural base and its exponential function — the key for the function is either e^x or exp. If you don't find a key on your calculator with either of these labels, use the inverse function key — probably Inv — followed by the ln key. The reason for this combination will become clear later in this section. Enter 1 in your calculator, and press the exponential key — the result will be the calculator's best estimate of $\exp(1)$, which equals e. (On some calculators, the correct procedure is the other way around: Press the exponential key and then 1 and Enter.)

Now we can see what is natural about the natural exponential function: Recall the formula for differentiating exponential functions:

$$\frac{d}{dt}\, b^t = L(b)\, b^t.$$

If we replace b by e and $L(b)$ by 1, we have

$$\frac{d}{dt}\, e^t = e^t.$$

Thus the derivative of exp is *itself*, not just proportional to itself. In particular, we don't need to find the limiting value of $(b^{\Delta t} - 1)/\Delta t$ to complete this derivative calculation.

Differentiating More General Exponential Functions

Have we really accomplished something by introducing e and concentrating on the one function $\exp(t) = e^t$? Suppose we consider functions $u(t) = e^{3t}$ or $v(t) = e^{5t}$ or

$w(t) = e^{2.35t}$. How do we differentiate these functions? In general, we need to know how to differentiate functions of the form e^{kt}, given that the derivative of e^t is e^t.

Exploration Activity 5

(a) Plot

$$f(t) = e^{kt}, \quad g(t) = \frac{f(t+0.001) - f(t)}{0.001}, \quad \text{and} \quad \frac{g(t)}{f(t)} \qquad \text{for } k = 2,\ 3,\ 4.$$

(b) Conjecture a general formula for

$$\frac{d}{dt}\,e^{kt}.$$

We already know that

$$\frac{d}{dt}\,e^{t} = e^{t}.$$

You should have found that

$$\frac{d}{dt}\,e^{2t} = 2e^{2t}, \quad \frac{d}{dt}\,e^{3t} = 3e^{3t}, \qquad \text{and} \qquad \frac{d}{dt}\,e^{4t} = 4e^{4t}.$$

This suggests the general formula

$$\frac{d}{dt}\,e^{kt} = ke^{kt}.$$

This is a special case of a much more general differentiation formula, the Chain Rule, which we take up in Chapter 4.

Checkpoint 3

Find the derivative of $e^{2.53t}$.

Now we establish a general formula for differentiation of exponential functions. Suppose A and k are constants. Then

$$\begin{aligned}
\frac{d}{dt}\,Ae^{kt} &= A\,\frac{d}{dt}\,e^{kt} && \text{(from the Constant Multiple Rule)} \\
&= A\,k\,e^{kt} && \text{(from the formula in Exploration Activity 5)} \\
&= k\,A\,e^{kt} && \text{(by commutativity of multiplication)}
\end{aligned}$$

Thus we have

$$\frac{d}{dt}\,Ae^{kt} = kAe^{kt}.$$

Example 1 Calculate the derivative of $5e^{3t}$.

Solution

$$\frac{d}{dt}5e^{3t} = 3 \cdot 5e^{3t} = 15e^{3t}$$

■

Symbolic Differentiation of Exponential Functions: Other Bases

We know how to differentiate e^{2t}, e^{3t}, and e^{5t}, but what about 2^t, 3^t, and 5^t? The formula

$$\frac{d}{dt}e^{kt} = k\,e^{kt}$$

is of help here as well.

Example 2 Differentiate 3^t.

Solution We begin by writing 3 in the form e^k. In fact, $3 = e^k$, where $k = \log_e 3$. (That's the definition of $\log_e 3$.) Now for this k we have

$$3^t = (e^k)^t = e^{kt}.$$

Thus

$$\begin{aligned}\frac{d}{dt}3^t &= \frac{d}{dt}e^{kt}\\ &= k\,e^{kt}\\ &= (\log_e 3)\,e^{kt}\\ &= (\log_e 3)\,3^t.\end{aligned}$$

■

Checkpoint 4

Give a reason for each step in Example 2.

There is nothing special about $b = 3$. In general,

$$\frac{d}{dt}b^t = (\log_e b)\,b^t.$$

Notice that we have now solved a problem we left dangling earlier: how to calculate $L(b)$ in the formula

$$\frac{d}{dt}b^t = L(b)\,b^t.$$

Checkpoint 5

(a) What's the answer—what is $L(b)$? In particular, what is $L(2)$? (Use your calculator. For reasons that we will explain very soon, the button on your calculator for "logarithm base e" is labeled "ln.")

(b) Look back at Checkpoint 4, and use what you know now to check your calculation.

In Chapter 1 we reviewed logarithms with an arbitrary base b. In this chapter we have already determined a special interest in the natural base e. The corresponding logarithm is also called "natural." Its abbreviated name is **ln**, which stands for "**l**ogarithm, **n**atural" (but which is read "natural logarithm"). Thus $\ln x = \log_e x$. At the risk of belaboring the obvious, we repeat the definition of logarithm for this important case.

Definition The **natural logarithm** is defined by

$$y = \ln x \quad \text{if and only if} \quad x = e^y.$$

Note, in particular, that if $y = 1$, then x must be e, and if x is e, then y must be 1. That is, $\ln e = 1$.

In this section we have developed formulas for differentiating exponential functions. In particular, we found that

$$\frac{d}{dt}\, b^t = (\ln b)\, b^t.$$

The logarithm in this formula, ln, is the *natural* logarithm, the one that has Euler's number e as its base. (An approximate value of e is 2.718281828.) When e is also the exponential base, we get the simpler formula

$$\frac{d}{dt}\, e^t = e^t.$$

This *natural* exponential function, the one that is its own derivative, is also called exp: $\exp(t) = e^t$. More generally, we found a whole family of functions that all have derivatives proportional to themselves:

$$\frac{d}{dt}\, A\, e^{kt} = k\, A\, e^{kt}.$$

Answers to Checkpoints

1. $\$869.26$
2. From Figure 2.34, $L(2)$ is between 0.68 and 0.70.
3. $2.53e^{2.53t}$
4. The reasons are: We chose k to make $3^t = e^{kt}$.

   The rule for differentiating exponential functions.

   $k = \log_e 3$.

   $e^{kt} = 3^t$.
5. $L(2) = \ln(2)$, which is approximately 0.693147.

Exercises 2.4

1. Suppose you invest $400 at an annual interest rate of 8.55% compounded annually. How much would you have in 20 years?
2. How much would you have to invest at 7.35% compounded annually so that you would have $20,000 after 10 years?
3. How much would you have to invest at 8.55% compounded annually so that you would have $30,000 after 20 years?

Rewrite each of the following equations in exponential form.

4. $\log_9 3 = \frac{1}{2}$
5. $\log_{10} 1000 = 3$
6. $\log_{10} 0.1 = -1$
7. $\log_2 4 = 2$
8. $\log_{10} 100 = 2$
9. $\log_{10} 0.01 = -2$

Find the value of each of the following expressions.

10. $\ln 1$
11. $\ln e$
12. $\ln e^3$
13. $\ln e^{2.7183}$
14. $\ln \dfrac{1}{e}$
15. $\ln \dfrac{1}{e^2}$

Solve each of the following equations for t.

16. $t^6 = 10$
17. $e^t = 10$
18. $10^t = 6$

Calculate each of the following derivatives.

19. $\dfrac{d}{dt} e^{-2t}$
20. $\dfrac{d}{dt} e^{0.07t}$
21. $\dfrac{d}{dt} 2^t$
22. $\dfrac{d}{dt} (2t - 5e^t)$
23. $\dfrac{d}{dt} 4t^3$
24. $\dfrac{d}{dt} e$
25. Express $2e^{3t}$ in the form cb^t.
26. Express $15e^{0.15t}$ in the form cb^t.
27. Express $2 \cdot 3^t$ in the form ce^{kt}.
28. Express $3 \cdot 10^t$ in the form ce^{kt}.

Find the derivative of each of the following functions.

29. $t^4 - 2t^3 + 2t^2 - t - 1$
30. $t^5 + t^3 - t^2 - 9 + e^{-t/3}$
31. $4e^4 - 3e^3 + 2e^2 - e + 7$
32. $13 - 26t + 6t^2 + e^t$
33. $2e^{3t} - 3t^2$
34. $\dfrac{1}{e^{2t}}$
35. $t^{10} + 10^t$
36. $\dfrac{2}{10^t}$

Find the slope of the graph of $y = e^t$ at each of the following points.

37. $(2, e^2)$
38. $(-2, e^{-2})$

Find the slope of the graph of $y = 2^t$ at each of the following points.

39. $(2, 4)$
40. $(-2, \frac{1}{4})$

Find the slope of the graph of $y = 10^t$ at each of the following points.

41. $(2, 100)$
42. $(-2, 0.01)$

Find the slope of the graph of $y = 3^t$ at each of the following points.

43. $(2, 9)$

44. $(-2, \frac{1}{9})$

45. You can use your calculator to determine the slope at $t = 0$ on the graph of $f(t) = b^t$ (for any particular value of b) by a procedure that does not depend on graphing. This slope is just the limiting value of the difference quotients

$$\frac{f(0 + \Delta t) - f(0)}{\Delta t}$$

as Δt shrinks to zero. If we replace $f(t)$ by b^t, we have

$$\frac{\Delta f}{\Delta t} = \frac{b^{\Delta t} - b^0}{\Delta t} = \frac{b^{\Delta t} - 1}{\Delta t}$$

so

$L(b)$ is the limiting value as $\Delta t \to 0$ of $\dfrac{b^{\Delta t} - 1}{\Delta t}$.

(a) Estimate this limiting value for $b = 2$ by filling in Table 2.3.

Table 2.3 Experimental determination of the limiting value for $L(2)$

Δt	$2^{\Delta t}$	$\frac{2^{\Delta t}-1}{\Delta t}$
0.01	1.00695555	0.695555
0.001		
0.0001		
0.00001		

(b) From this calculation, how many digits in $L(2)$ are you sure about?
(c) How does this fit with your calculations in Exploration Activities 2 and 3?
(d) Does it agree with our first graphical observation of $L(2)$ in Figure 2.26?

46. Repeat Exercise 45 with $\Delta t = -0.01$, -0.001, -0.0001, -0.00001. Do you get more evidence, less evidence, or about the same evidence about the value of $L(2)$? Do you get evidence for a different value of $L(2)$?

47. Repeat Exercise 45 for $b = 3$. What do you conclude about $L(3)$?

48. Repeat Exercise 45 for $b = 4$. What do you conclude about $L(4)$?

49. What happens if you extend Table 2.3 several lines further? Try it and see. Does this alter your opinion about the correct value for $L(2)$? Why or why not?

50. (a) Replace b by e and $L(b)$ by 1 in the expression

$L(b)$ is the limiting value as $\Delta t \to 0$ of $\dfrac{b^{\Delta t} - 1}{\Delta t}$.

The resulting statement gives us a definition of e: "e is the number such that . . . ." Finish the definition.
(b) Interpret this definition (as you did in Exploration Activity 3) as giving the slope of the graph of exp at a particular value of t. What value of t? What is the slope there?
(c) Fill in the blanks in the following characterization of exp: "Among all exponential functions, the natural one is the one that has slope ______ at $t =$ ______."

51. You now know the limiting value in

$L(b)$ is the limiting value as $\Delta t \to 0$ of $\dfrac{b^{\Delta t} - 1}{\Delta t}$.

What is it? What did the L stand for when we wrote the proportionality constant as $L(b)$? Fill in the blank in this updated version:

The limiting value as $\Delta t \to 0$ of $\dfrac{b^{\Delta t} - 1}{\Delta t}$ is ______.

52. The formula

$$\frac{d}{dt}\, b^t = (\log_e b)\, b^t$$

reveals, for an arbitrary constant k, the base b such that

$$\frac{d}{dt}\, b^t = k\, b^t.$$

How is b determined from k? For example, what is b if $k = 0.7$?

53. (a) Sketch the graph of the natural exponential function $y = e^t$.
(b) Use properties of exponents to show that, for every number t,

$$e^{-t} = \frac{1}{e^t}.$$

(c) On the same axes as you used in (a), sketch the graph of $y = e^{-t}$.
(d) Explain the symmetry the two graphs exhibit.
(e) Now sketch the graph of $y = e^t + e^{-t}$. Is this function even, odd, or neither? Confirm your answer algebraically.

54. (a) Find the derivative of $y = e^t + e^{-t}$.
(b) Graph the derivative of y.
(c) Is the derivative even, odd, or neither? Confirm your answer algebraically.

55. (a) Sketch the graph of $y = e^t - e^{-t}$. Is this function even, odd, or neither? Confirm your answer algebraically.
(b) Find the derivative of $y = e^t - e^{-t}$.
(c) Graph the derivative of y.
(d) Is the derivative even, odd, or neither? Confirm your answer algebraically.

2.5 Modeling Population Growth

A Difference Equation and a Differential Equation

We return to our examination of models of population growth. For our first model we make the assumption that populations tend to grow at a rate proportional to their numbers. In particular, this says that the more individuals there are in a population, the faster the population grows. However, "proportional" says something more specific about the rate of growth: For a given time step Δt, the mathematical interpretation of this biological principle is

$$\frac{\Delta P}{\Delta t} = kP,$$

where k is a constant, and P is the population at the start of a time interval of length Δt. Because of the differences, ΔP and Δt, and the assertion of equality appearing in this formula, it is called a **difference equation**.

We encountered a concrete example of such a difference equation in our discussion of interest rates and exponential growth. There, our P represented the bank balance (think *principal*), Δt was the time between interest compoundings (one year in our example), and k was the interest rate for the compounding period (0.05 in our original example). The upshot of our calculations was that the bank balance at time t could be represented by an exponential function with base $b = 1 + k$. In Project 3 at the end of this chapter you will find that all solutions of difference equations of the form

$$\frac{\Delta P}{\Delta t} = kP$$

are exponential functions, no matter what the time step or the proportionality constant.

For some populations—bears, for example—Δt is essentially constant. That is, there are fixed times at which reproduction is possible. If we assume that conditions are so favorable for our population that few or no deaths occur in the time frame of interest and that there are no migrations in or out of the population, then reproduction accounts for all the change in the population, and we may reasonably assume that Δt is constant, that it measures the smallest interval between reproduction times. On the other hand, there are biological populations—bacteria and humans, for example—in which reproduction can and does occur at any time. In these cases there is no smallest value Δt can assume.

Now let's examine this growth equation,

$$\frac{\Delta P}{\Delta t} = kP,$$

under conditions of continuous change. The equation says that no matter how small Δt is, the average rate of change of P is proportional to P. We need to state this a little more carefully: The proportionality constant k actually depends on the time step (but not on the population). For each fixed Δt, no matter how small Δt is, the average rate of change of P is proportional to P. If we let $\Delta t \to 0$, we have already seen that $\Delta P/\Delta t$ approaches the instantaneous rate of change of P.

Of the two factors on the right, only the k can change as Δt changes; P stands for the population at the start of a time interval, and this is independent of the length of the interval. Thus the instantaneous rate of change of P also must be proportional to P, i.e.,

$$\frac{dP}{dt} = KP,$$

where K is whatever k approaches as Δt approaches zero.

An instantaneous growth rate of slightly less than 2% is necessary if the average annual rate is to be 2%, because the instantaneous rate is being applied later in the year to a bigger population than was present at the start of the year. Thus, if $k = 0.02$ when $\Delta t = 1$, then k shrinks a little as Δt gets smaller, which means K will be a little smaller than 0.02 for the same population. In fact, to have a 2% yearly growth, K should be approximately 0.0198. In Exercise 23 we will work out how to find K when you know the average annual percentage growth.

Since the *differences* in

$$\frac{\Delta P}{\Delta t} = kP$$

have become *differentials* in

$$\frac{dP}{dt} = KP,$$

we call the latter a **differential equation**. We mentioned the term "differential" when we introduced the derivative, but we have yet to say what a differential is. The presence of a derivative—the apparent ratio of two differentials—in our equation suggests the term "derivational equation," but that term is *never* used.

To summarize, our model of natural biological growth is the difference equation if the reproductive times are discrete and the differential equation if reproduction is occurring continuously. As we consider influences that hinder or enhance natural growth rates, we will modify the models accordingly. Our immediate goal is to determine the implications of the differential equation, i.e., to solve the equation in order to find functions that can represent natural populations.

In Chapter 1 we raised some questions for which the answers had to be expressed in terms of functions rather than in terms of numbers or variables—for example, "What functions are additive?" or "What is the inverse of base 10 exponentiation?" Now we have posed another kind of problem that calls for an answer of the same type: Given an equation that describes an instantaneous rate of change, what ***functions*** satisfy that equation?

Suppose we set $k = 0.25$. What does it mean for a function to satisfy the differential equation

$$\frac{dP}{dt} = 0.25P\,?$$

Consider the function P_1 defined by

$$P_1(t) = 3e^{0.25\,t}.$$

This function is a solution, because

$$\frac{d}{dt}P_1(t) = 0.25 \cdot 3e^{0.25\,t} = 0.25P_1(t) \qquad \text{for all values of } t.$$

On the other hand, the function P_2 defined by

$$P_2(t) = t^2$$

does not satisfy the differential equation, since

$$\frac{d}{dt}P_2(t) = 2t$$

and

$$0.25P_2(t) = 0.25t^2$$

do not define the same function.

Checkpoint 1

Write down three more functions, in addition to P_1, that are solutions of $dP/dt = 0.25P$.

Slope Fields

In this section we will learn how to draw a picture of the differential equation that will help us understand what it means for a function to be a solution. We'll continue to look at the differential equation

$$\frac{dP}{dt} = 0.25P.$$

In Figure 2.36 we have drawn the graphs of a number of solutions. In Figure 2.37 we plot the same solutions but only short segments of each solution curve. The short segments still give us a good idea of what the solution curves look like, and we could draw in the rest of the curves if we wanted to.

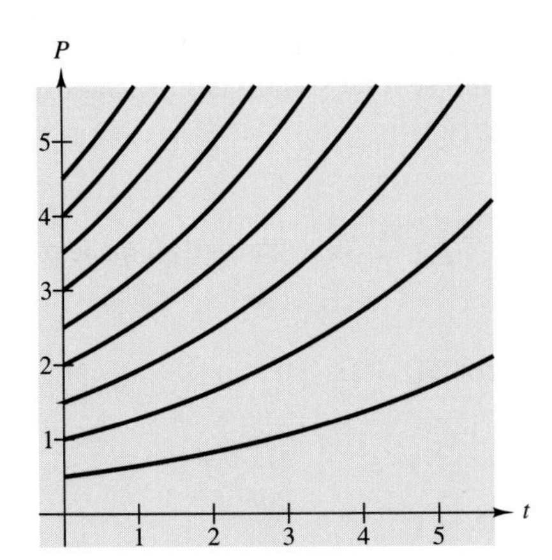

Figure 2.36 Solution curves for the differential equation $dP/dt = 0.25P$

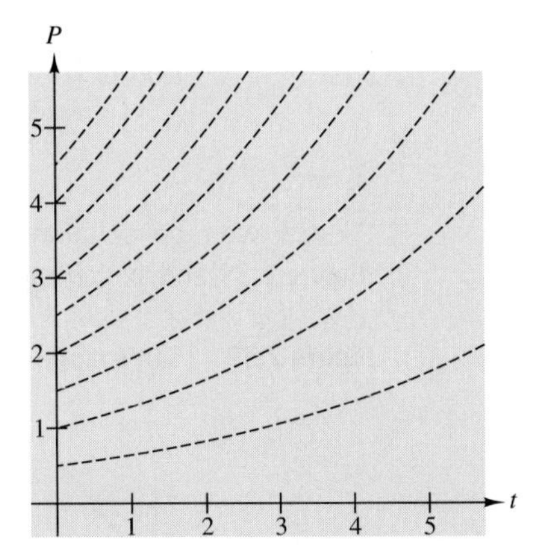

Figure 2.37 Segments of solution curves for the differential equation $dP/dt = 0.25P$

Notice that the curve segments in Figure 2.37 are nearly straight; i.e., they are essentially line segments. Here is the important point: **We know what the *slopes* of these line segments must be from the differential equation. We do not need to know formulas for the solutions in order to determine slopes.**

For this particular differential equation, we know that the slope of a very short piece of solution curve (a line segment) at any point (t, P) is $0.25P$. Thus, for example, the slope at $(1, 2)$ is $0.25 \times 2 = 0.5$, the slope at $(2, 6)$ is 1.5, and so on.

This gives us a way to represent the differential equation

$$\frac{dP}{dt} = 0.25P$$

graphically without reference to its solutions: At lots of points in the (t, P) plane we draw short line segments with slopes that match the equation. Figure 2.38 is such a picture: At each point (t, P) on a closely spaced grid we have drawn a small line segment whose slope is $0.25P$. The resulting picture is called a **slope field**.

Figure 2.38 Slope field for $\dfrac{dP}{dt} = 0.25P$

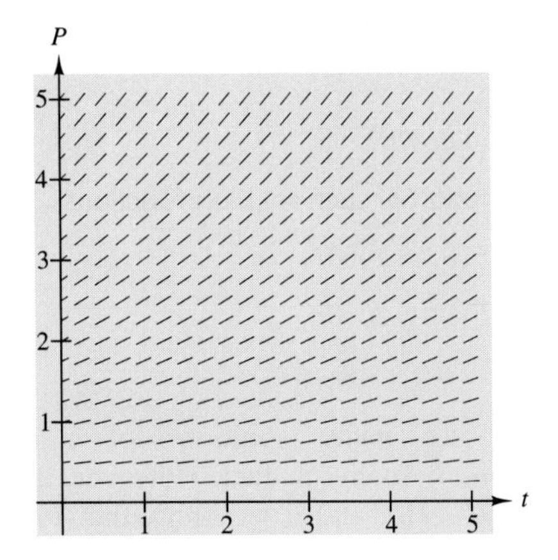

The selected points are only representative of the infinity of points in the (t, P) plane, but they show us a clear picture of the slope field described by the differential equation. Any solution of the differential equation is a function $P = P(t)$ whose graph passes through the slope field in the directions of the line segments it meets along the way. In Figure 2.39 we show the same field with several solutions superimposed.

Checkpoint 2

Sketch two more solution curves on the slope field in Figure 2.39. (Alternately, make a photocopy of Figure 2.39 and sketch the new curves on that.)

Figure 2.39 Same slope field with several solution curves

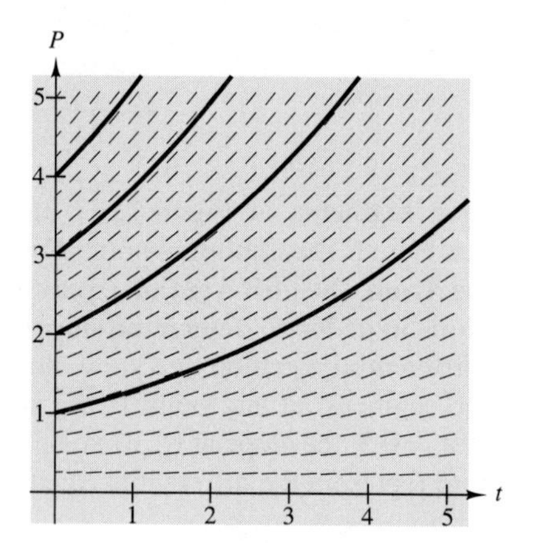

Differential-Equation-with-Initial-Value Problems

We now look at the question of how many solutions a differential equation might have.

Exploration Activity 1

You already have identified several solutions to the differential equation

$$\frac{dP}{dt} = 0.25P$$

(Checkpoint 1).

(a) How many solutions can you write down?

(b) Give a general form for the description of such solution functions.

(c) How many solutions also satisfy $P(0) = 1000$?

(d) How many solutions also satisfy $P(0) = 2000$?

In general, there are many solutions of a differential equation such as $dP/dt = 0.25P$—*infinitely* many, in fact. However, it is certainly plausible that if we choose an initial point (t_0, P_0), and if we know the slope of the solution at every point (t, P), then there must be only *one* solution that passes through (t_0, P_0). Thus, in hope of finding a uniquely determined solution — and because it's reasonable to assume we know a starting population — we turn our attention to a **differential-equation-with-initial-value problem**:

$$\frac{dP}{dt} = KP \quad \text{and} \quad P = P_0 \qquad \text{when } t = t_0.$$

Such a problem describes a unique function $P = P(t)$.

For the differential-equation-with-initial-value problem

$$\frac{dP}{dt} = 0.25P \quad \text{and} \quad P = 1000 \qquad \text{when } t = 0,$$

we have already done most of the work for finding a formula for the solution. Indeed, if we let $P = Ae^{0.25t}$, then we know from Section 2.4 that

$$\frac{dP}{dt} = \frac{d}{dt}Ae^{0.25t} = 0.25Ae^{0.25t} = 0.25P.$$

Thus, for every constant A, the function $P = Ae^{0.25t}$ is a solution of the differential equation

$$\frac{dP}{dt} = 0.25P.$$

Now we choose A so that $P = 1000$ when $t = 0$:

$$1000 = P(0) = Ae^0 = A.$$

Thus $P = 1000e^{0.25t}$ is the desired solution.

Checkpoint 3

Show that the general differential-equation-with-initial-value problem

$$\frac{dP}{dt} = KP \quad \text{and} \quad P = P_0 \qquad \text{when } t = 0$$

has the solution

$$P = P_0e^{Kt}.$$

In this section we introduced the difference equation $\Delta P/\Delta t = kP$ for modeling natural growth over discrete time intervals and looked at what happens as Δt approaches 0. In this case we obtain the differential equation $dP/dt = KP$ as a model for continuous natural growth.

In general, infinitely many functions are solutions of a given differential equation. We can visualize these solutions, even without formulas, by using the slope field. If we specify the value of a solution at one point, that also specifies a unique function in the family of solutions. Our knowledge of derivatives of exponential functions enables us to find formulas for solutions of all natural growth problems.

Answers to Checkpoints

1. $5e^{0.25t}$, $7e^{0.25t}$, $9e^{0.25t}$ or any three new functions of the form $Ae^{0.25t}$.

3. We know that functions of the form $P = Ae^{Kt}$ satisfy the differential equation. We want to select A so that $Ae^0 = P_0$. Since $e^0 = 1$, this means $A = P_0$ and $P = P_0e^{Kt}$.

Exercises 2.5

On each of the following slope fields, sketch three solution functions. (You may photocopy the page if you wish.)

1.

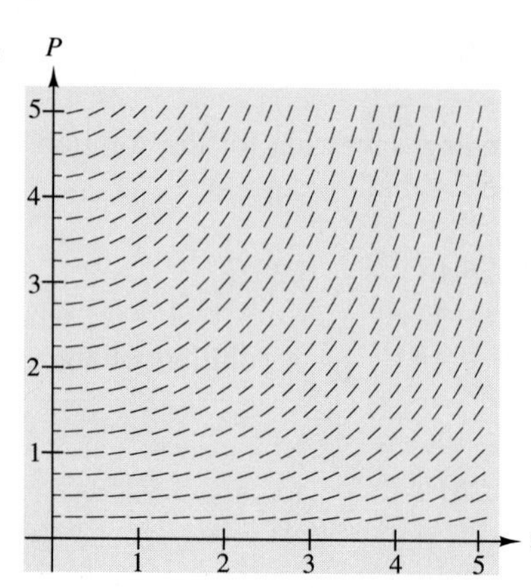

2.

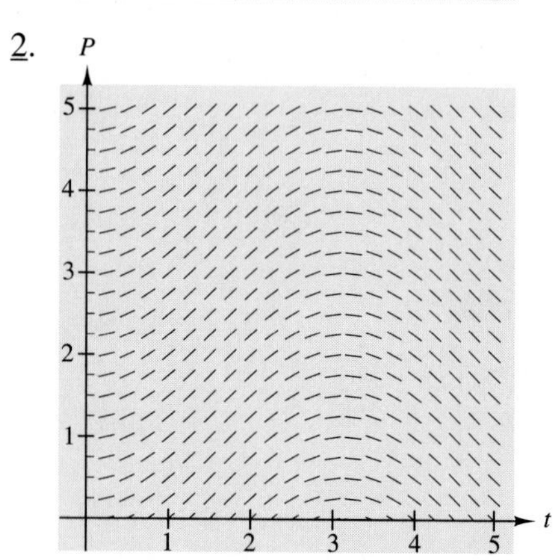

3.

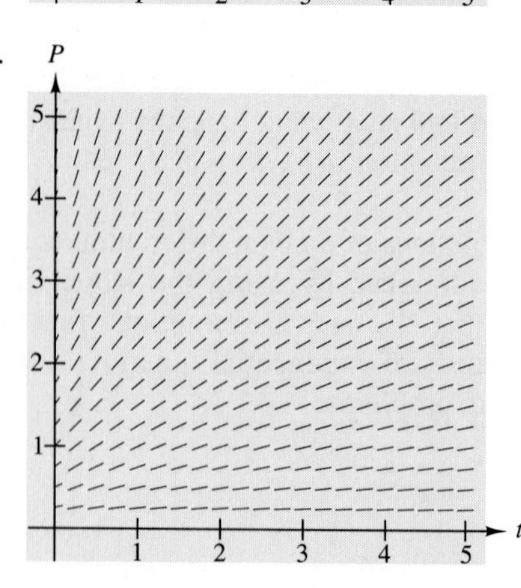

4.

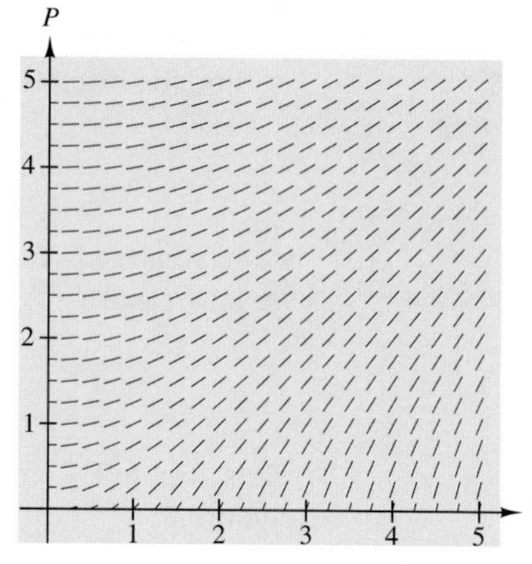

5.

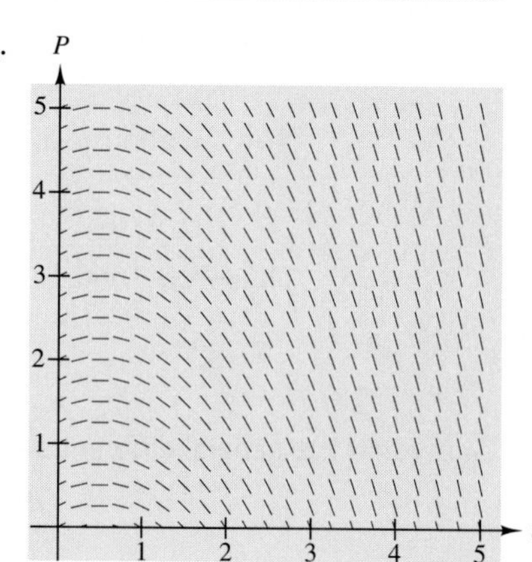

6.

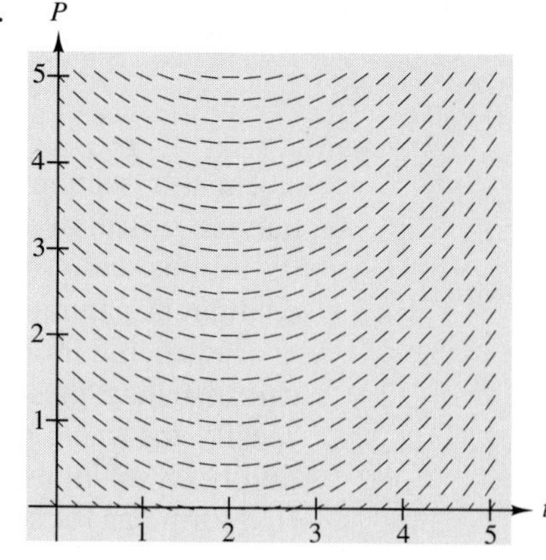

7. (a) Write down three functions that are solutions of the differential equation

$$\frac{dP}{dt} = 0.02P.$$

(b) Describe the infinite family of functions that satisfy the differential equation in part (a).
(c) Solve the differential-equation-with-initial-value problem

$$\frac{dP}{dt} = 0.02P \qquad \text{with } P = 5 \text{ at } t = 0.$$

(d) Solve the differential-equation-with-initial-value problem

$$\frac{dP}{dt} = 0.02P \qquad \text{with } P = 5 \text{ at } t = 2.$$

8. (a) Write down three solutions of the differential equation

$$\frac{dP}{dt} = 0.5P.$$

(b) Describe the infinite family of functions that satisfy the differential equation in part (a).
(c) Solve the differential-equation-with-initial-value problem

$$\frac{dP}{dt} = 0.5P \qquad \text{with } P = 2 \text{ at } t = 0.$$

(d) Solve the differential-equation-with-initial-value problem

$$\frac{dP}{dt} = 0.5P \qquad \text{with } P = 2 \text{ at } t = 5.$$

9. Solve the differential-equation-with-initial-value problem

$$\frac{dP}{dt} = 0.25P \qquad \text{and } P = 2000 \text{ when } t = 1.$$

10. Show that the general differential-equation-with-initial-value problem

$$\frac{dP}{dt} = KP \qquad \text{and } P = P_0 \text{ when } t = t_0$$

has the solution

$$P = P_0 e^{K(t-t_0)}.$$

Solve each of the following differential-equation-with-initial-value problems.

11. $\frac{dP}{dt} = 0.12P$ with $P = 500$ at $t = 0$

12. $\frac{dP}{dt} = -0.12P$ with $P = 500$ at $t = 0$

13. $\frac{dP}{dt} = 0.12P$ with $P = 500$ at $t = 2$

14. $\frac{dP}{dt} = -0.12P$ with $P = 500$ at $t = 2$

15. $\frac{dy}{dt} = 0.35y$ with $y = 100$ at $t = 0$

16. $\frac{dy}{dt} = -0.35y$ with $y = 100$ at $t = 0$

17. $\frac{dy}{dt} = 0.35y$ with $y = 100$ at $t = 1$

18. $\frac{dy}{dt} = -0.35y$ with $y = 100$ at $t = 1$

Solve each of the following equations for t.

19. $100 = 27e^{0.07t}$

20. $200 = 33e^{0.14t}$

21. $P = P_0 e^{kt}$

22. $y = Ae^{-\alpha t}$

23. The function $y = y_0 e^{kt}$ models continuous growth at a rate proportional to the amount present, with proportionality constant k.
(a) Find the constant k that produces 2% growth in one time period. (*Hint*: Set $t = 1$ and $y = 1.02y_0$. Solve for k.)
(b) Find the constant k that produces r% growth in one time period.

24. A town with a population of 10,240 at the start of a year has an average growth rate of 5% per year.
(a) What is the population at the end of that year?
(b) What is the population halfway through the next year?
(c) What is the population halfway through the third year?
(d) Explain your strategy for answering these questions.

25. (a) A biologist has a colony of bacteria that contains 1.5 biomass units when first weighed and

1.95 biomass units an hour later. What is the biomass of the colony t hours later?

(b) Her first measurement was made at 8:30 A.M. and her last just before she went home at 4:30 P.M. Estimate the last measurement.

26. "In recent years there has been a rise in the concentration of atmospheric methane of more than 1 percent per year. The increase is both rapid and significant because . . . methane is 20 times as effective as carbon dioxide in trapping heat."[2] How long will it take to increase the concentration of atmospheric methane by 50%? (The answer is not "50 years.")

27. Suppose you deposit \$1000 in a savings account that pays 8% interest compounded annually, and you leave it (and the interest) there indefinitely.
    (a) What is your balance one year after you made the deposit?
    (b) By what factor has your money increased?
    (c) What is your balance two years after the initial deposit?
    (d) What is your balance 20 years after the initial deposit?
    (e) What is your balance t years after the initial deposit?[3]

28. Answer the same questions as in the preceding problem
    (a) if the interest is compounded quarterly.
    (b) if the interest is compounded monthly.
    (c) if the interest is compounded daily.
    (d) The bank offering 8% compounded daily might advertise an "effective yield" of 8.33%. What does that mean?

29. Some banks advertise continuous compounding of interest. This means (or should mean) that the instantaneous rate of change of your balance is the advertised interest rate times the balance.
    (a) Answer the same questions as in Exercise 27 under the assumption of continuous compounding.
    (b) How much difference is there between daily compounding and continuous compounding? In particular, what is the effective yield in this case?

30. Relate your computations in the three preceding problems to the statement

    The limiting value of $(1 + k\Delta t)^{1/\Delta t}$ as $\Delta t \to 0$ is e^k.

    What specific values of k and Δt were you using?

31. The First National Bank offers $8\frac{1}{4}$% interest on deposits compounded daily, the Merchants and Farmers Bank offers $8\frac{3}{8}$% compounded quarterly, and the Central Savings and Loan offers $8\frac{1}{2}$% compounded annually.
    (a) Which bank would you choose if you planned to deposit a substantial amount of money in a long-term savings account?
    (b) Can you think of a reason to split your deposit between two (or more) of the banks?

32. An important concept in business and banking is the time required for an investment to double. A common rule of thumb, called the **rule of 72**, is that the doubling time in years is approximately 72 divided by the interest rate r (as a percentage). For example, at 8% interest, it takes about 9 years to double an investment.
    (a) If the interest is compounded continuously at r%, find the exact doubling time.
    (b) Make a table showing the doubling time in years, for $r = 2\%, 4\%, 8\%, 12\%$, and 18%, calculated three ways: from the exact formula in part (a), from the rule of 72, and from the similar **rule of 69**. Which rule of thumb comes closest to the exact answers?
    (c) Can you think of a reason why people in business would prefer the rule of 72 to the rule of 69?[4]

33. Table 2.4 contains data on the number of lawyers in a certain town and the total annual income of those lawyers. Assume that during the 5-year period represented by the table, the rate of inflation was 5% per year.
    (a) What statistic best describes the financial state of a typical lawyer in a given year? What formula computes it? Construct a table corresponding to this statistic.
    (b) When were lawyers most well off? Which year saw the greatest improvement? Which year saw the worst deterioration?

Table 2.4 Lawyers and their income

Year	*Number of Lawyers*	*Total Annual Income*
1982	40	$2,500,000
1983	45	$2,700,000
1984	50	$3,400,000
1985	40	$3,000,000
1986	50	$3,600,000

34. This spring, after cubs were born, the bear population of Big Bear Natural Wilderness Area was 251, half of them female. Each spring 65% of the females each produce two cubs, and 50% of all the cubs are female. Assuming no deaths, estimate the bear population in 3 years. Give your answer in "whole" bears.

35. (a) Make a table of doubling times for percentage growths from 1% per year to 20% per year.
(b) What percentage growth rate leads to doubling in 10 years?
(c) How does your table change if the time unit is a minute instead of a year?

36. Suppose a quantity grows at $n\%$ per year for a human lifetime. Explain why the factor by which the quantity grows is about 2^n.

2. From "Global Climatic Change" by R. A. Houghton and G. M. Woodwell, *Scientific American*, April 1989, pp. 36–44.
3. This problem and the four following it are adapted from D. A. Smith, *Interface: Calculus and the Computer*, 2nd Ed., Saunders College Publishing, Philadelphia, 1984.
4. This problem and the one following are adapted from *Calculus Problems for a New Century*, edited by Robert Fraga, MAA Notes Number 28, 1993.

2.6 Logarithms and Representation of Data

We turn our attention now to an aspect of growth problems that is not addressed directly by the formal manipulations of calculus: Now that we have a model for growth of populations, how can we tell if biological populations actually grow that way? We will see that the same exponential and logarithmic functions help us answer this and related questions. More important, they provide a set of tools that we will continue to use throughout the course.

Logarithmic Plots of Data: Natural Base

Early in this chapter we considered Thomas Malthus's concept of human population growth, which we subsequently identified as exponential. Indeed, we have seen that the natural assumption about growth of biological populations (growth rate proportional to the population) leads to exponential functions representing population sizes. How can we tell whether populations really grow that way?

We consider two case studies. The first uses data on fruit flies in an ideal laboratory setting,[5] and the other uses historical data on the human population of the earth. In each case we want to examine whether the data could have come from a population that is represented approximately by a function of the form $P = P_0 e^{kt}$.

Case Study 1 Table 2.5 lists numbers of fruit flies in a laboratory colony, under ideal conditions, for the first ten days of an experiment. Figure 2.40 is a plot of the same data.

5. We used these same data in Project 2 at the end of Chapter 1, where we considered the same question from a different point of view. Candor compels us to admit that we made up these data — they did not come from a "real" laboratory.

Table 2.5 Fruit fly data

Day Number	*Number of Flies*
0	111
1	122
2	134
3	147
4	161
5	177
6	195
7	214
8	235
9	258
10	283

Figure 2.40 Graph of fruit fly data

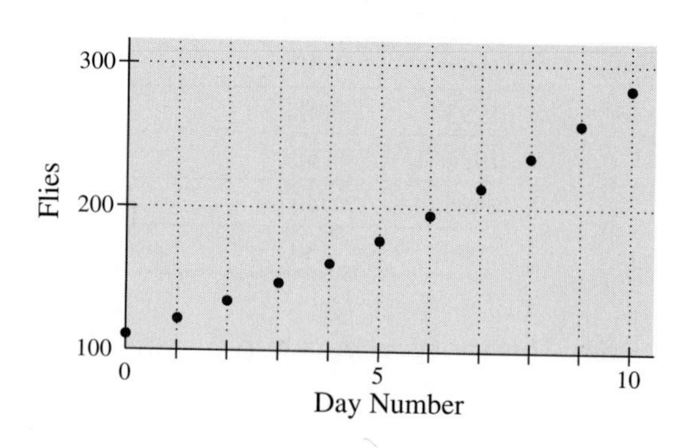

Case Study 2 Table 2.6 lists authoritative estimates of world population in the indicated years. Figure 2.41 is a plot of the same data.

Table 2.6 World population

Date (A.D.)	*Population (billions)*
1650	0.545
1750	0.728
1800	0.906
1850	1.171
1900	1.608
1950	2.517

Figure 2.41 Graph of world population

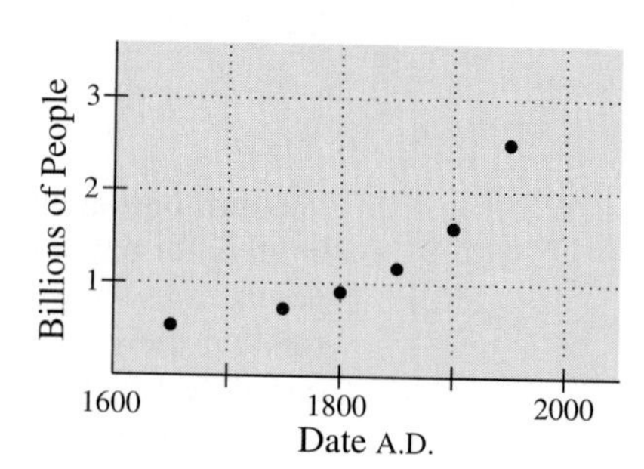

In Figures 2.40 and 2.41 you will have to imagine smooth curves connecting the data points. Could those curves be exponential? It's hard to be certain at this point—that's a question we have to investigate.

In your previous study of algebra, as well as in Chapter 1, you learned that taking logs is a good thing to do when confronted with problems involving exponents, and that's the kind of problem we have here. Recall the reason for this: "Taking logs" is defined by inversion (undoing) of exponentiation.

What happens if we take logs on both sides of the equation $P = P_0 e^{kt}$? The answer to this question depends on properties of logarithmic functions that are the most important reasons for studying these functions: *Logarithms turn products into sums and exponents into multipliers* (see Exercise 11). Because our exponent has base e, we will use base e logarithms as well. Here is the calculation:

$$\begin{aligned} \ln P &= \ln P_0 e^{kt} \\ &= \ln P_0 + \ln e^{kt} \\ &= \ln P_0 + kt \end{aligned}$$

Checkpoint 1

Use properties of logarithms to give reasons for each step in the calculation. Write a complete sentence for each reason.

The result of this calculation is that $\ln P$ (log of the population) turns out to be a constant ($\ln P_0$) plus another constant (k) times t. That is, the (natural) log of the population is a *linear* function of time! Now that's important, because it is easy to tell whether data points come from a linear function: Just plot them and see if they lie on a straight line.

Exploration Activity 1

(a) Turn back now to Table 2.5 for Case Study 1. Make a third column, headed "$\ln P$." Then use the [ln] key on your calculator to fill in this column.

(b) Plot $\ln P$ as a function of time for Case Study 1. Does the population grow exponentially?

(c) Turn back to Table 2.6 for Case Study 2. Make a third column, headed "$\ln P$." Then use the [ln] key on your calculator to fill in this column.

(d) Plot $\ln P$ as a function of time for Case Study 2. Does the population grow exponentially?

(e) If either population fails to grow exponentially, describe in your own words the way in which it fails to be exponential—for example, by growing too fast or too slow, or perhaps something else.

We show the logarithmic plots in Figures 2.42 and 2.43. We see that the fruit fly data indicate exponential growth; the plotted logarithm values seem to lie on a straight line. Check this by putting a straightedge along the plot. However, the plotted logarithm values for the world population data do not lie along a straight line; the growth is not exponential. Again, check this with a straightedge. In fact, the upward curve indicates that world population is growing too fast for exponential growth.

Figure 2.42 Graph of logarithms of the fruit fly data (Case Study 1)

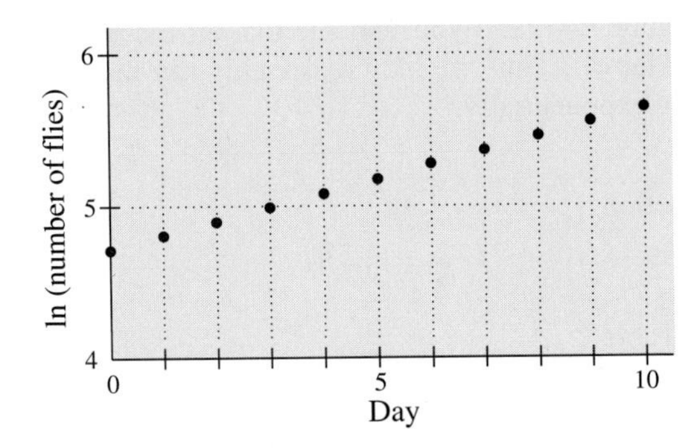

Figure 2.43 Graph of logarithms of the world population data (Case Study 2)

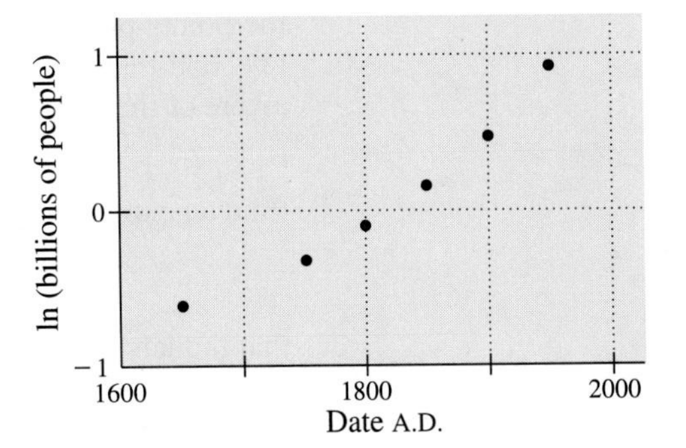

In the following example we show how to find a model function for data that appear to grow exponentially.

Example 1 Students in a psychology class planted a rumor that all university classes would be canceled on the university president's sixtieth birthday. After the rumor had run its course, they gathered the data in Table 2.7 on the number of students aware of the rumor at the end of each of six days. Is it reasonable to say that the number of students aware of the rumor grew exponentially?

Table 2.7 Spread of a rumor

Day	*No. of Students Aware*
1	6
2	11
3	25
4	45
5	70
6	130

Solution All we have to do is plot the natural logarithm of the number of students aware against the number of the day and see if the data points lie on a straight line. Figure 2.44 shows that plot.

Figure 2.44 Graph of the logarithm of the number of students aware

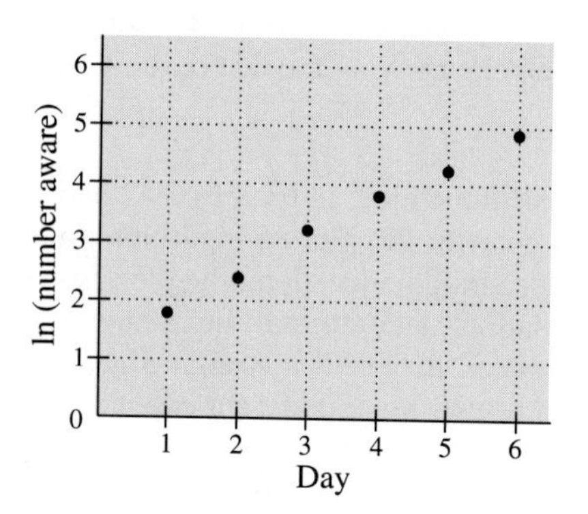

Now we have to decide whether the points lie on a straight line. Lay a straightedge along the points plotted in Figure 2.44, and you will see that all the points lie on a line except one, and that one is not far off. Thus we may reasonably say that the number of students aware of the rumor grew exponentially. ■

Example 2 Find an exponential function

$$N = N_0 e^{kt}$$

that models the data in Table 2.7.

Solution As we saw in Checkpoint 1, an equivalent equation is

$$\ln(N) = \ln(N_0) + kt,$$

so k should be the slope of the line on the ln N plot. If we take the line determined by the first and last points, this makes

$$k = \frac{\ln(130) - \ln(6)}{6 - 1} \approx 0.615.$$

Thus our model function should have the form

$$N = N_0 e^{0.615t}.$$

If we pick N_0 to fit the first data point, we have $6 = N_0 e^{0.615}$ or

$$N_0 = 6e^{-0.615} \approx 3.244.$$

This makes our model function

$$N = 3.244e^{0.615t}.$$

There are more systematic ways to pick the constants k and N_0 — we will return to this question later. ■

In Figure 2.45 we plot both the data and the graph of the model function.

Figure 2.45 Rumor data with model function

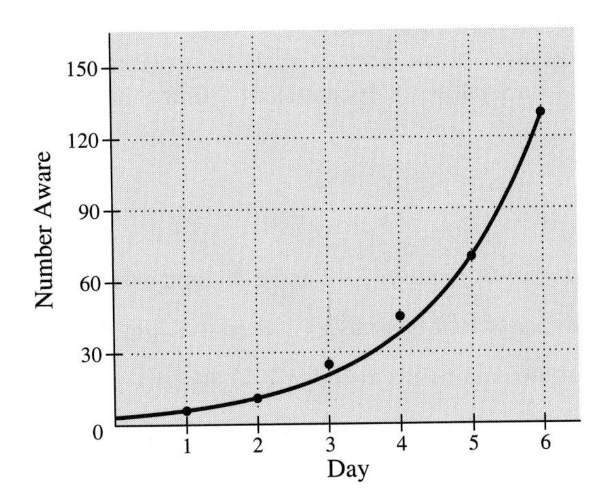

Logarithmic Plots of Data: Base 10

The graphs in Figures 2.42, 2.43, and 2.44, even though they serve their intended purpose, have a distinct disadvantage as pictures of our data: You can't look at these pictures and read off populations at particular dates. Contrast this with Figure 2.46 (repeated from Exercise 16 at the end of Chapter 1): That's also a picture of approximately exponential data — as we shall see — and the population figures can at least be estimated from the graph. This graph tracks a single cohort of students, ninth graders in 1972, through the years those students would have been in college and graduate school, counting those who eventually got degrees in the mathematical sciences.

Figure 2.46 The math pipeline

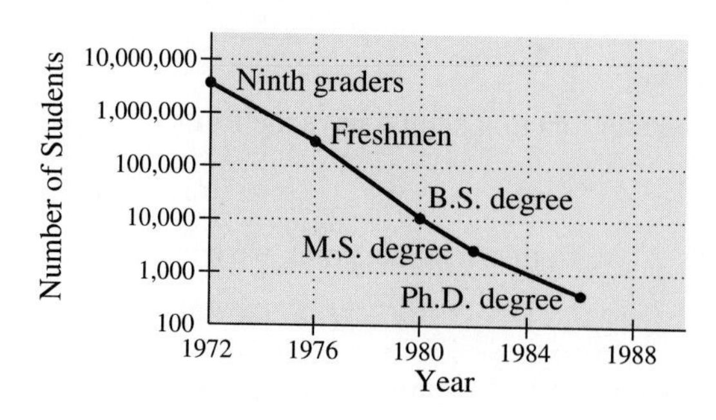

The secret of the graph in Figure 2.46 is its vertical scale. The numbers on that scale increase *exponentially* rather than linearly—each is 10 times the one below it. But those numbers, 10^2, 10^3, 10^4, 10^5, 10^6, and 10^7, are arranged on a *linear* scale, so the effect of graphing against that scale is to undo exponentiation, i.e., to graphically take a logarithm—with base 10. This kind of graphing is called **semilog** plotting. You can buy semilog graph paper, but, of course, it is much easier if your computer or calculator will do semilog plotting for you.

In Figures 2.47 and 2.48 we show semilog plots of the data from Case Studies 1 and 2. The data in both Case Study 1 and Case Study 2 vary by less than a factor of 10, so we set the top level of the vertical scale to be 10 times the bottom level in each case. The horizontal lines show the locations of 2 times the bottom level, 3 times the bottom level, and so on.

Checkpoint 2

(a) Compare these figures with Figures 2.42 and 2.43, and also with Tables 2.5 and 2.6.

(b) Verify that the number of fruit flies passes 200 on the right day.

(c) Check that each of the data points in Figure 2.48 appears to be at the right multiple of 0.3.

Figure 2.47 Semilog plot of fruit fly data (Case Study 1)

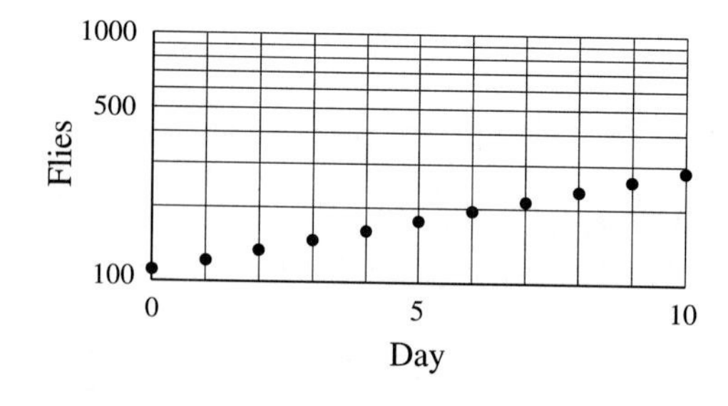

Figure 2.48 Semilog plot of world population (Case Study 2)

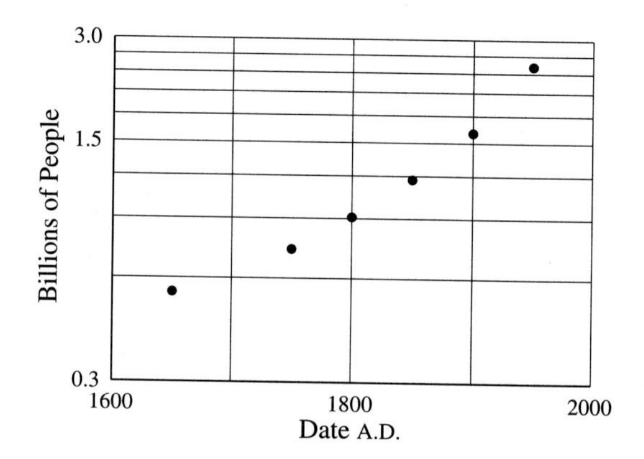

Semilog plotting of data has the effect of taking $\log_{10}$ (also called **common logarithm**) of each value of the dependent variable. Notice that the semilog plot for Case Study 1 is straight and the semilog plot for Case Study 2 is not. That's the same result we got by taking natural logarithms.

A word about the terminology: "Semi" means *half.* We are using a logarithmic scale in half the possible places, that is, on one out of two axes. That was true with our natural base logarithmic plotting as well, but standard usage assigns the name "semilog" only to base 10 logarithmic plotting. For purposes of finding model functions, it doesn't matter much which base you use. (See Exercise 13.) The primary virtue of base 10 is enabling direct visual estimates of function and data values.

The vertical scale in Figure 2.46 (the math pipeline) varies by many factors of 10, or **orders of magnitude**, as they are called. The only problem with reading values from the graph is that the horizontal lines for "2 times," "3 times," etc. have not been drawn in as we did in Figures 2.47 and 2.48. Figure 2.49 shows approximately the same pipeline data[6] plotted on a true semilog graph, with the appropriate number of orders of magnitude in the vertical direction and the logarithmically scaled reference lines within each order of magnitude.

Figure 2.49 The math pipeline

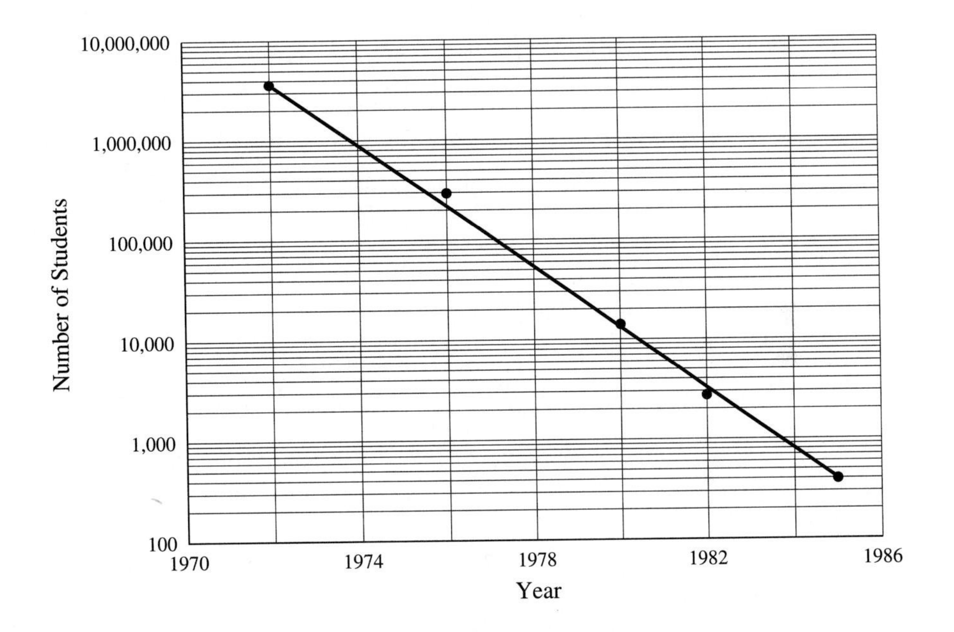

Checkpoint 3

(a) Compare each order of magnitude in Figure 2.49 with Figure 2.46, and label the lines that represent 1000, 10,000, 100,000, and 1,000,000.

---

6. The numbers we used in Figure 2.49 are based on a discussion on pages 35 and 36 of *A Challenge of Numbers: People in the Mathematical Sciences*, by B. L. Madison and T. A. Hart, National Academy Press, Washington, 1990. These numbers are not the actual reported figures in the indicated years; for example, there were only 11,000 bachelor's degrees in the mathematical sciences in 1980, not 14,000. However, the yield that year happened to be lower than in previous and subsequent years. Thus the near-linear relationship in Figure 2.49 is typical, if not completely accurate for the given starting year.

(b) Identify the horizontal lines in Figure 2.49 that represent 700, 20,000, 400,000, and 5,000,000.

(c) For each of the five data points, estimate the number of students directly from the graph.

Recall that the data points in Figure 2.49 track a single cohort of students, ninth graders in 1972, through the years those students would have been in college and graduate school, counting those who eventually got degrees in the mathematical sciences. The relationship on the semilog graph is nearly linear. Thus, if P represents the population in the math pipeline, then

$$\log_{10} P = \log_{10} P_0 + kt,$$

where $t = 0$ in the starting year (1972), and P_0 is the population in that year.

Exploration Activity 2

(a) Estimate the slope k of the linear function giving $\log_{10} P$ in Figure 2.49: Choose your own run, and calculate the corresponding rise.

(b) Solve the equation $\log_{10} P = \log_{10} P_0 + kt$ for P to express P as an exponential function (what base?).

(c) By what factor does P change in one year's time? Explain the following statement from *Everybody Counts*[7]: "[O]n average, we lose half the students from mathematics each year."

You should have found that k is approximately 0.30. When we exponentiate the equation $\log_{10} P = \log_{10} P_0 + kt$ to the base 10, we obtain

$$P = P_0 10^{kt}.$$

Thus the fraction left after any year is approximately

$$\frac{P_0 10^{k(t+1)}}{P_0 10^{kt}} = 10^k \approx 10^{0.30} \approx 0.50,$$

that is, approximately half the population is lost.

Power Functions and Log-Log Plotting

We turn now to data of another sort, namely, (simulated) data giving the position of a falling body at half-second time intervals. Table 2.8 contains a refinement of the falling body data given in Table 2.1 earlier in this chapter, where we explored a theoretical model for these data from elementary physics: $s = ct^2$, where s is distance fallen at time t, and c is a constant. We call such a function—a constant multiple of a constant power of the independent variable—a **power function**. This is to distinguish such functions from exponential functions, which have constant base and variable exponent. We can now address the question of whether the data actually fit a power function model.

7. From *Everybody Counts: A Report to the Nation on the Future of Mathematics Education*, National Academy Press, Washington, 1989.

Exploration Activity 3

(a) Take logs of both sides of the proposed functional relationship $s = ct^2$, and simplify as much as possible. (You decide what base to use for the logarithms.)

(b) The resulting equation says something is a linear function of something else. What is a linear function of what? What is the slope of that linear function?

(c) Propose a graphical test for deciding whether the data in Table 2.8 fits the theoretical model.

(d) Use a computer or graphing calculator to fill in the third and fourth columns of Table 2.8 (your choice of base), and do the graphing to carry out your test. What do you conclude about whether the data fit the model?

Table 2.8 Position of a falling object

Time (minutes)	*Distance (meters)*	*Log of Time*	*Log of Distance*
0.5	1.2		
1.0	5.0		
1.5	11.1		
2.0	19.4		
2.5	30.6		
3.0	44.1		
3.5	60.3		
4.0	78.0		
4.5	99.2		
5.0	122.8		
5.5	147.7		
6.0	175.8		
6.5	207.3		
7.0	240.0		
7.5	277.0		
8.0	312.8		
8.5	354.8		
9.0	396.1		
9.5	443.1		
10.0	489.0		

If we use the common logarithm, we have

$$\log s = \log c + 2 \log t.$$

Since $\log c$ is a constant, this equation has the form

$$y = \text{constant} + 2x,$$

where $y = \log s$ and $x = \log t$. Thus $\log s$ is a linear function of $\log t$ with slope 2.

We can test this model by plotting $\log s$ versus $\log t$ and see if the points lie along a line of slope 2. In Figure 2.50 we show such a plot with both scales treated logarithmically, as we did with only the vertical scale in the semilog plots. This is called a **log-log** plot or sometimes just a **log** plot.

The plotted points do seem to lie along a straight line. We can estimate the slope by checking the rise over the run for the extreme points

$$\frac{\log 489.0 - \log 1.2}{\log 10.0 - \log 0.5} \approx 2.01.$$

Thus it looks like the model fits the data.

Figure 2.50 Log-log plot of falling body data

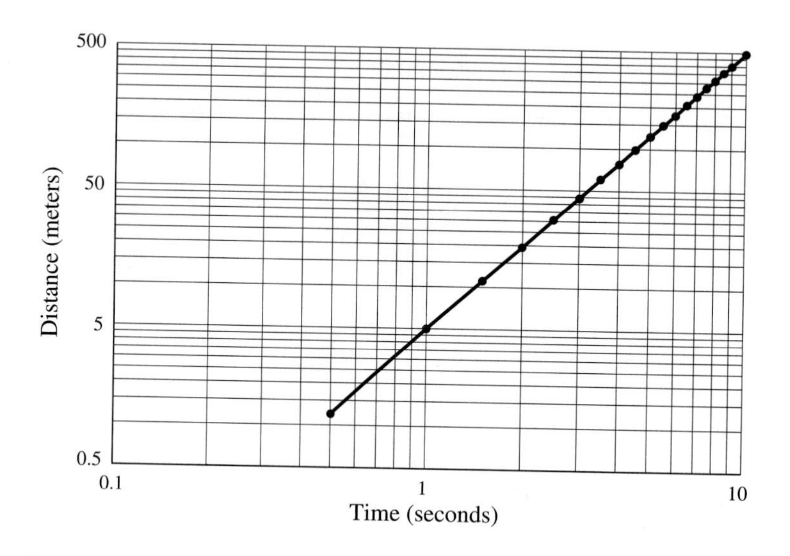

In the following example we show how to find a model function for data that appear to fit a power model.

Example 3 The Jurassic Toy Company produces seven models of their stuffed brachiosaurus. Each of the newer six models scales up the original successful model. Table 2.9 gives, for each model, the volume v of compressed stuffing required (in cubic inches) together with the scale factor x. Find a model formula for v as a function of x.

Table 2.9 Toy dinosaur stuffing

x	1	2	3	4	5	6	7
v	2.4	19.1	64.0	152	295	510	813

Solution To check whether a power model is reasonable, we look at a log-log plot of the data and decide whether the plotted points lie close to a straight line. Figure 2.51 shows the plot for these data. In this case the data seem to fall along a straight line, so a power model is reasonable.

Now how do we find an appropriate model of the form

$$v = cx^r?$$

If we have such a model, then

$$\log(v) = \log(c) + r\log(x).$$

Thus the straight line in the log-log plot should have slope r. If we use the first and last points to determine the slope, then our choice of r should be

$$\frac{\log(813) - \log(2.4)}{\log(7) - \log(1)} \approx 2.994.$$

It looks like 3 is a reasonable power.

Now our model has the form $v = cx^3$. If we use, say, the fourth point to determine c, we have

$$152 = c4^3 \qquad \text{or} \qquad c \approx 2.38.$$

Our model function is

$$v = 2.38x^3.$$

A plot of the data and the graph of the model function are shown in Figure 2.52. ■

Figure 2.51 Log-log plot of dinosaur stuffing data

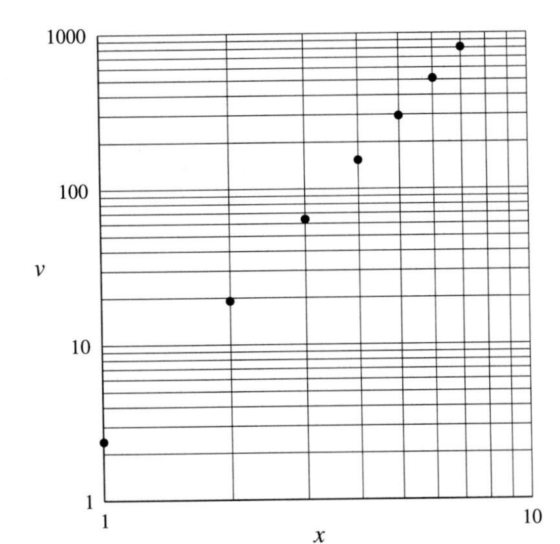

Figure 2.52 Plot of data and graph of the model function for Example 3

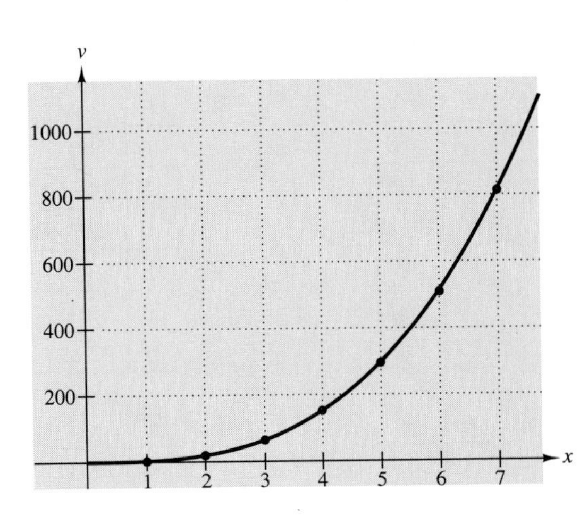

In this section we have studied semilog and log-log plots of data. If a semilog plot of the data lies approximately along a straight line, then it is reasonable to approximate the data with an exponential function. The slope of the approximating line may be used as the coefficient in the exponent of the model function—provided the base used is the same for the semilog plot and for the model function.

If a log-log plot of the data lies approximately along a straight line, then it is reasonable to approximate the data with a power function. The slope of the approximating straight line may be used for the power of the independent variable—and the power is the same, no matter what base is used for the slope calculation.

Answers to Checkpoints

1. We substitute the definition of P.

   The logarithm of a product is the sum of the logs.

   The natural logarithm and the natural exponential functions are inverses.

2. (a) Figures 2.47 and 2.48 show the same shapes for the scatter plots as Figures 2.42 and 2.43, respectively. The essential differences are in the spacing and meaning of the horizontal grid lines.

   (b) On day 6 the plotted point is just below the 200 line. The day 7 point is well above that line. Table 2.5 shows the actual numbers to be 195 and 214, respectively.

   (c) For example, the 1800 data point is at $3 \times 0.3 = 0.9$. The corresponding entry in Table 2.6 is 0.906.

3. (c) $3,400,000;\quad 220,000;\quad 15,000;\quad 2500;\quad 420$

Exercises 2.6

The data in each of the following tables approximately fit either an exponential function or a power function. In each case, decide which, and find a formula that fits the data. Each formula should have one of the forms $y = cb^t$, $y = ce^{kt}$, or $y = ct^k$, with explicit values for the constants.

1.

t	y
0	0.0
5	14.855
10	20.292
15	24.354
20	27.720
25	30.648

2.

t	y
0	1.0
5	4.452
10	7.339
15	12.101
20	19.950
25	32.893

3.

t	y
1	24.7
5	17.2
10	11.0
20	4.5
35	1.2
50	0.3

4.

t	y
1	25.0
5	7.5
10	4.4
20	2.6
35	1.7
50	1.3

5. (a) We now know three ways to plot data: Cartesian (evenly spaced grid in both directions), semilog (logarithmic scale in one direction), log-log (logarithmic scale in both directions). In each case, seeing a straight line tells you something about the data. What does it tell you in each of these cases?

   (b) What can you conclude about data if their plot is not straight in any of the three cases described in part (a)?

   (c) Consider the points $(0,0)$, $(1,3)$, $(2,6)$, $(3,9)$, and $(4,12)$, all of which have $y = 3x$. On how many of the three types of plots would these data appear as a straight line? Which ones, and why?

   (d) Consider the points $(0.1,-1)$, $(1,0)$, $(10,1)$, $(100,2)$, and $(1000,3)$, all of which satisfy $y = \log_{10} x$. What kind of plot of these data would appear to be a straight line? Why?

6. (a) The data in Table 2.10 come from the early U.S. censuses. Fill in the third column of the table, and decide whether the U.S. population grew exponentially during its first century. If so, explain why you think so. If not, explain the way in which the data deviate from being exponential.

   (b) Answer the same question for the time from the first census to the start of the Civil War. (Use a computer or graphing calculator to fill in the table and plot the data.)

7. The data in Table 2.11 are graphed in Figures 2.53, 2.54, and 2.55 in Cartesian, semilog, and log-log plots, not necessarily in that order.

   (a) Decide whether the data come from a linear function, a power function, an exponential function, or none of these.

   (b) Estimate $f(2)$.

   (c) Estimate $f'(2)$.

Table 2.10 U.S. population for the first 100 years

Year	*Population (millions)*	*ln of Population*
1790	3.929	
1800	5.308	
1810	7.240	
1820	9.638	
1830	12.866	
1840	17.069	
1850	23.192	
1860	31.443	
1870	38.558	
1880	50.156	
1890	62.948	

Table 2.11 Values of a function

t	$f(t)$
0.27	1.618
0.80	2.786
1.33	3.592
1.86	4.248
2.39	4.815
2.92	5.322
3.45	5.785
3.98	6.213
4.51	6.614
5.04	6.992
5.57	7.350

Figure 2.53 A plot of data in Table 2.11

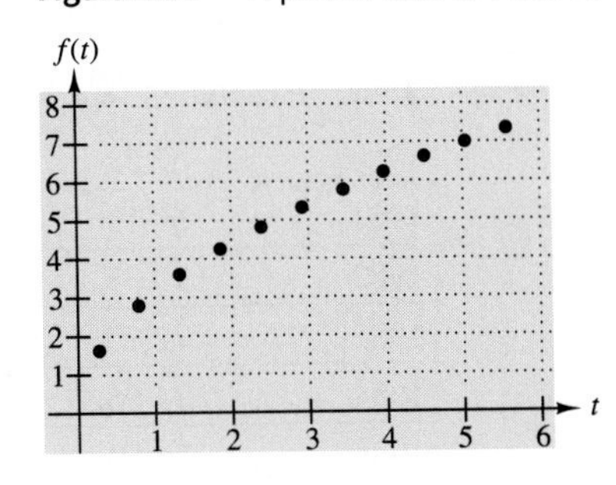

Figure 2.54 A plot of data in Table 2.11

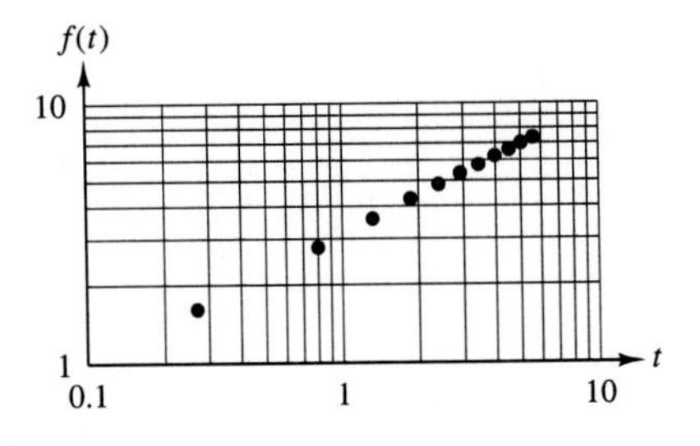

Figure 2.55 A plot of data in Table 2.11

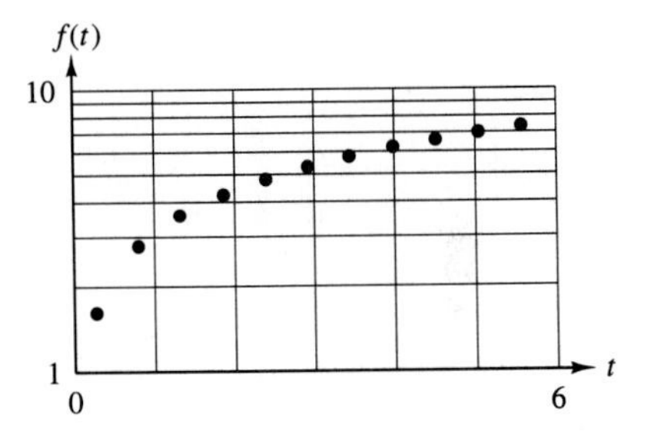

8. The data in Table 2.12 are graphed in Figures 2.56, 2.57, and 2.58 in Cartesian, semilog, and log-log plots, not necessarily in that order.
   (a) Decide whether the data come from a linear function, a power function, an exponential function, or none of these.
   (b) Find a formula for f as a function of t.

Table 2.12 Values of a function

t	$f(t)$
0.27	1.618
0.80	2.240
1.33	2.664
1.86	3.088
2.39	3.512
2.92	3.936
3.45	4.360
3.98	4.784
4.51	5.208
5.04	5.632
5.57	6.056

Figure 2.56 A plot of data in Table 2.12

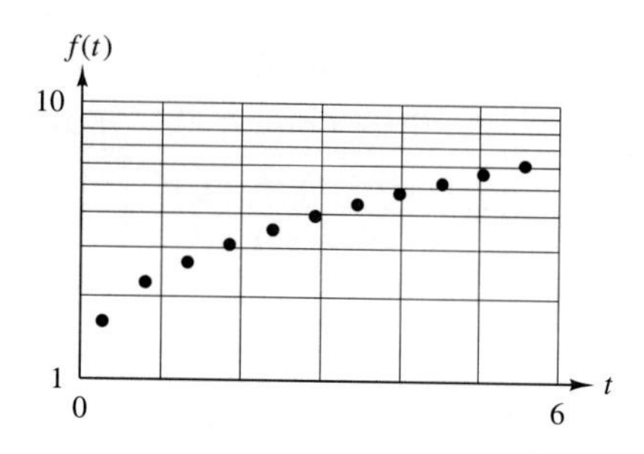

Figure 2.57 A plot of data in Table 2.12

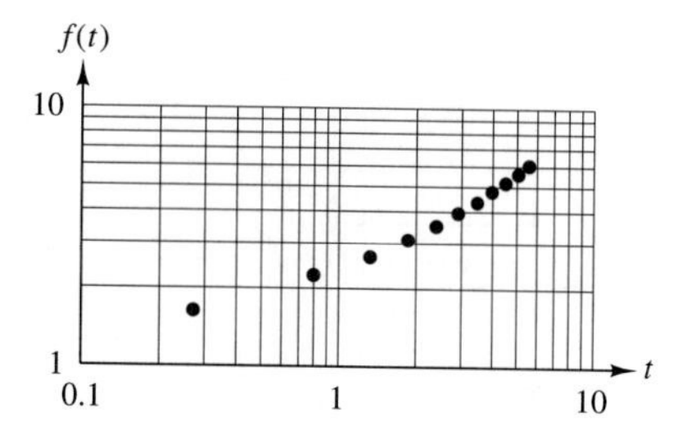

Figure 2.58 A plot of data in Table 2.12

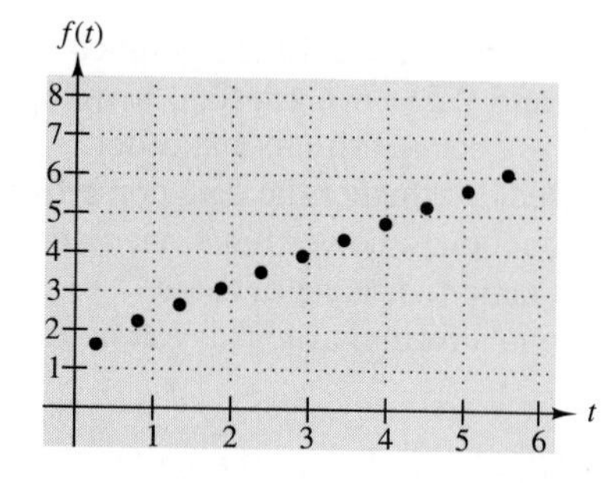

9. In our calculation of the linear relationship between $\ln P$ and t, we chose "ln" because the base of our exponential function was e. Show that this choice was unnecessary. That is, given $P = P_0 e^{kt}$, show that we could use logarithms with any base b and still find that "log of P is a linear function of t."

10. It may have occurred to you while working Exploration Activity 3 that you don't really need logarithms to tell whether the data in Table 2.8 fit $s = ct^2$ — taking square roots would do as well.
    (a) Verify this by adding a column for $\sqrt{s}$ to Table 2.8 and testing for linearity as you did in that activity.
    (b) Discuss the relative merits of these two procedures. Suppose you wanted to test whether $s = ct^p$ for some unknown power p. Could you do this by taking roots? Could you do it by taking logarithms? In the process, could you find out what p must be? Explain.

11. Use properties of exponentials and the definition of logarithms,

$$y = \log_b x \quad \text{if and only if} \quad x = b^y,$$

to establish each of the following properties of logarithms.
    (a) $\log_b(AB) = \log_b A + \log_b B$
    (b) $\log_b A^B = B \log_b A$
    (c) $\log_b \dfrac{A}{B} = \log_b A - \log_b B$

12. Suppose we model the rumor data in Example 1 with a function of the form $N = N_0 10^{mt}$.
    (a) Show that m is the slope of the line in the semilog plot. Calculate this slope m.
    (b) Show that $10^m = e^k$, where k is the slope of the $\ln N$ plot. Conclude that $10^{mt} = e^{kt}$ for all times t, and that this model gives the same function as in Example 1.

13. Show that if the data represent a function of the form $P = P_0 e^{kt}$, then $\log_b P$ will be a linear function of t no matter what the base b is.

14. Show that it would not have made a difference in Example 3 if we had used natural logarithms instead of common logarithms.

15. Table 2.13 gives the population of Mexico for each of the years 1980 through 1986. Decide whether the population was growing exponentially during those years. If so, find an exponential function that approximately fits these data. By what percentage did the population grow each year?

Table 2.13 Population of Mexico

Year	*Population (millions)*
1980	67.38
1981	69.13
1982	70.93
1983	72.77
1984	74.66
1985	76.60
1986	78.59

16. (a) In Table 2.14 we show the number of miles of railroad track in the United States in the late nineteenth century. Determine the growth pattern, and find a formula for a function that fits these data.
(b) If that pattern you found in part (a) had continued for another century, how many miles of track would there have been in the continental United States in 1990? What percentage of the area of the United States would have been covered by railroad beds in 1990?
(c) In what year would this growth pattern have led to covering the entire country with railroad bed?
(d) Suppose your estimate of the area of the United States is off by 10% one way or the other. How would that affect your answer to part (c)? Suppose your estimate of the width of a roadbed is off by a factor of 2 one way or the other. How would that affect your answer to part (c)?[8]

Table 2.14 Miles of railroad track in the United States

Year	*Miles of Track*
1860	30,626
1870	52,922
1880	93,262
1890	166,703

17. In Table 2.15 we show the U.S. Census data for the Houston, Texas, primary statistical metropolitan area.
(a) Show that the historic population data for Houston are approximated by an exponential function, and find a formula for the function.
(b) Find the doubling time for Houston's population.
(c) If the growth pattern continues, what will Houston's population be in 2000? In 2050?

Table 2.15 Houston area population

Census Date	*Population*
1850	18,632
1860	35,441
1870	48,986
1880	71,316
1890	86,224
1900	134,600
1910	185,654
1920	272,475
1930	455,570
1940	646,869
1950	947,500
1960	1,430,394
1970	1,999,316
1980	2,905,344

8. Problem suggested by R. H. Romer, "Covering the U.S.A.: Exponential Growth of Railroad Tracks," *The Physics Teacher* 28 (1990), pp. 46–47, and A. A. Bartlett, "A World Full of Oil," *The Physics Teacher* 28 (1990), pp. 540–541.

Chapter 2 Summary

The unifying theme of this chapter was our discussion of models of natural population growth and the mathematical concepts and tools necessary to understand them. The first of these concepts was the notion of *rate of change*.

We introduced difference quotients as average rates of change and then considered the limiting values of these rates as the time interval approached zero. This led to the important concept of instantaneous rate of change. In graphical terms, this was the slope of the graph when we zoomed in far enough for the curve to appear to be a straight line. The corresponding mathematical concept was the derivative — one of the two fundamental notions of all calculus.

In Section 2.3 we developed tools for calculating derivatives and used these tools to calculate derivatives of polynomials.

Next, we investigated the family of functions most closely associated with natural growth — the exponential functions b^t. We saw that these functions have derivatives that are constant multiples of themselves.

Then we returned to natural population growth. We saw that the appropriate mathematical model in the continuous case was a differential equation. The family of solutions of a differential equation can be pictured using slope fields. To identify a particular solution, we need to specify an initial value. Together the differential equation and the initial value constitute a differential-equation-with-initial-condition problem.

Finally, we developed graphical tools for deciding whether we could reasonably assume that a set of data could be modeled by an exponential function or by a power function. These tools were semilog plots and log-log plots, respectively. In addition, we saw how to determine a model function in each of these two cases.

Concepts

- Average rate of change
- Difference quotient
- Instantaneous rate of change
- Derivative
- Exponential function
- Natural exponential function
- Difference equation
- Differential equation
- Differential-equation-with-initial-condition problem
- Semilog plot
- Log-log plot

Applications

- Discrete natural growth
- Continuous natural growth
- Objects falling under the influence of gravity
- Finding model functions for data

Formulas

The natural base

$$e \approx 2.718282$$

The Power Rule

$$\frac{d}{dt}\, t^n = n\, t^{n-1}$$

The Constant Multiple Rule

$$\frac{d}{dt}\, cf(t) = c\, \frac{df}{dt}$$

The Sum Rule

$$\frac{d}{dt}\, [f(t) + g(t)] = \frac{df}{dt} + \frac{dg}{dt}$$

Derivatives of exponential functions

$$\frac{d}{dt}\, e^t = e^t$$

$$\frac{d}{dt}\, e^{kt} = k\, e^{kt} \quad \text{for any constant } k$$

$$\frac{d}{dt}\, A\, e^{kt} = kA\; e^{kt} \quad \text{for any constant } k$$

$$\frac{d}{dt}\, b^t = (\ln b)\, b^t \quad \text{for any constant exponential base } b$$

Chapter 2 Exercises

Given that $\log_b 2 = 0.6309$ and $\log_b 5 = 1.465$, find

1. $\log_b 10$
2. $\log_b \frac{5}{2}$
3. $\log_b \frac{1}{5}$
4. $\log_b \sqrt{b^3}$
5. $\log_b 5b$
6. $\log_b 8$

Solve each of the following equations for t.

7. $t^8 = 20$
8. $e^t = 20$
9. $10^t = 20$
10. $t^5 = 20$
11. $e^{3t} = 20$
12. $3^t = 20$

Solve each of the following equations for x.

13. $2^{-x} = 3^x$
14. $2^{-x} = 2^{x+2}$
15. $2^x = 3^{x+2}$
16. $2^{-x} = 3^{x+2}$
17. $3^x = 2^{x+2}$
18. $3^{-x} = 2^{x+2}$

Calculate each of the following derivatives.

19. $\frac{d}{dt}\, e^{2t}$
20. $\frac{d}{dt}\, e^{-t}$
21. $\frac{d}{dt}\, 3^t$
22. $\frac{d}{dt}\, e^{t/2}$
23. $\frac{d}{dt}\, e^{-4/3\,t}$
24. $\frac{d}{dt}\, \pi^t$
25. $\frac{d}{dt}\, (3t - 5)$
26. $\frac{d}{dt}\, 6\,t^2$
27. $\frac{d}{dt}\, 7$
28. $\frac{d}{dt}\, (3t - 5)^2$
29. $\frac{d}{dt}\, (t^6 + 6^t)$
30. $\frac{d}{dt}\, (2^t - 3^t)$

31. Express $17e^{0.07t}$ in the form cb^t.

32. Express $23 \cdot 2^t$ in the form ce^{kt}.

33. Express $0.07e^{17t}$ in the form cb^t.

34. Express $22 \cdot 3^t$ in the form ce^{kt}.

Find the derivative of each of the following functions.

35. $t^5 - 7e^{4t} + 2t^2 - 6e^{-t} + 1$
36. $t^5 + 2t^4 - 3t^2 - 8 + e^{-t/2}$
37. $4e^{4t} - e^{3t} + 2e^{2t} - e^t + 7$
38. $13e - 27t + 5t^2 + e^{3t}$

39. Manufacturing companies in Mexico owned by U.S. corporations are called *maquiladoras*. From 1993 to 1998, gross production of the *maquila* industry is predicted to grow at an average rate of 19% per year, and employment is predicted to grow at 9% per year.[9]

    (a) By what factor will *maquila* production grow over that five-year period?

    (b) By what factor will *maquila* employment grow over the same time?

40. Suppose you invest a sum of money in a savings account that pays 4.5% interest, compounded quarterly. Suppose further that, over the time the money is invested, there is an annual 3.5% inflation rate.

    (a) Would you expect the value of your investment to grow linearly or exponentially? Explain.

    (b) If you expect the growth to be exponential, what is the proportional growth rate? If you expect the growth to be linear, what is the growth rate?

41. A colony of bacteria living on a petri dish under optimal conditions doubles in size every 10 minutes. At noon on a certain day the petri dish is completely covered with bacteria. At what times (to the nearest hour, minute, and second) was the covered percentage of the dish

    (a) 50% ?

    (b) 25% ?

    (c) 5% ?

    (d) 1% ?[10]

On each of the following slope fields, sketch three solution functions. (You may photocopy the page if you wish.)

42.

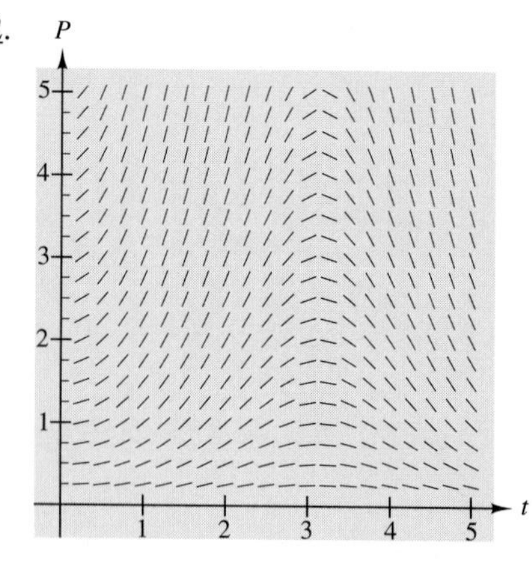

43.

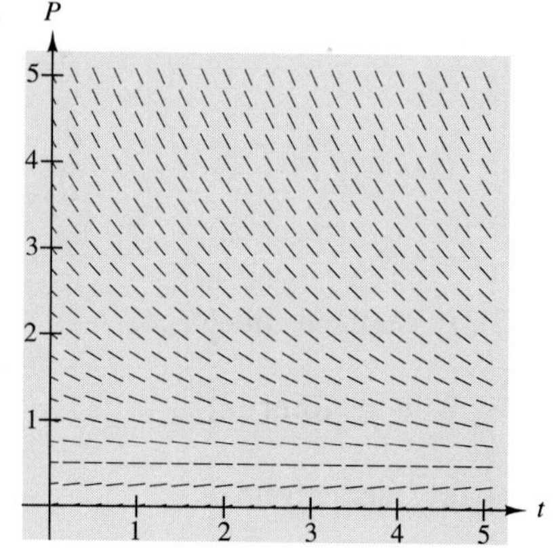

44. (a) Mosquitoes reproduce in 2-week cycles, and the typical female mosquito lays 400 eggs at a time. Assume that half the offspring are female and that all live to reproduce at the end of the next 2-week cycle. How many descendants could one female produce in a whole summer?

---

9. Source: "Maquila Industry Annual Review," as reported in Our World supplement to *USA Today*, September 30, 1994.
10. From *Calculus Problems for a New Century*, edited by Robert Fraga, MAA Notes Number 28, 1993.

(b) *USA Today* reported on its front page on July 26, 1993, that "One female mosquito today" could produce "16 million mosquitoes by Labor Day." A footnote added that "8 million are females and could, by [that same] Labor Day, lay 3.2 billion *more* eggs." Are either or both of those claims correct? Why or why not?

45. Max Wertheimer, founder of Gestalt psychology, was a friend and correspondent of physicist Albert Einstein. At the end of one of his letters, Wertheimer added a couple of brain teasers for Einstein to solve. Here is one of them:

> An amoeba propagates by simple division: It takes three minutes for each split. We put such an amoeba in a glass container with nutrient fluid, and it takes one hour until the vessel is full of amoebas. How long will it take to fill the vessel if we start not with one amoeba but two?

Solve it.[11]

46. A typical ream of paper (500 sheets) is about two inches thick, so we may assume that the thickness of a single sheet is 0.004 inch.

(a) If you fold a single sheet of paper 5 times, what will be the thickness of the folded paper? Do it, and check with a ruler, as carefully as you can.

(b) If you could fold a piece of paper 50 times, what would be the thickness of the folded paper? Compare your answer with some familiar distance of the same order of magnitude.

47. At the start of a (not very realistic) trip, an automobile is moving at 10 miles per hour. It accelerates at a constant rate of 10 miles per hour per hour for five hours. (Acceleration is the instantaneous rate of change of velocity.)

(a) Find an equation for the velocity of the car as a function of the number of hours after the start of the trip, and graph the velocity as a function of time.

(b) What is the slowest velocity of the car for $1 \leq t \leq 3$? Use that slowest velocity to estimate the distance the car travels between $t = 1$ and $t = 3$.

(c) What is the fastest velocity of the car for $1 \leq t \leq 3$? Use that fastest velocity to estimate the distance the car travels between $t = 1$ and $t = 3$. What have you learned from your two estimates?

(d) What is the slowest the car travels for $1 \leq t \leq 2$? What is the slowest the car travels for $2 \leq t \leq 3$? What is the minimum distance the car can travel in the second hour of the trip? What is the minimum distance the car can travel in the third hour of the trip? Use these results to estimate the distance the car travels between $t = 1$ and $t = 3$.

(e) What is the fastest the car travels for $1 \leq t \leq 2$? What is the fastest the car travels for $2 \leq t \leq 3$? What is the maximum distance the car can travel in the second hour of the trip? What is the maximum distance the car can travel in the third hour of the trip? Use these results to estimate the distance the car travels between $t = 1$ and $t = 3$.

(f) What have you learned from the two estimates in parts (d) and (e)? How does this compare with what you learned from parts (b) and (c)?

(g) Suggest a technique based on this procedure that might lead to an exact answer (or a very precise estimate) for the total distance the car travels between $t = 1$ and $t = 3$.[12]

---

11. Problem suggested by Agnes Wieschenberg, *UME Trends*, November 1993, p. 3.
12. Adapted from *Calculus Problems for a New Century*, edited by Robert Fraga, MAA Notes Number 28, 1993.

Chapter 2 Projects

1. **Spread of AIDS** Figure 2.59 shows data on the numbers of AIDS cases in the early years of the AIDS epidemic in the United States.

   (a) What is the significance of the approximately straight lines shown in the graph? (Look carefully at what is being plotted on the vertical axis.)

   (b) For reasons we will see later in the course (Chapter 5), many epidemics start out with roughly exponential growth in the number of cases; are the AIDS data exponential? Why or why not?

   (c) What sort of model (formula) would you expect to fit the "total" data in Figure 2.59?

   (d) If you had the numbers of cases in each year (instead of the graph), how would you test the data graphically to see what sort of model might fit? What would cause you to accept or reject each of the possible models you can think of?

   (e) Knowing what you know from Figure 2.59, what would you expect to see in each of your graphical tests?

Figure 2.59 Data on AIDS cases reported by the Centers for Disease Control[13] (Breakdown by race: 1: white; 2: black; 3: Hispanic; 4: unknown)

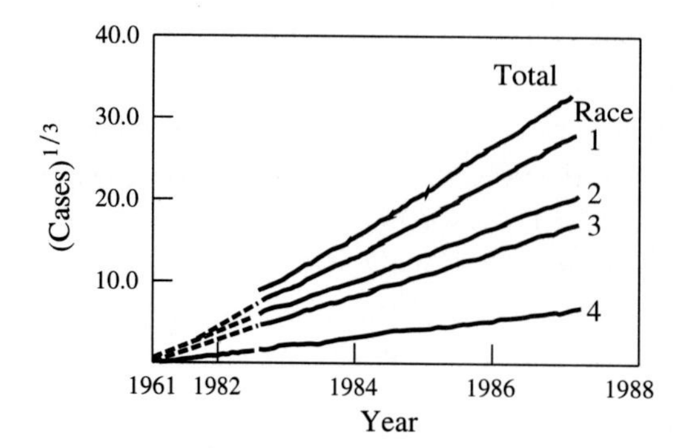

2. **Electromagnetic Field Exposure** Figure 2.60 shows electromagnetic field (EMF) exposure from three different sources as "functions" of distance from source.

   (a) Explain why none of the relationships shown is really a function.

   (b) We could, nevertheless, replace each of the three non-function graphs by an "average" function. Draw on Figure 2.60 what you think each of those average functions would look like.

   (c) The graphs for high-voltage transmission lines and for distribution lines each have a "horizontal" part and a "sloping" part. What does it mean for a part of the graph to be horizontal?

---

13. From "Modeling the AIDS Epidemic" by Allyn Jackson, *Notices of the American Mathematical Society* 36 (1989), pp. 981–983.

(d) On the type of graph shown, what does it mean for a part of the graph to be sloping and straight?

(e) For each of the sloping parts, estimate the slope, and then determine an appropriate model to fit that part of the data, using the graph type, the straightness, and the slope.

Figure 2.60 Electromagnetic field exposure from three sources[14]

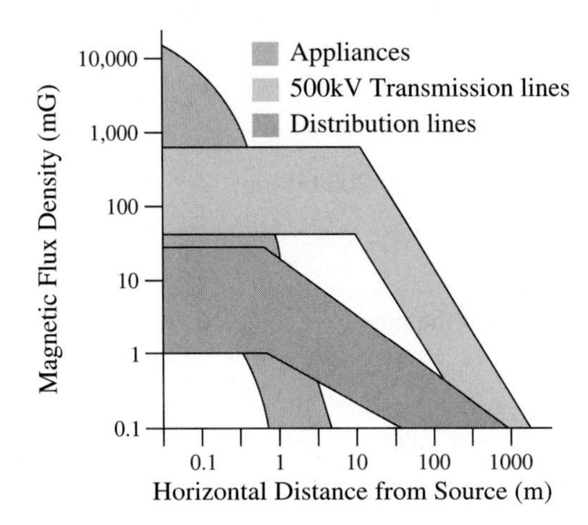

3. **Difference Equations** In this project we study the difference equation approach to modeling population growth. Recall that we started with a difference equation with a discrete time step Δt:

$$\frac{\Delta P}{\Delta t} = kP,$$

where k is a constant, and P is the population at the start of a time interval of length Δt. We have already studied an example of this difference equation in Section 2.4, where arithmetic calculations of compound interest led us to consider exponential functions in the first place. Indeed, compound interest and natural population growth are conceptually the *same idea*: Populations without constraints grow by adding "interest" (more individuals) to the population already present. In this project we pursue algebraically (as opposed to arithmetically) the same kind of "compound interest" calculations in order to find all possible solutions of the difference equation. Not surprisingly, the solutions turn out to be exponential functions.

We can rewrite equation the difference equation as

$$\Delta P = kP\Delta t.$$

In words, the change in population over a given time interval is *jointly* proportional to the population at the start of the interval and to the length of the interval. Since we are now assuming that the time interval remains constant, the equation becomes the simpler statement

14. From a "News and Comment" article on EMF exposure and cancer risk, *Science*, September 7, 1990, p. 1097. This figure is typical of data displays in the scientific literature.

that the change in population in one reproductive cycle is proportional to the population at the start of that reproductive cycle.

The equation

$$\Delta P = kP\Delta t$$

also can tell us something about *continuous* population growth, as long as we limit our attention to *average* change over discrete time intervals. For example, suppose we assume that human population is growing at 2% per year, and, at the start of 1988, world population was 5 billion. (This assumption is not consistent with the population data. We'll look at better models later.)

(a) What would have been the population at the start of 1989?

(b) Show that at the start of 1990 the population would have been 5.202 billion.

(c) Suppose the 2% annual rate continued for the rest of the century. What would the population be at the start of 2001?

(d) Still assuming the continued 2% rate, write a formula for the population any number of years after 1988?

(e) Suppose we set $t = 0$ at a time at which we know P, namely, at the start of 1988, when $P = P_0 = 5$ (billion). Then we saw in part (a) that P_1, the value of P at $t = 1$, is 5.1. We also saw that P_2 is 5.202. Fill in the following table to continue this process up to $t = 7$.

t	P	ΔP
0	5	0.1
1	5.1	0.102
2	5.202	
3		
4		
5		
6		
7		

(f) Fill in a similar table for a solution to the same difference equation but starting from $P(0) = 3.8$.

t	P	ΔP
0	3.8	
1		
2		
3		
4		
5		
6		
7		
8		

(g) Rewrite the equation

$$\Delta P = kP\Delta t$$

with $\Delta t = 1$, in the form

$$P_{n+1} - P_n = kP_n$$

at time $t = n$. Show that

$$P_{n+1} = (1+k)P_n \quad \text{for } n = 0, 1, 2, \cdots.$$

(This is called an **iterative** formula; it gives each value of P if you know the previous one. However, if you want to know P_{30}, you have to do 30 multiplications.) Write out the first several cases of this formula, starting at $n = 0$. Find a pattern from which you can deduce an explicit formula for P_n, one that would enable you to find P_{30} from P_0 and k with only a few keystrokes on your calculator.

(h) Use your formula, with $k = 0.02$, to check your answer to part (c) for the world population at the start of the twenty-first century.

(i) We can solve $\Delta P/\Delta t = kP$ algebraically by following the same pattern of steps you carried out in part (g). In order to proceed, we have to know a value of P (an "initial value") at some particular time t (the "initial time"). Since the time scale is under our control, we choose to make the initial time $t = 0$, and we denote the initial value by P_0. Thus we are really solving the **difference-equation-with-initial-value problem**

$$\frac{\Delta P}{\Delta t} = kP \qquad \text{and} \quad P = P_0 \quad \text{when } t = 0.$$

Rewrite the equation in the form

$$P_{n+1} - P_n = k\ P_n\ \Delta t,$$

and then in the form

$$P_{n+1} = P_n + k\ P_n\ \Delta t = (1 + k\Delta t)\ P_n.$$

Here is a table of the algebraic results:

n	t	P
0	0	P_0
1	Δt	$(1+k\Delta t)\ P_0$
2	$2\Delta t$	$(1+k\Delta t)\ P_1 = (1+k\Delta t)^2\ P_0$
3	$3\Delta t$	$(1+k\Delta t)\ P_2 = (1+k\Delta t)^3\ P_0$
4	$4\Delta t$	$(1+k\Delta t)\ P_3\ =$
5	$5\Delta t$	

Explain the algebraic step following each equals sign. Fill in the blanks in lines 4 and 5 of the table. What pattern do you see emerging? Explain why $P_n = a^n P_0$ for some constant a. What is a?

(j) Your explanation shows that given a starting population, each later population value can be expressed explicitly as a function of the step number n. What we really want as a solution to the difference-equation-with-initial-value problem

$$\frac{\Delta P}{\Delta t} = kP \quad \text{and} \quad P = P_0 \qquad \text{when } t = 0$$

is P expressed as a function of t. Show that $n = t/\Delta t$ and that $a^n = b^t$, where $b = a^{1/\Delta t}$. Then show that our desired form for the solution to the difference-equation-with-initial-value problem is

$$P(t) = P_0\, b^t \qquad \text{where } b = (1 + k\Delta t)^{1/\Delta t}.$$

(k) This spring, after cubs were born, the bear population of Black Bear Natural Wilderness Area was 317, half of them female. Each spring 70% of the females each produce two cubs, and 50% of all the cubs are female. Assuming no deaths, estimate the bear population in 5 years. Do your counting with whole bears, not with the solution we gave above, which may produce "fractional bears." (*Hint:* You may find it simpler to count the females year by year and then double to get the total population.)

(l) Now use your solution in part (j) to estimate the population in 5 years. How close are your two estimates?

CHAPTER

3

Initial Value Problems

In Chapter 2 we encountered the awkwardly named concept of **differential-equation-with-initial-value problem** *as the key to our study of natural population growth. Indeed, this concept is key to the study of many problems of growth and decay, of motion, and of all sorts of change. In this chapter we shorten the name to* **initial value problem**, *and we focus on this fundamental concept.*

After reviewing definitions and elementary examples, we will study two more problems that can be modeled as initial value problems: the change in temperature of a body cooling in the air and the speed of an object falling under the force of gravity. These two problems are prototypes for all the differential equation problems we will encounter in this course. That is, at the most basic level, every differential equation we see will be like the one or the other of these problems.

In the laboratory reading at the end of the chapter we extend our discussion of the falling body problem to include the possibility of air resistance.

Imagine yourself a script writer for your favorite TV detective. Here's the scene: On Monday morning, a wealthy industrialist is found murdered in his office in the climate-controlled building he owns (sorry, owned). The detective has a range of suspects: members of the janitorial crew, the industrialist's ex-wife, his nephew, and the firm's vice-president. Solving the case may depend critically on accurate determination of the *time* of the crime, since that may show that some (perhaps all but one?) of the suspects could not have been present. How can our detective determine the victim's time of death—and thereby narrow down this list of suspects?

We will soon put calculus to work to help our detective sort out the suspects. First, we review and formalize the concepts, terminology, and notation introduced in Chapter 2.

3.1 Differential Equations and Initial Values

We often identify functions by the differential equations they satisfy, so it is important to know what is meant by a function satisfying a differential equation.

Definitions Suppose y is a function of t, say, $y = f(t)$. We say the function **satisfies** an equation in t and y if $f(t)$ can be substituted for y in the equation to produce an identity for all time values t in some interval. A **solution** of a differential equation is a function that satisfies the equation.

Example 1 Show that the function $f(t) = \sqrt{1 - t^2}$ satisfies the equation

$$t^2 + y^2 = 1.$$

Solution We substitute $f(t)$ for y in the equation and simplify to see if the equation has become an identity. The largest interval of values for t for which this calculation makes sense is $-1 \leq t \leq 1$.

$$\begin{aligned} t^2 + \left(\sqrt{1 - t^2}\right)^2 &= 1 \\ t^2 + 1 - t^2 &= 1 \\ 1 &= 1 \end{aligned}$$

The last equation is an identity in t—even though no t remains—so $f(t)$ does indeed satisfy the equation. ■

Example 2 Show that the function $f(t) = 2t - 3$ is a solution of the differential equation

$$\frac{dy}{dt} = 2.$$

Solution Note that we are being asked to show — in possibly unfamiliar words — that a linear function of the form $mt + b$ has slope m, something we already know. Substitution of $f(t)$ in the left side of the equation means calculate its derivative:

$$\frac{d}{dt}(2t - 3) = 2.$$

Thus this substitution leads to the identity $2 = 2$ (for all values of t), so the linear function does indeed satisfy the differential equation. ■

Exploration Activity 1

Here are two similar-looking differential equations:

$$\frac{dy}{dt} = 2y$$

and

$$\frac{dy}{dt} = 2t$$

The slope fields that represent these two equations are shown in Figures 3.1 and 3.2, not necessarily in the same order. Match each slope field with the corresponding differential equation. Explain how you decided which is which.

Figure 3.1 A slope field

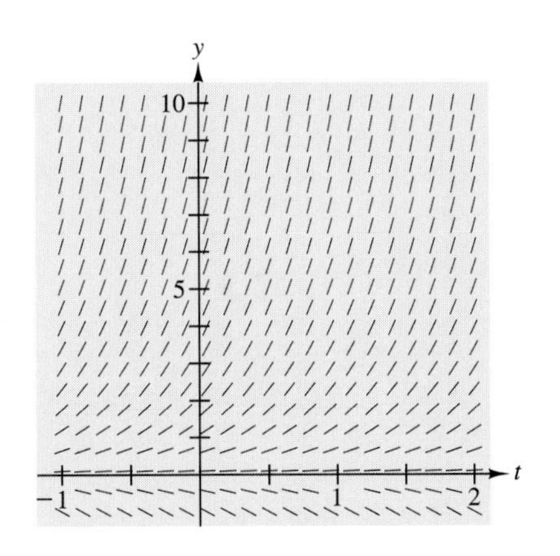

Figure 3.2 Another slope field

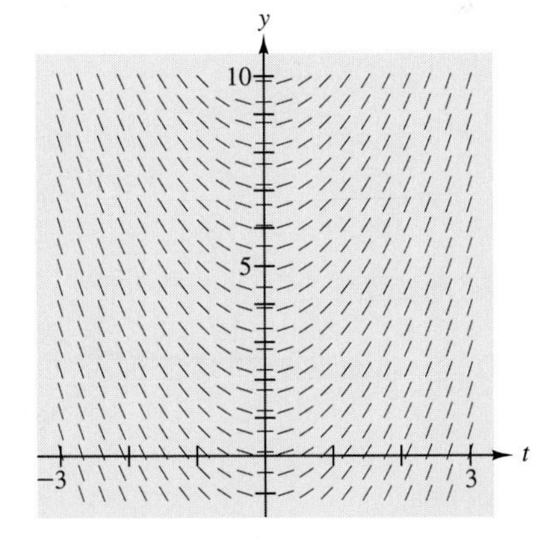

There are a number of ways to decide which slope field goes with which differential equation. For example, the first equation says that slopes will be positive when y (the dependent variable) is positive and slopes will be negative when y is negative. The second equation says that slopes will have the same sign as t, positive on the right and negative on the left. Thus the slope field for the first equation is Figure 3.1, and the slope field for the second equation is Figure 3.2.

Another way to decide is to notice that in Figure 3.1 all the slopes along any horizontal line are the same. Thus slope depends only on y, not on t, as in the first differential equation. Similarly, in Figure 3.2 the slopes along any vertical line are all the same, so slope depends only on t, not on y. Some other features you might have used: (1) whether slopes increase numerically in the t direction or in the y direction, or (2) whether the zero slopes occur where $y = 0$ or where $t = 0$.

Exploration Activity 2

(a) Our first differential equation from Exploration Activity 1

$$\frac{dy}{dt} = 2y,$$

states a problem: Find a function—perhaps many functions—having the property that the derivative is twice as large as the function itself. Give formulas for three different functions that have this property.

(b) Graph your three solutions in the same graphing window.

(c) Sketch your three solutions on the appropriate slope field from Exploration Activity 1. (You may copy the page if you don't want to draw in the book.)

(d) How many such functions do you think there are? That is, how many solutions do you think the differential equation has?

We saw such functions in the last chapter, for example, $f(t) = e^{2t}$, $g(t) = 3e^{2t}$, $h(t) = -7e^{2t}$. The graphs of these functions are shown in Figure 3.3.

Figure 3.3 Three solutions of $\dfrac{dy}{dt} = 2y$

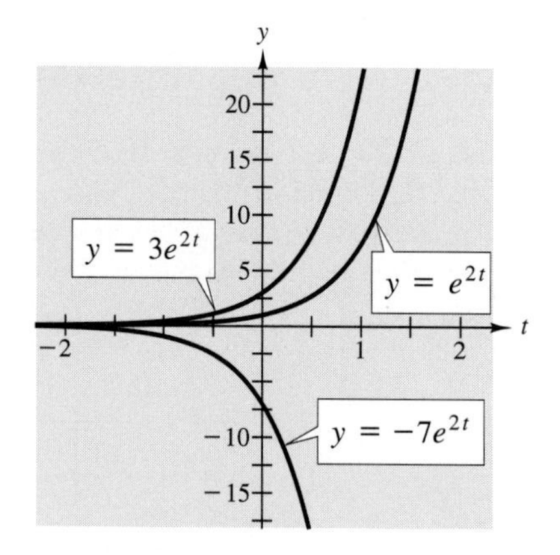

In general, if f is a function of the form

$$f(t) = Ce^{2t}$$

where C is any constant, then

$$f'(t) = \frac{d}{dt}\,Ce^{2t} = C\left(\frac{d}{dt}\,e^{2t}\right) = C(2e^{2t}) = 2Ce^{2t} = 2f(t).$$

Thus, if we substitute $f(t)$ for y and $f'(t)$ for dy/dt, we see that f satisfies the differ-ential equation $dy/dt = 2y$. Since there are infinitely many different choices for C, there are infinitely many different solutions of the differential equation.

Exploration Activity 3

(a) Give formulas for three different solutions of our second differential equation from Exploration Activity 1,

$$\frac{dy}{dt} = 2t.$$

(b) Graph your three solutions in a single graphing window.

(c) Sketch your three solutions on the appropriate slope field from Exploration Activity 1. (You may copy the page if necessary.)

(d) How many solutions do you think this differential equation has?

(e) How do the two differential equations (here and in Exploration Activity 2) differ from each other? How do their solutions differ?

The difference between the differential equations in Exploration Activities 2 and 3 is important. In Exploration Activity 3 it is the independent variable that appears on the right-hand side. Thus we are looking for a function whose derivative is $2t$. We know from Chapter 2 that one such function is t^2. Another is $t^2 + 3$. (See Figure 3.4.) In fact, if f is a function of the form

$$f(t) = t^2 + C,$$

where C is any constant, then f satisfies the differential equation $dy/dt = 2t$, because

$$f'(t) = 2t + 0 = 2t.$$

Thus, again, we find that the differential equation has infinitely many solutions. Some of the differences between the two families of solutions are described in our discussion of Exploration Activity 1. We also can describe the difference in terms of formulas: (1) If the slope of a function is always proportional to the function, then the function is a constant times an exponential function. (2) If the slope is always proportional to the independent variable, then the function is a constant plus a quadratic power function.

Checkpoint 1

(a) Find three different solutions of

$$\frac{dy}{dt} = -3t.$$

(b) Use your calculator to graph all your solutions in the same window.

(c) Describe the family of all solutions of the equation in part (a).

Figure 3.4 Two solutions of $\frac{dy}{dt} = 2t$

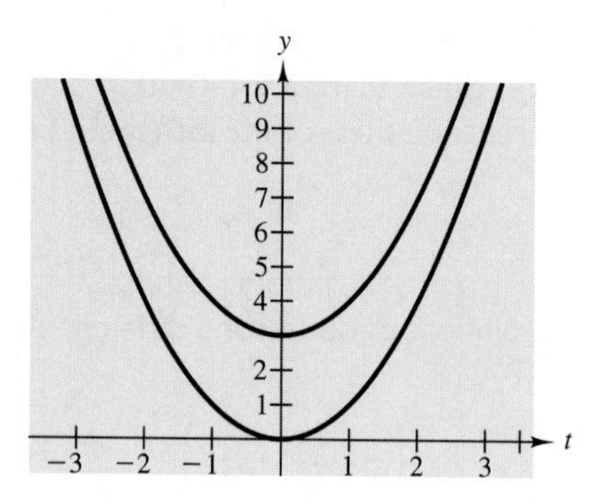

Checkpoint 2

(a) Find three different solutions of

$$\frac{dy}{dt} = -3y.$$

(b) Use your calculator to graph all your solutions in the same window.

(c) Describe the family of all solutions of the equation in part (a).

Checkpoint 3

Figure 3.5 shows the slope field for another differential equation. Sketch the graphs of four functions that are solutions of the differential equation. (You may copy the page if you don't want to make your sketches here.)

Figure 3.5 Another slope field

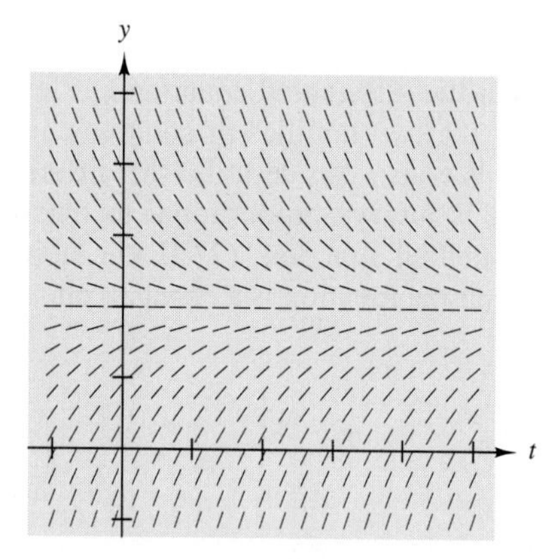

The slope fields in Figures 3.1 through 3.5 suggest that for each point in the (t, y) plane, there is exactly one solution curve passing through that point. Our solution formulas suggest the same thing: One (t, y) pair of numbers that must satisfy the formula is enough information to determine the constant C in each case. Thus, in order to specify a particular solution of the differential equation, all we have to do is specify its value at one point.

Exploration Activity 4

(a) Find the particular solution of the differential equation $dP/dt = 0.25P$ that satisfies the condition $P(0) = 2$.

(b) Find the particular solution of the same differential equation that satisfies the condition $P(1) = 2$.

(c) Use your calculator to graph your solution functions from parts (a) and (b).

(d) Sketch your solution functions on the slope field shown in Figure 3.6. (You may copy the page first if you wish.)

(e) Confirm that each of your solutions does indeed satisfy the corresponding condition stated in part (a) or part (b).

Figure 3.6 Slope field for $\dfrac{dP}{dt} = 0.25P$

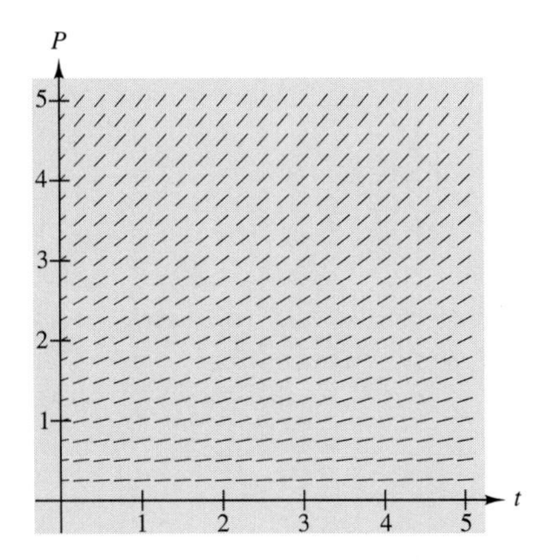

We know from Chapter 2 that the family of solutions of the differential equation

$$\frac{dP}{dt} = 0.25P$$

can be described by

$$P(t) = Ce^{0.25t}$$

where C can be any real number. (Compare this with Exploration Activity 2.) To find the particular solution such that

$$P(0) = 2$$

we set

$$2 = P(0) = Ce^{0.25 \cdot 0} = C \cdot 1 = C.$$

We find that $C = 2$, so the specific function

$$P_1(t) = 2e^{0.25t}$$

is the particular solution we seek.

Similarly, to find a solution of the form $P(t) = Ce^{0.25t}$ such that $P(1) = 2$, we set

$$2 = P(1) = Ce^{0.25},$$

and we find

$$C = 2e^{-0.25} \approx 1.55760.$$

It follows that our second solution (at least approximately) is

$$P_2(t) = 1.55760\, e^{0.25t}.$$

In Figure 3.7 we show the solutions $P_1(t)$ and $P_2(t)$ sketched on the slope field from Figure 3.6.

Figure 3.7 Slope field and solutions P_1 and P_2 of $\frac{dP}{dt} = 0.25P$

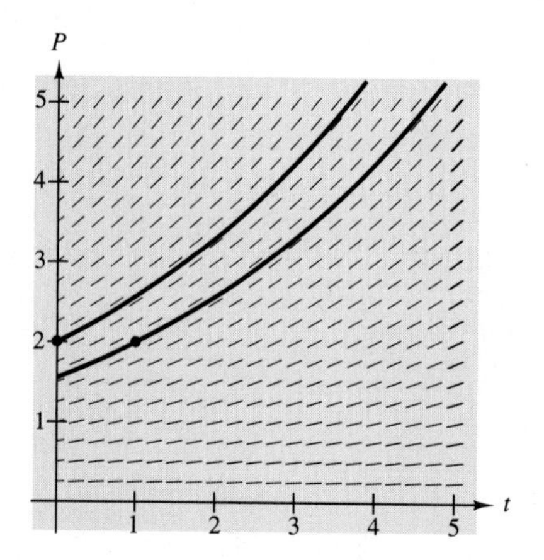

The number 2 that appears in parts (a) and (b) of Exploration Activity 4 is called an "initial value." The statement that P must have the value 2 at $t = 0$ (or at $t = 1$) is called an "initial condition." The entire problem stated in either part (a) or part (b) of Exploration Activity 4 is called an "initial value problem." We now give formal definitions of these terms.

Definitions An **initial value** is a specified value of a particular solution function at a specific number in the domain of the function. An **initial condition** is the specification of an initial value for a solution function. An **initial value problem** consists of a differential equation together with an initial condition. A **solution** of an initial value problem is a solution of the differential equation that also satisfies the initial condition.

For problems in which the independent variable is *time* t, it is often convenient to specify the initial value at $t = 0$, but that is not required—as you just saw in Exploration Activity 4.

We have seen several times now that an initial value problem has a unique solution. This is true for most differential equations—in particular, it is true for *all* the differential equations we consider in this course. For reference later in the text, we record this as a statement we assume to be true.

Fundamental Assumption In this book we assume that every initial value problem has a unique solution.

If we are looking for a formula for the solution of an initial value problem, our approach will be (1) to describe *all* solutions of the differential equation and (2) to determine the *unique* one that satisfies the initial condition.

Example 3 Solve the initial value problem

$$\frac{dy}{dt} = 2y \qquad \text{with} \qquad y(1) = 3.$$

Solution We know from Exploration Activity 2 that the solutions of the differential equation are functions of the form

$$y(t) = C\,e^{2t}.$$

We want to select the constant C so that

$$y(1) = 3.$$

This means that we must have

$$3 = y(1) = Ce^{2\cdot 1} = Ce^2$$

or

$$C = 3e^{-2} \approx 0.40601.$$

Therefore,

$$y(t) \approx 0.40601\,e^{2t}.$$

In Figure 3.8 we show the slope field for the differential equation $dy/dt = 2y$ and the particular solution for which $y(1) = 3$. ■

Checkpoint 4

Solve the initial value problem

$$\frac{dy}{dt} = -2y \qquad \text{with} \quad y(0) = 3.$$

Use your calculator to graph your solution function. Confirm that your solution does indeed satisfy the stated initial condition.

Figure 3.8 Slope field and solution for $\frac{dy}{dt} = 2y$, $y(1) = 3$

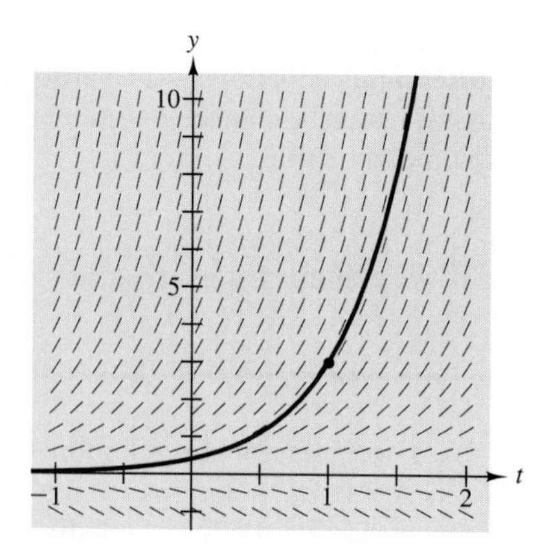

Example 4 Solve the initial value problem

$$\frac{dy}{dt} = 2t \qquad \text{with} \qquad y(1) = 3.$$

Solution We know from Exploration Activity 3 that the solutions of the differential equation have the form

$$y(t) = t^2 + C.$$

In order to satisfy the condition

$$y(1) = 3$$

we must have

$$3 = 1 + C$$

so

$$C = 2.$$

Thus the solution of the initial value problem is the function

$$y(t) = t^2 + 2.$$

In Figure 3.9 we show the slope field for the differential equation $dy/dt = 2t$ and the particular solution for which $y(1) = 3$. ■

Checkpoint 5

Solve the initial value problem

$$\frac{dy}{dt} = -3t \qquad \text{with} \quad y(0) = 3.$$

Use your calculator to graph your solution function. Confirm that your solution does indeed satisfy the stated condition.

Figure 3.9 Slope field and solution for $\dfrac{dy}{dt} = 2t,\ y(1) = 3$

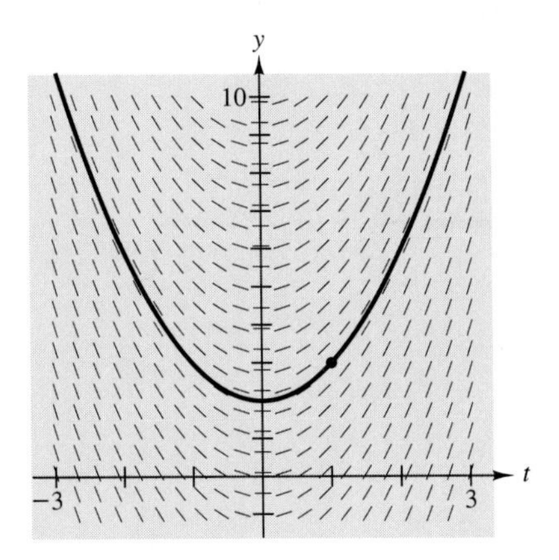

In this section we have seen that a differential equation can specify slope information about an unknown function either in terms of the dependent variable or in terms of the independent variable. Either way, such an equation will have an infinite number of solutions. With the specification of an initial condition, the number of possible solutions is reduced to one. We can solve initial value problems in a two-step process: First, find the entire family of solutions of the differential equation, which will involve an arbitrary (or not-yet-specified) constant in the formula. Then substitute the initial condition to determine the value of the arbitrary constant.

Answers to Checkpoints

1. (a) $y = -1.5t^2,\ \ y = -1.5t^2 + 2,\ \ y = -1.5t^2 - 1$

   (b)

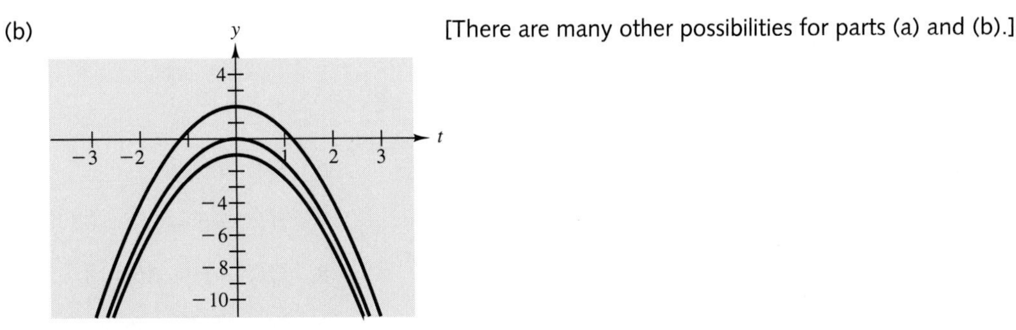

[There are many other possibilities for parts (a) and (b).]

   (c) All functions of the form $y = -1.5t^2 + C$.

2. (a) $y = e^{-3t},\ \ y = 1.5e^{-3t},\ \ y = -3e^{-3t}$

(b)

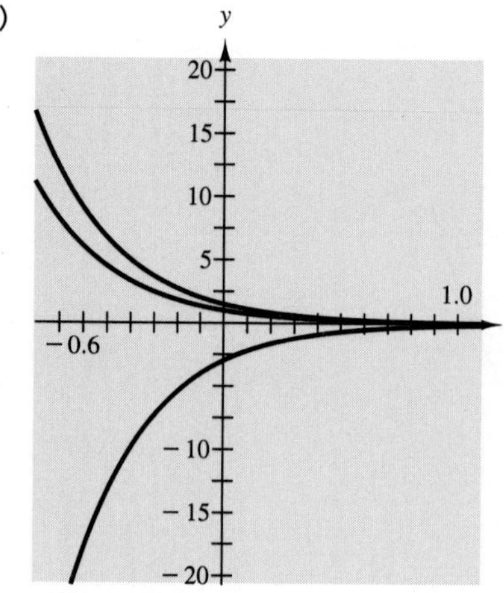

[There are many other possibilities for parts (a) and (b).]

(c) All functions of the form $y = Ce^{-3t}$.

3.

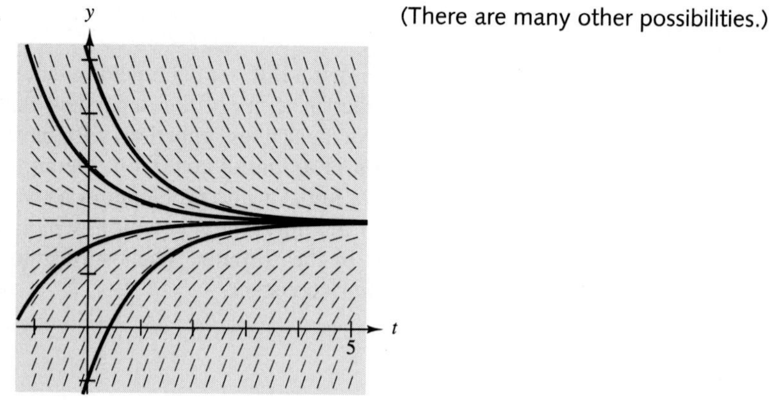

(There are many other possibilities.)

4. $y = 3e^{-2t}$

5. $y = -1.5t^2 + 3$

Exercises 3.1

Differentiate each of the following functions.

1. $f(t) = t^3$
2. $f(t) = e^{3t}$
3. $f(t) = 3 - e^t$
4. $f(t) = 1 - e^{3t}$
5. $f(t) = t - e^{3t}$
6. $f(t) = t^3 - e^t$
7. $f(t) = t^3 - 3^t$
8. $f(t) = t^3 - 3t^2 + 1$
9. $f(t) = e^{3t} - 3t^2 + 3t$
10. $f(t) = 3 + 7t - 2t^2 + 3t^3$

11. Which of the following functions are solutions of the differential equation $dy/dt = 4y$?
(i) $f(t) = e^{4t}$
(ii) $f(t) = e^{2t}$
(iii) $f(t) = 2t^2$
(iv) $f(t) = 5e^{4t}$
(v) $f(t) = e^{4t} + 5$
(vi) $f(t) = 2t^2 + 3$

12. Which of the following functions are solutions of the differential equation $dy/dt = 4t$?
(i) $f(t) = 2t^2$
(ii) $f(t) = 2t^2 + 3$
(iii) $f(t) = 5e^{4t}$
(iv) $f(t) = e^{4t}$
(v) $f(t) = 3t^2$
(vi) $f(t) = 2 + 2t^2$

13. Which of the following functions is the solution of the initial value problem $dy/dt = 4t$ with $y = 4$ at $t = 1$?
(i) $f(t) = 2t^2$
(ii) $f(t) = 2t^2 + 1$
(iii) $f(t) = 4e^{4t}$
(iv) $f(t) = 2t^2 + 2$
(v) $f(t) = 2t^2 + 4$
(vi) $f(t) = 3 + 2t^2$

14. Which of the following functions is the solution of the initial value problem $dy/dt = 4t$ with $y = 4$ at $t = 0$?
(i) $f(t) = 2t^2$
(ii) $f(t) = 2t^2 + 1$
(iii) $f(t) = 4e^{4t}$
(iv) $f(t) = 2t^2 + 2$
(v) $f(t) = 2t^2 + 4$
(vi) $f(t) = 3 + 2t^2$

15. Which of the following functions is the solution of the initial value problem $dy/dt = 4y$ with $y = 3$ at $t = 0$?
(i) $f(t) = e^{4t}$
(ii) $f(t) = 2e^{4t}$
(iii) $f(t) = 3 + 2t^2$
(iv) $f(t) = 3e^{4t}$
(v) $f(t) = 4e^{4t}$
(vi) $f(t) = 3t^2$

16. Which of the following functions is the solution of the initial value problem $dy/dt = 4y$ with $y = 4$ at $t = 0$?
(i) $f(t) = e^{4t}$
(ii) $f(t) = 2e^{4t}$
(iii) $f(t) = 3 + 2t^2$
(iv) $f(t) = 3e^{4t}$
(v) $f(t) = 4e^{4t}$
(vi) $f(t) = 3t^2$

Find three solutions of each of the following differential equations.

17. $\dfrac{dy}{dt} = 8y$

18. $\dfrac{dy}{dt} = 8t$

19. $\dfrac{dy}{dt} = 1.07y$

20. $\dfrac{dy}{dt} = 1.07t$

For each of the following differential equations, sketch the graphs of three solutions. (You may use your graphing calculator or computer and/or the solutions to Exercises 17–20.)

21. $\dfrac{dy}{dt} = 8y$

22. $\dfrac{dy}{dt} = 8t$

23. $\dfrac{dy}{dt} = 1.07y$

24. $\dfrac{dy}{dt} = 1.07t$

Find the solution of each of the following initial value problems.

25. $\dfrac{dy}{dt} = 8y$ with $y = 2$ at $t = 0$

26. $\dfrac{dy}{dt} = 8t$ with $y = 2$ at $t = 0$

27. $\dfrac{dy}{dt} = 1.07y$ with $y = 2.3$ at $t = 0$

28. $\dfrac{dy}{dt} = 1.07t$ with $y = 2.3$ at $t = 0$

For each of the following initial value problems, sketch the graph of the solution. (You may use your graphing calculator or computer and/or the solutions to Exercises 25–28.)

29. $\dfrac{dy}{dt} = 8y$ with $y = 2$ at $t = 0$

30. $\dfrac{dy}{dt} = 8t$ with $y = 2$ at $t = 0$

31. $\dfrac{dy}{dt} = 1.07y$ with $y = 2.3$ at $t = 0$

32. $\dfrac{dy}{dt} = 1.07t$ with $y = 2.3$ at $t = 0$

On each of the following slope fields, sketch three solution curves.

33.

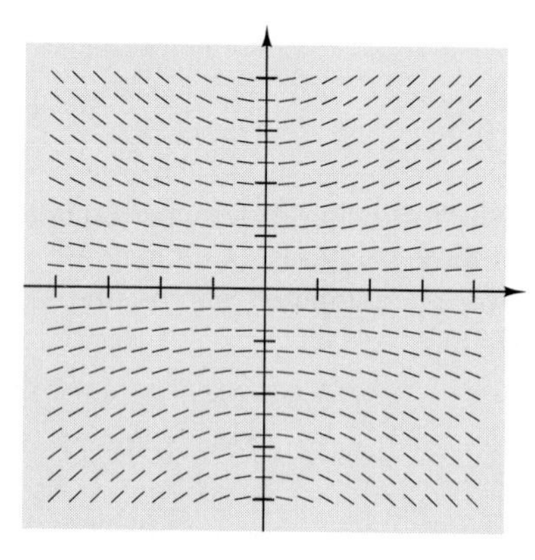

34.

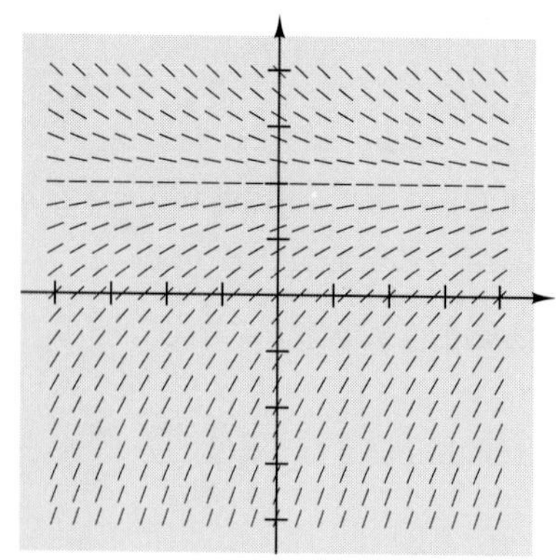

35.

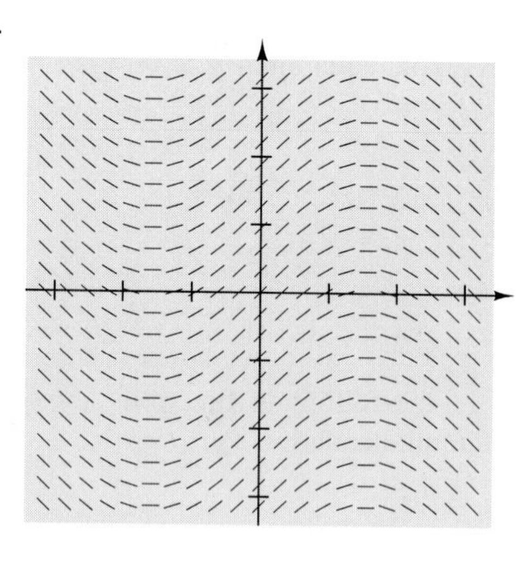

36.

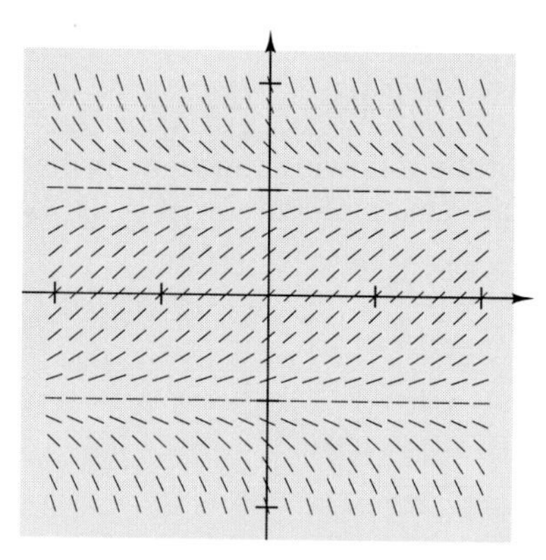

37. (a) Sketch a slope field for $dy/dt = 0.5t$.
    (b) Sketch a solution.
    (c) What feature of the slope field corresponds to the fact that y does not appear in the derivative?

38. The slope field in Exercise 35 comes from a differential equation of the form $dy/dt = g(t)$. What geometric characteristic of the field signals the fact that the right-hand side of the differential equation does not depend on y?

39. The slope fields in Exercises 34 and 36 come from differential equations of the form $dy/dt = g(y)$. What geometric characteristic of these field signals the fact that the right-hand side of each differential equation does not depend on t?

40. (a) Let $y = 2 - 2\,e^{-3t}$. Find dy/dt.
    (b) Show that $y = 2 - 2\,e^{-3t}$ is a solution of $dy/dt = 6 - 3y$.

41. Slope fields for three of the following differential equations are among the four displayed below. Match up the three differential equations with the corresponding slope fields.
(i) $y' = y^2 + 1$
(ii) $y' = t^2 + 1$
(iii) $y' = -y + 1$
(iv) $y' = -t - 1$

(a)

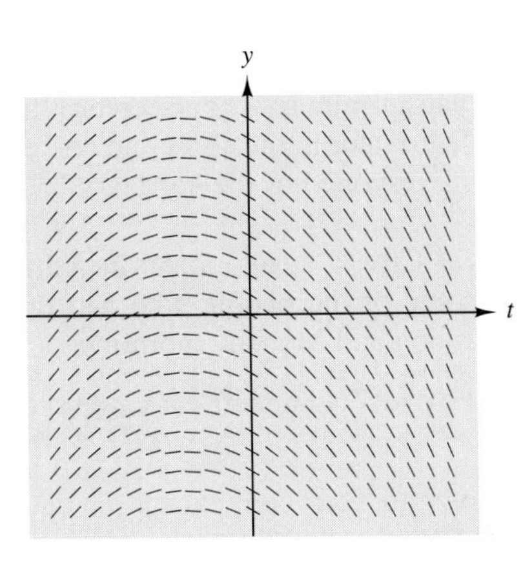

(b)

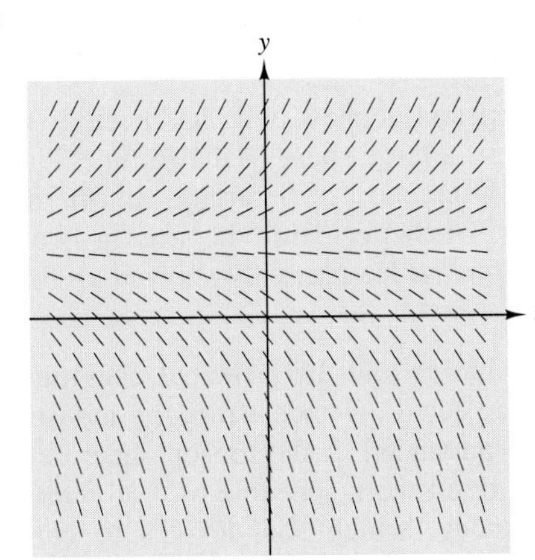

(c)

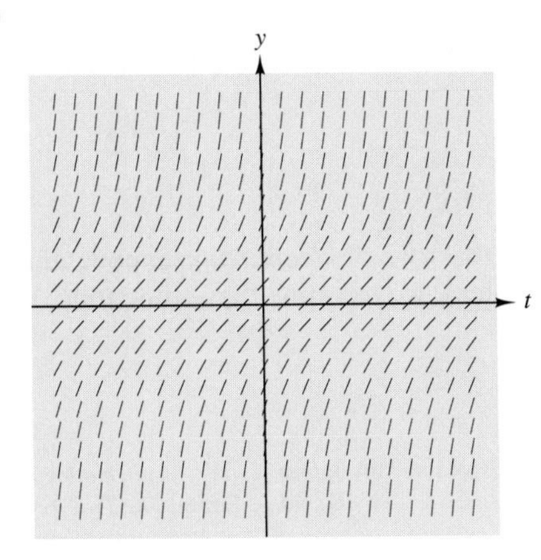

(d)

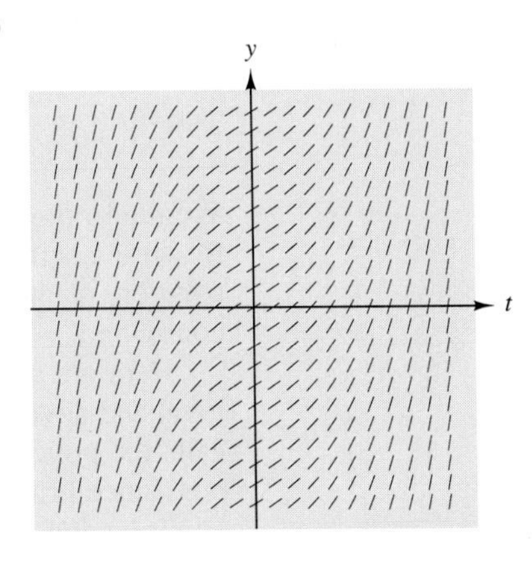

42. For the slope field in Exercise 33, does the right-hand side of the differential equation depend on y only, on t only, on both y and t, or on neither y nor t? Explain.

43. Table 3.1 shows the number of disintegrations per minute counted by a Geiger counter that is testing a sample of radioactive barium-137. The disintegration rate of a radioactive substance is proportional to the amount of the substance not yet disintegrated.
(a) Find a formula for a function that approximately fits the data.
(b) Find the *half-life* of barium-137, that is, the time it takes for a given sample to decay to half the original amount.[1]

Table 3.1 Disintegrations per minute for radioactive barium-137

Time (minutes)	***Counts per minute***
0	10,034
1	8105
2	5832
3	4553
4	3339
5	2648
6	2035

1. Problem adapted from the Core-Plus Mathematics Project.

3.2 An Initial Value Problem: A Cooling Body[2]

We return now to the problem with which we opened the chapter. Recall that our task is to write a script for a television detective who is confronted with a murder mystery, the solution of which may depend on determining the time at which the murder was committed.

Here are a few more details. The discovery of the industrialist's body is made by his loyal and devoted secretary when she reports for work as usual at 8:30 A.M. She promptly calls the police, and our hero is on the scene in minutes. The secretary has not seen her boss since Friday afternoon, and (we will learn later in the show) no one else is willing to admit having seen him over the weekend. Everyone but the secretary hated this man, so there are lots of suspects.

The office building is located in Crystal City, a development in Arlington, Virginia, just outside Washington, D.C. Here are some of the people who had or may have had access to the building between Friday afternoon and Monday morning:

- The janitorial crew that works an 11 P.M. to 7 A.M. shift each night, Sunday through Thursday.

- The industrialist's ex-wife, who, unknown to everyone (except the detective, who has ways of finding out), still has a key to the building. Her whereabouts over the weekend cannot be accounted for, but she is known to have reported for work at 9 A.M. Monday at her regular job—in Cleveland.

- The industrialist's trusted but incompetent nephew, who had been given a managerial job in the firm and who would inherit a large share of the fortune. His tennis partner confirms his presence on the tennis court most of Saturday. He (the nephew) says he went sailing alone all day Sunday. His wife claims he was in bed with her from midnight Sunday to about 9 A.M. Monday. He normally showed up for work between 10 A.M. and noon.

- The vice-president who is the real brains and the principal operating officer of the corporation. She is known to have rejected advances from the industrialist (when he was still married) and still to have advanced to the top on sheer determination and talent. She is also known to have a fixed work schedule of 6:30 A.M. to 6:30 P.M. six days a week—every day but Sunday, which she devotes to church and social activities. She was at work two hours before the secretary came in and discovered the body, but her usual practice was not to go near the owner's office, because he would always waste her time with off-the-wall ideas that would have been detrimental to the firm.

Decreasing Exponential Functions

A key feature of the solution will be the role of *decreasing* exponential functions. As preparation for solving the murder problem, we examine these functions first.

---

2. Based on "An Application of Newton's Law of Cooling," by James Hurley, *Mathematics Teacher* 67 (1974), pp. 141–142. See also "The Homicide Problem Revisited," by David A. Smith, *The Two-Year College Mathematics Journal* 9 (1978), pp. 141–145.

Exploration Activity 1

(a) Graph all three of the following exponential functions in a single graphing window.

$$y = e^{-1.7t} \qquad y = e^{-2.5t} \qquad y = e^{-3.7t}$$

(b) What point do the three graphs have in common?

(c) What happens to each of the three graphs as t becomes large?

(d) What can you say in general about functions of the form $y = e^{-kt}$, where k is a positive number?

(e) What connection do you find between the size of k and the rate of decrease of $y = e^{-kt}$?

In Figure 3.10 we graph the functions $y = e^{-1.7t}$, $y = e^{-2.5t}$, and $y = e^{-3.7t}$. Notice that all these functions have the value 1 at $t = 0$, so the shared point on all three graphs is $(0, 1)$. All three functions decrease to 0 as t increases to infinity. In fact, these properties are shared by all functions of the form $y = e^{-kt}$. The figure illustrates the fact that the larger the value of k, the faster the function decreases.

Figure 3.10 Graphs of functions of the form $y = e^{-kt}$

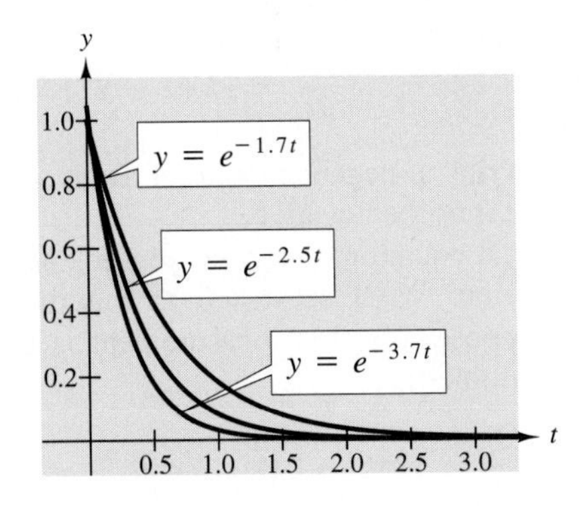

We already know how to differentiate exponential functions

$$\frac{d}{dt}\, e^{rt} = re^{rt}$$

for any constant r. In particular,

$$\frac{d}{dt}\, e^{-kt} = -ke^{-kt}$$

for all constants k.

Checkpoint 1

(a) Find the derivative of $e^{-1.7t}$.

(b) Graph $y = e^{-1.7t}$ together with its derivative.

Formulating the Initial Value Problem

Since the body of the murder victim has been cooling off since the time of death, our detective can determine how long the industrialist has been dead by examining the temperature of the body. In the remainder of this section we study an initial value problem that describes heat loss from a cooling body. Our analysis involves the following steps:

- We formulate the initial value problem.
- We solve the initial value problem by transforming it to one we have solved already.
- We interpret the mathematical solution in the context of the original problem.

Physics provides us a theoretical model (verifiable by simple experiments) called **Newton's Law of Cooling**:

> The rate of change in temperature of a cooling body is proportional to the *difference* between the temperature of the body and the surrounding temperature.

Our detective can determine the surrounding (room) temperature by checking the thermostat: 21°C.

Our next task is to write the physical law in mathematical notation. We'll let $T = T(t)$ represent the temperature of the cooling body in degrees Celsius at time t in hours. Then Newton's Law of Cooling may be written

$$\frac{dT}{dt} = -k\,(T - 21).$$

We have chosen to write a negative sign in front of the k because we know the proportionality constant must be negative.

That's our differential equation—except we don't know k yet. Now we need to find an initial value condition. We'll set $t = 0$ at the time the detective takes the first measurement. Let's suppose this first measurement is taken at 8:50 A.M. and is 30°C. Then our initial value problem is

$$\frac{dT}{dt} = -k(T - 21) \qquad \text{with } T = 30 \text{ at } t = 0.$$

Now we have to

1. Solve the initial value problem.
2. Determine k —to make the solution an explicit function of time.
3. Use the solution to figure out the time of death.

Exploration Activity 2

In Figures 3.11, 3.12, and 3.13 we show slope fields for $dT/dt = -k(T - 21)$, for $k = 1,\ 0.5,$ and 0.25.

(a) On each of the slope fields, sketch the graph of the solution that satisfies the initial condition $T(0) = 30$. (You may copy the page first.)

(b) What features do these three solution functions have in common?

(c) How does reducing the size of k affect the solution?

Figure 3.11 Slope field for $dT/dt = -(T - 21)$

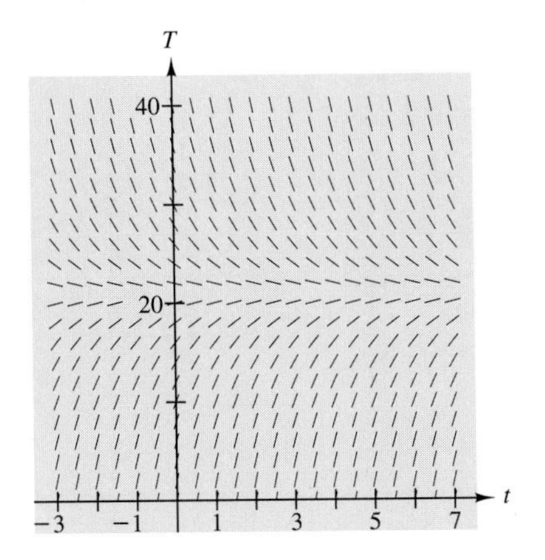

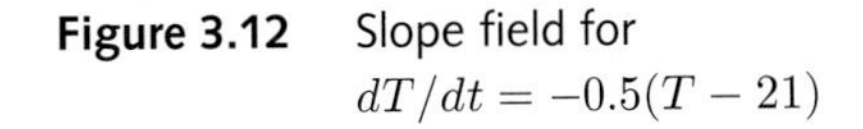

Figure 3.12 Slope field for $dT/dt = -0.5(T - 21)$

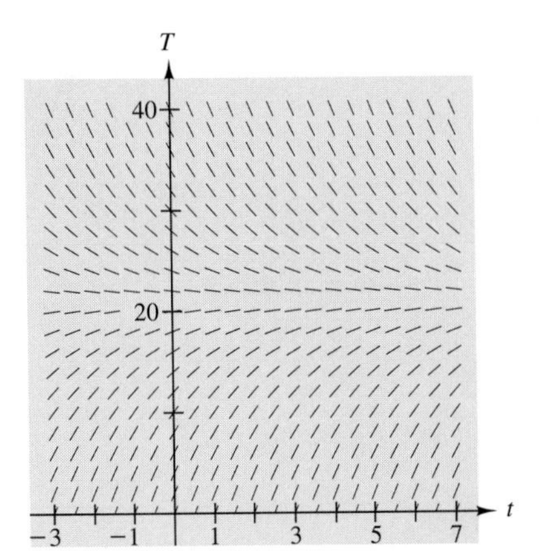

Figure 3.13 Slope field for $dT/dt = -0.25(T - 21)$

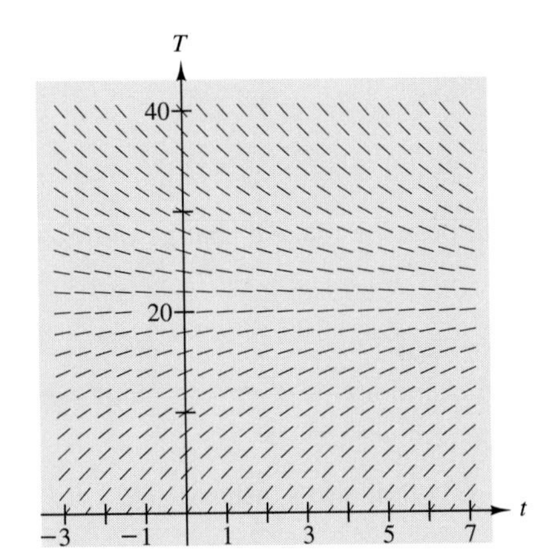

You should see in your sketches that all three temperature functions pass through the point $(0, 30)$ and all decrease toward a steady-state value of $T = 21$, the room temperature. The effect of decreasing k is to slow the rate of decrease. Thus the smaller k is, the longer it takes for the body to approach room temperature — and also the longer it has been since the time of death. We show the three solutions for these values of k in Figures 3.14, 3.15, and 3.16.

Figure 3.14 Solution for $k = 1$ **Figure 3.15** Solution for $k = 0.5$ **Figure 3.16** Solution for $k = 0.25$

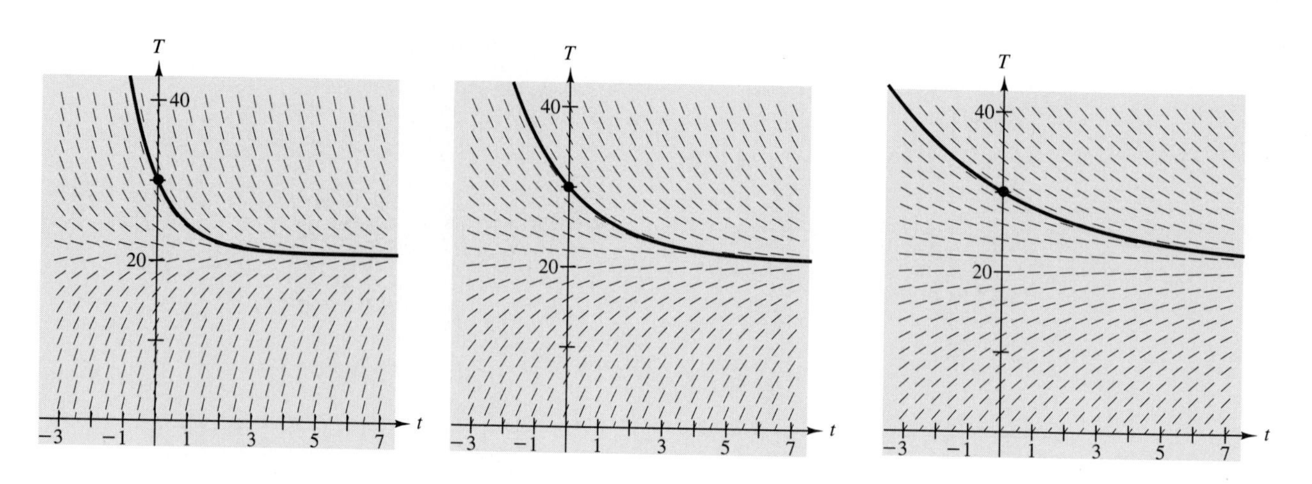

Solving the Initial Value Problem Symbolically

Now that we have seen a graphical interpretation of the solution of the initial value problem, we'll find a symbolic representation.

Exploration Activity 3

In the differential equation $dT/dt = -k(T - 21)$, suppose we make a change of dependent variable by setting $y = T - 21$.

(a) How are dy/dt and dT/dt related?

(b) If $T = T(t)$ is a solution of $dT/dt = -k(T - 21)$, and $y(t) = T(t) - 21$, find a differential equation that is satisfied by $y(t)$.

(c) If $T = T(t)$ is the solution of $dT/dt = -k(T - 21)$ that also satisfies $T = 30$ at $t = 0$, what initial value problem is satisfied by $y(t) = T(t) - 21$?

(d) Why is the initial value problem for y easier to solve than the initial value problem for T?

Our differential equation for temperature of the cooling body is not exactly like the problems considered in Chapter 2 because the rate of change is proportional to $T - 21$, not to T. That observation is what suggests the proposed change of dependent variable to $y = T - 21$. Now

$$\frac{dy}{dt} = \frac{d}{dt}(T - 21) = \frac{dT}{dt}.$$

Thus, by substituting $y = T - 21$ and $dy/dt = dT/dt$ into the original differential equation, we get a differential equation for the new unknown function y:

$$\frac{dy}{dt} = -ky.$$

Also, at $t = 0$, $T = 30$, so $y = T - 21 = 9$, which gives us an initial condition for the unknown function y. Thus the substitution $y = T - 21$ transforms the original initial value problem to a new initial value problem:

$$\frac{dy}{dt} = -ky \qquad \text{with} \quad y = 9 \text{ at } t = 0.$$

We illustrate the transformation from T to y in Figures 3.17 and 3.18. Figure 3.17 shows a slope field for the original differential equation. In Figure 3.18 we show the effect of the change of dependent variable, $y = T - 21$, which is to move the horizontal axis from $T = 0$ to the level of room temperature, $T = 21$, or $y = 0$. None of the slopes change because, at each value of t, $dy/dt = dT/dt$. Notice that the point $(0, 9)$ in Figure 3.18 corresponds to the point $(0, 30)$ in Figure 3.17.

Figure 3.17 Slope field for $dT/dt = -k(T - 21)$

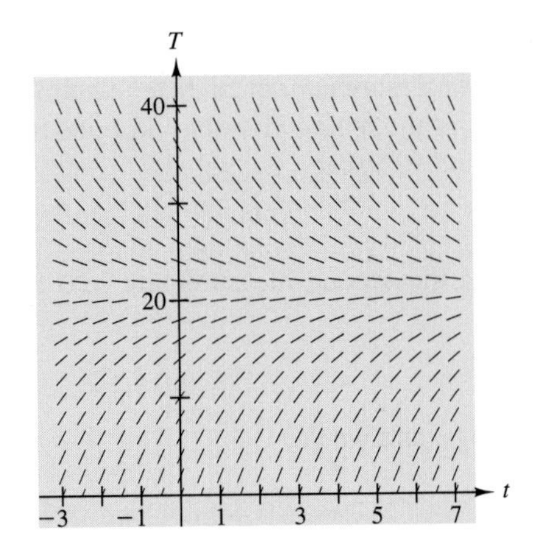

Figure 3.18 Effect of substituting $y = T - 21$

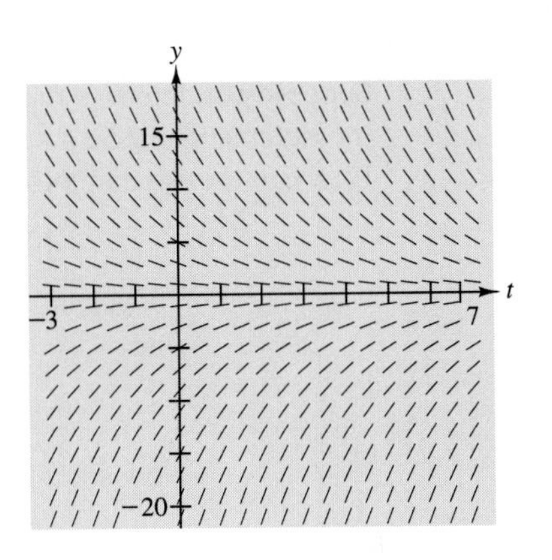

Our new initial value problem

$$\frac{dy}{dt} = -ky \qquad \text{with} \quad y = 9 \text{ at } t = 0$$

should look familiar: It's the natural growth equation—except our growth factor is negative. The problem has exactly the form we saw in Chapter 2:

$$\frac{dy}{dt} = ry \qquad \text{and} \quad y = y_0 \text{ when } t = 0$$

with $r = -k$. (In form, it is also the problem you studied in Exploration Activities 2 and 4 and Example 3 in the preceding section.) Thus the solution must have the form we saw in Chapter 2:

$$y = y_0 e^{rt}.$$

That is,

$$y = 9\, e^{-kt}.$$

Of course, y is not the quantity we wanted to know about—it was merely a computational convenience. But we can reverse the change of the dependent variable by substituting $y = T - 21$ and solving for T:

$$T = 21 + 9\,e^{-kt}.$$

Checkpoint 2

Use your calculator to graph the functions $y = 9e^{-kt}$ and $T = 21 + 9e^{-kt}$, both with $k = 0.5$. Sketch these solution functions on their respective slope fields in Figures 3.17 and 3.18. (If you prefer, you may copy this page instead of drawing in the book.)

Determining the Proportionality Constant

Recall that we don't actually know a value for k yet. Next, we will see how our detective might determine a value of k that fits the problem being solved.

Exploration Activity 4

Suppose our detective makes a second measurement of body temperature. He takes his second measurement at 9:50 A.M. and finds that the temperature has dropped to $28°C$; that is, $T = 28$ at $t = 1$. How can we use that information to find k? What is k?

When we substitute the second measurement into the equation $T = 21 + 9e^{-kt}$, we find

$$7 = 9e^{-k}.$$

Now we take natural logs, and we get

$$\ln 7 = \ln 9 + \ln e^{-k} = \ln 9 - k$$

or

$$k = \ln 9 - \ln 7 \approx 0.251.$$

The numerical value comes from a calculator, which any modern detective always has at hand. Figure 3.19 shows the slope field for $dT/dt = -k(T - 21)$, now drawn with the *known* value of k, along with the now fully known solution through $(0, 30)$.

Figure 3.19 Slope field and solution for $dT/dt = -k(T - 21)$ with $T(0) = 30$

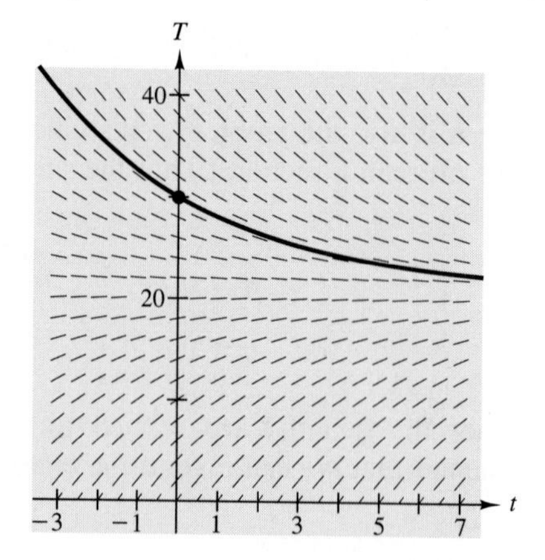

Interpreting the Result

We are now ready to use the solution of the initial value problem to estimate the time of death.

Exploration Activity 5

Use the solution function just obtained to determine the time of death. Once you know the time of death, make a conjecture about the likely murderer.

We have just finished step 2—finding the decay constant k—and now we are asking you to do step 3—to find the time on the clock when the decaying temperature function

$$T = 21 + 9e^{-kt}$$

had the value 37. The procedure is similar to step 2, except now we know k and want to solve for t. We substitute $T = 37$ in our formula, and we find

$$16 = 9e^{-kt}.$$

Taking logarithms again, and substituting our known value for k, we find

$$\ln 16 = \ln 9 - kt = \ln 9 - t\,(\ln 9 - \ln 7).$$

We solve for t and find the time of death to be

$$t = \frac{\ln 9 - \ln 16}{\ln 9 - \ln 7} = -2.29.$$

This still needs interpretation. The negative sign means the time of death was *before* the time of the first measurement at 8:50 A.M.—could it be otherwise? In fact, the time of death was 2 hours, 17 minutes before 8:50, or 6:33 A.M.

How does this help our detective sort out the suspects? Let's recall who they were: the industrialist's ex-wife, who reported for work at 9 A.M. Monday in Cleveland; his nephew, whose wife claims he was in bed with her from midnight Sunday to about 9 A.M. Monday; and the vice-president, who was at work two hours before the secretary discovered the body. We have found the time of death to be close enough to the time of discovery to make the rest of the details irrelevant.

Conclusion (of the calculation and of the TV show): The ex-wife didn't have time to commit the murder and drive to Cleveland, nor could she have gotten a plane that early. The nephew was probably in bed. So that leaves the vice-president, who was probably in the building at the right time. Or did the janitor do it?

In this section we have studied a cooling body problem as an initial value problem. We have seen that application of mathematics to a real problem involves at least these three steps:

Step 1: Translate the problem into mathematical form.

Step 2: Solve the mathematical problem.

Step 3: Interpret the mathematical solution.

Our particular problem had two interesting features that will reappear from time to time.

First, we encountered *decaying exponentials*, functions of time t of the form e^{-kt}, where the constant k is positive. These functions have the property that the functional values approach zero as t becomes large. Such functions appear often in models of *transient* effects, that is, things that die out as time goes on.

Second, we made a change of the dependent variable—replacing T by $y = T - 21$—to simplify our problem to one we had seen before. Simplification to an already-solved problem is a standard problem-solving technique. It makes it unnecessary to keep doing the same work over and over.

Answers to Checkpoints

1. (a) $\dfrac{d}{dt}\, e^{-1.7t} = -1.7e^{-1.7t}$

   (b) Graphs of $e^{-1.7t}$ and $-1.7e^{-1.7t}$:

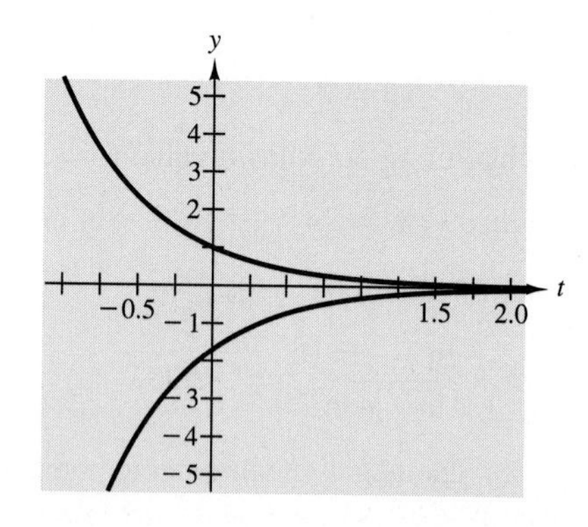

2.

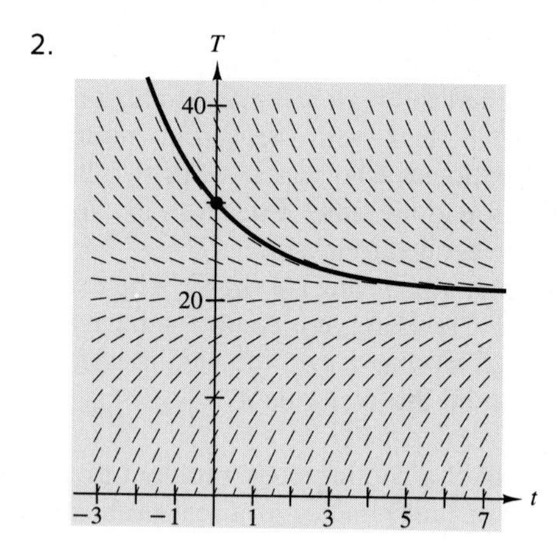

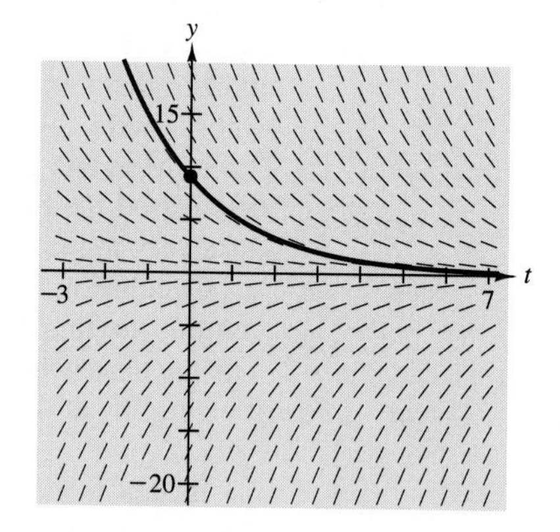

Exercises 3.2

Differentiate each of the following functions.

1. $f(t) = 3 - t^2$
2. $f(t) = e^{-3t}$
3. $f(t) = 3 - e^{-t}$
4. $f(t) = 1 - e^{-3t}$
5. $f(t) = t - e^{-3t}$
6. $f(t) = t^3 - e^{-t}$
7. $f(t) = t^4 - 4^t$
8. $f(t) = e^{-3t} - 3t^2 + 3t^3$

Find three solutions of each of the following differential equations.

9. $\frac{dy}{dt} = -8y$
10. $\frac{dy}{dt} = -8t$
11. $\frac{dy}{dt} = -1.07y$
12. $\frac{dy}{dt} = -1.07t$

Sketch the graphs of three solutions of each of the following differential equations. (You may use your graphing calculator or computer and/or the solutions to Exercises 9–12.)

13. $\frac{dy}{dt} = -8y$
14. $\frac{dy}{dt} = -8t$
15. $\frac{dy}{dt} = -1.07y$
16. $\frac{dy}{dt} = -1.07t$

Find the solution of each of the following initial value problems.

17. $\frac{dy}{dt} = -8y$ with $y = 20$ at $t = 0$
18. $\frac{dy}{dt} = -8t$ with $y = 20$ at $t = 0$
19. $\frac{dy}{dt} = -1.07y$ with $y = 2.3$ at $t = 0$
20. $\frac{dy}{dt} = -1.07t$ with $y = 2.3$ at $t = 0$
21. $\frac{dy}{dt} = -8(y - 10)$ with $y = 20$ at $t = 0$
22. $\frac{dy}{dt} = -8(t - 10)$ with $y = 20$ at $t = 0$
23. $\frac{dy}{dt} = -1.07(y - 10)$ with $y = 2.3$ at $t = 0$
24. $\frac{dy}{dt} = -1.07(t - 10)$ with $y = 2.3$ at $t = 0$

In Exercises 25–32, sketch the graph of the solution of the initial value problem. (You may use your graphing calculator or computer and/or the solutions to Exercises 17–24.)

25. $\frac{dy}{dt} = -8y$ with $y = 20$ at $t = 0$
26. $\frac{dy}{dt} = -8t$ with $y = 20$ at $t = 0$
27. $\frac{dy}{dt} = -1.07y$ with $y = 2.3$ at $t = 0$
28. $\frac{dy}{dt} = -1.07t$ with $y = 2.3$ at $t = 0$
29. $\frac{dy}{dt} = -8(y - 10)$ with $y = 20$ at $t = 0$
30. $\frac{dy}{dt} = -8(t - 10)$ with $y = 20$ at $t = 0$
31. $\frac{dy}{dt} = -1.07(y - 10)$ with $y = 2.3$ at $t = 0$
32. $\frac{dy}{dt} = -1.07(t - 10)$ with $y = 2.3$ at $t = 0$
33. If $z = 10 + 8e^{-kt}$ for all t and $z = 14$ at $t = 1$, find k.
34. If $u = 10 - 8e^{-kt}$ for all t and $u = 4$ at $t = 1$, find k.
35. If $z = 10 + 8e^{-0.693t}$ for all t, find t when $z = 12$.
36. If $u = 10 - 8e^{-0.288t}$ for all t, find t when $u = 1$.
37. In the solution of the cooling body problem, we found k to be $\ln 9 - \ln 7$.
    (a) Show that $-k = \ln(7/9)$.

(b) Show that the solution function T also can be expressed as

$$T = 21 + 9\left(\frac{7}{9}\right)^t.$$

(c) Recall where the numbers in this expression came from: 21 is the room temperature, 9 is the difference between body temperature at $t = 0$ and room temperature, and 7 is the difference between body temperature at $t = 1$ and room temperature. Interpret what this new formula for T says about change in the first hour.

(d) Interpret the new formula for T as saying that the "same thing" happens in *every* hour. What same thing?

38. You may think it unnatural to talk about room temperature and body temperature in degrees Celsius. Convert the room temperature, 21°C, and the body temperature at time of death, 37°C, to degrees Fahrenheit. Do these seem like reasonable numbers?

39. In Figure 3.20 we show a graph of $T = 21 + 9e^{-kt}$, with $k = 0.2513144$, along with a graph of normal body temperature, 37°C. Sketch your own graph (or mark it on this one, if you prefer) of the industrialist's body temperature before and after death.

Figure 3.20 Solution of the cooling body problem

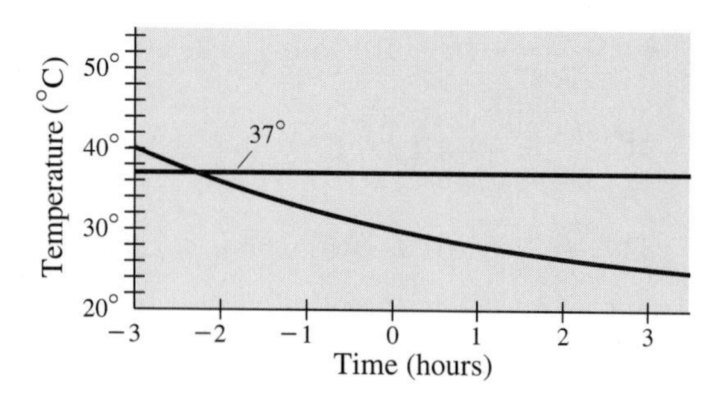

40. The graph of $T = 21 + 9e^{-kt}$ approaches a horizontal asymptote for large values of t.

(a) Extend your graph from Exercise 39 (or the one provided already in Figure 3.20) to show this asymptote and the approaching graph.

(b) What is the physical meaning of the approach of the temperature graph to this asymptote?

<u>41</u>. Suppose the detective's two measurements had been 26 and 25°C.

(a) What would the time of death have been?

(b) Whom would you have the detective arrest in this case?

42. In this problem we explore the possible complications created by setting $t = 0$ at time of death instead of at the time of first temperature mea-surement. We can still write the differential equation as

$$\frac{dT}{dt} = -k\,(T - 21),$$

and we can still transform the equation to

$$\frac{dy}{dt} = -ky.$$

Thus the solution for y still has the form

$$y = y_0 e^{rt}$$

where $r = -k$.

(a) What should y_0 be in this case?

(b) Suppose we call the time of first measurement t_1 so that the time of second measurement is $t_1 + 1$. Suppose the body temperatures at these two times are, as before, 30 and 28°C, respectively. In light of our previous calculation, what do you expect t_1 to turn out to be? What do you expect k to turn out to be?

(c) Use the solution for y and the two temperature measurements to write two equations in the two unknown quantities, k and t_1. Solve these simultaneous equations. Do you get the answers you expected?

43. A company is considering two ways to depreciate a piece of capital equipment that originally cost $14,000 and is worth $10,000 after one year:

Method I assumes the equipment depreciates at a rate proportional to the difference between its value and its scrap value of $400.

Method II assumes the equipment depreciates linearly, that is, at a constant rate.

(a) Using method I, find the value of the equipment at the end of 2 years and at the end of 3 years.

(b) Using method II, find the value of the equipment at the end of 2 years and at the end of 3 years.
(c) Which method produces faster depreciation? Explain.

44. Radioactive substances tend to decay at a rate proportional to the amount present at any given time. Let y_0 be the amount of a radioactive substance present at time $t = 0$, and let $y = y(t)$ be the amount present at any time t.
(a) Write an initial value problem whose solution is $y(t)$.
(b) Solve your initial value problem to find a formula for $y(t)$.
(c) The **half-life** of a radioactive substance is the time it takes for a given amount to decay to half that amount. Find an expression for half-life that depends only on the proportionality constant in your differential equation.

45. Superman has a violent reaction to green kryptonite, which fortunately has a half-life of only 15 hours as it decays into red kryptonite. It is no longer dangerous to Superman when 90% of the green kryptonite has decayed. If Superman is exposed to pure green kryptonite, for how long is he in danger?[3]

46. (a) What is the relation between the half-life H of a radioactive substance and the "third-life" T of the same substance?
(b) What is the relation between the doubling time D of a colony of rabbits and the tripling time T of the same colony?

3. This exercise and the next one are adapted from *Calculus Problems for a New Century*, edited by Robert Fraga, MAA Notes Number 28, 1993.

3.3 Another Initial Value Problem: A Falling Body

Suppose we dropped a marble from the window at the top of the Washington Monument. How fast would it be going when it hit the ground, about 535 feet below? Our approach to modeling this problem will involve *two* initial value problems—one for the velocity of the marble and another for the distance it travels. However, before we determine the form of these initial value problems, we will estimate the answer we seek by using the tabulated data from Chapter 2 on approximate position of a falling object for the first 10 seconds of fall.

Example 1 Table 3.2 (repeated here from Chapter 2) shows simulated positions s of a falling object as a function of time t. Estimate how fast the marble is falling just before it hits the ground.

Solution The tabulated distances are in meters, so we need to convert 535 feet to meters. A meter is about 3.28 feet, so the distance the marble falls is $535/3.28 = 163$ meters. Now, that distance lies between the table entries for 5.5 and 6.0 seconds, so a reasonable guess about its speed at impact would be the *average* speed for a falling object between times $t = 5.5$ and $t = 6.0$:

$$\frac{\text{change in position}}{\text{change in time}} = \frac{175.8 - 147.7}{6.0 - 5.5} = 56.2 \text{ m/s.}$$

In our original units, that would be 184 feet per second, or 126 miles per hour.

Table 3.2 Position of a falling object

Time (seconds)	*Distance (meters)*	*Time (seconds)*	*Distance (meters)*
0.5	1.2	5.5	147.7
1.0	5.0	6.0	175.8
1.5	11.1	6.5	207.3
2.0	19.4	7.0	240.0
2.5	30.6	7.5	276.9
3.0	44.1	8.0	312.8
3.5	60.3	8.5	354.8
4.0	78.0	9.0	396.1
4.5	99.2	9.5	443.1
5.0	122.8	10.0	489.0

Exploration Activity 1

(a) In Chapter 2 we showed that position s might be a function of the form ct^2. Explain why this would imply that velocity v is a linear function of t.

(b) How can you use the data in Table 3.2 to check whether the velocity is approximately linear? Carry out your check.

(c) Explain why if the velocity is linear, then the acceleration must be constant.

In Chapter 2 we learned that the derivative of $s = ct^2$ is $v = 2ct$ —and the time derivative (rate of change) of position is velocity. Similarly, acceleration is the derivative of velocity, and the derivative of a linear function is the constant slope of the graph of that function—in this case, $2c$.

As we saw in Example 1, we can *estimate* the velocity v of the falling object by calculating difference quotients of the data in Table 3.2. We have $s_0 = 0,\ s_1 = 1.2$, $s_2 = 5.0$, and so on. Thus the velocity $v(0)$ at time $t_0 = 0$ is approximately

$$\frac{\Delta s}{\Delta t} = \frac{s_1 - s_0}{t_1 - t_0} = \frac{1.2 - 0}{0.5 - 0} = 2.4 \text{ meters per second.}$$

Similarly,

$$v(0.5) \approx v_1 = \frac{s_2 - s_1}{t_2 - t_1} = \frac{5.0 - 1.2}{1.0 - 0.5} = 7.6 \text{ meters per second.}$$

In general,

$$v(t_i) \approx v_i = \frac{s_{i+1} - s_i}{t_{i+1} - t_i}$$

for each i from 0 to 19. In Figure 3.21 we show the results of these calculations.

Figure 3.21 Approximate velocity (in meters per second) of a falling object

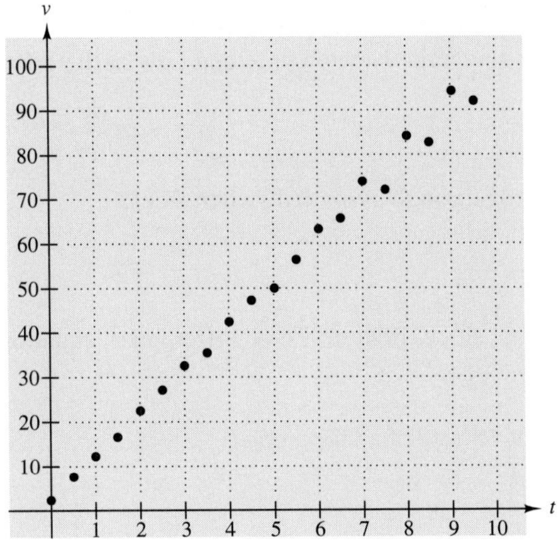

Checkpoint 1

What do you see in Figure 3.21 that tells you that the acceleration (the time derivative or rate of change of v) is approximately constant?

Checkpoint 2

(a) According to Table 3.2, when an object has fallen 200 meters, it has been falling between 6.0 and 6.5 seconds. Estimate its speed when it passes the 200-meter mark.

(b) Locate the points in Figure 3.21 that correspond to 6.0 and 6.5 seconds. Check that the velocity in part (a) is consistent with the plot of approximate velocities.

Gravity and Acceleration

According to **Newton's Second Law of Motion** (for an object of unchanging mass), the net force acting on an object is the product of its mass and its acceleration. In symbols, this says

$$F = ma$$

where F is the net force, m is the mass of the object, and a is the acceleration. Since acceleration is the rate of change of velocity, we also may write Newton's Second Law in the form

$$F = m\frac{dv}{dt}.$$

Our marble is subject to a gravitational force, called its "weight." Careful experiments have shown that as long as the falling object is near sea level, this force is approximately proportional to the mass of the object. The proportionality constant, usually abbreviated g, is approximately 32.17 feet per second per second.

Thus we may write Newton's Second Law as the differential equation

$$m\,g = m\,\frac{dv}{dt}$$

or, if we divide through by m and switch sides,

$$\frac{dv}{dt} = g.$$

We show the slope field for this differential equation—a particularly simple slope field—in Figure 3.22.

Figure 3.22 Slope field for velocity of a falling object

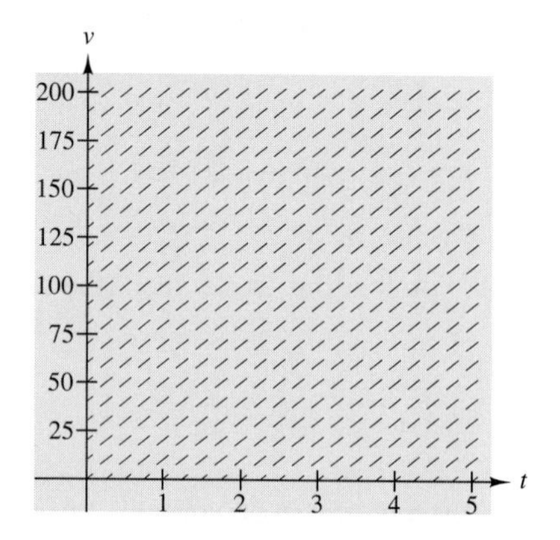

Checkpoint 3

The distance unit for the falling marble problem happens to be feet, but we will often use metric units as well—as we did in Example 1. Convert $g = 32.17$ feet per second per second to meters per second per second.

An Initial Value Problem for Velocity

Now that we have determined an appropriate differential equation, we'll look for the particular solution to model the fall of the marble.

Exploration Activity 2

(a) Find a family of solutions of the differential equation $dv/dt = g$. Check that your solution functions agree with the slope field in Figure 3.22.

(b) Find the solution of the initial value problem

$$\frac{dv}{dt} = g \qquad \text{with} \quad v(0) = 0.$$

(c) Find the solution of the initial value problem

$$\frac{dv}{dt} = g \qquad \text{with} \quad v(0) = 10.$$

Give a physical interpretation of this problem and its solution.

(d) Find the solution of the initial value problem

$$\frac{dv}{dt} = g \qquad \text{with} \quad v(0) = -10.$$

Give a physical interpretation of this problem and its solution.

The differential equation $dv/dt = g$ is the inverse of a problem we have studied already: If a function is linear, then its average rate of change is constant (the slope of the linear graph), so its instantaneous rate of change is also constant. Having studied the inverse problem, we know solutions to the differential equation: Any linear function with slope g, say, $v = gt + C$, satisfies the differential equation. Substituting $v = 0$ and $t = 0$ simultaneously, we see that the initial condition is satisfied if $C = 0$. Thus

$$v = gt$$

is one solution of the initial value problem

$$\frac{dv}{dt} = g \qquad \text{with} \quad v(0) = 0.$$

We assumed that each initial value problem has only one solution, so the solution must be $v = gt$. If our units are metric, this means that the velocity at time t is $9.807t$ meters per second. If the units are English, then $v(t)$ is $32.17t$ feet per second. Notice that by choosing acceleration to be positive in the downward direction, we also have made "down" the positive direction for velocity as well.

In general, we have $v(0) = g \cdot 0 + C$, so C is the initial velocity. If $v(0) = 10$, then $v = gt + 10$ at all times t until impact. The physical interpretation in this case is that the object was thrown downward with an initial speed of 10 units. Of course, it matters whether those units are feet per second or meters per second, but the formula works either way, as long as the proper value is given to g. Similarly, an initial velocity of -10 means the object is thrown upward with a speed of 10 units, and its velocity thereafter will be $v = gt - 10$.

For our falling marble (the case of zero initial speed), we now know velocity as a function of time. Recall that we asked "How fast would it be going when it hit the ground, about 535 feet below?" The formula $v = gt$ would answer this question if we knew how long it takes to fall 535 feet. But that's a question about the functional relationship between distance and time, which we don't know yet. However, we can read $v = gt$ as a differential equation relating distance and time—because $v = ds/dt$—and we can set up another initial value problem:

$$\frac{ds}{dt} = gt \qquad \text{with} \quad s = 0 \text{ at } t = 0.$$

In Figure 3.23 we show the slope field for this initial value problem.

Figure 3.23 Slope field for position of a falling object

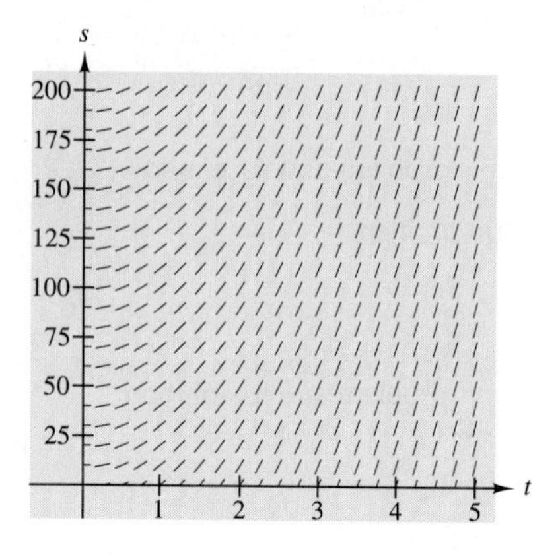

Exploration Activity 3

(a) Find a family of solutions of the differential equation $ds/dt = gt$.

(b) Find the solution of the initial value problem

$$\frac{ds}{dt} = gt \qquad \text{with} \quad s(0) = 0.$$

Sketch the solution on the slope field in Figure 3.23. (You may copy the page.)

(c) Find the solution of the initial value problem

$$\frac{ds}{dt} = gt \qquad \text{with} \quad s(0) = 10.$$

Sketch the solution on the slope field in Figure 3.23. Give a physical interpretation of this problem and its solution.

Again, we have studied the appropriate inverse problem already. In Chapter 2 we showed that any function of the form $s = ct^2$ has a derivative of the form $2ct$. The constant factor c was arbitrary, so we can set $2c = g$, and we will match the derivative in $ds/dt = gt$. Thus $c = \frac{1}{2}g$, and $s = \frac{1}{2}gt^2$ is one solution. As with the preceding initial value problem, we now know a whole family of solutions of the differential equation: $s = \frac{1}{2}gt^2 + C$, where C could be any constant. But when we substitute the initial condition for our falling marble, $s = 0$ at $t = 0$, we see that this C also has to be zero. Thus

$$s = \tfrac{1}{2}\, g\, t^2$$

is *a* solution of the initial value problem and therefore (by uniqueness) *the* solution.

More generally, for any solution of the differential equation, we see that $s(0) = 0 + C$, so the value of C is the initial position relative the selected coordinate system. For the falling marble problem, we found it convenient to place the origin of the coordinate system at the point from which the marble was dropped. Given that choice, the initial condition $s(0) = 10$ would mean that a second object was dropped 10 units (feet or meters) below the first object. If they were dropped simultaneously, the two curves you drew on Figure 3.23 would give their positions as functions of time.

Now that we have explicit formulas for position and velocity of the falling marble, we can determine the time of fall from the position formula and then the speed at impact from the velocity formula.

Example 2 How long does the marble fall?

Solution The time to fall 535 feet is the (positive) solution of

$$535 = \frac{1}{2}(32.17)t^2.$$

We have $t^2 = 33.26$, or $t = 5.77$ seconds. ■

Checkpoint 4

Finish the marble problem: How fast is the marble going just before it hits the ground? Express your answer first in feet per second. Then convert it to meters per second and miles per hour. Compare your answer to the estimate in Example 1.

Finally, we illustrate how all the pieces fit together in an extended example that treats a similar, but different, situation.

Example 3 Suppose the marble is thrown upward with a velocity of 10 feet per second from a height of 535 feet. How long will it take to hit the ground? How fast will it be traveling when it reaches the ground?

Solution We have assumed throughout the preceding discussion that down is the positive direction, so an upward initial velocity is negative. As we saw in Exploration Activity 2, the velocity function is

$$v(t) = gt - 10.$$

Now we have a new initial value problem for the distance traveled:

$$\frac{ds}{dt} = gt - 10 \qquad \text{with} \quad s = 0 \text{ at } t = 0.$$

We know that a function has derivative gt when it is of the form

$$\frac{1}{2}gt^2 + \text{constant}.$$

We also know that a function has the constant derivative 10 when it is of the form

$$10t + \text{constant}.$$

Thus, using our rule for the derivative of a difference, we find that any function of the form

$$s(t) = \tfrac{1}{2}gt^2 - 10t + K$$

(where K is a constant) satisfies the differential equation

$$\frac{ds}{dt} = gt - 10.$$

Our initial condition, $s(0) = 0$, requires that K be 0, so the unique solution of the initial value problem for distance fallen is

$$s(t) = \tfrac{1}{2}gt^2 - 10t.$$

We leave it to you to finish the example in the next checkpoint. ■

Checkpoint 5

(a) Find the particular value of t such that

$$535 = \tfrac{1}{2}gt^2 - 10t.$$

(b) What is the impact speed of the marble?

In this section we have studied a second kind of differential equation, that with the right-hand side involving the independent variable only. We started our discussion of falling bodies from the physical principle embodied in Newton's Second Law of Motion. That led us to the observation that a formula for distance fallen is just the end result of solving two simple initial value problems, one to get the velocity formula from acceleration and the other to get the position formula from the velocity formula. This discussion is continued in the Lab Reading on raindrops, where we bring in air resis-tance as another force acting on our falling object.

Answers to Checkpoints

1. The graph of v is approximately a straight line.
2. (a) 63 meters per second

   (b) The fourteenth point from the left was calculated by exactly the same quotient as in part (a).
3. 9.807 meters per second per second.
4. $32.17 \times 5.77 = 186$ feet per second $= 56.6$ meters per second $= 127$ miles per hour. The estimate in Example 1 is reasonably close.
5. (a) 6.09 seconds

   (b) Approximately 186 feet per second or about 127 miles per hour

Exercises 3.3

1. Suppose the marble is dropped from a height of 100 feet (approximately the height of a 10-story building).
   (a) What is its velocity (in feet per second) just before impact?
   (b) How long does it take to hit the ground?
2. Suppose the marble is dropped from a height of 4 feet (i.e., from arm's height above the ground).
   (a) What is its velocity (in feet per second) just before impact?
   (b) How long does it take to hit the ground?
3. (a) Suppose the marble is tossed straight up from a height of 535 feet with an initial speed of 40 feet per second. How does this affect the time of fall and velocity at impact?
   (b) Suppose it is tossed straight down at 40 feet per second. How does that affect the time of fall and velocity at impact?

Differentiate each of the following functions.

4. $f(t) = 3 + t - t^2$
5. $g(t) = 2t + e^{-1.7t}$
6. $h(t) = 2 - e^{1.7t}$
7. $\phi(t) = t^2 - e^{-2t}$
8. $F(t) = t^2 + e^{-3t}$
9. $G(t) = t + t^3 - 3e^{-t}$
10. $P(t) = 2t^4 - 2^t$
11. $\Phi(t) = 7e^{-3t} - 3t^2 + t^3$

Use your calculator to graph each of the following functions together with its derivative using your answer to the corresponding Exercise 4–11. In each case note any features of the derivative graph that correspond to what you observe about slopes on the graph of the original function. (This is a way to check your work on Exercises 4–11. If the features of your derivative graph *don't* match slope features on the original graph, that may mean you did something wrong in your algebraic calculation.)

12. $f(t) = 3 + t - t^2$
13. $g(t) = 2t + e^{-1.7t}$
14. $h(t) = 2 - e^{1.7t}$
15. $\phi(t) = t^2 - e^{-2t}$
16. $F(t) = t^2 + e^{-3t}$
17. $G(t) = t + t^3 - 3e^{-t}$
18. $P(t) = 2t^4 - 2^t$
19. $\Phi(t) = 7e^{-3t} - 3t^2 + t^3$

Solve each of the following initial value problems.

20. $\dfrac{dy}{dt} = 1 - 2t$ with $y = 3$ at $t = 0$
21. $\dfrac{dy}{dt} = 4 - 2y$ with $y = 1$ at $t = 0$
22. $\dfrac{dy}{dt} = 3t^2$ with $y = 5$ at $t = 2$
23. $\dfrac{dz}{dt} = 2 - 1.7e^{-1.7t}$ with $z = 1$ at $t = 0$
24. $\dfrac{dw}{dt} = -1.7e^{1.7t}$ with $w = 3$ at $t = 0$
25. $\dfrac{dY}{dt} = 2t + 2e^{-2t}$ with $Y = 3$ at $t = 0$
26. $\dfrac{dP}{dt} = 2t - 3e^{-3t}$ with $P = 1$ at $t = 0$
27. $\dfrac{dq}{dt} = 1 + 3t^2 + 3e^{-t}$ with $q = 1$ at $t = 0$
28. $\dfrac{du}{dt} = 8t^3 - (\ln 2)\, 2^t$ with $u = 3$ at $t = 0$
29. $\dfrac{dx}{dt} = -21e^{-3t} - 6t + 3t^2$ with $x = 3$ at $t = 0$
30. Suppose a baseball is thrown straight up, from a height of 5 feet above the ground, with an initial velocity of 70 feet per second. Ignoring air resistance and other nongravitational forces, find formulas for its velocity and height as functions of time, and then answer the following questions, not necessarily in the order given.
    (a) How high does the ball rise?
    (b) How fast is it going at the peak of its flight?
    (c) How long does it take to reach its peak height?
    (d) How fast is the ball going when it hits the ground?
    (e) What is the total elapsed time from release of the ball until it hits the ground?

31. Compare the impact speed for the marble dropped from 535 feet (Checkpoint 4) with that for the marble thrown upward from the same point at 10 feet per second (Checkpoint 5). Is this what you would expect? Explain.

32. Suppose a marble is thrown downward with a velocity of 10 feet per second from a height of 535 feet. How long will it take to hit the ground? How fast will it be traveling when it reaches the ground?

33. (a) Checkpoint 4 and Exercise 32 tell us that a marble thrown up at 10 feet per second from 535 feet hits the ground at the *same speed* as a marble thrown down at 10 feet per second from the same height. Explain this by showing that the marble thrown up is traveling at 10 feet per second when it passes the throwing point on the way down.

(b) In Figure 3.24 we show a portion of the slope field and solution for the position of a marble thrown at 10 feet per second. Was the marble thrown up or down? Explain your answer. Also explain the significance of the positive value of t at which the position curve crosses the t-axis.

Figure 3.24 Slope field for position of a falling object

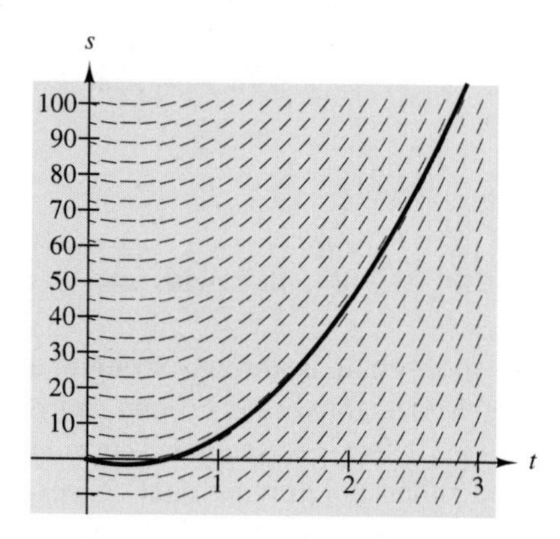

34. Suppose we check the answer to Checkpoint 5 by substituting $t = 6.09$ back into the equation

$$535 = \tfrac{1}{2}gt^2 - 10t.$$

On the right-hand side we get

$$\frac{1}{2} \cdot 32.17 \cdot (6.09)^2 - 10 \cdot 6.09 = 535.66,$$

which doesn't even round off to the expected 535. Explain what is going on in terms of what you learned in Chapter 1 about significant digits.

35. Suppose your car is capable of accelerating smoothly from a standing start at the rate of 1.4 feet per second per second.

(a) Find the acceleration, the velocity, and the distance traveled, all as functions of time, over the first minute of travel.

(b) What is the speedometer reading at the end of one minute?

(c) How far, in miles, does the car travel in that time?

36. If someone dropped a baseball out of the Washington Monument, do you think a professional catcher could catch it? Why or why not?

37. In Exercise 3 you modified the falling marble problem by giving the marble an initial velocity of 40 feet per second (up or down).

(a) Solve the problem for an arbitrary initial velocity v_0 (positive or negative) and an arbitrary starting height s_0.

(b) What are the time of fall and the velocity at impact? (Your answer will involve s_0 and v_0.)

38. Although many people know of the famous golf ball experiment, conducted during the *Apollo 11* mission to the moon, most people are unfamiliar with the experiments performed by the Frenchman, Forget Menot. Having calculated the acceleration due to gravitational force on the moon to be 5.4 feet per second per second, Forget threw a ball straight up with an initial velocity of 5 feet per second. Unfortunately, Forget neglected to bring his watch and was unable to estimate the total time it took for the ball to rise to its peak and return to the ground. Help Capitaine Menot by calculating this time and also by calculating the height of the ball at its peak.[4]

39. At a certain instant, just before lifting off, a plane is traveling down the runway at 285 kilometers per hour. The pilot suddenly realizes something is wrong and aborts the takeoff by cutting power and applying the brakes. Assume that the effect of this action is a deceleration proportional to time t. After 28 seconds the plane comes to a stop. How far does the plane travel after the takeoff is aborted?

4. Problem contributed by Sam Morris, Duke University.

40. Suppose a raindrop falls to the earth from a cloud 3000 feet high, and the only force acting on the raindrop is that of gravity. How long does it take the raindrop to fall? How fast is it going when it strikes the ground? Does this fit with your experience with rain?

Chapter 3 Summary

In this chapter we have considered two kinds of differential equations. The first kind expresses the derivative of an unknown function in terms of the dependent variable only, e.g., $dy/dt = 4y$. The second expresses the derivative as an explicit function of the independent variable alone, e.g., $dy/dt = 4t$. Every differential equation in this course will be of one or the other of these two types.

Either kind of differential equation can be expected to have infinitely many solutions. When we specify a single point on the solution curve (an initial condition), we usually single out a unique solution function. Since this will always be true for the differential equations in this course, we simply *assume* that **every initial value problem has a unique solution**.

We recalled at the beginning of the chapter the useful visual device called the "slope field." The most important role of the slope field is as a conceptual tool. You should imagine each differential equation being represented by such a field—that is, by a large number of direction markers whose slopes are determined by the differential equation. Solving the differential equation then corresponds to finding curves (graphs of functions) that pass through the slope field always following the direction markers. And solving an initial value problem means picking out the particular solution curve that goes through the right initial point.

In the remaining sections we studied applications that led naturally to the two different types of differential equation.

- Section 3.2 presented a scenario for application of Newton's Law of Cooling. The main step in determining the time of death of a murder victim was to find a solution of a differential equation in which the right-hand side involved only the unknown temperature function. Our technique for solution was to turn the problem into one we had already solved in Chapter 2.
- In Section 3.3, in the context of a falling marble, we saw that we could find a formula for velocity from the assumption of constant acceleration, and then we could find a formula for distance from the formula for velocity. In each case we were solving an initial value problem in which the right-hand side of the differential equation depended only on the independent variable, time.

Concepts

- Slope field
- Differential equation
- Initial condition
- Initial value problem
- Solution of a differential equation
- Substitution in a differential equation

Applications

- Newton's Law of Cooling
- Velocity and position of a freely falling object

Formulas

This summary repeats derivative formulas from Chapter 2, but with some reorganization and renaming.

The Constant Multiple Rule

$$\frac{d}{dt}\, Af(t) = A\,\frac{df}{dt}, \text{ for any constant } A$$

The Sum Rule

$$\frac{d}{dt}\,[f(t) + g(t)] = \frac{df}{dt} + \frac{dg}{dt}$$

The Difference Rule

$$\frac{d}{dt}\,[f(t) - g(t)] = \frac{df}{dt} - \frac{dg}{dt}$$

Derivatives of exponential functions

$$\frac{d}{dt}\, e^{kt} = k\, e^{kt}, \text{ for any constant } k$$

$$\frac{d}{dt}\, b^{t} = (\ln b)\, b^{t}, \text{ for any constant } b$$

Derivatives of power functions (the Power Rule)

$$\frac{d}{dt}\, t^{n} = n\, t^{n-1}$$

Derivatives of polynomial functions

Combine the Power Rule, the Constant Multiple Rule, and the Sum Rule.

Chapter 3 Exercises

Calculate the derivative of each of the following functions.

1. $f(t) = e^{3t}$
2. $g(t) = e^{-2t}$
3. $P(t) = 1.2^{t}$
4. $h(t) = 3t - 5t^2$
5. $\phi(t) = 6 - e^{3t}$
6. $q(t) = 2t + e^{-t}$
7. $F(t) = e^{-3t} + 6t^2$
8. $G(t) = 6 - e^{-2}$
9. $p(t) = 2^{3t}$
10. $H(t) = 3 + t + 5t^2$
11. $\psi(t) = e^{3t} - 6t + 7$
12. $r(t) = 2t^3 - e^{-t}$
13. $Q(t) = 7t^4 + 5t^3 - 4t^2 + 3t - 10$
14. $w(t) = t^3 + \sqrt{2}\, e^{2t}$

Each of the next four exercises shows a slope field. In each case sketch the graphs of at least three different solutions of the corresponding differential equation. (You may copy the page first.)

15. **Figure 3.25** Slope field for $dy/dt = \frac{1}{2}\, y$

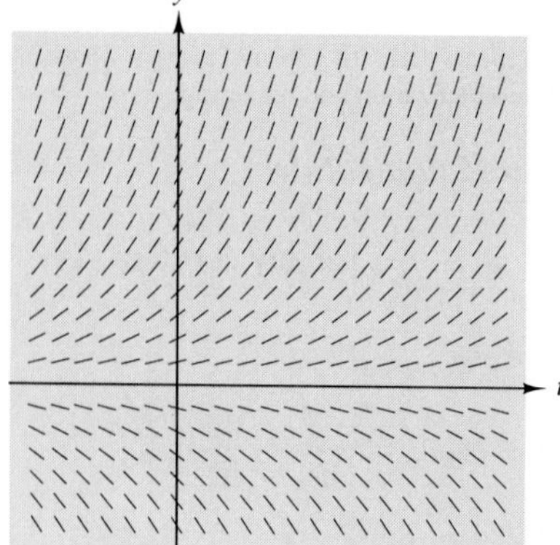

16. **Figure 3.26** Slope field for $dy/dt = \frac{1}{2}\, t$

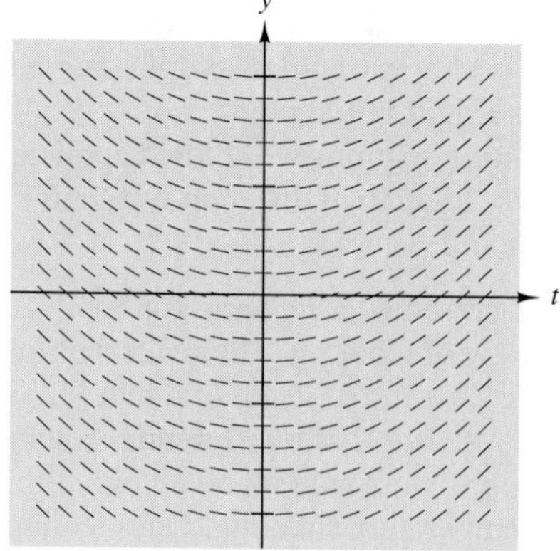

17. **Figure 3.27** Slope field for $dy/dt = \frac{3}{10}\,y\,(3-y)$

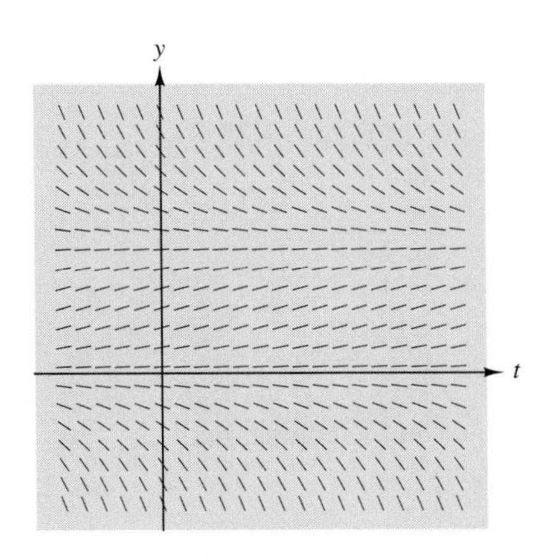

18. **Figure 3.28** Slope field for $dy/dt = \frac{3}{10}\,t\,(3-t)$

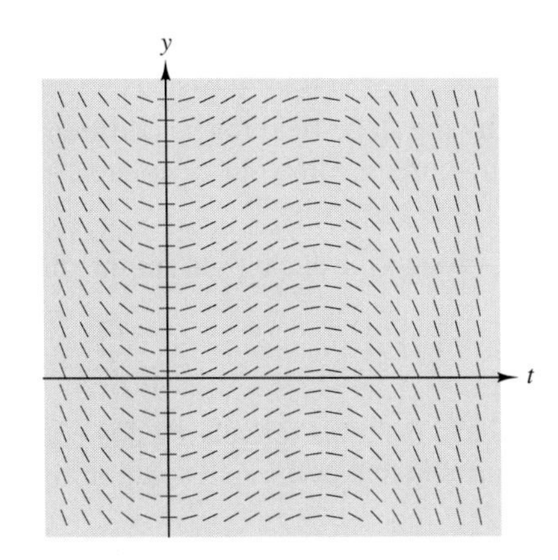

19. Solve the initial value problem

$$\frac{dy}{dt} = \frac{1}{2}y \qquad \text{with} \quad y(0) = 3.$$

Graph your solution with your calculator, and compare with the slope field in Figure 3.25.

20. Solve the initial value problem

$$\frac{dy}{dt} = \frac{1}{2}y \qquad \text{with} \quad y(0) = -1.$$

Graph your solution with your calculator, and compare with the slope field in Figure 3.25.

21. Solve the initial value problem

$$\frac{dy}{dt} = \frac{1}{2}t \qquad \text{with} \quad y(0) = 3.$$

Graph your solution with your calculator, and compare with the slope field in Figure 3.26.

22. Solve the initial value problem

$$\frac{dy}{dt} = \frac{1}{2}t \qquad \text{with} \quad y(0) = -1.$$

Graph your solution with your calculator, and compare with the slope field in Figure 3.26.

23. Solve the initial value problem

$$\frac{dy}{dt} = 0.9t - 0.3t^2 \qquad \text{with} \quad y(0) = 3.$$

Explain why the slope field for this differential equation is the one shown in Figure 3.28. Graph your solution with your calculator, and compare with the slope field.

24. Solve the initial value problem

$$\frac{dy}{dt} = 0.9t - 0.3t^2 \qquad \text{with} \quad y(0) = -1.$$

Explain why the slope field for this differential equation is the one shown in Figure 3.28. Graph your solution with your calculator, and compare with the slope field.

25. Figure 3.27 shows the slope field for the differential equation

$$\frac{dy}{dt} = \frac{3}{10}\,y\,(3 - y).$$

We won't have the tools for solving this equation symbolically until later in the course, but here is a formula that defines all solution functions that cross the y-axis:

$$y = \frac{3y_0}{y_0 + (3 - y_0)e^{-t/10}}$$

where $y_0 = y(0)$.

(a) Verify that the formula for y does indeed give the value y_0 when $t = 0$.

(b) What is the limiting behavior of y as t becomes large? What does that tell you about the vertical scale in Figure 3.27? What is the limiting behavior of dy/dt as t becomes large?

(c) Write down the solution of the initial value problem

$$\frac{dy}{dt} = \frac{3}{10}\,y\,(3 - y) \qquad \text{with} \quad y(0) = 1.$$

Use your calculator to graph the solution function, and compare the graph with Figure 3.27. Is the solution function consistent with the slope field? Explain.

(d) Write down the solution of the initial value problem

$$\frac{dy}{dt} = \frac{3}{10}\,y\,(3 - y) \qquad \text{with} \quad y(0) = 5.$$

Use your calculator to graph the solution function, and compare the graph with Figure 3.27. Is the solution function consistent with the slope field? Explain.

(e) Write down the solution of the initial value problem

$$\frac{dy}{dt} = \frac{3}{10}\,y\,(3 - y) \qquad \text{with} \quad y(0) = -1.$$

Use your calculator to graph the solution function, and compare the graph with Figure 3.27. Is the solution function consistent with the slope field? Is it consistent with your answers in part (b)? Explain.

26. Which of the following functions are solutions of the differential equation $dy/dt = 3y$?

(i) $f(t) = e^{4t}$ (ii) $f(t) = e^{3t}$ (iii) $f(t) = 3t^2$

(iv) $f(t) = 5e^{3t}$ (v) $f(t) = e^{3t} + 5$ (vi) $f(t) = 1.5t^2 + 3$

27. Which of the following functions are solutions of the differential equation $dy/dt = 3t$?

(i) $f(t) = 3t^2$ (ii) $f(t) = 1.5t^2 + 3$ (iii) $f(t) = 5e^{3t}$

(iv) $f(t) = e^{4t}$ (v) $f(t) = 0.5t^2$ (vi) $f(t) = 2 - 1.5t^2$

28. Which of the following functions is the solution of the initial value problem $dy/dt = 3t$ with $y = 4$ at $t = 1$?

(i) $f(t) = 3t^2$ (ii) $f(t) = 1.5t^2 + 4$ (iii) $f(t) = 4e^{3t}$

(iv) $f(t) = 1.5t^2 + 2.5$ (v) $f(t) = 3t^2 + 4$ (vi) $f(t) = 3 + 1.5t^2$

29. Which of the following functions is the solution of the initial value problem $dy/dt = 3t$ with $y = 4$ at $t = 0$?

(i) $f(t) = 3t^2$ (ii) $f(t) = 1.5t^2 + 4$ (iii) $f(t) = 4e^{3t}$

(iv) $f(t) = 1.5t^2 + 2.5$ (v) $f(t) = 3t^2 + 4$ (vi) $f(t) = 3 + 1.5t^2$

30. Which of the following functions is the solution of the initial value problem $dy/dt = 3y$ with $y = 3$ at $t = 0$?

(i) $f(t) = e^{3t}$ (ii) $f(t) = 2e^{3t}$ (iii) $f(t) = 3 + 1.5t^2$

(iv) $f(t) = 3e^{3t}$ (v) $f(t) = 4e^{3t}$ (vi) $f(t) = 3t^2$

31. Which of the following functions is the solution of the initial value problem $dy/dt = 3y$ with $y = 4$ at $t = 0$?

(i) $f(t) = e^{4t}$ (ii) $f(t) = 2e^{4t}$ (iii) $f(t) = 4 + 1.5t^2$

(iv) $f(t) = 3e^{3t}$ (v) $f(t) = 4e^{3t}$ (vi) $f(t) = 4t^2$

Find three solutions of each of the following differential equations.

32. $dy/dt = 5y$

33. $dy/dt = 5t$

34. $dy/dt = -1.7y$

35. $dy/dt = -1.7t$

For each of the following differential equations, sketch the graphs of three solutions. (You may use your graphing calculator or computer and/or the solutions to Exercises 32–35.)

36. $dy/dt = 5y$

37. $dy/dt = 5t$

38. $dy/dt = -1.7y$

39. $dy/dt = -1.7t$

Solve each of the following initial value problems.

40. $dy/dt = 5y$ with $y = 2$ at $t = 0$

41. $dy/dt = 5t$ with $y = 2$ at $t = 0$

42. $dy/dt = -1.7y$ with $y = 2.3$ at $t = 0$

43. $dy/dt = -1.7t$ with $y = 2.3$ at $t = 0$

For each of the following initial value problems, sketch the graph of the solution. (You may use your graphing calculator or computer and/or the solutions to Exercises 40–43.)

44. $dy/dt = 5y$ with $y = 2$ at $t = 0$

45. $dy/dt = 5t$ with $y = 2$ at $t = 0$

46. $dy/dt = -1.7y$ with $y = 2.3$ at $t = 0$

47. $dy/dt = -1.7t$ with $y = 2.3$ at $t = 0$

48. Suppose a wrench is dropped from a stationary blimp at a height of 2000 feet.

(a) How far does it travel in 2 seconds?

(b) How fast is it traveling when it hits the ground?

(c) How long does it take to hit the ground?

Describe a family of solutions of each of the following differential equations.

49. $\dfrac{dy}{dx} = -2(y-1)$

50. $\dfrac{dy}{dx} = 2(y-1)$

51. $\dfrac{dy}{dx} = -2(y+1)$

52. $\dfrac{dy}{dx} = 2(y+1)$

Solve each of the following initial value problems.

53. $\dfrac{dy}{dx} = -2(y-1),\ \ y(0) = 3$

54. $\dfrac{dy}{dx} = 2(y-1),\ \ y(0) = 3$

55. $\dfrac{dy}{dx} = -2(y+1),\ \ y(0) = 3$

56. $\dfrac{dy}{dx} = 2(y+1),\ \ y(0) = 3$

57. Exercises 49–52 can all be done with a single calculation. Decide how to do that — and do it. (*Hint*: Think about the solution of the cooling body problem. Your calculation here should be general enough to include that case as well.)

58. Table 3.3 shows world record holders in the men's 100-meter dash, together with the dates on which they set their records and their world-record times.

(a) Find a formula for a function that approximately fits the data.

(b) Predict the next time a world record will be set. (Think about the next smaller time that can be measured.)

(c) Does your formula allow the possibility that a time of 9.5 seconds for the 100-meter dash will ever be reached? If so, when? If not, what is the fastest time permitted by your formula?

Table 3.3 World record times in the 100-meter dash[5]

Record Holder	*Date*	*Time (seconds)*
Donald Lippincott	July 6, 1912	10.6
Charles Paddock	April 23, 1921	10.4
Percy Williams	August 9, 1930	10.3
Jesse Owens	June 20, 1936	10.2
Willie Williams	August 3, 1956	10.1
Armin Hary	June 21, 1960	10.0
Jim Hines	June 20, 1968	9.99
Jim Hines	October 14, 1968	9.95
Calvin Smith	July 3, 1983	9.93
Carl Lewis	September 24, 1988	9.92
Leroy Burrell	June 14, 1991	9.90
Carl Lewis	August 25, 1991	9.86
Leroy Burrell	July 6, 1994	9.85

59. Suppose there is some physical reason why no man can ever run the 100-meter dash in less than, say, 9.63 seconds. Then it should be harder to set a new record as that barrier time is approached. A reasonable conjecture might be that the rate of change of record times is proportional to how far the record time is from the barrier time. That is,

$$\frac{dT}{dt} = -k(T - 9.63)$$

where T is the record time at time t.

(a) What is the form of a function that satisfies this differential equation?

(b) How can you use a semilog plot to test whether the data in Table 3.3 is reasonably approximated by such a function? Carry out your test, and describe the result.

(c) Find a solution of the differential equation that reasonably fits the historic data. Use your formula to predict the next time a record will be set. How does this prediction compare with the one in the preceding exercise?

(d) By the time you read this a new record may have been set. If you know the date and time, compare that with your predictions in this exercise and the preceding one.

5. Source: Associated Press, July 7, 1994.

Chapter 3 Projects

1. **Glucose in the Bloodstream** A physician decides to give a patient an infusion of glucose at a rate of c grams per hour. The body of the patient simultaneously converts the glucose and removes it from the bloodstream at a rate proportional to the amount present in the bloodstream, say, at r grams per hour per gram of glucose present.

   (a) Explain why the amount $G = G(t)$ of glucose present at time t can be modeled by a differential equation of the form $dG/dt = c - rG$.

   (b) Given an initial amount G_0 of glucose in the bloodstream at time 0, find an explicit formula for the amount present at any time t. (*Hint*: Compare with Newton's Law of Cooling.)

   (c) What rate of infusion would keep the glucose level in the bloodstream constant?

   (d) A physician orders an infusion of 10 grams of glucose per hour for a patient. Laboratory technicians determine that the patient has 2 grams of glucose in his bloodstream and that his body will remove glucose from the bloodstream at a rate of 3 grams per hour per gram of glucose. How much glucose will be in the bloodstream t hours after the infusion is started? In particular, how much after 2 hours? How long will it take for the glucose level in the bloodstream to reach 3 grams?

   (e) Suppose a patient's bloodstream has 2 grams of glucose, and her physician wants to raise this amount to 3.5 grams in 3 hours. It is determined that her system removes glucose from the bloodstream at a rate of 4 grams per hour per gram of glucose. How fast should the physician order the glucose to be infused into the patient's body?

2. **RL Circuits** Figure 3.29 is a diagram of an electrical circuit with a resistance of R ohms, an inductance of L henries, and a battery (i.e., a constant voltage source) of V volts. If the current (in amperes) at time t (in seconds after closing the switch) is $i = i(t)$, then the voltage drop at the resistor is Ri, and the voltage drop at the inductor is $L\,di/dt$. According to Ohm's Law, the voltage input V must balance the voltage drops, that is,

$$L\,\frac{di}{dt} + R\,i = V.$$

   (a) Find the current as a function of time. (*Hint*: Compare with Newton's Law of Cooling.)

   (b) If the switch is left closed for a long time, what is the limiting value of the current?

Figure 3.29 An *RL* circuit

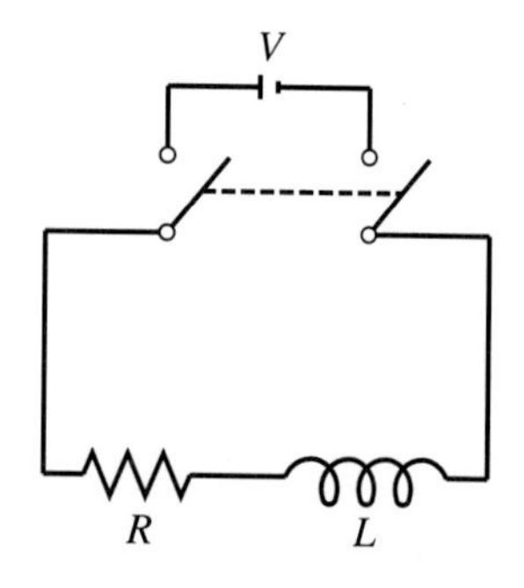

3. **A Falling Body with Air Resistance** Suppose we consider air resistance in our model for the falling marble. Then the total force F acting on the marble has two components, the gravitational force and the retarding force. If we write F_g for the gravitational force and F_r for the retarding force (air resistance), then $F = F_g + F_r$. For purposes of this project, we assume that F_r is proportional to the velocity, say, $F_r = -kv$.

   (a) Why is the coefficient of v negative?

   (b) Use Newton's Second Law of Motion to argue that v must satisfy a differential equation of the form $dv/dt = g - cv$ for some constant c. How is c related to k and the mass of the object?

   (c) Under the assumptions stated above, set up an initial value problem for the velocity of the marble dropped from the Washington Monument.

   (d) Assume further that the numerical values of k and of the mass are such that c turns out to be 0.08. Solve the initial value problem to find an explicit formula for velocity as a function of time. (*Hint*: Compare with Newton's Law of Cooling. In particular, note that the right-hand side of the differential equation is no longer constant but now depends on the dependent variable.)

   (e) Your explicit formula for velocity in part (d) also provides a differential equation for the position function $s = s(t)$. Set up and solve an initial value problem for the position function. (*Hint*: See the summary of derivative formulas at the end of the chapter.)

   (e) Under the assumptions stated above, how long would it take a marble dropped from the Washington Monument to reach the ground? How fast would it be going when it hit the ground?

4. **Brine Mixing** A tank contains 25 pounds of salt dissolved in 200 gallons of water. There is a pipe at the top of the tank to bring in additional solution and a pipe at the bottom of the tank to drain off solution. The tank itself contains a large stirrer that keeps the solution in the tank thoroughly mixed. Starting at time $t = 0$, water containing $\frac{1}{2}$ pound of salt per gallon enters the tank at the rate of 4 gallons per minute, and the well-stirred solution leaves the tank at the same rate.

   (a) If this process continues a long time, how much salt do you think will eventually be in the tank? Does your answer about the eventual salt content depend on how much salt there was in the tank at the start of the process? Why or why not?

   (b) Find a differential equation that describes the rate of change of the amount of salt in the tank. Supply an initial condition for your differential equation, and solve for an explicit formula that gives the amount of salt as a function of time. How much salt is in the tank after 10 minutes? after 2 hours? What does your formula tell you about how much salt there will be in the tank after a long time? Does your answer agree with what you determined in part (a)? Why or why not? [*Hints*: Let $S = S(t)$ be the amount of salt in the tank at time t, and let $C = C(t)$ be the concentration of the salt at time t. How are S and C related? The flow of salt out of the tank depends on the concentration; use the relationship between S and C to express that flow in terms of S.]

5. **Antifreeze Mixing** The cooling system in an old truck holds about 10 liters of coolant. While the engine is running, the system is being flushed by running tap water into a tap-in on the heater hose and simultaneously draining the thoroughly mixed fluid from the bottom of

the radiator. Water flows in at the same rate that the mixture flows out — about 2 liters per minute. The system initially contains 50% antifreeze.

(a) Let W be the amount of water in the system after t minutes. Find an initial value problem that determines W as a function of time.

(b) Solve the initial value problem to find W as a function of time.

(c) How long should water run into the system to ensure that 95% of the mixture is water?[6]

6. **Federal Pensions** In 1969 Congress added a cost-of-living escalator clause to the federal civilian and military pension programs: For each 3% rise in the consumer price index that lasted 3 months, benefits would increase 4%. The extra 1% "kicker" was to offset lost purchasing power during the 3-month qualifying period.

   (a) In 1976 Senator Thomas Eagleton predicted that if the formula were maintained for 25 years, the costs of all federal pension programs would quadruple. What rate of inflation would make that prediction correct?

   (b) In the same year, former Congressman Hastings Keith said, "One result of the staggering congressional practice is that taxpayers will have had to cough up an extra $300 billion by 1990 . . . just to cover the *unanticipated* retirement costs of federal workers." The omitted part of the quote stated an assumed inflation rate. What rate?[7]

7. **Radiocarbon Dating**[8] In 1960 Willard Libby won a Nobel prize in chemistry for his work in the 1940s and 1950s that included discovery of a radioactive isotope of carbon and development of the technique of radiocarbon dating to determine the ages of once-living objects.

   Ordinary carbon, found in abundance in all living things, has an atomic weight of 12; its unstable isotope has an atomic weight of 14 and is therefore called **carbon-**14 (**^{14}C**) or **radiocarbon**. This isotope is created in the upper atmosphere by interaction of cosmic ray neutrons with nitrogen. It is then oxidized to a radioactive form of carbon dioxide, which is mixed by winds with the stable carbon dioxide already present in the atmosphere. Because this process — formation of carbon-14, oxidation, mixing, and decay of the radioactive carbon dioxide back to nitrogen — has been going on for eons, the ratio of ^{14}C to ordinary carbon in the atmosphere has long since reached a steady state. That is, the proportion of radiocarbon in the air is constant.

   All air-breathing plants take in carbon dioxide with this constant proportion of carbon-14 (relative to carbon-12), and thus the carbon in their tissues has the same proportion of radiocarbon. Animals that eat these plants (e.g., humans) also incorporate carbon in their tissues with this constant proportion of radiocarbon. However, when plant or animal dies, it ceases to take in any more carbon (or anything else!). The radioactive carbon then in its tissues continues to decay, but the ordinary carbon does not, so the proportion of radiocarbon in its tissues decreases.

   Radioactivity is observed and measured by devices, such as the Geiger counter, which count events or disintegrations of the isotope that emit a subatomic particle. From a count of

---

6. Problem contributed by Lewis Blake, Duke University.
7. Adapted from A. A. Bartlett, *The Physics Teacher* 14 (1976), p. 485.
8. Adapted from *Calculus with Analytic Geometry*, by George F. Simmons, McGraw-Hill, New York, 1985, pp. 226–229, and from *Some Applications of Exponential and Logarithmic Functions*, by W. Thurmon Whitley, UMAP Module 444, COMAP, 1989.

disintegrations per minute (dpm), one can infer the proportion of ^{14}C in a given quantity of carbon.

Facts about the decay rate, coupled with observation of the actual proportion of carbon-14 present, suffice to determine the approximate time of death of organisms that lived many thousands of years ago:

Fact 1 Any radioactive isotope tends to decay at a rate proportional to the amount of that isotope present.

Fact 2 ^{14}C has a half-life of approximately 5730 years. That is, in 5730 years a given quantity of ^{14}C will have decreased to half as much.

Fact 3 The radiocarbon in living tissue decays at a rate of about 15.30 (measured) dpm per gram of contained carbon.

(a) Suppose $y = y(t)$ is a function whose value at time t is the proportion of carbon-14 in a once-living tissue, where $y_0 = y(0)$ is the proportion at time of death. Explain how Fact 1 leads to a formula for $y(t)$ that involves y_0 and the proportionality constant.

(b) Explain how knowledge of the half-life determines the proportionality constant, whether or not you know y_0. What is the proportionality constant for carbon-14? Does your answer depend on the unit in which time is measured? Explain why it does or does not.

(c) Invert the function you found in part (a) to find time as a function of amount of ^{14}C present.

(d) A fossilized bone of a man found in 1980 in western Pennsylvania contained approximately 17% of its original ^{14}C. Estimate the year the man died.

(e) A bone uncovered in Kenya was found to contain only 10% of its original ^{14}C. Approximately how long ago did death occur?

The objects listed in Table 3.4 were tested in 1950 for radioactivity, with the indicated results expressed in disintegrations per minute per gram of carbon.

Table 3.4 Objects measured for ^{14}C decay rate

Object	*dpm per Gram*
A chair leg from the tomb of Tutankhamen	10.14
A house beam from Babylon in the reign of Hammurabi	9.52
Giant sloth dung from Gypsum Cave in Nevada	4.17
A hardwood atlatl from Leonard Rock Shelter in Nevada	6.42

(f) Explain why knowing the rate of ^{14}C disintegration is enough to determine the proportion of ^{14}C present at time t. In particular, for the chair leg in Table 3.4, why is the ratio 10.14: 15.30 the same as the ratio $y(t): y_0$?

(g) Estimate the age of each object in Table 3.4.

(h) The objects in Table 3.5 have been dated by radiocarbon dating, with the indicated results. Estimate the measured disintegrations per minute per gram of carbon in each object, and explain your estimate.

Table 3.5 Objects dated by ^{14}C disintegrations

Object	*Age in Years*
Linen wrappings from the Dead Sea Scrolls	$1{,}917 \pm 200$
Charcoal from the Lascaux Caves in France	$15{,}516 \pm 900$
Charcoal from Stonehenge in England	$3{,}798 \pm 275$
Charcoal from the Crater Lake volcano eruption, Oregon	$6{,}453 \pm 250$

(i) Think about the uncertainties in the estimated ages. What do they say about possible uncertainties in the measured numbers of disintegrations?

8. **Immigration** The United States is a nation of immigrants—over its 200-year history, some $50,000,000$ people have come here to live from someplace else. How is it, then, that we could model our population growth in Chapter 2 and completely ignore immigration, not to mention wars, the Great Depression, and the postwar baby boom? We couldn't, of course; we found in Chapter 2 that the natural growth model is adequate from 1790 to 1860, but the effect of the Civil War was evident in the census data immediately after that.

A model that takes immigration into account might have the form

$$\frac{dP}{dt} = kP + (\text{immigration rate}).$$

Let's assume that the immigration rate is constant, say, that m people per year are added to the population through immigration. Then our differential equation is

$$\frac{dP}{dt} = kP + m$$

As with our previous population models, we can turn this into an initial value problem by declaring $t = 0$ to mean a time at which we know the population P_0.

(a) Show that the substitution $y = P + (m/k)$ turns the initial value problem into

$$\frac{dy}{dt} = ky \qquad \text{with} \quad y_0 = P_0 + \frac{m}{k} \text{ at } t = 0.$$

(b) Explain why the solution of this initial value problem is

$$y = \left(P_0 + \frac{m}{k}\right) e^{kt}.$$

Then show that the solution of our original problem is

$$P = \left(P_0 + \frac{m}{k}\right)e^{kt} - \frac{m}{k}.$$

Now we need to know both k and m. The growth constant k cannot be inferred directly from the data, but the immigration parameter m is observable: Every (legal) immigrant is counted. Suppose we take m to be the average immigration rate over a decade in which immigration was relatively constant, namely, the 1950s. For that decade, about $250,000$ immigrants arrived each year.[9] Then we can ask whether there is a

9. This is also the long-term average immigration rate. Source: Table 1 in the *U.S. Immigration and Naturalization Service Annual Report*, 1961, reproduced in M. T. Bennett, *American Immigration Policies: A History*, Public Affairs Press, Washington, 1963.

constant k such that the immigration model reasonably fits the census data over any substantial period of time.

If k really is constant, then we can estimate it by using a second measurement of population. Specifically, we can substitute $P = P_1$ at time $t = t_1$ to obtain an equation involving k alone. For example, we could take $t = 0$ in 1790, and let t_1 be 50 (which would represent 1840). From Table 1.4 (or Table 2.5), we see that P_0 is 3.929 (in millions) and P_1 is 17.069. Since we are measuring population in millions, our assumed value for m is 0.25.

(c) Substitute all these values into the solution from part (b) to show that

$$17.069 = \left(3.929 + \frac{0.25}{k}\right) e^{50k} - \frac{0.25}{k}.$$

(d) Check that $k = 0.001227$ is an approximate solution to this equation.

We have our proposed solution to the immigration problem:

$$P = \left(P_0 + \frac{m}{k}\right) e^{kt} - \frac{m}{k}$$

where $m = 0.25$ and $k = 0.001227$. Does this solution make any sense? There are at least two aspects of the solution that need to be checked. First, in general, it's a good idea to check whether the result of your calculations is an object that satisfies the mathematical conditions of the problem. No matter how good we are at calculation, we make mistakes some of the time. And answers corrupted by even small mistakes are of no particular value to anyone. Second, if the solution turns out to mathematically correct, we still need to know whether it actually fits the data for which it was proposed as a model.

(e) Check the initial condition by showing that $P(0) = P_0$.

(f) Check that the proposed solution function satisfies the original differential equation.

Next, we ask whether the solution of the proposed immigration model actually fits the census data for the U.S. population. (We repeat the data in Table 3.6.) The solution, rewritten in the form

$$P + \frac{m}{k} = \left(P_0 + \frac{m}{k}\right) e^{kt}$$

suggests that we should look at a semilog plot of the shifted data values

$$P_0 + \frac{m}{k}, P_1 + \frac{m}{k}, \ldots, P_{19} + \frac{m}{k}$$

versus time t.

(g) Explain why the semilog plot of shifted data should be a straight line if the function $P(t)$ fits the data.

(h) Fill in the $P + (m/k)$ columns in Table 3.6.

Table 3.6 U.S. Census Data, 1790–1980

t (years A.D.)	*P* (millions)	$P + (m/k)$	*t* (years A.D.)	*P* (millions)	$P + (m/k)$
1790	3.929		1890	62.948	
1800	5.308		1900	75.996	
1810	7.240		1910	91.972	
1820	9.638		1920	105.711	
1830	12.866		1930	122.775	
1840	17.069		1940	131.669	
1850	23.192		1950	150.697	
1860	31.443		1960	179.323	
1870	38.558		1970	203.185	
1880	50.156		1980	226.546	

(i) Figure 3.30 shows a semilog plot of $P + (m/k)$ versus t. Write a short discussion (not more than a page) of the extent to which the semilog plot confirms or refutes the reasonableness of the immigration model as a description of U.S. population growth.

Figure 3.30 Semilog plot of $P_i + (m/k)$ versus time

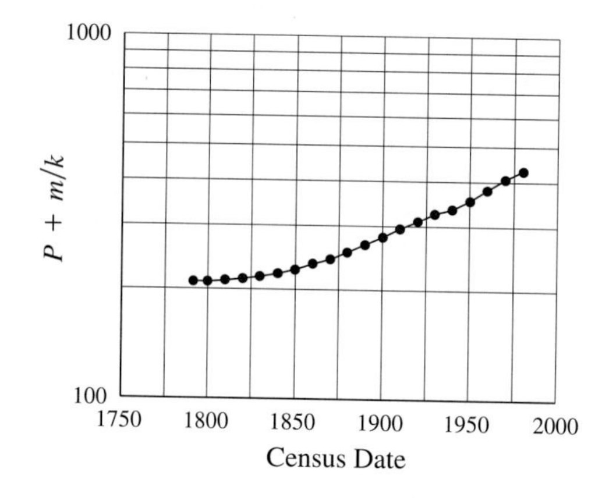

(j) How much difference would it have made in the immigration model if we assumed $m = 0.24$ instead of 0.25? What would k be in this case? (*Hint*: If you are not sure how to find k, see the next Project.)

9. **Solution of Nonlinear Equations** In Project 8 we encountered the problem of finding a value of k to satisfy the equation

$$17.069 = \left(3.929 + \frac{0.25}{k}\right) e^{50k} - \frac{0.25}{k}.$$

(a) Show that an equivalent equation is

$$(3.929\,k + 0.25)\,e^{50k} - 0.25 - 17.069\,k = 0.$$

(b) Show that $k = 0$ is a solution of the equation in part (a) but is not a solution of the original equation.

In this case 0 is a spurious solution, one that was introduced by the algebraic manip-ulation in part (a). Having a spurious solution is not usually a serious problem, as long as we are aware of it and are not misled by it. However, the form of the equation in part (a) suggests that any positive solution k must be very small. Specifically, the term involving e^{50k} is very large even for fairly small values of k, so there is no way that this term can be offset by the negative terms in the equation unless k is really tiny. This means that the solution we want must be very close to the solution we already know—and don't want. Parts (c) and (d) suggest two different ways to find the nonzero solution for k. If you succeed in part (c), you don't need to do part (d).

(c) You may be able to solve for k by graphing the function in part (a) and zooming in until you see a separation of the positive solution from the zero solution. Then either continue to zoom on the appropriate root, or use your Root or Solve key. Find the value of k to at least six decimal places, and compare this with the value we used in Project 8.

(d) If you had difficulty finding an appropriate window in part (c), you can spread out the values of the independent variable by changing to a new independent variable, $x = 100k$. Substitute $x/100$ each place k occurs, and show that the equation becomes

$$(0.03929\, x + 0.25)\, e^{x/2} - 0.25 - 0.17069\, x = 0.$$

Let $f(x)$ stand for the left-hand side of this equation. Figure 3.31 shows a starting view of the graph of f. Now use your graphing calculator to find a six-decimal-place approximation to a positive solution x. Then find the corresponding value of k, and compare this with the value we used in Project 8.

Figure 3.31 $f(x) = (0.03929\, x + 0.25)\, e^{x/2} - 0.25 - 0.17069\, x$

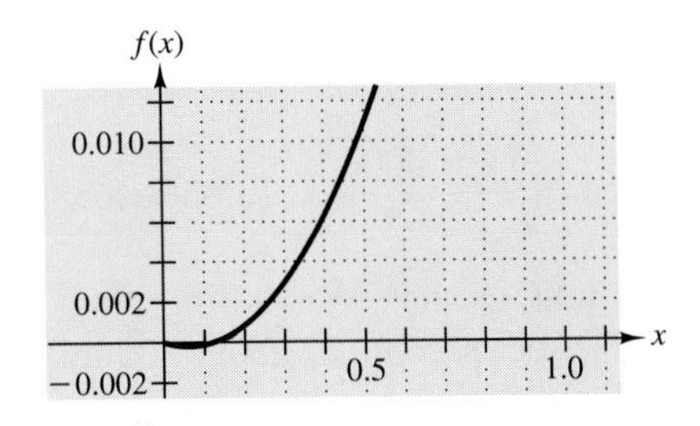

Chapter 3 Optional Lab Reading: Raindrops[10]

In Section 3.3 we studied a model for a dropped object falling under the influence of a constant gravitational force—and no other forces. In this reading we take a close look at possible assumptions about air resistance. First, we observe that ignoring air resistance is an unrealistic assumption for a familiar falling object: a raindrop. Then we look at two models that incorporate air resistance. We solve a differential equation that includes resistance proportional to velocity and examine a proposed solution for a differential

10. Adapted from Walter J. Meyer, "Falling Raindrops," pp. 101–111, in *Applications of Calculus*, edited by Phillip Straffin, MAA Notes No. 29, 1993. Our raindrop lab and much of this reading are based on Meyer's presentation.

equation in which resistance is proportional to the square of velocity. We don't have the tools yet for deriving symbolic solutions for the second equation, but we will develop those tools later in this text.

In a laboratory exercise associated with this chapter you may experiment with these models for falling raindrops. Much of that lab project is devoted to numerical exploration of the two models that incorporate air resistance.

Galileo's Model

Our discussion of the falling marble in Section 3.3 was based on **Galileo's Model**, the assumption that the only significant force acting on a falling object is the gravitational force. According to Newton's Second Law of Motion, such an object experiences a constant acceleration. In symbolic form, this model is the simplest initial value problem we have encountered so far:

$$\frac{dv}{dt} = g \qquad \text{with} \quad v(0) = 0.$$

Its solution, which can be expressed as "velocity is proportional to time," leads to our second simplest initial value problem:

$$v = \frac{ds}{dt} = gt \qquad \text{with} \quad s(0) = 0.$$

And the solution of this initial value problem tells us that the distance fallen is a quadratic function of time:

$$s = \tfrac{1}{2}gt^2.$$

In particular, if the raindrop falls, say, 3000 feet, then this last equation says that the time of fall would be

$$t = \sqrt{\frac{2 \times 3000}{32.2}} = 13.65 \text{ seconds.}$$

Then the velocity equation tells us that after falling 3000 feet, the raindrop would be going $32.2 \times 13.65 = 440$ feet per second, or about 300 miles per hour. Would you take a stroll in such a rain? Would you carry an umbrella? A raindrop striking the earth (or anything it encountered sooner) at 300 miles per hour just doesn't fit with our experience about raindrops. (You may have already examined this question in Exercise 40 in Section 3.3.)

Stokes's Law

For very small spherical droplets (those of diameter $D \leq 0.003$ inches) falling in still air, experience suggests that the drag force of the air (thus also the corresponding component of acceleration) is proportional to the velocity of the drop. Specifically, this model (called **Stokes's Law**) is expressed in the form of a differential equation by

$$\frac{dv}{dt} = g - kv.$$

Experimental evidence suggests that the constant k is approximately $0.329 \times 10^{-5}/D^2$ second^{-1} when the units are feet and seconds.

Problem 1

The differential equation for Stokes's Law implies the existence of a *terminal velocity*, a speed that the drops approach but cannot exceed.

(a) Calculate the terminal velocity v_{term} for drops small enough to obey Stokes's Law. In particular, what is the terminal velocity of a drizzle drop with $D = 0.003$ inches ($= 0.00025$ feet)?

(b) We will see shortly that drops this size reach their terminal velocities very quickly, so it is a reasonable approximation to assume that such drops are traveling at terminal velocity for the entire fall. With that assumption, how long does it take for a drizzle drop to fall 3000 feet?

Problem 2

(a) Stokes's Law, together with the assumption of zero starting velocity, constitutes an initial value problem. Solve this problem to find a formula for v as a function of t. *Hint*: Write the differential equation as

$$\frac{dv}{dt} = -k\left(v - \frac{g}{k}\right)$$

and make the change of dependent variable $y = v - g/k$.

(b) Does your formula confirm your calculation of v_{term} in Problem 1? How?

Problem 3

(a) Use the results from the two previous activities to find a formula for the ratio v/v_{term}.

(b) How long does it take a drizzle drop to reach 99% of terminal velocity?

(c) Was it reasonable to assume that these drops travel at terminal velocity for their entire fall? Why or why not?

The Velocity-Squared Model

For large raindrops, specifically, those with diameter $D \geq 0.05$ inches, the component of acceleration due to air resistance is found to be proportional to the square of velocity. That is, Stokes's Law is replaced by

$$\frac{dv}{dt} = g - cv^2.$$

In this case, experimental determination of the proportionality constant shows that c, in units of feet and seconds, is approximately $0.000460/D$ feet^{-1}. This model is called the **Velocity-Squared Model**.

Problem 4

(a) The Velocity-Squared Model, like Stokes's Law, implies a terminal velocity v_{term}. Use the Velocity-Squared differential equation to calculate v_{term} as a function of the diameter D. (Be careful with the units.)

(b) What is the terminal velocity of a raindrop of diameter 0.05 inches? How does that compare with the drizzle drop? If the raindrop falls 3000 feet at terminal velocity, how long does it fall?

(c) Answer the questions in part (b) for drops twice as large.

(d) Answer the questions in part (b) for drops four times as large.

(e) The fastest known times for raindrops to fall 3000 feet are between one and two minutes. Are your answers consistent with that fact?

We won't attempt a direct attack on the Velocity-Squared equation until later in the course. In the meantime, we can state what you will find later as the solution: The function $v(t)$ defined by

$$v(t) = \sqrt{\frac{g}{c}\,\frac{1 - e^{-2\sqrt{gc}\,t}}{1 + e^{-2\sqrt{gc}\,t}}}$$

satisfies the Velocity-Squared equation and has the right initial value.

Problem 5

(a) Does the formula just stated as the solution confirm your calculation of v_{term} in Problem 4?

(b) Use the solution formula to write a formula for the ratio v/v_{term}.

(c) How long does it take each of the raindrops in parts (b), (c), and (d) of Problem 4 to reach 99% of terminal velocity?

(d) Was it reasonable to assume in parts (b), (c), and (d) of Problem 4 that these drops travel at terminal velocity for their entire fall? Why or why not?

CHAPTER

4

Differential Calculus and Its Uses

This chapter, approximately at the center of our first course in calculus, has a title that might well be the title of the whole course. Here we will consolidate and build on what we have accomplished already, laying the groundwork for the rest of the course.

By addressing problems that appeared naturally as differential equations, we found a need for addressing the (easier) inverse problem of calculating derivatives of known functions. The problems in this chapter lead directly to the need for calculating derivatives—in many cases, derivatives of functions we have not yet differentiated. As we find out how to differentiate these new functions, we will expand our repertoire of derivative formulas and thereby enhance our ability to solve differential equations as well.

We begin by studying what derivatives tell us about the graphs of functions and what graphs tell us about derivatives. We will find an intimate connection with problems of optimization, that is, of finding the best (or worst) way to do something. That connection leads to the problem of solving nonlinear algebraic equations, not a difficult task with a graphing calculator. When we ask what the calculator's Solve key might be doing, we find an elegant and practical application of the derivative called Newton's Method.

Throughout the chapter we study prototypical applications—manufacturing costs, growth of energy consumption, physical properties of light. In each case we find that our mathematical models require us to expand our repertoire of derivative formulas. By the end of the chapter, our growing list of formulas will include the Product Rule, the General Power Rule, and the most important derivative formula of all: the Chain Rule.

4.1 Derivatives and Graphs

We know that local linearity ties the derivative of a function f to the local slope of the graph of f. We should not be surprised, then, that derivative calculations can tell us a great deal about the shape of the graph of a function. In this section we explore this connection, with particular emphasis on the use of the derivative to identify highest and lowest points on a graph.

Maximum and Minimum Values

We begin by examining the relationship between the graph of a function and the values of the derivative.

Exploration Activity 1

A child releases a helium-filled balloon on a still day. The balloon rises for a while, and then, because of a slow leak, it descends to the ground. The graph in Figure 4.1 plots the vertical velocity of the balloon as a function of time.

(a) At what time after release was the balloon at its highest point?

(b) Sketch an approximate graph of the height as a function of time.

Figure 4.1 Velocity of the balloon as a function of time

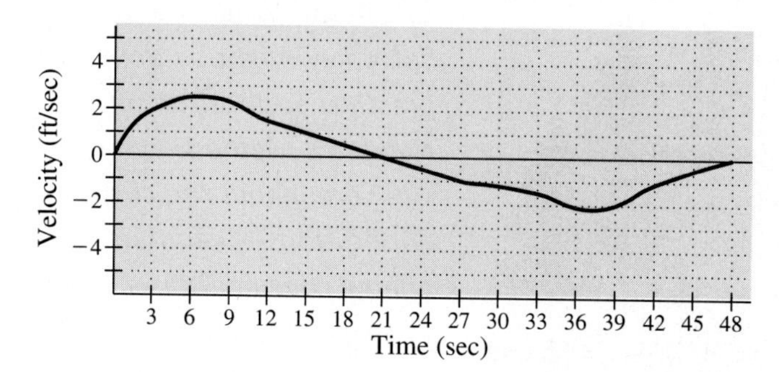

When the balloon is at its highest point, it stops rising and starts falling. Thus the velocity must be passing from positive values to negative values, i.e., the velocity must be zero. Checking our graph, we see that this happens at about 21 seconds after release.

In Figure 4.2 we sketch the height as a function of time. We know that the maximum height occurs when the velocity is zero. This correspondence is reflected in the geometry of the graphs in Figures 4.1 and 4.2. At its maximum, the height graph is horizontal. Thus the derivative — the velocity — must be zero.

Figure 4.2 Height of the balloon as a function of time

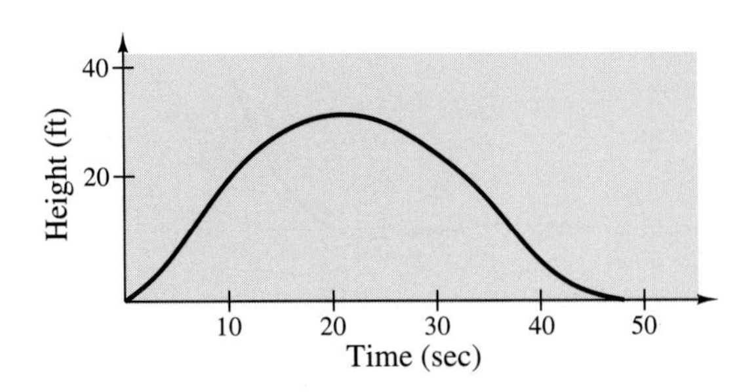

This correspondence between maximum values of a function and zeros of the derivative is one we will exploit often. For the moment, we'll illustrate this by describing how to sketch a graph by hand.

Example 1 Sketch the graph of the function $f(t) = t^4/4 - t^3 + t^2 - 1/6$ on the domain $-1 \le t \le 3$ by hand. (Suppose the batteries in your graphing calculator have run down!)

Solution We begin by calculating the derivative of f:

$$f'(t) = t^3 - 3t^2 + 2t = t(t-1)(t-2).$$

We can explore the sign of the derivative by examining the signs of the factors—sign changes can occur only where one of the factors changes sign. We see that the derivative is negative at $t = -1$ and remains negative until $t = 0$. At that point, the derivative switches over to positive values and remains positive until $t = 1$. Between $t = 1$ and $t = 2$, the derivative is negative again. Then at $t = 2$ the derivative becomes positive and remains that way.

Now we use this information to sketch the graph of f:

- At $t = -1$, the function value is $25/12$. As t increases, the graph decreases until it bottoms out at $(0, -1/6)$.
- Beyond $t = 0$, the graph starts to rise and continues upward until it reaches $(1, 1/12)$.
- At this point the graph turns downward again and decreases to $(2, -1/6)$.
- Here the graph turns upward and continues to rise for all $t > 2$.

Figure 4.3 shows a sketch of the graph of f drawn using this information. Check to see that this sketch reflects each point of the discussion.

Figure 4.3 Graph of $f(t) = t^4/4 - t^3 + t^2 - 1/6$

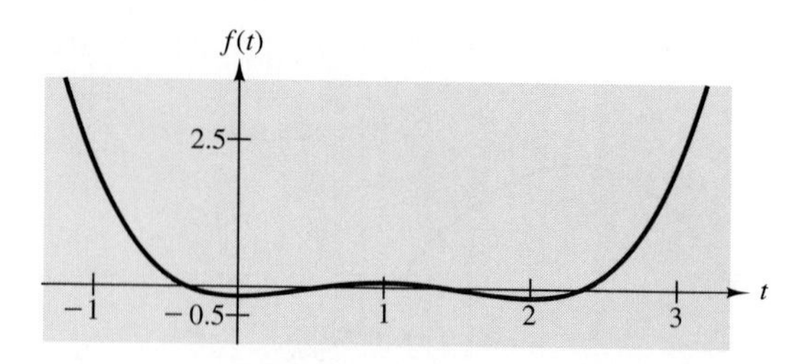

Now compare the graph of f with the graph of f' in Figure 4.4. Notice that two of the zeros of f' ($t = 0$ and $t = 2$) correspond to *minimum* values of f. The derivative must be zero at points where the graph stops increasing and starts decreasing—at maximum values—and it also must be zero at points where the graph stops decreasing and starts increasing—at minimum values. Thus the zeros of the derivative are helpful in locating both minimum and maximum values of the function.

Figure 4.4 $f'(t) = t^3 - 3t^2 + 2t$

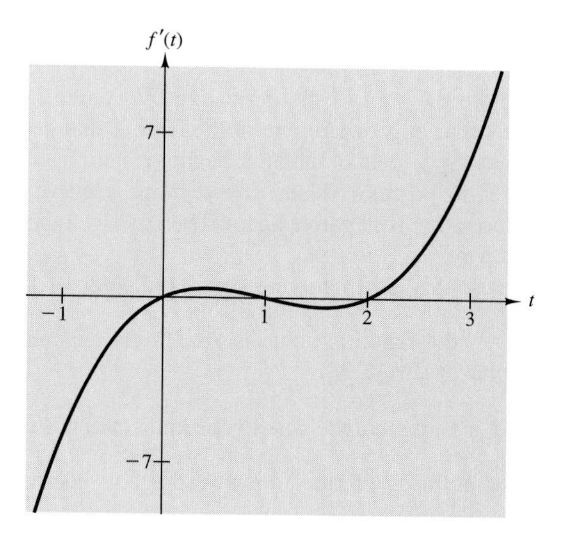

The remaining zero of the derivative at $t = 1$ corresponds to a maximum value. However, it is a maximum only relative to values at nearby points. The maximum value over the whole domain occurs at the two endpoints of the interval, where $f(t) = 25/12$. At these endpoints the derivative is not zero. These are maximum values only because we are not considering values of t less than -1 or greater than 3.

Definitions A function $f(t)$ has a **local maximum value** at $t = t_0$ if $f(t_0) \geq f(t)$ for all t close to t_0. The corresponding point on the graph of f is called a **local maximum point**. Similarly, the function $f(t)$ has a **local minimum value** at $t = t_0$ if $f(t_0) \leq f(t)$ for all t close to t_0. The corresponding point on the graph of f is called a **local minimum point**. Collectively, local maximum and minimum points are called **local extreme points**, and local maximum and minimum values are called **local extreme values**.

Figure 4.5 illustrates local extreme points that are not at the endpoints of a domain. Observe that this function clearly has values that are smaller (more negative) than the local minimum. If the domain continues to the right, it also may have values that are larger than the local maximum.

Figure 4.5 Local extreme points

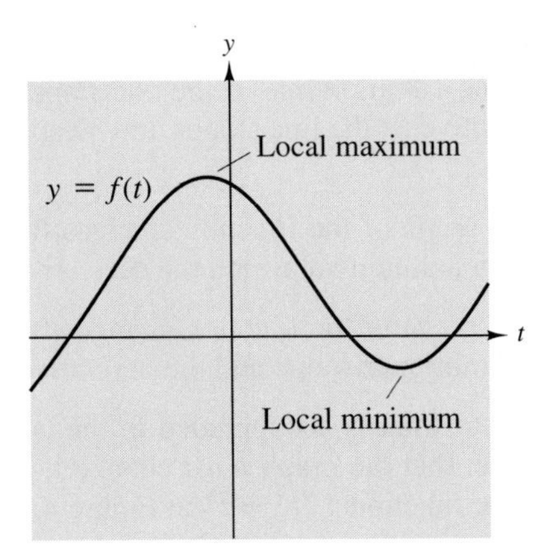

Example 2 Identify the local maxima and minima of $f(t) = t^4/4 - t^3 + t^2 - 1/6$ on the domain $-1 \leq t \leq 3$.

Solution Look again at the graph of $f(t) = t^4/4 - t^3 + t^2 - 1/6$ in Figure 4.6. The function has local minimum values at $t = 0$ and $t = 2$ and a local maximum value at $t = 1$, the three values of t at which $f'(t) = 0$. The function also has a local maximum value at the endpoints of the interval $t = -1$ and $t = 3$. At these points the derivative is not zero. These are local maximum points because $f(-1)$ and $f(3)$ are bigger than any nearby values of f *in the domain.*

Figure 4.6 Graph of $f(t) = t^4/4 - t^3 + t^2 - 1/6$

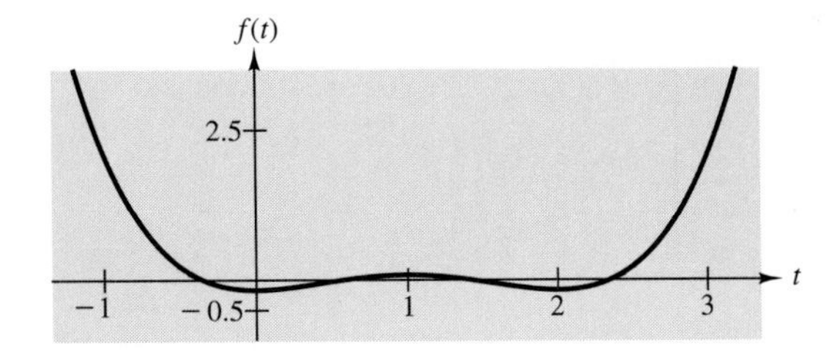

Checkpoint 1

Find the local maximum and minimum values of $f(t) = t^4/4 - t^3 + t^2 - 1/6$ on the domain $-\infty < t < \infty$.

Let's summarize what we have discovered about the relationship between the graph of a function f and values of its derivative f'.

- At points where the derivative of the function is positive, the graph of the function is increasing (from left to right).
- At points where the graph of the function is increasing, the value of the derivative is positive.
- At points where the derivative of the function is negative, the graph of the function is decreasing (from left to right).
- At points where the graph of the function is decreasing, the value of the derivative is negative.

These four statements reflect the local linearity of the graph. When we zoom in on a point of the graph, we see a straight line. If the line slopes upward, the graph is increasing and the derivative is positive. If the line slopes downward, the graph is decreasing and the derivative is negative.

- At points where the graph of the function changes from increasing to decreasing, the function has a local maximum value and the derivative is zero.
- At points where the graph of the function changes from decreasing to increasing, the function has a local minimum value and the derivative is zero.

It is important to note what is *not* included in the last two statements. A zero of the derivative does not mean that the graph *must* change its vertical direction. For example, consider the graph of the function $f(t) = t^3$ in Figure 4.7. The derivative of f at $t = 0$ is zero, but the graph goes from increasing to increasing. In other words, there is neither a local maximum nor a local minimum value for f at $t = 0$—even if we restrict our attention to just the vicinity of $t = 0$.

Figure 4.7 Graph of $f(t) = t^3$

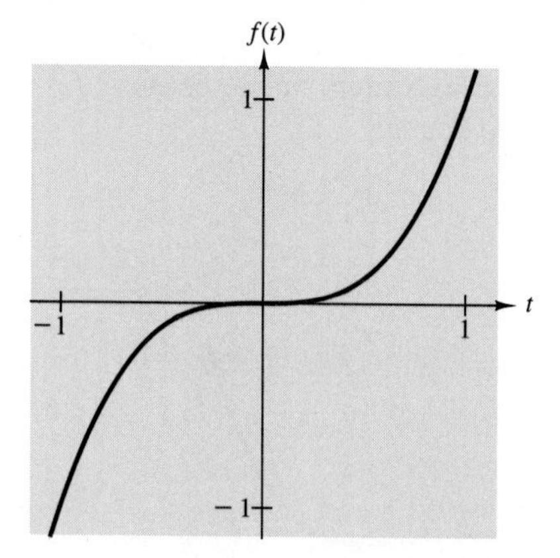

Now we concentrate on finding largest and smallest values of a function over a specified domain for the independent variable. This is the form in which differential calculus is usually applied to optimization problems. The local information provided by the derivative identifies candidates for the best or worst function values, and the endpoints (if any) identify additional candidates. Once we have a finite list of candidates, it is easy to check the value at each and choose the largest or smallest.

Example 3 Find the largest and smallest values of $f(t) = -t^3 + 12t$ on the domain $-4 \leq t \leq 3$.

Solution We display the graph of f in Figure 4.8. (Check the graph with your calculator.) It is clear from the graph that the smallest value occurs at one of the points where the derivative is zero, i.e., at a root of the equation $-3t^2 + 12 = 0$. These roots are $t = \pm 2$. The smallest value of f is clearly the local minimum that occurs at $t = -2$: $f(-2) = -32$. Note that the function also has a local minimum at $t = 3$, but $f(3)$ is clearly not the smallest (or largest) value of f.

Figure 4.8 Graph of $f(t) = -t^3 + 12t$

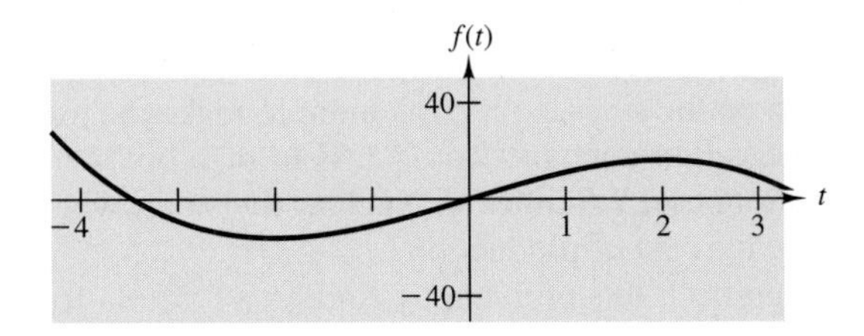

On the other hand, it is not clear from the graph whether the largest value occurs at $t = 2$ (the other zero of the derivative) or at $t = -4$ (the left-hand endpoint of the interval). In fact, $f(2)$ and $f(-4)$ are both equal to 16, which is the largest value of f on the given domain.

The largest and smallest values of a function and the corresponding points on the graph also get special names.

Definitions If $f(t_0)$ is greater than or equal to $f(t)$ for all values of t in the domain, then f has a **global maximum value** at $t = t_0$. Similarly, if $f(t_0)$ is less than or equal to $f(t)$ for all values of t in the domain, then f has a **global minimum value** at $t = t_0$. The corresponding points on the graph of f are called **global maximum points** and **global minimum points**, respectively. Collectively, global maximum and minimum points are called **global extreme points**, and global maximum and minimum values are called **global extreme values**.

We summarize what we have learned about the location of global extreme values of a function.

- A global extreme value (maximum or minimum) may occur at a point where the derivative is zero.
- A global extreme value (maximum or minimum) may occur at one of the boundary points of the domain.

Checkpoint 2

(a) Sketch the graph of the function $f(t) = t^4 - 4t^2 + 2$ over the domain $-2 \leq t \leq 2.1$.

(b) Find the maximum and minimum values of f on this domain.

Minimizing Production Cost

The Orlando Juice Company sells its product in distinctive rectangular paper boxes. The company would like to make the boxes as inexpensively as possible, without violating various design specifications. In particular, the boxes must be stronger on the top and bottom than on the sides, so it costs more to make the top and bottom.

The one-gallon container for O.J. is to be a box with a square top and bottom. If the top and bottom cost 2.5 times as much as the sides, what dimensions should the box have to minimize the cost of production?

We'll approach this problem by finding a formula for the cost of manufacturing a box of a given shape and then decide what shape has the lowest cost.

Figure 4.9 Orange juice box

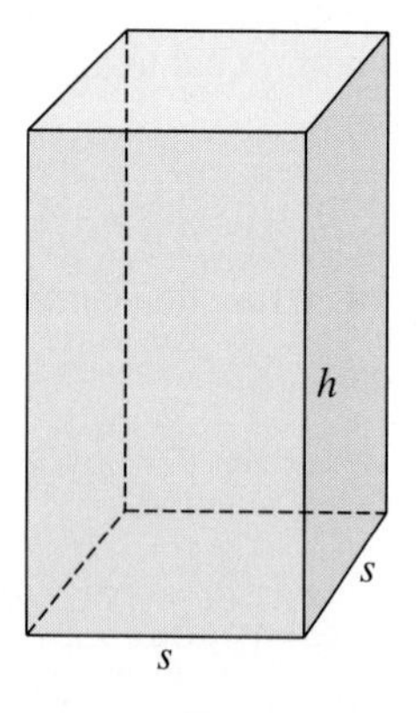

Suppose that both the square top and square bottom are s inches on a side and the height is h inches. (See Figure 4.9.) Then each of the four sides has surface area sh square inches, and each of the top and bottom has surface area s^2 square inches. If c is the cost per square inch of the sides, then the total cost is

$$C = 4shc + 2(2.5s^2)c.$$

We can express the cost in terms of s alone — by eliminating h. The volume of the box — one gallon — is 231 cubic inches. This volume is also s^2h, so $s^2h = 231$, and $h = 231/s^2$. Thus the cost of manufacturing the box is

$$C = 4\,\frac{231c}{s} + 2\big(2.5cs^2\big)$$
$$= c\left(\frac{924}{s} + 5s^2\right).$$

Now we want the value of s that minimizes C.

The value of c plays no role in the minimization calculation — just in what that minimum cost is. Thus we can concentrate on minimizing the weighted surface area function

$$S = \frac{924}{s} + 5s^2.$$

A graph of this function is shown in Figure 4.10.

Figure 4.10 Weighted surface area S versus side length s

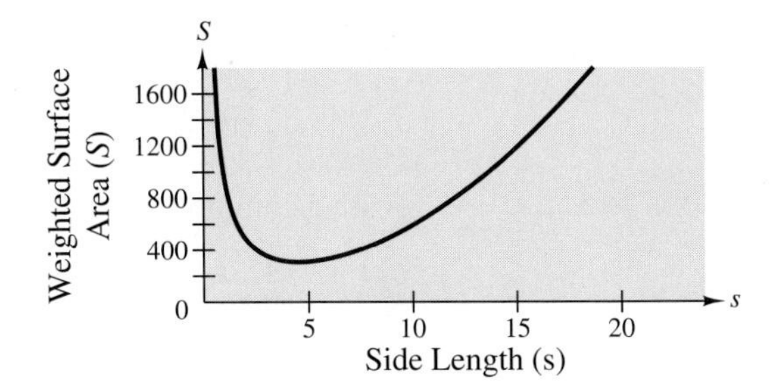

It is apparent that both large values of s (thin, wide boxes) and small values of s (tall, narrow boxes) yield large surface areas — and therefore large production costs. The minimum surface area seems to correspond to a value of s around 5 inches.

We could zoom in to try to obtain a better estimate of the optimal s, but the graph is very flat at this point. (Try it on your calculator.) Another approach is to calculate the derivative of the surface area function and find where this derivative is zero. The graph of dS/ds is plotted in Figure 4.11. Notice that this graph shows us that the optimal value of s is *not* 5.

Figure 4.11 The graph of dS/ds

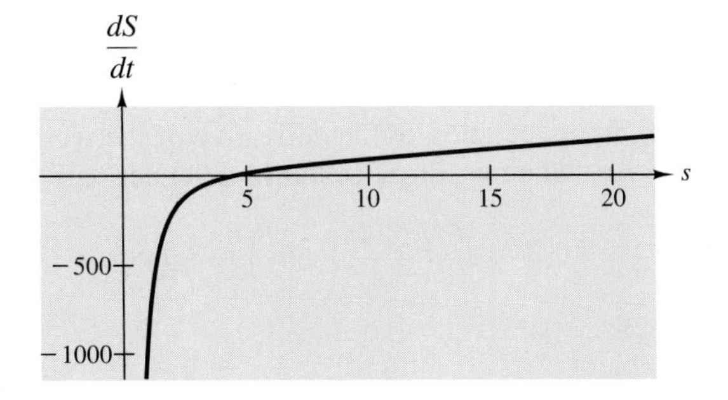

Here is the start of the derivative calculation:

$$\frac{d}{ds}\left(\frac{924}{s} + 5s^2\right) = 924\,\frac{d}{ds}\,\frac{1}{s} + 5\frac{d}{ds}s^2$$
$$= 924\,\frac{d}{ds}\,\frac{1}{s} + 10s.$$

To complete this calculation, we need a formula for the derivative of the reciprocal function:

$$\frac{d}{ds}\,\frac{1}{s}.$$

We digress briefly from the orange juice problem to work out the necessary formula.

The Derivative of the Reciprocal Function

If we set $u = 1/s$, then the difference quotient has the form

$$\frac{\Delta u}{\Delta s} = \frac{1/(s+\Delta s) - 1/s}{\Delta s}$$
$$= \frac{1}{\Delta s}\left(\frac{1}{s+\Delta s} - \frac{1}{s}\right).$$

If we rewrite $1/(s+\Delta s) - 1/s$ with a common denominator, we obtain

$$\frac{\Delta u}{\Delta s} = \frac{1}{\Delta s}\,\frac{s-(s+\Delta s)}{s(s+\Delta s)}$$
$$= \frac{1}{\Delta s}\,\frac{-\Delta s}{s(s+\Delta s)}$$
$$= -\frac{1}{s(s+\Delta s)}.$$

As Δs approaches zero, the difference quotient approaches $-1/s^2$. Thus the derivative of the reciprocal function is the negative of the reciprocal square function:

$$\frac{d}{ds}\,\frac{1}{s} = -\frac{1}{s^2}.$$

Completion of the Minimal Cost Calculation

We return to our calculation of the derivative of the weighted surface area function. Here was our progress when we digressed to find the derivative of the inverse function:

$$\frac{d}{ds}\left(\frac{924}{s} + 5s^2\right) = 924\,\frac{d}{ds}\,\frac{1}{s} + 5\,\frac{d}{ds}\,s^2$$
$$= 924\,\frac{d}{ds}\,\frac{1}{s} + 10s.$$

Now we continue:

$$\frac{d}{ds}\left(\frac{924}{s}+5s^2\right)=924\left(-\frac{1}{s^2}\right)+10s.$$

This is the function we graphed in Figure 4.11.

The minimum cost occurs when this derivative is zero:

$$-\frac{924}{s^2}+10s=0.$$

Solving for s, we find

$$s^3=92.4,$$

so $s \approx 4.52$ inches. The corresponding height $h=231/s^2$ is approximately 11.3 inches.

Exploration Activity 2

For the calculation just completed, the ratio of the cost of the top and bottom (per unit area) to the cost for the sides (per unit area) was 2.5. Let's call this ratio r. The Orlando Juice Company finds that r has been going up recently.

(a) Find a formula, with r as a parameter, for the length s that yields the least expensive box.

(b) The Sales Department has decreed that s must be restricted so that $4.25 \le s \le 6$. (Their research shows that customers don't buy boxes with values of s outside this range.) If the cost ratio r increases to 3.4, what size box should the company manufacture to minimize cost and stay within the Sales Department restrictions.

The formula for the total cost of a juice box with the top and bottom r times as expensive per unit area as the sides is

$$\begin{aligned} C &= 4\,\frac{231c}{s}+2\left(rcs^2\right) \\ &= c\left(\frac{924}{s}+2rs^2\right). \end{aligned}$$

Again, it is enough to minimize the formula for the weighted surface area,

$$S=\frac{924}{s}+2rs^2.$$

The optimal s is the one such that $dS/ds=0$. Since

$$\frac{dS}{ds}=-\frac{924}{s^2}+4rs,$$

the value of s that we want satisfies

$$-\frac{924}{s^2}+4rs=0.$$

Solving this equation, we find

$$s^3 = \frac{231}{r}$$

or

$$s = \left(\frac{231}{r}\right)^{1/3}.$$

Now, if we assume that r increases to 3.4, this formula gives $s = 4.08$. But this is outside the Sales Department's restrictions. It follows that there is no value of s in the interval $[4.25, 6]$ where dS/ds is zero. Thus the minimum value of S must occur at one or the other endpoint. Indeed, it occurs at $s = 4.25$. Figure 4.12 shows the graph of S versus s over this interval.

Figure 4.12 Graph of S versus s for $r = 3.4$

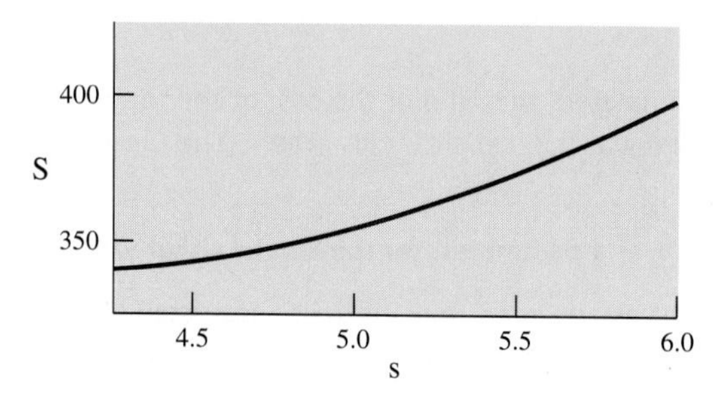

In this section we examined what we can learn about the graph of a function from a formula for the derivative. In particular, if the graph of f has a point (x_0, y_0) that is higher than any nearby points, then either x_0 is an endpoint of the domain or $f'(x_0) = 0$. Similarly, if the graph of f has a point (x_0, y_0) that is lower than any nearby points, then either x_0 is an endpoint of the domain or $f'(x_0) = 0$. Thus the natural candidates for maximum and minimum points of the graph are the endpoints of the domain and the values of the independent variable where the derivative is zero.

As part of our study of minimal cost of production, we also derived a differentiation formula for the reciprocal function. In fact, this formula turned out to be just like the Power Rule for nonnegative integer powers:

$$\frac{d}{dx}x^{-1} = (-1)\,x^{-2}.$$

Answers to Checkpoints

1. Local maximum at $t = 1$ and local minima at $t = 0$ and $t = 2$.

2. (a)

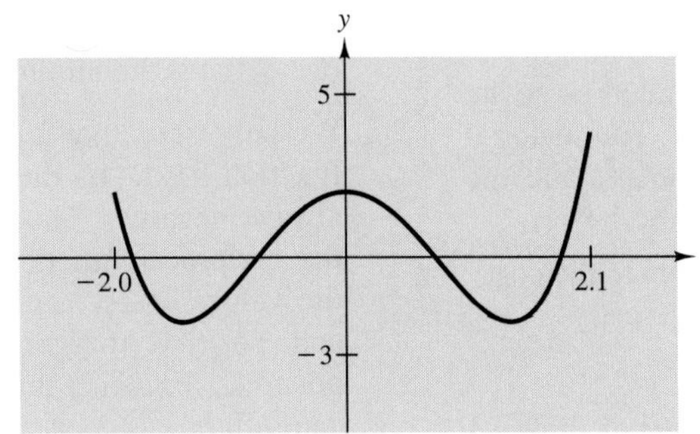

(b) The maximum value is $f(2.1) \approx 3.808$, and the minimum value is $f(\sqrt{2}) = f(-\sqrt{2}) = -2$.

Exercises 4.1

Calculate the derivative of each of the following functions.

1. $f(t) = e^{3t}$
2. $g(t) = e^{-3t}$
3. $h(t) = e^t - t^3$
4. $u(t) = t + 1$
5. $v(t) = t^2 + t + \dfrac{1}{t}$
6. $w(t) = t^3 + t^2 + t + 1$

For each of the following functions, (a) graph the function, (b) use your calculations in the corresponding Exercise from 1 through 6 to identify the range of values of t for which the graph is increasing, the range of values of t for which the graph is decreasing, and all local maxima and local minima.

7. $f(t) = e^{3t}$
8. $g(t) = e^{-3t}$
9. $h(t) = e^t - t^3$
10. $u(t) = t + 1$
11. $v(t) = t^2 + t + \dfrac{1}{t}$
12. $w(t) = t^3 + t^2 + t + 1$

13. Water is the only common liquid whose greatest density occurs at a temperature above its freezing point. (This phenomenon favors the survival of aquatic life by preventing ice from forming at the bottoms of lakes.) According to the *Handbook of Chemistry and Physics*, a mass of water that occupies one liter at 0°C occupies a volume of

$$1 + aT + bT^2 + cT^3 \text{ liters}$$

at T°C, where $0 \le T \le 30$, and where the coefficients are

$$a = -6.42 \times 10^{-5},$$
$$b = 8.51 \times 10^{-6}, \text{ and}$$
$$c = -6.79 \times 10^{-8}.$$

Find the temperature between 0 and 30°C at which the density of water is greatest.

14. Show that $3x^2 - 3x + 1$ is never negative. (Graphing the function with your calculator is not enough. You have to justify your answer.)

15. Find the number x between 0 and 4 where $f(x) = x^3 - 3x^2 + 9x + 5$ is the smallest. Explain your answer.

16. Let $P(x) = x^4 - 9x^3 + 20x^2 + 1$.
    (a) Find the maximum and minimum values of P on the interval $[0, 4]$.
    (b) Find the maximum and minimum values of P on the interval $[0, 5]$.

(c) Find the maximum and minimum values of P on the interval $[0, 6]$.

17. Suppose a can (a right circular cylinder) is to be made to hold 63 in^3 of Blastola Cola. The material for the sides and bottom costs 0.12¢/in^2, and the material for the easy-open top costs 0.5¢/in^2.

    (a) What height and radius will minimize the cost of the material?

    (b) What is the minimum cost?

18. Find formulas for the radius and height of the least expensive can in Exercise 17 if the top costs d¢/in^2 more than the sides and bottom. (In Exercise 17, d was 0.38.)

19. Suppose the value of d in Exercise 18 is 0.7; i.e., the easy-open top costs 0.92¢/in^2. Suppose also that design considerations restrict the radius to lie between 1.4 and 1.7 inches.

    (a) What should the height and radius of the can be to minimize the cost of the material?

    (b) What is the minimum cost?

20. We have seen that the derivative of the reciprocal function, $1/t$, is the function $-1/t^2$, which is always negative. That would seem to suggest that the reciprocal function is always decreasing. But the values of $1/t$ are negative when t is negative and positive (bigger than negative) when t is positive. Explain this apparent contradiction—carefully.

21. (a) Find the minimum value of the function

$$f(x) = 2|x - 1| - |x - 2| + \frac{1}{5}x^2.$$

    (b) Explain why the minimum does not occur at a value of x for which $f'(x) = 0$.

    (c) Graph the function.

4.2 Second Derivatives and Graphs

The Second Derivative

We get more information about the graph of a function if we differentiate the function twice. The derivative of a function $f(t)$ is a new function that we have denoted $f'(t)$ or

$$\frac{d}{dt} f(t).$$

For example, if

$$f(t) = t^3 - t,$$

then

$$\frac{d}{dt} f(t) = 3t^2 - 1.$$

This new function also has a derivative, which we may denote $f''(t)$ or

$$\frac{d}{dt}\left[\frac{d}{dt} f(t)\right].$$

For $f(t) = t^3 - t$, this new function is $6t$. It is common practice to discard the brackets in the second derivative notation and to replace the symbols dd with d^2 and $dtdt$ with

dt^2. This yields the notation

$$\frac{d^2}{dt^2}\,f(t).$$

For our example,

$$\frac{d^2}{dt^2}\,f(t) = 6t.$$

Definition The function

$$f''(t) \qquad \text{or} \qquad \frac{d^2}{dt^2}\,f(t)$$

obtained by differentiating f twice is called the **second derivative** of f.

Example 1 Calculate the second derivative of the function f defined by

$$f(t) = e^{3t} + t^3.$$

Solution Using the Sum, Exponential, and Power Rules, we calculate the first derivative:

$$\begin{aligned} f'(t) &= \frac{d}{dt}\,e^{3t} + \frac{d}{dt}\,t^3 \\ &= 3e^{3t} + 3t^2. \end{aligned}$$

Then we differentiate again, using the same rules:

$$\begin{aligned} f''(t) &= \frac{d}{dt}\,3e^{3t} + \frac{d}{dt}\,3t^2 \\ &= 9e^{3t} + 6t. \end{aligned}$$

We don't have to stop with first and second derivatives—we could talk of third derivatives, fourth derivatives, and so on. For now, we have no need for these higher derivatives,

$$\frac{d^3}{dt^3}\,f(t), \qquad \frac{d^4}{dt^4}\,f(t), \qquad \ldots,$$

and we confine our attention to first and second derivatives. However, higher derivatives will play an important role much later in the course.

Second derivatives are not entirely new in our development—we just didn't point them out before. When we were discussing falling bodies, we let $s(t)$ represent the distance an object had fallen at time t. The velocity $v(t)$ was the derivative of $s(t)$, and the acceleration $a(t)$ was the derivative of $v(t)$. That is,

$$a(t) = s''(t) = \frac{d^2 s}{dt^2}\,.$$

Graphs and the Second Derivative

Now let's think about how the graph of a function $f(t)$ and its *second* derivative $f''(t)$ are related. We know that the first derivative gives us information about slopes of small segments of the graph. The second derivative gives us information about the rate of change of these slopes. How is that information related to the shape of the graph?

Exploration Activity 1

(a) Use your graphing tool to plot the graph of f and the graph of f'' at the same time, where

$$f(t) = t^3 - t.$$

(b) Where is $f''(t)$ positive? What does the graph of f look like in this range?

(c) Where is $f''(t)$ negative? What does the graph of f look like in this range?

(d) Where is $f''(t)$ zero?

(e) Sketch the graph of f and then trace the graph with your finger. How does the curve change when you pass over the point where $f''(t)$ is zero?

(f) Repeat steps (a) to (e) for the function $g(t) = t^4 - 4t^2 + 2t$.

Your graphs in part (a) should combine our Figures 4.13 and 4.14. The second derivative f'' measures the rate of change of f', that is, the rate of change of *slope*. You should find that the slope of the function f is decreasing when t is negative and increasing when t is positive. If you imagine walking along the graph of f from left to right, "decreasing slope" means you are turning right, and "increasing slope" means you are turning left. We describe the graph in the first case as **concave down** and in the second case as **concave up**. Thus our function $f(t) = t^3 - t$ has a graph that is concave down for negative values of t and concave up for positive values of t. As you trace out the graph with your finger, you should find that you have to shift the direction of the heel of your hand when the concavity changes.

Figure 4.13 Graph of $f(t) = t^3 - t$

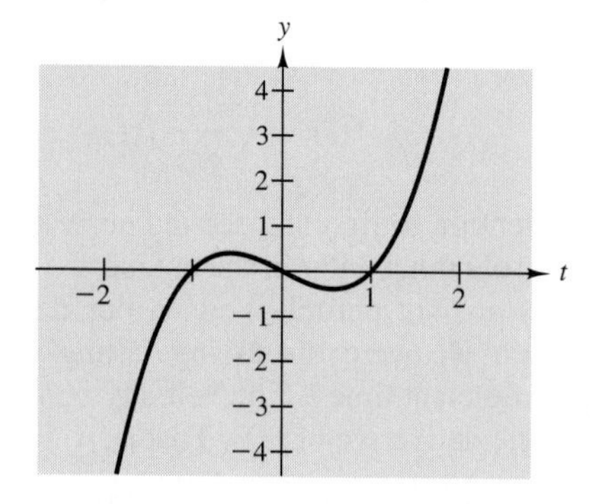

Figure 4.14 Graph of $f''(t) = 6t$

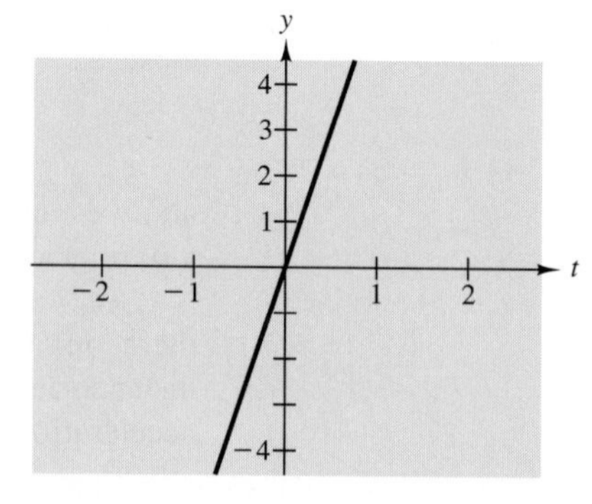

The second derivative of g is positive for $t < -1/2$, negative for $-1/2 < t < 1/2$, and positive again for $1/2 < t$. Thus the graph of g is concave up for $t < -1/2$, concave down for $-1/2 < t < 1/2$, and concave up again for $1/2 < t$. See Figures 4.15 and 4.16 for graphs of g and g''.

Figure 4.15 Graph of $g(t) = t^4 - 4t^2 + 2t$

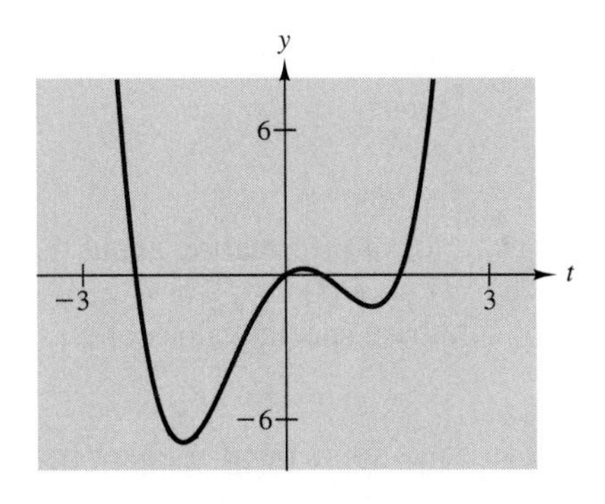

Figure 4.16 Graph of $g''(t) = 12t^2 - 8$

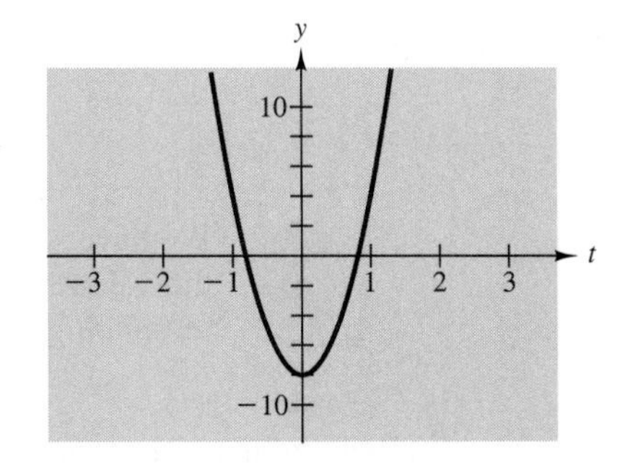

Checkpoint 1

In Figure 4.17 we show the graph of the function

$$h(t) = -t^3 + 12t.$$

(a) For what range of t does the graph appear to be concave up?

(b) For what range of t does the graph appear to be concave down?

(c) Calculate the second derivative $h''(t)$, and use algebra to check your answers to parts (a) and (b).

Figure 4.17 Graph of $y = -t^3 + 12t$

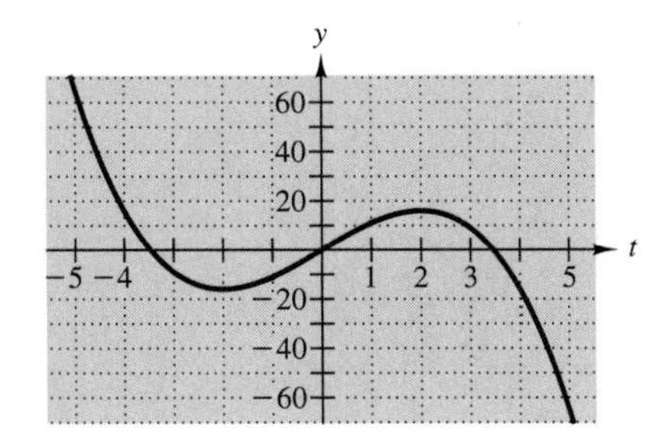

Finally, we note the interplay of first and second derivatives. For the function $f(t) = t^3 - t$ (see Exploration Activity 1), there are two places where the first derivative changes sign, and they are important for locating the local maximum and the local minimum on the graph of f. But we found them both by solving $f'(t) = 0$. In this case, it is pretty easy to tell which solution corresponds to the local maximum and which to the local minimum, because sketching the graph is easy. If the function were not so easily graphed — and we did not have a computer or graphing calculator at hand — how would we know which was which? Well, the maximum occurs where the graph is concave down (negative value of the second derivative) and the minimum where the graph is concave up (positive value of the second derivative). Thus the sign of the second derivative is sufficient to distinguish between peaks and valleys.

Checkpoint 2

(a) For the function $h(t) = -t^3 + 12t$ find the numbers t for which $h'(t) = 0$.

(b) Confirm that the second derivative identifies which of these inputs makes $h(t)$ a local maximum and which makes $h(t)$ a local minimum.

Terminology: Zeros, Critical Values, and Inflection Points

We have seen that we get a lot of information about the graph of a function f by finding the values of t at which $f(t)$ or $f'(t)$ or $f''(t)$ is zero. We summarize here the common terminology associated with these special values of t.

Definitions A **root** (or **solution**) of an equation in one variable t is a value of t that satisfies the equation. A **zero** of a function f is a root (or solution) of the equation $f(t) = 0$, that is, a value of the independent variable for which the corresponding function value is zero.

We have already seen that the zeros of a function tell us where the function might change from positive to negative values or vice versa. More specifically, they tell us where the graph either crosses or touches the horizontal axis. (See Figure 4.18.)

Figure 4.18 Crossing and touching the horizontal axis

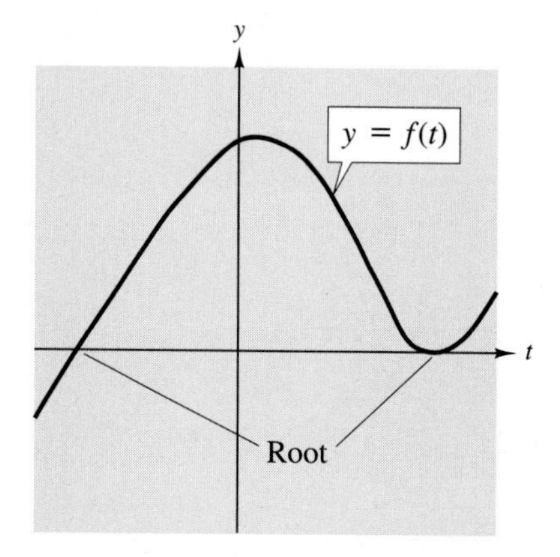

Also, we have seen that the zeros of f' [roots of $f'(t) = 0$] tell us where the graph of f might change from increasing to decreasing or vice versa. More specifically, they tell us where the graph of f might have a local maximum or a local minimum.

Definitions A number t at which $f'(t) = 0$ is called a **critical value** for f. A point on the graph of f at which $f'(t) = 0$ is called a **critical point**.

Now the zeros of f'' [roots of $f''(t) = 0$] tell us where the graph of f might change from concave down to concave up or vice versa. The points on the graph where a change of concavity occurs are given a special name.

Definition A point on the graph of a function at which a change of concavity occurs is called an **inflection point**.

The origin is both a critical point and an inflection point of the graph of $y = t^3$ (Figure 4.19). In particular, this is an example of a critical point that is neither a maximum nor a minimum point.

Figure 4.19 An inflection point

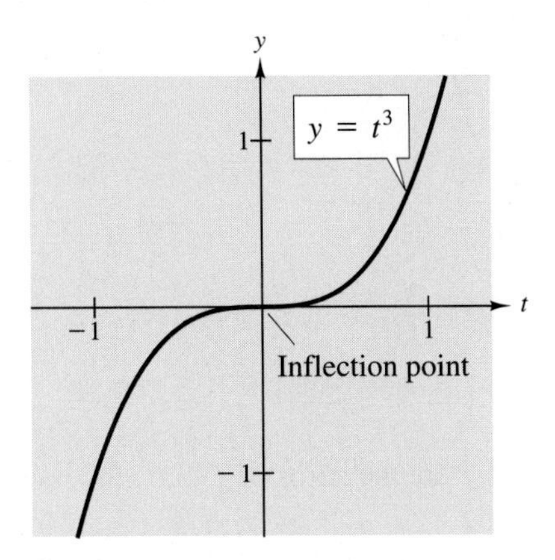

The origin is also an inflection point of the graph in Figure 4.20. In this case, there are two critical points, neither of them at the origin. (Where are they?)

Figure 4.20 An inflection point that is not at a critical value

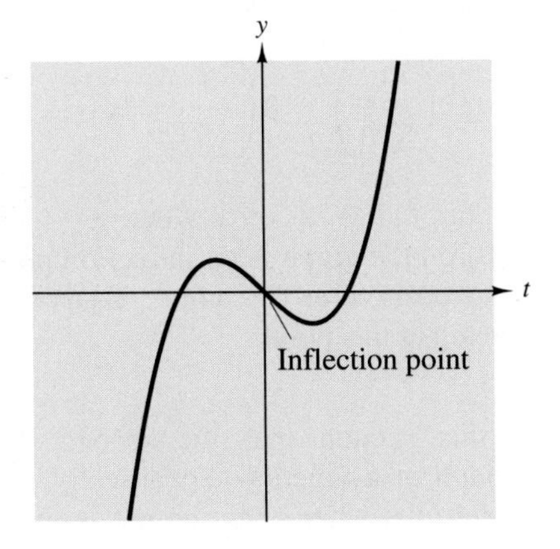

Example 2 Find the zeros, critical points, and inflection points for the function f defined by $f(t) = t^3 - t$. Determine what each tells us about the graph of f.

Solution The zeros are the roots of

$$t^3 - t = 0.$$

As

$$t^3 - t = t\,(t^2 - 1) = t\,(t-1)(t+1),$$

the zeros are $0,\ 1,$ and -1. (See Figure 4.21.) These are the numbers at which $f(t)$ changes sign, from negative to positive at -1, from positive to negative at 0, and from negative to positive at 1.

Figure 4.21 $f(t) = t^3 - t$

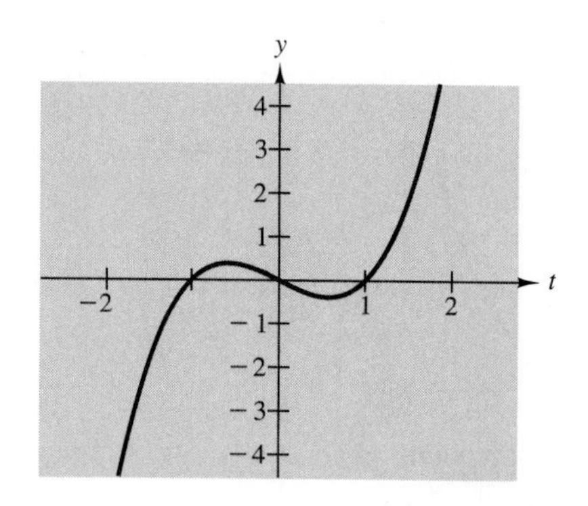

The critical values of f are the zeros of f', i.e., the roots of $3t^2 - 1 = 0$:

$$t = \frac{1}{\sqrt{3}} \quad \text{and} \quad t = -\frac{1}{\sqrt{3}}.$$

As we can see from the graph, f has a local maximum value at $t = -1/\sqrt{3}$ and a local minimum value at $t = 1/\sqrt{3}$. We can confirm this with the second derivative:

$$f''\left(-\frac{1}{\sqrt{3}}\right) = 6\left(-\frac{1}{\sqrt{3}}\right) = -\frac{6}{\sqrt{3}} < 0$$

and

$$f''\left(\frac{1}{\sqrt{3}}\right) = 6\left(\frac{1}{\sqrt{3}}\right) = \frac{6}{\sqrt{3}} > 0.$$

The second derivative of f has a zero when $6t = 0$, that is, when $t = 0$. The corresponding point $(0, 0)$ on the graph is an inflection point, since the second derivative changes from negative to positive at this point. Equivalently, the graph changes from concave down to concave up at this point. ■

We have seen in this section that the second derivative gives us additional information about the graph of a function. Specifically, positive and negative values of the second derivative tell us where the graph is concave up or concave down, respectively. If t is a critical value of f and the graph is concave up at the corresponding point on the graph, that point must be a local minimum point. If t is a critical value and the graph is concave down, the point must be a local maximum. Where the second derivative is 0, there may be an inflection point. If the second derivative changes sign, then this point *is* an inflection point.

Answers to Checkpoints

1. The graph is concave up for $t < 0$ and concave down for $t > 0$. $h''(t) = -6t$, which is positive for $t < 0$ and negative for $t > 0$.
2. (a) $h'(t) = -3t^2 + 12 = 0$ for $t = \pm 2$.

   (b) $h''(2) < 0$, which identifies a local maximum on the graph of h (concave down). $h''(-2) > 0$, which identifies a local minimum on the graph of h (concave up).

Exercises 4.2

Calculate the second derivative of each of the following functions.

1. $f(t) = e^{3t}$
2. $g(t) = e^{-3t}$
3. $h(t) = e^t - t^3$
4. $u(t) = t + 1$
5. $v(t) = t^2 + t + 1$
6. $w(t) = t^3 + t^2 + t + 1$

For each of the following functions, use your calculations in the corresponding Exercise from 1 through 6 to identify the range of values of t for which the graph is concave up and the range of values for which the graph is concave down.

7. $f(t) = e^{3t}$
8. $g(t) = e^{-3t}$
9. $h(t) = e^t - t^3$
10. $u(t) = t + 1$
11. $v(t) = t^2 + t + 1$
12. $w(t) = t^3 + t^2 + t + 1$
13. Figure 4.22 shows a graph of the polynomial function $f(t) = t^4 - 18t^2 - 10t + 39$.
    (a) Approximately where is $f(t) = 0$?
    (b) Approximately where is $f'(t) = 0$? On the graph of f (or a copy), mark these points.
    (c) On what intervals is f' positive?
    (d) On what intervals is f' negative?
    (e) Sketch a graph of f'.
    (f) Confirm your results in parts (b) through (e) by calculating f' and graphing it on your calculator.

Figure 4.22 $f(t) = t^4 - 18t^2 - 10t + 39$

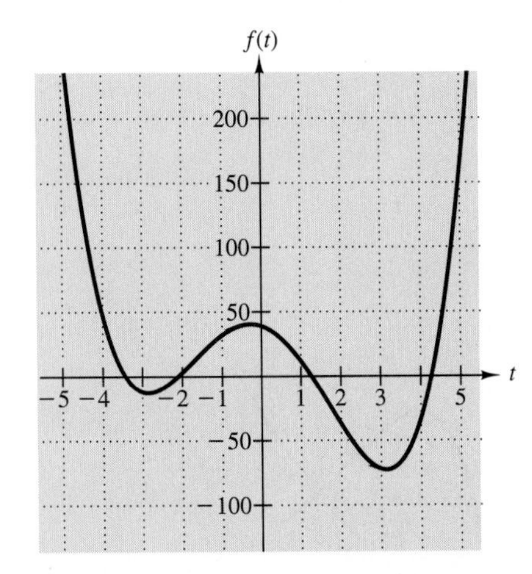

14. (a) For the function graphed in Figure 4.22, where (approximately) is $f''(t) = 0$? (If you wish, you can determine algebraically exactly where f'' is zero.)
    (b) On the graph of f (or a copy), mark the points where $f''(t) = 0$.
    (c) On what intervals is $f''(t)$ positive?
    (d) On what intervals is $f''(t)$ negative?
    (e) Sketch a graph of $f''(t)$.

(f) Confirm your results in parts (b) through (e) by calculating f'' and graphing it on your calculator.

15. (a) Find the zeros of the function f defined by the formula

$$f(t) = t^3 - 36t.$$

(b) Calculate f', and find the zeros of f'.

(c) Calculate f'', and find the value of f'' at each zero of f'.

(d) For each zero of f', decide whether the corresponding point on the graph is a local maximum, a local minimum, or neither.

(e) Decide whether the zero of f'' represents an inflection point.

(f) Sketch a graph of f.

16. The graphs of three functions appear in Figure 4.23. Identify which is the graph of f, which is the graph of f', and which is the graph of f''.

Figure 4.23 Identify the function f and its first two derivatives

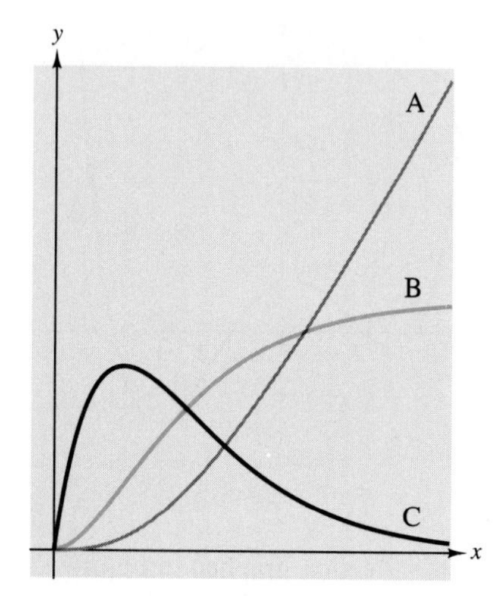

17. Imagine the graph of a polynomial of degree 3 or higher. Imagine that this graph is a map of a relatively flat road. Imagine that you are riding a bicycle along this road in the positive x-direction.
    (a) Describe what you experience as you pass through a point at which the second derivative of the polynomial is zero.
    (b) What are you doing along a stretch of the road on which the second derivative is positive?
    (c) What are you doing along a stretch of the road on which the second derivative is negative?

18. Sketch the graph of a function that is increasing at an increasing rate.

19. Sketch the graph of a function that is increasing at a decreasing rate.

20. Sketch the graph of a function that is decreasing at an increasing rate.

21. Sketch the graph of a function that is decreasing at a decreasing rate.

22. Figure 4.24 shows the graph of the *derivative* $f'(x)$ of an unknown function $f(x)$.
    (a) From the figures $\boxed{A}$–$\boxed{I}$ following, choose the one that shows the graph of $f(x)$.
    (b) From the figures $\boxed{A}$–$\boxed{I}$ following, choose the one that shows the graph of $f''(x)$.
    (c) Justify each of your choices in parts (a) and (b).

Figure 4.24 Graph of $f'(x)$

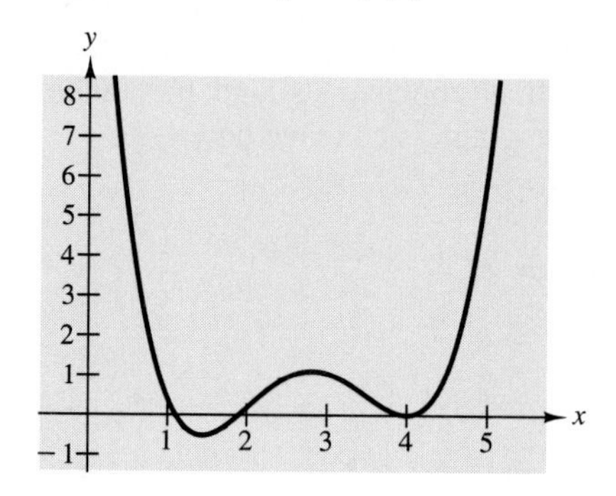

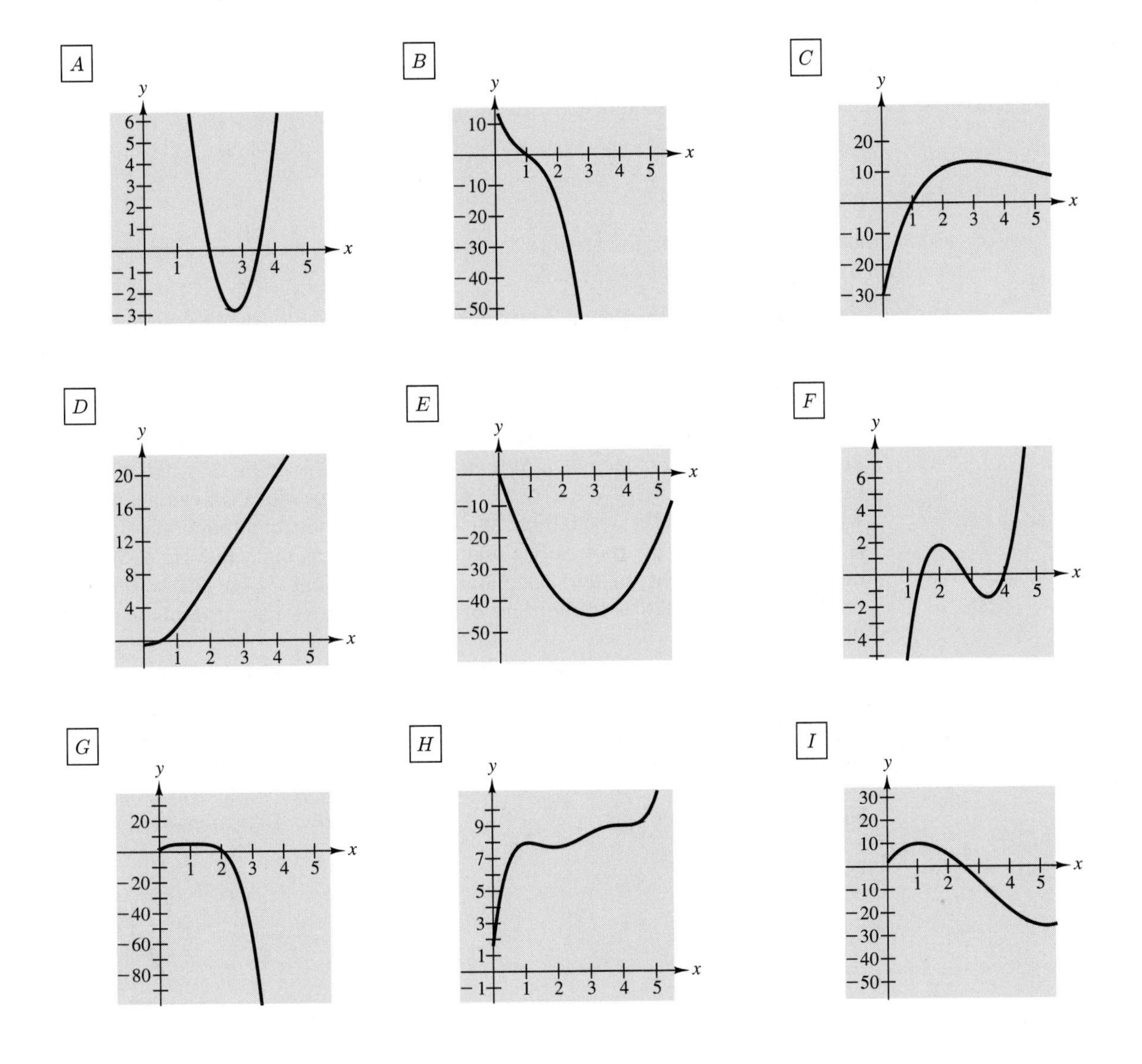

4.3 Solving Nonlinear Equations by Linearization: Newton's Method

We have seen that analysis of the graph of a function leads naturally to the problem of finding zeros. This is the same problem as finding roots or solutions of equations in one variable — not a hard problem if you have a graphing calculator with a Root or Solve key.

Exploration Activity 1

(a) Use your calculator to graph the function f defined by

$$f(t) = e^{3t} - t^2.$$

(b) Zoom in on the root near -0.5.

(c) Find numbers b and c such that you are sure $b < \text{root} < c$ and that the distance between b and c is less than 0.01.

(d) Use the Root or Solve procedure on your calculator to find an approximation to the root that is accurate to 10 significant digits.

Here's what happens with one particular graphing calculator. *The numbers you get and the keys you press may be different.* We show a first graph of f and a zoom-in graph in Figures 4.25 and 4.26. We trace the graph to see how close we can get to the root, and we find that $f(-0.489) = -0.009$. We see from the negative function value that -0.489 is just to the left of the root, so we may take -0.489 as our value for b. When we move one step to the right on our calculator graph, we find $f(-0.479) = 0.008$, which is positive. Thus -0.479 is to the right of the root, and we may use this value for c. Note that $c - b = 0.001$.

Now we press the Root key, and the calculator asks us for three inputs: First, a *lower bound*—we use b. Second, an *upper bound*—we use c. Finally, a *guess*—we use c again. Almost instantly, the calculator tells us that the root is approximately -0.4839075718, and the value of f at that input is -1×10^{-14}.

Figure 4.25 Graph of $f(t) = e^{3t} - t^2$

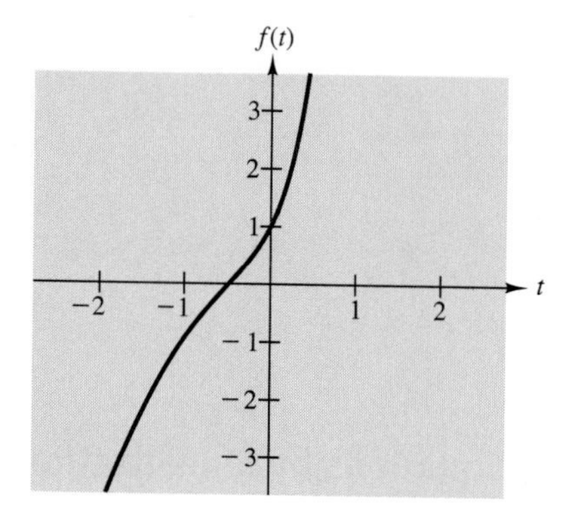

Figure 4.26 First zoom-in

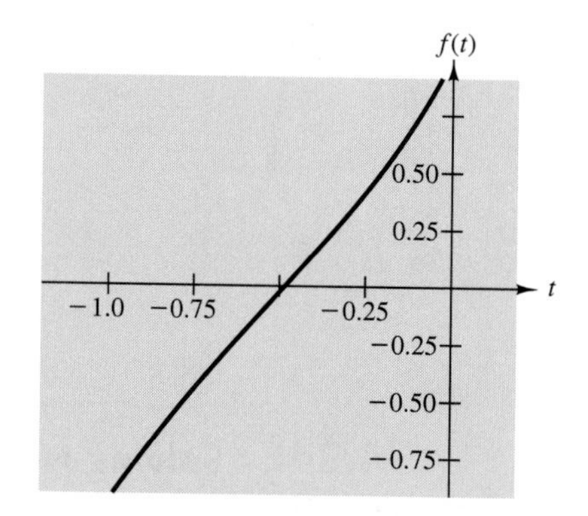

How did our calculator find a 10-decimal-place solution of a nonlinear equation? (If you think that's not an interesting question, consider this: Somebody studying this text—maybe you—could turn out to be an electronic engineer designing the next generation of calculators. Even if that's not your goal, you should know something about how your tools work.) Figure 4.26 suggests an answer: A small segment of the graph—in particular, a segment close to the root—is nearly a straight line. That is, our non-linear function is *locally linear*, so we can treat it like a linear function if the problem is sufficiently local.

Definitions An equation is **linear** if it can be put in the linear form $mt + b = 0$, where t is the variable and m and b are constants. Similarly, a function f is **linear** if the rule determining its values can be put in the linear form $f(t) = mt + b$. An equation is **nonlinear** if it cannot be put in the linear form, and a function f is **nonlinear** if the rule determining its values cannot be put in the form $f(t) = mt + b$.

For example, the equation $3t + 7 = 13$ is linear because it can be put in the form $3t - 6 = 0$. Similarly, the function f defined by $f(t) = 3t - 6$ is linear. On the other hand, the equation $e^{3t} - t^2 = 0$ is not linear, and neither is the function f defined by $f(t) = e^{3t} - t^2$. The graph of a linear function is a straight line. The graph of a nonlinear function (see Figure 4.25) is something other than a straight line.

Here is one possibility for what our calculator did in Exploration Activity 1. We identified the points $(-0.489, -0.009)$ and $(-0.479, 0.008)$ on the graph of f when we entered lower and upper bounds. These points are close to—and bracket—the point $(r, 0)$ where the graph of f crosses the t-axis. If the segment of the graph connecting these bounding points is sufficiently linear, then we can locate $(r, 0)$ as a point on the line segment connecting the bounding points. See Figure 4.27.

Figure 4.27 Local linearity near a root

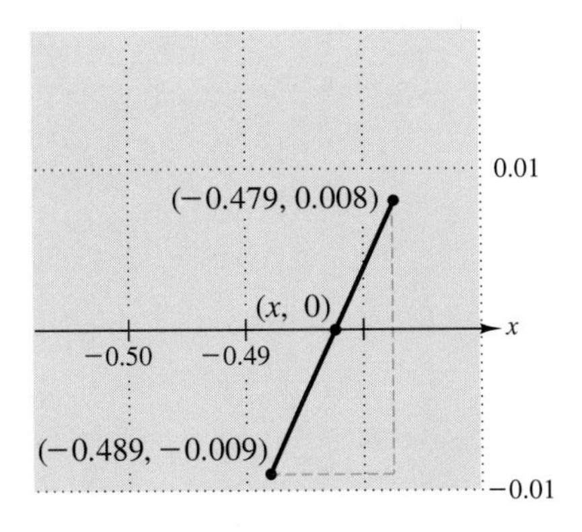

The slope of the line segment can be calculated in two ways as "rise over run." First, we calculate the slope using the upper right point, $(-0.479, 0.008)$, and the lower left point, $(-0.489, -0.009)$:

$$\frac{0.008 - (-0.009)}{-0.479 - (-0.489)} = 1.7.$$

Now we calculate the slope using the upper right point $(-0.479, 0.008)$ and the point of intersection with the t-axis $(r, 0)$:

$$\frac{0.008 - 0}{-0.479 - r}.$$

Since these two calculations of the slope must be the same, we may set them equal:

$$\frac{0.008 - 0}{-0.479 - r} = 1.7.$$

When we clear the fraction and solve for r (check the details), we get

$$r = -0.479 - \frac{0.008}{1.7} = -0.48370\ldots .$$

That's the right answer to 3 decimal places—the error is about 0.0002—but not to 10 places.

In principle, our calculator could repeat the process, using -0.4837 as the next guess. However, we are going to look next at a slight variation of this idea. Instead of using the slope of the line segment connecting two bracketing points, we will use the slope *at* our current guess, calculated as a value of the derivative.

Newton's Method

Here's our plan for finding a root r of the equation $f(t) = 0$: We start close to $t = r$, say, at $t = t_0$. We find the slope of the graph at the point $(t_0, f(t_0))$ as $f'(t_0)$. Our next guess will be the number t_1 at which a line of slope $f'(t_0)$ crosses the horizontal axis. That number won't necessarily be the solution of $f(t) = 0$ that we seek, but it will usually be much closer than t_0 was. Then we start over, replacing t_0 by t_1. The key to the success of this idea is the "much closer" that we achieve at each step, which usually makes it unnecessary to do more than a few steps.

In our example, $f(t) = e^{3t} - t^2$, so $f'(t) = 3e^{3t} - 2t$. We start at the point $(-0.479, 0.008)$, so our slope (see Figure 4.28) is

$$f'(-0.479) = 3e^{3(-0.479)} - 2(-0.479) = 1.670918828.$$

Figure 4.28 First approximation step

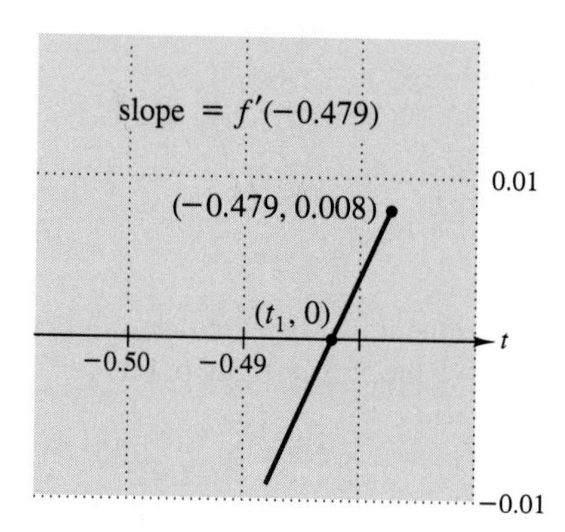

Now our calculation proceeds very much as before. Again, we equate two expressions for the slope: the value, 1.670918828, of the derivative and the slope of the line from the upper right point $(-0.479, 0.008)$ to the point $(t_1, 0)$ of intersection with the t-axis. Equating these two slope calculations, we find

$$1.670918828 = \frac{0.008 - 0}{-0.479 - t_1}.$$

When we clear the fraction and solve for t_1, we get

$$t_1 = -0.479 - \frac{0.008}{1.670918828} = -0.48378778\ldots .$$

This is a little better than before—the error is about 0.00012—but it's still not correct to 10 places. On the other hand, if we start again from t_1 and calculate the next guess t_2, we get

$$\begin{aligned} t_2 &= t_1 - \frac{f(t_1)}{f'(t_1)} \\ &= -0.4837\ldots - \frac{0.00020008\ldots}{1.67032768\ldots} \\ &= -0.4839075714. \end{aligned}$$

Now we're getting somewhere. The error, when we compare this answer with that from the calculator's Root key, is about 4×10^{-10}, so we're already at 9-place accuracy, with just two steps.

Let's outline the calculation in general. The slope of the graph of f at $t = t_0$ is $f'(t_0)$. The point $(t_1, 0)$ is also on the line through (t_0, y_0) with the same slope. Thus the slope also has the form "rise over run" or $(0 - y_0)/(t_1 - t_0)$, where y_0 represents $f(t_0)$. Now we set our two expressions for slope equal:

$$f'(t_0) = -\frac{y_0}{t_1 - t_0}.$$

And we solve for t_1:

$$\begin{aligned} t_1 - t_0 &= -\frac{y_0}{f'(t_0)}, \\ t_1 &= t_0 - \frac{y_0}{f'(t_0)}. \end{aligned}$$

This equation tells us how to get from t_0 to t_1: Subtract the ratio of the function value at t_0 to the slope at t_0. Look at Figure 4.29, and explain the calculation to yourself in these terms: *The run is the rise divided by the slope*. Be sure you understand where the minus sign comes from.

Figure 4.29 First two steps in Newton's Method for solving $f(t) = 0$

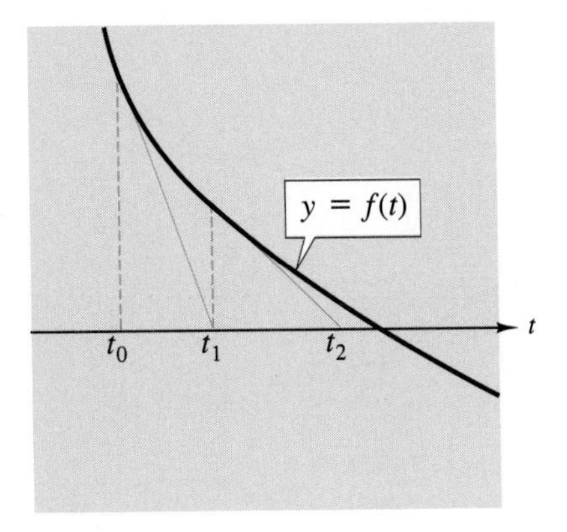

Now we get t_2 from t_1 in exactly the same way. Then t_3 from t_2, and so on. In fact, what we have here is an *iterative* formula for generating the numbers $t_1, t_2, t_3, \ldots$ from the starting point t_0:

$$t_{n+1} = t_n - \frac{f(t_n)}{f'(t_n)}.$$

This formula is generally known as **Newton's Method** for solving equations.

Example 1 Find a root of the equation $e^{-t} = t$.

Solution We start by setting $f(t) = e^{-t} - t$. A graph of this function (Figure 4.30) shows that there is a solution a little bigger than 0.5. Suppose we begin with a guess that is not very close, say, $t_0 = 0.1$. Then

$$y_0 = f(t_0) \approx 0.804837.$$

Now, $f'(t) = -e^{-t} - 1$ for all t, so

$$f'(t_0) = -e^{-0.1} - 1 \approx -1.904837.$$

Then

$$t_1 = t_0 - \frac{y_0}{f'(t_0)} = 0.1 - \frac{0.804837}{-1.904837} \approx 0.522523.$$

The next two steps, calculated in the same way, give

$$t_2 \approx 0.566778$$

and

$$t_3 \approx 0.567143.$$

Since $f(t_3) \approx 4.55 \times 10^{-7}$, it is likely that all six decimal places shown for t_3 are correct. Indeed, since the slope near the root is about -2, the difference between t_3 and the exact root (a run) should be about half the difference between $f(t_3)$ and zero (the corresponding rise). Run equals rise over slope.

Figure 4.30 $y = e^{-t} - t$

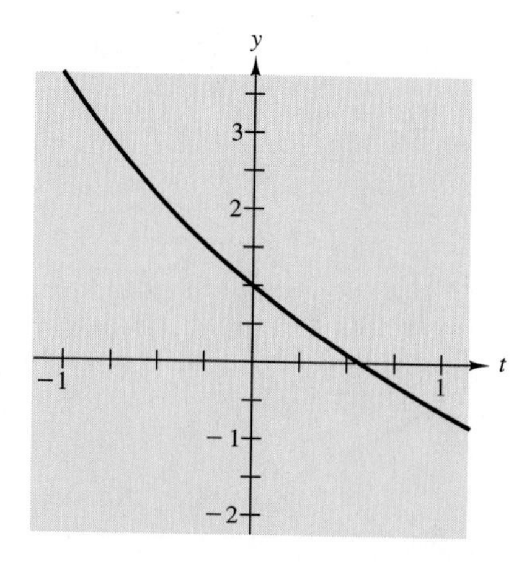

Our next example illustrates a lengthier calculation with Newton's Method. We will solve an equation $f(t) = 0$ without first zooming in to find a good starting guess.

Example 2 Find the smallest (left-most) solution of $t^4 - 18t^2 - 10t + 39 = 0$ without zooming in.

Solution Figure 4.31 shows the graph of the function $f(t) = t^4 - 18t^2 - 10t + 39$. From the graph we see that the solution is between -4 and -3. We could use either -4 or -3 as a starting point—let's use -4. We have

$$f'(t) = 4t^3 - 36t - 10.$$

Table 4.1 shows our calculations. Check the calculation of t_1 with your own calculator.

Figure 4.31 $f(t) = t^4 - 18t^2 - 10t + 39$

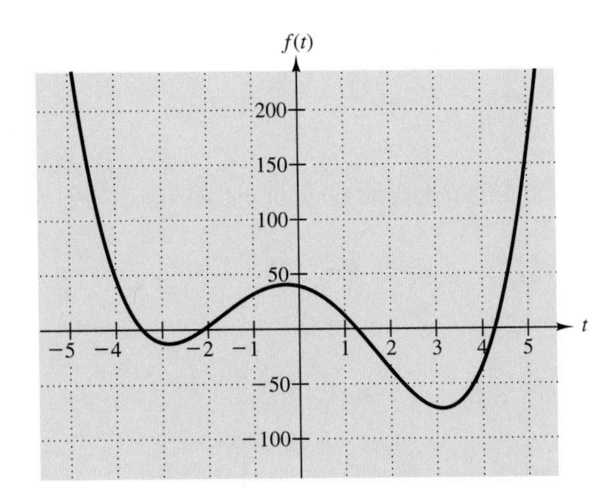

Table 4.1 Newton's Method calculation of a root

n	t_n	$f(t_n)$	$f'(t_n)$	$f(t_n)/f'(t_n)$
0	−4.0	47.0	−122.0	−0.385246
1	−3.614754	10.683527	−68.796815	−0.155291
2	−3.459463	1.402967	−51.069143	−0.027472
3	−3.431991	0.040318	−48.144002	−0.000837
4	−3.431154	0.000058	−48.055859	−0.000001
5	−3.431153	0.000058	−48.055859	−0.0000002

This is as far as we can go with six-decimal-place accuracy. The change in t in the last column is too small to affect the sixth place, which is why neither f nor f' changed in the last step. We can safely conclude that the solution we seek is $t = -3.43115$ to five-place accuracy. The next digit remains in doubt because of our rounding. ■

Notice how rapidly the changes in the last column got small. This only hints at how effective Newton's Method can be—a matter you may take up in a lab related to this chapter.

Checkpoint 1

Use Newton's Method to find four approximations to the zero of $f(t) = t^4 - 2t^2 + 5t - 3$ that lies between 0 and 2.

The real point of this section is not so much *how to find roots*—you already know how to do that with your calculator. Rather, the point is the continuing theme of *local linearity* of the well-behaved functions that we use for most of our models. We used this concept to define the derivative in Chapter 2 and to visualize differential equations via slope fields in Chapter 3. We have used it here to explain the principle of a fast, accurate, automatic root-finding method—which might be what your Solve or Root key does. We will use local linearity again in the next section to derive a new differentiation formula. And we will use it in the next chapter to generate solutions of initial value problems. That will not be your last opportunity to put this important concept to work—the *essence* of differential calculus is linear approximation of small segments of curves.

Answers to Checkpoint

1. The approximations you obtain depend on your initial value. For $t_0 = 1$, the first four approximations are

n	t_n
1	0.8
2	0.7663201
3	0.76579371
4	0.76579360

Exercises 4.3

1. (a) Use Newton's Method to find all the zeros of the function in Example 2:

$$f(t) = t^4 - 18t^2 - 10t + 39.$$

   (b) Check your answers by calculating the value of $f(t)$ at each zero.
   (c) Check your answers by zooming in on the graph.
   (d) Check your answers by using the Solve or Root program on your calculator.

2. (a) Use Newton's Method to find all three solutions of the equation $3x^2 = e^x$.
   (b) Check your answers by zooming in on the graph.
   (c) Check your answers by using the Solve or Root program on your calculator.

3. Let $f(t) = t^3 + 8t^2 + t - 6$.
   (a) Calculate the values of f at $t = -10,\ t = -5,\ t = 0,$ and $t = 2$.
   (b) Why do your calculations in part (a) show that f has at least three zeros?
   (c) How do you know that f does not have more than three zeros?
   (d) Use Newton's Method to find all zeros of f.
   (e) Check your answers by using the Solve or Root program on your calculator.

4. Let $f(t) = e^t + e^{-t} - 2t^2$.
   (a) Find the four zeros of f.
   (b) Find the three zeros of f'.
   (c) Find the smallest value of f and the two values of t where f has its smallest value.
   (d) Find the largest value of f between $t = -1$ and $t = 1$.

(e) Use the information from parts (a) through (d) to sketch the graph of f. Use your calculator to check your graph.

5. (a) Use Newton's Method to find the root of $t^2 - e^{-3t} = 0$.
(b) Check your answers by using the Solve or Root program on your calculator.

Use Newton's Method to find (approximately) all solutions of each of the following equations. Check your answers by using the Solve or Root program on your calculator.

6. $x^3 + 5x = 10$
7. $x^3 + 5x = e^x$
8. $x^3 + 5x = -e^x$
9. $x^3 - 5x = 10$
10. $x^3 - 5x = e^x$
11. $x^3 - 5x = -e^x$
12. Let $f(t) = t^4 - 10t^2 + e^{2t}$.
(a) Find all the zeros of $f(t)$. (There are four.)
(b) Find all the zeros of $f'(t)$. [There are three— how must they be related to the zeros of $f(t)$?]
(c) Find all the zeros of $f''(t)$. [There are two— how must they be related to the zeros of $f'(t)$?]
(d) Sketch the graphs of $f(t)$, $f'(t)$, and $f''(t)$. (Use your calculator to check.)
(e) Find all local maximum points and all local minimum points of the graph of f.
(f) Find all inflection points of the graph of f.
13. If Newton's Method is applied to $f(x) = 3x + 4$, and x_0 is chosen to be 15, what will x_{17} be?
14. Find at least one solution to the equation $4x^3 - x^4 = 30$, or explain why no such solution exists.
15. (a) Suppose you are using Newton's Method to find a zero of the function f whose graph is shown in Figure 4.32. If your starting point is at $x_0 = 2.5$, label where (approximately) x_2 will be. (You may copy the page first.)
(b) Suppose you try again with Newton's Method for the same function f. If your starting point is at $x_0 = 5$, label where (approximately) x_2 will be.

Figure 4.32 Start Newton's Method (a) at 2.5 and (b) at 5

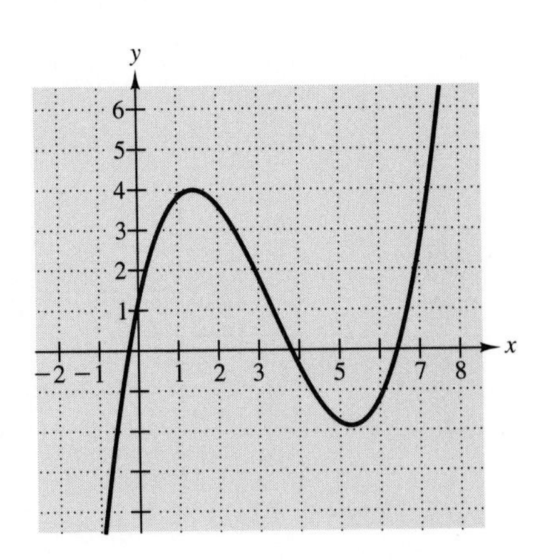

16. Figure 4.33 shows a slope field for $dy/dt = 0.3y(3 - y)$ for t ranging from -2 to 6 and y also ranging from -2 to 6. Figure 4.34 shows a slope field for $dy/dt = 0.3t(3 - t)$ with both t and y ranging from -2 to 5. You may copy this page, if necessary, for drawing on the slope fields.
(a) On Figure 4.33 sketch the solution of $dy/dt = 0.3y(3 - y)$ for which $y(0) = 0.5$.
(b) On Figure 4.34 sketch the solution of $dy/dt = 0.3t(3 - t)$ for which $y(0) = 0.5$.
(c) Solve the initial value problem: $dy/dt = 0.3(3t - t^2)$ with $y(0) = 0.5$.
(d) Your graph in part (b) should show that the solution in part (c) has a root near $t = 5$. Take that as a starting guess, and find the next approximation to the root from Newton's Method.

Figure 4.33 $dy/dt = 0.3y(3 - y)$

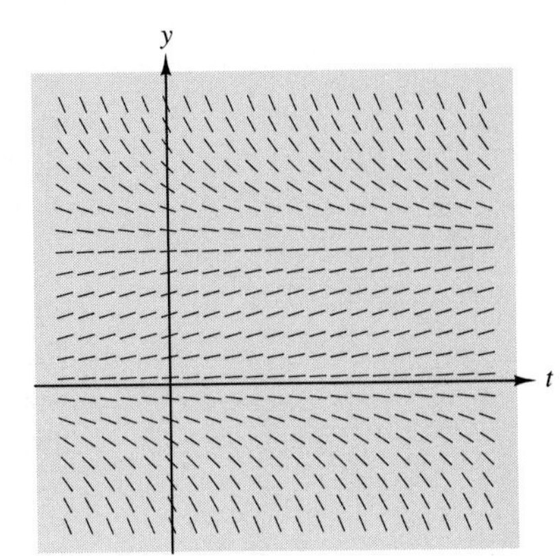

Figure 4.34 $dy/dt = 0.3t(3 - t)$

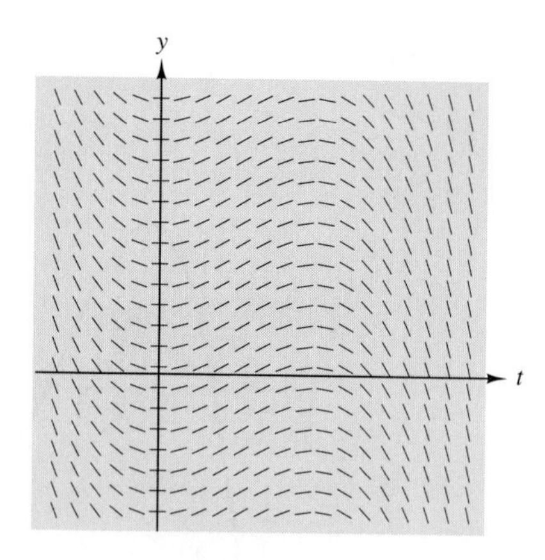

17. The square root of a number A can be calculated by solving the equation $x^2 - A = 0$ numerically.

    (a) Show that application of Newton's Method to this equation leads to a simple repeated averaging scheme. What is being averaged at each step?

    (b) Start with $x_0 = 1$, and compute $\sqrt{7}$ by this scheme. Compare your result with that obtained from the square root key on your calculator. Could your calculator actually be doing repeated averaging?

4.4 The Product Rule

The first three sections of this chapter have given us some new reasons to be interested in calculating derivatives: They provide us important information about and interpretations of the graphs of functions, and they play an important role in root finding via Newton's Method. So far we know how to calculate derivatives of polynomial functions and of exponential functions. But that hardly accounts for all the functions of interest. In the remainder of the chapter we concentrate on expanding our set of tools for computing derivatives. We start with a problem that shows the need for our first step in this development.

The Growth Rate of Energy Consumption[1]

The population of the United States (in millions) from 1982 to 1990 is closely approximated by an exponential growth formula of the form

$$P = P_0 e^{kt},$$

where time t is measured in years from 1982 (i.e., $t = 0$ in that year), $k = 0.00917$, and $P_0 = 232$. That is, a population of 232 million in 1982 grew at a rate of just under 1% per year for the next eight years. The per capita energy consumption over the same period (in millions of BTUs per person) is roughly approximated by a linear formula of the form

$$E = mt + b,$$

1. Based on data from the U.S. Commerce Department. Problem suggested by John Goebel, North Carolina School of Science and Mathematics.

where $m = 3.4$ and $b = 302$. Thus the total energy use T is approximated by the *product* of these two functions:

$$T = PE = (mt + b)P_0 e^{kt}.$$

Suppose we want to estimate the growth rate of energy consumption at some time in the 1980s. How can we calculate dT/dt? We know how to differentiate the population and per-capita-energy functions separately, but how does that help us find the derivative of their product?

A General Rule for Differentiating Products

To see that the problem is a real one, let's practice with some products of functions for which we already know the derivative of each factor *and* the product.

Exploration Activity 1

(a) Suppose $g(x) = x$, and $h(x)$ also equals x. Let $f(x) = g(x)h(x)$; thus $f(x) = x^2$. What are the derivatives of $f(x)$, $g(x)$, and $h(x)$?

(b) Suppose $g(x) = x$ and $h(x) = x^2$. Let $f(x) = g(x)h(x)$; thus $f(x) = x^3$. What are the derivatives of $f(x)$, $g(x)$, and $h(x)$?

(c) Suppose $g(x) = e^{2x}$ and $h(x) = e^{3x}$. Let $f(x) = g(x)h(x)$. What is $f(x)$ as an explicit function of x? What are the derivatives of $f(x)$, $g(x)$, and $h(x)$?

(d) Try to give a formula describing $f'(x)$ in terms of $g'(x)$ and $h'(x)$ that is valid for each of parts (a), (b), and (c).

For part (b), $f(x) = g(x)h(x)$, where $g(x) = x$ and $h(x) = x^2$. Since $f(x) = x^3$, we know $f'(x) = 3x^2$. On the other hand, $g'(x) = 1$ and $h'(x) = 2x$. Thus it is clear that $f'(x)$ is *not* $g'(x)h'(x)$, i.e., the obvious candidate for a product formula is wrong. Parts (a) and (c) show the same thing. We will see shortly that the correct formula is

$$f'(x) = g(x)h'(x) + g'(x)h(x).$$

For example, in part (b),

$$g(x)h'(x) + g'(x)h(x) = 1 \cdot x^2 + x \cdot 2x = 3x^2.$$

Let's try this out for part (c). Here, $f(x) = g(x)h(x)$, where $g(x) = e^{2x}$ and $h(x) = e^{3x}$. For this example,

$$g(x)h'(x) + g'(x)h(x) = e^{2x} \cdot 3e^{3x} + 2e^{2x} \cdot e^{3x} = 5e^{5x}.$$

On the other hand, $f(x) = e^{5x}$ and $f'(x) = 5e^{5x}$.

Now let's see why this rule holds in general. Suppose $f(x) = g(x)h(x)$, where g and h are any functions that have derivatives. We attempt to find $f'(x)$—a rate of growth—as approximately a rise divided by a run. From a starting x, a run of Δx changes the independent variable to $x + \Delta x$. This also changes the value of f from $f(x) = g(x)h(x)$ to $f(x + \Delta x)$, which is equal to $g(x + \Delta x)h(x + \Delta x)$.

To understand how changes in g and h affect the product f, we approximate those changes by local linearity. We have

$$\text{Rise} \approx \text{slope} \times \text{run}$$

or

$$g(x+\Delta x) - g(x) \approx g'(x)\Delta x$$

and

$$h(x+\Delta x) - h(x) \approx h'(x)\Delta x.$$

We solve these approximate equalities for the new values of g and h:

$$g(x+\Delta x) \approx g(x) + g'(x)\Delta x$$

and

$$h(x+\Delta x) \approx h(x) + h'(x)\Delta x.$$

Now we multiply the last two expressions to get an approximate description of the new value of f:

$$\begin{aligned} f(x+\Delta x) &= g(x+\Delta x)h(x+\Delta x) \\ &\approx [g(x) + g'(x)\Delta x][h(x) + h'(x)\Delta x] \\ &= g(x)h(x) + g(x)h'(x)\Delta x + h(x)g'(x)\Delta x + g'(x)h'(x)(\Delta x)^2 \\ &= f(x) + g(x)h'(x)\Delta x + h(x)g'(x)\Delta x + g'(x)h'(x)(\Delta x)^2. \end{aligned}$$

Next we approximate the rise in f by subtracting $f(x)$ from both sides:

$$f(x+\Delta x) - f(x) \approx g(x)h'(x)\Delta x + h(x)g'(x)\Delta x + g'(x)h'(x)(\Delta x)^2.$$

From the rise, we can calculate the rate of growth by dividing both sides by the run:

$$\frac{f(x+\Delta x) - f(x)}{\Delta x} \approx g(x)h'(x) + h(x)g'(x) + g'(x)h'(x)\Delta x.$$

Finally, we consider what happens as the run Δx shrinks to zero. On the left, we get $f'(x)$. On the right, nothing happens to the first two terms, but the third term approaches zero. And the approximation becomes *exact*:

$$f'(x) = g(x)h'(x) + h(x)g'(x).$$

The Product Rule (Functional Notation) If $f(x) = g(x)h(x)$, then

$$f'(x) = g(x)h'(x) + h(x)g'(x).$$

The Product Rule (Variable Notation) If $u = g(x)$, $v = h(x)$, and $w = g(x)h(x)$, then

$$\frac{dw}{dx} = u\frac{dv}{dx} + v\frac{du}{dx}.$$

In words, the Product Rule says: "The derivative of a product is the first factor times the derivative of the second plus the second factor times the derivative of the first."

More on the Growth Rate of Energy Consumption

We can use our energy consumption example to interpret the Product Rule. Recall that U.S. population in millions is approximated by

$$P = P(t) = P_0 e^{kt},$$

where time t is measured in years from 1982, $k = 0.00917$, and $P_0 = 232$. Per capita energy consumption (in millions of BTUs per person) is approximated by

$$E = E(t) = mt + b,$$

where $m = 3.4$ and $b = 302$. Thus total energy use $T = T(t)$ is approximated by the product

$$T = PE = (mt + b)P_0 e^{kt}.$$

We wanted to calculate how fast total energy use in the United States is growing.

We know the growth rate of population,

$$\frac{dP}{dt} = kP_0 e^{kt},$$

and of per capita energy consumption,

$$\frac{dE}{dt} = m.$$

The Product Rule then tells us that

$$\begin{aligned} \frac{dT}{dt} &= P\,\frac{dE}{dt} + E\,\frac{dP}{dt} \\ &= \left(P_0 e^{kt}\right)m + (mt + b)\left(kP_0 e^{kt}\right) \\ &= P_0(kmt + kb + m)e^{kt}. \end{aligned}$$

When we substitute numeric values for k, m, and P_0, we find an explicit formula for the derivative:

$$\frac{dT}{dt} = (7.23t + 1430)e^{0.00917t}$$

If we want to know, for example, the growth rate of U.S. energy consumption in 1985, we substitute $t = 3$ to get about 1.5 billion BTUs per year. (Check the details.)

More important than specific numbers is the formula

$$\frac{dT}{dt} = P\,\frac{dE}{dt} + E\,\frac{dP}{dt},$$

which tells us that the energy consumption growth rate has two components. The first, $P\ dE/dt$, represents population times growth rate of per capita consumption. The second, $E\ dP/dt$, is per capita consumption times growth rate of the population.

Example 1 Find the derivative of $w = x^3 e^{4x}$.

Solution If we set $u = x^3$ and $v = e^{4x}$, we have $du/dx = 3x^2$ and $dv/dx = 4e^{4x}$. Thus

$$\begin{aligned}\frac{dw}{dx} &= u\,\frac{dv}{dx} + v\,\frac{du}{dx}\\ &= x^3\left(4\,e^{4x}\right) + e^{4x}\left(3x^2\right)\\ &= (4x^3 + 3x^2)e^{4x}.\end{aligned}$$

Checkpoint 1

Find the derivative of each of the following functions.

(a) $f(x) = x^2 e^{3x}$

(b) $f(x) = (0.3x + 2.5)e^{x/2} - 7x$

In this section, by examining locally linear change in the factors, we have discovered how to find the rate of change of a product. This is summarized in the **Product Rule**: If $f(x) = g(x)h(x)$, then

$$f'(x) = g(x)h'(x) + h(x)g'(x).$$

In variable notation, if $w = f(x)$, $u = g(x)$, and $v = h(x)$, then

$$\frac{dw}{dx} = u\,\frac{dv}{dx} + v\,\frac{du}{dx}.$$

Each of these formulations tells us that the growth rate of a product is made up of two terms, and each term is one of the factor functions times the growth rate of the other.

Answers to Checkpoint

1. (a) $(2x + 3x^2)e^{3x}$ (b) $(0.15x + 1.55)e^{x/2} - 7$

Exercises 4.4

Calculate the derivative of each of the following functions.

1. $t^2 e^t$
2. $(t^3 + t)e^{-3t}$
3. $(t - 3)e^{2t}$
4. $x^3 e^x$
5. $(x^2 - x)e^{2x}$
6. $(x + 2)e^{-x}$
7. $2^x\,(x^2 + 1)$
8. $7t^3 - 5t^2 + 13t - \sqrt{2}$
9. $\left(\frac{3}{5}\right)^t$

Calculate the second derivative of each of the following functions.

10. $t^2 e^t$

11. $(t^3 + t)e^{-3t}$
12. $(t - 3)e^{2t}$
13. $x^3 e^x$
14. $(x^2 - x)e^{2x}$
15. $(x + 2)e^{-x}$
16. $2^x\,(x^2 + 1)$
17. $7t^3 - 5t^2 + 13t - \sqrt{2}$
18. $\left(\frac{3}{5}\right)^t$

Calculate each of the following derivatives.

19. $\frac{d}{dx}\, x\, e^x$
20. $\frac{d^2}{dx^2}\, x\, e^x$
21. $\frac{d}{dy}\,(y + 1)e^{-2y}$
22. Given $y = (1 + t^2)(t^3 - 3t^2 + 1)$, calculate dy/dt two ways:
    (a) using the Product Rule.
    (b) multiplying the two factors before differentiating.
    Verify that the two answers you get are equivalent.
23. (a) Calculate dy/dt, where $y = (1 + 2t)e^{-3t}$.
    (b) Check your formula for the derivative at three different values of t by calculating $\Delta y/\Delta t$ for an appropriately small Δt.
24. Use Newton's Method to find the smaller root of $f(x) = 0$, where

$$f(x) \;=\; (0.3x + 2.5)e^{x/2} - 7x.$$

    (a) Start by verifying that $f(0.4)$ and $f(0.5)$ have different signs. Why does this imply that the root lies between $x = 0.4$ and $x = 0.5$?
    (b) Set $x_0 = 0.4$, and verify the entries in Table 4.2 using your result from Checkpoint 1. Fill in the rest of the table.

Table 4.2 Newton's Method calculation

n	x_n	$f(x_n)$	$f'(x_n)$	$f(x_n)/f'(x_n)$
0	0.4	0.400075	–5.033542	–0.079482
1	0.479482			
2				
3				

25. When you cough, your windpipe contracts. The velocity v at which air comes out depends on the radius r of your windpipe. If R is the normal (rest) radius of your windpipe, then (for each possible radius r), $v = a\,(R - r)\,r^2$, where a is a constant.
    (a) What value of r maximizes the velocity?
    (b) What is that maximum velocity?
    Both answers may involve either or both of the constants a and R.
26. A patient's reaction $R(x)$ to a drug dose of size x is given by a formula of the form

$$R(x) = A\,x^2(B - x),$$

    where A and B are positive constants. The *sensitivity* of the patient's body to a dose of size x is defined to be $R'(x)$.
    (a) What do you think the domain of x is? What is the physical meaning of the constant B? Of the constant A?
    (b) For what value of x is R a maximum? (Your answer should contain the constant B.)
    (c) What is the maximum value of R?
    (d) For what value of x is the sensitivity a maximum?
    (e) Why is it called "sensitivity"?[2]
27. Your local pizza delivery uses square boxes that are made from rectangular pieces of corrugated cardboard, each 47 by 90 cm. The box is made by cutting out six small squares, three from each of the 90-cm edges, one square at each corner and one in the middle of the edge. The result is then folded into a box in the obvious fashion. (If it isn't obvious, order one with sausage and mushrooms to investigate. If you order from Domino's, you will get an octagonal box that is made from a single rectangular piece of cardboard. If you ignore the clever diagonal cuts, the basic structure of removing squares is still there, but you will have to use a little imagination to see it.) How does one design a box in this fashion to obtain the largest volume?
28. Figure 4.35 repeats the figure from Chapter 2 that depicts Thomas Malthus's view of the eventual population crisis, with population growing exponentially and food supply growing linearly. Suppose we give Malthus the benefit of the doubt on both growth rates and try to determine when the

food supply would become inadequate. We set $t = 0$ in 1800, and we take one unit on the food scale to mean just enough food to feed the entire population in that year. The unit of population is the population in 1800; i.e., $P(0) = 1$. We interpret the food function $F(t)$ to mean *potential* for producing food at a given time. If we were to zoom in on the graph, we might find that a reasonable initial value is $F(0) = 5$.

Figure 4.35 Malthus's view of population growth

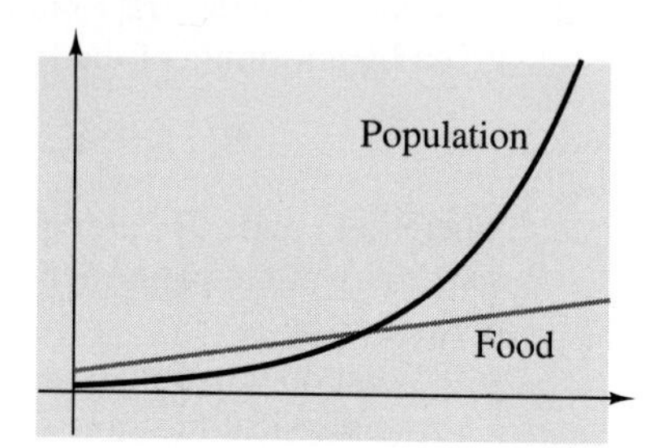

Suppose the population grows at a rate of 2% per year (compounded continuously), and the slope of the food function is 0.1. In what year would the potential food supply be just adequate for the population? What would happen after that? In what year would the potential food surplus (relative to the population) be greatest? (Our numbers here are fictional — they should not be taken seriously.)

29. Which would you think is the larger term in growth rate of energy consumption — population times growth rate of per capita consumption or per capita consumption times population growth rate? Check to see how much each term contributed to the 1985 growth rate of 1.5 billion BTUs per year.

2. This problem and the next are adapted from *Calculus Problems for a New Century*, edited by Robert Fraga, MAA Notes Number 28, 1993.

4.5 The Chain Rule

Optimization in Nature[3]

Do the following experiment: Look at a mirror, and focus on a small object that you see in the mirror. You see that object because there are light rays traveling from it to the mirror and then to your eye. Light rays normally travel in straight lines, but they can be bent — for example, by reflection in mirrors. Now visualize three points in space: the object, your eye, and the *image* of the object in the mirror (see Figure 4.36). The ray from the object to its image in the mirror makes an angle with the mirror that we label α in Figure 4.36. The ray from the image to your eye makes an angle we call β. As you are looking in the mirror, estimate the sizes of α and β. Vary your position in the room with respect to the object, and for each new position, estimate α and β again. How do you think α and β are related?

3. Much of Sections 4.5 and 4.6 is based on "Somewhere Within the Rainbow," by Steven Janke, in *Applications of Calculus*, edited by Philip Straffin, MAA Notes, 1993, and in part on *Five Applications of Max-Min Theory from Calculus*, by W. Thurmon Whitley, UMAP Module 341, COMAP, 1979.

Figure 4.36 Reflection of a light ray in a mirror

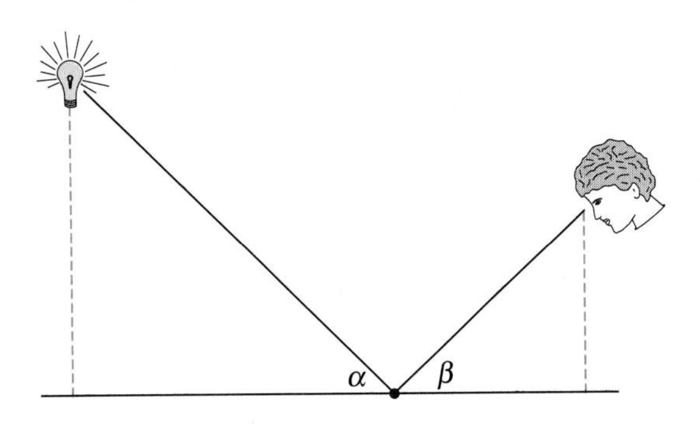

Now try this experiment: Fill a glass with water, and place a pencil in the water at an angle (not vertical). Notice the apparent bend in the pencil at the surface of the water? You know the water is not actually bending the pencil, so it must be bending the light rays from the pencil to your eye. More precisely, a light ray from a point on the pencil (say, the bottom end) travels through the water along a straight line to the surface, and then it changes direction to travel along another straight line to your eye.

We illustrate this situation in the left half of Figure 4.37 with the actual and apparent positions of the pencil. The heavy line is the route taken by a light ray from the end of the pencil to your eye. This bending phenomenon is called **refraction**, and it occurs whenever a light ray passes from one medium to another in which the speed of light is different. In the right half of Figure 4.37 we have labeled the angles the light ray makes with the vertical α (in water) and β (in air). How do you think the angles α and β in Figure 4.37 are related? The question is much harder for refraction than for reflection—don't worry if you can't answer it yet.

In both our experiments, the angle we have labeled α is called the **angle of incidence**. For reflection, this means the angle at which the light ray meets the mirror surface. For refraction, it means the *complement* of the angle at which the light ray meets the water-air surface.

A physics book would use the complementary angles in both cases, that is, angles with the perpendicular to the surface. For the reflection experiment, we thought you would find it easier to estimate the angles with the surface. Whether you use angles with the surface or their complements doesn't really matter. We can arrive at the same conclusions either way.

The angles we have labeled β are called, respectively, **angle of reflection** and **angle of refraction**. In the case of reflection, you probably have a pretty good idea from our simple experiment how α and β are related, even if you have never thought about it before. In the case of refraction, it is unlikely that you can even guess the relationship—unless you remember it from a physics course.

Figure 4.37 Apparent bending of pencil in water due to refraction

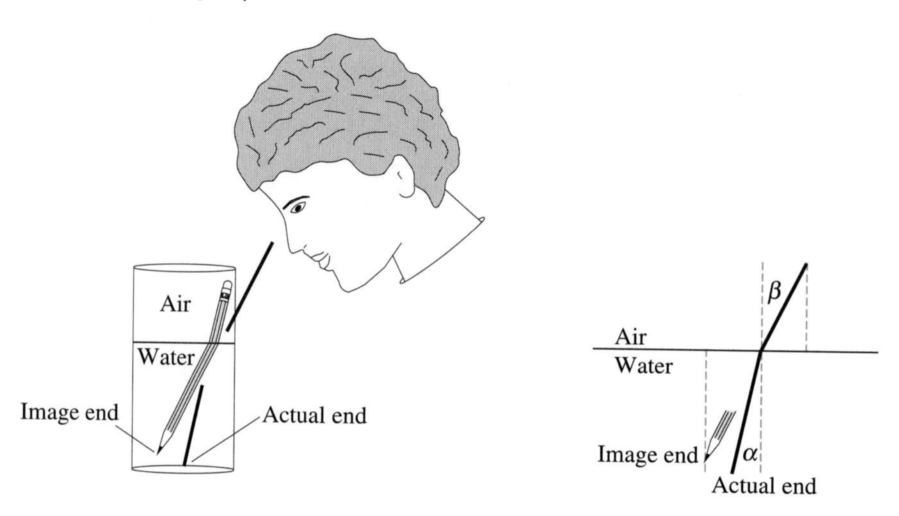

Experiments with light rays, confirmed many times over a period of at least 300 years, suggest that light obeys a simple and plausible optimization law:

> **Fermat's Principle** Light follows a path that minimizes total travel time.

This principle enables us to calculate the bending of light, whether by mirrors, water-air interfaces, lenses, prisms, or rainbows. All we have to do is write down a formula for total travel time as a function of something, and find the value of "something" that gives the minimum value of the function. As we have seen earlier in the chapter, one way to do this is to calculate a derivative and find its zeros — the critical values of the total-time function. For the case of reflection, we will carry out that process in the next section. The case of refraction we leave to a project at the end of the chapter.

In the next section we will find that the total distance function for reflection has the form

$$L(x) = \sqrt{p^2 + x^2} + \sqrt{q^2 + (D - x)^2},$$

where x measures the position of the reflection point on the mirror, and p, q, and D are constants. Some features of this function — for example, sums and squares — are familiar. But this function has two features that we have not encountered in our previous studies of derivatives: (1) square roots and (2) a composite structure. What we mean by "composite" is this: We have not just the square of x but the square of $D - x$ to deal with. And we have not just the square root of x but square roots of polynomials in x.

We devote the rest of this section to learning these two things:

- how to differentiate the square root function, and
- how to differentiate a composite function.

The Derivative of the Square Root Function

Let's write $y = \sqrt{x}$. The difference quotient that approximates dy/dx is

$$\frac{\Delta y}{\Delta x} = \frac{\sqrt{x + \Delta x} - \sqrt{x}}{\Delta x}.$$

It's not obvious how to simplify this expression—in particular, how to find a factor of Δx in the numerator. We need to use algebra to get a more useful form—but the most useful manipulation might not occur to you immediately.

The difference of square roots is the source of the difficulty. We could make that go away if we rationalize the numerator. It's the same idea you learned in high school for rationalizing a denominator: Multiply and divide by the sum of the square roots. Since we are multiplying and dividing by the same quantity, we are not changing the value of the expression. Let's see what happens:

$$\begin{aligned}\frac{\Delta y}{\Delta x} &= \frac{\sqrt{x + \Delta x} - \sqrt{x}}{\Delta x} \cdot \frac{\sqrt{x + \Delta x} + \sqrt{x}}{\sqrt{x + \Delta x} + \sqrt{x}} \\ &= \frac{(\sqrt{x + \Delta x})^2 - (\sqrt{x})^2}{\Delta x\,(\sqrt{x + \Delta x} + \sqrt{x})} \\ &= \frac{(x + \Delta x) - x}{\Delta x\,(\sqrt{x + \Delta x} + \sqrt{x})} \\ &= \frac{1}{\sqrt{x + \Delta x} + \sqrt{x}}.\end{aligned}$$

Exploration Activity 1

(a) Give a reason for each step in this calculation.

(b) Finish the calculation of dy/dx by determining the limiting value as Δx approaches 0.

To get from the first line to the second, we multiplied the two fractions by multiplying numerators and denominators, and we used the identity

$$(a - b)(a + b) = a^2 - b^2.$$

To get from the second line to the third, we used the fact that the squaring function is the inverse of the square root function. To get from the third line to the fourth, we simplified the numerator to Δx and canceled this Δx with the one in the denominator.

As $\Delta x \to 0$, $\sqrt{x + \Delta x} + \sqrt{x}$ approaches the limiting value $2\sqrt{x}$. Thus $1/(\sqrt{x + \Delta x} + \sqrt{x})$ approaches the limiting value $1/2\sqrt{x}$. This shows that

$$\frac{d}{dx}\sqrt{x} = \frac{1}{2\sqrt{x}}.$$

Example 1 Calculate the derivative of $\sqrt{x}\, e^x$.

Solution By the Product Rule,

$$\frac{d}{dx}\sqrt{x}\, e^x = \left(\frac{d}{dx}\sqrt{x}\right)e^x + \sqrt{x}\left(\frac{d}{dx}e^x\right).$$

Now, using our new formula for the derivative of the square root function, we find

$$\frac{d}{dx}\sqrt{x}\, e^x = \frac{1}{2\sqrt{x}}\, e^x + \sqrt{x}\, e^x.$$

■

Exploration Activity 2

(a) Calculate the derivative of $x\sqrt{x}$ by using a difference quotient and a limiting value, as in the calculation preceding Exploration Activity 1.

(b) Calculate the same derivative by using the Product Rule and your formula for the derivative of the square root function.

(c) Compare the results of parts (a) and (b), both in terms of algebraic equivalence of the two answers and in terms of labor required to get the answers.

For this derivative, the difference quotient is

$$\frac{(x+\Delta x)\sqrt{x+\Delta x} - x\sqrt{x}}{\Delta x}.$$

If we multiply both numerator and denominator by $(x+\Delta x)\sqrt{x+\Delta x} + x\sqrt{x}$, we have

$$\frac{\left[(x+\Delta x)\sqrt{x+\Delta x}\right]^2 - \left(x\sqrt{x}\right)^2}{\Delta x\left[(x+\Delta x)\sqrt{x+\Delta x} + x\sqrt{x}\right]}.$$

This simplifies to

$$\frac{(x+\Delta x)^3 - x^3}{\Delta x\left[(x+\Delta x)\sqrt{x+\Delta x} + x\sqrt{x}\right]}.$$

When we expand the numerator and cancel, we have

$$\frac{3x^2\Delta x + 3x\,\Delta x^2 + \Delta x^3}{\Delta x\left[(x+\Delta x)\sqrt{x+\Delta x} + x\sqrt{x}\right]}.$$

Next, we cancel the Δx in the denominator with a factor of Δx in the numerator to obtain

$$\frac{3x^2 + 3x\,\Delta x + \Delta x^2}{(x+\Delta x)\sqrt{x+\Delta x} + x\sqrt{x}}.$$

Now the numerator of this approaches $3x^2$ as $\Delta x \to 0$, and the denominator approaches $x\sqrt{x} + x\sqrt{x} = 2x\sqrt{x}$. Thus the limiting value of the difference quotient is

$3x^2/(2x\sqrt{x})$ or

$$\frac{3}{2}\sqrt{x}\,.$$

If we use the Product Rule and the formula for the derivative of the square root function, we find

$$\frac{d}{dx}\,x\sqrt{x} = 1 \cdot \sqrt{x} + x \cdot \frac{1}{2\sqrt{x}} = \frac{3}{2}\sqrt{x}\,.$$

Both approaches give the same formula for the derivative, but the second approach is certainly easier.

Differentiating a Function of a Function: The Chain Rule

Now that we have a formula for differentiating the square root function, we need to extend that to a formula for the derivative of $\sqrt{u}$, where u is itself a function of x. Actually, square root has little to do with this step—this is a problem that we will encounter over and over with many different functions. If we write $y = \sqrt{u}$, then our problem takes this form: y is a function of u, and u is a function of x. Therefore, y is a function of x also. How do we find dy/dx when we know dy/du and du/dx ?

Let's relate this question to something we have done before. In Chapter 2 we saw that, for any constant k,

$$\frac{d}{dx}\,e^{kx} = k\,e^{kx}.$$

If we set $u = kx$ and $y = e^u$, then this calculation takes the form

$$\begin{aligned}\frac{dy}{dx} &= e^{kx} \cdot k \\ &= e^{u}\,\frac{du}{dx} \\ &= \frac{dy}{du}\,\frac{du}{dx}.\end{aligned}$$

We show next that the formula

$$\frac{dy}{dx} = \frac{dy}{du}\,\frac{du}{dx}$$

holds for *any* function, not just the exponential function, and for *any* dependence of u on x. In words, this says that the rate of change of y as a function of x is the rate of change of y as a function of u times the rate of change of u as a function of x.

Suppose we fix a number x at which we want to know dy/dx, and we compute an approximating difference quotient for a small increment Δx. We write simply u for the value of u at x and $u + \Delta u$ for the value at $x + \Delta x$. That is, Δu is the corresponding increment in the intermediate variable. Similarly, we write y for the value of the outer variable at x and $y + \Delta y$ for the value at $x + \Delta x$, so Δy is the corresponding increment in the outer variable. Then dy/dx is approximated by $\Delta y/\Delta x$, and simple algebra tells us that

$$\frac{\Delta y}{\Delta x} = \frac{\Delta y}{\Delta u} \frac{\Delta u}{\Delta x}.$$

We may not know yet what du is — or why it appears to cancel in the derivative for-mula — but Δu is an ordinary numerical quantity, so, whenever it is not zero, it is subject to the algebraic cancellation law.

The two factors $\Delta y/\Delta u$ and $\Delta u/\Delta x$ in the last equation approximate, respectively, the rate of change of y with respect to u and the rate of change of u with respect to x. Furthermore, the approximations to instantaneous rates of change all get better as the increment in x shrinks to zero, so, when we take limiting values of all three quotients, we find

$$\frac{dy}{dx} = \frac{dy}{du} \frac{du}{dx}$$

as predicted.

Simple as it is, this equation is perhaps the most important formula of differential calculus, because so many other formulas and calculations depend on it. Important results have names — the name of this one is the **Chain Rule**. It is called that because it tells us how to differentiate chains of functions, i.e., how to find dy/dx when y is a function of u and u is a function of x.

The Chain Rule If y is a function of u and u is a function of x, then

$$\frac{dy}{dx} = \frac{dy}{du} \frac{du}{dx}.$$

In functional notation, if $u = g(x)$ and $y = f(u) = f(g(x))$, then

$$\frac{d}{dx} f(g(x)) = f'(g(x))\, g'(x).$$

Example 2 Calculate the derivative of $\sqrt{p^2 + x^2}$, where p is a constant.

Solution We set $u = p^2 + x^2$ and $y = \sqrt{u} = \sqrt{p^2 + x^2}$. Then $du/dx = 2x$ and $dy/du = 1/2\sqrt{u}$, so

$$\frac{dy}{dx} = \frac{dy}{du} \frac{du}{dx} = \frac{1}{2\sqrt{u}} 2x = \frac{x}{\sqrt{p^2 + x^2}}.$$

■

Checkpoint 1

If $y = \sqrt{e^x + x}$, find dy/dx.

In this section we developed two new differentiation formulas, one specific and the other general. The specific formula is for the square root function. You may have noticed that this is consistent with the Power Rule, previously known to us only for nonnegative integer powers and the -1 power:

$$\frac{d}{dx}\,x^{1/2} = \frac{d}{dx}\,\sqrt{x} = \frac{1}{2\sqrt{x}} = \frac{1}{2}\,x^{-1/2}.$$

The new general formula is the most important of the differentiation rules, the Chain Rule:

$$\frac{dy}{dx} = \frac{dy}{du}\,\frac{du}{dx}.$$

Answer to Checkpoint

1. $\dfrac{e^x + 1}{2\sqrt{e^x + x}}$

Exercises 4.5

Use the Chain Rule and the rule for differentiating the square root function to calculate the derivative of each of the following functions.

1. $\sqrt{2t+3}$
2. $\sqrt{2t^2+3}$
3. $\sqrt{t^3-t}$

Use the Chain Rule and the Product Rule to calculate the derivative of each of the following functions.

4. e^{4t-6}
5. e^{t^2}
6. e^{t+4}
7. $t\,e^{4t-6}$
8. $t\,e^{t^2}$
9. $t\,e^{t+4}$
10. t^2e^{4t-6}
11. $t^2e^{t^2}$
12. t^2e^{t+4}
13. $\sqrt{t}\,e^{4t-6}$
14. $\sqrt{t}\,e^{t^2}$
15. $\sqrt{t}\,e^{t+4}$

For each of the following functions, find dy/dx.

16. $y = xe^{-x}$
17. $y = e^{(x^2+x)}$
18. $y = (2x^3+2x)^2$
19. $y = v^2 + v$, where $v = \sqrt{u^2+1}$ and $u = x^3 - x$
20. $y = \sqrt{x^5+x^2+1}$
21. If $z = y^3 + y + 2$ and $y = 3x + 1$, find $\dfrac{dz}{dx}$.

Suppose f and g are functions such that $f'(2) = 4$, $g'(2) = -3$, $f(2) = -1$, $g(2) = 1$, $f'(1) = 2$, and $g'(-1) = 5$. For each of the following functions, find the value of the derivative at $x = 2$.[4]

22. $s(x) = f(x) + g(x)$
23. $p(x) = f(x)g(x)$
24. $h(x) = f(g(x))$
25. $k(x) = g(f(x))$
26. Suppose $y = f(u(t))$, $f'(1) = 2$, $f'(3) = 8$, $f'(5) = 13$, $u(1) = 3$, $u(3) = 20$, $u'(1) = 5$, and $u'(3) = 51$. Find dy/dt at $t = 1$.
27. Use the Chain Rule to calculate the derivative of $(1+3x)^4$. Check your answer by expanding $(1+3x)^4$ and differentiating term by term.
28. Suppose $z^2 = x^2 + y^2$, where x, y, and z are all functions of t. Find dz/dt in terms of x, y, dx/dt, and dy/dt.
29. In an early laboratory you studied the logistic growth equation $dP/dt = cP(M-P)$ as a mod-el for growth of fruit flies. Among other things, you

located the approximate size of the population at the time when it is growing most rapidly.

(a) Explain why the population must be growing most rapidly at a time at which the second derivative of P is zero.

(b) Differentiate both sides of the differential equation to find an expression for the second derivative. (The Product Rule works for differentiating the right-hand side, but you can make the computation easier if you rewrite the expression in another form first.) Be careful with your differentiation—you are differentiating with respect to t, not P, so the Chain Rule must come into play every time you run into the unknown function P.

(c) What does your expression for the second derivative tell you about the population size when the growth rate is maximal? How does this fit with your data from the laboratory?

30. When a drug is injected into the bloodstream, its concentration $C(t)$ at t minutes after injection is given by a formula of the form

$$C(t) = K\,\frac{e^{-bt} - e^{-at}}{a - b},$$

where K, a, and b are positive constants, and $a > b$. When does the maximum concentration occur?[5]

31. What is the cone of largest volume that can be formed by rotating a right triangle of fixed hypotenuse h around one of its legs? What is the volume of that cone?

32. Develop the following argument to establish the Product Rule by using the Chain Rule and without resorting to difference quotient calculations.

(a) Let $x = x(t)$ and $y = y(t)$ be any two functions. Expand the square $(x + y)^2$ to find an equal expression in terms of x^2, y^2, and xy.

(b) Solve your equation for xy.

(c) The other side of the equation now contains only squares of unknown functions, which you can differentiate with the aid of the Chain Rule. Thus, when you differentiate both sides, you should find an expression for the derivative of xy with respect to t. With any luck (skill?), your expression will turn out to be the Product Rule.

33. (a) Write down formulas for two even functions. Sketch the graph of each of these functions. Sketch the graph of the derivative of each of these functions.

(b) Write down formulas for two odd functions. Sketch the graph of each of these functions. Sketch the graph of the derivative of each of these functions.

(c) What conclusion(s) do you draw from parts (a) and (b)?

34. If $y = f(u)$ and $u = g(x)$, then $y = f(g(x))$. The function $f(g(x))$ is called the **composite** of f and g. For example, if $f(u) = u^2$ and $g(x) = x^3 + x$, then $f(g(x)) = (x^3 + x)^2$. What can be said about the oddness or evenness of the composite of f and g if

(a) f and g are both odd?

(b) f and g are both even?

(c) f is odd and g is even?

(d) f is even and g is odd?

Give an example for each of the cases in parts (a) through (d). Verify that the composite in each case has the oddness or evenness property you said it should have.[6]

35. (a) Explain in geometric terms (i.e., by using symmetries and slopes) why the derivative of an odd function is an even function and why the derivative of an even function is an odd function. (See Example 3 in Section 4.1, Exploration Activity 1 in Section 4.2, and Exercise 33 for examples that illustrate both these statements.)

(b) Explain in algebraic terms (i.e., by using difference quotients) why the derivative of an odd function is an even function and why the derivative of an even function is an odd function.

(c) Explain in calculus terms (i.e., by using the Chain Rule) why the derivative of an odd function is an even function and why the derivative of an even function is an odd function.

---

4. Exercises 22–25 are adapted from *Calculus Problems for a New Century*, edited by Robert Fraga, MAA Notes Number 28, 1993.
5. Adapted from *Five Applications of Max-Min Theory from Calculus*, by W. Thurmon Whitley, UMAP Module 341, COMAP, 1979.
6. Adapted from *Calculus Problems for a New Century*, edited by Robert Fraga, MAA Notes Number 28, 1993.

4.6 Analysis of Reflection

With the necessary differentiation tools in hand, we return to our investigation of the reflection of light. In Figure 4.38 we have labeled the object, the viewer's eye, and the image of the object in the mirror by P, Q, and R, respectively. For fixed positions of the mirror, the object being viewed, and the eye, the quantities that are constant are the distance p from object to mirror, the distance q from eye to mirror, and the distance D between the projections of object and eye on the mirror. In principle, until we know the relationship between α and β, the point R could be anywhere along the mirror. We label its distance from the left edge of the figure by x, and we imagine that x can take any value from 0 to D. Our question, then, is what value of x produces the minimal travel time for a ray from P to R to Q? When we can answer that question, we will know how α and β are related.

Figure 4.38 Reflection of light ray in a mirror

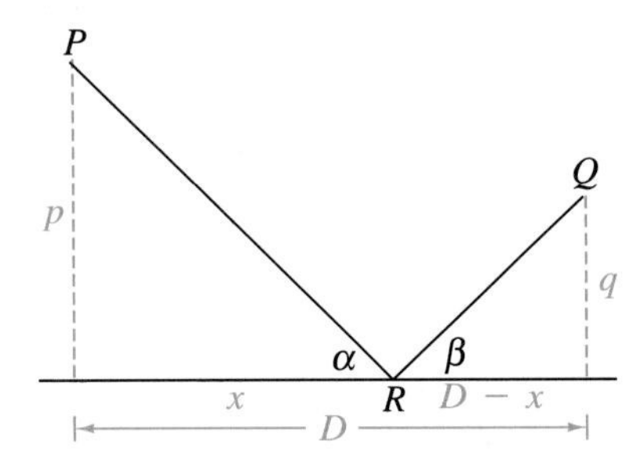

Since the speed of light in air is constant, minimizing the time of travel for the light ray is equivalent to minimizing the distance traveled, i.e., the sum of the distances from P to R and from R to Q. Both distances depend on x, of course, so their sum does also.

Exploration Activity 1

(a) Figure 4.38 contains two triangles. We have explained the labeling of three of the six sides. Explain the fourth distance label, $D - x$, in the figure.

(b) Calculate the lengths of the two remaining sides, and label them accordingly.

(c) Find a formula for the length of path function $L(x)$ for any x between 0 and D.

Since the distance of the image of the object in the mirror from the left edge of the figure is x, the remaining distance to the right side is $D - x$. Both triangles are right triangles and the remaining sides are the hypotenuses. By the Pythagorean Theorem, the distance from P to R is

$$\sqrt{p^2 + x^2},$$

and the distance from Q to R is

$$\sqrt{q^2 + (D - x)^2}.$$

Thus the length of the total path $L(x)$ is just the sum of these two lengths:

$$L(x) = \sqrt{p^2 + x^2} + \sqrt{q^2 + (D - x)^2}.$$

In Figure 4.39 we show a graph of the function $L(x)$ for typical values of p, q, and D, along with a graph of its derivative. The graphs confirm that $L(x)$ has a unique minimum, that the minimum occurs somewhere between $x = 0$ and $x = D$, and that $L'(x) = 0$ at the minimum point. Thus all we have to do is calculate $L'(x)$, set the result equal to zero, and solve for x.

Figure 4.39 $L(x)$ and its derivative

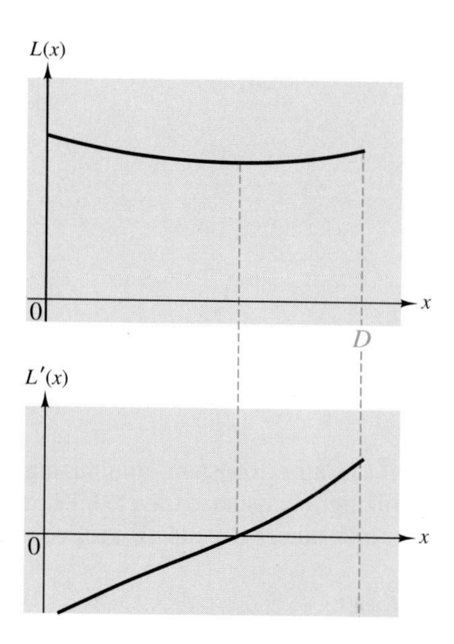

Figure 4.39 also has a message about the limitations of graphical and numerical methods for solving the type of problem we are studying now. First, in order to draw the graph of L, we had to specify values of the constants p, q, and D. If we made different choices, would the graph look the same? We want to know the relationship between α and β regardless of the values of the constants.

Second, as we observed in Section 4.1, we would have a problem finding the leveling-off point on the graph of L by zooming in on it. That portion of the graph is already very flat, and when we zoom in, we will see nothing but a horizontal line long before the scale is fine enough to tell us anything about the lowest point. We could zoom in instead on the graph of L' to find where it crosses the x-axis—but only if we know a formula for L', or at least some way to compute its values. Similarly, we could apply Newton's Method to the equation $L'(x) = 0$, but for that we need to be able to compute values of both L' and its derivative, L''. Thus, even with graphical and numerical tools at hand, we still have to come to grips with the problem of calculating L'.

Exploration Activity 2

Calculate

(a) $\dfrac{d}{dx}\left[q^2 + (D-x)^2\right]$

(b) $\dfrac{d}{dx}\sqrt{q^2 + (D-x)^2}$

(c) $\dfrac{d}{dx}L(x)$

If we apply the Sum Rule and note that q^2 is a constant, we find

$$\frac{d}{dx}\left[q^2 + (D-x)^2\right] = \frac{d}{dx}(D-x)^2.$$

Now, $(D-x)^2$ is a function of a function to which we can apply the Chain Rule. Specifically, if we write $v = w^2$ and $w = D - x$, then $v = (D-x)^2$. The Chain Rule tells us that

$$\frac{dv}{dx} = \frac{dv}{dw}\frac{dw}{dx} = 2w(-1) = -2(D-x).$$

And since adding the constant q^2 does not change the derivative,

$$\frac{d}{dx}\left[q^2 + (D-x)^2\right] = -2(D-x).$$

Now, if we write $u = q^2 + (D-x)^2$ and $z = \sqrt{u}$, then

$$\frac{dz}{dx} = \frac{dz}{du}\frac{du}{dx} = \frac{1}{2\sqrt{u}}\left[-2(D-x)\right] = \frac{x-D}{\sqrt{q^2 + (D-x)^2}}.$$

In the preceding section we calculated the derivative of $y = \sqrt{p^2 + x^2}$:

$$\frac{d}{dx}\sqrt{p^2 + x^2} = \frac{x}{\sqrt{p^2 + x^2}}$$

Finally, we find that

$$\frac{d}{dx}L(x) = \frac{dy}{dx} + \frac{dz}{dx} = \frac{x}{\sqrt{p^2 + x^2}} + \frac{x-D}{\sqrt{q^2 + (D-x)^2}}.$$

Notice what happened here: From two very simple formulas—the Power Rule and the Chain Rule—the first very specific and the second very general, we differentiated a very complicated formula, one we could never have dealt with directly by using difference quotients.

Now what did we learn about reflection? Well, $L'(x) = 0$ is equivalent to

$$\frac{x}{\sqrt{p^2 + x^2}} = \frac{D-x}{\sqrt{q^2 + (D-x)^2}}.$$

This still looks like a substantial algebraic challenge, but let's keep our goal in focus. We wanted to determine the relationship between α and β, and the two quotients in the preceding equation have a direct connection with those two angles. In Figure 4.40 we see that each numerator in the equation is the length of an adjacent side for one of the angles, and each denominator is the length of the corresponding hypotenuse. Thus the equation reduces to a very simple statement about the angles:

$$\cos \alpha = \cos \beta$$

And since both angles are between 0 and 90 degrees, the only way they can have the same cosine is to be the same angle. Conclusion:

The angle of incidence equals the angle of reflection.

This conclusion probably comes as no surprise to you. Indeed, you probably figured it out from our mirror experiment.

Figure 4.40 Reflection of light ray in a mirror

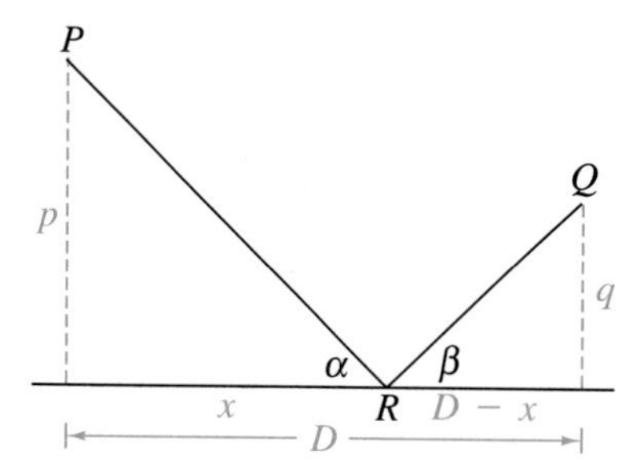

In this section we have put our new differentiation formulas to work to see that a physical optimization law—Fermat's Principle—implies equality of angles of incidence and reflection. This is not a terribly profound conclusion, and it could be deduced without any calculus at all (see Exercise 4). However, we also have laid the groundwork for addressing more difficult questions, such as the refraction question we posed at the start of the chapter. That issue is pursued in Project 2 at the end of the chapter.

Exercises 4.6

1. We show typical graphs of $L(x)$ and $L'(x)$ again in Figure 4.41, repeated from Figure 4.39.
   (a) It is clear from this figure that $L(x)$ also has a maximum value on the interval $0 \leq x \leq D$—and it does not occur at a point where $L'(x) = 0$. Where does it occur?
   (b) Recall that

$$L(x) = \sqrt{p^2 + x^2} + \sqrt{q^2 + (D - x)^2}$$

   for any x between 0 and D. Does your answer to part (a) depend on the relative sizes of p and q? If so, how?

Figure 4.41 $L(x)$ and $L'(x)$

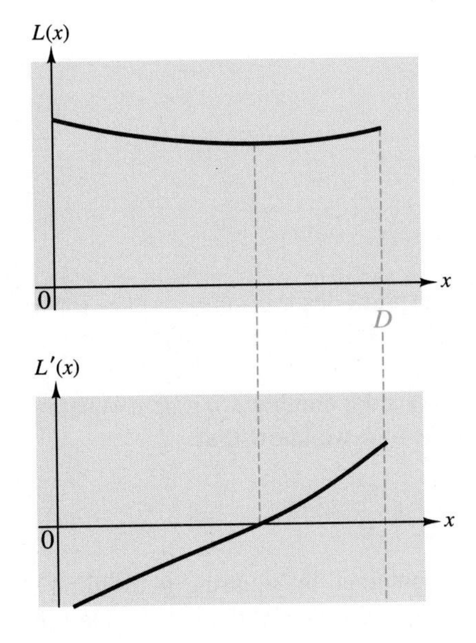

2. A square plot of ground 100 feet on a side has corners labeled A, B, C, and D clockwise. Pipe is to be laid in a straight line from A to a point P on the side BC and thence to C. (P could be one of the corners B or C.) The cost of laying the pipe is \$20 a foot if it goes through the lot (because it must be laid underground) and \$10 a foot if it is laid along one of the sides of the square. What is the most economical way to lay the pipe?[7]

3. Lee and Dana are in a rowboat half a mile from the nearest point on a relatively straight shoreline when Dana suddenly becomes ill. The closest place to find a telephone to call for help is at a seaside restaurant 1 mile from the nearest point on shore. Lee plans to row to a point on the shore and then jog to the restaurant. Lee can jog at 6 miles per hour and row at 1.5 miles per hour.
   (a) How long would it take to get to the restaurant if Lee rows to the nearest point on shore?
   (b) How long would it take if Lee rows directly to the restaurant?
   (c) What point on the shore would make the trip to restaurant as quick as possible? How long would it take to get to the restaurant?
   (d) How fast would Lee have to row so that the quickest route would be to row directly to the restaurant?

4. Make a copy of Figure 4.40. Add to the figure the point P' at which the point P *appears* to be (behind the mirror) when your eye is at Q. Use a shortest-distance-between-two-points argument and elementary geometry to explain why α and β must be the same angle.

5. A mirror is placed flat on the ground 7 meters from the base of a tree. When a person whose eyes are 180 centimeters from the ground stands 1 meter beyond the mirror, he or she can see the top of the tree in the mirror. How tall is the tree?

---

7. This problem and the next one are adapted from *Calculus Problems for a New Century*, edited by Robert Fraga, MAA Notes Number 28, 1993.

4.7 Derivatives of Functions Defined Implicitly

In Figure 4.42 we graph the ellipse with equation

$$\frac{x^2}{4} + \frac{y^2}{9} = 1.$$

Suppose we want to know the slope of the tangent line to the ellipse at the point (x_0, y_0), where $x_0 = 1$ and $y_0 = 3\sqrt{3}/2 \approx 2.598$.

Figure 4.42 Graph of $\frac{x^2}{4} + \frac{y^2}{9} = 1$

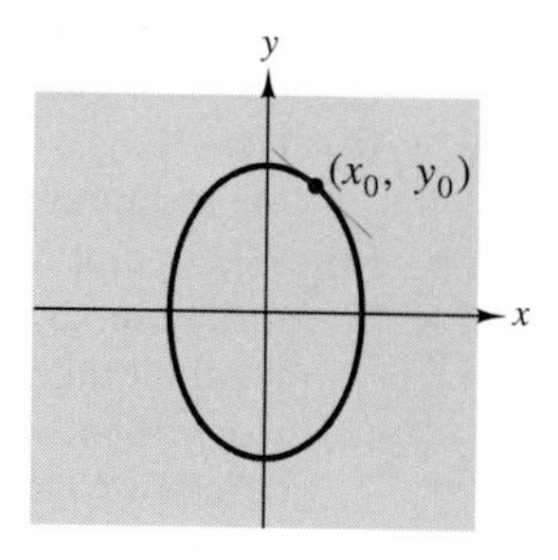

One approach would be to find an explicit description $y = f(x)$ for the top half of the ellipse, differentiate that function, and evaluate the derivative at $x = 1$. You will carry out that computation in Checkpoint 1.

However, before you do that, let's consider another approach. Whatever the functional relation between y and x, for each value of x we know that

$$\frac{x^2}{4} + \frac{y^2}{9} = 1.$$

Thus the function of x on the left side, obtained by squaring y, dividing by 9, and adding $x^2/4$, must be the same function as the one on the right side; i.e., the function on the left must be the constant function 1.

Now we differentiate both the left-hand and the right-hand functions. On the left, we use the Sum Rule, the Power Rule, and the Chain Rule to obtain

$$\frac{d}{dx}\left(\frac{x^2}{4} + \frac{y^2}{9}\right) = \frac{d}{dx}\frac{x^2}{4} + \frac{d}{dx}\frac{y^2}{9} = \frac{x}{2} + \frac{2}{9}y\frac{dy}{dx}.$$

On the right, the derivative of the constant 1 is 0. Since the two derivatives are equal, we may write

$$\frac{x}{2} + \frac{2}{9}y\frac{dy}{dx} = 0.$$

Solving for dy/dx, we obtain

$$\frac{dy}{dx} = \frac{-x/2}{\frac{2}{9}y} = -\frac{9x}{4y}.$$

In particular, at $x_0 = 1$ and $y_0 = 3\sqrt{3}/2$, we find

$$\frac{dy}{dx} = -\frac{9}{6\sqrt{3}} = -\frac{\sqrt{3}}{2} \approx -0.866.$$

The ellipse in Figure 4.42 is the combined graphs of two functions of the form $y = y(x)$, one for the top half of the ellipse and one for the bottom half. Both these functions are **defined implicitly** by the equation of the ellipse:

$$\frac{x^2}{4} + \frac{y^2}{9} = 1$$

Our technique for calculating dy/dx directly from the implicit definition applies to both functions.

Checkpoint 1

(a) Show that, for the top half of the ellipse,

$$y = 3\sqrt{1 - \frac{x^2}{4}}.$$

(b) Calculate dy/dx explicitly, substitute $x = 1$, and show that your answer agrees with the one we obtained by our differentiation of the implicit definition.

Checkpoint 2

In Figure 4.43 we show the graph of the equation

$$y^2 - x^2 = 1.$$

(a) Use the method of implicit differentiation to find the slope of the tangent line to the curve at $x = 1$ and $y = \sqrt{2}$.

(b) Use the explicit method of Checkpoint 1 to find the slope of the tangent line to the curve $(1, \sqrt{2})$.

Figure 4.43 Graph of $y^2 - x^2 = 1$

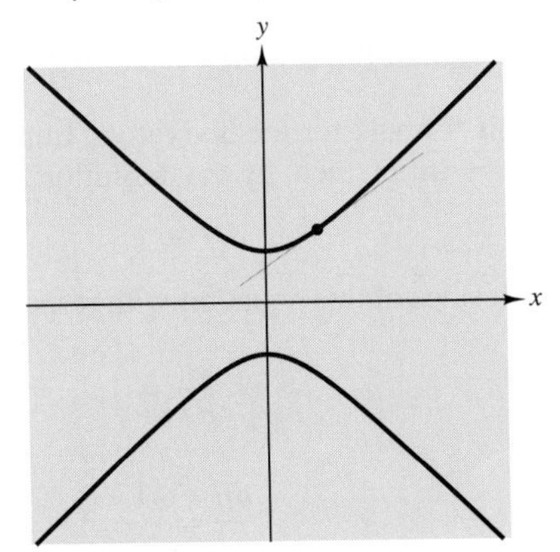

The Derivative of the Natural Logarithm

We have formulas for the derivatives of all but one of the functions we have encountered so far—all except the natural logarithm. We will find a formula for this derivative in this section. Before we do, let's return to the definition of the derivative and look at difference quotients for $\ln(x)$.

In Figure 4.44 we graph the difference quotient function

$$y = \frac{\ln(x+0.001) - \ln(x)}{0.001},$$

which should closely approximate the derivative of $\ln(x)$. Do you recognize this function?

Figure 4.44 Graph of the difference quotient for $\ln(x)$

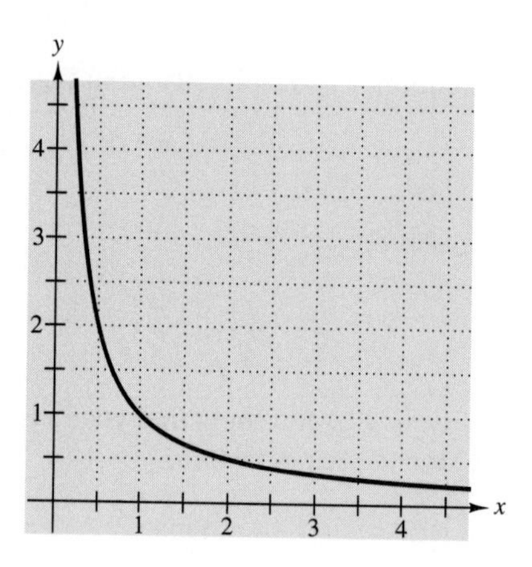

Exploration Activity 1

(a) Use the graph to estimate the difference quotient for $x = 1/4$, $x = 1/2$, $x = 1$, $x = 2$, and $x = 4$.

(b) On the basis of your estimates in (a), make a conjecture about the derivative of $\ln(x)$.

From the graph in Figure 4.44, the difference quotient is approximately 4 at $x = 1/4$, 2 at $x = 1/2$, 1 at $x = 1$, $1/2$ at $x = 2$, and $1/4$ at $x = 4$. It looks like the derivative of $\ln(x)$ is probably $1/x$.

Now let's verify this formula for the derivative. Implicit differentiation turns out to be just what we need. If $y = \ln(x)$, then, by the definition of the natural logarithm,

$$e^y = x.$$

If we differentiate both sides of this equation with respect to x, we obtain

$$e^y \frac{dy}{dx} = 1,$$

which leads to

$$\frac{dy}{dx} = \frac{1}{e^y}.$$

But, since $e^y = x$, we find

$$\frac{dy}{dx} = \frac{1}{x}.$$

What we have shown is that

$$\frac{d}{dx}\ln(x) = \frac{1}{x}.$$

Since the natural logarithm is only defined for positive numbers x, this formula is correct for all $x > 0$.

Checkpoint 3

(a) Use the formula just derived for the derivative of $\ln(x)$ and the Chain Rule to calculate

$$\frac{d}{dx}\ln(x^2).$$

(b) Use a property of logarithms to simplify the calculation in (a) to one that does not require the Chain Rule.

Implicit differentiation is a useful way to obtain differentiation formulas. In Example 1 we use this method to derive again the formula we know already for the derivative of the square root function.

Example 1 Apply the implicit differentiation technique to the equation

$$y^2 = x$$

to obtain the formula

$$\frac{dy}{dx} = \frac{1}{2\sqrt{x}}.$$

Solution Differentiating both sides of the equation $y^2 = x$ with respect to x, we obtain

$$2y\,\frac{dy}{dx} = 1.$$

Solving for dy/dx, we get

$$\frac{dy}{dx} = \frac{1}{2y}.$$

Since $y = \sqrt{x}$, we may substitute to obtain

$$\frac{dy}{dx} = \frac{1}{2\sqrt{x}}.$$

■

Derivatives of Inverse Functions

We have just seen two instances of the relationship between the derivatives of inverse functions—for the pair

$$y = \ln(x) \qquad x = e^y$$

and for the pair

$$y = \sqrt{x} \qquad x = y^2.$$

Since we will use this relationship again to determine other derivatives of inverses, it is worth our while to look at derivatives of inverse functions in general.

Suppose f and g are an inverse pair of functions: If $y = f(x)$, then $x = g(y)$. When we apply one after the other, we get

$$g(f(x)) = x\,.$$

Now we apply the Chain Rule to this equation to get

$$g'(f(x))f'(x) = 1$$

or

$$f'(x) = \frac{1}{g'(f(x))}\,.$$

In differential notation, this becomes

$$\frac{dy}{dx} = \frac{1}{dx/dy}\,.$$

Thus we see another instance (like the Chain Rule) in which differentials appear to behave just like numbers. In words, this formula says that derivatives of inverse functions are reciprocals of each other.

Example 2 Use the inverse function derivative formula to give another derivation of the formula

$$\frac{d}{dx}\ln(x) = \frac{1}{x}.$$

Solution If $y = \ln(x)$, then $x = e^y$. By the inverse derivative formula, we have

$$\frac{dy}{dx} = \frac{1}{dx/dy} = \frac{1}{e^y} = \frac{1}{x}.$$

■

We have accomplished three things in this section:

- a specific formula for the derivative of the natural logarithm function:

$$\frac{d}{dx}\ln(x) = \frac{1}{x} \qquad \text{for } x > 0;$$

- a general formula for derivatives of inverse functions:

$$\frac{dy}{dx} = \frac{1}{dx/dy};$$

- the technique of implicit differentiation.

If y is defined implicitly as a function of x by an equation relating y and x, we can find an expression for the derivative dy/dx (in terms of both x and y) by differentiating both sides of the equation with respect to x and solving the resulting equation for dy/dx.

Answers to Checkpoints

1. (a) Solve the given equation for y: $y = \pm 3\sqrt{1 - x^2/4}$. The positive solution corresponds to the top half of the ellipse.

   (b) $\dfrac{dy}{dx} = 3\,\dfrac{1}{2}\,\dfrac{1}{\sqrt{1 - x^2/4}}\,\dfrac{d}{dx}(1 - x^2/4) = \dfrac{3}{2}\,\dfrac{1}{\sqrt{1 - x^2/4}}\,\dfrac{-2x}{4} = -\dfrac{3}{4}\,\dfrac{x}{\sqrt{1 - x^2/4}}$.

   At $x = 1$, $\dfrac{dy}{dx} = -\dfrac{3}{4}\,\dfrac{1}{\sqrt{3/4}} = -\sqrt{3/4} = -\dfrac{\sqrt{3}}{2}$.

2. (a) $2y\,\dfrac{dy}{dx} - 2x = 0$, so $\dfrac{dy}{dx} = \dfrac{x}{y}$. At $x = 1$ and $y = \sqrt{2}$, $\dfrac{dy}{dx} = \dfrac{1}{\sqrt{2}}$.

   (b) $y = \sqrt{1 + x^2}$, so $\dfrac{dy}{dx} = \dfrac{1}{2}\,\dfrac{1}{\sqrt{1 + x^2}}\,2x = \dfrac{x}{\sqrt{1 + x^2}}$.

   At $x = 1$, $\dfrac{dy}{dx} = \dfrac{1}{\sqrt{2}}$.

3. (a) $\dfrac{d}{dx}\ln(x^2) = \dfrac{1}{x^2}\,(2x) = \dfrac{2}{x}$.

   (b) $\dfrac{d}{dx}\ln(x^2) = \dfrac{d}{dx}\,2\ln x = \dfrac{2}{x}$.

Exercises 4.7

1. (a) Differentiate implicitly to find the slope of the tangent line to the curve $x = y^2$ at $x = 1$ and $y = 1$.

   (b) Calculate dx/dy for the equation $x = y^2$, and use the derivative to find the slope of the tangent line to this curve at $(1, 1)$.

   (c) Find the slope of the tangent line to the curve $x = y^2$ at $(1, 1)$ by first solving for y as an explicit function of x.

2. Find the slope of the tangent line to the curve $x^3 + y^3 = 28$ at $(1, 3)$.

3. Find the slope of the tangent line to the curve $x^2 - y^2 = 5$ at $(3, 2)$.

Calculate the derivative of each of the following functions.

4. $\ln(x + 2)$

5. $\ln(x^2 + 2)$

6. $x^2 + 2 \ln x$

7. $x \ln x$

8. $x^2 \ln(x + 2)$

9. $x \ln(x^2 + 2)$

For each of the following functions, find $\frac{dy}{dx}$.

10. $y = \ln(x^2 + 1)$

11. $y = 2^x(x^2 + 1)$

12. $y = e^{x^2}$

13. $y = \sqrt{x^3 + 1}$

14. $y = \ln(5x)$

15. $y = e^{\ln(5x)}$

16. $y = \dfrac{1}{x + 5}$

17. $y = \dfrac{1}{\sqrt{x + 5}}$

18. $y = \dfrac{1}{\ln(x + 5)}$

19. $y = \dfrac{1}{x^2 + 5}$

20. $y = \dfrac{1}{\sqrt{x^2 + 5}}$

21. $y = \dfrac{1}{\ln(x^2 + 5)}$

22. Find (approximately) a number x that solves the equation $x = 5 \ln x$.

23. Use the inverse function derivative formula to give another derivation of the formula

$$\frac{d}{dx}\sqrt{x} = \frac{1}{2\sqrt{x}}.$$

24. (a) Use implicit differentiation to obtain a formula for the derivative of $x^{1/3}$.

    (b) Use the inverse function derivative formula to confirm your result in part (a).

25. We began our discussion of implicit differentiation by finding a slope formula for points on the ellipse $x^2/4 + y^2/9 = 1$:

$$\frac{dy}{dx} = -\frac{9x}{4y}.$$

We applied the formula only to points on the upper half of the ellipse.

   (a) Find the slope at the point on the lower half of the ellipse at which $x = 1$.

   (b) Find an explicit formula for slopes at all points on the lower half of the ellipse.

26. In Checkpoint 2 you found slopes on the upper half of the hyperbola $y^2 - x^2 = 1$.

   (a) Find the slope at the point on the lower branch of the hyperbola at which $x = 1$.

   (b) Find an explicit formula for slopes at all points on the lower branch of the hyperbola.

4.8 The General Power Rule

We need one more general rule for differentiation. We know how to differentiate power functions such as $f(x) = x^2$ and $g(x) = x^9$—the derivatives are $f'(x) = 2x$ and $g'(x) = 9x^8$. We can summarize what we know about derivatives of power functions:

$$\frac{d}{dx}\, x^n = nx^{n-1}.$$

Here, n can be 0 or 1 or 2 or any nonnegative integer. The cases of $n = 1/2$ and $n = -1$, both discussed earlier in this chapter, also fit this formula:

$$\frac{d}{dx}\, x^{1/2} = \frac{d}{dx}\, \sqrt{x} = \frac{1}{2\sqrt{x}} = \frac{1}{2}\, \frac{1}{\sqrt{x}} = \frac{1}{2}\, x^{-1/2},$$

$$\frac{d}{dx}\, x^{-1} = \frac{d}{dx}\, \frac{1}{x} = -\frac{1}{x^2} = (-1)x^{-2}.$$

What about the derivative of $y = x^{7/3}$? Does it fit the form of the Power Rule? To answer this, we use again implicit differentiation. The 7/3 power function satisfies the equation

$$y^3 = x^7.$$

Now we can differentiate both sides with respect to x—because both exponents are positive integers:

$$3y^2\, \frac{dy}{dx} = 7x^6.$$

When we solve for dy/dx, we get

$$\frac{dy}{dx} = \frac{7x^6}{3y^2}.$$

If we substitute $y = x^{7/3}$, we get

$$\frac{dy}{dx} = \frac{7x^6}{3(x^{7/3})^2}$$

or

$$\frac{dy}{dx} = \frac{7}{3}\, x^{6-14/3} = \frac{7}{3}\, x^{4/3}.$$

Note that if we set $n = 7/3$, this fits the pattern we have already seen:

$$\frac{d}{dx}\, x^n = nx^{n-1}.$$

This Power Rule formula holds for any rational power n. Indeed, it holds for any power function, whether n is rational or not, although it may not be clear at this point what an irrational power means.

Before we discuss how to establish a General Power Rule, we give another example that shows how we may use the Power Rule and the Chain Rule in combination to calculate a rather complicated derivative. This example is much like the calculations we did in Section 4.6 when we examined the reflection property of light.

Example 1 Differentiate the function $(e^{3x} + 5x)^{7/3}$.

Solution Using both the Power Rule and the Chain Rule, we find

$$\frac{d}{dx}(e^{3x} + 5x)^{7/3} = \frac{7}{3}(e^{3x} + 5x)^{4/3} \frac{d}{dx}(e^{3x} + 5x).$$

When we finish calculating the second factor, we have

$$\frac{d}{dx}(e^{3x} + 5x)^{7/3} = \frac{7}{3}(e^{3x} + 5x)^{4/3}(3e^{3x} + 5).$$

The Power Rule for Rational Exponents

We begin by showing that the Power Rule holds for positive rational exponents, i.e., exponents of the form $r = p/q$, where p and q are positive integers. We have just worked out the case with $p = 7$ and $q = 3$. Now we show that the same argument works for any p and q.

We introduce a name for the rth power function, say, $u = t^r$. Then u is also equal to $t^{p/q}$. If we raise both sides of this equation to the qth power, we have

$$u^q = t^p.$$

On the right in this equation we have a function of t whose derivative we know from the Power Rule, since p is a positive integer. On the left we have the same function of t (because of the equality) written in a different form, namely, as the qth power of some other function u. We know exactly what function u is, but that's not important at the moment.

From the positive integer version of the Power Rule, we know that

$$\frac{d}{du}\, u^q = q\, u^{q-1}.$$

Thus the Chain Rule—with u^q as a function of u and u as a function of t —tells us that

$$\frac{d}{dt}\, u^q = \frac{d}{du}\, u^q\, \frac{du}{dt} = q\, u^{q-1} \frac{du}{dt}.$$

This gives us the derivative of the left-hand side of $u^q = t^p$, which must equal the derivative of the right-hand side, so

$$q\, u^{q-1} \frac{du}{dt} = p\, t^{p-1}.$$

Now u is the function whose derivative we wanted to know, so we solve this equation for du/dt and use algebra to simplify:

$$
\begin{aligned}
\frac{du}{dt} &= \frac{p\,t^{p-1}}{q\,u^{q-1}} \\
&= \frac{p}{q}\,\frac{t^{p-1}}{u^{q-1}} \\
&= \frac{p}{q}\,\frac{t^{p-1}}{(t^{p/q})^{q-1}} \\
&= \frac{p}{q}\,\frac{t^{p-1}}{t^{(q-1)p/q}} \\
&= \frac{p}{q}\,\frac{t^{p-1}}{t^{p-p/q}} \\
&= \frac{p}{q}\,t^{p-1-p+p/q} \\
&= \frac{p}{q}\,t^{-1+p/q} \\
&= r\,t^{r-1}.
\end{aligned}
$$

Exploration Activity 1

In the first step we solved for du/dt by dividing both sides of the previous equation by qu^{q-1}. Give reasons for the rest of the steps in the argument.

These steps are justified by properties of fractions, by the rules of exponents, and by substitution. For example, in the step from the second line to the third we substitute $t^{p/q}$ for u, and the next step is justified by the fact that $(a^b)^c = a^{bc}$. Next, we multiply out the exponent in the denominator. The fifth line follows from

$$\frac{1}{t^{p-p/q}} = t^{-(p-p/q)}$$

and the exponent of a product is the sum of the exponents. We substitute r for p/q to obtain the final result.

Exploration Activity 2 shows how to extend the Power Rule to negative rational exponents.

Exploration Activity 2

Combine the previous result—the Power Rule for positive rational powers—with the Power Rule for the -1 power to derive the Power Rule for negative rational powers. Notation is important for making sense of this task. Suppose you want to differentiate t^r, where r is a *negative* rational number. Then $r = -s$, where s is *positive* (in fact, s is the absolute value of r), so we know the Power Rule for t^s. If we write $u = t^r = t^{-s}$, then we can take reciprocals on both sides and write $u^{-1} = t^s$.

We differentiate both sides of $u^{-1} = t^s$ with respect to t. On the left side we use the result of the Power Rule for the -1 power and the Chain Rule:

$$\frac{d}{dt}u^{-1} = -u^{-2}\frac{du}{dt}.$$

On the right side we use the Power Rule for positive rational powers:

$$\frac{d}{dt}t^s = st^{s-1}.$$

Equating the two sides, we obtain

$$-u^{-2}\frac{du}{dt} = st^{s-1}$$

or

$$\frac{du}{dt} = -su^2t^{s-1}$$

When we substitute t^{-s} for u, we get

$$\frac{du}{dt} = -s(t^{-s})^2t^{s-1} = -st^{-2s+s-1} = -st^{-s-1}.$$

Now, since $r = -s$,

$$\frac{du}{dt} = rt^{r-1}.$$

We have now established that the Power Rule

$$\frac{d}{dt}t^r = rt^{r-1}$$

is correct for every power function with a rational exponent r.

Power Functions with Irrational Exponents

We turn now to the question of what might be meant by t^π or $t^{\sqrt{2}}$ or $t^{\sqrt{\pi}}$ or $t^{-\pi/4}$. When we can answer this question, we also will see that we can calculate the derivatives of such functions with rules that are mostly known to us already. Even though the techniques will turn out to be different from the algebraic calculations we have done already, the answer will be the same—the Power Rule.

We can sneak up on t^π by considering a sequence of rational approximations to π, namely, the successive decimal approximations to π:

$$3, \quad 3.1, \quad 3.14, \quad 3.141, \quad 3.1415, \quad \ldots.$$

Each number in this list is rational, i.e., is the ratio of two integers. Thus, at least for positive values of t, it makes sense to use each of these numbers as an exponent on t. For example, $t^{3.14}$ means the same thing as $t^{314/100}$, which can be interpreted as the 100th root of the 314th power of t. The next number in the list, $t^{3.141}$, means the same things as $t^{3141/1000}$, or the 1000th root of the 3141th power. The restriction that t be positive is necessary to make sense out of the general root step in the interpretations just given.

Now that we understand what is meant by an exponent that is a decimal fraction, we consider the graphs of

$$t^3, \quad t^{3.1}, \quad t^{3.14}, \quad t^{3.141}, \quad \text{and} \quad t^{3.1415}.$$

See Figure 4.45, in which the first four of these functions are graphed with colored lines and the last with a black line. Because each exponent is slightly larger than the one before, the order on the graph is bottom to top. In fact, on the interval shown in Figure 4.45, only the first two of these power functions are visually distinct from the last.

Figure 4.45 Successive power functions approximating t^{π}

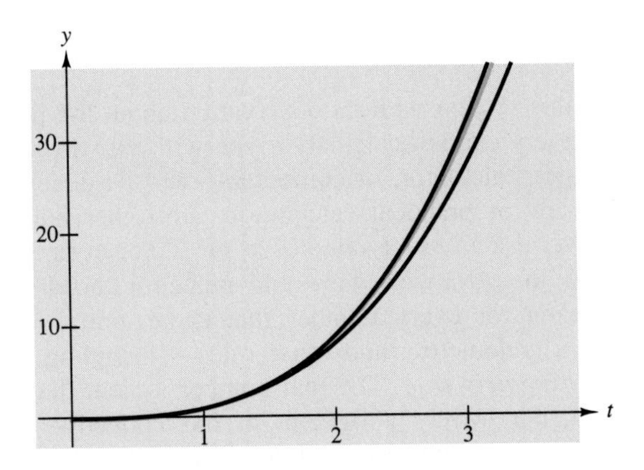

The power functions—the five shown and all the later ones determined by longer decimal approximations to π—*are* distinct, but the differences between them quickly get smaller and smaller, which suggests that there is a limiting function that the power functions approach as the powers get closer to π. That limiting function is what we call t^{π}. If we asked the computer to add the graph of t^{π} to Figure 4.45, we would see no change in the figure—the limiting function is visually indistinguishable from the power function shown already as a black line.

Exploration Activity 3

Write out formulas for the derivatives of each of the following functions:

$$t^3, \quad t^{3.1}, \quad t^{3.14}, \quad t^{3.141}, \quad \text{and} \quad t^{3.1415}.$$

You know how to do this because each of the exponents is a rational number. Now use Figure 4.45 to explain in your own words why

$$\frac{d}{dt} t^{\pi} = \pi\, t^{\pi-1}.$$

We know from the Power Rule for rational exponents that the functions in the list have derivatives $3t^2$, $3.1t^{2.1}$, $3.14t^{2.14}$, $3.141t^{2.141}$, and $3.1415t^{2.1415}$. From Figure 4.45 we can see that the slope of the approximating functions approaches the slope of the limiting function. That is, the derivative of t^{π} is the limiting value of $3t^2$, $3.1t^{2.1}$,

$3.14t^{2.14}$, $3.141t^{2.141}$, $3.1415t^{2.1415}$, ... In short,

$$\frac{d}{dt}t^{\pi} = \pi t^{\pi-1}.$$

Checkpoint 1

Calculate $\dfrac{d}{dt}(t^2+1)^{\sqrt{2}}$.

Symbols

We close this section with an observation about the power and convenience of abstract symbols. Every explicit calculation you will ever do with numbers—whether with pencil and paper, calculator, or computer—will be done with *rational numbers*. From the point of view of practical calculation, any other numbers are fictional. In particular, we can never use an exact value of π, or 2^{π}, or even $2^{3.1415}$ in an arithmetic calculation. But think of the extra work we would create for ourselves if we had to write out a rational approximation for every number that came along. Worse than that, all our familiar algebraic, trigonometric, and other rules—including those of calculus—would be, at best, *approximately true*. The real number system that is the basis of this course is not *real* at all, but merely a figment of our collective imagination. On the other hand, irrational numbers enable us to do exact calculations with circumferences and diameters of circles, with angles and sides of triangles, and with many other important relationships. We have just seen that they enable us to talk about power functions without worrying about whether base, exponent, or value is rational—and the key derivative formula remains the same for all exponents, rational or not.

In this section we have established the **General Power Rule**:

$$\frac{d}{dx}x^r = r\,x^{r-1}.$$

The formula is correct throughout the domain of the function x^{r-1}. For positive integer values of r, the formula is correct for all values of x. For negative integer values of r, the domain consists of all x except zero. For other rational and all irrational values of r, the rth power function is defined for positive numbers x, and the General Power Rule works for these functions as well.

Answer to Checkpoint

1. $2\sqrt{2}\,t\,(t^2+1)^{\sqrt{2}-1}$

Exercises 4.8

Calculate the derivative of each of the following functions.

1. $x^{5/2}$
2. $x^{-5/2}$
3. $x^{5/2} + x^2$
4. $\sqrt{x^2 + 3}$
5. $(x^2 + 3)^{5/2}$
6. $(\ln x)^2$
7. $x^{4/3} \ln x$
8. $(\ln x + x)^{-1/3}$
9. $x\,(p^2 + x^2)^{-1/2}$
10. $(1 - \sqrt{x}) \ln x$
11. $x^{-2/3} \ln x$
12. $(1 + x^2)^{1/3}$
13. Use implicit differentiation to show that the Power Rule holds for the derivative of $x^{5/3}$.
14. Use implicit differentiation to show that the Power Rule holds for the derivative of $x^{9/2}$.
15. Write an operational definition of $t^{3.1415}$ in terms of integer powers and roots.
16. (a) Calculate the derivatives of the two functions e^x and x^e.
    (b) Use your graphing calculator or a computer graphing program to plot the graphs of e^x and x^e on the interval from 0 to 4.
    (c) Do the two graphs in (b) intersect? Do they cross?
    (d) Plot the graphs of the derivatives of the two functions over the same interval. Find the values of x where these two graphs cross.
    (e) What is the geometric significance for the graphs of e^x and x^e of values of x where the derivative graphs cross?
17. A rectangular computer chip is made of ceramic material with circuitry placed in an interior rectangle. Two parallel edges of the circuitry are each 1 millimeter from the corresponding edge of the chip, and the other two edges are each 2 millimeters from the corresponding edges of the chip. How should a chip whose area is 200 square millimeters be designed in order to maximize the area of the circuitry it can accommodate?
18. In a recent (fictional) study, scientists formulated the following equation to represent the amount i of information retained from one day of studying:

$$i = k\,s\,c^{3/2},$$

where s is the number of hours slept, c is the number of hours spent cramming, and k is a proportionality constant. Assuming that the average freshman prepares for finals by cramming every hour that she or he is not sleeping, how much should a student sleep in the course of a day to maximize the knowledge retained? Justify that your answer yields a maximum.

19. A truck traveling on a flat interstate highway at a constant rate of 50 mph gets 4 miles to the gallon. Fuel costs \$1.15 per gallon. For each mile per hour increase in speed, the truck loses a tenth of a mile per gallon in its mileage. Drivers get \$27.50 per hour in wages, and fixed costs for running the truck amount to \$12.33 per hour. What constant speed should a dispatcher require on a straight run through 260 miles of Kansas interstate to minimize the total cost of operating the truck?[8]
20. We gave meaning to the expression t^{π} as a function by treating it as a limiting function obtained from successive rational approximations to π. Since we knew the Power Rule applied to rational powers of t, we were able to show (Exploration Activity 3) that the Power Rule also applies to this irrational-power function. With the aid of the Chain Rule and the rule for differentiating the natural logarithm function, you can show that the Power Rule applies to all power functions:
    (a) Explain why t^r must be the same thing as $e^{r \ln t}$ for every pair of positive real numbers t and r, regardless of what's variable and what's constant.
    (b) Now suppose t is the independent variable and r is a constant. Explain why the function $y = t^r$ is the same as $y = e^u$, where $u = r \ln t$.

(c) Use the Chain Rule to find dy/dt, and simplify to show that the result is the same as the Power Rule.

21. Let f be the function defined by $f(x) = x^x$.
(a) Sketch the graph of f.
(b) Find a formula for $f'(x)$. (*Hint*: $y = x^x$ satisfies the equation $\ln y = x \ln x$.)
(c) Find the smallest value of x^x.

22. In Figure 4.46 we show again the graphs of $L(x)$, the length of path for a reflected ray as a function of where the ray would hit the mirror, and its derivative $L'(x)$.

Figure 4.46 $L(x)$ and $L'(x)$

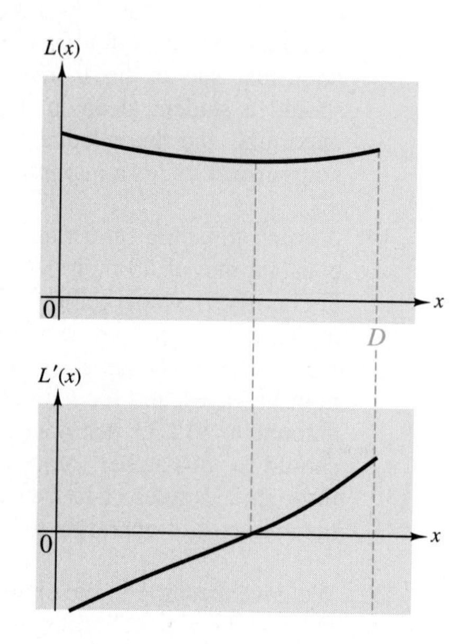

Here again are the formulas for $L(x)$ and $L'(x)$:

$$L(x) = \sqrt{p^2 + x^2} + \sqrt{q^2 + (D - x)^2}$$

and

$$L'(x) = \frac{x}{\sqrt{p^2 + x^2}} + \frac{x - D}{\sqrt{q^2 + (D - x)^2}}$$

for any x between 0 and D.

(a) Find the second derivative of $L(x)$, and show that it is always positive. (*Hint*: See Exercise 9. The p-term and the q-term of L' each produce two terms in the derivative. Combine each pair by using a common denominator, and you should find that each numerator must be positive.)

(b) Explain why the unique solution of $L'(x) = 0$ must be the x-coordinate of a minimum point.

23. (a) Derive a Quotient Rule for differentiation. Let $f(t)$ and $g(t)$ be any two functions, and find a way to express the derivative of $f(t)/g(t)$ in terms of the derivatives of $f(t)$ and $g(t)$. [*Hint*: Write the quotient as $f(t)\,g(t)^{-1}$.]

(b) Use your rule to find the derivative of $(t^2 + 1)/e^t$.

(c) Write the function in part (b) as $(t^2 + 1)\,e^{-t}$, and confirm that you get the same answer for the derivative by using the Product Rule.

Calculate the derivative of each of the following functions.

24. $\dfrac{t^{2/3}}{1 + t^2}$

25. $\dfrac{1 + t^2}{t^{2/3}}$

26. $\dfrac{\ln t}{1 + t^{1/3}}$

---

8. Adapted from *Calculus Problems for a New Century*, edited by Robert Fraga, MAA Notes Number 28, 1993.

4.9 Differentials and Leibniz Notation

It's time to answer the question, "What's a differential?" Perhaps you have forgotten the question. It came up in Chapter 2 when we introduced the notation dy/dt for the instantaneous rate of change of y with respect to t. The numerator dy and the denominator dt were called *differentials* to justify the standard name "differential equation" for an equation containing a derivative of the unknown function. In fact, the portion of calculus that pertains to the study of derivatives is universally known as **differential calculus**, so we really have an obligation to tell you why.

The concept of differential and the quotient notation for the derivative are part of the contribution to the development of calculus made by Leibniz[9]—a small but very important part. As we will see from time to time, notation can have power. Good notation is a powerful tool enabling us to develop concepts and to solve problems. Bad notation can seriously hinder our efforts to do either.

We have already seen an example of the use of Leibniz notation in the development of a concept: the Chain Rule. The notation immediately suggests both the correct statement of the rule and a means for showing that it is indeed correct. Our other notation for derivatives, the prime notation that traces its heritage to Newton, would not have served us nearly so well.

To illustrate the power of the Leibniz notation for problem solving, we provide a small problem you can solve now—with some effort—with nothing but elementary geometry and perhaps a calculator. Later, we will show how the problem can be solved with a quick pencil-and-paper calculation that uses Leibniz notation.

Exploration Activity 1

Estimate the total volume of the earth's crust. You may assume the earth is a sphere with a radius of 4000 miles, and the average depth of the crust is 20 miles.

The volume of a sphere of radius r is $\frac{4}{3}\pi r^3$. If we assume that the radius of the earth is 4000 miles and that the core is a sphere with radius 20 miles smaller, then the volume of the crust is

$$\frac{4}{3}\pi(4000)^3 - \frac{4}{3}\pi(3980)^3 \text{ cubic miles}$$

or approximately

$$4.0 \times 10^9 \text{ cubic miles.}$$

As you may have guessed by now, the concepts of differential and derivative are closely related, but they are not the same thing. To explain the difference, we once again call on the fundamental concept for this course:

$$\text{Slope} = \frac{\text{rise}}{\text{run}}.$$

9. Gottfried Wilhelm Leibniz (1646–1716), German philosopher and mathematician. He and Isaac Newton (1642–1727) independently assembled the key ideas of calculus into a coherent theory, Newton several years earlier, but without publishing until considerably later. This resulted in a bitter controversy over priority between the two men of genius and their followers. Today they are accorded equal status as codiscoverers of calculus.

This equation is meaningful, of course, only if there is a run. That is, it describes the calculation of average rate of change (slope), but it does not directly explain the calculation of instantaneous rate of change (derivative), except through an approaching process in which the run shrinks to zero.

Zooming In—Again

In Figures 4.47, 4.48, 4.49, and 4.50 we recall the zooming-in process by which we observed what happens when the run shrinks to zero: At most points on the graphs of most functions, a sufficiently small segment of the curve looks like a straight line, and the slope of that line is what we called the instantaneous slope of the function, i.e., the derivative. The figures here are like our zooming figures in Chapter 2, with these exceptions: (1) The point at which we are doing the zooming is at the left edge of each picture, not in the center, and (2) we have added to each figure a colored line tangent to the graph at the left edge of the figure.

Figure 4.47 $\Delta t = 1, \ \Delta y = 3$

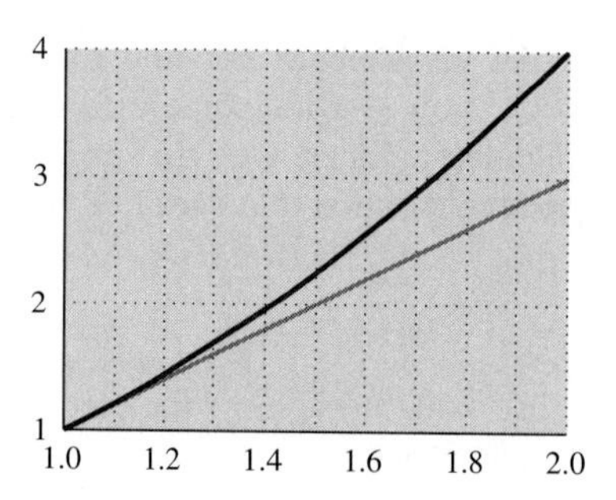

Figure 4.48 $\Delta t = 0.1, \ \Delta y = 0.21$

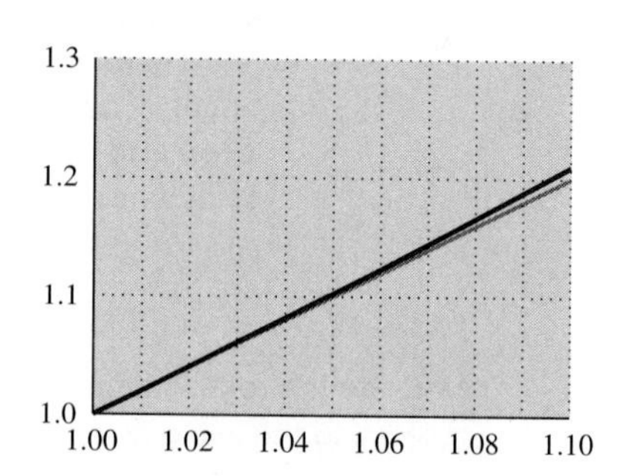

Figure 4.49 $\Delta t = 0.01, \ \Delta y = 0.02$

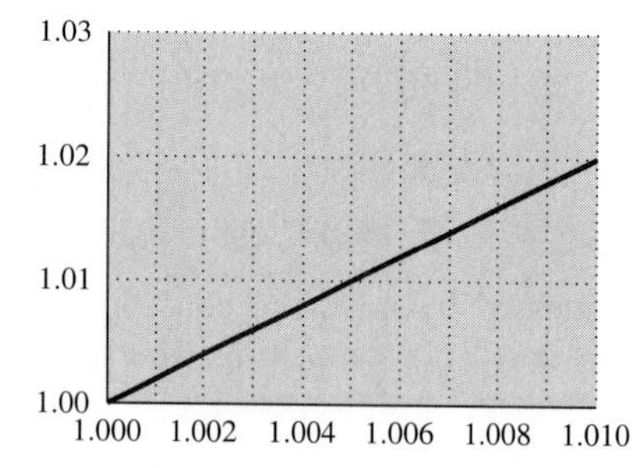

Figure 4.50 $\Delta t = 0.001, \ \Delta y = 0.002$

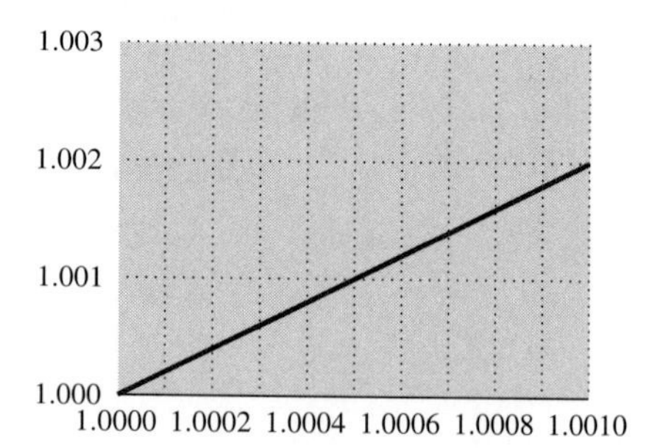

Observe what happens as we zoom in: The curve eventually appears to coincide with the tangent line. Indeed, the line whose slope we called the derivative *is* the tangent line at the point at which we are calculating instantaneous rate of change. Conclusion: At any particular point on the graph of a smooth function $y = y(t)$, the derivative dy/dt is the slope of the tangent line at that point.

As a slope, dy/dt is also a rise over run. What rise, and what run? Since the tangent line is a line, we can take any run we like and calculate the corresponding rise — the ratio of rise to run is constant. Suppose we name the run dt (even though it already has the name Δt) and the rise dy. Then we will have identified two quantities, dy and dt, whose ratio is indeed the derivative. That's it!

Definitions The **differential of** t, calculated at a particular point on the graph of $y = y(t)$, is a (possibly small) change in t, and the **differential of** y is a corresponding change in y — not on the graph of $y = y(t)$ but on the tangent line at the point in question.

Figure 4.51 Rise to the curve (Δy) and rise to the tangent line (dy) for a given run Δt (or dt)

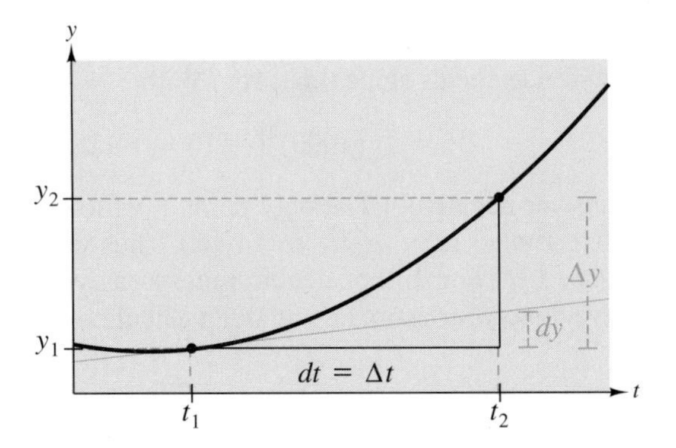

We summarize the definitions and the paragraph preceding them in Figure 4.51, in which we deliberately exaggerate the size of the change in t to show clearly the distinction between the curve and its tangent line. Observe the two different rises in the figure:

Δy is the rise *to the curve* after a run of Δt.

dy is the rise *to the tangent line* after a run of dt.

When we take dt and Δt to be the same run, Δy and dy are clearly different, and it follows that the slopes $\Delta y/\Delta t$ (between two points on the curve) and dy/dt (between two points on the tangent line) are also different.

However, we know the figure is an exaggeration. Our zoom-in figures show what happens when we consider a very small run: The rise to the curve and the rise to the tangent line are approximately equal. In one sense, we already knew that. We introduced the derivative as the limiting value of difference quotients (slopes between points on the curve) when the run shrinks to zero. Thus, for a very small run, the slope between points is approximately the slope at a point, i.e., the derivative. If the slopes are approximately equal, the rises also must be.

In another sense, we have a new idea in the approximate equality of the two rises:

$$dy \approx \Delta y.$$

This says that we can approximate an actual change in y values—a rise on the curve—by something that may be easier to compute, a rise on a straight line. Indeed, for the line, if we know the slope and the run, then we know the rise. The idea is not entirely new—we used this concept in Section 4.4 to derive the Product Rule.

An Application of the Differential

To illustrate the point of approximating change on a curve by change along a tangent line, we now solve Exploration Activity 1 by using a differential. Recall that we wanted to estimate the volume of the earth's crust, knowing that the radius of the earth is about 4000 miles and the thickness of the crust is about 20 miles. As we have seen, you can do this with the formula for volume of a sphere: $V = \frac{4}{3}\pi r^3$. All you have to do is calculate $V(4000) - V(3980)$, a ΔV! But because 20 is small relative to 4000, we can estimate the actual rise in V by a differential, dV. The calculation by hand is a little simpler if we take the run in the negative direction, i.e., from 4000 to 3980. Thus the run is $dr = -20$. We need the rate of change of V with respect to r, but that's easy: $dV/dr = 4\pi r^2$. Thus $dV = 4\pi r^2 dr$—rise equals slope times run. With $r = 4000$ and $dr = -20$, we get

$$dV = 4\pi(4000)^2(-20) = -128 \times 10^7 \times \pi.$$

(Why is the answer negative?) The only point at which we might need some help from a calculator is to estimate 128π —it's about 400. Thus we estimate the earth's crust to have a volume of 4×10^9 (four billion) cubic miles—as you saw in Exploration Activity 1. Could you have done your estimate without a calculator?

Checkpoint 1

(a) Calculate the differential dy for the function $y = 2xe^x$.

(b) Use the differential to estimate the change in y as x increases from 1 to 1.1.

(c) Compare your estimate with the actual change in y.

In this section we have given a name—differential—and a notation—dy—to an idea we have seen frequently, that of linear approximation to change in nonlinear functions.

We also have seen that differentials can be used for quick estimates of changes. Suppose the dependent variable y is a function of the independent variable t, say,

$$y = f(t).$$

The differential dt represents a (possibly small) change in t from a particular value, say, t_1. The corresponding differential dy is the corresponding change in y along the line tangent to the graph of $y = f(t)$ at $(t_1, f(t_1))$. Since the slope of this line is $f'(t_1)$, we have

$$dy = f'(t_1)\,dt.$$

Thus we get a quick estimate of change in y as slope $\times$ run.

Answers to Checkpoint

1. (a) $dy = (2xe^x + 2e^x)dx$
   (b) 1.09
   (c) The actual change is 1.17. The error in the estimate is just over 7%.

Exercises 4.9

1. (a) In the course of Exploration Activity 1 we also computed the area of the surface of the earth. Where did we do that? What is it?
   (b) How is the volume of the crust related to the area of the surface? Is this reasonable? Why or why not?
2. Suppose we were able to determine that the (average) radius of the earth is actually 3959 miles, plus or minus 1 mile.
   (a) Calculate the area of the earth's surface.
   (b) How far off might this calculation be if our radius is actually off by 1 mile?
   (c) Find some region (city, state, country, continent) whose area is approximately equal to the possible error.
   (d) Calculate the percentage error. Is this error small or large?

Calculate the differential dy of each of the following functions.

3. $y = x^{5/2}$
4. $y = x^{-5/2}$
5. $y = x^{5/2} + x^2$
6. $y = \sqrt{x^2 + 3}$
7. $y = (x^2 + 3)^{5/2}$
8. $y = (\ln x)^2$
9. $y = x^{4/3} \ln x$
10. $y = (\ln x + x)^{-1/3}$
11. $y = x\,(p^2 + x^2)^{-1/2}$
12. $y = \ln(x + 2)$
13. $y = \ln(x^2 + 2)$
14. $y = x^2 + 2 \ln x$
15. $y = x \ln x$
16. $y = x^2 \ln(x + 2)$
17. $y = x \ln(x^2 + 2)$
18. $y = \ln(x^2 + 1)$
19. $y = 2^x(x^2 + 1)$
20. $y = e^{x^2}$
21. $y = \sqrt{x^3 + 1}$
22. $y = \ln(5x)$
23. $y = e^{\ln(5x)}$
24. There is a technical problem with the way we described "differential" in that the word has two definitions depending on whether we are talking about independent or dependent variables. The same variable sometimes plays both roles in a single context—for example, the intermediate variable u that we introduced in our discussion of the Chain Rule. How are we to know that these possibly different meanings are actually the same?
   (a) First consider the case of the function $y = t$, that is, the function for which independent and dependent variables always have the same value. Sketch the graph of this function. Explain why $dy = dt$ in this case.
   (b) Now consider a situation in which y is a function of u and u is a function of t. Thus u is both an independent and a dependent variable. Explain why the Chain Rule requires that du means the same thing whether u is viewed as independent or dependent.

Chapter 4 Summary

Much of this chapter has been devoted to developing the body of computational formulas that make calculus a CALCULUS, i.e., a system of calculation. Most of us don't calculate just for the fun of it — we calculate to solve problems. Thus our development of calculational tools has been embedded in the process of solving meaningful problems.

We have now developed all the general rules for differentiation: the **Constant Multiple Rule**, the **Sum Rule**, the **General Power Rule**, the **Product Rule**, the **Inverse Function Rule**, and the **Chain Rule**. The last four were new in this chapter.

Early in the chapter we introduced second derivatives and considered the relationships between the graph of a function and the values of its first two derivatives. This led us to discover the importance of points at which the first derivative is zero — **critical points** — and at which the second derivative is zero — **inflection points** (usually). This brought us to the problem of finding numeric solutions of equations in one variable. Modern electronic tools solve this problem for us, but differential calculus is the key to understanding how they do it. In particular, we saw that local linearity leads to an elegant way to solve an equation quickly: **Newton's Method**.

To calculate critical points, to use Newton's Method, or to carry out many other problem-solving procedures, one must have formulas for computing derivatives. The focus of the rest of the chapter was on developing those formulas. For example, in the context of computing the rate of change of U.S. energy consumption, we developed the **Product Rule** and gave an interpretation of its components.

Next we turned to a mathematical analysis of the physical observation that light rays follow a path which minimizes travel time. In particular, analysis of the angle of reflection led us naturally to the most important derivative formula of all, the **Chain Rule**. When combined with the formulas known already, plus a direct calculation of the derivative of the square root function, the Chain Rule provided the tool we needed for optimizing the reflection travel-time function.

At the end of the chapter we examined the meaning of **differential** as a free-standing object. We found that we could interpret the differential of a dependent variable to mean rise to a tangent line, as opposed to difference of a dependent variable, which is the rise to a point on the graph. For small values of the run, the tangent line and the curve are close together, so differential and difference approximate each other. We used that approximation to estimate the volume of the earth's crust — modeled as a difference of the volumes of two concentric spheres — and we saw that the corresponding differential was easier to calculate.

Concepts

- Second derivative
- Local extreme values and extreme points
- Global extreme values and extreme points
- Zeros and roots
- Critical values and points
- Inflection points
- Newton's Method
- Product Rule
- Chain Rule
- Inverse Function Rule
- General Power Rule
- Implicit differentiation
- Derivative of the natural logarithm function
- Differential
- Differential approximation

Applications

- Minimal production cost
- Growth rate of energy consumption
- Path of reflected light
- Volume of the Earth's crust

Formulas (cumulative listing)

Newton's Method

Starting from an approximation t_0 to a zero of f, generate

$$t_{n+1} = t_n - \frac{f(t_n)}{f'(t_n)}, \quad \text{for } n = 0, 1, 2, \ldots .$$

Exponential Functions

$$\frac{d}{dt} e^t = e^t, \text{ where } e \text{ is the natural base, } 2.71828 \ldots$$

$$\frac{d}{dt} e^{kt} = k\, e^{kt}, \text{ for any constant } k.$$

$$\frac{d}{dt} b^t = (\ln b)\, b^t, \text{ for any constant } b.$$

Power Rule

$$\frac{d}{dt} t^r = r\, t^{r-1}, \text{ for any real constant } r.$$

Natural logarithm function

$$\frac{d}{dt} \ln t = \frac{1}{t}$$

Constant Multiple Rule

$$\frac{d}{dt} A f(t) = A \frac{d}{dt} f(t), \text{ where } A \text{ is any constant.}$$

Sum Rule

$$\frac{d}{dt}[f(t) + g(t)] = \frac{d}{dt} f(t) + \frac{d}{dt} g(t)$$

Product Rule

$$\frac{d}{dt}[g(t)\, h(t)] = g(t) \frac{d}{dt} h(t) + h(t) \frac{d}{dt} g(t)$$

Inverse Function Rule

$$\frac{dy}{dt} = \frac{1}{dt/dy}$$

Chain Rule

$$\frac{dy}{dt} = \frac{dy}{du} \frac{du}{dt}$$

Special case of the Chain Rule

$$\frac{d}{dt} f(kt) = k \frac{d}{du} f(u),$$

where $u = kt$ and k is any constant.

Combinations of the Chain Rule with specific function rules

$$\frac{d}{dt} f(t)^r = r\, f(t)^{r-1} \frac{d}{dt} f(t), \text{ for any real constant } r.$$

$$\frac{d}{dt} e^{f(t)} = e^{f(t)} \frac{d}{dt} f(t)$$

$$\frac{d}{dt} \ln f(t) = \frac{1}{f(t)} \frac{d}{dt} f(t)$$

Chapter 4 Exercises

Calculate each of the following derivatives.

1. $\frac{d}{dt}(6t^5 - 4t^3 + 7)$

2. $\frac{d}{dw} \frac{1}{w^4}$

3. $\frac{d}{dt} t^{3/5}$

4. $\frac{d}{dt} e^{t^2}$

5. $\frac{d}{dt} \sqrt{t^3 + 1}$

6. $\frac{d^2}{dt^2} \sqrt{t^3 + 1}$

7. $\dfrac{d}{dy}\sqrt{y+1}$

8. $\dfrac{d}{dx}\sqrt{x^2+1}$

9. $\dfrac{d}{dx}(x \ln x + 1)^{3/2}$

10. $\dfrac{d}{dw}(7w^7 - 3w^3 + 8)$

11. $\dfrac{d}{dx}\dfrac{1}{(x+1)^2}$

12. $\dfrac{d^2}{dx^2}\dfrac{1}{(x+1)^2}$

13. $\dfrac{d}{dx}\ln(x^2+2)$

14. $\dfrac{d}{dt}\dfrac{(\ln t)^3}{t}$

15. $\dfrac{d}{dt}\ln 5t^3$

16. $\dfrac{d}{dt}\dfrac{e^{-t}}{1+t^2}$

17. $\dfrac{d}{dx}\dfrac{\ln x + x}{\sqrt{x}}$

18. $\dfrac{d}{dy}\dfrac{e^{3y}}{\ln(2y+1)}$

19. If $f(x) = x$, find $f'(5)$.

Find the derivative of each of the following functions.

20. $f(x) = x^3 + 4x^2 - 2x - 7$

21. $g(t) = t - e^{-3t}$

Find the second derivative of each of the following functions.

22. $f(x) = x^3 + 4x^2 - 2x - 7$

23. $g(t) = t - e^{-3t}$

Use the ideas developed in Sections 4.1 and 4.2 (as many as apply) to sketch the graph of each of the following functions (see Exercises 20–23). Compare your hand-drawn graphs with graphs drawn by calculator or computer. (Do your own first!)

24. $f(x) = x^3 + 4x^2 - 2x - 7$

25. $g(t) = t - e^{-3t}$

26. (a) Find a function of t whose derivative is t.
    (b) Find a function whose derivative is t^2.
    (c) Find a function whose derivative is t^3.
    (d) Find a function whose derivative is t^n, where n is any positive integer. Verify your result by showing that the derivative of your answer function really is t^n.
    (e) Does your answer to part (d) work if $n = 0$? Why or why not?

27. In the preceding exercise we used the phrase "a function whose derivative is" several times. Such a function is called an **antiderivative**, and the process of finding it (i.e., of undoing differentiation) is called **antidifferentiation**.
    (a) Find three different antiderivatives of t^7.
    (b) Formulate a general principle that enables you to find infinitely many antiderivatives of a given function, once you know one of them.

28. Find an antiderivative of the function e^{kx}, where k is any nonzero constant, and x is the independent variable. (*Hint*: Guess the form of such a function, differentiate it, and compare the derivative to e^{kx}. Then adjust your guess.)

29. Suppose you know antiderivatives for the functions $f(t)$ and $g(t)$. How would you find an antiderivative for $f(t) + g(t)$? Use a differentiation rule to justify your answer.

30. Suppose you know an antiderivative for the function $f(t)$. How would you find an antiderivative for $cf(t)$, where c is a constant? Use a differentiation rule to justify your answer.

Find an antiderivative of each of the following functions.

31. $f(x) = x^3 + 4x^2 - 2x - 7$

32. $g(t) = t - e^{-3t}$

33. For what point (or points) on the graph of $y = 3 - x^2$ does the tangent line go through the point $(2, 0)$?

34. A ship is moving directly away from shore at a speed of 15 knots. A person on the ship is walking toward the bow at a speed of 3 knots.

    (a) How fast is this person moving away from shore?

    (b) What differentiation rule does this problem illustrate?[10]

35. Take a circle of radius r, and cut out a sector subtended by an angle θ, producing a PacMan-like figure. Join the cut edges together (the upper and lower jaws of PacMan) to form a cone. What angle θ produces the largest volume of such a cone?

36. What is the area of the largest triangle that can be formed in the first quadrant by the x-axis, the y-axis, and a tangent to the graph of $y = e^{-x}$?

37. In Figure 4.52 we repeat the velocity graph from Section 4.1 for the leaky helium balloon. How high does the balloon go before it starts to descend? Explain your strategy for answering this question.

Figure 4.52 Velocity of a balloon rising and falling

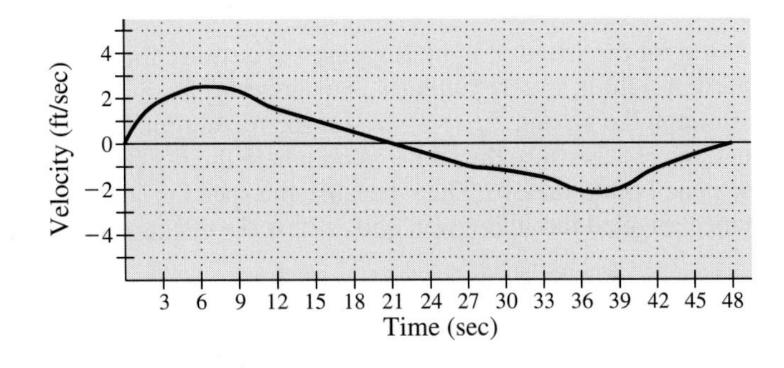

10. This problem and the next two are adapted from *Calculus Problems for a New Century*, edited by Robert Fraga, MAA Notes Number 28, 1993.

Chapter 4 Projects

1. **Air Traffic Control** American Flight 1003 is traveling from Minneapolis to New Orleans, and United Flight 366 is traveling from Los Angeles to New York. Both flights are at 33,000 feet, and the flight paths intersect over Ottumwa, Iowa. At 1:30 P.M. (Central time), the American flight is 32 nautical miles (horizontally) from Ottumwa and is approaching it on a heading of 171 degrees at a rate of 405 knots. The United flight is 44 nautical miles from Ottumwa and is approaching it on a heading of 81 degrees at a rate of 465 knots.

   (a) At this instant, how fast is the distance between the planes decreasing?

   (b) How close will the planes come to each other? Will they violate the FAA's minimum separation requirement of 5 nautical miles? Will they collide if Air Traffic Control does not take action to separate them?

   (c) What time will it be at the time of closest approach? Is there enough time for ATC to take appropriate action?

2. **Refraction**[11] In this project we use Fermat's Principle to analyze the refraction of light. Review Section 4.6 for background information. To help you get started, we repeat the refraction picture (Figure 4.53) with some added labels.

Figure 4.53 Angle of incidence and angle of refraction

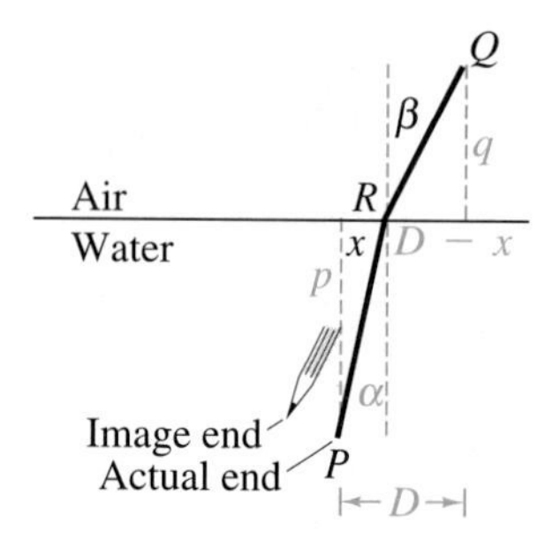

In contrast to reflection, with refraction there isn't just one speed of light, but two—one speed in air and one in water. In general, light has a different speed for each medium through which it travels.

The symbol c is commonly used for the speed of light. Suppose we write c_a for the speed of light in air and c_w for the speed of light in water. Similarly, we can write $T_a = T_a(x)$ for the travel time of the light ray through the air and $T_w = T_w(x)$ for the travel time through water. The travel times are functions of x because they depend on where the ray passes from water to air on its way from the pencil tip to the eye. Then, because the speed is constant in each medium, $c_w T_w$ is the distance traveled through water, and $c_a T_a$ is the distance traveled through air.

11. Based on "Somewhere Within the Rainbow," by Steven Janke, in *Applications of Calculus*, edited by Philip Straffin, MAA Notes, 1993, and in part on *Five Applications of Max-Min Theory from Calculus*, by W. Thurmon Whitley, UMAP Module 341, COMAP, 1979.

(a) Show that the total travel time, $T(x)$, is given by

$$T(x) = \frac{\sqrt{p^2 + x^2}}{c_w} + \frac{\sqrt{q^2 + (D - x)^2}}{c_a}$$

for any x between 0 and D.

Since $T(x)$ is similar in form to the reflection function $L(x)$, its graph is not very different from the one shown in Figure 4.39. Minimizing the total travel time for a refracted ray is essentially the same problem as minimizing the travel time for a reflected ray. The formula has two added symbols, but $1/c_w$ and $1/c_a$ are constant factors in their respective terms.

(b) Calculate $T'(x)$.

(c) Show that $T(x)$ is minimized when

$$\frac{\sin \alpha}{c_w} = \frac{\sin \beta}{c_a}.$$

(d) Use (b) to justify the classic description of the angle of refraction:

Snell's Law. *The ratio of the sine of the angle of incidence to the sine of the angle of refraction is a constant.*

(e) Show that the second derivative of T is always positive, and therefore that any solution of $T'(x) = 0$ is the x-coordinate of a minimum point.

(f) Suppose a light ray passes through a plate of glass, such as a window pane, entering at an angle α and leaving at an angle γ. Show that $\alpha = \gamma$. (*Hint*: Draw a picture that shows the path of the ray entering, traversing, and leaving the plate. Use Snell's Law.)

3. **Oil Consumption**[12] In Section 4.9 we determined that the volume of the earth's crust is about 4×10^9 cubic miles. Here we explore the somewhat fanciful question of when we might run out of oil in a "best case" scenario, namely, if the entire crust were oil.

(a) The rate of consumption of oil grew steadily for a century at 7.04% per year. Thus the consumption rate has a formula of the form $r(t) = r_0 e^{kt}$, for some constants k and r_0. Take the time unit to be one day (not one year), and find k.

(b) In 1988, the rate of consumption of oil world wide was $9,280,000$ cubic meters per day. Convert this to appropriate units to be an initial rate r_0. Suppose the consumption rate *stopped* growing in 1988 and continued at the level r_0 forever. How long would it take to use up 4×10^9 cubic miles of oil? Compare this with the estimated age of the earth.

(c) Now suppose the consumption rate continued to grow exponentially for as long as possible. That is, suppose the formula $r(t) = r_0 e^{kt}$ continues to describe the consumption rate beyond 1988. Find an antiderivative of $r(t)$, that is, a function for which $r(t)$ is the growth rate. Find the particular antiderivative that has the value r_0 at time $t = 0$ in 1988. At any time t beyond $t = 0$, this function describes the total consumption since 1988. How long would it take to use up 4 billion cubic miles of oil if the historic increase of use continues? Contrast this with the answer to part (b).

12. Based on A. A. Bartlett, "A World Full of Oil," *The Physics Teacher*, 28 (1990), pp. 540–541.

(d) What fraction of the volume of the earth's crust do you think could possibly be oil? How do your answers to parts (b) and (c) change if your estimate of the possible resource is correct?

4. **Pricing and Production** We quote the scenario from Whitley[13]:

> Several years ago, the Boeing Aircraft Company was faced with the problem of determining the selling price for a new model jet airliner. The basic problem was to find the price per aircraft which would maximize the company's profit. In this particular case, Boeing had one competitor, which had a similar plane. It was understood that the companies would charge the same price, since any price adjustment by one company would automatically be met by the other. Thus, the price would not affect the relative shares of the market. It could, however, have a significant impact on the total size of the market.

We denote by $N(p)$ the total number of airliners that would be sold at price p (in millions of dollars) by Boeing and its competitor. We write $C(X)$ for the total cost (also in millions of dollars) to Boeing of manufacturing X airliners. Analysts at Boeing made market predictions and cost estimates and arrived at the following expressions for these functions:

$$\begin{aligned} N(p) &= -78p^2 + 655p - 1125 \\ C(X) &= 50 + 1.5X + 8X^{3/4}. \end{aligned}$$

(a) Find the range of prices p for which $N(p) \geq 0$.

(b) If P denotes the total profit (in millions of dollars) to Boeing and h is the fraction (between 0 and 1) of the market to be won by Boeing, show that

$$P = phN(p) - C(hN(p)).$$

(c) Calculate $P'(p)$. (*Hint*: Use the Chain Rule.)

(d) Show that the critical price p that maximizes profit P must satisfy the condition

$$p + \frac{N(p)}{N'(p)} = C'(X) \qquad \text{if } N'(p) \neq 0.$$

Also show that if $N'(p) = 0$, then $P'(p) \neq 0$ unless h is zero. Thus, if Boeing has any market share at all, then this condition must hold at the critical price.

(e) Calculate $N'(p)$ and $C'(X)$.

(f) Substitute your formulas for $N(p)$, $N'(p)$, and $C'(X)$, along with $X = hN(p)$, into the condition in part (d) to find a condition on the critical price that (almost) involves only p. You can't quite get rid of h, but in principle, one could set various values of market share and solve for the critical price. However, you may find that the equation to be solved is sufficiently complicated that you wouldn't even want to try solving it by Newton's Method. We'll try another tack first.

(g) Calculate $C'(X)$ for $X = 60$, 80, 100, and 120. You should find that C' doesn't change much over a wide range of production numbers. Why is that? (*Hint*: Calculate *its* rate of change, and explain why it is small when X is large.)

---

13. Based on pages 6–10 of *Five Applications of Max-Min Theory from Calculus*, by W. Thurmon Whitley, UMAP Module 31, COMAP, 1979. Whitley's treatment is in turn based on "Pricing, Investment, and Games of Strategy" by Georges Brigham, in *Management Sciences, Models and Techniques*, C. W. Churchman and M. Verhulst (eds.), Pergamon Press, 1960.

(h) The implication of (g) is that the right-hand side of the condition in part (d) is almost constant. Select an "average" value to replace the right-hand side. Then clear the fractions to rewrite the condition in part (d) as a quadratic equation in p. Now you can solve for the (approximately) optimal price with your calculator.

(i) The equation in part (h) has two solutions. Explain why the profit-maximizing price is about five million dollars per plane, regardless of the number made or the market share.

(j) Find the maximum value of the total market function $N(p)$. What price produces the largest number of sales? How many planes will be sold at this price?

(k) How many planes can be sold by both companies at the optimal price you arrived at in part (i)?

(l) If Boeing makes 70 planes at the optimal price, what will their profit be? How many planes will the competitor make, and what will Boeing's market share be?

(m) For each Boeing production level X from 50 to 200 (in steps of 10), calculate $C'(X)$, solve the condition in part (d) for the optimal price, and find the total production for both companies, the Boeing market share, and the Boeing profit. From your tabulated results, advise the company on their pricing and production strategies.

CHAPTER

5

Applications of Iteration

We begin this chapter where we left off in Chapter 4, looking at the differential as **rise over run***. In particular, we will see that the differential approximation is also the heart of an iterative numerical method for solving initial value problems:* **Euler's Method***.*

We have already used iterative methods to solve different types of problems. For example, in Chapter 2, we saw that we could record the growing balance in an interest-bearing bank account by writing down an iterative formula that calculated each year's balance from the previous year's balance. (Replacing the iterative formula by an explicit formula led us to exponential functions.) In a laboratory exercise associated with Chapter 3 (Raindrops), you may have encountered the numerical method for solving initial value problems that we will here identify as Euler's Method. And in Chapter 4 we saw that algebraic equations in one unknown can be solved by an iteration called Newton's Method. The idea of iteration—repeating the same type of step over and over—is a powerful concept that we will use many more times in this course. Most iterations would not be practical to carry out with pencil and paper, but the concept is ideally suited to programmable calculators and computers. Indeed, this concept is at the heart of most applications of electronic computing technology.

After we formalize Euler's Method, we extend the method to handle several dependent variables at once, and we use this version to study a model for the spread of epidemics. We will see that this model enables us to predict the behavior of future epidemics—and to plan inoculation campaigns to prevent them.

We then turn to the study of price evolution in an economy in which prices adjust so that supply equals demand. Plausible simplifying assumptions, when incorporated into a differential equation model, lead to a problem we have already seen several times—and therefore know how to solve. The same assumptions, when incorporated into a difference equation model—an iteration—lead to the need for new solution techniques.

5.1 Euler's Method

In an early lab you may have considered a model for constrained population growth—growth for which there is a maximum sustainable population. We repeat the model here in a more symbolic form. In this model, t denotes time, and $P = P(t)$ denotes the unknown population function. The initial value problem has the form

$$\frac{dP}{dt} = cP(M - P) \qquad \text{with } P = P_0 \text{ at } t = t_0,$$

where

$$\begin{aligned} M &= \text{maximum sustainable population,} \\ P_0 &= \text{initial population, and} \\ c &= \text{a proportionality constant.} \end{aligned}$$

In the lab we found both numerical and graphical approximations to the solution of such an equation. A typical solution has the S-form displayed in Figure 5.1.

Figure 5.1 Typical solution curve

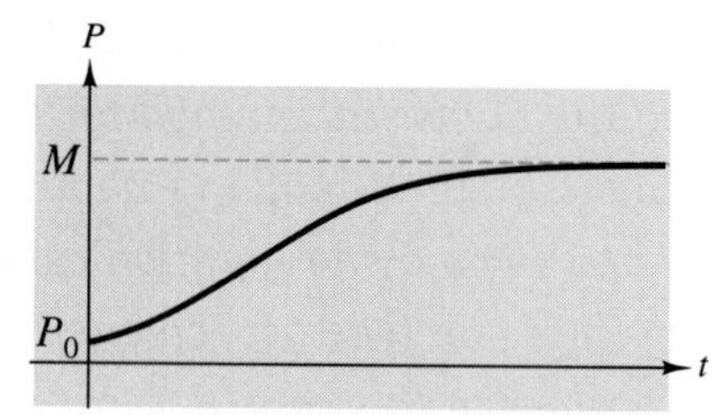

Our strategy for generating approximate values of P was this: We selected a time step Δt, so we had a sequence of time values

$$\begin{aligned} t_0 &= 0, \\ t_1 &= 0 + \Delta t = \Delta t, \\ t_2 &= t_1 + \Delta t = 2\,\Delta t, \end{aligned}$$

and so on. Starting from P_0, we wanted to compute

$$P_1 = P(t_1) = P_0 + \Delta P_0,$$

then

$$P_2 = P(t_2) = P_1 + \Delta P_1,$$

and so on. That is, to get from each P to the next one, we needed to compute a *rise*—a vertical change from one point on the solution curve to another. Because we had no way to do that exactly, we computed the rise to the tangent line as *slope* times *run*:

$$\begin{aligned} dP &= cP(M - P)dt \\ &= cP(M - P)\Delta t, \end{aligned}$$

since $dt = \Delta t$ for the independent variable. Then we found each *new* P as *old* P *plus* dP

or *old P plus slope times run*. In symbols, that gave us the iterative formula

$$P_{k+1} = P_k + cP_k(M - P_k)\Delta t, \qquad k = 0,\ 1,\ 2,\ \ldots\ .$$

That's **Euler's Method** for solving an initial value problem: Starting from a known point, we move to the next one by computing the rise as slope times run, where the slope at the known point is given by the differential equation and the run is chosen by the problem solver.

In general, if the slope at every point is given by a differential equation of the form

$$\frac{dP}{dt} = f(P)$$

and a starting point P_0 is known, then Euler's Method computes values of an approximate solution function by the iteration

$$P_{k+1} = P_k + f(P_k)\Delta t, \qquad k = 0,\ 1,\ 2,\ \ldots\ .$$

If the right-hand side of the differential equation is a function of t rather than of P, say, $g(t)$, then the Euler iteration takes the form

$$P_{k+1} = P_k + g(t_k)\Delta t, \qquad k = 0,\ 1,\ 2,\ \ldots\ .$$

The computation of the *new* P is always *old* P *plus slope times run*, regardless of how the slope is calculated.

Example 1 Calculate approximate values of $P(0.1)$, $P(0.2)$, and $P(0.3)$, where $P(t)$ is the solution of the initial value problem

$$\frac{dP}{dt} = P + 1, \qquad P(0) = P_0 = 2.$$

Solution We calculate the first three steps of the Euler iteration with $\Delta t = 0.1$:

$$P(0.1) \approx P_1 = P_0 + (P_0 + 1)\Delta t = 2 + 2(0.1) = 2.2$$
$$P(0.2) \approx P_2 = P_1 + (P_1 + 1)\Delta t = 2.2 + (3.2)(0.1) = 2.52$$
$$P(0.3) \approx P_3 = P_2 + (P_2 + 1)\Delta t = 2.52 + (3.52)(0.1) = 2.872$$

■

In this section we have reviewed and formalized a numerical method for approximating the solution of an initial value problem—Euler's Method. As long as the run is sufficiently small, Euler's Method succeeds because $dP \approx \Delta P$. Thus the method can be viewed as repeated application of the differential approximation. In the next section we apply this approximation scheme to the study of epidemics.

Exercises 5.1

In Exercises 1–3, use Euler's Method to generate approximate values P_1, P_2, and P_3 of the solution $P(t)$ of the initial value problem

$$\frac{dP}{dt} = P^2, \qquad P(0) = 1,$$

with the indicated time step.

1. $\Delta t = 0.1$
2. $\Delta t = 0.01$
3. $\Delta t = 0.001$

In Exercises 4–6, use Euler's Method to generate approximate values P_1, P_2, and P_3 of the solution $P(t)$ of the initial value problem

$$\frac{dP}{dt} = P^2 + 1, \qquad P(0) = 2,$$

with the indicated time step.

4. $\Delta t = 0.1$
5. $\Delta t = 0.01$
6. $\Delta t = 0.001$

In Exercises 7–9, use Euler's Method to generate approximate values P_1, P_2, and P_3 of the solution $P(t)$ of the initial value problem

$$\frac{dP}{dt} = t^2, \qquad P(0) = 1,$$

with the indicated time step.

7. $\Delta t = 0.1$
8. $\Delta t = 0.01$
9. $\Delta t = 0.001$
10. Find the *exact* solution of the initial value problem

$$\frac{dP}{dt} = t^2, \qquad P(0) = 1.$$

[*Hint*: Determine all the functions that have derivative t^2, and let $P(t)$ be the one that satisfies the initial condition.]

11. Evaluate your solution function from Exercise 10 at $t = 0.1$, 0.2, and 0.3. Compare these exact values with the approximate values from Euler's Method in Exercise 7. How good are the approximations?
12. Evaluate your solution function from Exercise 10 at $t = 0.01$, 0.02, and 0.03. Compare these exact values with the approximate values from Euler's Method in Exercise 8. How good are the approximations?
13. Evaluate your solution function from Exercise 10 at $t = 0.001$, 0.002, and 0.003. Compare these exact values with the approximate values from Euler's Method in Exercise 9. How good are the approximations?

In Exercises 14–16, use Euler's Method to generate approximate values P_1, P_2, and P_3 of the solution $P(t)$ of the initial value problem

$$\frac{dP}{dt} = t^2 + 1, \qquad P(0) = 2,$$

with the indicated time step.

14. $\Delta t = 0.1$
15. $\Delta t = 0.01$
16. $\Delta t = 0.001$
17. Find the *exact* solution of the initial value problem

$$\frac{dP}{dt} = t^2 + 1, \qquad P(0) = 2.$$

(*Hint*: See Exercise 10.)

18. Evaluate your solution function from Exercise 17 at $t = 0.1$, 0.2, and 0.3. Compare these exact values with the approximate values from Euler's Method in Exercise 14. How good are the approximations?
19. Evaluate your solution function from Exercise 17 at $t = 0.01$, 0.02, and 0.03. Compare these exact values with the approximate values from Euler's Method in Exercise 15. How good are the approximations?
20. Evaluate your solution function from Exercise 17 at $t = 0.001$, 0.002, and 0.003. Compare these exact values with the approximate values from Euler's Method in Exercise 16. How good are the approximations?

5.2 Modeling Epidemics

Hong Kong Flu

During the winter of 1968–1969, the United States was swept by a virulent new strain of influenza, named "Hong Kong flu" for its place of discovery. We will study the spread of the disease through a single urban population, that of New York City. The available data (Figure 5.2 and Table 5.1) consist of weekly totals of excess pneumonia-influenza deaths, that is, the number of such deaths in excess of the average number to be expected from other sources.

Table 5.1 Flu-related deaths[1]

Week	*Excess Deaths*
1	14
2	28
3	50
4	66
5	156
6	190
7	156
8	108
9	68
10	77
11	33
12	65
13	24

Figure 5.2 Flu-related deaths

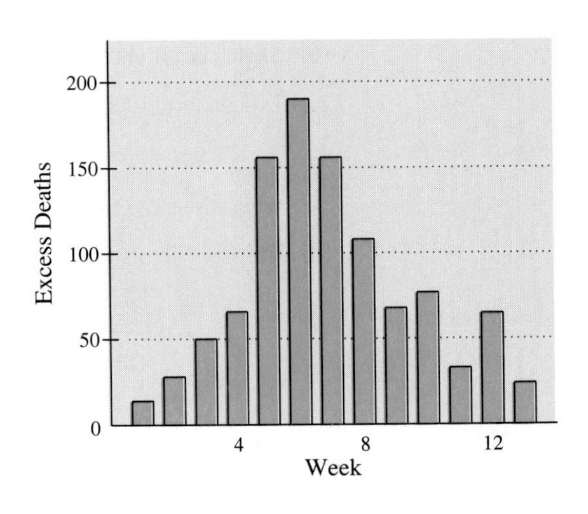

Relatively few flu sufferers died from the disease or its complications. However, we may reasonably assume that the number of excess deaths in a week was proportional to the number of new cases of flu in some earlier week, say, three weeks earlier. Thus the figures in Table 5.1 reflect (proportionally) the rise and subsequent decline in the number of new cases of Hong Kong flu. We want to model the spread of such a disease so that we can predict what might happen with similar epidemics in the future.

At any given time during a flu epidemic, we want to know the number of people who are infected. We also want to know the number who have been infected and have recovered, because these people now have an immunity to the disease. (As a matter of convenience, we include in the recovered group the relative handful who do not recover but die—they too can no longer contract the disease.) If we ignore movement into and out of the infected area, then the remainder of the population is still susceptible to

1. Source: Centers for Disease Control, as quoted in "Mathematical Models for Urban-Suburban Spread of Disease," by Geoffrey M. Davis, a paper submitted for Graduation with Distinction from Duke University, April 11, 1988.

the disease. Thus, at any time, the fixed total population (approximately 7,900,000 in the case of New York City) may be divided into three distinct groups: those who are *infected*, those who have *recovered*, and those who are still *susceptible*.

The Initial Value Problem

For a new disease, which Hong Kong flu was in 1968, the initial number of infected individuals is quite small, and everyone else is initially susceptible.

Exploration Activity 1

Let $S = S(t)$ stand for the number of susceptible individuals as a function of time, $R = R(t)$ the number of recovered individuals, and $I = I(t)$ the number of infected individuals.

(a) Under the assumptions we have made, how do you think S should vary with time? How should R vary with time? How should I vary with time?

(b) Sketch what you think the graph of each of these functions looks like.

(c) Explain why, at each time t,

$$S(t) + R(t) + I(t) = N,$$

where N is the total population.

(d) List some factors that you think might govern the rates of change of S, I, and R.

Our assumptions suggest that, at the start of the epidemic, $S \approx N$, with I having a small positive value, and $R = 0$. As time progresses and susceptible individuals become infected, S should decrease and I increase. Also, as infected individuals recover, R should increase. Eventually, the epidemic should peak, and I should start to decline, perhaps eventually reaching zero. Our graphs incorporating these features will appear later in the section. Keep track of yours for comparison.

At each time t, we have assigned every individual to one of the three groups, so the sum of their numbers, $S + R + I$, must always equal the total population. This is one reason why it is convenient to include those who have died in the recovered group—we don't have to account for them separately.

We will investigate a model that accounts for some, but probably not all, of the factors you listed in part (d) of Exploration Activity 1. We ignore minor changes in the population by births, travel, deaths unrelated to flu, and perhaps others you thought of.

The only way an individual is removed from the susceptible group is by becoming infected with the disease. We assume that the rate of change of S with respect to time depends on the number of individuals in the susceptible category, the number of individuals in the infected category, and the amount of contact there is between them. Suppose that each infected individual has a fixed number β of contacts per day that are sufficient to spread the disease. Not all these contacts are with individuals susceptible to the disease. If we assume a homogeneous mixing of the population, the fraction of these contacts that are with susceptibles is $S(t)/N$. Thus, on average, each infected individual generates $\beta S(t)/N$ new infected individuals per day. Accounting for the new cases of the disease generated by all the infected individuals, we have

$$\frac{dS}{dt} = -\frac{\beta}{N} I(t)\, S(t),$$

where time is measured in days.

Checkpoint 1

Explain carefully how each component of this differential equation follows from the text preceding. In particular, why is the factor of $I(t)$ present? Where did the negative sign come from?

The quotient β/N appears so often in what follows that we find it useful to introduce a single symbol for it. We let

$$\alpha = \frac{\beta}{N}\,.$$

Using this notation, we can simplify our differential equation to

$$\frac{dS}{dt} = -\,\alpha\, I(t)\, S(t).$$

Now we turn to the rate of change of $I(t)$, for which we need to consider both the movement of individuals from the susceptible group into the infected group (as we have just done) and the movement of individuals from the infected group to the recovered group. We assume that a fixed fraction λ of the infected group will recover during any given day. Thus

$$\frac{dI}{dt} = \alpha\, I(t)\, \mathrm{S}(t) - \lambda\, I(t).$$

Checkpoint 2

Explain carefully how each component of this equation follows from the text preceding. In particular, why are there two terms? Why is it reasonable that the rate of flow from the infected population to the recovered population should depend only on $I(t)$? Where did the minus sign come from?

Exploration Activity 2

(a) Using an argument similar to those above, explain why

$$\frac{dR}{dt} = \lambda I(t).$$

(b) Explain why the differential equations for S, I, and R are compatible with the formula

$$S(t) + R(t) + I(t) = N.$$

Every individual leaving the infected group (including those who die) is entering the recovered group, so the growth rate of R is the negative of the second term in the rate of change of I. One can quibble about whether the recovery-rate coefficient λ is the right thing to be using—in either differential equation. However, the number who die is such a small fraction of the infected population that there is no need to account for their transition rate separately.

If you add the three differential equations together, you will find that

$$\frac{dS}{dt} + \frac{dI}{dt} + \frac{dR}{dt} = 0,$$

which is consistent with the total population remaining constant.

We now have the three quantities of interest, $S(t)$, $I(t)$, and $R(t)$, governed by a system of three differential equations. Each equation has an initial condition, that is, a (presumably) known value at the start of the epidemic.

Checkpoint 3

Explain why reasonable initial values for the Hong Kong flu epidemic in New York are

$$S(0) = 7{,}900{,}000$$
$$I(0) = 10$$
$$R(0) = 0.$$

We must determine the constants β and λ experimentally; i.e., we select them so that the resulting functions fit the data on hand. For the New York flu epidemic, reasonable values turn out to be $\beta = 0.6$ and $\lambda = 0.34$. (This value of λ corresponds to the observation that the average flu sufferer remains infectious for about three days—see Exercise 2.) In a lab associated with this chapter, you may experiment with other values to see what effect they have on the solution functions.

Here is a summary of our problem: We want to find functions $S(t)$, $I(t)$, and $R(t)$ that satisfy these differential equations and initial conditions:

$$\frac{dS}{dt} = -\alpha\, I(t)\, S(t) \qquad S(0) = 7{,}900{,}000,$$

$$\frac{dI}{dt} = \alpha\, I(t)\, S(t) - \lambda\, I(t) \qquad I(0) = 10,$$

$$\frac{dR}{dt} = \lambda\, I(t) \qquad R(0) = 0.$$

This looks like *three* initial value problems, but it is really *only one*. That's because each of three differential equations involves at least one of the other unknown functions.

The Euler's Method Approximation

As we did with initial value problems involving a single differential equation, we can approximate a solution by using Euler's Method. What is new here is that at each time step t_k we have three approximate quantities—S_k, I_k, and R_k—and we have to alter each one by adding *its* approximate rise. For example,

$$S_{k+1} = S_k - \alpha I_k S_k \Delta t.$$

Checkpoint 4

(a) Why is the approximate rise equal to $-\alpha I_k S_k \Delta t$? (In this case the rise is actually a drop.)

(b) Study the differential equations for I and R, and write the other two iterative formulas.

Checkpoint 5

Assume that $\beta = 0.6$ and $\lambda = 0.34$. Then $\alpha = \beta/N = 7.6 \times 10^{-8}$. With time measured in days, and $\Delta t = 1$ day, use your formulas from Checkpoint 4 and ours for S_{k+1} to

(a) check the entries in Table 5.2,

(b) fill in the missing entries in Table 5.2.

It would be extremely tedious to extend Table 5.2 to, say, 13 weeks—unless we wrote a program to carry out the iteration. In Figure 5.3 we show a computer-generated solution for each of the three population groups.

Table 5.2 Approximate values of S, I, and R for five days of the New York flu epidemic

k	t_k	S_k	I_k	R_k
0	0	7,900,000	10	0
1	1	7,899,994	13	3
2	2	7,899,986	16	8
3	3			
4	4			
5	5			

Figure 5.3 Susceptible, infected, and recovered populations generated by the theoretical model with $\beta = 0.6$ and $\lambda = 0.34$

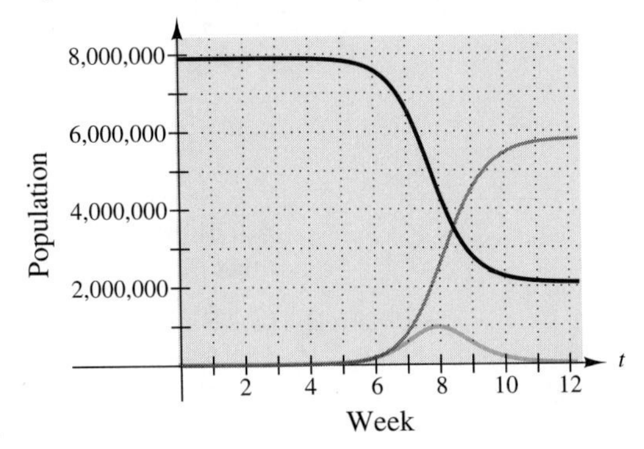

Checkpoint 6

Figure 5.3 shows the graphs of the Euler approximations to $S = S(t)$, $I = I(t)$, and $R = R(t)$.

(a) Decide which curve is which, and label the curves accordingly.

(b) How do these curves compare with the ones you drew in Exploration Activity 1?

In this section we have constructed the S-I-R model for the spread of epidemics, and we have illustrated its use with the 1968–1969 Hong Kong flu epidemic in New York City. The model consists of three linked differential equations. We observed that we can generate approximate numerical solutions of these equations by Euler's Method—that the method works as well for three dependent variables as it does for one. As evidence that the model is reasonable for a flu epidemic, we note that the infected population graph in Figure 5.3 has roughly the same shape as the presumably proportional excess death data in Figure 5.2.

Answers to Checkpoints

1. The left-hand side is the rate of change of the susceptible population. The negative sign reflects the fact that S is always decreasing. The rate of decrease of S is the rate at which susceptibles are becoming infected. That in turn is the number $I(t)$ already infected times the number of new infecteds each day generated by each infected individual.

2. The first term on the right is explained in Checkpoint 1. The negative term expresses the rate at which $I(t)$ declines by infecteds recovering or dying. This rate is essentially the number of infecteds recovering per day (only a tiny percentage die), and that is the fraction λ recovering per day times the number of infecteds.

3. $S(0)$ is essentially the entire population. $I(0)$ represents the trace of infection initially present—it could be any positive number that is very small relative to $S(0)$. $R(0)$ must be 0, because no one can be recovered before they are infected.

4. (a) The approximate rise is the approximate slope times the run. The run is Δt, and the approximate rise is $S'(t_k) \approx -\alpha I(t_k)S(t_k) \approx -\alpha I_k S_k$.
   (b) $I_{k+1} = I_k + (\alpha I_k S_k - \lambda I_k)\Delta t$, $R_{k+1} = R_k + \lambda I_k \Delta t$.

5. (b) For $k = 5$, you should have $S = 7{,}899{,}950$, $I = 32$, $R = 28$.

6. (a) The graph of $S(t)$ starts in the upper left corner and ends just above 2 million. $R(t)$ starts at 0 and ends just below 6 million. $I(t)$ starts near 0, peaks at almost a million in week 8, and declines to 0.

Exercises 5.2

1. Explain how the differential equation

$$\frac{dR}{dt} = \lambda\, I(t)$$

could have been derived from the differential equations for S and I, plus the fact that $S + I + R$ must be constant.

2. (a) Suppose $\lambda = 1/4$. Explain why the average length of time that an individual remains infected is approximately four days.

(b) For a general λ, what is the average number of days that an individual remains infected?

3. Observe from Figure 5.3 and Checkpoint 5 that for some time at the beginning of the epidemic, the susceptible population remains nearly constant.
   (a) Explain why this means

$$\frac{dI}{dt} \approx (\beta - \lambda)I(t)$$

   during that time.
   (b) From your observation in part (a), explain why the infected population tends to grow exponentially at the start of the epidemic, provided β is greater than λ.
   (c) What happens if β is less than λ? Interpret your answer in terms of the meanings of β and λ.

4. Many important epidemics are not at all like flu epidemics; the most important one afflicting the world at this writing is the AIDS epidemic. In Exercise 3 you showed that a flulike epidemic should grow exponentially at first. Table 5.3 shows the data from the Centers for Disease Control on reported cases for the first four years of the AIDS epidemic in the United States. Show that these data are *not* growing exponentially.

Table 5.3 The start of the AIDS epidemic

Date	*Cumulative Cases Reported*
Sep. 1981	129
Feb. 1982	257
July 1982	514
Jan. 1983	1,029
Aug. 1983	2,057
Apr. 1984	4,115
Feb. 1985	8,229
Jan. 1986	16,458

5. Figure 5.4 is a representation of roughly the same data (the line labeled "TOTAL") as those shown in Table 5.3, with an added breakdown of the total by race. The graph is *not* a log-log or semilog graph; look at what the vertical axis represents. Explain carefully what the graph shows about the data. (See also Project 1 in Chapter 2.)

Figure 5.4 Data on AIDS cases by race: (1) white, (2) black, (3) Hispanic, (4) unknown[2]

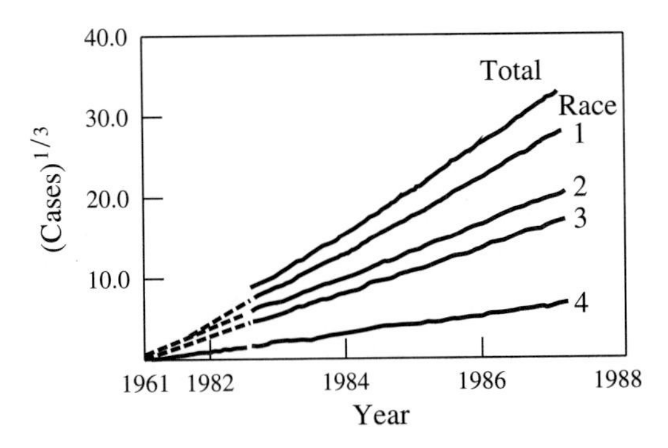

6. Exercise 4 tells us that the early growth of the AIDS epidemic was not exponential, and Exercise 5 suggests that the growth was that of a power function.
   (a) Construct a log-log plot of the data in Table 5.3, measuring time t in months from the start of 1981.
   (b) Find a function of the form $f(t) = at^b$ that fits the data reasonably well.
   (c) How well does the power b in your function agree with Exercise 5? Explain.

7. (a) Find and list the assumptions we made about the New York flu epidemic. How reasonable are these assumptions?
   (b) What other assumptions might we make that would be appropriate for a flu epidemic? How would these change the system of differential equations?
   (c) Think of some other infectious disease, and determine which of the assumptions in the flu model would not apply to it. How would you modify the assumptions to construct a model for the spread of your selected disease in the form of an initial value problem?

---

2. This figure is adapted from "Modeling the AIDS Epidemic," by Allyn Jackson, in the October 1989 issue of *Notices of the American Mathematical Society*. This is a highly readable source, with references, that reports on what is known about reasons for the non-exponential behavior of the epidemic.

5.3 Estimating and Using Parameters[3]

In the preceding section we took it for granted that the parameters β and λ could be estimated somehow, and therefore it would be possible to generate numerical solutions of the differential equations, as we did in constructing Figure 5.3. You may have already explained (in Exercise 2 of the preceding section) why λ should be the reciprocal of the average number of days of infection. Indeed, for many contagious diseases, the infection time is approximately the same for most infecteds and is known by observation. There is no *direct* way to observe β, but there is an indirect way.

Consider the ratio of β to λ:

$$\begin{aligned}\frac{\beta}{\lambda} &= \beta \cdot \frac{1}{\lambda} \\ &= \text{the number of close contacts per day per infected} \\ &\quad \times \text{the number of days infected} \\ &= \text{the number of close contacts per infected individual.}\end{aligned}$$

We call this ratio the **contact number**, and we write $c = \beta/\lambda$. The contact number c is a combined characteristic of the population and of the disease. In similar populations, it measures the relative contagiousness of the disease, because it tells us indirectly how many of the contacts are close enough to actually spread the disease. We now use calculus to show that c can be estimated after the epidemic has run its course. Then β can be calculated as $c\lambda$.

Finding the Contact Number

Here again are our differential equations for S and I, which we write in a form that displays β explicitly, since we need to find β before we can find α:

$$\begin{aligned}\frac{dS}{dt} &= -\frac{\beta}{N}\, I(t)\, S(t) \\ \frac{dI}{dt} &= \frac{\beta}{N}\, I(t)\, \mathrm{S}(t) - \lambda\, I(t).\end{aligned}$$

We observe about these two equations that the most complicated term in both would cancel and leave something simpler if we were to divide the second equation by the first—provided we can figure out what it means to divide the derivatives on the left. But the Chain Rule solves that puzzle for us: The number of infecteds at any given time is implicitly a function of the number of susceptibles, so *the rate of change of I with respect to S* makes sense, and the Chain Rule tells us that

$$\frac{dI}{dt} = \frac{dI}{dS}\,\frac{dS}{dt}\,.$$

3. Based in part on information from a preliminary version of *Calculus Using Mathematica*, by K. D. Stroyan, Academic Press, New York, 1994. Some of our information comes from "Herd Immunity," a student project handout by A. L. Miller and K. D. Stroyan.

Thus,
$$\frac{dI/dt}{dS/dt} = \frac{dI}{dS}.$$

Put another way, the Chain Rule says that if it looks like differentials cancel, they really do. Now we can divide the differential equations as we suggested and simplify:

$$\begin{aligned}\frac{dI}{dS} &= \frac{(\beta/N)\,I\,S - \lambda\,I}{-\,(\beta/N)\,I\,S} \\ &= -1 + \frac{\lambda N}{\beta S} \\ &= -1 + \frac{N}{cS} \\ &= -1 + \frac{1}{c(S/N)}.\end{aligned}$$

Checkpoint 1

Write a reason for each step of this calculation.

The ratio S/N that turned up in the last step is the *fraction* of the population that is susceptible. We can simplify further by introducing variables, say, i and s, to stand for the infected and susceptible fractions of the population: $i = I/N$ and $s = S/N$. Then

$$\frac{di}{ds} = \frac{dI}{dS}$$

and
$$\frac{di}{ds} = -1 + \frac{1}{cs}.$$

This is a differential equation that determines (except for dependence on an initial condition) the infected fraction as a function of the susceptible fraction. Three features of the equation are particularly worth noting:

1. The only parameter that appears is c, the one we are trying to determine.
2. The equation is *independent of time*. That is, whatever we learn about the relationship between i and s, it must be true for the entire duration of the epidemic.
3. The right-hand side is an explicit function of s, the *independent* variable.

The conclusion we can draw from observation 3 is that $i = i(s)$ is a function whose derivative is $-1 + 1/cs$. Thus, to solve this differential equation, all we have to do is find a function that has this derivative. Of course, c will appear in the answer (observation 1). Since the relationship between i and s will be independent of time (observation 2), we can write it down at *two different* times at which we know i and s. The equality of these two expressions will give us an equation we can solve for c.

So what's a function whose derivative is $-1 + 1/cs$? Well, it has two terms, so we can take it a term at a time. The first term, -1, is the derivative of $-s$.

We can write the second term as

$$\frac{1}{c} \cdot \frac{1}{s},$$

a constant times the *reciprocal* of s, so a function that has the right derivative will be the same constant times a function whose derivative is the reciprocal of s. We know that such a function is $\ln(s)$:

$$\frac{d}{ds} \ln s = \frac{1}{s}.$$

Checkpoint 2

(a) Show that $i = -s + \frac{1}{c} \ln s$ is a solution of

$$\frac{di}{ds} = -1 + \frac{1}{cs}.$$

(b) Show that any solution of the differential equation in part (a) must have the form

$$i = -s + \frac{1}{c} \ln s + k$$

where k is a constant. (Recall that any two functions that have the same derivative on a given interval must have a difference that is constant.)

The immediate implication we can draw from the solution above is

$$i + s - \frac{1}{c} \ln s = k,$$

i.e., the expression on the left is constant for all values of time. Now we need to know the values of i and s at two different times in order to find an equation that involves c alone.

The times at which we can estimate i and s with some precision are just before the epidemic ($t = 0$) and after it has run its course ($t = \infty$). We write i_0 and s_0 for the starting fractions and i_∞ and s_∞ for the ending fractions—that is, the limiting values of i and s as t becomes large. Then we substitute these values into the equation

$$i + s - \frac{1}{c} \ln s = k$$

to get

$$i_0 + s_0 - \frac{1}{c} \ln s_0 = i_\infty + s_\infty - \frac{1}{c} \ln s_\infty.$$

The meaning of "the epidemic has run its course" is that there are essentially no infecteds left in the population; i.e., $i_\infty = 0$. (See Figure 5.5, which repeats Figure 5.3). Furthermore, the typical epidemic starts with a very small number of infecteds being introduced into a largely susceptible population, so, relative to s_0, we have $i_0 \approx 0$

(again, see Figure 5.5). These observations allow us to simplify to the approximate equality:

$$s_0 - \frac{1}{c}\ln s_0 \approx s_\infty - \frac{1}{c}\ln s_\infty.$$

Figure 5.5 Susceptible, infected, and recovered populations

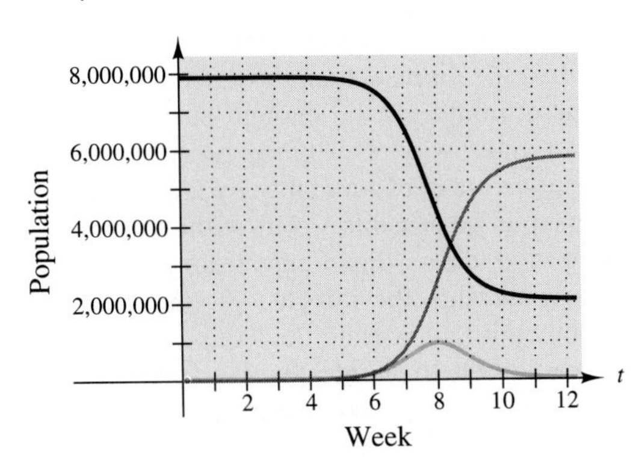

Checkpoint 3

Solve this approximate equation for the contact number c to show that

$$c \approx \frac{\ln s_\infty - \ln s_0}{s_\infty - s_0}.$$

Exploration Activity 1

(a) For the 1968–1969 flu epidemic in New York City, what was s_0?

(b) Use Figure 5.5 to estimate the number R_∞ of recovered individuals and then the fraction r_∞ of recovered individuals at the end of the epidemic.

(c) Find s_∞ from r_∞.

(d) Now finish the calculation of c from the approximate equation above, and then calculate β. Does your result agree with our use of $\beta = 0.6$ in Figure 5.5?

We assumed at the outset that essentially everyone in New York was susceptible, so the starting fraction was 1. (If you take the initial infecteds into account, the initial fraction still rounds to 1.) In particular, this means that $\ln s_0$ is 0.

Figure 5.5 suggests that R_∞ is about 5.8 million. (This is also the total number of people who got infected—a number that would be known to public health officials if the reporting of cases is thorough.) Thus r_∞ is about 0.73. The number of remaining susceptibles cannot be counted directly, so it must be computed from the known total number of cases. In particular, as $I_\infty = 0$, we have $S_\infty = N - R_\infty = 2.1$ million. Thus $s_\infty = 0.27$.

Now our approximate formula for c gives $\ln 0.27/(0.27-1) = 1.79$. We multiply by λ to get $\beta = 0.6$. This has to agree with the parameter values we used to draw Figure 5.5 because that figure is the only evidence we have about r_∞ or s_∞. The point of the calculation is that public health officials would have enough information to at least estimate c and β.

How Effective Is Inoculation?

For a disease like flu—one that confers future immunity on its sufferers—if almost everyone has had the disease, then those who have not had the disease are protected from getting it: There would not be enough susceptibles left in the population to allow an epidemic to get under way. This group protection is called **herd immunity**. The point of inoculation is to create herd immunity by stimulating the antibodies that confer immunity in as many people as possible—but without actually giving those people the disease. Thus inoculation creates a direct path from the susceptible group to the recovered group without passing through the infected group. And a large-scale inoculation program to head off an impending epidemic does this rapidly enough to artificially lower the initial susceptible population to a safe level—safe enough that if a trace level of infection enters the population, a few people may get sick, but no epidemic will develop.

At least that's how it is supposed to work. In order to be sure it *will* work, public health officials need to know what fraction of the susceptible population must be inoculated. That's the question we address now.

In Exercise 3 of the preceding section you may have observed that the time derivative of I would be initially *negative* if β is less than λ—and if I decreases with time, then clearly no epidemic develops.

Checkpoint 4

Explain why no epidemic occurs if the contact number c is less than 1.

Exercise 3 and Checkpoint 4 are based on a simplification of

$$\frac{dI}{dt} = \alpha\, I(t)\, S(t) - \lambda\, I(t)$$

that got the susceptible population out of the picture. Now we consider how the susceptible population itself contributes to the increase or decrease in the infected population. Let's use this differential equation to relate the rate of change of the number of infecteds to the *fraction* of susceptibles:

$$\begin{aligned}\frac{dI}{dt} &= \frac{\beta}{N}\, I(t)\, S(t) - \lambda\, I(t)\\ &= \beta I(t)\, s(t) - \lambda\, I(t)\\ &= [\beta\, s(t) - \lambda] I(t)\\ &= \lambda\, [cs(t) - 1]\, I(t).\end{aligned}$$

Checkpoint 5

Write a reason for each step in the preceding calculation.

Thus we see that to ensure a negative derivative for I, we don't have to have $c < 1$—which is likely to be impossible—it's enough to have $cs(t) < 1$. In particular, to have I decreasing from time zero, it's enough to have $cs_0 < 1$, or $s_0 < 1/c$.

Now we have a handle on how to determine an inoculation strategy: Determine the contact number c for a given disease from historical data, take its reciprocal, and reduce the susceptible fraction of the population below that level. "*The* contact number" is a little misleading because, for a given epidemic, this number depends on both the disease and the population. However, if we can determine a worst case or a likely case value of c, then we can determine what we have to do to protect against that level of contagiousness.

Example 1 In the United States from 1912 to 1928, the contact number for measles was 12.8. The reciprocal of this number is about 0.078. If we assume that c is still 12.8, then reducing the susceptible population to 7% of the total population of any given area should confer herd immunity on the entire population. This requires vaccination of *more* than 93% of the population, because the measles vaccine is only 95% effective; that is, 5% of those inoculated do not acquire immunity. ■

Checkpoint 6

What fraction of a population in the United States must be vaccinated to ensure herd immunity against measles?

In this section we have shown how calculus can be used, in conjunction with data available to public health officials, to determine the parameters in an S-I-R model for a specific disease and a specific population. We also have seen how knowledge of the contact number can help us determine what percentage of the population must be inoculated in order to prevent an epidemic.

Answers to Checkpoints

1. We divided the formula for I' by the formula for S'.
   We divided the denominator into each term of the numerator and simplified.
   We replaced λ/β by $1/c$.
   We replaced N/S by $1/(S/N)$.
2. (a) $\frac{d}{ds}\left(-s + \frac{1}{c}\ln s\right) = -1 + \frac{1}{c}\frac{1}{s} = -1 + \frac{1}{cs}$
   (b) If $f(s)$ and $g(s)$ are functions with the same derivative, then
   $$f(s) = g(s) + k$$
   for some constant k. If $g(s) = -s + (\ln s)/c$ and f is any other solution, then $f(s) = -s + (\ln s)/c + k$ for some constant k.
3. When we multiply both sides by c, we get
   $$c\, s_0 - \ln s_0 \approx c\, s_\infty - \ln s_\infty.$$

Now we collect the terms involving c:

$$c\,s_0 - c\,s_\infty \approx \ln s_0 - \ln s_\infty.$$

Factoring out c on the left, and dividing by the other factor, we get

$$c \approx \frac{\ln s_0 - \ln s_\infty}{s_0 - s_\infty} = \frac{\ln s_\infty - \ln s_0}{s_\infty - s_0}.$$

4. $c = \beta/\lambda$, so if $c < 1$, then $\beta < \lambda$.
5. The first line restates the differential equation for I.
   We replaced S/N by s.
   We factored $I(t)$ out of both terms.
   We factored λ out of both terms and replaced β/λ by c.
6. 98%

Exercises 5.3

Table 5.4 lists recorded contact numbers in the United States for several different diseases. This information is used in the first 12 exercises.[4]

Table 5.4 Historical contact numbers

Disease	*Year(s)*	c
Whooping cough	1943	17.3
Chicken pox	1943	11.3
Scarlet fever	1908–1917	8.5
Mumps	1943	8.1
Diphtheria	1943–1947	7.4
Poliomyelitis	1955	4.9

Assuming 100% effectiveness of the vaccine for each of these diseases, what percentage of the population would have to be vaccinated against each disease to ensure herd immunity?

1. Whooping cough
2. Chicken pox
3. Scarlet fever
4. Mumps
5. Diphtheria
6. Poliomyelitis

Assuming 95% effectiveness of the vaccine for each of these diseases, what percentage of the population would have to be vaccinated against each disease to ensure herd immunity?

7. Whooping cough
8. Chicken pox
9. Scarlet fever
10. Mumps
11. Diphtheria
12. Poliomyelitis
13. Rubella (German measles) had a contact number of 7.7 in West Germany in 1972 and 7.0 in England and Wales in 1979. The rubella vaccine, like the measles vaccine, is 95% effective. What fraction of the population must be inoculated to ensure herd immunity from rubella?
14. Laws in the United States require children to have a rubella immunization prior to entering school. The typical vaccine is called MMR, for measles-mumps-rubella; that is, it is a combined vaccine for all three. It is estimated that 98% of children entering school have been vaccinated. Is this fraction high enough to ensure herd immunity for any or all of these diseases?
15. The contact number for measles in England and Wales in 1956–1959 was 15.6. With a 95% effective vaccine, what percentage of the population would have to be vaccinated to ensure herd immunity?
16. The contact numbers for measles in Nigeria in 1960–1968 was 17.0. With a 95% effective vac-

cine, what percentage of the population would have to be vaccinated to ensure herd immunity?

17. (a) Use the equations

$$\frac{dS}{dt} = -\frac{\beta}{N}\,I(t)\,S(t) \quad \text{and} \quad \frac{dR}{dt} = \lambda\,I(t)$$

to show that

$$\frac{ds}{dr} = -cs,$$

where $s = S/N$ is the fraction of the population that is susceptible, $r = R/N$ is the fraction of the population that is recovered, and c is the contact number.

(b) Add an appropriate initial condition, and solve the initial value problem to find the relationship between s and r.

(c) Check your answer against the data in Figure 5.6 (copied from Figure 5.3) at the end of the flu epidemic. Calculate c from the given values of β and λ, and estimate s_∞ and r_∞ from the figure. Then check that these numbers satisfy the relationship you have derived from the initial value problem.

Figure 5.6 Susceptible, infected, and recovered populations

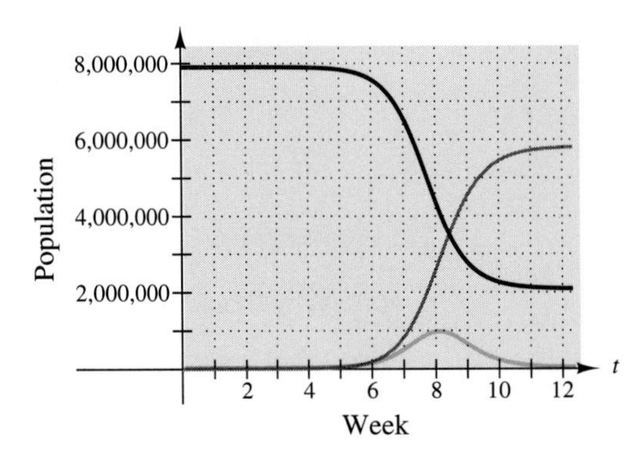

18. At the start of this section we divided the equation

$$\frac{dI}{dt} = \frac{\beta}{N}\,I(t)\,S(t) - \lambda\,I(t)$$

by the equation

$$\frac{dS}{dt} = -\frac{\beta}{N}\,I(t)\,S(t)$$

to achieve a simplification. Explain why it would not have been particularly useful to *add* these two equations.

4. These exercises and the next four are based on "Herd Immunity," a student project handout by A. L. Miller and K. D. Stroyan.

5.4 Modeling the Evolution of Prices in a Simple Economy[5]

Continuous Price Evolution

We begin our study of price evolution with some assumptions about the economic mechanism whereby prices shift to balance supply and demand. First, we consider a very simple market economy in which there is just one good being produced and bought. Moreover, we assume that transactions are being made at all times—i.e., continuously—and that buyers and sellers respond immediately to changes in the price. All these are unrealistic assumptions—we'll add more realistic features later. For the moment, our goal is to describe this situation mathematically.

5. Adapted from *An Introduction to Mathematical Models in Economic Dynamics*, by David Clements, Polygonal Publishing, Washington, NJ, 1984.

We start with demand D—in bushels, say, for a good such as peaches. We assume that D is a function of the price p. Lacking any specific knowledge of demand functions, we make the simplest possible assumption—that demand is a *linear* function of price:

$$D(p) = a + bp$$

where $b < 0$.

Checkpoint 1

(a) Why should b be negative?

(b) What does a represent? Should it be positive or negative?

We assume that the supply Q—also in bushels, in the case of peaches—has the same simple form

$$Q(p) = c + dp$$

where d is positive.

Checkpoint 2

(a) Why should d be positive?

(b) Should we expect c to be positive or negative? Does it depend on the commodity?

The **equilibrium price** p^* is the price at which $Q = D$. Figure 5.7 shows that there is such a price, namely, the p-coordinate of the point of intersection of the two lines.

Figure 5.7 Model supply and demand functions and the equilibrium price

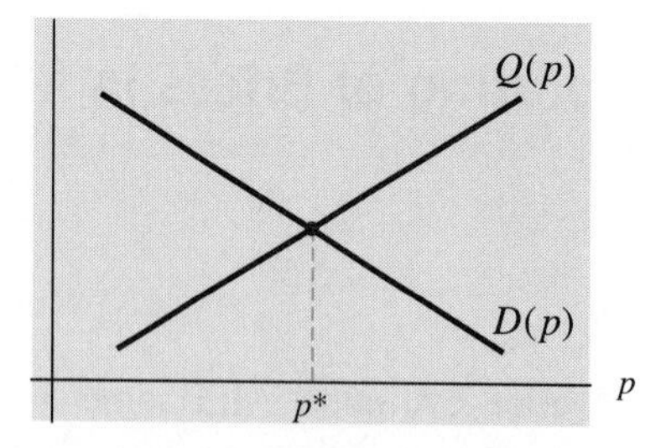

Given a starting price at which supply and demand *do not* balance, we want to know whether the price moves toward equilibrium. That is, we want to introduce time t as the independent variable and see how supply, demand, and price vary as functions of time. To do this, we have to make some assumption about the way the price reacts to an imbalance between supply and demand.

It is reasonable to assume that the rate of change of price as a function of time, dp/dt, depends on the **excess demand**, that is,

$$E(p) = D(p) - Q(p).$$

Specifically, we assume that dp/dt is proportional to $E(p)$:

$$\frac{dp}{dt} = \gamma\, E(p)$$

where γ is positive. (Why is γ positive?)

Now we can combine our assumptions to formulate a differential equation for the unknown price function p.

$$\begin{aligned}\frac{dp}{dt} &= \gamma\,[D(p) - Q(p)]\\ &= \gamma\,[a + bp - (c + dp)]\\ &= -\gamma(d-b)p + \gamma(a-c)\end{aligned}$$

Checkpoint 3

Give a reason for each step in the preceding calculation.

Now we can describe the price p, a function of time t, as the solution of an initial value problem:

$$\frac{dp}{dt} = -\gamma(d-b)p + \gamma(a-c) \qquad \text{with } p(0) = p_0,$$

where p_0 is the price at the initial time.

Example 1 If $a = 12$, $b = -1.5$, $c = -1$, $d = 1$, $\gamma = 0.17$, and $p_0 = 5$, then the differential equation simplifies to

$$\frac{dp}{dt} = -0.17(1 + 1.5)p + 0.17(12 + 1) = -0.425p + 2.21,$$

so the initial value problem becomes

$$\frac{dp}{dt} = -0.425p + 2.21 \qquad \text{with } p(0) = 5.$$

■

Checkpoint 4

(a) Show that the unique solution of the initial value problem in Example 1 is

$$p(t) = 5.2 - 0.2e^{-0.425t}.$$

(b) What is p^* in this case?

Exploration Activity 1

(a) Use the differential equation

$$\frac{dp}{dt} = -\gamma(d-b)p + \gamma(a-c)$$

to describe p^* in terms of a, b, c, and d. (*Hint*: What is the rate of change when $p = p^*$?)

(b) Does your general description of p^* in part (a) agree with your calculation of the specific case in Checkpoint 4?

When the price is at equilibrium, the excess demand is 0, so the rate of change of price also must be 0. When you set the derivative equal to 0 and p equal to p^*, you get

$$0 = -\gamma(d-b)p^* + \gamma(a-c),$$

from which it follows that

$$p^* = \frac{a-c}{d-b}.$$

For the numbers in Example 1, this formula gives $p^* = (12+1)/(1.5+1)$, or 5.2, which is also the limiting value of the function in Checkpoint 4.

We may simplify the appearance of the differential equation

$$\frac{dp}{dt} = -\gamma(d-b)p + \gamma(a-c)$$

by introducing a single symbol, say, α, for the constant $\gamma(d-b)$ and another single symbol, say, β, for the constant $\gamma(a-c)$. Then the initial value problem becomes

$$\frac{dp}{dt} = -\alpha p + \beta \qquad \text{with } p(0) = p_0.$$

We have seen this problem many times now. In particular, in Chapter 3 you learned how to solve it by scaling the dependent variable. That may be what you did for the particular case in Checkpoint 4. Now we ask you to tackle the general case. As we have seen on other occasions, the calculation is actually less messy with letters than with numbers as the coefficients.

Exploration Activity 2

(a) Solve the initial value problem

$$\frac{dp}{dt} = -\alpha p + \beta \qquad \text{with } p(0) = p_0.$$

[You may find it useful to begin by rewriting the right-hand side of the differential equation in the form $-\alpha\,(p - \beta/\alpha)$.]

(b) Substitute the definitions of α and β to express the solution function in terms of the constants a, b, c, d, γ, and p_0.

(c) Use the result of Exploration Activity 1 to show that the solution function can be written in the form

$$p(t) = (p_0 - p^*)e^{-\gamma(d-b)t} + p^*.$$

(d) Check that this general solution agrees with your calculation of the particular solution in Checkpoint 4.

(e) What does the graph of the function in part (c) look like? How does the graph change for different starting values p_0? Make a rough sketch showing several solutions starting from different initial points.

If we substitute $y = p - \beta/\alpha$ in the differential equation, it becomes an exponential decay equation whose solution is

$$y = y_0 e^{-\alpha t}.$$

Now, when we make the reverse substitution, $p = y + \beta/\alpha$, and solve for p, we find

$$p = \left(p_0 - \frac{\beta}{\alpha}\right)e^{-\alpha t} + \frac{\beta}{\alpha}.$$

When we replace α and β by their definitions, the solution becomes

$$p = \left(p_0 - \frac{a-c}{d-b}\right)e^{-\gamma(d-b)t} + \frac{a-c}{d-b}.$$

We recognize the quotient that appears twice in this formula as the expression we found for p^* in Exploration Activity 1. Thus the answer can be written in the form

$$p = (p_0 - p^*)e^{-\gamma(d-b)t} + p^*$$

as desired.

Since γ and d are positive and b is negative, the exponential part of this formula must approach 0 as t grows, and therefore, $p(t)$ approaches p^*. If $p_0 > p^*$, then $p(t)$ is always bigger than p^*. If $p_0 < p^*$, then $p(t)$ is always smaller than p^*. Your sketch should show several graphs of functions that start above or below $p = p^*$ and approach that horizontal line with an exponential decay.

Discrete Price Evolution

You might object to the continuous model for prices because suppliers cannot react instantaneously to changes in prices. Indeed, for many commodities, goods are exchanged only at regular intervals, say, once a day or once a week at market. We turn now to a model that incorporates discrete trading times but retains the other assumptions of the continuous model.

We assume that trading times occur at regular time intervals of length Δt. (There is no reason to assume that Δt is small. For example, in the case of a farm product such as peaches, the product might be brought to market once a week during a certain season.) Thus the trading times are $t_0 = 0,\ t_1 = \Delta t,\ t_2 = 2 \cdot \Delta t,\ \ldots$. We denote the price at trading time t_k by p_k. Similarly, the demand and supply at this trading time are denoted

D_k and Q_k, respectively. The linearity assumptions on the demand and supply take the form

$$D_k = a + bp_k \qquad \text{and} \qquad Q_k = c + dp_{k-1}.$$

Notice the p_{k-1} in the formula for Q_k: The suppliers base the amount to bring to the market at time t_k on the price at the *previous* market time, p_{k-1}.

Now we impose the condition that $D_k = Q_k$. This is a *market clearing* assumption: Everything must be sold. From $D_k = a + bp_k$ and $Q_k = c + dp_{k-1}$, we have

$$a + bp_k = c + dp_{k-1}.$$

Hence,

$$p_k = \frac{c-a}{b} + \frac{d}{b}\, p_{k-1}.$$

This iterative formula allows us to substitute a starting value p_0 on the right to find p_1, then substitute p_1 to find p_2, and so on, through the entire price evolution, that is, for as many trading times as the market stays in business.

We have encountered iterative formulas for generating sequences of numbers on other occasions: discrete natural growth at the end of Chapter 2, Newton's Method for solving equations at the end of Chapter 4, and Euler's Method for solving initial value problems earlier in this chapter.

Our discussion of natural growth in Chapter 2 had both discrete and continuous models based on the same assumption about biological growth—growth rate proportional to the population. We found that the discrete model (with a very small time step) closely approximated the continuous model, and the same could be said about the solutions—both of which were exponential functions.

Here we have both discrete and continuous initial value problems to model price evolution that are based on the same assumptions about supply and demand. Could we have a similar connection between the discrete and continuous price models? In particular, will the solutions of the difference equation

$$p_k = \frac{c-a}{b} + \frac{d}{b}\, p_{k-1}$$

display the decaying exponential behavior of the solutions

$$p(t) = (p_0 - p^*)e^{-\gamma(d-b)t} + p^*$$

to the differential equation

$$\frac{dp}{dt} = -\gamma(d-b)p + \gamma(a-c)\,?$$

We can attempt to answer this question by actually solving the difference equation, that is, by finding an explicit formula for p_k in terms of the starting price p_0 and the constants a, b, c, and d. This is a different problem from that of solving the discrete natural growth equation, so we will have to develop a somewhat different technique.

Geometric Sums

The question before us is how to solve the iterative formula

$$p_k = \frac{c-a}{b} + \frac{d}{b}\, p_{k-1}$$

for an explicit formula for p_k in terms of the starting price p_0 and the constants a, b, c, and d. We start as we did with the continuous price evolution model, introducing single symbols for the complicated constants: say, $\alpha = (c-a)/b$ and $\beta = d/b$. Then the iterative formula assumes the simpler form

$$p_k = \alpha + \beta p_{k-1}.$$

Now we write out the first few cases for $k = 1$, 2, and 3:

$$p_1 = \alpha + \beta p_0$$

$$p_2 = \alpha + \beta p_1 = \alpha + \beta\,(\alpha + \beta p_0) = \alpha + \beta\alpha + \beta^2 p_0$$

$$p_3 = \alpha + \beta p_2 = \alpha + \beta\,(\alpha + \beta\alpha + \beta^2 p_0) = \alpha + \beta\,\alpha + \beta^2\alpha + \beta^3 p_0\,.$$

We see a pattern emerging. In fact,

$$p_k = \alpha\,(1 + \beta + \beta^2 + \cdots + \beta^{k-1}) + \beta^k p_0$$

for each positive integer k.

This equation is a solution of the difference equation $p_k = \alpha + \beta p_{k-1}$ in the sense that it expresses each price p_k in terms of the starting price p_0 and the constants α and β. However, it is not an *explicit* solution because of the $\cdots$ (ellipsis) in the middle of the sum. Our next task is to evaluate the sum that appears in parentheses.

For the time being, let us suppose the values of β and k are fixed, and let's give a name to the sum, say,

$$S = 1 + \beta + \beta^2 + \cdots + \beta^{k-1}.$$

If we multiply both sides of this equation by β, we get

$$\beta S = \beta + \beta^2 + \cdots + \beta^{k-1} + \beta^k.$$

Next, we subtract this equation from the previous one to get

$$S - \beta S = 1 - \beta^k.$$

Finally, we solve for S to get an explicit expression involving β and k—but no ellipsis:

$$S = \frac{1-\beta^k}{1-\beta}\,.$$

Here we must assume that β is not equal to 1, of course. When we replace S by its definition (the sum), we get

$$1 + \beta + \beta^2 + \cdots + \beta^{k-1} = \frac{1-\beta^k}{1-\beta}\,.$$

The sum on the left-hand side of this equation is called a **geometric sum**. It is characterized by the fact that each term is β times the previous term, where β is a constant. We have just derived a formula for evaluating geometric sums.

Checkpoint 5

Check the geometric sum formula:

(a) Start with $\beta = 1/2$ and $k = 4$; do both sides give you the same number?

(b) Try $\beta = 2$ and $k = 5$.

(c) Choose two other combinations of β and k, and check that the formula is correct for your choices.

This is not the first time we have seen something growing in proportion to its size at the previous step. We have already associated the names "exponential" and "geometric" with such quantities. Indeed, the terms of the geometric sum

$$1 + \beta + \beta^2 + \cdots + \beta^{k-1}$$

are values of the discrete exponential function with base β. What is new here, thrust on us by the need to solve a new discrete initial value problem, is the idea of *summing* such terms.

Checkpoint 6

(a) Substitute the geometric sum formula

$$1 + \beta + \beta^2 + \cdots + \beta^{k-1} = \frac{1 - \beta^k}{1 - \beta}$$

into the solution formula for price at the kth trading session

$$p_k = \alpha\,(1 + \beta + \beta^2 + \cdots + \beta^{k-1}) + \beta^k p_0$$

to find an explicit solution for p_k in terms of α, β, and p_0.

(b) Now replace α and β by their definitions in terms of our original constants a, b, c, and d to get an explicit solution of the difference equation

$$p_k = \frac{c - a}{b} + \frac{d}{b}\, p_{k-1}.$$

Simplify the result.

A Comparison of the Continuous and Discrete Models

In Figures 5.8 and 5.9 we show the solution

$$p(t) = (p_0 - p^*)e^{-\gamma(d-b)t} + p^*$$

to the continuous initial value problem

$$\frac{dp}{dt} = -\gamma(d - b)p + \gamma(a - c) \qquad \text{with } p(0) = p_0$$

and the solution you just worked out to the discrete initial value problem

$$p_k = \frac{c-a}{b} + \frac{d}{b}\,p_{k-1}.$$

Both figures use the numerical values from Example 1 and Checkpoint 4: $p_0 = 5$, $a = 12$, $b = -1.5$, $c = -1$, $d = 1$, and (for the continuous case) $\gamma = 0.17$. For this particular selection of constants, the two models lead to similar evolutions of prices toward the equilibrium price p^* at which supply and demand exactly balance. (Compare the prices on even-numbered trading days.) For any reasonable selection of constants, the continuous solution always has the same general shape that you see in Figure 5.8. However, the discrete model has some features—in addition to the odd-even oscillation seen here—that the continuous model does not exhibit. You may have an opportunity to explore these features in a laboratory experiment.

Figure 5.8 Solution of the continuous price model for $a = 12$, $b = -1.5$, $c = -1$, $d = 1$, $\gamma = 0.17$

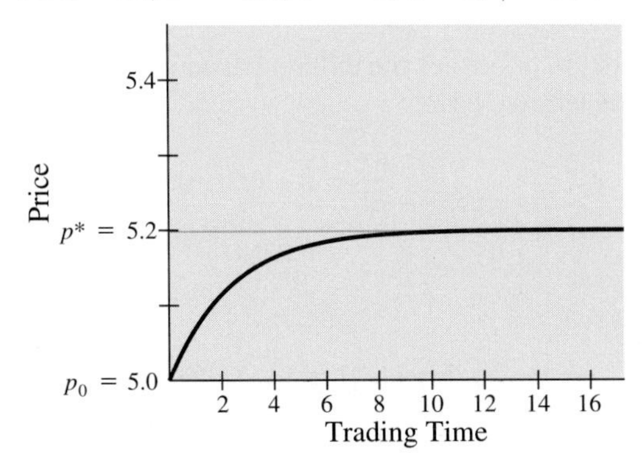

Figure 5.9 Solution of the discrete price model for $a = 12$, $b = -1.5$, $c = -1$, $d = 1$

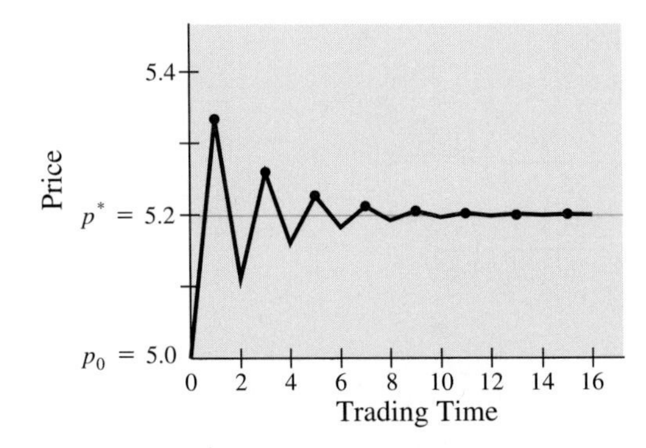

In this section we have studied both continuous and discrete models of price evolution. We saw that the continuous model leads to a familiar type of differential equation with a familiar solution—exponential decay. The discrete model is more complicated. Here we encountered the technique of summing geometric sums—a technique of great importance in this course and in mathematics in general.

Answers to Checkpoints

1. (a) Higher price should lead to lower demand.

   (b) a is the demand when the good is free, which will certainly be positive.

2. (a) Higher price should induce suppliers to bring more to market.

   (b) c is the amount supplied when the price is 0, which one might think should be 0. However, c could be negative if the price has to reach a certain positive threshold before suppliers will bring anything to market. On the other hand, it could also be positive if the suppliers are determined to keep the market open even when they have to give away the goods.

3. We substituted the definition of $E(p)$.
   We substituted the formulas for $D(p)$ and $Q(p)$.
   We collected like terms (those with p and those without) and multiplied through by γ.

4. (a) First, we verify the initial condition:

$$p(0) = 5.2 - 0.2 \cdot 1 = 5.$$

   Next, we show that $p(t)$ satisfies the differential equation by calculating each side of the equation separately. On the left, we have

$$\begin{aligned} \frac{dp}{dt} &= 0 - 0.2(-0.425)e^{-0.425t} \\ &= 0.085e^{-0.425t}. \end{aligned}$$

   On the right,

$$\begin{aligned} -0.425p + 2.21 &= -0.425\left(5.2 - 0.2e^{-0.425t}\right) + 2.21 \\ &= 0.085e^{-0.425t} - 2.21 + 2.21 \\ &= 0.085e^{-0.425t}. \end{aligned}$$

   Thus the given function p satisfies both the differential equation and the initial condition.

   (b) 5.2

5. (a) Both sides give $15/8$.

   (b) Both sides give 31.

6. (a) $p_k = \dfrac{\alpha}{1-\beta} + \left(p_0 - \dfrac{\alpha}{1-\beta}\right)\beta^k$

   (b) $p_k = \dfrac{c-a}{b-d} + \left(p_0 - \dfrac{c-a}{b-d}\right)\left(\dfrac{d}{b}\right)^k$

Exercises 5.4

Calculate each of the following sums exactly (i.e., do not compute decimal approximations).

1. $1 + 3 + 9 + 27 + \cdots + 3^{10}$
2. $1 + \dfrac{1}{3} + \dfrac{1}{9} + \dfrac{1}{27} + \cdots + \left(\dfrac{1}{3}\right)^{10}$
3. $5 + 25 + 125 + \cdots + 5^9$ (Watch the first term!)
4. $3^6 + 3^7 + 3^8 + \cdots + 3^{12}$
5. $1 + \dfrac{1}{4} + \dfrac{1}{4^2} + \dfrac{1}{4^3} + \cdots + \dfrac{1}{4^8}$
6. $1 + 2 + 4 + 8 + 16$

7. $1 + \frac{1}{2} + \frac{1}{4} + \frac{1}{8} + \frac{1}{16}$

8. $1 + \frac{1}{2} + \frac{1}{4} + \frac{1}{8} + \frac{1}{16} + \cdots + \frac{1}{1024}$

9. $1 + \frac{1}{2} + \frac{1}{4} + \frac{1}{8} + \frac{1}{16} + \cdots + \frac{1}{2^{12}}$

10. $\frac{1}{3^4} + \frac{1}{3^5} + \frac{1}{3^6} + \frac{1}{3^7} + \cdots + \frac{1}{3^{12}}$

11. $\frac{7}{13} + \left(\frac{7}{13}\right)^2 + \left(\frac{7}{13}\right)^3 + \cdots + \left(\frac{7}{13}\right)^{20}$

12. $\frac{2}{3} + \left(\frac{2}{3}\right)^2 + \left(\frac{2}{3}\right)^3 + \cdots + \left(\frac{2}{3}\right)^{14}$

13. $\frac{13}{7} + \left(\frac{13}{7}\right)^2 + \left(\frac{13}{7}\right)^3 + \cdots + \left(\frac{13}{7}\right)^{25}$

14. The NCAA Championship basketball tournament is a single elimination tournament that starts with 64 teams. All teams play in the first round, and the winners progress to the second round. This pattern is repeated until only one team is left undefeated.
    (a) Express the number of games played as a geometric sum, and calculate this number.
    (b) Describe a simpler method for determining the number of games played in the tournament.

15. (a) Suppose the supply and demand for a given product are given by $Q = 2 + 9p$ and $D = 40 - 10p$, respectively, and that the market for the product fits the continuous price evolution model

$$\frac{dp}{dt} = -\gamma(d - b)p + \gamma(a - c), \ \ p(0) = p_0$$

with $\gamma = 0.17$. If the initial price is 10% above the equilibrium price, at what time will the price be 5% above equilibrium?[6]

    (b) Answer the same question as in part (a) if the market instead fits the discrete price evolution model

$$p_k = \frac{c - a}{b} + \frac{d}{b}\, p_{k-1}.$$

16. Use your graphing tool to graph the solution function in Exercise 15(a). Trace the solution to confirm your numerical answer.

17. Use your graphing tool to graph the solution function in Exercise 15(b). Trace the solution to confirm your numerical answer.

18. In Figure 5.9 we saw that an iterative formula could be represented graphically by plotting p_k as a function of k. Figure 5.10 shows another popular way to represent an iteration: a **web diagram.** (Some calculators and many computer systems can draw such diagrams.) The evolution of the iterative sequence—in this case, the sequence of prices generated by the discrete price model—is represented by plotting p_k against p_{k-1}. The specific iteration shown in Figure 5.10 is the same as the one shown in Figure 5.9.

The web diagram can be constructed without knowing an explicit solution of the iteration. Here's how: First, draw the graphs of $y = x$ and of $y = f(x)$, where $p_k = f(p_{k-1})$— these are the colored lines in Figure 5.10. Start at (p_0, p_1). Draw a horizontal line to $y = x$, which takes you to the point (p_1, p_1). Next, move vertically (up or down) to the graph of $y = f(x)$, which you meet at the point (p_1, p_2). Now start over, alternating horizontal moves to $y = x$ and vertical moves to $y = f(x)$.

Figure 5.10 Web diagram for the iteration $p_k = \frac{c - a}{b} + \frac{d}{b}\, p_{k-1}$

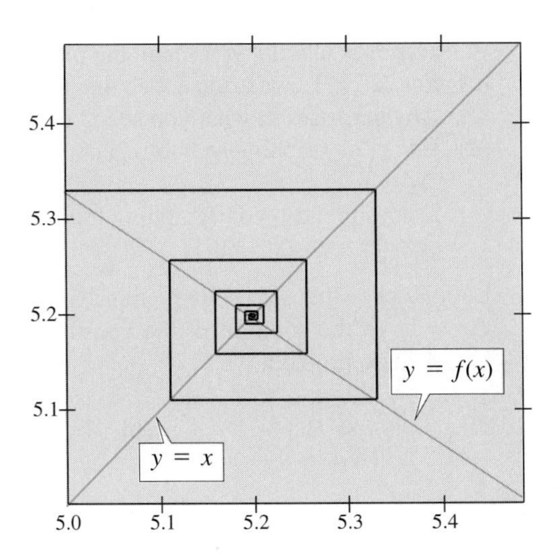

(a) Explain the construction. Why does the sequence of points visited contain all the points of the form (p_{k-1}, p_k)? Where are those points in the diagram?

(b) What property of the graph of f in Figure 5.10 ensures that the web will spiral in to the point (p^*, p^*)?

(c) The iteration shown in Figure 5.10 has $b = -1.5$ and $d = 1$. What would the web look like if we changed d to 1.5? to 1.8? Draw each to be sure of your conclusions.

19. (a) Construct a web diagram for the iteration in Exercise 15(b).

(b) Check your diagram against your graph from Exercise 17 to make sure the two are consistent as representations of the same iteration.

20. We constructed Figure 5.8 from an explicit solution of the initial value problem

$$\frac{dp}{dt} = -\gamma(d-b)p + \gamma(a-c), \qquad p(0) = p_0$$

with $a = 12$, $b = -1.5$, $c = -1$, $d = 1$, $\gamma = 0.17$, and $p_0 = 5$. But we also could have solved the initial value problem approximately by Euler's Method and plotted the results, either by a function plot or a web diagram.

(a) Explain why the iterative formula for Euler's Method in this case is

$$p_k = p_{k-1} + (2.21 - 0.425 p_{k-1})\Delta t.$$

(b) For $\Delta t = 1$, plot p_k as a function of k. Any surprises in what you see in the plot?

(c) For $\Delta t = 1$, construct a web diagram for p_k. Any surprises in what you see in the plot?

(d) Use your graphing tool to repeat part (b) with $\Delta t = 0.1$.

(e) If possible, use your graphing tool to repeat part (c) with $\Delta t = 0.1$.

21. Let P_k be the kth Euler approximation, with $\Delta t = 1$, to the solution of the constrained growth initial value problem

$$\frac{dP}{dt} = P(1-P) \qquad \text{with } P_0 = 0.1.$$

(a) Plot P_k as a function of k. Compare the result with Figure 5.1. Any surprise in what you see?

(b) Construct a web diagram for P_k. Describe the similarities to and differences from the web diagrams for discrete and continuous price evolution models.

22. Let $S_k = 1 + \beta + \beta^2 + \cdots + \beta^k$. We know an explicit formula for these sums, but they also can be represented by an iterative formula.

(a) Explain why $S_k = S_{k-1} + \beta^k$. What is S_0?

(b) For $\beta = 1/3$, plot S_k as a function of k. What is the long-term behavior of S_k as k becomes large?

(c) For $\beta = -1/3$, plot S_k as a function of k. What is the long-term behavior of S_k as k becomes large?

(d) If you tried to construct a web diagram for S_k (any β) where would you run into a problem?

23. Suppose that 70 cents of every dollar spent in the United States is spent *again* in the United States. (Economists call this the **multiplier effect.**) If the federal government pumps an extra billion dollars into the economy, how much total spending in the United States occurs as a result?

24. Our derivation of the geometric sum formula required the assumption that $\beta \neq 1$. But the geometric sum clearly has a value when $\beta = 1$, a value that you can easily write down without an ellipsis. What is

$$1^0 + 1^1 + 1^2 + \cdots + 1^{k-1}$$

in general ?

6. Both parts of this exercise are adapted from *An Introduction to Mathematical Models in Economic Dynamics*, by David Clements, Polygonal Publishing Co., Washington, NJ, 1984.

Chapter 5 Summary

We began this chapter with a formalization of Euler's Method, an iterative method we had seen informally in an early laboratory. Euler's Method is a tool for generating numerical and graphical solutions to initial value problems. With a computer or graphing calculator at hand, we can solve an initial value problem, whether or not we know how to find a formula for the solution.

Of course, Euler's Method has nothing to do with epidemics — it's a *general tool*. But we demonstrated the power of this tool by applying it to the S-I-R epidemic model, which has (1) three interrelated dependent variables and (2) no possibility of a simple formula solution for any of the variables as functions of time. The effectiveness of Euler's Method is shown in the simultaneous solutions for all three functions shown in Figure 5.3. In particular, the infected population graph has roughly the same shape as the (presumably proportional) excess death data shown in Figure 5.2.

Iteration also played a prominent role in the discrete model in Section 5.4. Here we studied price evolution in a single-product market, given very simple assumptions about supply and demand functions and a requirement that prices adjust so that supply equals demand. We had two purposes in doing this: (1) to illustrate the very beginnings of a calculus-based study of economic theory and (2) to show that discrete and continuous models based on the same underlying assumptions can lead to very different conclusions — even when each approximates the other.

The process of solving the difference equation for discrete price evolution bears some similarity to Project 3 at the end of Chapter 2, in which you may have solved the discrete natural growth problem. But there is a difference: Here we have not just powers of a growth factor but *sums* of those powers. Thus we found it necessary to digress briefly to work out a formula for calculating geometric sums.

Concepts

- Euler's method for approximating the solution of an initial value problem
- linked systems of differential equations
- geometric sums

Applications

- the S-I-R model for spread of an epidemic
- a continuous model for price evolution
- a discrete model for price evolution

Formulas

Euler's Method for approximation of the solution of an initial value problem

$$\frac{dy}{dt} = f(y) \qquad \text{with } y(0) = y_0$$

on the interval $0 \le t \le b$ using n steps:

$$y_0 = y(0),$$
$$y_{k+1} = y_k + f(y_k)\Delta t,$$

where $\Delta t = b/n$.

Geometric Sums

$$1 + \beta + \beta^2 + \cdots + \beta^{k-1} = \frac{1-\beta^k}{1-\beta} \qquad \text{for } \beta \neq 1.$$

Chapter 5 Exercises

Calculate each of the following derivatives.

1. $\dfrac{d}{dt}\, 2\ln t$
2. $\dfrac{d}{dt}\ln\left(t^2\right)$
3. $\dfrac{d}{dt}\ln 2t$
4. $\dfrac{d}{dt}\ln(t+2)$
5. $\dfrac{d}{dt}(\ln t)^2$
6. $\dfrac{d}{dt}\, e^{3t}\ln t$

7. $\frac{d}{dt}\sqrt{1+t^2}\ln 3t$

Calculate each of the following second derivatives.

8. $\frac{d^2}{dt^2}2\ln t$

9. $\frac{d^2}{dt^2}\ln(t^2)$

10. $\frac{d^2}{dt^2}\ln 2t$

11. $\frac{d^2}{dt^2}\ln(t+2)$

12. $\frac{d^2}{dt^2}(\ln t)^2$

13. $\frac{d^2}{dt^2}e^{3t}\ln t$

14. There was no flu vaccine in 1968, but there is now.
   (a) How would you incorporate the effect of a vigorous inoculation program into the flu epidemic model? Assume that the program is *not* complete before the first trace infection appears in the population but that it continues as the disease is spreading.
   (b) How would you expect this change to affect the solution of the initial value problem?[7]

15. Before vigorous worldwide attempts to wipe out smallpox, its contact number in India was 5.2, relatively low among the contagious diseases considered in this chapter. Smallpox vaccinations began in 1958, but the disease persisted until 1977, almost 20 years. What explanations would you offer for this?

16. (a) In your view, why do we still have measles and rubella but not polio or smallpox?
   (b) About 1970, there were some problems with the measles vaccines. How are those problems related to outbreaks of measles in the late 1980s and to vigorous revaccination campaigns on college campuses?
   (c) If measles and rubella vaccines are given together and have the same level of effectiveness, how would you explain the facts that the incidence of rubella has been steadily declining in recent years and the incidence of measles has not?

17. Here again is the formula for calculating a geometric sum with k terms, each term being β times the previous term:

$$1+\beta+\beta^2+\cdots+\beta^{k-1}=\frac{1-\beta^k}{1-\beta}.$$

   (a) Explain why, if $|\beta|<1$, the sum has a finite limiting value as k becomes infinitely large. Explain why that limiting value is $1/(1-\beta)$.
   (b) If $\beta>1$ or $\beta<-1$, what is the limiting sum (if any) as k becomes infinitely large?
   (c) If $\beta=1$, what is the limiting sum (if any) as k becomes infinitely large?
   (d) If $\beta=-1$, what is the limiting sum (if any) as k becomes infinitely large?

7. This exercise and the next two are based in part on "Herd Immunity," a student project handout by A. L. Miller and K. D. Stroyan.

18. (a) Use the formula from part (a) of the preceding exercise to explain why $0.3333\ldots$ is the same number as $1/3$.

    (b) What fraction is the same number as $0.353535\ldots$?

    (c) What fraction is the same number as $0.199519951995\ldots$?

19. (a) Find the *exact* sum:

$$1+\frac{6}{7}+\left(\frac{6}{7}\right)^2+\left(\frac{6}{7}\right)^3+\cdots+\left(\frac{6}{7}\right)^{40}.$$

    (b) Find a *decimal approximation* to the sum in part (a) with 7SD accuracy.

    (c) Find the *exact* sum with infinitely many terms:

$$1+\frac{6}{7}+\left(\frac{6}{7}\right)^2+\left(\frac{6}{7}\right)^3+\cdots.$$

20. Zeno of Elea, a Greek philosopher of the fifth century B.C., constructed several paradoxes to show the impossibility of motion. Here is one of them:

> You cannot walk across the room because, to do so, you would first have to walk half-way across the room, then half the remaining distance, half of that distance, and so on. As you have to cover half the remaining distance an infinite number of times, you will never complete the trip. Therefore, motion is an illusion. (If you are having any doubts about the wisdom of this argument, get up and walk across the room.)

    (a) Using the width of the room as a unit of distance, how far do you walk in the first step of this process? In the first two? The first three?

    (b) Without any further adding of fractions (but using your formula for calculating a geometric sum), how far do you walk in the first 27 steps?

    (c) What happens when the number of steps becomes infinite? And just how is it that you can complete an infinite number of these steps in a finite time?

21. Suppose you drop a ball from a height of one meter, and the ball bounces to a height of three-quarters of a meter. (This is about the right coefficient of restitution for a Superball.) You let the ball continue to bounce until it comes to rest.

    (a) How high does it rise on the second bounce? On the third?

    (b) Estimate the total distance that the ball travels up and down.

    (c) Estimate the total time the ball is bouncing before it comes to rest.

    (d) If you have a reasonably bouncy ball handy, you can check to see how close the theoretical model is to reality. First, determine what number should replace "three-quarters," and solve the exercise with a coefficient of restitution that matches your real ball. Compare the time result with your observation of total time.

22. In Chapter 1 we stated without explanation the following formula for the monthly payment on a new car:

$$p=\frac{(P-D)\,r\,(1+r)^n}{(1+r)^n-1},$$

where p is the monthly payment, P is the price of the car, D is the down payment, r is the monthly interest rate (as a decimal fraction), and n is the number of months required to pay back the loan. Explain the formula. [*Hints*: $P - D$ is the amount borrowed. The total amount paid back is np. The factor by which the outstanding balance grows each month (before a payment is made) is $1 + r$.]

23. A physician decides to give a patient an infusion of glucose at a rate of 10 grams per hour. The body of the patient simultaneously converts the glucose and removes it from the bloodstream at 3 grams per hour per gram of glucose present. There are 2 grams of glucose in the bloodstream at time 0.

    (a) Explain why the amount $G = G(t)$ of glucose present at time t can be modeled by the differential equation $dG/dt = 10 - 3G$.

    (b) Find an explicit formula for the amount present at any time t.

    (c) At what time is the glucose level in the patient's bloodstream rising fastest? What is that fastest rate of increase?

    (d) Suppose Euler's Method is used to generate approximate values of the glucose level, G_1, G_2, G_3, and so on, at times t_1, t_2, t_3, and so on, that are 15 minutes apart. What will the calculated approximate level of glucose be one hour after the infusion starts?

    (e) Use your formula from part (b) to calculate the glucose level at the end of an hour. How does your approximation in part (d) compare with the glucose level calculated from the formula?

Chapter 5 Projects

1. **Rocketing Away: The Ultimate Escape** Suppose we fire a projectile directly away from the surface of the earth—straight up. What initial velocity must it have so that it *will not return*?

   Since most of the motion will be far from earth, we may reasonably ignore air resistance. But, for the same reason, we must account for the fact that the force due to gravity diminishes as we move away. Newton's Law of Gravitation says that the force of gravity on a mass m at a distance x from the center of the earth is proportional to the mass and inversely proportional to the square of the distance. In symbols, this force is

$$F = -\frac{k\,m}{x^2}$$

   where the negative sign represents the convention that the positive direction for distance is away from the earth.

   (a) Use Newton's Second Law of Motion to show that while traveling away from the earth, the gravitational acceleration is given by

$$\frac{dv}{dt} = -\frac{k}{x^2}.$$

   (b) Use the fact that the radius of the earth is approximately 3960 miles to show that k is approximately 95,040 miles cubed per second squared.

(c) Use the Chain Rule to show that

$$\frac{dv}{dt} = \frac{dv}{dx}\,v.$$

(d) What is the sign of dv/dt ? What is the sign of dx/dt when the projectile is rising from earth? When it is falling toward earth? What is the value of dx/dt when the projectile stops rising and starts falling? Why is the velocity v not a function of distance x if the projectile both rises and falls?

(e) Explain why the equation in part (a), which is correct for the rising phase of the projectile, cannot be correct beyond the point at which the projectile stops rising and starts falling.

(f) Show that as long as the projectile is rising,

$$\frac{d}{dx}\left(\frac{1}{2}\,v^2\right) = \frac{d}{dx}\left(\frac{k}{x}\right).$$

[*Hint*: Compute each side of the equation directly, and use the equations in parts (a) and (c).]

Since $v^2/2$ and k/x are functions of x that have the same derivative, we know that

$$\frac{1}{2}\,v^2 = \frac{k}{x} + C.$$

(g) Substitute $v = v_0$ and $x = 3960$ (at the surface of the earth) to find C.

(h) Show that as long as the projectile is rising,

$$v^2 = 2k\left(\frac{1}{x} - \frac{1}{3{,}960}\right) + v_0^2.$$

In order for the projectile to escape from the earth's gravity, we can never have $v = 0$ (why?). Although the projectile may go far into space, if v_0 is not large enough, v will eventually decrease to 0. We illustrate this situation in Figure 5.11, which shows the graph of v^2 (with v in miles per second) as a function of x (in miles) for two different values of v_0, one small enough to allow v to reach zero and the other possibly large enough to allow escape.

Figure 5.11 Squared velocity of the projectile versus distance from the center of the earth for two values of v_0

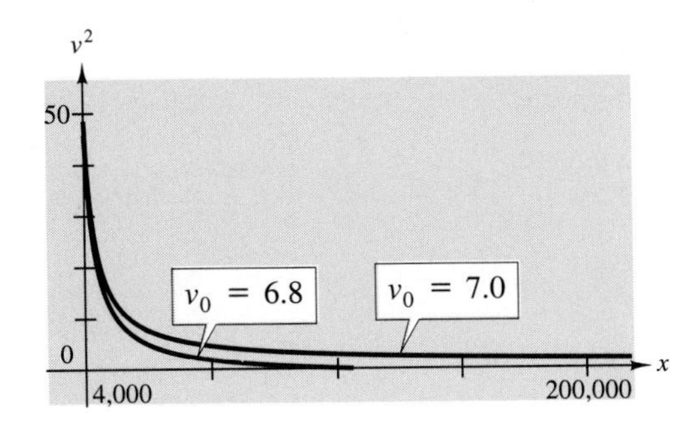

There is a critical value for v_0 — let's call it v_0^* — such that v will eventually decrease to zero if v_0 is less than v_0^* but will never reach zero if v_0 is greater than v_0^*.

(i) Show that v_0^* is approximately 6.93 miles per second or about 25,000 mph. Conclude that if v_0 is greater than 25,000 mph, then the projectile will never return.

(j) If $v_0 = 6.8$ miles per second, find the distance of the projectile from the center of the earth when the velocity reaches zero. Does your answer agree with Figure 5.11? What happens to the projectile after v reaches zero?

It is not easy to launch a projectile at a speed of 25,000 mph from the earth's surface. That's one reason why space probes are launched with rockets, not cannons. As long as a rocket has fuel to burn, it continues to accelerate. When its fuel is spent, it becomes a projectile whose initial velocity v_0 is its speed at the time the rocket motor shuts down and whose initial position is the height achieved at that time.

(k) Show that, if the rocket engine ceases to burn at a distance D from the center of the earth, then, in order to escape the earth's gravity, the velocity at that point must exceed $\sqrt{2k/D}$.

(l) What is the escape velocity if the rocket becomes a projectile 200 miles above the surface of the earth?

(m) Make a quick sketch of your idea of the graph of escape velocity as a function of D. Then fill in the following table, and plot these points (by computer, by calculator, or on graph paper) to see if your sketch is right.

D (miles)	*Escape Velocity (miles per second)*
4,000	
10,000	
20,000	
30,000	
50,000	
75,000	
100,000	
150,000	

(n) Think about the television pictures you have seen of rockets being launched from Cape Canaveral. At the moment of launch, and for a brief time after, the rocket appears to hesitate and only gradually pick up speed, in spite of the tremendous force generated by its engines. Which of the following do you think is the *major* cause of that hesitation? Explain your answer.

A. Air resistance
B. Gravity
C. Release of the hold-down clamps
D. Inertial resistance
E. Cross winds
F. Distraction of the crew
G. The lunar cycle
H. Atlantic Ocean tides
I. All of the above
J. None of the above

2. **Burning Rocket Fuel: The Effect of Decreasing Mass** We have in our storeroom a rocket of mass 3000 kilograms that will carry 12,000 kilograms of fuel. This fuel is burned at a constant rate and is consumed in 30 seconds. The exhaust gases exit the rear of the rocket at a speed of 10,000 mph or 2.78 miles per second. At the end of the 30-second burn, will the rocket have enough velocity to escape the earth's gravity? (See Project 1.)

   We need an equation of motion for the rocket that takes into consideration the changing mass of the rocket. To model this situation, we use a common procedure in physics and other sciences—one that fits naturally with our development of rates of change. To find a formula for the velocity v of the rocket, we consider the average rate of change of v over a small time interval Δt. Then we let the time interval shrink to zero to find an expression for the derivative of v, i.e., a differential equation. A rocket that starts from a launch pad has initial velocity zero. Thus we can expect to find a unique solution for our initial value problem. And if it turns out that we cannot find a *formula* for the unique solution, we can resort to numerical solution by Euler's Method.

   We assume for now that exhaust gases represent the only force on the rocket. Thus we ignore gravity, at least until the rocket becomes a projectile. If we find that the rocket will not reach escape velocity even when we ignore gravity, then it certainly won't escape when we consider gravity. On the other hand, if the rocket appears to reach escape velocity when we ignore gravity, then we have to consider whether gravity makes a significant difference.

   We can find the equation of motion from the physical principle called Conservation of Momentum. The **momentum** of a system is the product of its velocity and mass or, if the system consists of several parts moving at different velocities, the sum of those products. The **Principle of Conservation of Momentum** states that *in the absence of an external force, the total momentum of a system remains constant.*

   At time t after liftoff of the rocket, we let $v(t)$ be the velocity of the rocket and $m(t)$ the mass of the rocket, including the remaining fuel. The momentum of the rocket and fuel at time t is then $v(t)\,m(t)$. At a slightly later time $t + \Delta t$, what was the system at time t now consists of two parts:

   - the rocket with the remaining unburned fuel, and
   - the fuel expelled in the time interval from t to $t + \Delta t$.

   We will calculate the momentum of each of these parts and then add the momenta of the parts to get a new expression for the momentum of the total system.

   Let $\Delta m = m(t + \Delta t) - m(t)$ and $\Delta v = v(t + \Delta t) - v(t)$.

   (a) What is the sign of Δm?

   (b) What is the mass of the fuel burned between time t and time $t + \Delta t$?

   (c) Show that the momentum of the rocket at time $t + \Delta t$ is

   $$[v(t) + \Delta v][m(t) - |\Delta m|].$$

   Why does one factor have a plus sign and the other a minus sign?

   (d) Figure 5.12 shows the rocket moving in one direction and the exhaust gases moving in the other. If γ is the constant speed at which exhaust gases are expelled from the rocket, show that the momentum of the expelled gases is

   $$[v(t) - \gamma]|\Delta m|.$$

(e) Show that

$$m\,v = (v + \Delta v)(m - |\Delta m|) + (v - \gamma)|\Delta m|,$$

where we have abbreviated $m(t)$ and $v(t)$ to m and v, respectively.

(f) Show that

$$\frac{\Delta v}{\Delta t} = \frac{\Delta v}{\Delta t}\frac{|\Delta m|}{m} + \frac{\gamma}{m}\frac{|\Delta m|}{\Delta t}.$$

Figure 5.12 Motion of rocket and expelled gases

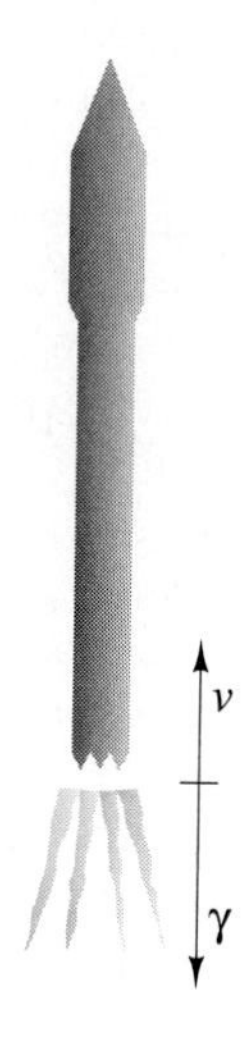

(g) What does $\dfrac{\Delta v}{\Delta t}$ approach as Δt approaches zero?

(h) What does $\dfrac{|\Delta m|}{m}$ approach as Δt approaches zero?

(i) Why is $|\Delta m| = -\Delta m$? What does $\dfrac{|\Delta m|}{\Delta t}$ approach as Δt approaches zero?

(j) Explain why

$$\frac{dv}{dt} = -\frac{\gamma}{m}\frac{dm}{dt}.$$

(k) For the rocket described in the first paragraph of this project, show that dm/dt is constant and $m(t) = 15{,}000 - 400t$.

(l) Show that

$$\frac{dv}{dt} = \frac{1112}{15{,}000 - 400t} \qquad \text{and } v(0) = 0.$$

(m) Show that the unique solution of the initial value problem is

$$v(t) = 2.78 \ln\left(\frac{15{,}000}{15{,}000 - 400\,t}\right).$$

(n) Find the velocity of the rocket after the 30-second burn, both in miles per second and miles per hour.

Your calculation should show that $v(30)$ is not as large as escape velocity from the surface of the earth (see Project 1), but we have not yet answered the question of whether our rocket can escape the earth's gravity. Since we are not considering a projectile shot from the surface of the earth, but farther out, we need to know how far away from the surface of the earth the rocket is when it has consumed all of its fuel.

An exact answer, i.e., a determination of the distance function from the velocity function, requires solution of another antidifferentiation problem, but in this case a simple estimate will suffice. The distance traveled by the rocket is certainly less than $4.47 \times 30 \approx 134$ miles. (Why?) The rocket is then at a distance less than $3960 + 134$ miles from the center of the earth. This is not far enough from the center of the earth to make a significant change in the escape velocity. (Use the result of Project 1 to check this!) Thus the rocket is still not going fast enough to escape.

There is a good reason why spacecraft are usually not launched with single-stage rockets, like the one we just considered, but rather with rockets that are *multistage*. In an accompanying laboratory project, you may have the opportunity to design a two-stage rocket that can achieve escape velocity, even when we take gravity into consideration, and still carry a payload.

Chapter 5 Optional Lab Reading: Projectiles

Winning the Peace: A Parable for Our Time

Relations between the two nations sharing the Island of Paradise have become increasingly belligerent over the last few months. Schwartz, the crazed dictator of South Paradise, is determined to inflame the patriotic passions of his people and lead them in an attack on North Paradise in order to gain control of the entire island.

Schwartz's daughter, Maria, is a playwright who is much beloved by her fellow South Paradisians. Her new play, Chaos in Paradise, *opened two weeks ago in Belmo, the capital of North Paradise. The next morning the headlines of the papers in South Paradise read*

"CHAOS" PANNED IN BELMO ! ! !
"UTTER MADNESS . . . CRAZINESS" REVIEWER STATES
SCHWARTZ DECLARES STATE OF EMERGENCY,
CLOSES BORDER WITH NORTH PARADISE,
URGES CITIZENS TO AVENGE INSULT TO MARIA

As Schwartz has been whipping up war fury, all communication with the North has been broken off.

However, the review in the Belmo paper was actually quite favorable: "This is a delightful farce with beautifully balanced examples of utter madness. The author has the touch of divine craziness that enables us both to laugh at our failings and view others in a new light. Don't miss it!"

Kept in ignorance of the favorable reception of Maria's play, the Southerners are preparing for war. The Northerners have indignantly mobilized to meet the attack. Fighting seems inevitable.

A small group of Northerners has decided on a daring plan to avoid the bloodshed. They have one old 81*-millimeter mortar shell and a mortar launcher. Under cover of darkness, they plan to paddle an inflatable boat around the border and carry the shell and launcher on a seldom-used trail to the top of a hill overlooking the Southern capital of Ergo. They have removed most of the explosive from the shell and have replaced it with copies of the Belmo review. At dawn the shell will be launched; the small explosive charge will burst the shell at a height of* 100 *meters and allow the copies of the review to flutter down and enlighten the war-mad Ergons.*

The hill is 400 meters above the plain on which Ergo is situated and 600 meters (horizontal distance) away from the center of Ergo. If the muzzle velocity of the shell is 100 meters per second, at what angle should the brave peacemakers aim the launcher to have the shell travel to a point 100 meters above the capital?

The small explosive device is activated by a timed fuse that is started when the shell is launched. For what length of time should the fuse be set so that the shell will burst when at the desired point, 100 meters over Ergo?

The Initial Value Problems: Symbolic Solution

We now assume the role of members of the peace-making band and attempt to solve this two-part problem. We start by attempting to describe the position of the shell as a function of time since launch. In other words, we try to find coordinate functions, $x = x(t)$ and $y = y(t)$, that describe the curve traced out by the shell.

We choose a coordinate system (see Figure 5.13) with origin at the launch site, the x-axis horizontal and pointing in the direction of Ergo, and the y-axis vertical with the positive direction upward. As usual in problems of motion, we obtain the description of the motion from Newton's Second Law.

Figure 5.13 Mortar shell trajectory over Ergo

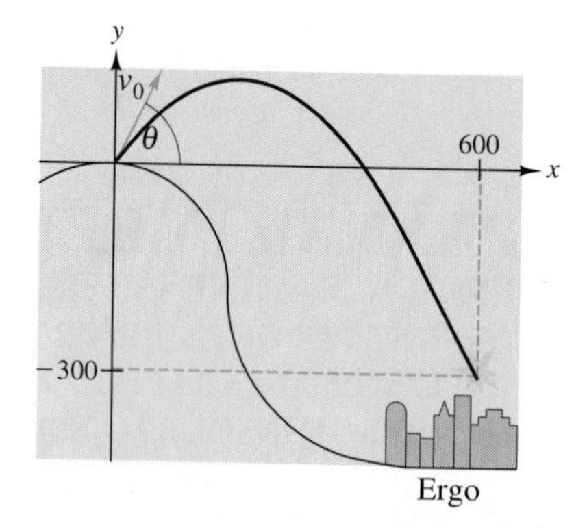

We call the mass of the shell m, the horizontal and vertical velocities v_x and v_y, respectively, and the corresponding accelerations a_x and a_y. We start by assuming that the only force acting on the projectile in flight is gravity, and we defer until later a discussion of what difference (if any) air resistance might make. Thus

$$ma_x = 0$$

and

$$ma_y = -mg.$$

Acceleration is the derivative of velocity, so — after dividing through each equation by m — we may write

$$\frac{dv_x}{dt} = 0$$

and

$$\frac{dv_y}{dt} = -g.$$

To determine v_x and v_y, we need initial conditions for these quantities. If the shell is launched at a speed of 100 meters per second and at an angle of θ with the horizontal, we have $v_x(0) = 100 \cos\theta$ and $v_y(0) = 100 \sin\theta$. (Again, see Figure 5.13.) Thus we have two initial value problems:

$$\frac{dv_x}{dt} = 0 \qquad \text{with } v_x(0) = 100\cos\theta$$

and

$$\frac{dv_y}{dt} = -g \qquad \text{with } v_y(0) = 100\sin\theta.$$

Exploration Activity 1

(a) Show that

$$v_x(t) = 100\cos\theta$$

and

$$v_y(t) = 100\sin\theta - gt.$$

(b) Show that

$$x(t) = 100(\cos\theta)t$$

and

$$y(t) = 100(\sin\theta)t - \frac{gt^2}{2}.$$

What did you use for initial conditions, and why?

Setting the Launch Angle and Fuse Timer

Now that we have explicit formulas for x and y as functions of t, our problem reduces to finding an angle θ and a time t such that $x(t) = 600$ and $y(t) = -300$. (Why are these the right numbers?) If we substitute those values into the formulas in Exploration Activity 1, we get

$$600 = 100(\cos\theta)t$$

and

$$-300 = 100(\sin\theta)t - \frac{gt^2}{2}.$$

This is a pair of equations in the unknowns t and θ, just the quantities we want to determine for placing the mortar and setting the fuse. Note that the symbol θ has been a *constant* throughout this calculation: The mortar is going to be placed and fired only once. However, it is still an *unknown*, because we haven't determined yet what value or values will meet our requirements for passing through a particular target point. The symbol g is also a constant, but its value is *known*. We have left g in symbolic form up to this point because it is easier to write g than to write 9.807.

Exploration Activity 2

(a) Solve the two equations for t and θ.

(b) We get only one try with the mortar, and it absolutely has to be right—a war hangs in the balance. Check your work by substituting your (θ, t) pair or pairs into the equations for $x(t)$ and $y(t)$ to see if placing the mortar at the angle θ and setting the timer for t seconds will send the shell to the right spot.

(c) It is not unusual for nonlinear equations (singly or in a system) to have multiple solutions. Did you find more than one solution for θ? Is there more than one real solution to the peacemakers' problem? Explain carefully your reasons for accepting or rejecting each possible solution.

Air Resistance: The Velocity-Squared Model in the Plane

Your solution to Exploration Activity 2 may or may not be adequate to prevent a war. This solution started with the assumption that air resistance was not a factor—indeed, that *nothing* affected the path of the projectile other than gravity. In the accompanying laboratory project you can explore what happens when air resistance is considered—and then see the effect of firing into a headwind. Here and in the next section we consider what differences these factors make in our formulas.

In the Chapter 3 Lab Reading we saw that air resistance on a fast-moving object is likely to be proportional to the square of the speed of the object. Lacking any better information at this point, we will accept that on faith. For objects traveling in a straight line, this was a relatively easy model to use. But our mortar shell does not travel in a straight line, so we have to consider how to resolve air resistance into its horizontal and vertical components—that is, how it affects motion in both the x- and y-directions.

Velocity is a **vector** quantity; that is, it has both a *direction* and a *magnitude*. Its direction is the direction of travel at any instant, i.e., the tangent direction to the curve traced out by the projectile. The magnitude of velocity is what we call **speed**—how fast the projectile is traveling at a given instant. Look again at Figure 5.13, in which we represented the *initial* velocity v_0 by an arrow drawn tangent to the path of motion at time $t = 0$. This is the usual representation of vector quantities: an arrow pointing the direction and with a length that gives the magnitude. It makes sense to square *speed*—which is just a number—but it doesn't make sense to square velocity. (What would it mean to square a direction-magnitude combination—or an arrow?)

The drag force is also a vector quantity. Its direction is *opposite* to the direction of motion (therefore, opposite to v), and its magnitude is proportional to the square of the speed. Since the acceleration induced by the drag force is proportional to that force (Newton's Second Law again), the same statements are true about the drag component of acceleration: Its direction is opposite to v, and its magnitude is proportional to the square of the speed. We illustrate the direction statement in Figure 5.14, which shows a typical segment of the path of motion, with a typical velocity vector v (at some specific, but unspecified, time t) and the corresponding typical drag acceleration vector a_{drag}.

Figure 5.14 Typical drag acceleration directed opposite to velocity

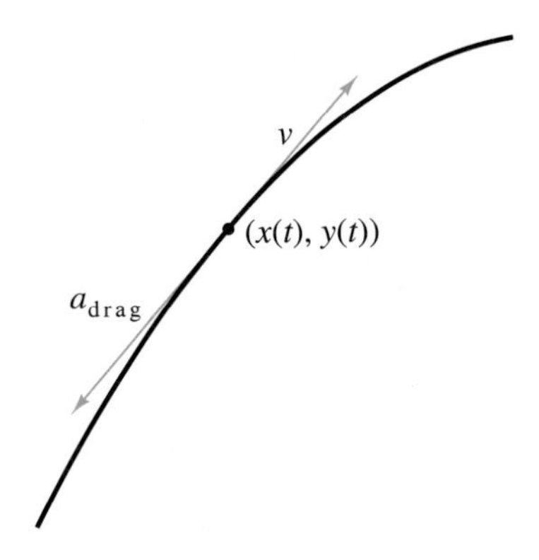

Since a_{drag} is directed along the same line as v, it must be a numerical multiple of v—with a multiplier that can change as a function of time. Since a_{drag} is directed opposite to v, the multiplier must be negative. Since the magnitude of a_{drag} must be proportional to the square of the speed, the multiplier must itself be proportional to the speed.

The conventional notation for magnitude of a vector quantity is the same as absolute value—two vertical bars. Thus the speed is written $|v|$, and a_{drag} can then be expressed as

$$a_{\text{drag}} = -c\ |v|\ v,$$

where c is a positive constant. This expression has all the right properties for a_{drag}: It is a numerical multiple of v with multiplier $-c\,|v|$, which is negative and proportional to the speed $|v|$. And the magnitude of a_{drag} is proportional to the square of the speed:

$$|a_{\text{drag}}| = c\,|v|^2.$$

In Figure 5.15 we show the resolution of v and a_{drag} into their horizontal and vertical components. The horizontal and vertical components of v are the functions we have already named v_x and v_y, respectively. The equation $a_{\text{drag}} = -c\ |v|\ v$ and the similar triangles in Figure 5.15 show that the horizontal and vertical components of drag acceleration must be, respectively, $-c\,|v|v_x$ and $-c\,|v|v_y$. Furthermore, we see from Figure 5.15 that the speed can be expressed in terms of the component functions of v via the Pythagorean Theorem:

$$|v| = \sqrt{v_x^2 + v_y^2}\ .$$

Figure 5.15 Resolution of velocity and drag acceleration into components

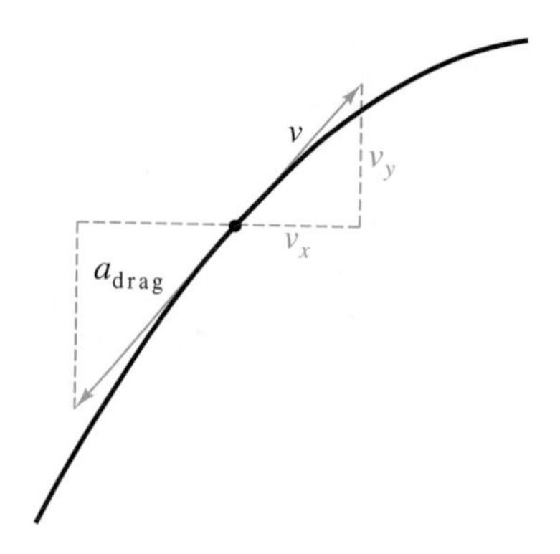

Now we can put all the pieces together and state how we will modify the gravity-only initial value problems,

$$\frac{dv_x}{dt} = 0 \qquad \text{with } v_x(0) = 100\cos\theta$$

and

$$\frac{dv_y}{dt} = -g \qquad \text{with } v_y(0) = 100\sin\theta,$$

to account for the drag force resulting from air resistance. When we add the drag components of acceleration, these become

$$\frac{dv_x}{dt} = -c\sqrt{v_x^2 + v_y^2}\, v_x \qquad \text{with } v_x(0) = 100\cos\theta$$

and

$$\frac{dv_y}{dt} = -g - c\sqrt{v_x^2 + v_y^2}\, v_y \qquad \text{with } v_y(0) = 100\sin\theta.$$

The gravity-only equations were simple to solve and completely separated into horizontal and vertical components. The drag equations are not only complicated but also inextricably linked: Each component's derivative involves the *other* unknown component function. There is no hope of solving the new initial value problems symbolically. However, we have already seen—with the S-I-R epidemic model—that Euler's Method will work quite well on complicated and linked initial value problems. Thus, it is possible to pursue the peacemakers' solution in the lab.

Wind

Now we add to our model of the projectile another physical reality that may confront our Paradisian peacemakers: wind. In particular, let's think about the effect of a 10 meter per second headwind on the equations we just derived.

You may find it easier to think about this problem in terms of something more familiar, such as riding a bicycle. You know that it is harder to ride with wind in your face than it is to ride the same road on a still day.

Exploration Activity 3

(a) Suppose you are riding at a speed of 12 meters per second into a 5 meter per second headwind. How fast would you have to ride on a still day to feel the same force of air resistance?

(b) Suppose you are riding at a speed of v meters per second into a 5 meter per second headwind. How fast would you have to ride on a still day to feel the same force of air resistance?

(c) Suppose that when you ride on a still day at a speed of v meters per second, the drag force of air resistance is kv^2. What is the drag force of air resistance if there is a 5 meter per second headwind?

(d) If you consider the direction in which you are riding to be positive, is the force of air resistance positive or negative?

(e) Now we return to the projectile problem, for which we assume that the projectile is traveling into a horizontal headwind of 10 meters per second. Explain why only the occurrences of v_x in the acceleration equations need to be changed to account for the headwind.

(f) How should the accelerations without wind in horizontal and vertical directions be changed to account for the headwind? Write down new versions of the initial value problems whose solutions give the horizontal and vertical velocities when the mortar is fired into a 10 meter per second wind.

You will need this revised initial value problem in the laboratory when we modify the peacemaking problem to account for the wind.

CHAPTER 6

Periodic Motion

Our investigation of population growth, cooling, falling bodies, electric circuits, optics, epidemics, price evolution, and other phenomena have led to models based on exponential, logarithmic, and algebraic functions. We turn now to phenomena that exhibit **repetitive** *behavior—for which none of the functions used so far is an especially good model. Exponential and logarithmic functions never repeat any of their values. Some of the algebraic functions do—after passing through peaks or valleys—but not in a regular, repeating pattern.*

Where do we see repetitive phenomena? Our hearts (if healthy) beat regularly, returning to the same state every second or so. The sun rises and sets in a regular cycle of about 24 hours. A long-play record spins at 33 revolutions per minute—that is, it returns to the same state 33 times every minute. The discs in our compact disc players also spin, but not at a constant speed. The electric current provided by our local utility (if properly regulated) alternates in direction 60 times every second. The planets, asteroids, and some comets travel around the Sun in repeating orbits. The hands of our analog clocks and watches traverse a complete circle every minute, hour, or half-day. And all the operations in our computers are timed by internal clocks that cycle millions of times every second.

To model such repetitive processes, we need to identify and study basic repetitive functions, the ones that will play the role of building blocks—as do power functions in the study of falling bodies and exponential functions in the study of biological growth.

6.1 Circular Functions: Sine and Cosine

We obtain the simplest repeating functions by considering the simplest repeating process we know: going around in circles.

Functions Generated by Circular Motion

Imagine a circular running track of radius one hundred meters, and imagine an xy-coordinate system placed on the track with the center of the circle at the origin. (See Figure 6.1.) Suppose you start at the point with coordinates $x = 100$ and $y = 0$ and run in a counterclockwise direction with a speed of five meters per second. How can we describe your position as a function of the angle θ between your line to the origin and the x-axis? How can we describe your position as a function of time?

Figure 6.1 Position of runner on a track of radius 100 meters

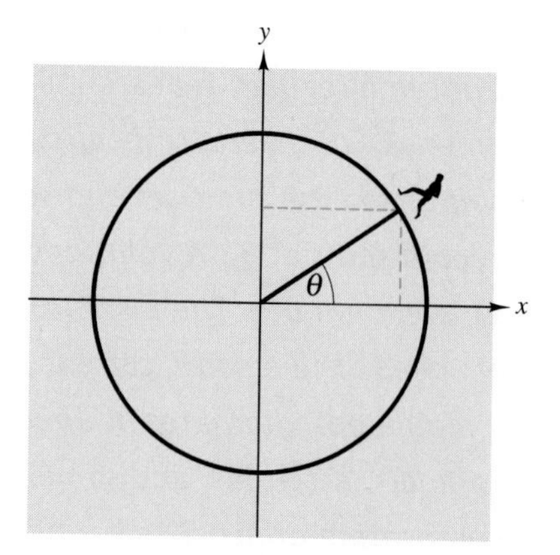

Consider first the question of describing your position in terms of θ. We'll measure the angle in radians, so one lap corresponds to 2π radians, a quarter lap to $\pi/2$ radians, and so on. To locate your position in the plane, we need to specify both the x- and y-coordinates. The functions that do this should be old friends—or at least passing acquaintances. If you have traveled through an angle of θ radians, then you are located at the point with x-coordinate $100 \cos \theta$ meters and y-coordinate $100 \sin \theta$ meters. That is,

$$x = 100 \cos \theta,$$

and

$$y = 100 \sin \theta.$$

Checkpoint 1

(a) Make your own sketch of Figure 6.1, and label the distances that give the x- and y-coordinates of the runner. Also label the 100-meter radius.

(b) Explain why x and y are related to θ by cosine and sine functions, respectively. (Look at an appropriate right triangle in the figure.)

(c) Explain why, as you continue running laps, both your x-coordinate and your y-coordinate oscillate between 100 meters and -100 meters.

In Figure 6.2 we show another circle of radius 100 with separate graphs of the x- and y-coordinates that result from motion around the circle. Put the index finger of your left hand on the circle at the point $(100, 0)$ and the index finger of your right hand on the graph of the y-coordinate at the point $(0, 0)$. As you move your left finger along the circle, let your right finger trace out the graph of y as a function of θ.

Figure 6.2 Generation of sine and cosine functions from motion around a circle

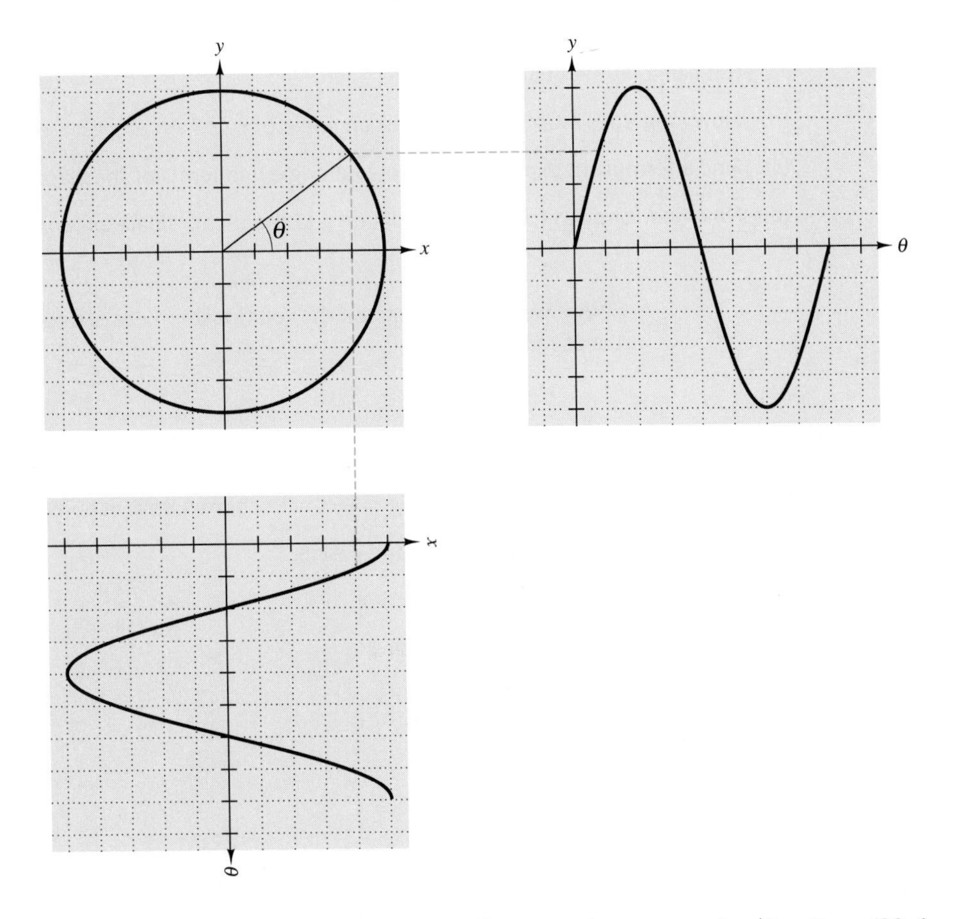

Notice that the axes for the x-coordinate graph are rotated $\pi/2$ radians (90 degrees) clockwise from the usual position. Turn the page counterclockwise to be sure you understand what that part of the figure shows. Now place your left index finger at the point $(0, 100)$ on the x-coordinate graph and your right index finger on the circle at the point $(100, 0)$. As you move your left finger along the circle, let your right finger trace out the graph of x as a function of θ.

Checkpoint 2

You have just experienced the generation of the sine and cosine functions from motion around a circle.

(a) What changes if the radius of the circle is 27 instead of 100? Give the new formulas for the x- and y-coordinates.

(b) What if the radius is 1? Give the new formulas for the x- and y-coordinates.

Now we turn to the question of your location on the track as a function of time in seconds. Let $t = 0$ correspond to your starting position at $x = 100$ and $y = 0$. Answer the following questions and find the formulas requested.

Checkpoint 3

(a) How long does it take you to run one lap at a speed of 5 meters per second? How long does it take to run half a lap? A quarter of a lap?

(b) How long does it take to sweep out an angle of $\pi/2$? An angle of $\pi/3$?

(c) Find a formula to express the swept-out angle θ in terms of the elapsed time t.

(d) Now find formulas for the x-coordinate and the y-coordinate as functions of time.

Now we'll concentrate on the relationship between the sine and cosine functions and a circle of radius one.

Exploration Activity 1

(a) What circle is traveled by a moving object with coordinate functions $x = \cos\theta$ and $y = \sin\theta$?

Explain the following facts about sine and cosine.

(b) For every angle θ, $\cos(\theta + 2\pi) = \cos\theta$.

(c) For every angle θ, $\sin(\theta + 2\pi) = \sin\theta$.

(d) For every angle θ, $\sin^2\theta + \cos^2\theta = 1$.

(e) For every angle θ, $|\sin\theta| \leq 1$.

(f) For every angle θ, $|\cos\theta| \leq 1$.

The formulas in part (a) of Exploration Activity 1 describe a circle of radius 1 centered at the origin. This circle is called the **unit circle**. We can interpret the properties in parts (b) and (c) in the following way: No matter where you start with θ, when you increase θ by 2π—that is, go once around the circle—you will return to the same function values for cosine (the x-coordinate) and sine (the y-coordinate) as when you started. Part (d) follows from a Cartesian formula for the same circle: $x^2 + y^2 = 1$. Parts (e) and (f) assert that the x- and y-coordinates of points on the unit circle always stay between -1 and 1.

In Figure 6.2 we graphed both the sine and cosine functions on the θ-interval from 0 to 2π. The repeating property tells us that we could construct the rest of the graph for each of these functions by copying over and over the part we have done already. Try it.

Checkpoint 4

(a) Copy from Figure 6.2 the graph of $y = \sin\theta$ from 0 to 2π. Then replicate this graph as many times as necessary to construct the graph of $y = \sin\theta$ from -3π to 7π.

(b) Repeat part (a) for the graph of $x = \cos\theta$.

(c) Check your work by graphing these functions with your calculator.

Period and Frequency

A function that repeats over and over, as do sine and cosine, is called **periodic**, and the horizontal length of the shortest pattern that repeats is called the **period**. We also can think of the period as being the time required to complete a single cycle (if time is the independent variable). Thus sine and cosine are periodic functions with period 2π.

A concept closely related to period is **frequency**, which is the rate at which periods are being completed. Thus, if the unit of time is seconds and a periodic function has period 5 seconds, then its frequency is $1/5$ cycles per second. In general, frequency and period are reciprocals of each other. The unit for frequency is cycles per unit of time, and the unit for period is units of time per cycle.

Exploration Activity 2

(a) Read the second paragraph on the introductory page of this chapter again.

(b) Complete Table 6.1: Next to each of the repeating phenomena, write the period and frequency (in compatible time units) of whatever function describes the phenomenon.

Table 6.1 Periodic phenomena

Phenomenon	*Period*	*Frequency*
Normal heart beat		
LP record		
Rotation of the earth on its axis		
Earth orbiting the sun		
Standard alternating current		
Second hand on a clock		
Minute hand on a clock		
Hour hand on a clock		

Derivatives of Sine and Cosine

To model repeating phenomena by sine and cosine functions, we need to know the rates of change of these two functions. Thus it is time to calculate those derivatives.

We start with some pictorial evidence of what to look for. In Figure 6.3 we show a graph of the sine function and in Figure 6.4 a graph of an approximate derivative, calculated as

$$\text{approximate derivative of } \sin t = \frac{\sin(t+0.001) - \sin t}{0.001}.$$

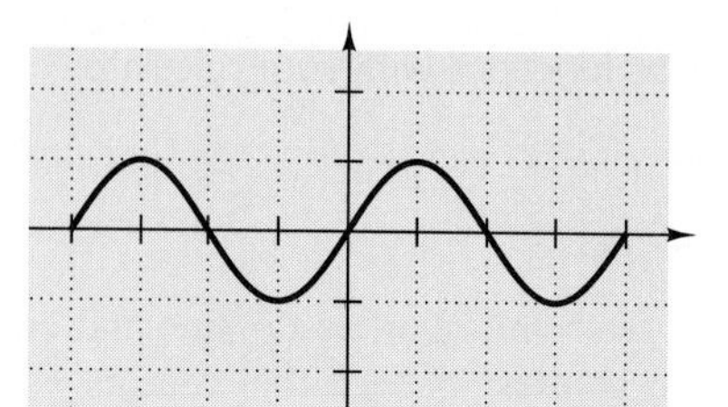

Figure 6.3 The sine function

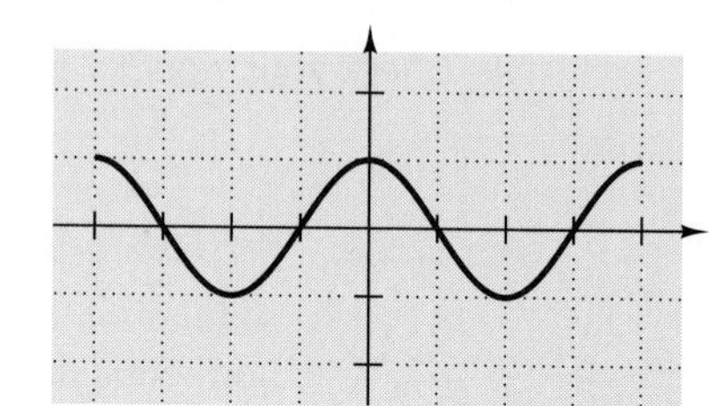

Figure 6.4 Approximate derivative of $\sin t$

Exploration Activity 3

(a) Label the axes and indicate the horizontal and vertical units on the left-hand graph in Figure 6.3. (You may copy this page if you do not want to write in the book.)

(b) The scales are the same for Figure 6.4; label it also.

(c) Why does the expression

$$\frac{\sin(t+0.001) - \sin t}{0.001}$$

approximate the derivative of $\sin t$?

(d) What do you see in the right-hand graph? What function do you think is the derivative of $\sin t$?

In both part (a) and part (b) the origin is at the center of the figure, with the horizontal axis representing t and the vertical axis y. The horizontal separation between grid lines is $\pi/2$ units, and the vertical separation is 1 unit. If $y = \sin t$, then the difference quotient $\Delta y/\Delta t$ that approximates dy/dt is

$$\frac{\sin(t+\Delta t) - \sin t}{\Delta t}.$$

The formula in part (c) is this difference quotient with $\Delta t = 0.001$, a small enough number to give a very good approximation to the derivative function. If you decided that the derivative looks a lot like $\cos t$, that's a good guess.

In Figures 6.5 and 6.6 we do the same thing with the cosine function that we just did with the sine function. The picture on the right is the graph of

$$\text{Approximate derivative of } \cos t = \frac{\cos(t+0.001) - \cos t}{0.001}.$$

Figure 6.5 The cosine function

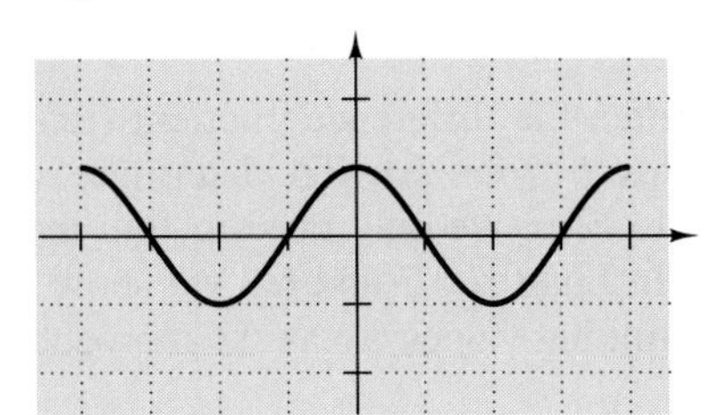

Figure 6.6 Approximate derivative of $\cos t$

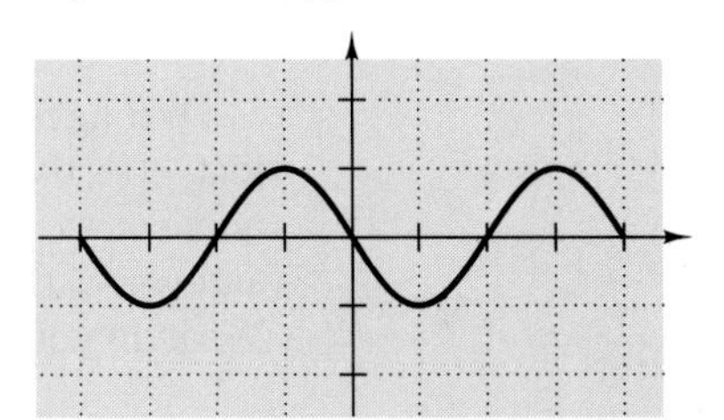

Checkpoint 5

(a) Label the axes and indicate the units in Figures 6.5 and 6.6. (The units are the same as in Figures 6.3 and 6.4.)

(b) What function do you think is the derivative of the cosine?

To actually calculate the derivative of $\sin t$, we examine algebraically the difference quotient

$$\frac{\sin(t+\Delta t) - \sin t}{\Delta t}$$

for small Δt. We know from Figure 6.4 what this expression looks like for varying t and a particular choice of Δt, and you have used this evidence to conjecture what the derivative is. However, there isn't any obvious way to use algebra to verify the conjecture, i.e., to find a factor of Δt in the numerator to cancel with the one in the denominator.

There is, however, a way to make progress with algebra, if we write the numerator in some other form. For this purpose, we recall the trigonometric identity for sine of a sum:

$$\sin(\alpha + \beta) = \sin\alpha\,\cos\beta + \cos\alpha\,\sin\beta.$$

Letting $\alpha = t$ and $\beta = \Delta t$, we may write

$$\sin(t+\Delta t) = \sin t\,\cos\Delta t + \cos t\,\sin\Delta t.$$

When we substitute this expression in the numerator of the difference quotient, we obtain

$$\frac{\sin(t+\Delta t)-\sin t}{\Delta t}=\frac{\sin t\cos\Delta t+\cos t\sin\Delta t\ -\ \sin t}{\Delta t}.$$

Now we regroup the terms on the right:

$$\frac{\sin(t+\Delta t)-\sin t}{\Delta t}=\frac{\sin\Delta t}{\Delta t}\cos t\ +\ \sin t\,\frac{\cos\Delta t-1}{\Delta t}.$$

We have now rewritten the difference quotient (left side of the equation) in a form that effectively separates t from Δt. To determine what happens to the difference quotient as Δt shrinks to zero, we only need to find the limiting values of $(\sin\Delta t)/\Delta t$ and $(\cos\Delta t - 1)/\Delta t$. In Figures 6.7 and 6.8 we show graphs of these functions of Δt. What do you think the limiting values are as Δt approaches 0?

Figure 6.7 Graph of $y=(\sin\Delta t)/\Delta t$

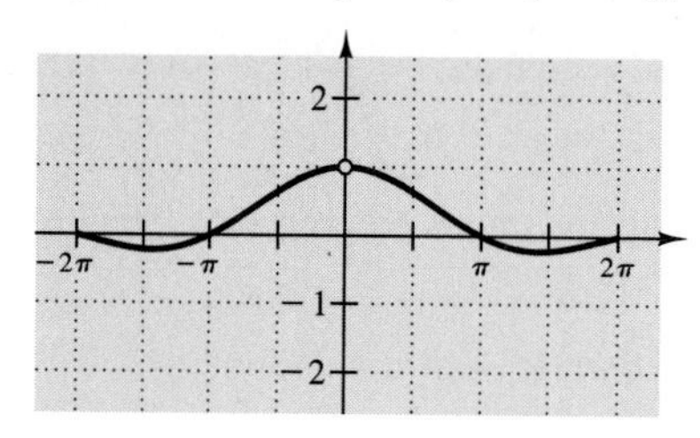

Figure 6.8 Graph of $y=(\cos\Delta t-1)/\Delta t$

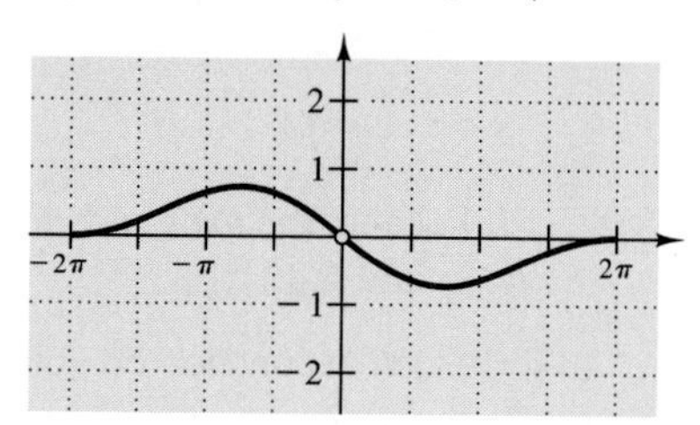

Exploration Activity 4

(a) Check your perception by using your calculator to fill in the missing entries in Table 6.2.

Table 6.2 Limiting values for $(\sin\Delta t)/\Delta t$ and $(\cos\Delta t-1)/\Delta t$ as $\Delta t\to 0$

Δt	$\sin\Delta t$	$\dfrac{\sin\Delta t}{\Delta t}$	$\cos\Delta t-1$	$\dfrac{\cos\Delta t-1}{\Delta t}$
1.0	0.84147	0.84147	−0.45970	−0.45970
0.1	0.099833	0.99833	−0.0049958	−0.049958
0.01				
0.001				
0.0001				
0.00001				

(b) What is the limiting value of $(\sin\Delta t)/\Delta t$ as Δt approaches 0?

(c) What is the limiting value of $(\cos\Delta t-1)/\Delta t$ as Δt approaches 0?

(d) Use your answers to (b) and (c), together with the equation

$$\frac{\sin(t+\Delta t)-\sin t}{\Delta t}=\frac{\sin\Delta t}{\Delta t}\cos t\ +\ \sin t\,\frac{\cos\Delta t-1}{\Delta t}$$

to show that the derivative of the sine function is what you already knew it had to be.

In the second and third columns of Table 6.2, you should see a definite pattern of digits. The last entry in the third column is 0.999999999983—very close to 1, which is the answer to part (b). A similar pattern appears in the cosine columns. Your calculator probably reports the last entry as -5×10^{-6}—very close to 0, which is the answer to part (c). In part (d), the first fraction on the right-hand side approaches 1, and the second approaches 0, so the difference quotient approaches $\cos t$ as Δt approaches 0.

We could find the derivative of $\cos t$ in much the same way (see Exercise 37), but it is easier to use what we already know about the derivative of $\sin t$. The two functions are related by this pair of identities:

$$\cos \alpha = \sin\left(\frac{\pi}{2} - \alpha\right)$$

and

$$\cos\left(\frac{\pi}{2} - \alpha\right) = \sin \alpha.$$

Exploration Activity 5

Use the two identities above, your answer to Exploration Activity 4, and your other rules for differentiation to find (or confirm) a formula for the derivative of the cosine function.

Since $\cos x = \sin(\pi/2 - x)$, we have

$$\frac{d}{dx} \cos x = \frac{d}{dx} \sin(\pi/2 - x).$$

Using the Chain Rule, we may rewrite the right-hand side

$$\begin{aligned}\frac{d}{dx} \cos x &= \cos(\pi/2 - x)\frac{d}{dx}(\pi/2 - x) \\ &= -\cos(\pi/2 - x).\end{aligned}$$

Since $\cos(\pi/2 - x) = \sin x$, we have

$$\frac{d}{dx} \cos x = -\sin x.$$

In this section we reviewed the two fundamental periodic functions—sine and cosine—and we discovered the differentiation formulas

$$\frac{d}{dt} \sin t = \cos t$$

and

$$\frac{d}{dt} \cos t = -\sin t.$$

Answers to Checkpoints

1. (a)

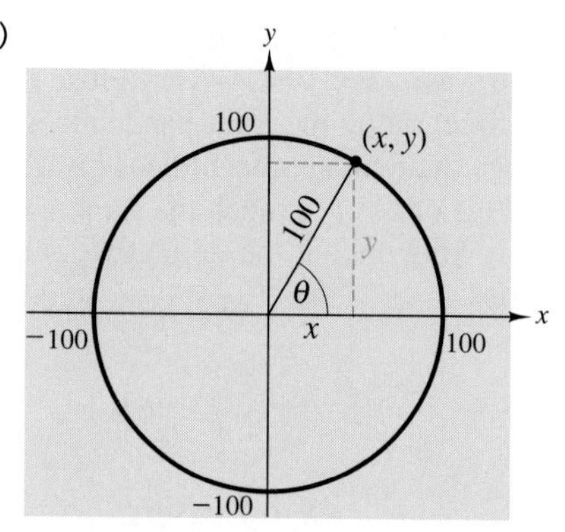

   (b) $x/100 = \cos\theta$ and $y/100 = \sin\theta$, so $x = 100\cos\theta$ and $y = 100\sin\theta$.

   (c) For all points on the circle, $-100 \le x \le 100$ and $-100 \le y \le 100$. Because you keep returning to the starting point, both coordinates oscillate between these extremes.

2. (a) $x = 27\cos\theta$ and $y = 27\sin\theta$

   (b) $x = \cos\theta$ and $y = \sin\theta$

3. (a) $40\pi \approx 126$ seconds, or 2 minutes, 6 seconds
   $20\pi \approx 63$ seconds, or 1 minute, 3 seconds
   $10\pi \approx 31$ seconds

   (b) $10\pi \approx 31$ seconds
   $20\pi/3 \approx 21$ seconds

   (c) $\theta = t/20$

   (d) $x = 100\cos(t/20)$ and $y = 100\sin(t/20)$

4. (a)

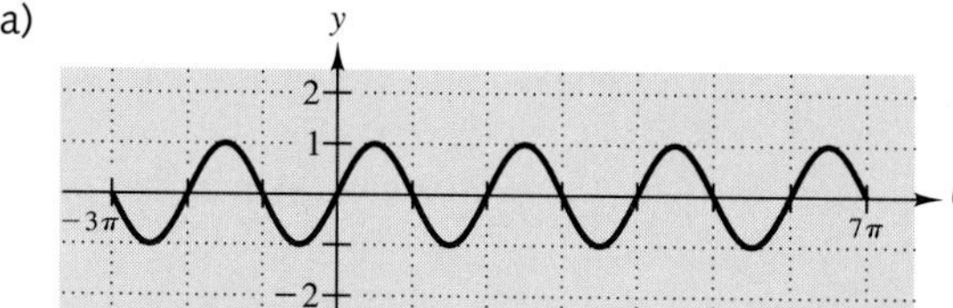

   (b)

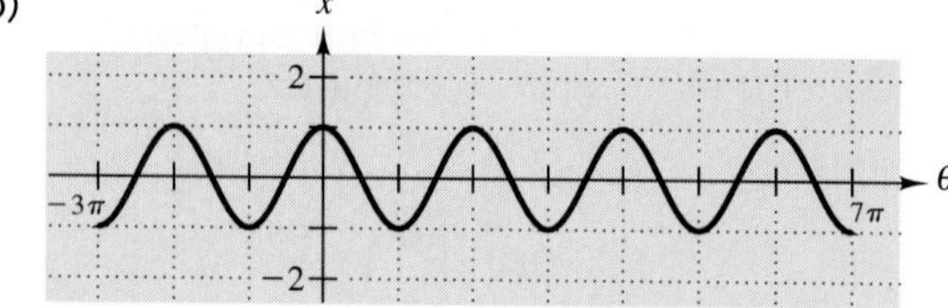

5. (a) See discussion of Exploration Activity 3 and answer to Checkpoint 4(b) with θ replaced by t.

   (b) The graph looks like that of $-\sin t$.

Exercises 6.1

Calculate each of the following derivatives.

1. $\dfrac{d}{dt}\, e^{2t} \sin t$
2. $\dfrac{d}{dt}\, e^{t} \cos 2t$
3. $\dfrac{d}{dt}\, \ln t \sin 2t$
4. $\dfrac{d^2}{dt^2}\, \sin t$
5. $\dfrac{d^2}{dt^2}\, \cos t$
6. $\dfrac{d}{dy}\, \sin y \cos y$
7. $\dfrac{d^2}{dt^2}\, \sin 2t$
8. $\dfrac{d^2}{dt^2}\, \sin^2 t$
9. $\dfrac{d^2}{dt^2}\, e^{2t} \sin t$

Find the derivative of each of the following functions.

10. $y = \sin 2t - 3 \cos t$
11. $y = \cos^2 t$
12. $y = \cos(t^2)$
13. $y = (\sin 3t)(\cos 2t)$
14. $z = \dfrac{2}{t-1}$
15. $z = \dfrac{1}{1+7t}$
16. $y = x^2 + 3x + 6$
17. $y = 5^x$
18. $y = x^3 \sin 2x$

For each of the following functions, (a) sketch the graph, and (b) sketch the graph of the derivative.

19. $y = \sin 2t - 3 \cos t$
20. $y = \cos^2 t$
21. $y = \cos(t^2)$
22. $y = (\sin 3t)(\cos 2t)$
23. $z = \dfrac{2}{t-1}$
24. $z = \dfrac{1}{1+7t}$
25. $y = x^2 + 3x + 6$
26. $y = 5^x$
27. $y = x^3 \sin 2x$

Find all antiderivatives of each of the following functions.

28. e^{2t}
29. t^n, where n is a positive integer
30. $\dfrac{1}{t^2}$
31. $\cos t$
32. $\dfrac{1}{t}$
33. $\sin 2t$
34. $\dfrac{2}{t-1}$
35. $\dfrac{1}{1+7t}$
36. Put your calculator in parametric mode, and graph the parametric equations

$$x = \cos t \qquad \text{and} \qquad y = \sin t.$$

Explain how the resulting graph relates to Figure 6.2.

37. Use a trigonometric identity for the cosine of a sum to rewrite the difference quotient

$$\frac{\cos(t + \Delta t) - \cos t}{\Delta t}$$

as we did for the difference quotient for sin t. Use what you already know about limiting values to show (another way) that

$$\frac{d}{dt} \cos t = -\sin t.$$

38. Given the initial value problem

$$\frac{dP}{dt} = 2 - \cos(\pi P) \qquad \text{with } P(0) = 1,$$

use Euler's Method with $\Delta t = 0.2$ to approximate $P(1)$.

39. Find the maximum and minimum values of

$$y = \sin x + \cos x$$

on the interval $[0, \pi]$.

40. (a) Show that sine is an odd function and cosine is an even function.
    (b) How do the symmetries of these functions fit with the information in Exercise 35 in Section 4.5?
41. Give a reason why $\cos x$ is not a polynomial.[1]
42. Determine the smallest positive number x for which the function

$$f(x) = -4 \sin\left(4x + \frac{\pi}{6}\right)$$

has
(a) the value zero.
(b) the maximum value of $f(x)$.
(c) the minimum value of $f(x)$.

43. (See Exercise 23 in Section 4.8) Recall that the tangent and secant functions are defined by

$$\tan t = \frac{\sin t}{\cos t} \quad \text{and} \quad \sec t = \frac{1}{\cos t}.$$

(a) Show that

$$\frac{d}{dt} \tan t = \sec^2 t.$$

(b) Show that

$$\frac{d}{dt} \sec t = \sec t \tan t.$$

44. Recall that the cotangent and cosecant functions are defined by

$$\cot t = \frac{\cos t}{\sin t} \quad \text{and} \quad \csc t = \frac{1}{\sin t}.$$

Find formulas for the derivatives of
(a) the cotangent function.
(b) the cosecant function.

45. Define a function f by $f(x) = (x + \sin x)/\cos x$ for $-\pi/2 < x < \pi/2$.
(a) Is $f(x)$ an even function, an odd function, or neither? Justify your answer.
(b) Find $f'(x)$.
(c) Find an equation of the line tangent to the graph of f at the point where $x = 0$.[2]

1. This exercise and the next are from *Calculus Problems for a New Century*, edited by Robert Fraga, MAA Notes Number 28, 1993.
2. From *Calculus Problems for a New Century*, edited by Robert Fraga, MAA Notes Number 28, 1993.

6.2 Spring Motion

The Spring-Mass System

The sine and cosine functions are the keys to describing all repetitive or periodic behavior. For our next illustration of this fact, we consider the motion of a mass bobbing up and down on the end of a spring. That may not sound like an important problem—unless you have an old car in which the shock absorbers are really shot—but it serves nicely as a prototypical oscillation, one that we can easily observe and measure. It's much harder to observe the crystals in our watches, the electric current in our walls, or the silicon circuits in our computers. And study of the bouncing mass on a spring is the beginning of analyses of phenomena such as vibrations of a bridge or the flutter of an airplane wing.

For all the examples just mentioned, we want to know such things as the frequencies and amplitudes of the vibrations. As we saw in the preceding section, **frequency** is the number of cycles per unit time. **Amplitude** is the maximum amount of displacement from zero in either direction.

We begin our study of the spring-mass system with a physical law—a property of springs that has been observed and measured in many physics labs.

Hooke's Law The force required to stretch a spring a distance s from its natural length is proportional to s.

In Exercise 18 at the end of Chapter 1 we asked you to express Hooke's Law by a simple formula. The law asserts that the force function has the form $F(s) = ks$ for some constant k, called the **spring constant** for the given spring. To find the spring constant, we can hang a known weight on the spring and measure the amount of displacement. For example, if a 5-pound weight stretches a spring 6 inches, then (in English units of feet and pounds), $5 = k \cdot (1/2)$, so the spring constant k is 10 pounds per foot.

Weight means the force exerted on an object by gravity. By Newton's Second Law, this force is the mass of the object times the acceleration g due to gravity. Thus weight and mass are related by

$$w = mg.$$

In Figure 6.9 we show two views of the same spring, one at its natural length (no force being applied) and the other stretched by hanging an object of mass m on it. We assume the spring-mass system is in equilibrium in the second view; i.e., the object is not moving. Thus the force the spring exerts on the object exactly balances the gravitational force the object exerts on the spring, namely, mg. By Hooke's Law, this force is also ks_0, so the mass, spring constant, and equilibrium displacement are related by

$$mg = ks_0.$$

Now we consider what happens if we stretch the spring an additional distance x, as in Figure 6.10. (Think of rescaling the vertical axis additively so that the new origin $x = 0$ is at $s = s_0$.) The system consisting of just the mass and spring is no longer in equilibrium. If we let go, there will be an imbalance of forces that will cause the object to move. On the one hand, there is the gravitational force mg—positive because our positive direction is downward. On the other hand, there is the restoring force of the spring, which, according to Hooke's Law, is $-ks = -k(x + s_0)$, negative because this force opposes increase in s. Thus the total force on the object is $mg - kx - ks_0$, which simplifies (because $mg = ks_0$) to just $-kx$.

Figure 6.9 Stretching a spring from its natural length with mass m

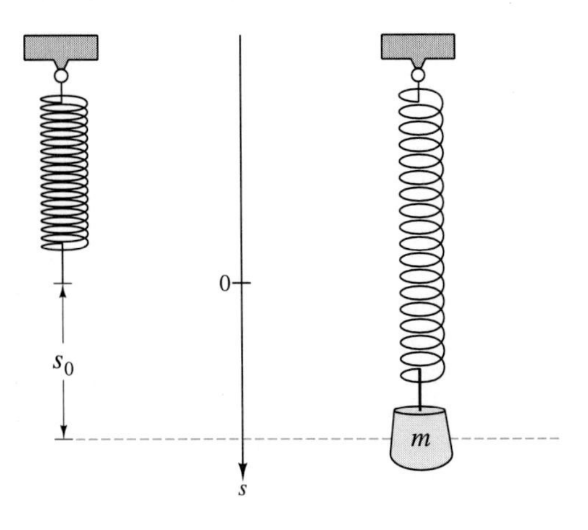

Figure 6.10 Stretching a spring an additional distance x

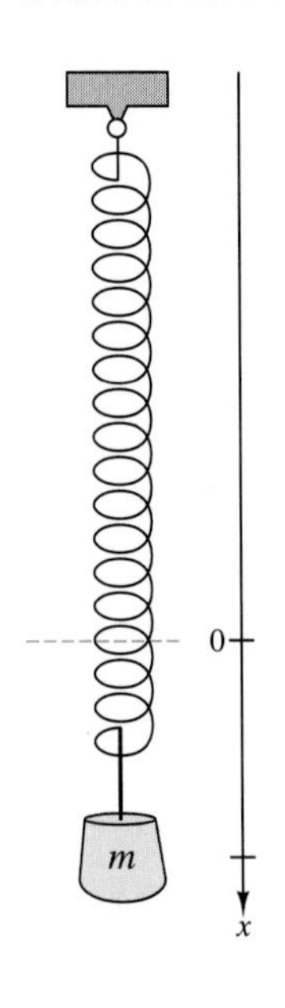

The Initial Value Problem

By Newton's Second Law of Motion, the total force on the moving object of mass m is also its mass times its acceleration. Thus we have a relationship between the displacement x and its second derivative:

$$m\,\frac{d^2x}{dt^2} = -kx.$$

Some features of this equation should look familiar, and some should appear new. First, under the influence of gravity, our bouncing mass is a falling body—but with a second force acting on it, the spring. When the falling body had no other force acting on it, our differential equation model expressed the second derivative of the position function as a constant. Here the second derivative is proportional to the position function.

Second, as was the case with the falling body, we should expect to undo differentiation twice to find x as a function of t. But unlike the falling body model, it is not at all clear how to do that. We also might expect that two "undifferentiation" steps will need two initial conditions, probably an initial velocity and an initial position.

Finally, note that the phrase "proportional to the unknown function" has appeared before—in natural population growth. What's different here is that it is the second derivative, not the first, that is proportional to the unknown function. Of course, an exponential function, say, $f(t) = e^{rt}$, has a second derivative that is proportional to the function itself, but it seems unlikely that such a function could tell us anything about bouncing up and down on a spring. Let's rule out exponential functions as solutions to the differential equation right away.

Exploration Activity 1

(a) Calculate the second derivative of

$$f(t) = e^{rt}.$$

(b) Show that f'' is proportional to f and that the proportionality constant must be positive, no matter what sign r has—as long as r is not 0.

(c) Explain why f cannot be a solution of equation

$$m\,\frac{d^2x}{dt^2} = -kx.$$

Each differentiation of e^{rt} inserts a factor of r, so the second derivative is r^2e^{rt}. This function is proportional to e^{rt} with proportionality constant r^2, which is always positive. The differential equation for motion of the spring-mass system has the second derivative of x proportional to x itself with proportionality constant $-k/m$, which must be negative. Thus no exponential function can be a solution.

We have just eliminated a promising candidate—promising mathematically, but not physically—for a solution to the differential equation

$$m\,\frac{d^2x}{dt^2} = -kx.$$

We have asserted—with little supporting evidence—that sines and cosines should have something to do with oscillations of the spring-mass system. As you may have found in Exercises 2 and 5 of Section 6.1, the second derivative of $\sin x$ is $-\sin x$, and the second derivative of $\cos x$ is $-\cos x$. Thus the sine and cosine functions have second derivatives that are proportional to those functions, with proportionality constant -1—at least the sign is right.

Before we pursue further how sines and cosines can help us describe the motion of the spring-mass system, we will study the differential equation from a numerical point of view.

A Numerical Solution: Euler's Method

If we formulate our problem as an initial value problem, then we have some hope that we can generate a numerical or graphical solution, which may in turn provide some evidence about the role to be played by sine and cosine functions.

In addition to the equation of motion,

$$m\,\frac{d^2x}{dt^2} = -kx,$$

we know something about how the motion starts. In particular, if we pull the mass to an initial displacement x_0 before we let go, then we know that $x(0) = x_0$. And if we simply let go, that is, if we don't throw the mass either up or down, then the initial velocity, $x'(0)$, must be zero. Thus we can state our initial value problem—after solving for the second derivative—in the following way:

$$\frac{d^2x}{dt^2} = -\frac{k}{m}\,x, \qquad \text{with} \quad x(0) = x_0 \quad \text{and} \quad x'(0) = 0.$$

Euler's Method, as we have used it so far, seems to apply only to first-order equations, that is, to equations in which the first derivative is expressed by a formula involving the independent or dependent variables. But, as we have seen with the flu epidemic model in the preceding chapter, the method can be used with a system of first-order differential equations, provided we have an initial condition for each of the unknown functions. We can use that knowledge here if we can write a single second-order differential equation as a system of two first-order equations.

We have two strong hints in our previous work about where to look for the second unknown function. First, we already have two initial conditions, one for the function x we are trying to find and the other for its derivative x'. Perhaps the second function should be the velocity! That would fit with the second hint, which comes from the falling body problem, in which our actual approach to solving the equation $s''(t) = g$ was to solve two first-order problems, one for finding velocity from acceleration and the other for finding position from velocity.

Suppose $x(t)$ is the desired function for the position of the bouncing mass at time t, i.e., the solution of the initial value problem. If we write $v(t) = x'(t)$, then x and v must satisfy the system of first-order equations

$$\frac{dx}{dt} = v$$

and

$$\frac{dv}{dt} = -\frac{k}{m}x$$

with the initial conditions

$$x(0) = x_0$$

and

$$v(0) = 0.$$

Be sure you know where each of these four equations comes from.

Recall the basic idea of Euler's Method: We step forward in time in steps of equal length Δt, so $t_0 = 0$, $t_1 = \Delta t$, $t_2 = 2(\Delta t)$, and so on. Initially, we know starting values for x and v from $x(0) = x_0$ and $v(0) = 0$, and we can calculate starting slopes for the two functions by substituting these values in $x' = v$ and $v' = -(k/m)x$. Given slopes, we can calculate a rise for each function as slope × run. Adding each rise to the corresponding starting value gives the next value for each function. And then we start over: From known function values, we use the differential equations to calculate slopes, use slopes to calculate rises, and add rises to find the next values.

Of course, as we saw in Section 5.1, the rises we are calculating are really dx and dv—rises to tangent lines—instead of what we really want to know, Δx and Δv—rises to the unknown curves. Euler's Method is only as good as the approximations (see Section 4.9) $dx \approx \Delta x$ and $dv \approx \Delta v$.

Verify that the discussion in the last two paragraphs leads to the forward-stepping formulas

$$x_{k+1} = x_k + v_k\,\Delta t$$

$$v_{k+1} = v_k - \frac{k}{m}x_k\,\Delta t$$

with

$$x_0 = \text{the starting displacement}$$

$$v_0 = 0.$$

In Figures 6.11 and 6.12 we show solutions for x and v generated from these formulas for a starting displacement x_0 of one unit and $k/m = 1$.

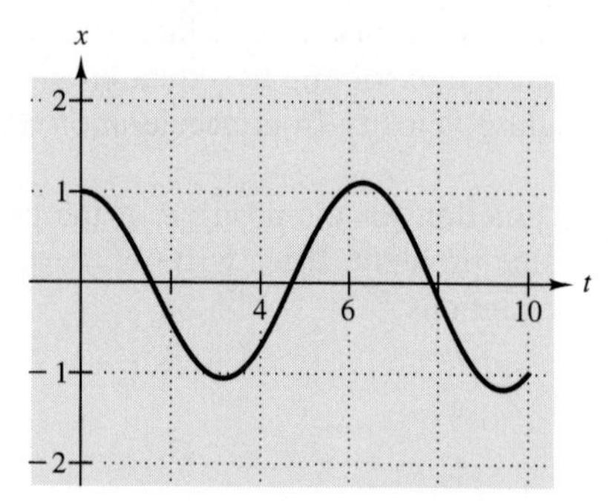

Figure 6.11 Numerical solution for position $x(t)$ of the mass bouncing on the spring

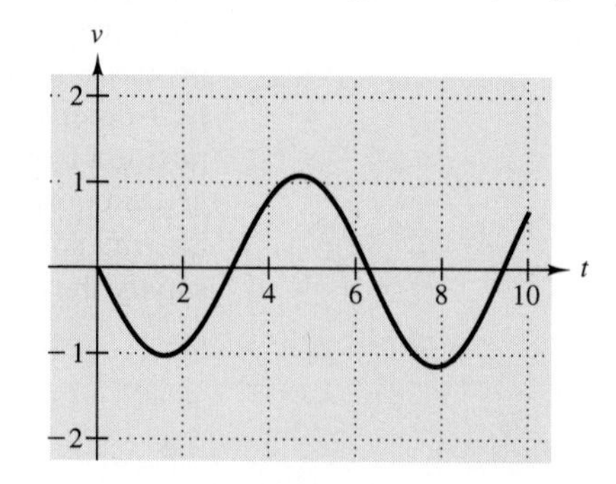

Figure 6.12 Numerical solution for velocity $v(t)$ of the mass bouncing on the spring

Exploration Activity 2

(a) Examine Figures 6.11 and 6.12 carefully. Do the curves behave the way you would expect from what you know about the physics of the spring-mass system?

(b) Are the curves believable as a solution to the initial value problem

$$\frac{dx}{dt} = v, \text{ with } x(0) = 1,$$

$$\frac{dv}{dt} = -x, \text{ with } v(0) = 0?$$

Why or why not?

(c) How do you think the numerically generated graphs of $x = x(t)$ and $v = v(t)$ are related to the sine and/or cosine functions? Does the relationship fit with what you know about derivatives of these functions?

The position curve in Figure 6.11 starts, as it should, at one unit away from equilibrium position and oscillates back and forth between that position and one unit away in the opposite direction. The velocity curve in Figure 6.12 starts at zero, as it should, and has a similar oscillation. In fact, the velocity appears to be 0 whenever the position is 1 or -1, which is where the moving mass changes directions. And the velocity appears to be 1 or -1 wherever x is 0. That is, the fastest speeds occur where the moving mass is passing its equilibrium position.

By comparison with Figures 6.5 and 6.6, we see that Figures 6.11 and 6.12 look a lot like graphs of $x = \cos t$ and its derivative, $x' = -\sin t$. And that pair of functions clearly solves the initial value problem.

A Symbolic Solution

We are ready now to find symbolic solutions of the differential equation

$$\frac{d^2x}{dt^2} = -\frac{k}{m}x,$$

where k and m are arbitrary positive constants. Then we can use the initial values for x and x' to complete our search for formulas for the solutions we generated graphically in Figures 6.11 and 6.12. As usual, we attack this problem by first solving an easier inverse problem. That is, we calculate derivatives of functions that would appear to have the right form to satisfy the differential equation.

In the next exploration activity we use ω for the constant multiplier of t. The symbol ω is a lower-case Greek omega—it is a conventional symbol for a period-altering scale factor in trigonometric functions.

Exploration Activity 3

(a) Show that

$$\frac{d^2}{dt^2}\sin\omega t = -\omega^2 \sin\omega t.$$

(b) Show that

$$\frac{d^2}{dt^2}\cos\omega t = -\omega^2 \cos\omega t.$$

(c) Explain why $x(t) = \sin\omega t$ and $x(t) = \cos\omega t$ are both solutions of the differential equation

$$\frac{d^2x}{dt^2} = -\omega^2 x.$$

(d) Explain why any function of the form $x(t) = A\sin\omega t + B\cos\omega t$ is a solution of the same differential equation.

By the Chain Rule, the derivative of sin ωt is ω cos ωt. When we differentiate again, using both the Chain Rule and the Constant Multiple Rule, we find that the second derivative of sin ωt is $-\omega^2$ sin ωt, as asserted in part (a). The calculation for part (b) is similar. These calculations show that the effect of differentiating either sin ωt or cos ωt twice is to multiply the original function by $-\omega^2$. That is, each of these functions satisfies the differential equation in part (c). For part (d), we can combine these results in another calculation using the Sum and Constant Multiple Rules:

$$\begin{aligned}\frac{d^2x}{dt^2} &= \frac{d^2}{dt^2}(A\sin\omega t + B\cos\omega t)\\ &= A\,\frac{d^2}{dt^2}\sin\omega t + B\,\frac{d^2}{dt^2}\cos\omega t\\ &= A\,(-\omega^2\sin\omega t) + B\,(-\omega^2\cos\omega t)\\ &= -\omega^2(A\sin\omega t + B\cos\omega t)\\ &= -\omega^2 x.\end{aligned}$$

Notice the parallel in Exploration Activity 3 with what we did in Chapter 2. There we looked for functions to satisfy an equation of the form

first derivative is proportional to the function

and found that the solutions were all of the form Ae^{kt}. In particular, the scale factor for the independent variable was the proportionality constant in the differential equation. Now we are solving

second derivative is *negatively* proportional to the function

and finding a similar result with sines and cosines. In particular, the scale factor for the independent variable is the square root of the absolute value of the proportionality constant.

Second derivative is *positively* proportional to the function

is a different, but also similar, situation—see Exercises 45–47 at the end of this chapter.

Our original differential equation

$$\frac{d^2x}{dt^2} = -\frac{k}{m}\,x$$

has precisely the form

$$\frac{d^2x}{dt^2} = -\omega^2 x$$

if we identify ω^2 with k/m; that is, if we set $\omega = \sqrt{k/m}$. And we can do that because we know that both k and m are positive. Thus, according to Exploration Activity 3, any function of the form

$$x(t) = A\sin\sqrt{\frac{k}{m}}\,t + B\cos\sqrt{\frac{k}{m}}\,t$$

is a solution of

$$\frac{d^2x}{dt^2} = -\frac{k}{m}\,x.$$

Now we need to determine what combination of coefficients A and B—if any—will give us one function that also satisfies the initial conditions for displacement and velocity. We start by substituting $t = 0$ and $x = x_0$:

$$x_0 = A\sin 0 + B\cos 0.$$

Since $\sin 0 = 0$ and $\cos 0 = 1$, this equation tells us that B must be the starting displacement x_0.

Since A dropped out of the equation when we substituted $t = 0$, we know nothing yet about the value of A. But we have another initial condition! In order to use it, we need a formula for $v(t)$, which you can provide by differentiating the formula for $x(t)$.

Checkpoint 1

(a) Find a formula for $v(t)$.

(b) Substitute the initial values for t and v in your formula to find A.

Conclusion: The unique solution of the spring-mass initial value problem,

$$\frac{d^2x}{dt^2} = -\frac{k}{m}\,x \qquad \text{with } x(0) = x_0 \quad \text{and} \quad x'(0) = 0$$

is

$$x(t) = x_0\cos\sqrt{\frac{k}{m}}\,t.$$

In order to generate a numerical solution for Figures 6.11 and 6.12, we had to give numerical values for the starting displacement and for the ratio of spring constant to mass. We chose both those numbers to be 1, so the solution we generated was just $x = \cos t$. The velocity function was then the derivative of $\cos t$, that is, $v = -\sin t$. And that was exactly what we concluded in our discussion of Exploration Activity 2.

In this section we have set up and solved an initial value problem that represents the repetitive motion of a simple spring-mass system. We first studied the problem numerically and graphically. In the course of that study we saw that we could apply Euler's Method by turning our second-order differential equation into two first-order equations, one for position and one for velocity. The graphical solutions for the two functions resembled graphs of cosine and sine functions, respectively. We then solved the problem symbolically and found that the position function is indeed a constant multiple of a cosine function, and therefore, the velocity function is a constant multiple of a sine function. Furthermore, we can now relate the frequency of the oscillation to the spring constant and the mass of the moving object.

Answers to Checkpoint

1. (a) $v(t) = A\sqrt{\frac{k}{m}}\cos\sqrt{\frac{k}{m}}\,t - B\sqrt{\frac{k}{m}}\sin\sqrt{\frac{k}{m}}\,t$ (b) $A = 0$

Exercises 6.2

Calculate each of the following derivatives.

1. $\frac{d}{dt}(10t^6 - 4t^3 + \pi)$
2. $\frac{d}{dx}\frac{\sin x}{x}$
3. $\frac{d}{dx}(\cos x)^3$
4. $\frac{d}{dt}\ln\sqrt{3t-9}$
5. $\frac{d}{d\theta}\frac{\sin\theta}{\cos 2\theta}$
6. $\frac{d}{dw}\frac{1}{w^3}$
7. $\frac{d}{dt}t^{7/5}$
8. $\frac{d}{dt}te^{-t}$
9. $\frac{d^2}{dt^2}te^{-t}$

Find the derivative of each of the following functions.

10. $y = \ln(x^{10} - 3x)^{1/3}$
11. $f(x) = \frac{x^2}{1+2x}$
12. $y = \frac{\cos x + x^2}{x^3}$
13. $y = \frac{\sin x}{\cos x + 1}$
14. $y = \ln(e^x)$
15. $f(x) = 3e^{-4x}$
16. $f(x) = \sin(x^2)$
17. $y = y(x)$, where $y^3 + x^2 = 3x + 7$

For each of the following functions, (a) sketch the graph, and (b) sketch the graph of the derivative.

18. $y = \ln(x^{10} - 3x)^{1/3}$
19. $f(x) = \frac{x^2}{1+2x}$
20. $y = \frac{\cos x + x^2}{x^3}$
21. $y = \frac{\sin x}{\cos x + 1}$
22. $y = \ln(e^x)$
23. $f(x) = 3e^{-4x}$
24. $f(x) = \sin(x^2)$
25. $y = y(x)$, where $y^3 + x^2 = 3x + 7$

Solve each of the following initial value problems.

26. $\frac{dy}{dt} = 1.3y, \quad y(0) = -0.5$
27. $\frac{dy}{dt} = -1.3y, \quad y(0) = 0.5$
28. $\frac{d^2y}{dt^2} = -1.3y, \quad y(0) = -0.5, \quad y'(0) = 0$

29. $\dfrac{d^2y}{dt^2} = -1.3y, \ y(0) = 0, \ y'(0) = -0.5$

30. Solve the following initial value problem:

$$\frac{d^2y}{dt^2} = 1.3y \quad \text{with } y(0) = -0.5 \text{ and } y'(0) = 0.$$

(*Hint*: Find two exponential functions, e^{rt} and e^{st}, that satisfy the differential equation. Then look for a solution of the initial value problem of the form $Ae^{rt} + Be^{st}$.)

31. (a) What is the frequency of the oscillation of a spring-mass system with mass m and spring constant k?
(b) What is the period?

32. The frequency of household electric current is 60 hertz. (One **hertz** is one cycle per second.) If this current is described by a formula of the form $x(t) = A \sin \omega t + B \cos \omega t$, and time is measured in seconds, what is ω?

33. The clock speed of a certain personal computer is 33 megahertz. If the oscillation is described by a formula of the form $x(t) = A \sin \omega t + B \cos \omega t$ and time is measured in microseconds, what is ω? (The prefix **mega-** means one million, and the prefix **micro-** means one-millionth.)

34. If the mass attached to a spring is thrown (vertically up or down) instead of let go, all that changes in our initial value problem is that the initial velocity is no longer zero.
(a) The initial velocity can be either positive or negative. Which sign corresponds to throwing the mass up and which down?
(b) Find values for the constants A and B such that the function

$$x(t) = A \sin \sqrt{\frac{k}{m}}\, t + B \cos \sqrt{\frac{k}{m}}\, t$$

satisfies the initial value problem

$$\frac{d^2x}{dt^2} = -\frac{k}{m}\, x$$

with $\quad x(0) = 1 \quad$ and $\quad \dfrac{dx}{dt}(0) = -0.5.$

35. Interpret the initial value problem

$$\frac{d^2y}{dt^2} = -1.07y$$

with $y(0) = -0.5$ and $y'(0) = 0.5$ as representing the motion of a spring-mass system. What are the mass and the spring constant? What is the initial state of the system?

36. A mass of 4 kilograms is suspended from a spring attached to the ceiling that has a spring constant of 9 kilograms per second squared. The mass is pulled downward 1 meter below its equilibrium position and then released from this point with an initial velocity upward of 1.5 meters per second.
(a) Write down the differential equation and initial conditions satisfied by the function $x(t)$ that gives displacement of the mass from equilibrium as a function of time.
(b) Solve the initial value problem to find $x(t)$.
(c) At what times will the mass be closest to the ceiling?

37. (a) Sketch the graph of $\cos 2t$ for $0 \le t \le 2\pi$.
(b) Sketch the graph of $\cos^2 t$ for $0 \le t \le 2\pi$.
(c) Use the identities $\cos 2t = \cos^2 t - \sin^2 t$ and $\cos^2 t + \sin^2 t = 1$ to show that

$$\cos^2 t = \frac{1 + \cos 2t}{2}$$

for all t.
(d) Find all the points of inflection on the curve $y = \cos^2 t$, and mark them on your graph in part (b). Why must these points also be points of inflection for the curve $y = \sin^2 t$?
(e) Sketch the graph of $\sin^2 t$ for $0 \le t \le 2\pi$.
(f) Show that

$$\sin^2 t = \frac{1 - \cos 2t}{2}$$

for all t.

38. An airplane is flying directly overhead at an altitude of 20,000 feet. One minute later, the line of sight to the airplane makes an angle of 49 degrees with the ground. What is the approximate speed of the airplane in miles per hour?

39. In the Chapter 5 Lab Reading we derived coordinate formulas for the position of a projectile fired at an angle θ at 100 meters per second, under the

assumption of negligible air resistance. Here are those formulas:

$$x = 100(\cos\theta)t$$

and

$$y = 100(\sin\theta)t - \frac{gt^2}{2}.$$

(a) Solve the first equation for t in terms of x, and substitute the result in the second equation to find an equation in x and y for the flight path of the projectile. What do you conclude about the shape of the path?
(b) Suppose (unlike our scenario in the lab reading) that the target point is at the same altitude as the launch point. Express the horizontal distance to the target point in terms of the firing angle θ.
(c) What angle θ gives the maximum possible horizontal distance? [You may need to use a trigonometric identity to simplify your formula in part (b) before trying to answer this question. The maximum horizontal distance a projectile can achieve is called the **range** of the projectile.]

40. A paper in the journal *Statistics and Medicine*[3] reported on efforts to find baseline models for normal biological rhythms, such as the sleep-wake cycle, hormone levels, and core body temperature. The goal of such research is to be able to diagnose conditions such as depression by measuring disturbances in these rhythms. The principal example in the paper is based on temperature data gathered over a 5-day period from a healthy 28-year-old female. An embedded temperature probe recorded core body temperature every 2 minutes. The researchers used every tenth data point, so their time unit was 20 minutes. On that scale, they found the primary rhythm to be modeled quite well by a function of the form

$$f(t) = \mu + A\cos(2\pi\omega t) + B\sin(2\pi\omega t),$$

where $\mu = 37.5$, $\omega = 0.0138$, $A = -0.103$, and $B = -0.429$.
(a) Graph the model function.
(b) What is the physiological significance of μ ?
(c) What is the physiological significance of ω ?

3. J. B. Greenhouse, R. E. Kass, and R. S. Tsay, "Fitting Nonlinear Models with ARMA Errors to Biogical Rhythm Data," *Statistics and Medicine* 6 (1987), pp. 167–183.

6.3 Pendulum Motion

We don't make much use of pendulum clocks (sometimes called grandfather clocks) anymore, but pendulum motion provides another example of a repetitive phenomenon that we can easily observe and measure. We will see it has some features that are similar to spring motion and some that are different.

Exploration Activity 1

Stop reading at the end of this paragraph, and make yourself a pendulum. String and a heavy object (such as a doorknob) work well. If the only doorknob you have is still attached to your door, substitute a coffee mug (empty) or your key case. If nothing else is at hand, take off your shoe and suspend it from a shoelace. Hold the top of the string fixed, and pull the object at the bottom (called the bob) to one side at an angle of approximately 30 degrees, hold it still, then let it go. The pendulum will swing back and forth, with the maximum angle from the vertical decreasing with time until the swinging stops.

(a) What causes the pendulum to stop?

(b) The length of time it takes the pendulum to swing out and back is the **period** of the pendulum. Does the period seem to depend on the initial angle? Try various starting angles to see.

(c) The length of the string from your hand to the bob is called the **length** of the pendulum. Move your hand up and down the string to vary the length. How does the period of the pendulum seem to depend on the length of the pendulum?

The most important forces acting on the bob are the gravitational force that makes it move in the first place and the force exerted by the string to keep it moving along a circular path. As we will see shortly, neither of these forces can make the bob stop moving. However, the significant forces also may include friction where you are holding the string and resistance from the air through which the bob is passing. These (and possibly other) retarding forces eventually cause the pendulum to stop swinging.

You probably found very little dependence of the period on the initial angle, but strong dependence on the length. In particular, the longer the string, the longer the period. However, if you timed periods carefully, you found that the relationship between period and length is not one of proportionality. In this section we will determine, at least approximately, how to describe this relationship between period and length.

An Initial Value Problem

We will construct a model to describe how the angle of the pendulum varies as a function of time. Then we will use this model to describe how the period of the pendulum varies as a function of the length of the pendulum. In particular, suppose a bob of mass m (say, 50 grams) is attached to the end of a string of length L (say, one meter). Assume the bob is pulled to one side so that the string makes an angle of 30 degrees with the vertical. The bob is held still and released.

To obtain a reasonably simple model for pendulum motion, we make a number of simplifying assumptions:

- Relative to the mass m of the bob, the mass of the string is negligible.
- Compared with the length of the string, the diameter of the bob is negligible.
- At the stable point of swing (the pivot), friction and other retarding forces are negligible.
- The retarding force of air resistance is negligible.

You may well question some or all of these assumptions—particularly if your pendulum is a shoe on a shoestring. Compare these assumptions with Exploration Activity 1. Under these assumptions, is there anything that can make the pendulum come to rest?

Holding our skepticism in abeyance, let's see where these assumptions lead us. Let $s = s(t)$ be the distance along the arc from the lowest point to the current position of the bob, with displacement to the right considered positive. (See Figure 6.13. Identify and label s in the figure.) Let $\theta = \theta(t)$ be the corresponding angle with respect to the vertical. For a particular pendulum, L is fixed and s varies with θ.

Figure 6.13 Tangential and radial components of gravitational force on the pendulum bob

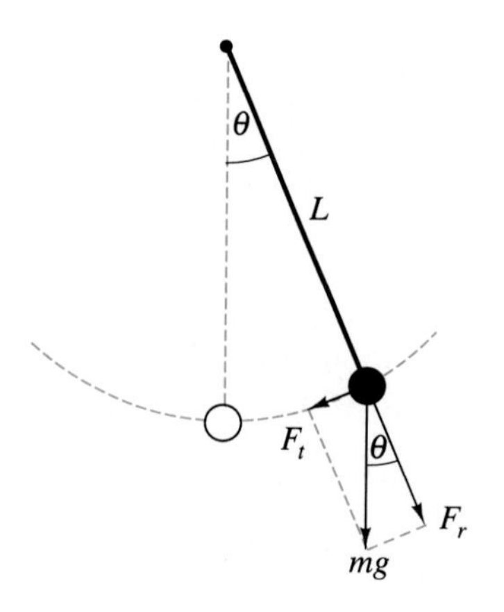

Our goal is to model θ as a function of time. In order to do this, we will need to find the relationship between s and θ.

Checkpoint 1

(a) What is s, the distance along a circular arc of radius L, when $\theta = 2\pi$?

(b) What is s when $\theta = \pi$?

(c) What is s when $\theta = \pi/2$?

(d) Explain why, for any angle θ, $s = L\,\theta$.

As usual for problems of motion, we will obtain a differential equation (for θ as a function of t) by applying Newton's Second Law. However, in this two-dimensional setting, all three of velocity, acceleration, and force are vector quantities; i.e., they have both magnitude and direction, and this has to be taken into account.

The direction of velocity is tangent to the curve of motion—the string prevents any motion in the perpendicular (or radial) direction. As $s = L\,\theta$, the magnitude of the velocity (i.e., the speed) ds/dt is $L\,(d\theta/dt)$.

You will learn in your physics class (or in the third semester of this course) that both the total force and the total acceleration may be broken down into tangential and radial parts and that Newton's Law may be applied to each of these parts separately. We study the total force first.

There are two forces on the bob—tension in the string and the force of gravity. The tension is entirely along the string, i.e., radial. Thus the tangential part of the tension is zero. In Figure 6.13 we show how the gravitational force mg is broken down into its tangential and radial parts. The magnitudes of the tangential force F_t and the radial force F_r should be such that they fit together as the sides of a rectangle of which the diagonal is the gravitational force.

Checkpoint 2

Use Figure 6.13 to explain why

$$F_t = -mg \sin\theta.$$

Where does the factor of $\sin\theta$ come from? Where does the negative sign come from?

Since the tangential part of the tension is zero, the total tangential force is $-mg\sin\theta$. You will learn in physics (or later in this course) that the tangential part of the acceleration is the second derivative of position or the first derivative of the speed. We are ready to apply Newton's Second Law,

$$\text{mass} \times \text{acceleration} = \text{force},$$

to the tangential parts of acceleration and force. For our pendulum,

$$m\,\frac{d^2 s}{dt^2} = -mg \sin\theta.$$

When we cancel the common factor m, we get

$$\frac{d^2 s}{dt^2} = -g \sin\theta.$$

There are two features of this differential equation that we need to think about—and then to do something about: (1) The variable s appears on the left side of the equation, while the variable θ appears on the right. Each is an unknown function of t. (2) The derivative on the left-hand side, as was the case with the spring-mass system, is a second derivative. The first of these features is easily dealt with, as we will see now. The second we will treat as we did with the spring.

Checkpoint 3

Show that $\dfrac{d^2 s}{dt^2} = L\,\dfrac{d^2\theta}{dt^2}$.

If we substitute $L\,\dfrac{d^2\theta}{dt^2}$ for $\dfrac{d^2 s}{dt^2}$ in

$$\frac{d^2 s}{dt^2} = -g \sin\theta$$

and divide through by L, we obtain

$$\frac{d^2\theta}{dt^2} = -\frac{g}{L}\sin\theta.$$

Now we have a differential equation with a single dependent variable θ.

This equation is similar to the spring equation

$$\frac{d^2x}{dt^2} = -\frac{k}{m}x,$$

with θ replacing x, g replacing k, and L replacing m. But an important difference between the two is the presence of the sine function in pendulum equation. Recall that, for the spring, trigonometric functions turned up only in the solutions.

For our assumed conditions—starting at a 30-degree angle to vertical and letting the bob drop—our initial conditions are

$$\theta = \frac{\pi}{6} \quad \text{when } t = 0$$

and

$$\frac{d\theta}{dt} = 0 \quad \text{when } t = 0.$$

Checkpoint 4

Write a description in words of the physical meaning of each of the initial conditions.

In summary, to describe the motion of the pendulum, we seek the function $\theta(t)$ that satisfies the second-order initial value problem

$$\frac{d^2\theta}{dt^2} = -\frac{g}{L}\sin\theta \qquad \text{with } \theta(0) = \frac{\pi}{6} \quad \text{and} \quad \theta'(0) = 0.$$

Now, we know this problem must have a solution, because we see the pendulum move. But the presence of $\sin\theta$ in the differential equation makes it impossible to give a simple formula that describes the solution function $\theta(t)$). Instead, we will take two other approaches to describing this function.

A Numerical Solution: Euler's Method

We will first find a numerical solution to the pendulum problem as we did for the spring-mass problem. We know now how to apply Euler's Method to a second-order initial value problem. The fact that the differential equation is slightly different for the pendulum has little effect on our ability to apply this method.

Checkpoint 5

Suppose $\theta(t)$ is the unknown solution function for position of the pendulum bob at time t, and let $v = \theta'(t)$. Explain why θ and v must satisfy the system of first-order equations

$$\frac{d\theta}{dt} = v$$

and

$$\frac{dv}{dt} = -\frac{g}{L}\sin\theta,$$

with the initial conditions

$$\theta(0) = \frac{\pi}{6}$$

and

$$v(0) = 0.$$

Now we apply Euler's Method to the system of first-order differential equations in Checkpoint 5.

Checkpoint 6

Explain why the forward-stepping formulas for Euler's Method are

$$\theta_{k+1} = \theta_k + v_k\,\Delta t$$

$$v_{k+1} = v_k - \frac{g}{L}\,\text{sin}(\theta_k)\,\Delta t$$

with

$$\theta_0 = \frac{\pi}{6}$$

$$v_0 = 0.$$

In Figure 6.14 we show solutions for θ and v generated from these formulas for a pendulum of length one meter. In a lab associated with this chapter you may experiment with Euler's Method to see what it tells us about the relationship between period and length of the pendulum.

Figure 6.14 Numerical solutions for the position and velocity of the pendulum bob

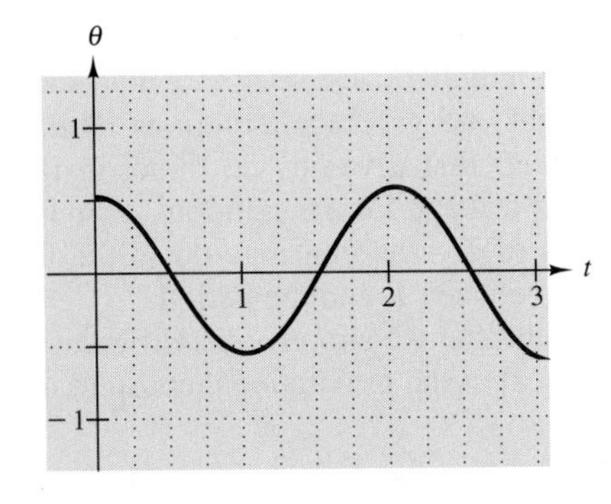

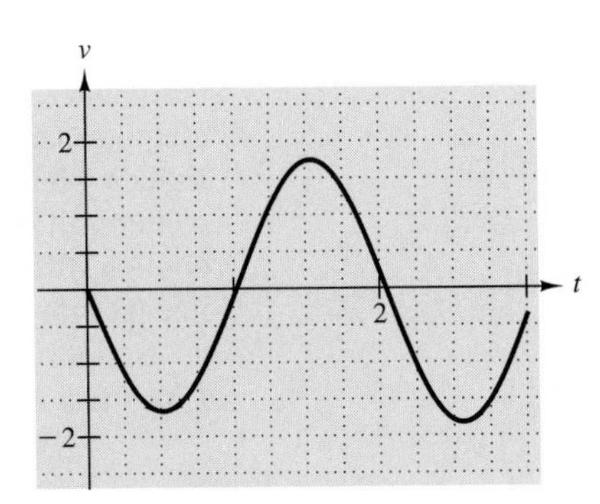

Exploration Activity 2

(a) Examine Figure 6.14 carefully. Do the curves behave the way you would expect from what you know about the physics of the pendulum?

(b) Are the curves believable as a solution to the initial value problem

$$\frac{d\theta}{dt} = v \qquad \text{with } \theta(0) = \frac{\pi}{6},$$

$$\frac{dv}{dt} = -\frac{g}{L}\sin\theta \qquad \text{with } v(0) = 0?$$

(c) Estimate the period of a one-meter pendulum from Figure 6.14. (Time is measured in seconds.)

(d) Why doesn't this period depend on the mass of the bob?

We note that the θ curve starts just above 0.5, which is about right for $\pi/6$, and the velocity graph starts at 0, as it should. The angle θ oscillates from its maximum value through 0, on to about $-\pi/6$, and back again. We see that the speeds are greatest when the angle is 0, and the speed is 0 when θ is at its extremes. All this appears to fit both the physics of the pendulum—if there is no friction—as well as the mathematics of the initial value problem.

The initial value problem makes no mention of the mass of the bob. Thus, if the physics is right and the model is right, the period should be independent of mass.

A Simpler Initial Value Problem: Symbolic Solution

Let's recall the problem we set out to solve: to determine the relationship between the period of a pendulum and its length. We modeled the motion of an idealized pendulum by the second-order differential equation

$$\frac{d^2\theta}{dt^2} = -\frac{g}{L}\sin\theta$$

where θ is the angle the pendulum makes with the vertical at time t. We saw that, for a particular choice of the length L, and for particular starting values of θ and $d\theta/dt$, we could generate numerical and graphical solutions of the differential equation by using Euler's Method. However, that gives us no direct insight into the symbolic form of a solution. It does show us a shape for the solution—see Figure 6.14. And, if repeated with enough different values of L—as you may do in the laboratory—it can give us some insight into the relation between L and the period.

Suppose for a moment that we simplify the differential equation by choosing a length L that is numerically equal to the gravitational constant g. Then the equation reduces to

$$\frac{d^2\theta}{dt^2} = -\sin\theta.$$

Does either $\theta = \sin t$ or $\theta = \cos t$ satisfy this equation? Not quite! For example, if $\theta = \sin t$, then

$$\frac{d^2\theta}{dt^2} = -\sin t = -\theta,$$

and the result is *not* $-\sin\theta$.

But we're also not so far off, either. Indeed, you showed in Exploration Activity 4 in Section 6.1 that, for small values of θ, sin θ and θ are approximately equal. Thus, instead of making a direct assault on symbolic solution of

$$\frac{d^2\theta}{dt^2} = -\frac{g}{L}\sin\theta,$$

we can replace the differential equation with one that is similar but has an easily described family of solutions. If we replace $\sin\theta$ by θ in the equation, we obtain

$$\frac{d^2\theta}{dt^2} = -\frac{g}{L}\,\theta.$$

Can we justify this replacement? For $\theta = \pi/6 \approx 0.5236$, we have $\sin\theta = 0.5$. So the substitution of θ for $\sin\theta$ is not a great approximation for θ near $\pi/6$, but it becomes better for smaller values of θ. Again, see your table in Section 6.1, Exploration Activity 4.

If we shift our attention to the differential equation

$$\frac{d^2\theta}{dt^2} = -\frac{g}{L}\,\theta,$$

what can we say about the family of solutions? We already know what they are! Indeed, this equation is just an alias for

$$\frac{d^2x}{dt^2} = -\frac{k}{m}\,x$$

with the names of the variables changed to protect the innocent. Specifically, x has been replaced by θ, k by g, and m by L. In the exercises at the end of this section we ask you to recall what you already know about solutions of this differential equation and then to answer the question about relationship between period and length of the pendulum.

In this section we have modeled the motion of a frictionless pendulum. We saw that the model is strikingly similar to—but still a little different from—our model of a spring-mass system in the preceding section. We took two approaches to finding approximate solutions of the initial value problem. First, we generated numerical and graphical solutions by Euler's Method, just as we did for the spring-mass system. Here, the difference between the two models is of no great significance. Not surprisingly, the graphical solutions look very similar. Then we used the approximation $\sin\theta \approx \theta$ to turn the pendulum model into the equivalent of a spring-mass model. For the latter, we already know solutions in terms of sine and cosine functions, so we can use these symbolic solutions for spring-mass as approximate solutions for the pendulum. As we have already worked out these solutions once, we leave the rest of the details for you to complete. In the process, you will answer—at least approximately—the question we posed at the start of the section: What is the relationship between the length of a pendulum and its period?

Answers to Checkpoints

1. (a) $2\pi L$ (b) πL (c) $\pi L/2$

(d) The radian measure of a central angle of a circle is the ratio of the length of the subtended arc to the radius of the circle.

2. θ is also an acute angle in the other triangle, opposite the side of length $|F_t|$. Thus $\sin\theta = |F_t|/mg$. When θ is positive, F_t is directed in the negative θ-direction, and when θ is negative, F_t is directed in the positive θ-direction. Thus the sign of F_t is opposite that of θ and hence also opposite that of $\sin\theta$. It follows that $F_t = -mg\sin\theta$.

3. From $s = L\theta$, where s and θ are functions of t and L is constant, it follows that $s' = L\theta'$ and $s'' = L\theta''$.

4. The first initial condition states that the starting position of the pendulum bob is $\pi/6$ radians or 30 degrees. The second initial condition states that the starting angular velocity is 0.

5. When we substitute the first equation, $v = \theta'$, into the second, we get $\theta'' = -(g/L)\sin\theta$, which is the equation of motion for the pendulum. The initial condition for v matches that for θ', so this first-order system is equivalent to the second-order initial value problem that defines θ and θ'.

6. Each stepping formula has the form new value = old value + slope × run, where the run is Δt and the slope is given by the appropriate differential equation. Also, the starting values match the initial conditions.

Exercises 6.3

Calculate each of the following derivatives.

1. $\dfrac{d}{dt}(t^6 - 4t^3 + 1)^2$

2. $\dfrac{d}{dx}\dfrac{x}{\sin x}$

3. $\dfrac{d}{dx}(\cos x)^{-3}$

4. $\dfrac{d}{dt}\sin\sqrt{3t-9}$

5. $\dfrac{d}{d\theta}\dfrac{\sin 2\theta}{\cos\theta}$

6. $\dfrac{d}{dw}\dfrac{1}{w^{-3}}$

7. $\dfrac{d}{dt}t^{-7/5}$

8. $\dfrac{d}{dt}t^2e^{-t}$

9. $\dfrac{d^2}{dt^2}t^2e^{-t}$

Find the derivative of each of the following functions.

10. $y = \ln(\sin x)^{1/3}$

11. $f(x) = \dfrac{x}{1+2x^2}$

12. $y = \dfrac{\sin x + x^2}{x}$

13. $y = \dfrac{\cos x}{\cos x - 1}$

14. $y = e^{\sin x}$

15. $f(x) = -3e^{4x}$

16. $f(x) = \sin(e^x)$

17. $y = y(x)$, where $y + \cos y = 3x + 7$

For each of the following functions, (a) sketch the graph, and (b) sketch the graph of the derivative.

18. $y = \ln(\sin x)^{1/3}$

19. $f(x) = \dfrac{x}{1+2x^2}$

20. $y = \dfrac{\sin x + x^2}{x}$

21. $y = \dfrac{\cos x}{\cos x - 1}$

22. $y = e^{\sin x}$

23. $f(x) = -3e^{4x}$

24. $f(x) = \sin(e^x)$

25. Sketch the graph of $y = y(x)$, where

$$y + \cos y = 3x + 7.$$

(*Hint*: Graph x as a function of y, and invert.)

Solve each of the following initial value problems.

26. $\dfrac{dy}{dt} = -0.2y, \quad y(0) = -0.5$

27. $\dfrac{dy}{dt} = 0.2y, \quad y(0) = 0.5$

28. $\dfrac{d^2y}{dt^2} = -0.2y, \quad y(0) = -0.5, \quad y'(0) = 0$

29. $\dfrac{d^2y}{dt^2} = -0.2y, \quad y(0) = 0, \quad y'(0) = -0.5$

30. Let f be the function defined by $f(x) = e^{\sin x}$.
    (a) Sketch the graph of f.
    (b) Sketch the graph of f'.
    (c) Estimate $f'(\pi)$.
    (d) Calculate $f'(\pi)$ exactly.
    (e) Find an equation of the tangent line to the graph of f at the point $(\pi, f(\pi))$.

31. Answer the following questions about the function $f(x) = e^{\sin x}$ in Exercise 30. Give a reason for each answer. ("My calculator said so" is not an adequate reason.)
    (a) Is this function periodic?
    (b) Does it grow without bound like an exponential function?
    (c) Is it even, odd, or neither?
    (d) Does it have any negative values?
    (e) Does it have any zeros?
    (f) Where are its maximum values located?
    (g) Where are its minimum values located?

32. Let g be the function defined by $g(x) = \sin(e^x)$.
    (a) Sketch the graph of g.
    (b) Sketch the graph of g'.
    (c) Estimate $g'(0)$.
    (d) Calculate $g'(0)$ exactly.
    (e) Find an equation of the tangent line to the graph of g at the point $(0, g(0))$.

33. Answer the following questions about the function $g(x) = \sin(e^x)$ in Exercise 32. Give a reason for each answer. ("My calculator said so" is not an adequate reason.)
    (a) Is this function periodic?
    (b) Does it grow without bound like an exponential function?
    (c) Is it even, odd, or neither?
    (d) Does it have any negative values?
    (e) Does it have any zeros?
    (f) Where are its maximum values located?
    (g) Where are its minimum values located?

34. Show that every function in the family

$$A \sin \sqrt{\frac{g}{L}}\, t + B \cos \sqrt{\frac{g}{L}}\, t$$

is a solution of the differential equation

$$\frac{d^2\theta}{dt^2} = -\frac{g}{L}\,\theta.$$

    (b) For each such function, explain why B is the value of the function at $t = 0$.
    (c) How is A related to the value of the first derivative at $t = 0$?

35. Find values for the constants A and B such that the function

$$\theta(t) = A \sin \sqrt{\frac{g}{L}}\, t + B \cos \sqrt{\frac{g}{L}}\, t$$

satisfies the initial value problem

$$\frac{d^2\theta}{dt^2} = -\frac{g}{L}\,\theta \qquad \text{with } \theta(0) = \frac{\pi}{6} \quad \text{and} \quad \theta'(0) = 0.$$

36. We know that the period of $\sin t$ is 2π. That is, 2π is the smallest number T such that $\sin t$ goes through a complete swing from 0 to 1 to -1 and back to 0 again in the interval $[0, T]$.
    (a) What is the period of $\sin 7t$? of $\sin 0.2t$?
    (b) What is the frequency of each of the functions in part (a)?

37. (a) What is the period of the function $\sin \omega t$? of $\cos \omega t$?
    (b) Find the period of the solution

$$\theta(t) = A \sin \sqrt{\frac{g}{L}}\, t + B \cos \sqrt{\frac{g}{L}}\, t$$

    to the initial value problem in Exercise 35, and simplify.
    (c) How is the period related to the length L?
    (d) Does your answer agree with Figure 6.15, in which we have plotted period as a function of length? Explain.

Figure 6.15 Graph of period T versus length L for the approximate pendulum equation $\frac{d^2\theta}{dt^2} = -\frac{g}{L}\,\theta$

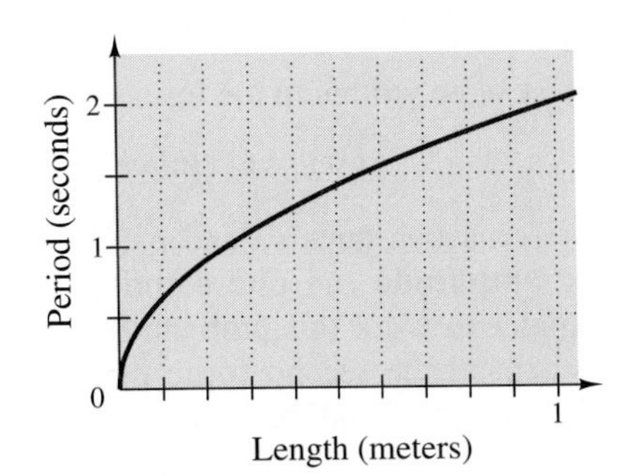

38. (a) Show that for a given spring, the period of a spring-mass system is determined by the mass in approximately the same way that the period of a pendulum is determined by the length. Why is the analogy only approximate?
    (b) What feature of the pendulum problem is analogous to spring constant? Why didn't we study the dependence of period on this feature?

39. Suppose the same mass m is attached, one at a time, to a variety of different springs with different spring constants. How do the periods of the different spring-mass systems depend on the spring constants?

40. If the pendulum bob is thrown (along the circle of motion) instead of being dropped, all that changes is that the initial velocity is no longer zero.
    (a) The initial velocity can be either positive or negative. Which sign corresponds to throwing the bob up and which down?
    (b) Find values for the constants A and B such that the function

$$\theta(t) = A \sin \sqrt{\frac{g}{L}}\, t + B \cos \sqrt{\frac{g}{L}}\, t$$

satisfies the initial value problem

$$\frac{d^2\theta}{dt^2} = -\frac{g}{L}\,\theta,$$

$$\text{with} \quad \theta(0) = \frac{\pi}{6}, \quad \text{and} \quad \theta'(0) = -0.5.$$

41. Figure 6.16 shows the solution of Exercise 40 for a pendulum of length one meter. The figure suggests that the solution is just a sinusoidal function, shifted horizontally. Show that every function of the form

$$\theta(t) = A \sin \omega t + B \cos \omega t$$

also can be written in the form

$$\theta(t) = C \sin(\omega t + \delta)$$

for some constants C and δ (called, respectively, the **amplitude** and **phase shift**). You may find it simpler to work this problem backwards—that is, to expand the second form, using the formula for sine of a sum, and then match coefficients with the first form. That will show how C and δ must be related to A and B.

Figure 6.16 Graph of the solution of Exercise 40

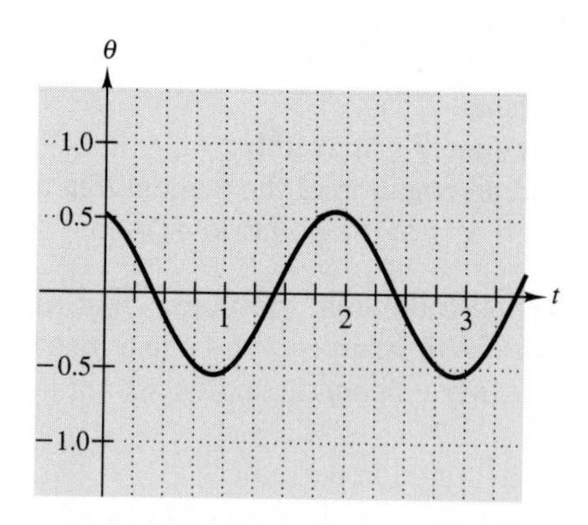

42. In Figure 6.17 we show the long-term average monthly temperatures (in degrees Fahrenheit) in Durham, N.C. Each average temperature is plotted as a dot. We have superimposed on the plotted data the graph of an approximating function of the form $f(t) = b + C \sin(\omega t + \delta)$, where time t is measured in months. The constant term b is an additive scaling of the vertical axis to center the data around an average value.
    (a) Estimate from the graph the values of b (average temperature for the year), C (amplitude), ω (related to the period), and δ (phase shift).
    (b) Check your answers by substituting into your function several integers between 0 and 12. If you don't get values close to those in Figure 6.17, make the necessary adjustments in your function until you do.

Figure 6.17 Average monthly temperatures in Durham, with a superimposed sine function

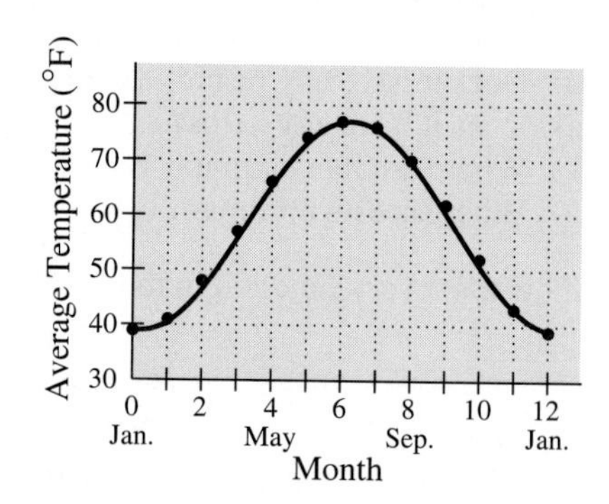

43. The long-term average monthly temperatures (in degrees Fahrenheit) for Kalamazoo, Michigan, are shown in Table 6.3. Find a function of the form

$f(t) = b + C\ \sin(\omega t + \delta)$ that approximates these data. Explain your strategy for finding b, C, ω, and δ from the data.

44. In Exercise 40 of Section 6.2 we considered a model of body temperature (in degrees Celsius) based on data from a healthy 28-year-old female. With a time unit of 20 minutes, the data could be modeled by a function of the form

$$f(t) = \mu + A\cos(2\pi\omega t) + B\sin(2\pi\omega t)$$

where $\mu = 37.5$, $\omega = 0.0138$, $A = -0.103$, and $B = -0.429$.

(a) Convert the model function to the form

$$f(t) = \mu + C\cos(2\pi\omega t - \delta).$$

(b) What is the physiological significance of C?

Table 6.3 Average monthly temperatures in Kalamazoo

Month	*Temperature*
January	24°
February	27°
March	36°
April	49°
May	60°
June	69°
July	73°
August	72°
September	65°
October	54°
November	40°
December	29°

6.4 Derivative Calculations

From our introduction of the derivative in Chapter 2, we have gradually expanded our repertoire of formulas for calculating derivatives of functions defined by formulas. Along the way, we have frequently asked you, in checkpoints and exercises, to calculate derivatives. Such calculation is not the main point of this course—indeed, relatively inexpensive electronic devices can calculate the derivative of any formula-defined function quickly, easily, and accurately. However, it is still important that you develop an appropriate level of competence in pencil-and-paper calculation of derivatives. There are several reasons for this:

- You need to know when it is quicker and easier to do your own calculation than to use an electronic device—or to interpret its output. Accurate and confident judgment about proper use of tools comes only from lots of practice.
- You need to be able to recognize wrong results from your computer or calculator—usually the result of wrong entry of the problem. Practice with hand calculation will help you develop a sense of the types of components that can turn up in a given derivative calculation.
- In later chapters we will confront the much harder—and more important—problem of finding antiderivatives. We will see that skill in derivative calculation is an essential part of this process.

In this section we review the important formulas that have appeared in our development thus far, and we illustrate how several different formulas may be needed in a single calculation.

Example 1 Find the derivative of the polynomial function $7x^5 - 4x^3 + 2$.

Solution
$$\frac{d}{dx}(7x^5 - 4x^3 + 2) = 35x^4 - 12x^2$$

At this point you can probably differentiate polynomials in a one-line calculation, as we did here. However, if you need to write down intermediate steps to be confident of your accuracy, please do so. Check carefully to see where we used the Sum Rule, Constant Multiple Rule, and Power Rule.

Checkpoint 1

Find the derivative of $5x^6 - 2x^4 + 3x^2 - 7$.

Example 2 Calculate $\frac{d}{du}\sqrt{5u^4 - u + 1}$.

Solution The first step is to notice that the function to be differentiated is a square root—a $\frac{1}{2}$-power—of something. Thus we combine the Power Rule and the Chain Rule:

$$\begin{aligned}\frac{d}{du}\sqrt{5u^4 - u + 1} &= \frac{1}{2}\,\frac{1}{\sqrt{5u^4 - u + 1}}\,\frac{d}{du}(5u^4 - u + 1)\\ &= \frac{20u^3 - 1}{2\sqrt{5u^4 - u + 1}}.\end{aligned}$$

The "something" under the square root sign is the polynomial function $5u^4 - u + 1$, so the Chain Rule tells us to include a factor that is the derivative of this function. In the second step we calculated that derivative and multiplied the fractions.

In Figure 6.18 we show the graphs of both $\sqrt{5u^4 - u + 1}$ and its derivative, as calculated in the preceding solution. Observe that the derivative is negative where the original function is decreasing, 0 when $\sqrt{5u^4 - u + 1}$ reaches its minimum, and positive beyond that. This is a simple graphical check of a symbolic calculation—one that you can always carry out with your calculator.

Figure 6.18 Graphs of $\sqrt{5u^4 - u + 1}$ and its derivative

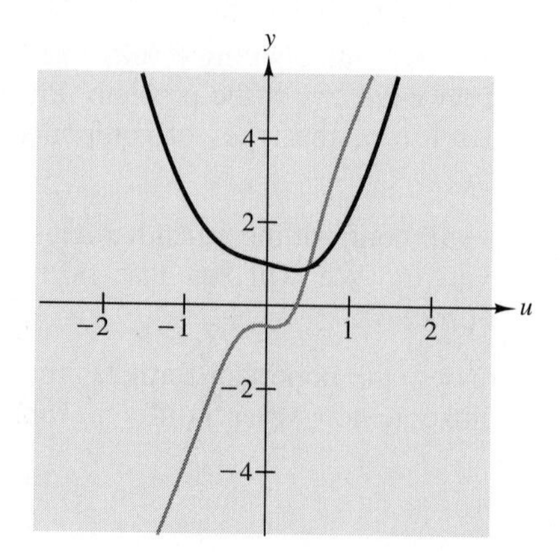

Checkpoint 2

(a) Find the derivative of $(5u^4 - u + 1)^{3/2}$.

(b) Graph both the function and its derivative. Confirm that your derivative formula has appropriate graphical properties to be the derivative of the given function.

Example 3 Differentiate $(\ln t)^4$.

Solution

$$\frac{d}{dt}(\ln t)^4 = 4\,(\ln t)^3 \frac{1}{t}\,.$$

Note the similarity to the preceding example. Again, we used the Power Rule and the Chain Rule, but this time the derivative of the inside function required the Logarithmic Rule. In Figure 6.19 we show the graphs of both $(\ln t)^4$ and its derivative, $4\,(\ln t)^3/t$. Again, the derivative is negative where the original function is decreasing, 0 when $(\ln t)^4$ reaches its minimum, and positive to the right of the minimum.

Figure 6.19 Graphs of $(\ln t)^4$ and $4\,(\ln t)^3/t$

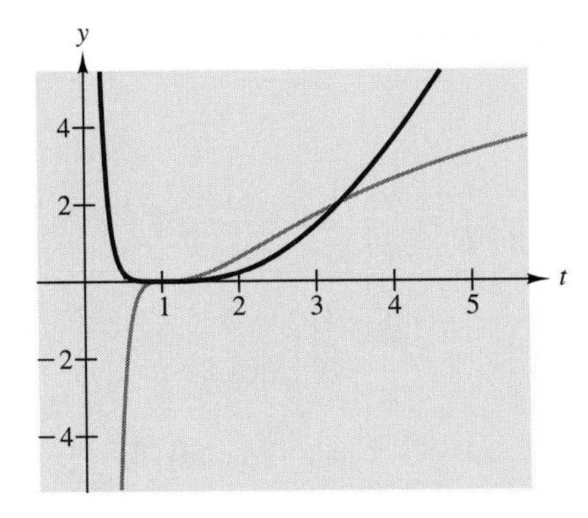

Checkpoint 3

(a) Find the derivative of $\ln(t^4)$.

(b) Graph both the function and its derivative. Confirm that your derivative formula has appropriate graphical properties to be the derivative of the given function.

Example 4 Calculate the derivative of $e^{2t}\cos 3t$.

Solution

$$\begin{aligned}\frac{d}{dt}\left(e^{2t}\cos 3t\right) &= e^{2t}\frac{d}{dt}\cos 3t + \cos 3t\,\frac{d}{dt}e^{2t}\\ &= 3e^{2t}(-\sin 3t) + 2\,e^{2t}\cos 3t\\ &= e^{2t}(2\cos 3t - 3\sin 3t).\end{aligned}$$

The formula for the function in Example 4, reading from outside in, is a product. Thus the Product Rule comes into play first. That rule in turn calls for a derivative of each factor, so the Cosine and Exponential Rules are used at the second step. But we have cosine of $3t$ and exponential of $2t$, so the Chain Rule requires a factor of 3 in the first term and a factor of 2 in the second term. The final step is an algebraic simplification—not part of the differentiation, but often a good idea for making the result more useful.

Checkpoint 4

Check our work in Example 4 graphically.

Checkpoint 5

(a) Differentiate $e^{2u}(\ln u)^4$.

(b) Check your result in part (a) graphically.

Example 5 Find the derivative of $\dfrac{e^{2\theta}}{\cos 3\theta}$.

Solution

$$\begin{aligned}
\frac{d}{d\theta}\left(\frac{e^{2\theta}}{\cos 3\theta}\right) &= \frac{d}{d\theta}\left(e^{2\theta}\,\frac{1}{\cos 3\theta}\right) \\
&= e^{2\theta}\,\frac{d}{d\theta}\left(\frac{1}{\cos 3\theta}\right) + \frac{d}{d\theta}\left(e^{2\theta}\right)\frac{1}{\cos 3\theta} \\
&= e^{2\theta}(-1)(\cos 3\theta)^{-2}\frac{d}{d\theta}\cos 3\theta + 2e^{2\theta}\frac{1}{\cos 3\theta} \\
&= e^{2\theta}(-1)(\cos 3\theta)^{-2}3(-\sin 3\theta) + 2e^{2\theta}\frac{1}{\cos 3\theta} \\
&= \frac{e^{2\theta}}{\cos 3\theta}\left(3\,\frac{\sin 3\theta}{\cos 3\theta} + 2\right).
\end{aligned}$$

■

Checkpoint 6

(a) Give reasons—as many as necessary—for each line in the preceding solution.

(b) Check the result of Example 5 graphically.

Checkpoint 7

Find

$$\frac{d}{ds}\ln(5s + 6).$$

In this section we have seen how differentiation rules for sums, constant factors, powers, composites, exponential, logarithmic, and trigonometric functions can be combined to differentiate rather complicated functions. We also have seen a simple graphical check on the accuracy of our symbolic calculations.

Answers to Checkpoints

1. $30x^5 - 8x^3 + 6x$

2. (a) $\dfrac{3}{2}\sqrt{5u^4 - u + 1}\,(20u - 1)$

   (b) See discussion after Example 2.

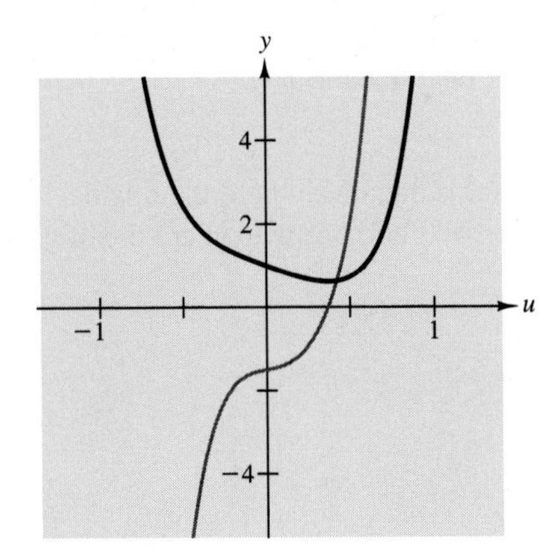

3. (a) $4/t$

   (b)

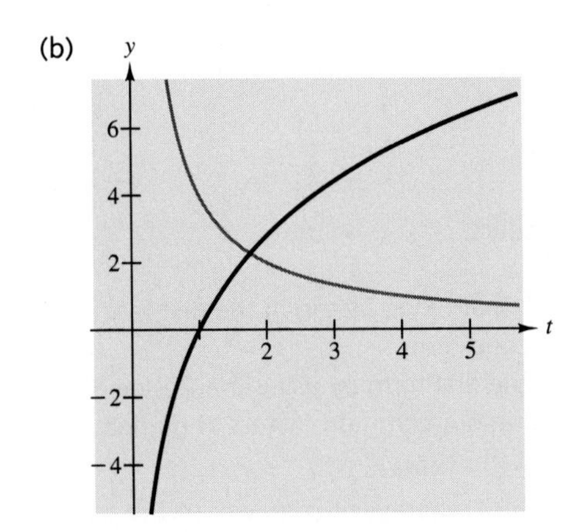

4.

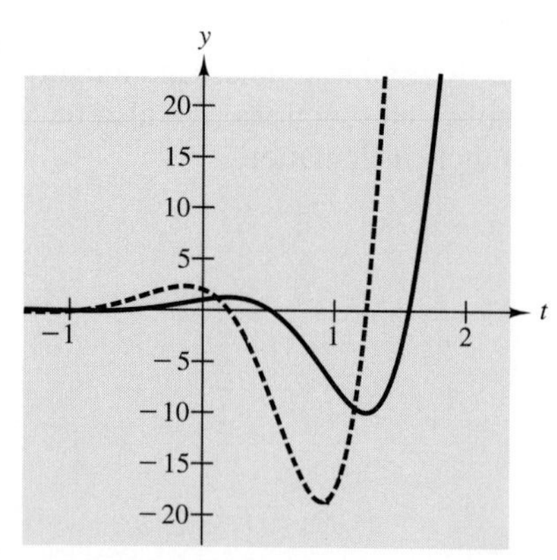

The figure shows the original function in black and the calculated derivative in color. Both formulas have the form "exponential times trig function"—with the same exponential and same frequency—so their graphs should have similar shapes. Also, the dashed curve crosses the t-axis where the solid curve levels off, near 0.2 and 1.2.

5. (a) $e^{2u}(\ln u)^3(4/u + 2\ln u)$

   (b) The graph of the formula in part (a) is negative where the original function is decreasing, 0 where the original function levels off at $u = 1$, and positive where the original function is increasing.

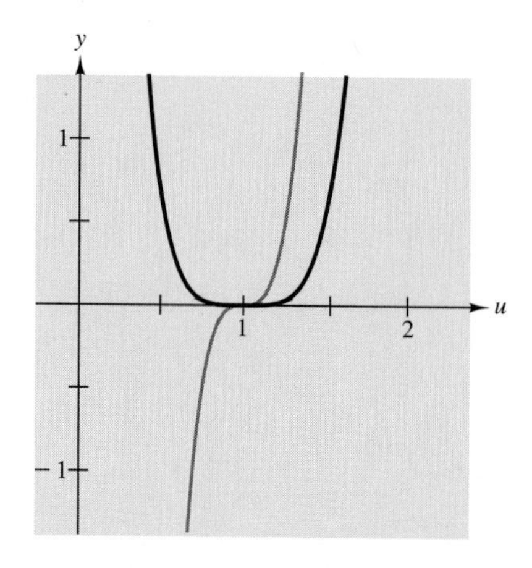

6. (a) We rewrote the quotient as a product.
   We applied the Product Rule.
   We applied the Power Rule (with Chain Rule factor) in the first term and the Exponential Rule (with Chain Rule factor) in the second term.
   We completed the calculation of the first term by using the Cosine Rule with Chain Rule factor.
   We simplified the result by factoring out common factors of numerator and denominator. (There are many correct simplifications of this expression.)

   (b) Both graphs have vertical asymptotes wherever $\cos 3\theta = 0$, so the graphs have many interesting pieces. We show two pieces of each graph in the figure. Note the correspondence of positive, negative, and zero values of the derivative with increasing, decreasing, and relative extreme points on the graph of the original function.

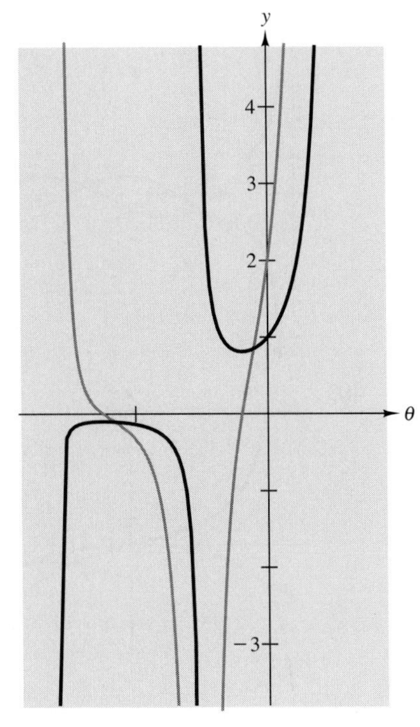

7. $\dfrac{5}{5s + 6}$

Exercises 6.4

Calculate each of the following derivatives.

1. $\dfrac{d}{dt}\ e^{-t}\sin 5t$

2. $\dfrac{d}{dt}\ e^{-2t}\cos 5t$

3. $\dfrac{d}{dt}\ (\ln t^2)\cos 2t$

4. $\dfrac{d}{dt}\ 3^{2t}$

5. $\dfrac{d^2}{dt^2}\ 3^{2t}$

6. $\dfrac{d}{dz}\ \dfrac{e^{2z}}{\sin 5z}$

7. $\dfrac{d^2}{dt^2}\sin 2t$

8. $\dfrac{d^2}{dt^2}\sin^2 t$

9. $\dfrac{d^2}{dt^2}\ e^{-t}\sin 5t$

10. $\dfrac{d^2}{dt^2}\cos 2t$

11. $\dfrac{d^2}{dt^2}\cos^2 t$

12. $\dfrac{d}{du}\sin^2 u\cos u$

Find the derivative of each of the following functions.

13. $y = 3\sin t\cos 3t$

14. $u = \cos^3 2t$

15. $v = \cos(2t^3)$

16. $y = \sin 2t - \cos 5t$

17. $z = \dfrac{1}{(2t+1)^2}$

18. $z = \dfrac{7}{(3+2t)^3}$

19. $y = \dfrac{\sin 2u}{\cos 5u}$

20. $z = (2t+1)^{12}$

21. $z = 7\ln(3+2t)$

22. $y = x^6 + 4x^2 + 6x^{-2}$

23. $y = 2e^{5x}$

24. $y = x^2 3^x$

For each of the following functions, (a) graph the function, (b) graph the derivative from your calculation in Exercises 13–24, and (c) explain how appropriate properties of the derivative graph confirm your symbolic calculation.

25. $y = 3\sin t\cos 3t$

26. $u = \cos^3 2t$

27. $v = \cos(2t^3)$

28. $y = \sin 2t - \cos 5t$

29. $z = \dfrac{1}{(2t+1)^2}$

30. $z = \dfrac{7}{(3+2t)^3}$

31. $y = \dfrac{\sin 2u}{\cos 5u}$

32. $z = (2t+1)^{12}$

33. $z = 7\ln(3+2t)$

34. $y = x^6 + 4x^2 + 6x^{-2}$

35. $y = 2e^{5x}$

36. $y = x^2 3^x$

In each of the following exercises, graphs of two functions are shown. In each case, one of the functions is the derivative of the other. Decide which function is f and which is f'. Give as many reasons as you can for your choice.

37.

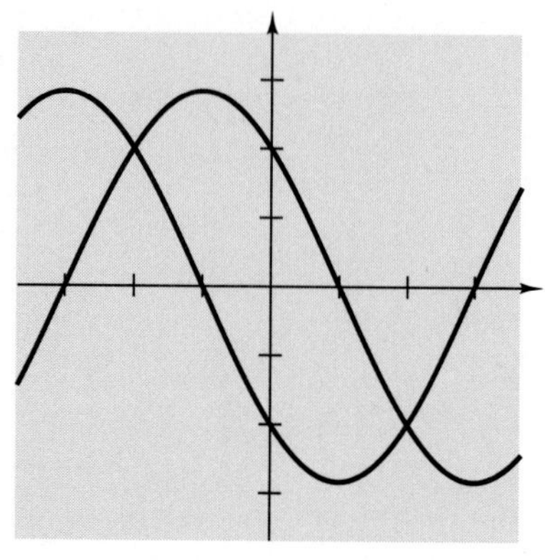

38.

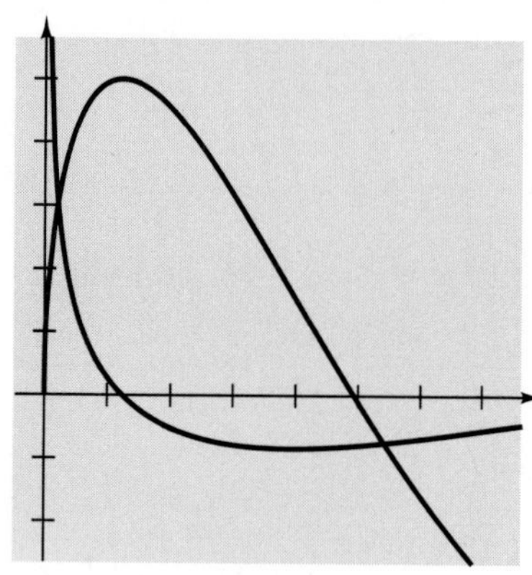

39.

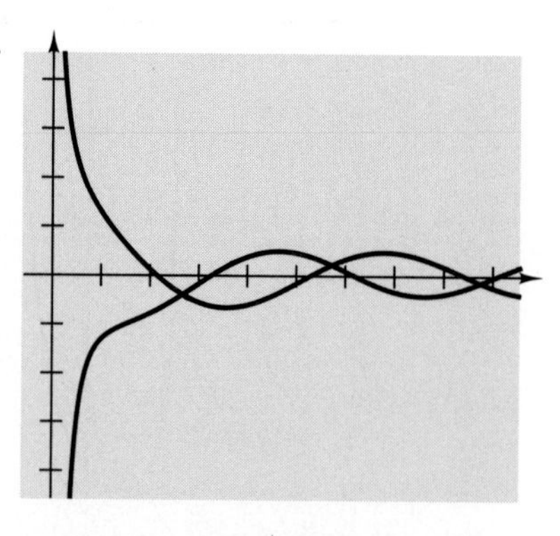

40.

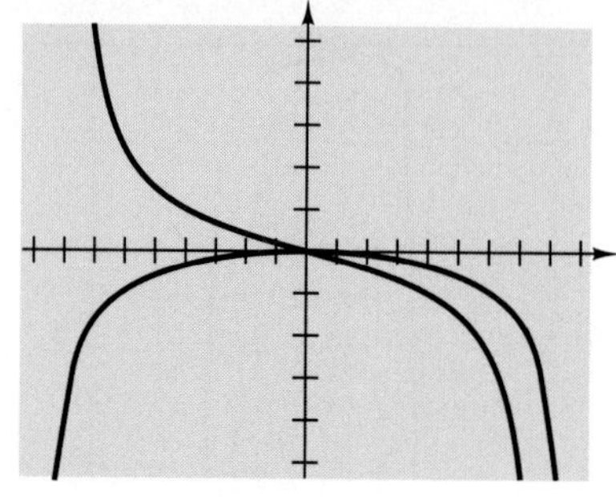

Chapter 6 Summary

In Section 6.1 we reviewed the basic properties of the sine and cosine functions, with emphasis on the origin of these functions in motion around a circle. We found, first graphically and then by studying limiting values of difference quotients, that sine and cosine have very simple derivative formulas.

Next, we turned to a physical problem as a prototype for problems whose solutions are periodic: the spring-mass system. We saw that Newton's Second Law of Motion leads to a description of the motion as

> acceleration is negatively proportional to displacement.

With appropriate initial conditions, this statement becomes a second-order initial value problem.

Our first approach to solving this problem was to rewrite it as a system of two first-order problems, using velocity as an intermediate variable. That turned our problem into one that we could solve numerically, just as we solved the S-I-R epidemic model in Chapter 5 by Euler's Method. The numerical solution looked like a cosine, which fit with our calculation of derivatives of sine and cosine. Then we found a symbolic solution, which was indeed a cosine function, with period determined by the spring constant and mass, and amplitude determined by the initial displacement.

In Section 6.3 we took up motion of a simple pendulum as a second study of repeating phenomena. In particular, we tried to describe the relationship between period and length. We found three different ways to address this question:

- We asked you to do physical measurements with casual experiments.
- We modeled the pendulum with a second-order initial value problem. Even with simplifying assumptions, this problem turned out to be one that we could not solve symbolically. Nevertheless, as we did for the spring equation, we used Euler's Method to generate numerical and graphical solutions.
- We replaced the differential equation with a simpler one and found solutions of this approximating differential equation the same way we solved the spring equation. These solutions have the attractive feature that their period is easy to calculate as a function of the length of the pendulum, thus giving at least an approximate answer to our primary question.

Finally, in Section 6.4 we consolidated all our derivative formulas from Chapter 2 on and illustrated how they could be used to differentiate virtually any function defined by formulas whose components are algebraic, exponential, logarithmic, or trigonometric functions. Furthermore, we saw that we could easily employ a graphing calculator to check the results of such calculations.

Concepts

- Trigonometric functions
- Period
- Frequency
- Amplitude
- Phase shift
- Reduction of second-order differential equations to first-order systems

Applications

- Motion of spring-mass systems
- Pendulum motion
- Numerical solution of second-order differential equations

Formulas (cumulative listing)

Exponential Functions

$\frac{d}{dt}\, e^t = e^t$, where e is the natural base, $2.71828\ldots$

$\frac{d}{dt}\, e^{kt} = ke^{kt}$, for any constant k

$\frac{d}{dt}\, b^t = (\ln b)\, b^t$, for any constant b

Power Rule

$$\frac{d}{dt} t^r = rt^{r-1}, \text{ for any real constant } r$$

Natural logarithm function

$$\frac{d}{dt} \ln t = \frac{1}{t}$$

Trigonometric functions

$$\frac{d}{dt} \sin t = \cos t$$

$$\frac{d}{dt} \cos t = -\sin t$$

Constant Multiple Rule

$$\frac{d}{dt} Af(t) = A \frac{d}{dt} f(t), \text{ where } A \text{ is any constant}$$

Sum Rule

$$\frac{d}{dt}[f(t) + g(t)] = \frac{d}{dt} f(t) + \frac{d}{dt} g(t)$$

Product Rule

$$\frac{d}{dt}[g(t)\, h(t)] = g(t) \frac{d}{dt} h(t) + h(t) \frac{d}{dt} g(t)$$

Chain Rule

$$\frac{dy}{dt} = \frac{dy}{du} \frac{du}{dt}$$

Special case of the Chain Rule

$$\frac{d}{dt} f(kt) = k \frac{d}{du} f(u), \text{ where } u = kt \text{ and } k \text{ is constant}$$

Combinations of Chain Rule with specific function rules

$$\frac{d}{dt} f(t)^r = rf(t)^{r-1} \frac{d}{dt} f(t), \text{ for any real constant } r$$

$$\frac{d}{dt} e^{f(t)} = e^{f(t)} \frac{d}{dt} f(t)$$

$$\frac{d}{dt} \ln f(t) = \frac{1}{f(t)} \frac{d}{dt} f(t)$$

$$\frac{d}{dt} \sin f(t) = \cos f(t) \frac{d}{dt} f(t)$$

$$\frac{d}{dt} \cos f(t) = -\sin f(t) \frac{d}{dt} f(t)$$

Chapter 6 Exercises

Calculate each of the following derivatives.

1. $\dfrac{d}{dt}\, e^{-3t} \sin 2t$

2. $\dfrac{d}{dt}\, e^{-t} \cos 3t$

3. $\dfrac{d}{dt}\, (\ln t)^2 \cos 2t$

4. $\dfrac{d^2}{dt^2} \cos 3t$

5. $\dfrac{d^2}{dt^2} \sin^3 t$

6. $\dfrac{d^2}{dt^2}\, e^{-3t} \sin 2t$

7. $\dfrac{d^2}{dt^2} \sin 3t$

8. $\dfrac{d^2}{dt^2} \cos^3 t$

9. $\dfrac{d}{dx} \sin x \cos^2 x$

Find the derivative of each of the following functions.

10. $y = \sin t - 3 \cos 3t$

11. $y = \cos^2 3t$

12. $y = \cos(3t^2)$

13. $y = (\sin 2t)(\cos 5t)$

14. $z = \dfrac{1}{2t+1}$

15. $z = \dfrac{7}{3+2t}$

16. $y = x^7 + 3x^2 + 6$

17. $y = 2^{5x}$

18. $y = x^2 \sin(3x)$

For each of the following functions, (a) graph the function, (b) graph the derivative from your calculation in Exercises 10–18, and (c) explain how appropriate properties of the derivative graph confirm your symbolic calculation.

19. $y = \sin t - 3\cos 3t$

20. $y = \cos^2 3t$

21. $y = \cos(3t^2)$

22. $y = (\sin 2t)(\cos 5t)$

23. $z = \dfrac{1}{2t+1}$

24. $z = \dfrac{7}{3+2t}$

25. $y = x^7 + 3x^2 + 6$

26. $y = 2^{5x}$

27. $y = x^2 \sin(3x)$

Find all antiderivatives of each of the following functions.

28. e^{-2t}

29. t^3

30. t^{-3}

31. $\dfrac{1}{t^{-2}}$

32. $\cos 3t$

33. t^{-1}

34. $\sin 2\pi t$

35. $\dfrac{1}{2t+1}$

36. $\dfrac{7}{3+2t}$

Solve each of the following initial value problems.

37. $\dfrac{dy}{dt} = -1.07y, \quad y(0) = -0.5$

38. $\dfrac{dy}{dt} = 1.07y, \quad y(0) = 0.5$

39. $\dfrac{d^2y}{dt^2} = -1.07y, \quad y(0) = 0.5, \quad y'(0) = -0.5$

40. $\dfrac{d^2y}{dt^2} = -1.07y, \quad y(0) = -0.5, \quad y'(0) = 0.5$

41. (See Exercise 44 in Section 6.1.) According to Poiseuille's Law, if blood flows into a straight blood vessel of radius r branching off another straight blood vessel of radius R at an angle α, the total resistance T of the blood in the branching vessel is given by

$$T = C\left(\frac{a - b\cot\alpha}{R^4} + \frac{b\csc\alpha}{r^4}\right),$$

where a, b, C, r, and R are constants, and $r < R$. Show that when $\cos\alpha = (r/R)^4$, the total resistance is minimized.

42. What is the cone of largest volume that can be formed by rotating a right triangle of fixed hypotenuse h around one of its legs? What is the volume of that cone? (*Note*: This was also Exercise 31 in Section 4.5. At that point you might have used the length of one leg of the

triangle as your independent variable. Try the problem again using one angle of the triangle as your independent variable. Do you get the same answer both ways? Which choice leads to an easier calculation?)

43. Suppose you walk counterclockwise around the perimeter of the square with corners at $(\pm 1, \pm 1)$, starting at the point $(1, 0)$. Let $(x(t), y(t))$ be your position on the square after you have walked t units.

    (a) Find formulas for $x(t)$ and $y(t)$ as functions of t.

    (b) Sketch the graph of $x(t)$ as a function of t.

    (c) Sketch the graph of $y(t)$ as a function of t.

    (d) Are the functions $x(t)$ and $y(t)$ periodic? If so, what is the period of each? Explain.[4]

44. If we begin with $x_0 = 0$, then Newton's Method fails to find a root for one of the following functions. Which function—and why?

    (i) $f(x) = \sin x$

    (ii) $f(x) = \cos x$

    (iii) $f(x) = 2e^x - 1$

    (iv) $f(x) = e^{-x} - x$

45. We discovered in Chapter 2 that solutions of $y' = ky$ are all of the form $y = Ae^{kt}$, and we discovered in this chapter that solutions of $y'' = -\omega^2 y$ are all of the form

$$A \sin \omega t + B \cos \omega t.$$

    Now we seek solutions of $y'' = \omega^2 y$.

    (a) Show that $e^{\omega t}$ and $e^{-\omega t}$ are both solutions of $y'' = \omega^2 y$.

    (b) Show that all functions of the form $Ae^{\omega t} + Be^{-\omega t}$ are solutions.

    (c) If $\omega = 2$, find a solution that satisfies the initial conditions $y(0) = 1$ and $y'(0) = 0.5$.

46. The solutions of $y'' = \omega^2 y$ can be written in another form that is more suggestive of the parallel with trigonometric functions (also called **circular** functions). This other form uses so-called **hyperbolic** functions, the **hyperbolic sine** (abbreviated **sinh**) and **hyperbolic cosine** (abbreviated **cosh**). These functions may be defined from exponential functions by the following formulas:

$$\sinh \omega t = \frac{1}{2}\left(e^{\omega t} - e^{-\omega t}\right)$$

    and

$$\cosh \omega t = \frac{1}{2}\left(e^{\omega t} + e^{-\omega t}\right).$$

    (a) How are the first derivatives of these hyperbolic functions related to the functions themselves?

4. Problem suggested by John Frampton, Northeastern University.

(b) Explain why you already know that sinh ωt and cosh ωt are solutions of

$$\frac{d^2y}{dt^2} = \omega^2 y.$$

(c) Explain why every function of the form A sinh $\omega t + B$ cosh ωt is also a solution of the same differential equation.

(d) Why must B be the value of such a function at $t = 0$?

(e) How is A related to the value of the first derivative at $t = 0$?

(f) Compare what you did in parts (c) through (e) with what you did in Exercise 34 of Section 6.3.

(g) If $\omega = 2$, find a solution expressed in hyperbolic functions that satisfies the initial conditions $y(0) = 1$ and $y'(0) = 0.5$. Is this the same solution you found in the preceding exercise? Why or why not?

47. You showed in Exploration Activity 1 of Section 6.1 that the circular functions

$$x = \cos \omega t \text{ and } y = \sin \omega t$$

satisfy the equation

$$x^2 + y^2 = 1$$

—the equation of a circle—which is not surprising, given the origins of these functions in a circle.

(a) Find a similar equation satisfied by the hyperbolic functions

$$x = \cosh \omega t \text{ and } y = \sinh \omega t.$$

(*Hint*: Calculate the squares of the hyperbolic sine and the hyperbolic cosine directly from their definitions in the preceding problem.)

(b) Why are these functions called hyperbolic?

48. (a) Make a sketch, on a single set of coordinate axes, of the graphs of $y = e^x$ and $y = e^{-x}$.

(b) The function $y = \cosh x$ is the average of the two functions whose graphs you just sketched. Sketch the average of the two graphs to find the graph of $y = \cosh x$.

(c) Construct the graph of $y = \sinh x$ in a similar manner, as the average of two exponential graphs.

(d) Check your work on a computer or graphing calculator.

49. In a Chapter 3 Lab Reading we saw that the velocity of a falling object (with air resistance proportional to the square of velocity) could be modeled by the initial value problem

$$\frac{dv}{dt} = g - cv^2 \qquad \text{with } v(0) = 0.$$

Show that

$$v(t) = \sqrt{\frac{g}{c}}\,\frac{1 - e^{-2\sqrt{gc}\,t}}{1 + e^{-2\sqrt{gc}\,t}}$$

is the solution of this initial value problem. (*Note*: This is a hard calculation. Keep careful track of your algebraic steps.)

50. (a) Show that the function defined in Exercise 49 also may be written in the form

$$v(t) = \sqrt{\frac{g}{c}}\,\frac{e^{\sqrt{gc}\,t} - e^{-\sqrt{gc}\,t}}{e^{\sqrt{gc}\,t} + e^{-\sqrt{gc}\,t}}.$$

(b) Use the definitions of hyperbolic sine and hyperbolic cosine (Exercise 46) to rewrite $v(t)$ in the form

$$v(t) = \sqrt{\frac{g}{c}}\,\frac{\sinh(\sqrt{gc}\,t)}{\cosh(\sqrt{gc}\,t)}.$$

(c) Define the hyperbolic tangent and hyperbolic secant functions in the obvious way:

$$\tanh u = \frac{\sinh u}{\cosh u}$$

and

$$\operatorname{sech} u = \frac{1}{\cosh u}.$$

Use the result of Exercise 47 to show that

$$1 - \tanh^2 u = \operatorname{sech}^2 u.$$

(d) Use the result of Exercise 46 to show that

$$\frac{d}{du}\tanh u = \operatorname{sech}^2 u.$$

(e) Combine parts (b), (c), and (d) to show that

$$v(t) = \sqrt{\frac{g}{c}}\tanh\sqrt{gc}\,t$$

is the solution of the initial value problem in Exercise 49. (Note that once the right concepts and formulas are in place, the computation is much easier than the one in Exercise 49.)

CHAPTER

7

Solutions of Initial Value Problems

We began our study of calculus in Chapter 2 by exploring the implications of the assumption that a biological population would grow at a rate proportional to its size. In brief, we found that the population would grow exponentially. You also may have studied numerical and graphical solutions of a model that also considered limited resources for the growing population. We noted at the time that we lacked the tools to find a symbolic expression for the solution function being graphed by your computer or calculator. Now we have those tools. In particular, building on ideas developed in Chapters 3, 4, and 5, we now will develop a more general approach to solution of initial value problems, and then we will apply this approach to the limited growth problem. In the process, we will find a need to expand our knowledge of techniques for antidifferentiation. We will explore the extent to which our population models describe the historical growth of human populations, specifically the populations of the United States and of the world.

Next, we turn our attention to discrete models, specifically to the discrete limited-growth model. This model has a very close connection to the continuous model, because it is the result of applying Euler's method to the continuous model. However, the discrete model has solutions that are—in some cases—wildly different from those of the continuous model. That observation will lead us into the modern mathematical theory of chaos, most of which has been developed since the mid-1970's. We will see that, even when we are studying discrete phenomena, calculus provides important tools for the analysis of those phenomena.

7.1 Separation of Variables

In this section we develop a systematic method for finding symbolic solutions of differential equations—even when we cannot guess the form. We begin with a review of our population growth models.

Natural Versus Constrained Growth

In Chapter 2 we studied growth of a biological population under the assumption that the growth rate would be proportional to the population. If reproduction took place continuously, then the growth rate assumption was

$$\frac{dP}{dt} = kP,$$

where P is a function of continuous time t and k is the proportionality constant. We found that all solutions of this natural-growth equation have the form

$$P(t) = P_0 e^{kt},$$

where e is the natural base and P_0 is the population at time $t = 0$. These continuous solutions also have the form

$$P(t) = P_0 b^t,$$

where $b = e^k$. The advantage of the natural base form is that it displays explicitly the role of the constant k. In short, our conclusion from Chapter 2 is that unfettered natural growth is exponential.

Of course, most population growth is not unfettered, even in the short run, and none is unfettered forever. In an early lab experiment, you may have studied the nonexponential form of solutions to a fruit fly model under the added assumption of a maximum number of flies that could be supported by the environment. At the time, we lacked the tools to solve the initial value problem symbolically, that is, to find a formula for $P(t)$. Now we have those tools.

Figure 7.1 shows a numerical solution as it might have been generated in the fruit fly lab, but with a smaller starting population of only 10 flies. Figure 7.2 shows a corresponding natural-growth solution of the form $P(t) = P_0 e^{kt}$, which would represent the growth of the same population if there were no constraints on space and food supply. Observe the similarity of the two curves for small time values (the first 25 days or so) and their obvious differences as the maximum supportable population becomes a fac-tor in the left-hand graph. Our immediate objective is to find a formula that describes curves such as the one in Figure 7.1.

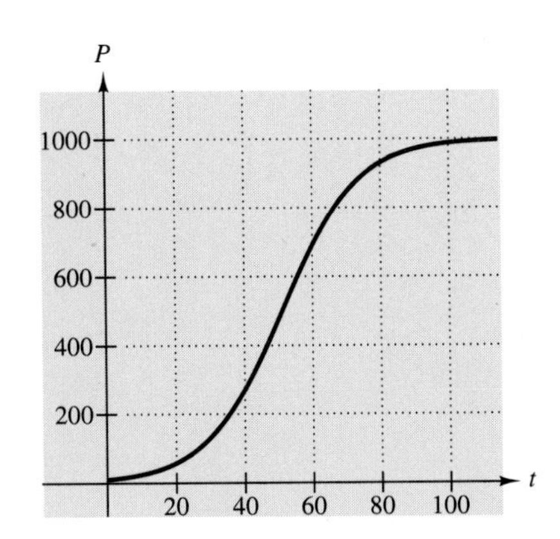

Figure 7.1 Euler solution of $dP/dt = 0.00009P\,(1000 - P)$, $P(0) = 10$

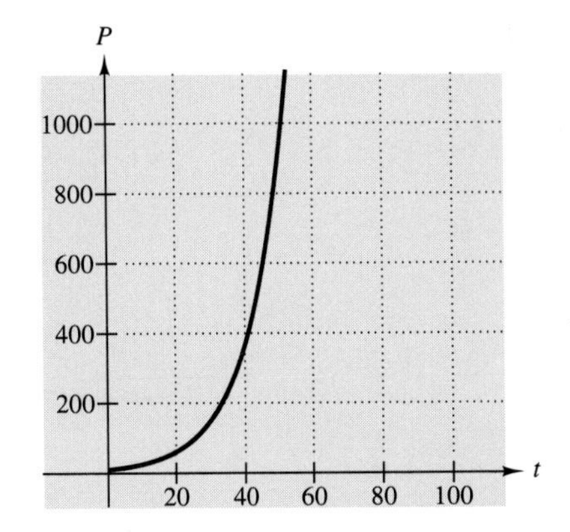

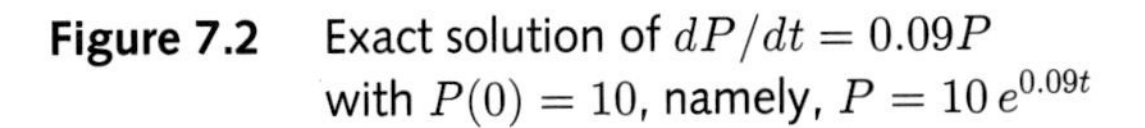

Figure 7.2 Exact solution of $dP/dt = 0.09P$ with $P(0) = 10$, namely, $P = 10\,e^{0.09t}$

We recall the assumption we made in the fruit fly lab about the growth rate: that it would be jointly proportional to the population size P (accounting for natural growth) and to the amount of remaining space, $M - P$, where M is the maximum supportable population. In symbols,

$$\frac{dP}{dt} = kP(M - P),$$

where k is the proportionality constant.

Exploration Activity 1

Explain how the factor $M - P$ affects the growth rate (a) when P is small and (b) when P is large, i.e., close to M. Your explanation should agree with what you see in Figure 7.1.

When P is small relative to M, the rate of growth dP/dt is approximately kPM. Since M is constant, this means that P grows almost exponentially, i.e., $P \approx P_0 e^{kMt}$. When P is close to M, the factor $M - P$ is close to zero and $dP/dt \approx 0$. This means that P is approximately constant, which corresponds to the leveling off of the curve in Figure 7.1.

As with the natural growth model, the constrained growth model

$$\frac{dP}{dt} = kP(M - P)$$

is a differential equation that has infinitely many solutions. Figure 7.3 shows the corresponding slope field, in which the slope at each point is computed from the differential equation. The complete model for constrained growth is an initial value problem consisting of $dP/dt = kP(M - P)$ and an initial condition, $P(0) = P_0$. That is, we assume we know the population at some particular time, which we then designate time zero. Our fundamental assumption continues to be that any initial value problem has a unique solution. Thus, once we have found any solution at all, it must be the only one.

Figure 7.3 Slope field for $dP/dt = 0.00009P\,(1000 - P)$

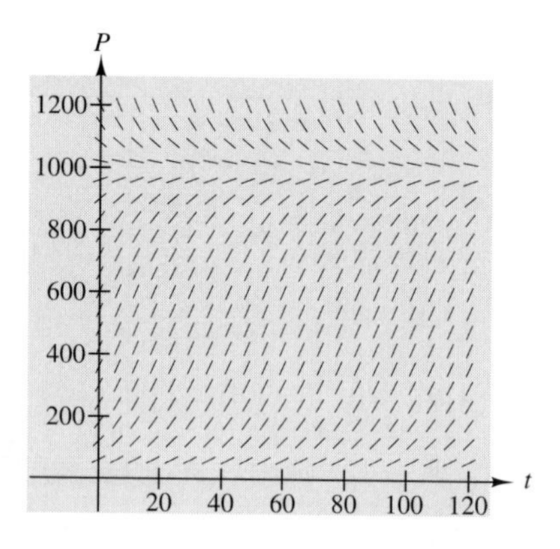

Checkpoint 1

(a) Use the slope field in Figure 7.3 to sketch the unique solution for which $P(0) = 10$. This should resemble the graph in Figure 7.1.

(b) Sketch the unique solution for which $P(0) = 111$. This should look like the solution generated in the fruit fly lab.

(c) Add a few more solutions with starting points of your own choosing.

(d) What happens if $P(0) = 1000$ (i.e., the starting point is on the horizontal line in Figure 7.3)? What happens if $P(0)$ is greater than 1000?

Symbolic Solutions

Our problem now is to find a symbolic solution for the constrained-growth equation

$$\frac{dP}{dt} = kP(M - P)$$

with the initial condition

$$P(0) = P_0.$$

But we are going to start by finding another way to solve the natural-growth equation

$$\frac{dP}{dt} = kP,$$

one that also will apply to constrained growth.

Our initial solution to the natural-growth equation, back in Chapter 2, amounted to guessing the answer and checking by direct calculation that the answer is correct. That won't work with constrained growth, though, because we have no information about a likely form for the answer.

So consider the natural-growth equation

$$\frac{dP}{dt} = kP$$

with the initial condition

$$P(0) = P_0.$$

The derivative dP/dt may be considered as the quotient of the differentials dP and dt. Recall that the differential of the dependent variable P is just

$$dP = \frac{dP}{dt}\,dt.$$

Now, if we multiply both sides of the differential equation $dP/dt = kP$ by the differential dt, we obtain

$$dP = kP\,dt.$$

Dividing both sides of this equation by P, we obtain

$$\frac{dP}{P} = k\,dt.$$

The right side of this equation is the differential of

$$kt + C,$$

where C can be any constant. We'll try to write the left side, dP/P, as a differential as well.

For the moment, think of P as an independent variable. Then dP/P is the differential of any function with derivative $1/P$. One such function is $\ln(P)$. (We know that P is positive, so we do not need to write this as $\ln|P|$.) But is $\ln(P)$ an antidifferential for dP/P when we consider t as the independent variable? Let's check. We know that

$$d\ln(P) = \frac{d}{dt}\ln P\,dt.$$

By the Chain Rule,

$$\frac{d}{dt}\ln P = \frac{1}{P}\frac{dP}{dt}.$$

So

$$\begin{aligned} d\ln P &= \frac{1}{P}\frac{dP}{dt}\,dt \\ &= \frac{1}{P}\,dP. \end{aligned}$$

Here is the important point: For antidifferential calculations, the Chain Rule shows that it doesn't make any difference whether we think of P as an independent variable or a dependent variable.

So for the function P that we are looking for, the differential of $\ln P$ equals $k\,dt$. Thus $\ln P$ must be one of the functions

$$kt + C.$$

We may use the initial condition to determine which constant C we need. Since $P(0) = P_0$, we know that

$$\ln P_0 = 0 + C,$$

or

$$C = \ln P_0.$$

Thus

$$\ln P = kt + \ln P_0.$$

Now we use algebra to solve for P. We get rid of the logarithms by taking exponentials on both sides:

$$P = e^{\ln P} = e^{(kt + \ln P_0)} = e^{kt}\, e^{\ln P_0} = P_0\, e^{kt}.$$

Not surprisingly, we get the same answer we found back in Chapter 2.

This may seem like a lot of work to find out something we already knew. But think about what we accomplished: We solved $dP/dt = kP$ without having to guess the answer—or even the form of the answer. Thus there is some hope for using this technique, called **separation of variables**, to solve differential equations and initial value problems in situations where there is no hope of guessing the answer—for example, the restricted growth model

$$\frac{dP}{dt} = kP(M - P).$$

As long as we were operating in the "guess the answer and check" mode, there was little doubt about correctness of the answer, because checking it was part of the process of finding it. Now that we are about to move beyond answers we can guess, we also have to pay attention to establishing the correctness of our answers. In most cases where "answer" means the result of antidifferentiation, this is not difficult, because checking can be done by the easier of the two inverse processes, differentiation. However, you need to be aware that the resulting equation will probably have to be manipulated algebraically to make it look like the original differential equation.

Pretending for a moment that we do not already know that $P = P_0 e^{kt}$ is a correct solution to $dP/dt = kP$, we illustrate the checking process with a short calculation:

$$P(t) = P_0 e^{kt}$$

$$\frac{dP}{dt} = kP_0 e^{kt}$$

$$\frac{dP}{dt} = kP$$

Checkpoint 2

(a) Give a reason for each step in the calculation.

(b) Check that the initial condition is satisfied by the proposed solution.

In the following Exploration Activity, we ask you to apply the separation of variables technique to another differential-equation-with-initial-value problem.

Exploration Activity 2

(a) Use the separation of variables technique to solve the initial value problem:

$$\frac{dP}{dt} = kP^2, \qquad P(0) = P_0.$$

(b) Check your answer to part (a) by showing that your solution function satisfies both parts of the initial value problem.

We start by writing the differential equation in differential form with the variables separated:

$$\frac{dP}{P^2} = k\,dt.$$

Now we antidifferentiate on both sides:

$$\text{An antidifferential of } \frac{dP}{P^2} = \text{some antidifferential of } k\,dt.$$

On the right, as before, we get $kt + C$. On the left, we use the Power Rule to find an antidifferential:

$$\frac{d}{dt}\left(-\frac{1}{P}\right) = \frac{d}{dt}(-P^{-1}) = -(-1)P^{-2} = \frac{1}{P^2}.$$

Thus

$$d\left(-\frac{1}{P}\right) = \frac{1}{P^2}\,dP.$$

So now we know

$$-\frac{1}{P} = kt + C.$$

Next, we determine C from the initial condition. When we set $t = 0$ and $P = P_0$, we find that $C = -1/P_0$. Thus

$$-\frac{1}{P} = kt - \frac{1}{P_0}.$$

The rest of the problem is algebra; we have to solve for P as a function of t. If we take reciprocals of both sides of our equation, we get

$$-P = \frac{1}{kt - \dfrac{1}{P_0}} = \frac{P_0}{kP_0t - 1}.$$

Now we change the sign on both sides to get our explicit formula for P:

$$P = \frac{P_0}{1 - kP_0t}.$$

Let's check this calculation. In order to differentiate P, we first write it in the form $P = P_0(1 - kP_0t)^{-1}$. Then, using the Power Rule and the Chain Rule, we find

$$\frac{dP}{dt} = -P_0(1 - kP_0t)^{-2}(-kP_0) = \frac{kP_0^2}{(1 - kP_0t)^2} = kP^2,$$

so our solution function does indeed satisfy the differential equation. Finally, substituting $t = 0$ in the formula for the solution function, we see that $P = P_0$ when $t = 0$.

Checkpoint 3

Solve the initial value problem

$$\frac{dP}{dt} = 2P^2 \qquad \text{with } P(0) = 1.$$

In this section we considered the constrained-growth model for population growth and realized that our previous guess-and-check method of attacking initial value problems would not work. Instead, we used the separation-of-variables approach. This approach requires rewriting the differential equation as an equality of differentials with all occurrences of the dependent variable on one side and all occurrences of the independent variable on the other. We look for antidifferentials of the two sides. When we find two such antidifferentials, we know that they must differ by a constant. At this point we have reduced our problem to an algebraic computation. In the next section we will apply this technique to the constrained-growth model.

Answers to Checkpoints

1. (d) If $P(0) = 1000$, then the solution is the constant function $P(t) = 1000$. If $P(0) > 1000$, then $P(t)$ decreases toward 1000.
2. (a) The reasons for the steps are:

   This is the proposed solution.

   We used the Exponential Rule for differentiation.

   We substituted the function definition for P.

   (b) When we substitute $t = 0$, the exponential factor reduces to 1 and we verify that $P(0) = P_0$.
3. $P = \dfrac{1}{1 - 2t}$.

Exercises 7.1

Solve each of the following initial value problems.

1. $\dfrac{dP}{dt} = 2P$ with $P(0) = 1$
2. $\dfrac{dP}{dt} = 2P$ with $P(0) = -1$
3. $\dfrac{dP}{dt} = 2P$ with $P(0) = 0$
4. $\dfrac{dP}{dt} = 2P^2$ with $P(0) = 2$
5. $\dfrac{dP}{dt} = 2P^2$ with $P(0) = -1$

6. $\dfrac{dP}{dt} = 2P^3$ with $P(0) = 0.5$

7. $\dfrac{dP}{dt} = 2P^3$ with $P(0) = 1$

8. $\dfrac{dP}{dt} = 2P^3$ with $P(0) = -1$

9. $\dfrac{dP}{dt} = 2P^3$ with $P(0) = 0$

For each of Exercises 10–20, (a) find a symbolic solution to the initial value problem, and (b) find the largest interval on which the solution is valid.

10. $\dfrac{dy}{dt} = \dfrac{1}{t}$ with $y(1) = 0$

11. $\dfrac{dy}{dt} = \dfrac{1}{t}$ with $y(1) = 1$

12. $\dfrac{dy}{dt} = e^{-y}$ with $y(1) = 0$

13. $\dfrac{dy}{dt} = e^{y}$ with $y(-1) = 0$

14. $\dfrac{dy}{dt} = e^{-y/2}$ with $y(1) = 0$

15. $\dfrac{dy}{dt} = e^{y/2}$ with $y(-1) = 0$

16. $\dfrac{dy}{dt} = e^{-y/2}$ with $y(2) = 0$

17. $\dfrac{dy}{dt} = e^{-t}$ with $y(0) = 1$

18. $\dfrac{dy}{dt} = e^{t}$ with $y(0) = 1$

19. $\dfrac{dy}{dt} = \dfrac{1}{y}$ with $y(0) = 1$

20. $\dfrac{dy}{dt} = \dfrac{1}{y}$ with $y(0) = -1$

21. The initial value problem in Exploration Activity 2,

$$\frac{dP}{dt} = kP^2, \quad P(0) = P_0,$$

is another population model of some interest; we will return to it in the Optional Lab Reading at the end of the chapter. In what important ways does the solution of this problem differ from the solution, $P(t) = P_0 e^{kt}$, to the apparently similar initial value problem,

$$\frac{dP}{dt} = kP, \quad P(0) = P_0?$$

7.2 The Logistic Growth Differential Equation

Here, again, is the differential equation we set out to solve at the start of this chapter:

$$\frac{dP}{dt} = kP(M - P),$$

where k and M are constants. The problem becomes an initial value problem if we know the population at some particular time, say, time zero. This equation is called the **logistic growth differential equation**. The graphs of its solutions are called **logistic curves**. You saw one in Figure 7.1, and you drew several yourself in Checkpoint 1 in the preceding section.

A New Antidifferentiation Problem

We begin the solution process just as we did for the natural-growth equation, by separating the variables:

$$\frac{dP}{P(M-P)} = k\,dt.$$

Now we want to antidifferentiate both sides. When we do that we will have rewritten the problem as

$$\text{An antidifferential of } \frac{dP}{P\,(M-P)} = \text{some antidifferential of } k\,dt.$$

This presents us with a new problem: The antidifferentiation on the right is easy—and just like the natural-growth case—but the one on the left is different from anything we have seen before.

Our immediate problem is not a population problem, or even a differential equation problem. The question we have to answer is, "What function of P has differential $dP/[P(M-P)]$?" That's the same as asking what function has derivative $1/[P(M-P)]$. And we don't have a ready answer, because nothing like this has turned up yet in our derivative formulas.

To emphasize the fact that the question in the last paragraph is unrelated to population growth—and because we might want to use the answer in some other context later—let's reformulate the question as

$$\text{"What function of a variable } x \text{ has derivative } \frac{1}{x(c-x)}\text{?"}$$

You may think of x as representing P and c as representing M if you wish.

We want to rewrite this problem in a form we can deal with more easily, i.e., use algebra to transform the problem into one where we can recognize what to do. The question we need to ask is: "What kind of algebraic calculation could have produced

$$\frac{1}{x(c-x)}\text{?"}$$

This is a little like playing "Jeopardy":

$$\text{"The answer is } \frac{1}{x(c-x)}\text{; what is the question?"}$$

Exploration Activity 1

Write down your own thoughts about this question. You may think of more than one way to answer the question. That's good, because we may need more than one way.

The first thoughts that usually appear in response to

$$\text{"The answer is } \frac{1}{x(c-x)}\text{; what is the question?"}$$

are (1) multiplication of algebraic fractions has answers that look like this, and (2) addition of algebraic fractions has answers that look like this.

If multiplication is what we are looking for, finding the direct problem is easy:

$$\frac{1}{x} \cdot \frac{1}{c-x} = \frac{1}{x(c-x)} .$$

That's somewhat promising, because each of the factors on the left is a (-1)th power of a linear expression in x. However, simple products don't turn up that often in derivatives—except in the Chain Rule—and the left side of the last equation doesn't look like a Chain Rule calculation. The Product Rule leads to a *sum* of products, and we don't have that either. Thus, while

$$\frac{1}{x} \cdot \frac{1}{c-x}$$

is a correct way to rewrite the function we want to antidifferentiate, it doesn't seem to lead to a solution.

Addition of fractions may be more useful for purposes of antidifferentiation because the Sum Rule for derivatives says the right thing: The derivative of a sum is the sum of the derivatives. As we have seen, it follows immediately that the same is true for antiderivatives. However, the inverse algebraic problem is not quite so obvious as was

$$\frac{1}{x} \cdot \frac{1}{c-x} = \frac{1}{x(c-x)}.$$

We want something like

$$\frac{A}{x} + \frac{B}{c-x} = \frac{1}{x(c-x)} ,$$

where the numerators A and B are yet to be determined.

Exploration Activity 2

(a) Find numbers A and B so that

$$\frac{A}{x} + \frac{B}{2-x} = \frac{1}{x(2-x)}.$$

(b) Repeat the calculation in (a) with 2 replaced by an arbitrary constant c, i.e., find numbers A and B so that

$$\frac{A}{x} + \frac{B}{c-x} = \frac{1}{x(c-x)}.$$

(This time your expressions for A and B will involve c.)

Let's consider part (b) first. Since we have a form for the inverse of adding fractions, we apply the direct process (addition) to the form, and try to match the answer to the expression with which we started. If we write the sum $A/x + B/(c-x)$ with a common denominator, we obtain

$$\frac{A(c-x)+Bx}{x(c-x)}.$$

In order for this to equal

$$\frac{1}{x(c-x)},$$

the numerators must match. Thus we must have

$$A(c-x)+Bx=1 \text{ for all values of } x.$$

To find algebraic equations that will pin down A and B, we may substitute any values we want for x. Particularly nice values to substitute are $x=0$ and $x=c$. For $x=0$, we obtain $Ac=1$, or $A=1/c$. And for $x=c$, we have $Bc=1$, so B is also $1/c$. In particular, in part (a) you should have found that both A and B are $1/2$.

Checkpoint 1

(a) We have used just two values of x to determine A and B. Substitute $A=B=1/c$ in

$$A(c-x)+Bx=1$$

and show that the result is an identity for all values of x.

(b) With the same values of A and B, show that

$$\frac{A}{x}+\frac{B}{c-x}=\frac{1}{x(c-x)}$$

is an identity for all values of x for which it makes sense.

Now we are ready to solve the antidifferentiation problem. Using the Sum and Constant Multiple Rules, we can write

$$\begin{aligned}\text{An antidifferential of } \frac{dx}{x(c-x)} &= \text{an antidifferential of } \left(\frac{1/c}{x}+\frac{1/c}{c-x}\right)dx \\ &= \text{the sum of antidifferentials of } \frac{1}{c}\frac{dx}{x} \text{ and } \frac{1}{c}\frac{dx}{c-x}.\end{aligned}$$

Exploration Activity 3

Explain why, when x is restricted to the interval between 0 and c, an antidifferential of $dx/[x(c-x)]$ is

$$\frac{1}{c}\ln\frac{x}{c-x}.$$

For positive values of x, we know that $\ln x$ is an antidifferential for $1/x$. Similarly, for $x<c$, we know $-\ln(c-x)$ is an antidifferential for $1/(c-x)$. (Here the negative sign in front of the logarithm is necessary to balance the negative sign generated by the Chain Rule in differentiation.) Putting these two antidifferentials together, we obtain

$$\frac{1}{c}\ln x - \frac{1}{c}\ln(c-x)$$

as an antidifferential for $dx/[x(c-x)]$. Now we may factor out the $1/c$ and use the fact that a difference of logarithms is the log of the quotient to rewrite the antidifferential as

$$\frac{1}{c}\ln\frac{x}{c-x}.$$

Checkpoint 2

(a) Write $1/(1-x^2)$ in the form $A/(1+x) + B/(1-x)$.

(b) Find an antidifferential for

$$\frac{dx}{1-x^2}$$

for $-1 < x < 1$.

Solution of the Logistic Growth Equation

Now we apply our antidifferential calculation to the logistic growth equation. Here is the differential equation again with variables separated and both sides ready to antidifferentiate:

$$\text{An antidifferential of } \frac{dP}{P(M-P)} = \text{some antidifferential of } k\,dt.$$

The left-hand side has exactly the form of the left-hand side of Exploration Activity 3—and we already knew how to antidifferentiate the right-hand side—so we are ready for the antidifferentiation step:

$$\frac{1}{M}\ln\frac{P}{M-P} = kt + C.$$

We have now solved the differential equation, in the sense that we have an equation without derivatives that relates the variables P and t. However, we don't have a completely satisfactory form of the solution yet, because we can't read off P as a function of t. The rest of the solution process consists of the necessary algebra to determine C from the initial condition and to solve for P.

First, we substitute $P = P_0$ and $t = 0$ to satisfy the initial condition:

$$\frac{1}{M}\ln\frac{P_0}{M-P_0} = C.$$

Next, we substitute this value of C back into

$$\frac{1}{M}\ln\frac{P}{M-P} = kt + C$$

to get

$$\frac{1}{M}\ln\frac{P}{M-P} = kt + \frac{1}{M}\ln\frac{P_0}{M-P_0}.$$

We proceed now to a sequence of transformations of this equation that lead to a solution for P as a function of t.

$$\begin{aligned}
\ln \frac{P}{M-P} &= Mkt + \ln \frac{P_0}{M-P_0} \\
\ln \frac{P}{M-P} - \ln \frac{P_0}{M-P_0} &= Mkt \\
\ln \frac{(M-P_0)P}{P_0(M-P)} &= Mkt \\
\frac{(M-P_0)P}{P_0(M-P)} &= e^{Mkt} \\
(M-P_0)P &= e^{Mkt}P_0(M-P) \\
(M-P_0)P &= P_0Me^{Mkt} - P_0e^{Mkt}P \\
(M-P_0+P_0e^{Mkt})P &= P_0Me^{Mkt} \\
P &= \frac{P_0Me^{Mkt}}{M-P_0+P_0e^{Mkt}}
\end{aligned}$$

Checkpoint 3

Provide a reason for each step in the calculation.

No wonder we couldn't guess the answer! But there it is—an explicit formula for population as a function of time, given a starting population P_0 and an assumption of logistic growth with maximum sustainable population M. Is it right? For $k = 0.00009$, $M = 1000$, and $P_0 = 10$ (the data for Figure 7.1), our formula becomes (check our arithmetic)

$$P = \frac{10{,}000e^{0.09t}}{990 + 10e^{0.09t}},$$

which should have a graph very similar to Figure 7.1. We repeat that graph in Figure 7.4 and show the graph of our formula in Figure 7.5—you be the judge.

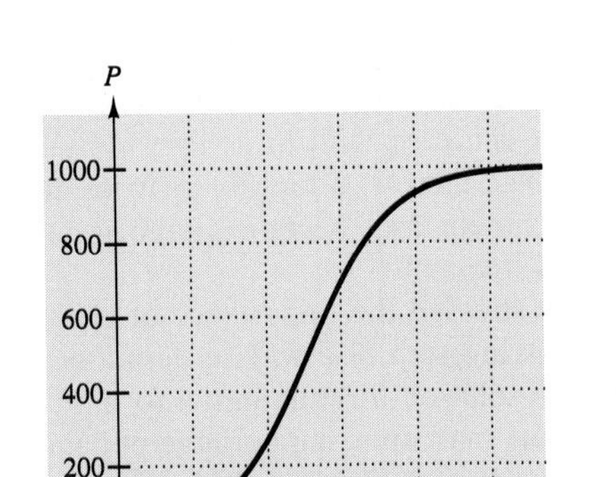

Figure 7.4 Numerical solution of the logistic equation, repeated from Figure 7.1

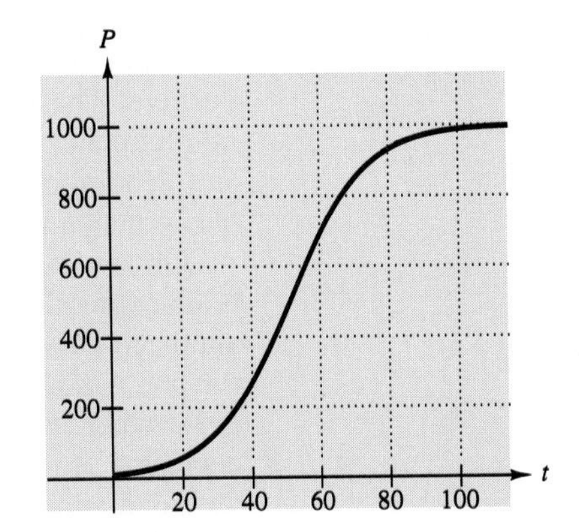

Figure 7.5 Exact solution of the logistic equation, graphed from $P = 10{,}000\ e^{0.09t}/(990 + 10\ e^{0.09t})$

Interlude: Exponential Approach to Stability

We pause for a moment to reflect on a theme that has already occurred in several apparently different contexts.

- In radioactive decay problems, we found that a function representing the quantity of a substance present at time t had the form Ae^{-kt}, an exponential with a negative exponent, and its values therefore decreased to zero as t became large.
- With Newton's Law of Cooling (Chapter 3), the solution function representing temperature had two terms: a constant, which was the ambient temperature, and an exponential that resembled radioactive decay. Thus, as time became large, the exponential term decreased to zero, and the temperature decreased to the ambient temperature.
- For falling bodies with air resistance proportional to velocity (Project 3 in Chapter 3), the computation was similar to Newton's Law of Cooling, but the factor multiplying the exponential term was negative. Thus the solution for velocity as a function of time,

$$v = \frac{g}{c}\,(1 - e^{-kt}),$$

  had a constant term representing terminal velocity and a decreasing exponential term representing how much faster the object had to go to reach terminal velocity.
- The RL circuit with a battery (Project 2 in Chapter 3) had a current function of exactly the same form, where the stable value being approached was the steady state current after the initial transient phase died away.

With the logistic model, we see a similar phenomenon, an approach as time goes on toward a stable population level, namely, the maximum supportable population M. We also see exponential functions in the solution

$$P = \frac{P_0 M e^{Mkt}}{M - P_0 + P_0 e^{Mkt}},$$

but the approach toward M is not immediately evident in this algebraic form of the solution—in fact, the coefficient of each exponential is Mk, a quantity that is clearly positive. These exponentials don't decrease with time, they *grow*—and at a very rapid rate (exponentially!). However, since there is a rapidly growing term in both numerator and denominator, the eventual behavior as t becomes large is indeterminate—another phenomenon we have encountered frequently.

On several occasions we have stressed the importance of changing the form of algebraic expressions as a problem-solving step or a way to gain a new insight. We use this approach here to mold our logistic solution formula into a form in which we can see that the population really approaches the maximum supportable population. The exponentials grow when we think they should decrease? Let's change them into exponentials that decrease! How? Multiply and divide by the appropriate decreasing exponential. Here's the calculation:

$$\begin{aligned} P &= \frac{P_0 M e^{Mkt}}{M - P_0 + P_0 e^{Mkt}} \cdot \frac{e^{-Mkt}}{e^{-Mkt}} \\ &= \frac{P_0 M e^{Mkt} \cdot e^{-Mkt}}{(M - P_0)\, e^{-Mkt} + P_0 e^{Mkt} \cdot e^{-Mkt}} \\ &= \frac{P_0 M}{(M - P_0)\, e^{-Mkt} + P_0} \end{aligned}$$

Thus exactly the same function also can be described in a more revealing form:

$$P = \frac{P_0 M}{(M - P_0)\, e^{-Mkt} + P_0}.$$

The only exponential in this equation has a negative exponent, so it decreases to zero as time becomes large. And, as that term in the denominator approaches zero, P approaches $(P_0 M)/P_0 = M$. Hence we see in the algebraic representation of the solution the approaching behavior we had seen previously in the numerical and graphical representations.

Checkpoint 4

(a) Carry out the same algebraic transformation in the specific case represented by

$$P = \frac{10{,}000 e^{0.09t}}{990 + 10 e^{0.09t}},$$

to get a formula for the fruit fly population that involves $e^{-0.09t}$ as the only exponential.

(b) Check your work by verifying directly that your function has the value 10 when t is 0 and approaches 1000 as t becomes large.

Population Growth in the United States

Is the logistic model of any real interest? Sure, some biologists might be interested in fruit flies, but are there any other populations that actually behave that way?

We began Chapter 2 with Thomas Malthus's observation, written about 1800, of the inevitable dire consequences of natural (exponential) growth of the human population of the earth. Some 40 years later, the Belgian mathematician P. F. Verhulst modified Malthus's prediction with the observation that the (human) population of any given region —including the whole Earth—could not exceed some "maximum capacity" M for that region. He may have been the first to propose the logistic equation

$$\frac{dP}{dt} = kP(M - P)$$

as a model for constrained natural growth, and he definitely had human populations in mind.

However, Verhulst was unable to test his theory or to compare it with Malthus's for two reasons: (1) In his day, there were no reliable estimates of world population at any time, past or present, so he had no way of estimating k, or P_0, or M. (2) Even if he knew, or could guess, reasonable values for the parameters, his formula for population would have to give essentially the same results as an exponential formula, since the population was clearly still very small relative to M.

Frustrated by his inability to say anything meaningful about world population—other than to suggest a reason why Malthus might be wrong—Verhulst turned to the only available source of good information about some large population: the U.S. Census. By 1840, there were six censuses, the results of which are shown in Table 7.1. Verhulst used these data to make a prediction, based on the logistic model, of the U.S. population 100 years later; he was off by only 1%. He didn't live to see how his prediction worked out, of course, so he probably died frustrated. His accomplishment was remarkable—perhaps also a bit lucky—because it took into account nothing but natural biological reproduction and the existence of some unknown maximum population—nothing about immigration, wars, depression, or anything else that affected our population in either direction over that 100-year period.

In a classroom or laboratory project associated with this chapter, you may have an opportunity to use the data in Table 7.1 to replicate Verhulst's prediction of U.S. population in 1940. To give you some feel for the likelihood of success, we show in Figure 7.6 a plot of the actual census data from 1790 to 1940; the first six points were those available to Verhulst.

Table 7.1 U.S. Census data

Date (A.D.)	*Population (millions)*
1790	3.929
1800	5.308
1810	7.240
1820	9.638
1830	12.866
1840	17.069

Figure 7.6 Census data on population of the United States (in millions), 1790–1940

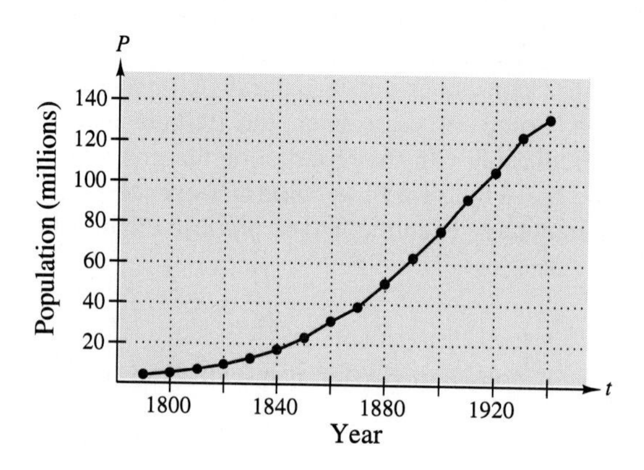

In this section we derived symbolic representations for solutions of the logistic growth differential equation:

$$\frac{dP}{dt} = kP(M - P).$$

In the process we developed a new approach to finding antiderivatives—an approach that uses the technique of rewriting quotients of polynomials as sums of simpler quotients. One form of our symbolic representation exhibits again the role of negative exponentials in expressions that approach a limiting value as t becomes large.

Answers to Checkpoints

2. (a) $\frac{1/2}{1+x} + \frac{1/2}{1-x}$, (b) $\frac{1}{2} \ln \frac{1+x}{1-x} = \ln \sqrt{\frac{1+x}{1-x}}$.

3. The reasons are:

   This is the original equation.

   We subtracted $\ln[P_0/(M - P_0)]$ from each side.

   The difference of logs is the log of the quotient.

   We exponentiated both sides of the equation.

   We multiplied across by $P_0(M - P_0)$.

   We expanded the product on the right-hand side.

   We gathered the terms with P as a factor together on the left-hand side and factored P out.

   We divided both sides by $M - P_0 + P_0e^{Mkt}$.

4. (a) $\frac{10,000}{10 + 990e^{-0.09t}}$

Exercises 7.2

In the following exercises, let f be the function given by $f(t) = e^{-2t}$, and let g be the function given by $g(t) = e^{2t}$. Sketch the graph of the indicated function over the domain $-2 \le t \le 2$.

1. f
2. f'
3. g
4. g'
5. $f + g$
6. $f - g$

Sketch the solution of each of the following initial value problems.

7. $\dfrac{dP}{dt} = 2P(4 - P)$ with $P(0) = 1$
8. $\dfrac{dP}{dt} = 2P(4 - P)$ with $P(0) = 4$
9. $\dfrac{dP}{dt} = 2P(4 - P)$ with $P(0) = 5$

In Exercises 10–15, find numbers A and B so that the relation holds for all appropriate values of x.

10. $\dfrac{A}{x} + \dfrac{B}{5 - x} = \dfrac{1}{x(5 - x)}$
11. $\dfrac{A}{x} + \dfrac{B}{5 + x} = \dfrac{1}{x(5 + x)}$
12. $\dfrac{A}{2 + x} + \dfrac{B}{2 - x} = \dfrac{1}{4 - x^2}$
13. $\dfrac{A}{x} + \dfrac{B}{x + 4} = \dfrac{1}{3x(x + 4)}$
14. $\dfrac{A}{x} + \dfrac{B}{4 - x} = \dfrac{1}{3x(4 - x)}$
15. $\dfrac{A}{3 - 2x} + \dfrac{B}{3 + 2x} = \dfrac{1}{9 - 4x^2}$

Find an antiderivative for each of the following functions.

16. $\dfrac{1}{x(5 + x)}$ for $x > 0$
17. $\dfrac{1}{x(5 - x)}$ for $0 < x < 5$
18. $\dfrac{1}{4 - x^2}$ for $-2 < x < 2$
19. $\dfrac{1}{3x(x + 4)}$ for $x > 0$
20. $\dfrac{1}{3x(4 - x)}$ for $0 < x < 4$
21. $\dfrac{1}{9 - 4x^2}$ for $-\dfrac{3}{2} < x < \dfrac{3}{2}$
22. In Exploration Activity 3 you showed that

$$\text{an antidifferential of } \frac{dx}{x(c - x)} = \frac{1}{c}\ln x - \frac{1}{c}\ln(c - x)$$

when x is restricted to range between 0 and c. Show that the restriction on the range of x can be dropped—that is, we can allow x to be negative or greater than c—if we write the formula as

$$\text{an antidifferential of } \frac{dx}{x(c - x)} = \frac{1}{c}\ln|x| - \frac{1}{c}\ln|c - x|.$$

23. (a) Use Exercise 22 to show that

$$\ln\frac{x}{x - 1}$$

is an antiderivative for

$$\frac{1}{x(1 - x)} \text{ for } x > 1.$$

(b) Use Exercise 22 to show that

$$\ln\frac{-x}{1 - x}$$

is an antiderivative for

$$\frac{1}{x(1 - x)} \text{ for } x < 0.$$

For each of the following functions, find an antiderivative for the indicated domain.

24. $\dfrac{1}{x(5 + x)}$ for $-5 < x < 0$
25. $\dfrac{1}{x(5 + x)}$ for $x < -5$

26. $\dfrac{1}{x(5-x)}$ for $x < 0$

27. $\dfrac{1}{x(5-x)}$ for $x > 5$

28. $\dfrac{1}{4-x^2}$ for $x > 2$

29. $\dfrac{1}{4-x^2}$ for $x < -2$

30. $\dfrac{1}{3x(x+4)}$ for $-4 < x < 0$

31. $\dfrac{1}{3x(x+4)}$ for $x < -4$

32. $\dfrac{1}{9-4x^2}$ for $x < -\dfrac{3}{2}$

33. $\dfrac{1}{9-4x^2}$ for $x > \dfrac{3}{2}$

34. Assume that the population of the United States from 1790 to 1940 is modeled by the logistic growth equation

$$\frac{dP}{dy} = kP(M-P)$$

for appropriate choices of k and M [*Note*: You may have done parts (b) through (d) of this exercise as Exercise 29 in Section 4.5.]

(a) Estimate from Figure 7.6 the time of most rapid population growth in the United States (up to 1940) and the population at that time.

(b) Explain why the population must be growing most rapidly at a time at which the second derivative of P is zero.

(c) Differentiate both sides of the differential equation to find an expression for the second derivative. (The Product Rule works for differentiating the right-hand side, but you can make the computation easier if you rewrite the expression in another form first.) Be careful with your differentiation—you are differentiating with respect to t, not P, so the Chain Rule must come into play every time you run into the unknown function P.

(d) What does your expression for the second derivative tell you about the population size when the growth rate is maximal?

(e) Estimate the maximum supportable population for the Verhulst model. How does your estimate compare with the population now?

(f) Do you think the Verhulst model will hold up for U.S. population beyond 1940? Why or why not?

7.3 Discrete Logistic Growth

Discrete Versus Continuous Population Models

Much of Section 7.2 centered around the logistic growth model

$$\frac{dP}{dt} = kP(M-P),$$

where P is a population at time t, M is the maximum population supported by the environment, and k is a proportionality constant. This model represents imposition of a logistic constraint (the maximum population) on natural or exponential growth. Thus, while the population P is small relative to M, the growth rate is roughly proportional to P, and the growth is roughly exponential. But when P becomes a significant fraction of M, the decreasing factor $M - P$ forces the growth rate to slow down. Eventually, the growth rate approaches zero, and the population approaches M.

In this section we take the logistic model as a point of departure for moving in a quite different direction from preceding sections. We observe that all our analysis of populations (since Chapter 2) has assumed that $P(t)$ is a continuous function of the continuous time variable t. In reality, we know that populations do not change continuously but rather in discrete amounts at discrete times. If a population is sufficiently large, then modeling its size with a continuous function may lead to reasonable and accurate results. On the other hand, many situations require that we use a discrete model for population.

We have frequently treated derivatives and difference quotients (for small values of Δt) as though they were interchangeable. Why are we now making an issue of this discrete versus continuous dichotomy? Well, recall that we encountered in Section 5.4 a situation in which discrete and continuous models for a single phenomenon behaved quite differently. The setting was that of determining the time-dependent behavior of prices for a given commodity under a market-clearing assumption. We saw that the continuous model always converged to a stable equilibrium price. On the other hand, the discrete model (explored in a laboratory assignment) allowed for an overreaction by suppliers that produced wild swings away from the equilibrium price. In technical jargon, if the proportionality constant representing supplier reaction is too large, we say that the equilibrium price is "unstable." Later in this section we will discover similar behavior of the discrete logistic model — and worse.

Rescaling to Maximum Supportable Population

For convenience in what follows we now rescale our population measurements to represent *fractions* of the maximum supportable population. Thus we introduce a new dependent variable p, which is defined in terms of population P by

$$p = \frac{P}{M}.$$

This has the effect of making the maximum supportable population equivalent to one unit of population. The new dependent variable p takes only values between 0 and 1. For example, $p = 0.25$ means that the population is one-quarter of the long-term sustainable population.

Exploration Activity 1

In this activity you will recast what we know about the dependent variable P in terms of the rescaled variable

$$p = \frac{P}{M}.$$

(a) We know that P is a solution of the differential equation

$$\frac{dP}{dt} = kP(M - P).$$

Find a similar differential equation satisfied by p.

(b) We know that P is of the form

$$P = \frac{P_0\, M}{(M - P_0)\, e^{-Mkt} + P_0}.$$

Find a similar form for solutions p of your differential equation in part (a).

(c) If $p_0 \neq 0$, show that any solution p approaches 1 as t becomes large.

Since $P = Mp$, we know that $dP/dt = M\, dp/dt$. Substituting both relations in the differential equation

$$\frac{dP}{dt} = kP(M - P),$$

we obtain

$$M\,\frac{dp}{dt} = kMp(M - Mp).$$

Dividing both sides of this equation by M, we obtain

$$\frac{dp}{dt} = kMp\,\frac{(M - Mp)}{M} = kMp\,(1 - p).$$

For convenience in what follows we substitute a single symbol κ for kM:

$$\frac{dp}{dt} = \kappa p\,(1 - p).$$

(The symbol κ is the lower-case Greek kappa.)

Suppose p is a solution of

$$\frac{dp}{dt} = \kappa p\,(1 - p),$$

and P is the corresponding solution of the original logistic equation, i.e., $P = Mp$. Then

$$P = \frac{P_0\, M}{(M - P_0)\, e^{-Mkt} + P_0}.$$

Since $P = Mp$ for all t, this is true, in particular, for $t = t_0$, i.e., $P_0 = Mp_0$. Using this and the relation $\kappa = kM$, we obtain

$$Mp = \frac{(Mp_0)\, M}{(M - Mp_0)\, e^{-\kappa t} + Mp_0}.$$

Dividing both sides by M, we obtain

$$p = \frac{Mp_0}{(M - Mp_0)\, e^{-\kappa t} + Mp_0}.$$

When we divide both the numerator and the denominator on the right by M, we obtain

$$p = \frac{p_0}{(1 - p_0)\, e^{-\kappa t} + p_0}.$$

Finally, the negative exponential $e^{-\kappa t}$ approaches 0 as t becomes large. Thus, if p_0 is not zero, the functional values of p approach $p_0/p_0 = 1$ as t becomes large.

Exploration Activity 1 describes what we know about the solutions of the *continuous* logistic differential equation

$$\frac{dp}{dt} = \kappa p\,(1 - p).$$

Now we turn to the corresponding *discrete* logistic model. If times are measured in discrete steps of length Δt, then this model is

$$\frac{\Delta p}{\Delta t} = \kappa p\,(1 - p).$$

We emphasize that in this discussion we are not necessarily considering small time steps, but just some fixed step length. It turns out to be convenient to let this time step be our unit of time; i.e., we assume that $\Delta t = 1$. With this assumption, our discrete model has the form

$$\Delta p = \kappa p\,(1 - p).$$

Now we write $\Delta p = p_{n+1} - p_n$ and rewrite $\Delta p = \kappa p\ (1 - p)$ in the more explicit form

$$p_{n+1} = p_n + \kappa p_n(1 - p_n) \qquad \text{for } n = 0, 1, 2, \ldots .$$

We may expand the product on the right and regroup to rewrite this iteration formula as

$$p_{n+1} = (\kappa + 1)p_n - \kappa p_n^2 \qquad \text{for } n = 0, 1, 2, \ldots .$$

A Rescaling of Convenience

We want to investigate how the iterations p_n respond to changes in the initial condition p_0 and, more important, how the iterations respond to changes in the responsiveness of the system, i.e., to changes in the factor κ. It turns out that these investigations will be simpler if we make still another change in scale. We write

$$x_n = \frac{\kappa}{\kappa + 1}\, p_n$$

for each value of n.

Checkpoint 1

The following calculation transforms

$$p_{n+1} = (\kappa + 1)p_n - \kappa p_n^2$$

into an equivalent equation with x as the dependent variable. Write a reason for each step.

$$\begin{aligned}
x_{n+1} &= \frac{\kappa}{\kappa+1}\, p_{n+1} \\
&= \frac{\kappa}{\kappa+1}\left[(\kappa+1)p_n - \kappa p_n^2\right] \\
&= (\kappa+1)\left(\frac{\kappa}{\kappa+1}\, p_n\right) - \frac{\kappa^2}{\kappa+1}\, p_n^2 \\
&= (\kappa+1)\left(\frac{\kappa}{\kappa+1}\, p_n\right) - (\kappa+1)\left(\frac{\kappa}{\kappa+1}\right)^2 p_n^2 \\
&= (\kappa+1)x_n - (\kappa+1)x_n^2 \\
&= (\kappa+1)x_n(1-x_n)
\end{aligned}$$

Finally, we replace the coefficient $\kappa + 1$ with a new (constant) symbol r to get

$$x_{n+1} = r\,x_n(1-x_n) \qquad \text{for } n = 0, 1, 2, \dots .$$

This is our convenient form of the discrete logistic equation.

We have changed the equation considerably. However, x still could represent a scaled population, and r is still a measure of how vigorously the system reacts. As before, our interest will be in examining how the solutions of this equation respond to changes in the initial condition and, especially, to changes in the factor r.

Function Iteration

To investigate the behavior of the discrete logistic equation, we make use of the function

$$F(x) = r\,x\,(1-x)$$

to generate a sequence of values for x_n starting with $x = x_0$. With this definition, we may rewrite the iteration formula

$$x_{n+1} = r\,x_n(1-x_n) \qquad \text{for } n = 0, 1, 2, \dots ,$$

as

$$x_{n+1} = F(x_n) \qquad \text{for } n = 0, 1, 2, \dots .$$

Thus the value of x_1 is $F(x_0)$, the value of x_2 is $F(x_1)$, and so on. We continue to turn each $F(x)$ back into an x to generate a new $F(x)$. We can find x_n for any value of n by starting with x_0 and iterating $F(x)$ a total of n times. We used this process of iteration extensively in Chapter 5—with Euler's Method for solving initial value problems and in the discrete price model. We also used it in Chapter 4 with Newton's Method for solving equations.

Exploration Activity 2

(a) In the following table we have started a list of values of x generated by carrying out the iteration

$$x_{n+1} = F(x_n) = r\,x_n(1-x_n) \qquad \text{for } n = 0, 1, 2, \dots .$$

with $r = 1.5$ and a starting value of $x_0 = 0.8$. Check the values given, and fill in the blank entries. What limiting value do the x's seem to be approaching?

n	x	n	x	n	x
0	0.8	5		10	
1	0.24	6		11	
2	0.2736	7	0.33064	12	
3	0.29811	8			
4		9			

(b) Suppose the sequence of x_n's generated by

$$x_{n+1} = r\,x_n(1 - x_n) \qquad \text{for } n = 0, 1, 2, \ldots$$

approaches a limiting value x. Explain why x must be a solution of the equation $x = F(x)$.

(c) For $r = 1.5$, solve the equation $x = F(x)$ for the limiting value of x. Does your result agree with your calculations in part (a)?

If the values x_n approach the limiting value x, then the values x_{n+1} also approach the same limiting value. (Both x_n and x_{n+1} represent steps in the same iteration.) Thus the iteration equation approaches

$$x = rx(1 - x)$$

or

$$x = F(x).$$

Note that this is true for any value of r. For the particular value $r = 1.5$, this equation becomes

$$x = 1.5x(1 - x)$$

or

$$x(1.5x - 0.5) = 0.$$

Thus this equation has two solutions: $x = 0$ and $x = 1/3$. The second of these is the limiting value you should have seen in part (a).

Now we'll look at a different value of r.

Exploration Activity 3

(a) Generate a sequence of x_n's with $r = 1$ and a starting value x_0 between 0 and 1. What does this sequence appear to be approaching? (*Warning*: The convergence of this sequence may be slow.)

(b) Choose another starting point in the same range and repeat part (a). What does the sequence appear to be approaching now?

(c) Solve the equation $x = F(x)$ in this case (i.e., with $r = 1$). Is your solution consistent with your computations? Why or why not?

For $r = 1$, any limiting value x must satisfy $x = x(1 - x)$. This is equivalent to $x^2 = 0$ or $x = 0$. Your sequences in parts (a) and (b) should both have approached 0. (If you take a starting value smaller than 0 or larger than 1, the sequence will not have a limiting value.)

Now we consider solutions of the equation $x = rx(1 - x)$ for general r.

Exploration Activity 4

(a) Without specifying the value of r (i.e., leaving r as a parameter), solve the equation

$$x = rx(1 - x)$$

for x. This equation is quadratic in x, so in general it should have two solutions. What are they?

(b) For what value or values of r is there only one solution?

(c) Are there any values of r for which the equation has no (real) solutions? Why or why not?

(d) Check that your general solution in terms of r agrees with the specific calculations you did in the two previous Exploration Activities.

We rewrite the equation in the form $rx^2 + x - rx = 0$. When we factor out x, we obtain $x[rx + (1 - r)] = 0$. Thus either $x = 0$ or $rx + (1 - r) = 0$. If $r \neq 0$, we have two possibilities:

$$\text{(i)} \quad x = 0$$

or

$$\text{(ii)} \quad x = \frac{r - 1}{r} = 1 - \frac{1}{r}.$$

If $r \neq 1$, these possibilities are distinct. If $r = 1$, then the second solution reduces to $x = 0$, and there is only one solution. If $r = 0$, the original equation reduces to $x = 0$, and there is again only the one solution.

We investigated the case $r = 1.5$ in Exploration Activity 2; here, the values x_n approached the limiting value $x = (r - 1)/r = 0.5/1.5 = 1/3$. In Exploration Activity 3 we investigated the case $r = 1$ and saw the limiting value $x = 0$.

Convergent, Periodic, and Chaotic Iterations

The sequence generated by the iteration

$$x_{n+1} = r\,x_n\,(1 - x_n) \qquad \text{for } n = 0, 1, 2, \ldots$$

is a function: x_n is a function of n. This function has a graph. We will examine such graphs for a variety of values of r and x_0. We begin with the case $r = 1.5$ and $x_0 = 0.8$. In Figure 7.7 the function graph consists of just isolated points; we have connected the points with line segments to make it easier to follow the progression. Next to the graph we tabulate some of the numbers used in drawing it.

Figure 7.7 Logistic iteration, $r = 1.5$, $x_0 = 0.8$

n	x
0	0.8
1	0.24
2	0.2736
3	0.29812
4	0.31386
5	0.32303
6	0.32802
7	0.33064
8	0.33197
9	0.33265
10	0.33299

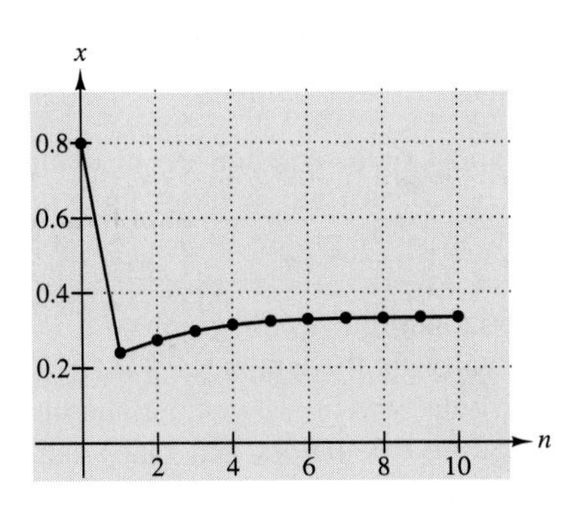

Also, we want to compare the discrete sequence with the corresponding solution for the continuous model. For each choice of r and x_0, there are corresponding values of κ and p_0 for the continuous model

$$\frac{dp}{dt} = \kappa p\,(1 - p) \qquad \text{with } p(0) = p_0.$$

We know the solution to this initial value problem is

$$p = \frac{p_0}{(1 - p_0)\,e^{-\kappa t} + p_0}.$$

Since $x = \kappa/(\kappa + 1)\,p$, the corresponding values for the continuous model are

$$x(t) = \frac{\kappa}{\kappa + 1}p(t).$$

In Figure 7.8 we show the graph of this continuous solution along with the discrete solution.

Figure 7.8 Logistic iteration, $r = 1.5$, $x_0 = 0.8$

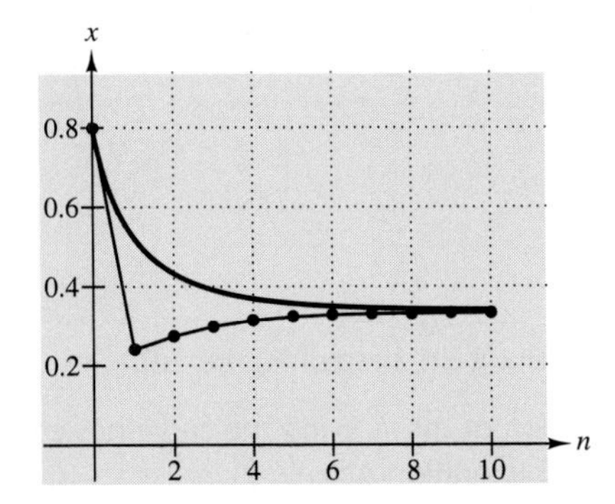

In Figure 7.8 we see that even if discrete and continuous models converge to the same number, the two solutions may not look anything alike. We have started with a population well above the maximum supportable population, so the continuous solution decreases exponentially to that level. The discrete solution, on the other hand, starts in the same place with the same slope and undershoots the limit on the very first step. From there on it tracks what would be the continuous solution starting from $x = 0.24$.

As we learned in Exercise 18 in Section 5.4, it is often helpful to use another graphical representation of an iteration, the so-called **web diagram**. The web diagram can be constructed without knowing an explicit solution of the iteration. Here's how: First, draw the graphs of $y = x$ and of $y = F(x)$. Start at (x_0, x_0) and draw a vertical line (up or down) to the graph of $y = F(x)$, which takes you to the point (x_0, x_1). Now draw a horizontal line to the graph of $y = x$, which takes you to the point (x_1, x_1). Again move vertically to the graph of $y = F(x)$, which you meet at the point (x_1, x_2), and then horizontally to (x_2, x_2). Continue the process.

Some calculators and many computer programs enable you to draw web diagrams yourself. Check your calculator manual or ask your instructor. Experimentation is the essence of this approach to understanding iteration. You will find it much more valuable to see the web diagrams being drawn than to look at our static figures.

Here is the process for the logistic iteration with $r = 1.5$. We first sketch a graph of $y = F(x)$, as shown in Figure 7.9. Then we add the line $y = x$. Iterating $F(x)$ means that we take a starting value for x ($x_0 = 0.8$ in Figure 7.9) and then move vertically to the curve to find $F(x)$. Each time we iterate $F(x)$, we must change $F(x)$ back into x for the next iteration. As we continue this process, we see that the lines in the web are tending toward one of the points of intersection of the curve with the line, i.e., a solution of the equation $x = rx(1 - x)$.

Figure 7.9 Web diagram for the logistic iteration, $r = 1.5$, $x_0 = 0.8$

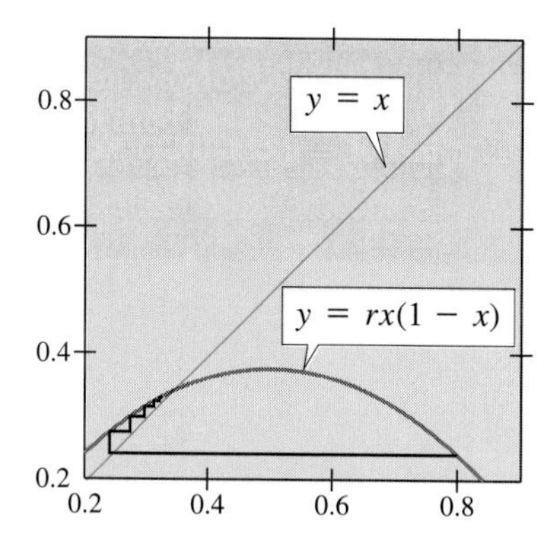

Checkpoint 2

(a) Identify the values x_1, x_2, and x_3 along the base of Figure 7.9.

(b) Explain why the point of intersection of the line and the curve has coordinates (x, x), where x is the same limiting value you found in Exploration Activity 2.

We now examine both these representations — function graphs and web diagrams — for the logistic iteration with a range of values of r. Because $r = \kappa + 1$, and κ is positive, only values of r larger than 1 are meaningful for population models. (Other values of r are explored in Exercises 9 and 10.) Our figures show discrete and continuous plots and the corresponding web diagrams for $r = 2.5$, 3.2, 3.5, and 3.9. These figures illustrate the facts that (1) the continuous solution always converges, whether the discrete one does or not, and (2) the discrete solution doesn't necessarily track the continuous one, even when both converge.

Figure 7.10 Logistic iteration, $r = 2.5$, $x_0 = 0.05$

n	x
0	0.05
1	0.11875
2	0.26262
3	0.48294
4	0.62427
5	0.58639
6	0.60634
7	0.59673
8	0.60161
9	0.59919
10	0.60040

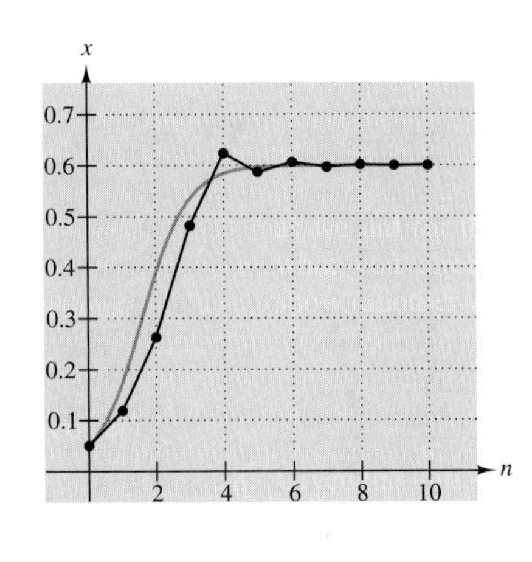

Figures 7.10 and 7.11, with a larger value for r than the one used in Figures 7.8 and 7.9, show another way the discrete and continuous solutions can converge to the same limiting value ($x = 0.6$) and yet look quite different. The starting point is a reasonable one for a logistic population model, and the continuous solution shows the familiar sigmoid shape that rises steadily toward the limiting value. The discrete solution undershoots the continuous solution for three steps, then overshoots on the next step. From then on, successive values (see the table) are alternately above and below the eventual limit.

Figure 7.11 Web diagram for logistic iteration, $r = 2.5$, $x_0 = 0.05$

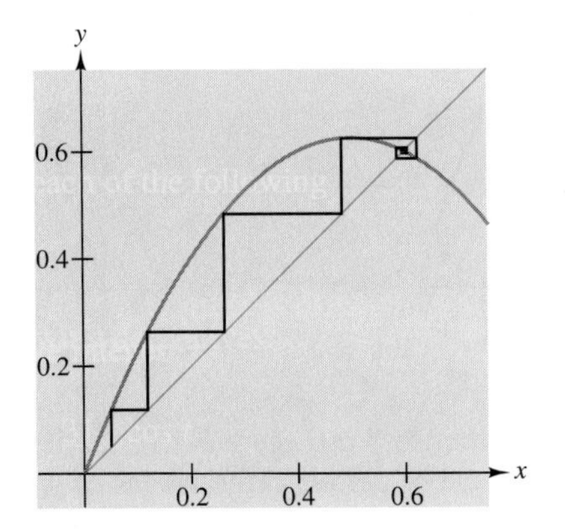

In Figure 7.12 we continue to increase r, this time to 3.2. Now we notice something very peculiar. The discrete sequence does not close in on any one number; instead, within only a few iterations it starts to oscillate back and forth between 0.799... and 0.513.... Indeed, no matter what starting value we supply for x_0, the iterations eventually fall into this same pattern. This limiting behavior is called a **cycle of period** 2, or simply a **2-cycle**.

Figure 7.12 Logistic iteration, $r = 3.2$, $x_0 = 0.02$

n	x
0	0.02
1	0.06272
2	0.18812
3	0.48873
4	0.79959
5	0.51278
6	0.79948
7	0.51300
8	0.79946
9	0.51304
10	0.79946

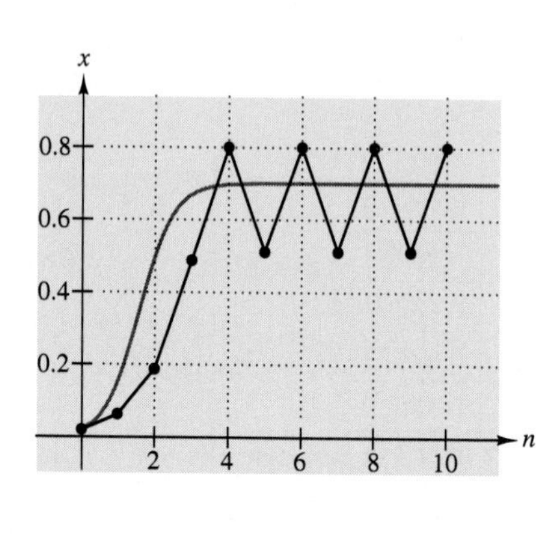

This behavior shows up clearly in the web diagram in Figure 7.13. Instead of settling down onto a single point, the diagram settles into a closed rectangular pattern. In Figure 7.14 we repeat the web diagram showing only the limiting 2-cycle.

Figure 7.13 Web diagram for the logistic iteration, $r = 3.2$, $x_0 = 0.02$

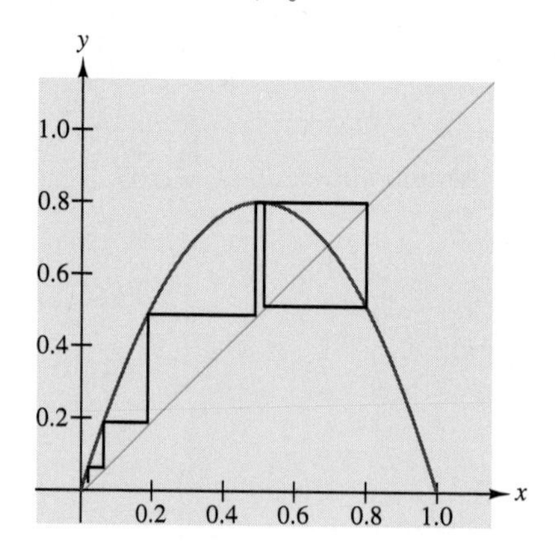

Figure 7.14 Limiting 2-cycle for $r = 3.2$

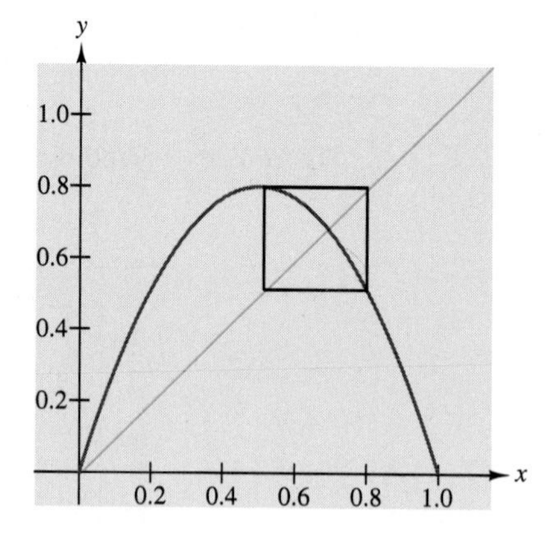

As we increase r further to 3.5 (Figure 7.15), the iterations settle into an even more complicated pattern, a cycle of period 4, the values of which we can read from the table: 0.5008..., 0.8750..., 0.3828...., and 0.8269.... In order to show the limiting 4-cycle clearly, we have extended the computation from 10 steps to 20.

Figure 7.15 Logistic iteration, $r = 3.5$, $x_0 = 0.02$

n	x
0	0.02
1	0.0686
2	0.22363
3	0.60767
4	0.83443
5	0.48355
⋮	⋮
13	0.50066
14	0.87500
15	0.38282
16	0.82694
17	0.50089
18	0.87500
19	0.38282
20	0.82694

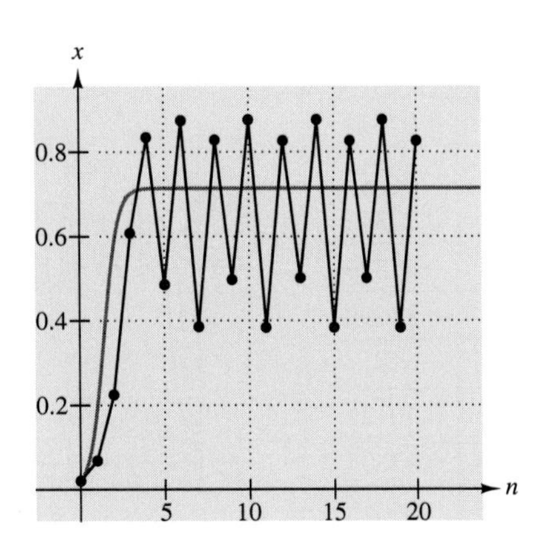

Again, this shows up clearly in the web diagram in Figure 7.16. Now the limiting figure cycles among four values rather than the two we saw with $r = 3.2$. In Figure 7.17 we show only the limiting 4-cycle.

Figure 7.16 Web diagram for the logistic iteration, $r = 3.5$, $x_0 = 0.02$

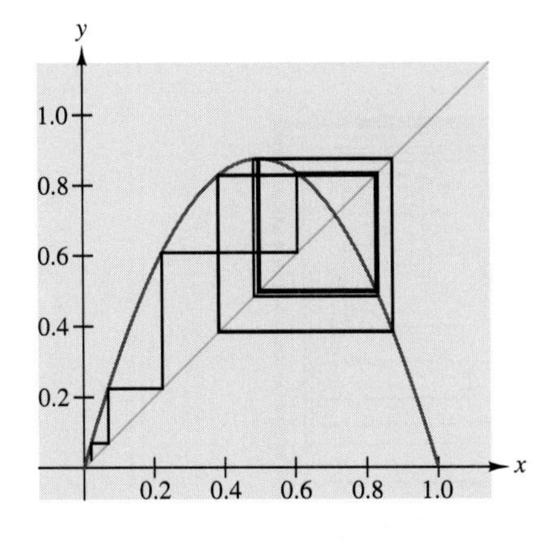

Figure 7.17 Limiting 4-cycle for $r = 3.5$

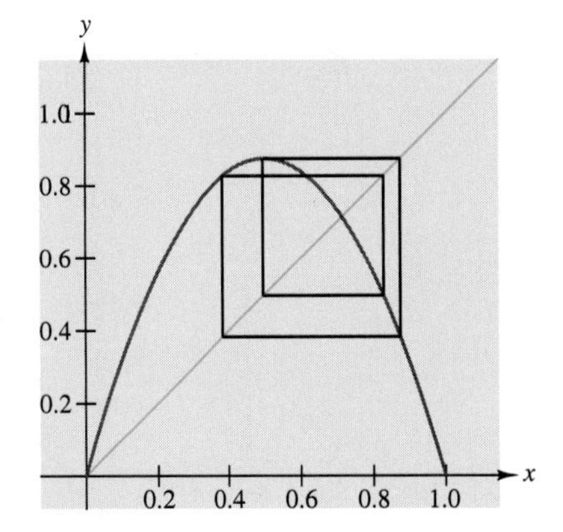

When we try even larger values of r (as in Figure 7.18), the behavior becomes increasingly complex. For $r = 3.9$, the sequence exhibits no discernible pattern, no matter what starting value is supplied and no matter how far we extend the computation (Figure 7.18 has 40 steps). The values of x seem to jump around at random. If we carried the computation far enough, the values would seem to cover most of the range from 0 to 1. This is your first look at behavior now known by the technical term **chaos**.[1]

1. For a fascinating and readable account that goes far beyond our brief treatment of this subject, see James Gleick's book, *Chaos: Making a New Science*, Viking, 1987.

Figure 7.18 Logistic iteration, $r = 3.9$, $x_0 = 0.02$

n	x
0	0.02
1	0.07644
2	0.27533
3	0.77814
4	0.67329
5	0.85788
⋮	⋮
13	0.40121
14	0.93694
15	0.23043
16	0.69160
17	0.83183
18	0.54557
19	0.96690
20	0.12482

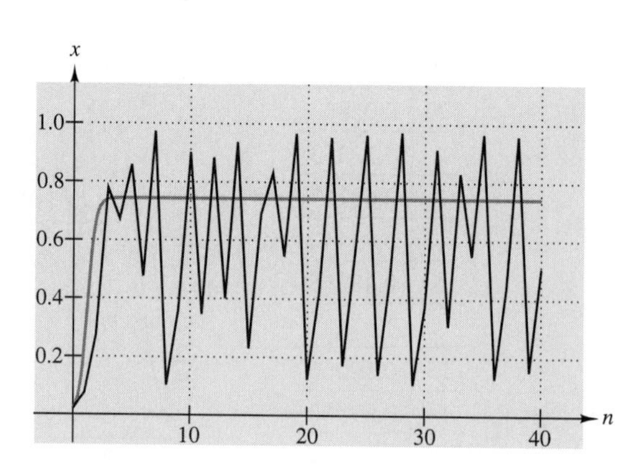

The web diagram for this iteration reflects the chaotic nature of the iteration. If we let the iteration run, we would eventually fill up most of the space in Figure 7.19 with black.

Figure 7.19 Web diagram for $r = 3.9$ and $x_0 = 0.02$

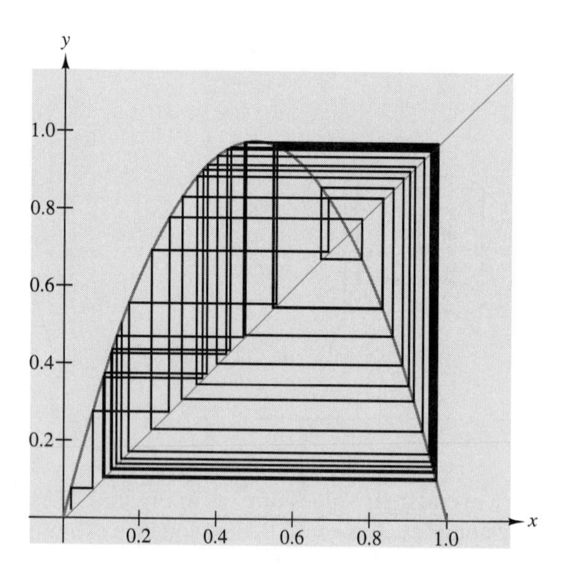

What is going on here? The continuous logistic function always converges. Why does iteration of the discrete logistic function yield such erratic results?

Bifurcation: An Application of the Derivative

Why do we find, when r increases past 3, that iteration does not approach the x-value where $y = F(x)$ and $y = x$ intersect? Suppose we consider such an r-value, for example, $r = 3.2$. If we start exactly at the point of intersection ($x = 2.2/3.2 \approx 0.688$ for $r = 3.2$), then $x = F(x)$, and the iteration will never move from that point. However, if we start anywhere else, the iteration for $r = 3.2$ will always approach the same 2-cycle.

We illustrate this in Figure 7.20 with another web diagram, this time for $r = 3.2$ and $x_0 = 0.9$.

Figure 7.20 Logistic iteration, $r = 3.2$, $x_0 = 0.9$

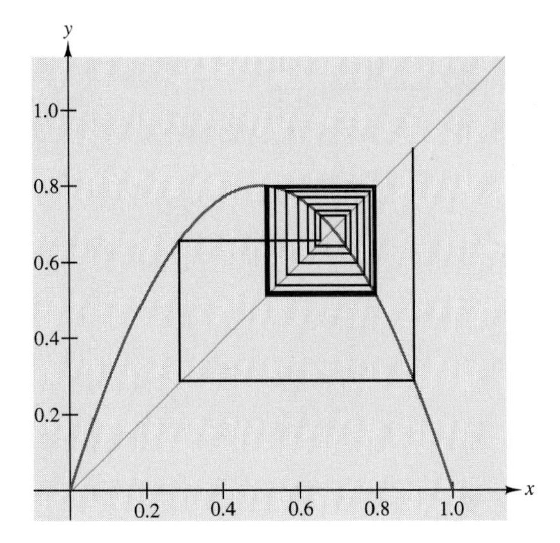

Trace the steps in Figure 7.20 with your finger, starting from the outermost ($x = 0.9$). You will see that the iteration quickly moves close to the intersection point and then gradually farther away. But it doesn't get much farther away. You can see only about 16 distinct x-coordinates in Figure 7.20, but the iteration was run for 30 steps. The box that contains most of the iterates is the 2-cycle we observed in Figure 7.14. Its lower left corner is at about $(0.513, 0.513)$, and its upper right corner is at about $(0.799, 0.799)$.

Why does the iteration process move away from the intersection point when $r = 3.2$? More generally, exactly where (in changing values of r) does the process stop converging to the intersection point and start moving toward the 2-cycle? And what feature of the curve causes this change?

As you can see in Figures 7.21, 7.22, and 7.23, increasing r makes the graph of $F(x)$ steeper.

Figure 7.21 Graphs of $y = rx(1-x)$ and $y = x$ for $r = 1.5$

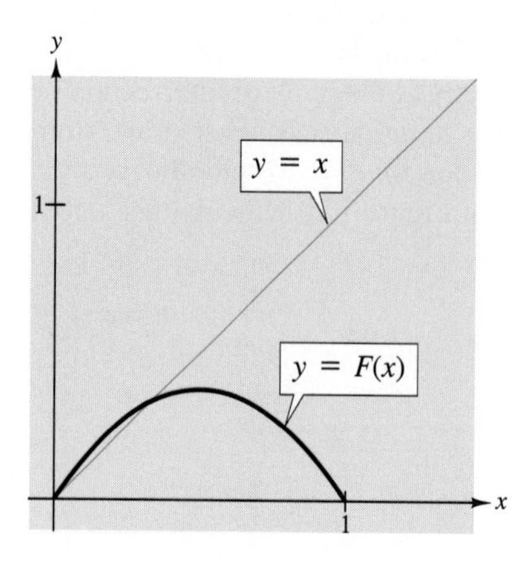

Figure 7.22 Graphs of $y = rx(1-x)$ and $y = x$ for $r = 2.5$

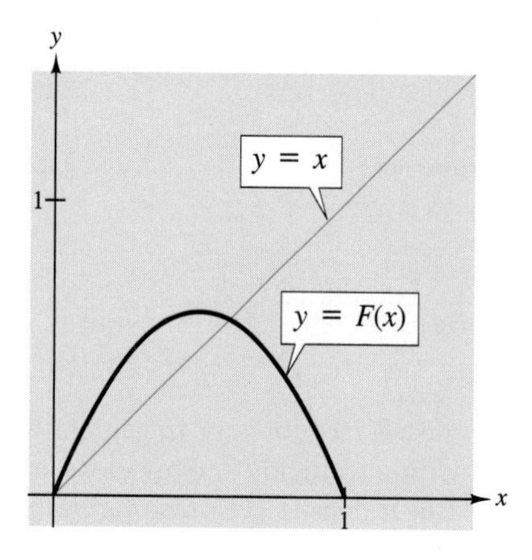

Figure 7.23 Graphs of $y = rx(1-x)$ and $y = x$ for $r = 3.2$

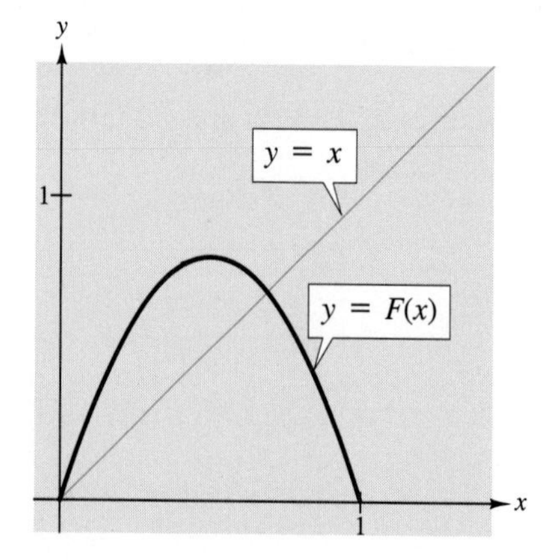

As the graph of $y = F(x)$ gets steeper, vertical distances in the web diagram get larger, relative to horizontal distances at the corresponding steps. In Figure 7.24 we repeat the approach to the 2-cycle for $r = 3.2$ and $x_0 = 0.02$.

Figure 7.24 Web diagram for $r = 3.2$ and $x_0 = 0.02$

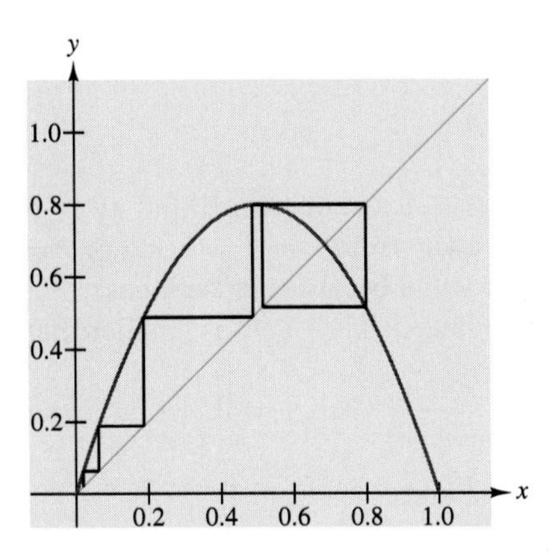

As you follow the web diagram toward the 2-cycle, the vertical distance from the curve to the line is always greater than the horizontal distance from the line to the curve on any one iteration. This characteristic of the curve near where it intersects $y = x$ carries the web away from the point of intersection until it settles into a 2-cycle. The curve falls away from the line $y = x$ too quickly for the iterations to approach the intersection point.

In Figure 7.25 examine again the web diagram for $r = 2.5$ and $x_0 = 0.05$. You will see a more gentle sloping of the curve near the intersection with $y = x$, one that ensures that each vertical step is shorter than the corresponding horizontal step, which brings each iteration closer to the intersection point.

Figure 7.25 Web diagram for $r = 2.5$ and $x_0 = 0.05$

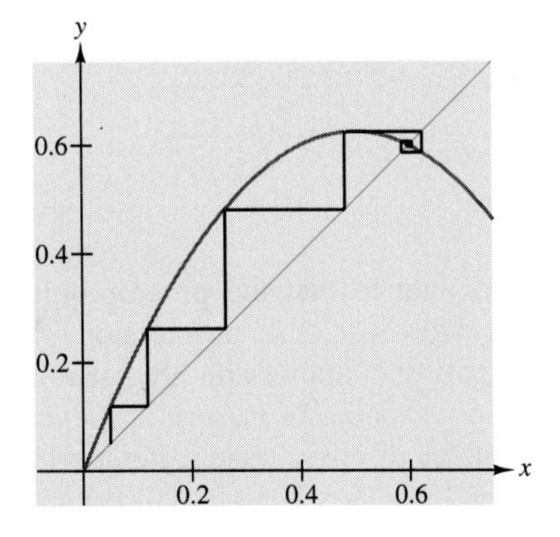

Checkpoint 3

The observation in the preceding paragraph leads us to suspect that the existence of a 2-cycle is related to the derivative of $F(x)$ at the intersection with $y = x$, specifically at $x = 1 - 1/r$.

(a) Find the derivative of F, and evaluate $F'(1 - 1/r)$.

(b) What is the value of this intersection slope when $r = 1$? When $r = 3$?

(c) What is the range of values of the intersection slope when r is between 1 and 3? When r is greater than 3?

When $F'(1 - 1/r) < -1$, the curve will fall away from the line $y = x$ so steeply that the iteration will move away from $x = 1 - 1/r$. For $r = 1.5$ and 2.5, the iteration process yields a single limiting value because, in each case, $F'(1 - 1/r)$ is greater than -1 (i.e., closer to zero). For $r = 3.2$, $F'(1 - 1/r)$ is smaller than -1 (i.e., farther from zero), and a 2-cycle results.

Since $F'(1 - 1/r) = -1$ when $r = 3$, let us examine the web diagram for $r = 3$ (Figure 7.26).

Figure 7.26 Logistic iteration, $r = 3.0$, $x_0 = 0.9$

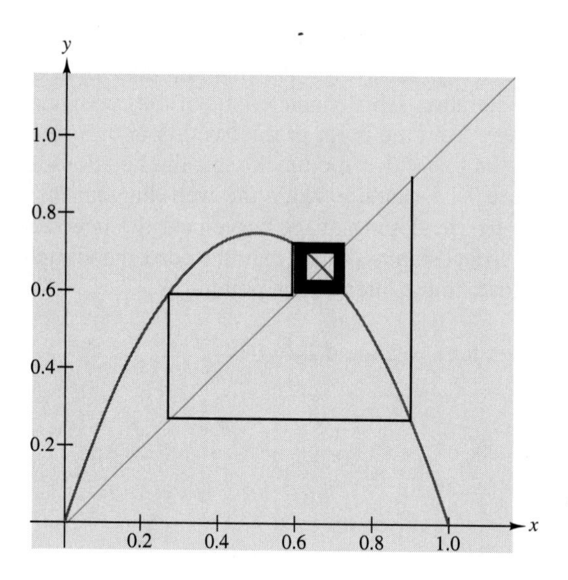

Starting with $x = 0.9$, notice that the process quickly carries the x-values toward $x = 1 - 1/r = 0.666...$. However, the progression slows down considerably as the process goes on. One gets the impression that the iteration sequence cannot decide whether to move in on $x = 0.666...$ or to form a 2-cycle. In fact, this tension is exactly what is occurring. For values of r less than 3, iteration leads us to $x = 1 - 1/r$. As soon as we move a little beyond $r = 3$, a 2-cycle will occur. The value $r = 3$ where iteration starts to yield 2-cycles is known as a **bifurcation point**—"bifurcation" is a fancy word for splitting or branching at a two-way fork.

As we allow r to increase from 3 to 4, we notice that the 2-cycles give way to 4-cycles. A 4-cycle is produced as each x-value in the 2-cycle splits, or bifurcates. These in turn give way to 8-cycles, then 16-cycles, and so on until we reach r-values for which no discernible pattern exists, which is the condition known as chaos. You may have an opportunity to explore these ideas further in a laboratory activity.

In this section we examined discrete logistic growth and compared it to the continuous model. After several changes of scale, we represented the logistic iteration by $x_{k+1} = F(x_k)$, where $F(x) = rx(1 - x)$. Here the factor r represents the responsiveness of the system. We found that we could distinguish different types of behavior of the iterations by studying the slope of the graph of F at its intersection with the graph of $y = x$. For values of $r < 3$, the iteration converges to the value x such that $x = F(x)$. For values of r slightly larger than 3, the iterations settle into a 2-cycle, a pattern of cycling between two distinct values. As we increase the value of r, we see next a 4-cycle, then an 8-cycle. As r is increased still more, we see chaotic behavior.

Answers to Checkpoints

1. The reasons for the steps are

   This is the given scaling.

   We substituted $p_n + \kappa p_n(1 - p_n)$ for p_{n+1}.

   We used the distributive law.

   We multiplied the numerator and denominator of the second term on the right by $\kappa + 1$.

   We substituted x_n for $[\kappa/(\kappa + 1)]\, p_n$.

   We factored $(\kappa + 1)x_n$ out of both terms.

2. 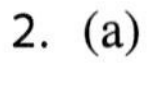

   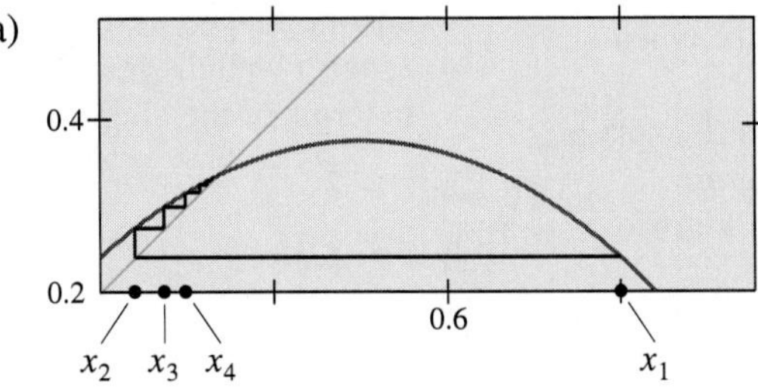

   (b) See the discussion following Exploration Activity 2.

3. (a) $F'(x) = r(1 - 2x)$, $F'(1 - 1/r) = 2 - r$.

   (b) $F'(1 - 1/r) = 1$ when $r = 1$, and $F'(1 - 1/r) = -1$ when $r = 3$.

   (c) $-1 < F'(1 - 1/r) < 1$ for $1 < r < 3$, and $F'(1 - 1/r) < -1$ for $r > 3$.

Exercises 7.3

In Exercises 1–8, for the specified value of r, experiment with the logistic iteration

$$x_{n+1} = rx_n(1 - x_n), \quad \text{for } n = 0, 1, 2, \dots,$$

with starting x-values of 0.1 and 0.9. For Exercises 7 and 8 you will need at least 40 iterations.
(a) Decide whether the sequence converges to a limit point, a 2-cycle, a 4-cycle, an 8-cycle, or is in the chaotic range.
(b) Make a web diagram and explain your conclusion in terms of the diagram.

1. $r = 1.7$
2. $r = 2.3$
3. $r = 2.95$
4. $r = 3.05$
5. $r = 3.45$
6. $r = 3.55$
7. $r = 3.568$
8. $r = 3.6$

9. (a) Experiment with the logistic iteration

$$x_{n+1} = rx_n(1 - x_n), \quad \text{for } n = 0, 1, 2, \dots,$$

with $r = 0.5$. Try a starting x-value of 0.2. Where are the sequence values headed? Try other starting values, such as 0.5 and 0.886. What happens? What do you conclude about the eventual behavior of the sequence when $r = 0.5$?
(b) Make a web diagram for this case, and explain your conclusion in terms of the diagram.
(c) In what way does this case relate to logistic population growth?

10. (a) Experiment with the logistic iteration

$$x_{n+1} = rx_n(1 - x_n), \quad \text{for } n = 0, 1, 2, \dots,$$

with $r = 1$. Try a starting x-value of 0.2. Where are the sequence values headed? Try other starting values, such as 0.5 and 0.886. What happens? What do you conclude about the eventual behavior of the sequence when $r = 1$?
(b) Make a web diagram for this case, and explain your conclusion in terms of the diagram.
(c) In what way does this case relate to logistic population growth?

11. (a) The relationship between population p (as a fraction of supportable population) and "scaled population" x was defined by

$$x = \frac{\kappa}{\kappa + 1}\, p,$$

where $r = \kappa + 1$. What values of p correspond to the solutions for x in Exploration Activity 4?
(b) The values of p in part (a) represent all the population levels to which the discrete logistic model might possibly converge, if it converges at all. Do they agree with your intuition about the long term behavior of the model? Why or why not?
(c) Do the values of p to which the discrete model might converge agree with the long term behavior of the continuous model, as determined in Section 2? Why or why not?

In Exercises 12–15, experiment with the iteration

$$x_{n+1} = 0.3r\sin(3x_n), \quad \text{for } n = 0, 1, 2, \dots$$

with the given value of r. Use starting x-values of 0.1 and 0.9.
(a) Decide whether the sequence converges to a limit point, a 2-cycle, a 4-cycle, an 8-cycle, or is in the chaotic range.
(b) Make a web diagram, and explain your conclusion in terms of the diagram.

12. $r = 2$
13. $r = 2.9$
14. $r = 2.95$
15. $r = 3.2$

16. Use Newton's Method to find the positive solution of the equation

$$x = 0.3r\sin(3x)$$

for $r = 2$. Compare this calculation with your calculations in Exercise 12.

17. Consider the iteration

$$x_{k+1} = F(x_k)$$

where $F(x) = rx(1 - x^2)$.
(a) For a general $r > 1$, find the positive solution of $x = F(x)$.
(b) For $r = 1.5$ and $x_0 = 0.1$, find the limiting value of the iteration $x_1,\ x_2,\ x_3,\ \dots$. Compare this to your answer in (a).

18. Consider the iteration

$$x_{k+1} = F(x_k)$$

where $F(x) = rx(1 - x^2)$.
(a) Find the value of r at the first bifurcation point of the iteration.
(b) Verify your answer to (a) by experimenting with the iteration for values of r that are larger and smaller than your answer.

19. Consider the iteration

$$x_{k+1} = F(x_k)$$

where $F(x) = rx(1 - x^2)$ and $r = 2.1$.
(a) Sketch the web diagram for $x_0 = 0.9$.
(b) Investigate the iteration for $x_0 = 0.1$. You will need to consider 40 or more iterations. Sketch the web diagram and explain what is going on.

Chapter 7 Summary

We began this chapter by picking up where we left off in Chapter 2 with the study of population growth. We saw there that unfettered natural growth is exponential and obtained simple symbolic solutions for the corresponding differential equations. For the more complicated logistic models, we could generate numerical and graphical solutions, even though, at that time, we could not find symbolic solutions. Our first objective in this chapter was to develop the tools we had previously lacked for finding those symbolic solutions.

The most important of these tools is the idea of **separation of variables** in a differential equation. The success of the separation-of-variables technique rests on Leibniz's powerful notation of differentials and the Chain Rule. The Chain Rule assures us that, after separating variables, it is legitimate to antidifferentiate both sides of an equation, even though the antidifferentiation on one side is with respect to the dependent variable and on the other side with respect to the independent variable.

The last section of the chapter was a study of the discrete logistic model, a reasonable model for many biological populations that reproduce at discrete times. We had already seen in Section 5.4, by comparing models for price evolution, that discrete and continuous models with the same assumptions about growth rates could lead to very different results. In particular, the continuous price evolution model was always stable, in the sense that its long-term behavior was to approach a market-clearing price. On the other hand, the discrete model might display the same stability, but it could also be unstable — for example, if suppliers were too responsive to previous prices and demands. Similarly, the discrete logistic model sometimes mimics the stability of the continuous logistic model, but if the growth-rate constant is too large, a fantastic variety of other eventual behaviors can appear. Some of these behaviors are stable cycles, and some are chaotic in the sense that no detectable pattern ever appears. Furthermore, tiny variations in the rate constant can lead to vastly different behavior.

Concepts

- Separation of variables to solve first-order differential equations
- Undoing addition of algebraic fractions
- Sequence iteration
- Cyclic behavior of iterated sequences
- Web diagrams
- Bifurcation
- Chaos

Applications

- Continuous logistic growth
- Population growth of the United States
- Discrete logistic growth

Formulas

- $\dfrac{1}{x(c-x)} = \dfrac{1}{c}\left[\dfrac{1}{x} + \dfrac{1}{c-x}\right]$
- An antidifferential for $\dfrac{dx}{x(c-x)}$ is $\dfrac{\ln|x|}{|c-x|}$.

- The solution of the logistic growth differential equation

$$\frac{dP}{dt} = kP(M - P) \quad \text{with} \quad P(0) = P_0$$

is
$$P = \frac{P_0 M}{(M - P_0)e^{-Mkt} + P_0}.$$

Chapter 7 Exercises

Each differential equation we have studied so far has been of one of the following types:

$$\frac{dy}{dt} = \text{an expression involving only } y \text{ (the dependent variable)}$$

or
$$\frac{dy}{dt} = \text{an expression involving only } t \text{ (the independent variable).}$$

For each of the following differential equations, decide which type of problem it is, separate the variables, and find all solutions.

1. $\dfrac{dy}{dt} = \dfrac{1}{y}$

2. $\dfrac{dy}{dt} = \dfrac{1}{t}$

3. $\dfrac{dy}{dt} = e^{-t}$

4. $\dfrac{dy}{dt} = e^{-y}$

5. $\dfrac{dy}{dt} = y^3$

6. $\dfrac{dy}{dt} = t^3$

Solve each of the following initial value problems.

7. $\dfrac{dy}{dt} = \dfrac{1}{y}, \quad y(0) = 2$

8. $\dfrac{dy}{dt} = \dfrac{1}{t}, \quad y(1) = 2$

9. $\dfrac{dy}{dt} = e^{-t}, \quad y(0) = 2$

10. $\dfrac{dy}{dt} = e^{-y}, \quad y(0) = \ln 2$

11. $\dfrac{dy}{dt} = y^3, \quad y(1) = 1$

12. $\dfrac{dy}{dt} = t^3, \quad y(1) = 1$

13. Checkpoint 1 in Section 7.1 and the accompanying slope field (Figure 7.3) made it clear that the logistic growth model is biologically meaningful even if P_0 is greater than M. This might be the case if there were a sudden wave of immigration—for example, large numbers of refugees from a civil war in a neighboring country—or an overstocking of a wildlife refuge, or a concentration of wildlife in a habitat that is shrinking because of development. Find the general solution of the logistic growth model when P_0 is greater than M. (*Hint*: Review Exercise 19 in Section 7.2.)

14. The logistic initial value problem,

$$\frac{dP}{dt} = kP(M - P), \qquad P(0) = P_0 ,$$

is mathematically meaningful for every starting point P_0 — negative, zero, less than M, equal to M, and greater than M. Expand on the preceding exercise to find the general solution in every case.

15. In Project 3 in Chapter 3 we modeled the acceleration of a falling body by the initial value problem

$$\frac{dv}{dt} = g - cv, \qquad v(0) = 0.$$

By use of a scaling argument, we saw that the unique solution of this initial value problem is

$$v = \frac{g}{c}\left(1 - e^{-ct}\right).$$

(a) Find this solution another way, by separating the variables in

$$\frac{dv}{dt} = g - cv.$$

(b) Write the left-hand side of $v = (g/c)\left(1 - e^{-ct}\right)$ as ds/dt, and find position s (distance fallen) as an explicit function of t. (You may have already done this in Project 3 in Chapter 3.)

16. Recall that in our derivation of the discrete logistic equation in the form

$$x_{n+1} = r\, x_n(1 - x_n),$$

x is related to p by

$$x_n = \frac{\kappa}{\kappa + 1}\, p_n, \qquad \text{with } r = \kappa + 1.$$

If p ranges between 0 and 1, what is the range of the variable x? In particular, what is the range of x if $\kappa = 0.05$? if $\kappa = 0.5$? if $\kappa = 2$?

17. Consider the iteration

$$x_{k+1} = F(x_k)$$

where $F(x) = rx\sqrt{1 - x^2}$ and $r > 1$.

(a) Find the value of r at the first bifurcation point of the iteration.

(b) Verify your answer to (a) by experimenting with the iteration for values of r that are larger and smaller than your answer.

In Exercises 18–21, experiment with the iteration

$$x_{n+1} = 0.3r \sin(3x_n), \qquad \text{for } n = 0, 1, 2, \ldots .$$

with the given value of r. Use starting x-values of 0.1 and 0.9.

(a) Decide whether the sequence converges to a limit point, a 2-cycle, a 4-cycle, an 8-cycle, or is in the chaotic range.

(b) Make a web diagram, and explain your conclusion in terms of the diagram.

18. $r = 1.8$ 19. $r = 1.86$ 20. $r = 1.88$ 21. $r = 2$

22. You may have noticed that we never checked our solution of the logistic growth equation. We leave this to you, with the warning that this is a serious challenge to your algebraic skills. You will find it easiest to work with the form

$$P = \frac{P_0 M}{(M - P_0)\, e^{-Mkt} + P_0}.$$

First do the easy part: Check that $P = P_0$ when $t = 0$. Then verify that the function $P(t)$ satisfies the growth equation

$$\frac{dP}{dt} = kP(M - P).$$

[*Hint*: Calculate dP/dt and $kP(M - P)$ separately. Then simplify each until you can see that the two expressions are the same.]

23. Figure 7.27 shows web diagrams for iterations of the function $F(x) = 0.3\, r \sin 3x$ with $r = 2$ and with $r = 2.85$. The diagram on the left shows convergence to a limiting value; that on the right a limit 2-cycle. Find the value of r at which the first bifurcation occurs. [*Hint*: Let x^* be the x-coordinate of the point of intersection of the graphs of $y = F(x)$ and $y = x$. You know already that the first bifurcation occurs at a value of r such that $F'(x^*) = -1$. The pair of conditions $x = F(x)$ and $F'(x) = -1$ is a set of two equations in the two unknowns x and r. If you solve those simultaneous equations, you will find both the intersection coordinate x^* and the bifurcation value r^*. It is easy to eliminate r from the two equations, because r appears linearly in both equations. Then you can solve for x^* numerically, say, by Newton's Method or with your calculator Solve function. Finally, you can substitute x^* into either equation (or both!) to find r^*.]

Figure 7.27 Web diagrams for iterations of $F(x) = 0.3\, r \sin 3x$ with $r = 2$ (*left*) and $r = 2.85$ (*right*)

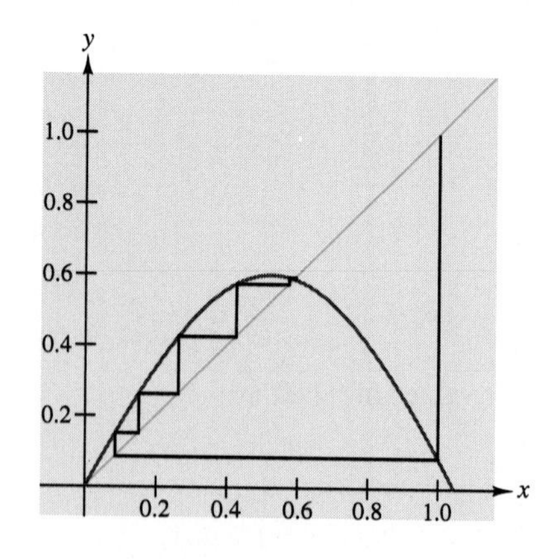

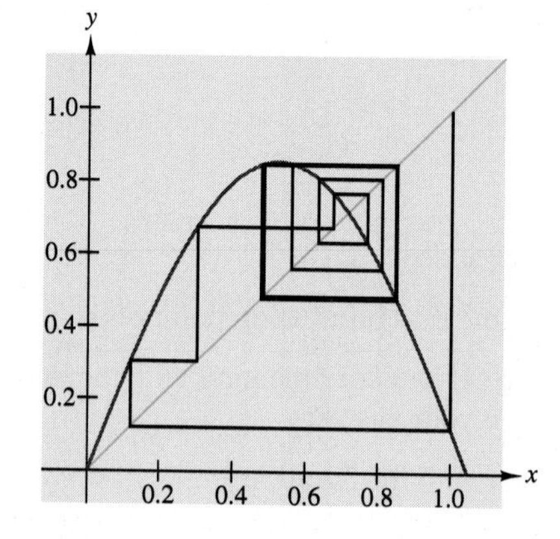

24. Figure 7.28 shows web diagrams for iterations of the function

$$F(x) = r\,x\,(1-x)(2-x)(3-x)$$

with $r = 0.5$ and with $r = 0.8$. The diagram on the left shows convergence to a limiting value; that on the right a limit 2-cycle. Find the value of r at which the first bifurcation occurs.

Figure 7.28 Web diagrams for iterations of $F(x) = r\,x(1-x)(2-x)(3-x)$ with $r = 0.5$ (*left*) and $r = 0.8$ (*right*)

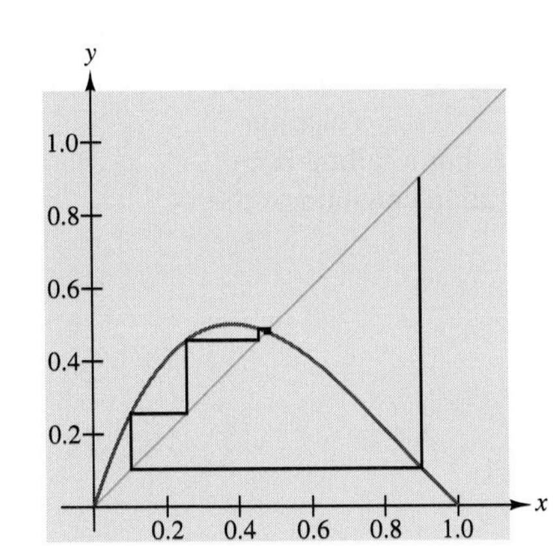

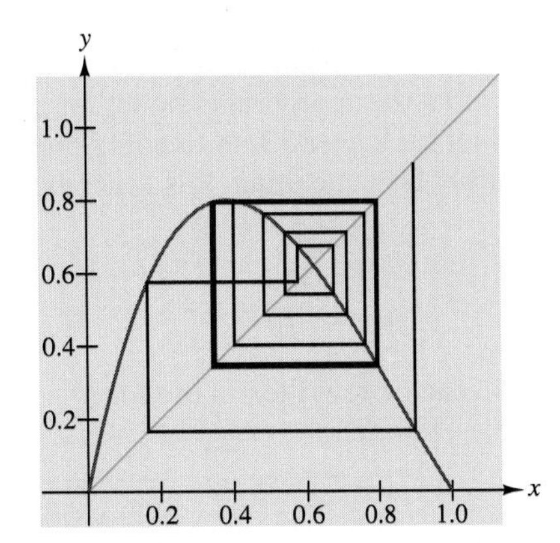

Chapter 7 Projects

1. **Testing for Logistic Growth**

(a) Devise a graphical test by which you could tell quickly (with help from a computer or calculator) whether given data satisfies a logistic growth pattern. (*Hint*: Suppose the variables are t and P; use the equation

$$\frac{1}{M}\ln\frac{P}{M-P} = kt + C$$

to argue that an appropriate semilog graph must give a straight line.) You need to estimate M from the data. Here are some possibilities: (i) If the data reveal the location of the maximal growth rate, use Exercise 34 in Section 7.2. (ii) If the data appear to have leveled off, use that fact. (iii) Otherwise, conduct computer experiments, varying M to get your line as straight as possible.

(b) In an early laboratory project you may have met the Project CALC mascot, a golden retriever named Sassy. You found that her recorded weight for the first seven weeks of her life appeared to grow exponentially, but that projection along that exponential curve led to a ridiculously large weight by the time she was six months old. Could it be that puppy weight grows logistically? Use your test from part (a) to determine whether the data in Table 7.2 fit a logistic growth model. The table contains more complete data than you saw before. [*Warning*: If you suspect that your last data value may be indistinguishable from M, you have to omit it when applying your test (why?).]

Table 7.2 Sassy's weight

Age (days)	0	10	20	30	40	50	60	70	101	115	150	195	230	332	436
Weight (pounds)	3.25	4.25	5.5	7	9	11.5	15	19	30	37	54	65	70	75	77

2. **Falling Bodies with Air Resistance: Sky Diving** Pilots and parachutists generally assume that the resisting force of air on an airplane wing or a falling human is proportional to the square of the velocity,[2] i.e., that the appropriate model for air resistance is the Velocity-Squared Model discussed in the Chapter 3 Lab Reading on Raindrops. For a falling body, since speed and velocity mean the same thing, this leads to an acceleration equation of the form

$$\frac{dv}{dt} = g - cv^2.$$

(a) Show that this equation can be rewritten in the form

$$\frac{dv}{(\sqrt{g} - \sqrt{c}v)(\sqrt{g} + \sqrt{c}v)} = dt.$$

(b) Solve the equation from part (a) for v as a function of t, assuming that $v_0 = 0$, using the same approach we used to solve the logistic growth equation.

(c) Checking the result is hard—we asked you to do it in Exercise 49 at the end of Chapter 6, although your solution may have a different form from the one given there. However, it is not difficult to check for the correct initial and terminal velocities. Find the terminal velocity directly from $dv/dt = g - cv^2$, and then verify that your velocity function approaches the correct terminal velocity. (If it doesn't, check the details of your calculation to see what went wrong.)

(d) Your answer to part (b), a formula for v as a function of t, is also a differential equation whose solution (with the right initial condition) gives position as a function of time. The required antidifferentiation is a little beyond our reach—until Chapter 9. Instead of asking you to tackle it at this point, we give you the answer and ask you to confirm that it agrees with your answer for $v(t)$. Given the formula

$$s(t) = \frac{1}{c} \ln \left(\frac{e^{\sqrt{gc}t} + e^{-\sqrt{gc}t}}{2} \right),$$

show that $s(0) = 0$ and that ds/dt matches the velocity you calculated in part (b). [If not, check the details of both part (b) and this part to see what went wrong.]

(e) Ms. Jennifer Phillips of Orange, MA, an experienced sky diver, reports the following free fall terminal velocities: 170 mph in fetal position, 140 mph in a nose dive, and 120 mph in a horizontal position with arms and legs spread out.[3] For each of these cases, find the constant c in equation $dv/dt = g - cv^2$, and then find explicit formulas for velocity and

2. W. K. Kershner, *The Advanced Pilot's Flight Manual*, 3rd ed., Iowa State University Press, 1970, p. 10.
3. We are indebted to Professor Richard Palmer of the Duke University Physics Department for assistance with the falling-body problem and, in particular, for providing the reference in the preceding footnote and the data in this problem.

position at time t after leaving the plane. In each case, how long does it take to reach 95% of terminal velocity, and how far has she fallen at that time? (Be careful with units.)

(f) Suppose the sky diver's air resistance is proportional to her velocity rather than to the square of her velocity. That is, suppose the formulas from Exercise 15 apply to sky diving. Answer the questions in part (e) under this assumption. Can you find something in your answers for time and distance (to reach 95% of terminal velocity) that might suggest one model is more reasonable than the other? Why or why not?

3. **Law of Mass Action** Consider a chemical reaction in which two substances combine to form a third substance. If this results from collision of the molecules of the two substances, then it is reasonable to suppose that the rate of formation of the new compound is proportional to the number of collisions per unit time and that this number is jointly proportional to amounts of the original substances not yet transformed. (Reactions that behave this way are said to satisfy the **Law of Mass Action**.) Suppose that y grams of the new compound are made up of py grams of the first reactant and qy grams of the second, where $p + q = 1$. Suppose further that the initial amounts of the two substances were P grams of the first and Q grams of the second.

(a) Explain why the Law of Mass Action says that the amount y of the new substance at time t is the solution of the initial value problem

$$\frac{dy}{dt} = k(P - py)(Q - qy), \qquad y(0) = 0.$$

(b) Use the same approach we used for the logistic growth equation to solve this initial value problem. At some point you will need to assume something about the relative sizes of the parameters. Assume that $Pq \neq Qp$. (For real chemicals, it is very unlikely that products like these could ever be exactly equal.)

(c) What can you say about the limiting value of y as t becomes large? Your answer may depend on which of Pq and Qp is the larger. Does your answer make sense in terms of the physical limitations of the reaction? Could you predict the answer directly from the differential equation?

Chapter 7 Optional Lab Reading: The Coalition Model of Population Growth

In 1960, Heinz von Foerster and two colleagues proposed a new approach to modeling world population,[4] rather different from either the Malthus or Verhulst models, but equally simplistic. They argued from a historical perspective that improvements in technology and in mass communication have had the effect of molding the world population into a more and more effective coalition in a vast game against nature, rendering natural environmental hazards less effective, improving living conditions, and

4. Heinz von Foerster, Patricia M. Mora, and Larry W. Amiot, "Doomsday: Friday, 13 November, A.D. 2026," *Science* 132 (1960), pp. 1291–1295. The authors were in the Department of Electrical Engineering at the University of Illinois, Urbana. Some of the material in this Lab Reading is based on a paper by David A. Smith, "Human Population Growth: Stability or Explosion?" *Mathematics Magazine* 50 (1977), pp. 186–197.

extending the average life span. It follows (they said) that the net production rate (i.e., the ratio of the growth rate to the population), instead of being constant (as in the Malthus model), might actually be an increasing function of the population.

They expected that the production rate would be a very slowly increasing function, and they tentatively proposed a function of the form kP^r, where the parameters k and r would be determined by the historical data, with r probably very small. Of course, it could turn out that no combination of k and r would give a reasonable fit to the historical data, in which case the tentative proposal would have to be abandoned. We (and the von Foerster group) have an advantage over Malthus and Verhulst, in that more or less reliable estimates of world population for the last 350 years are now available, and there are educated guesses for population figures going back much further.

In a laboratory assignment associated with this chapter, you will determine whether either the Malthus (natural growth) or the von Foerster (coalition) model can possibly fit the historical data and, if either does, to estimate the relevant parameters. Here we will determine what sort of functions can satisfy the differential equation resulting from the von Foerster assumption.

An Initial Value Problem and Its Solution

In the form of a differential equation, the von Foerster proposal is

$$\frac{\frac{dP}{dt}}{P} = kP^r$$

or

$$\frac{dP}{dt} = kP^{r+1}.$$

Since a starting time and corresponding population can be determined from the historical data, we have another initial value problem to solve. If r happens to be zero, this is exactly the Malthus model for population growth, and we know already that the solution is an exponential function. However, we are assuming (with von Foerster) that r is not zero, even if it turns out to be very small. Thus we have to solve a differential equation that is more like the one you solved in Exploration Activity 2 of Section 7.1. Indeed, when we separate the variables and antidifferentiate, we get

$$\text{An antidifferential of } \frac{dP}{P^{r+1}} = \text{some antidifferential of } k\,dt.$$

The antidifferential on the left, which is the same as the antidifferential of $P^{-r-1}dP$, brings up the general question of antidifferentiating powers of the variable, i.e., the Power Rule for antidifferentiation.

In words, if we differentiate the nth power of the independent variable, the result is obtained by multiplying by the old power (n) and decreasing the power by 1. Antidifferentiation is the inverse process: Increase the power by 1, and divide by the new power. This would not work if the power being antidifferentiated were -1, of course, but -1 can't turn up as the result of a Power Rule differentiation (why not?).

In symbols, here is the **Power Rule for antidifferentiation**:

$$\text{An antidifferential of } x^n\,dx \text{ is } \frac{x^{n+1}}{n+1}, \text{ if } n \neq -1.$$

$$\text{For } x \neq 0, \text{ an antidifferential of } x^{-1}\,dx \text{ is } \ln|x|.$$

Problem 1

(a) Carry out the antidifferentiation on both sides of

$$\text{An antidifferential of } \frac{dP}{P^{r+1}} = \text{some antidifferential of } k\,dt.$$

How do we know that the first of the two formulas above applies to this problem and the second does not?

(b) Show that the solution equation can be written in the form

$$P^r = \frac{1}{rk(T-t)},$$

where T is a new constant that depends on k and on the antidifferentiation constant. [*Hint*: For the case of $r = 1$, this is the same as Exploration Activity 2 of Section 7.1. But the form of the equation above is not the same as the answer we worked out in that activity. Try to put our earlier answer in the new form by factoring kP_0 out of the denominator. Then apply the same steps to your result from part (a) of this activity to see that everything works the same way when r is not necessarily 1.]

(c) Since P must be positive, T must be greater than t. What happens to P as t approaches T? Make a rough sketch of P as a function of t. (*Hint*: First sketch P^r as a function of t; then think about how the graph will change when you take an rth root of the dependent variable.)

For reasons you should have determined by now, the time T is called "doomsday." If there is indeed such a time (i.e., if the historical data actually fit the coalition model, which is a question you will address in the laboratory), it is clearly important to know how far away that time is. If we are talking about extremely rapid population growth 300 years from now, that's not a serious problem for us, and there is plenty of time for the people of the world to figure out what to do about it. If doomsday, as predicted by the historical data, is right around the corner, then world population growth is a much more serious problem than Malthus ever imagined.[5]

Just to give the matter a little perspective, consider this: When the von Foerster paper was published in 1960, there were just over 3 billion people in the world, and it had taken 30 years to go from 2 billion to 3 billion. It took roughly 17 years to add the next billion and 9 more years to add the fifth billion. Is that pattern likely to continue for another 300 years?

5. Whatever you think about infinite population in finite time, we recommend that you read *The Population Explosion* by Paul R. Ehrlich and Anne H. Ehrlich (Simon and Schuster, 1990). Without any assumptions leading to infinite population, the Ehrlichs explain why overpopulation is already having disastrous consequences.

Adding a Logistic Constraint to the Coalition Model[6]

Your graph of world population P as a function of time (Problem 1), according to the coalition model, should show P becoming infinitely large as t approaches T. At the same time, it should show dP/dt also becoming infinitely large. A reasonable person might object that neither of those phenomena is physically or biologically possible. Indeed, Verhulst already made the argument for the first: We may not know what the number is, but the Earth must have a finite maximum supportable population. Why can't the growth rate of the population be arbitrarily large?

If we accept Verhulst's concept of a logistic limit and incorporate it into our model for population growth, can we still achieve a reasonable fit to the historical data? If so, what conclusion can we draw about the eventual size of the population?

This question was addressed by Arthur Austin and John Brewer[7] in a paper that first appeared a decade after the von Foerster paper. To explain their approach, we need to rewrite the logistic growth equation in another form. Here is the original equation again:

$$\frac{dP}{dt} = kP(M - P);$$

if we factor M out of the last factor, we can write the equation as

$$\frac{dP}{dt} = mP\left(1 - \frac{P}{M}\right),$$

where $m = kM$. Now we see more clearly the relationship of the logistic model to the natural growth model: When P is small relative to M, the last factor is approximately 1, and m is approximately the production rate, i.e., the ratio of growth rate to population. As P increases, the factor $1 - P/M$ decreases, so m is also the maximal production rate.

Austin and Brewer suggested another production rate that would grow like the coalition production rate for relatively small values of the population and still account for the existence of a maximal production rate, m. The form of their production rate for small P is

$$m\left[1 - \exp\left(-\frac{kP^r}{m}\right)\right].$$

As P becomes large, the exponential term approaches zero, so this expression approaches m. Following Problem 2 we show that the expression matches the coalition rate when P is small.

Problem 2

(a) Show that the tangent line to the graph of $y = e^x$ at the point $(0, 1)$ has the equation $y = 1 + x$.

(b) Draw a picture that shows both the exponential graph and its tangent line at $(0, 1)$.

---

6. This section is optional reading for the interested student. It is not necessary for the World Population Laboratory Activity.

7. A. L. Austin and J. W. Brewer, "World Population Growth and Related Technical Problems," *IEEE Spectrum* 7 (Dec., 1970), pp. 43–54, 8 (Mar., 1971), p. 10; reprinted in *Technological Forecasting and Social Change* 3 (1971), pp. 23–49. A quite different approach to fitting the historical data and still having an eventual leveling off was studied by William Squire in "Taming Hyperbolic Growth," *Technological Forecasting and Social Change* 10 (1977), pp. 15–20. Squire's paper can be understood with the mathematical background you have achieved at this point in the course.

The implication of Problem 2 is that e^x is approximately equal to $1 + x$ for values of x close to zero. Thus, for small values of P (relative to its eventual maximum), if we set $x = -kP^r/m$, we can write

$$m\left[1 - \exp\left(-\frac{kP^r}{m}\right)\right] \approx m\left[1 - \left(1 - \frac{kP^r}{m}\right)\right] = m\left(\frac{kP^r}{m}\right) = kP^r.$$

This shows that the proposed production rate has both the desired characteristics — a match with the coalition rate for small P and an approach to m for large P.

Next, Austin and Brewer incorporated the logistic factor $1 - P/M$ into their growth rate. As we have seen, this factor is approximately 1 for small values of P and approximately 0 for large values of P. Combining all the factors, we get the final form of the Austin-Brewer model for population growth:

$$\frac{dP}{dt} = m\left[1 - \exp\left(-\frac{kP^r}{m}\right)\right]\left(1 - \frac{P}{M}\right)P.$$

The combination of just these factors in the growth rate — still ignoring all other factors that affect population growth — gives us a differential equation of such complexity that we have little hope of solving it in the sense of finding a formula for the solution function. We can still separate the variables, as we did with all the other examples, but we have no tools for carrying out the antidifferentiation.

On the other hand, the differential equation

$$\frac{dP}{dt} = m\left[1 - \exp\left(-\frac{kP^r}{m}\right)\right]\left(1 - \frac{P}{M}\right)P$$

defines a perfectly reasonable slope field, and numerical methods for generating solution curves through slope fields are usually not daunted by complexity of the formula defining the slopes. Figure 7.29 shows the slope field defined by the Austin-Brewer model and a solution curve that starts from a population of three billion in 1960. It happens that this curve fits the historical data approximately as well as the coalition model does. Its parameters are $k = 5 \times 10^{-12}$, $r = 1.0$, $m = 0.1$ (10% per year maximum production), and $M = 50 \times 10^9$ (50 billion maximum supportable population).

Figure 7.29 Slope field for the Austin-Brewer model and solution for $P(1960) = 3$ billion

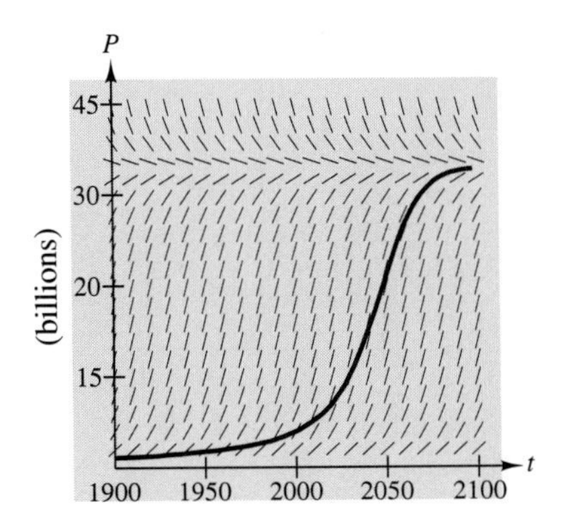

The message of the Austin-Brewer model is that historical population data doesn't necessarily predict infinite population in finite time; an equally good fit can be obtained with finite limits on both the production rate and the size of the population. At the doomsday predicted by the coalition model, the Austin-Brewer model predicts "only" about 10 billion — merely double the 1987 population. But by then the population is growing by at least a billion a year until it gets close to the maximum and starts leveling off. Can we take any comfort in that message if the population still might rise to 50 billion in the 21st century — 10 times the 1987 world population?

CHAPTER

8

The Fundamental Theorem of Calculus

To this point we have concentrated on just one central concept: **instantaneous rate of change** *(the derivative) of a continuously varying function. We have seen some important problems—for example, optimization—for which we need derivatives of known functions, and others—for example, initial value problems—for which we need antiderivatives of known functions. What we needed to know about antiderivatives we learned by studying derivatives.*

In this chapter we take up the second key concept of calculus: **continuous accumulation.** *"Accumulation" is just a fancy word for adding to what you have already. Indeed, it is not an entirely new idea—it appeared, for example, in growth of populations and of interest-bearing accounts. In fact, continuous accumulation can be viewed as a continuously varying function, and thus it has an instantaneous rate of change. That will be the connecting link between our two central concepts. And when the two rather different concepts are connected by that link, we will learn something very important: the Fundamental Theorem of Calculus.*

Before we can link the second concept to the first, we need to know what the second concept is. We know what it means to add things together—but what does it mean to accumulate continuously? We answer that question in a familiar context that involves adding a lot of things together: averaging.

8.1 Averaging Continuous Functions: The Definite Integral

Average Temperature

The weather report in your morning newspaper may contain a statement like, "The average temperature yesterday was 56." Temperature at a particular weather station is a continuously changing quantity. At every time t, there is a definite temperature, namely, the current reading on the weather station's thermometer. What does it mean to average a continuous function?

This is an important question for your gas and electric utility companies and your local heating oil supplier. In fact, the average temperature is the basis for running calculations of energy use based on heating and cooling degree days. By constantly monitoring energy use—at least the part determined by the weather—the utilities can anticipate short-run demand, and the oil supplier will know when to show up to deliver oil.

Exploration Activity 1

Figure 8.1 shows the continuously recorded temperature at Carlisle, PA, for part of January 1993. Estimate the average temperature for Wednesday, January 19.

Figure 8.1 Carlisle, PA, temperatures (°F), January 1993

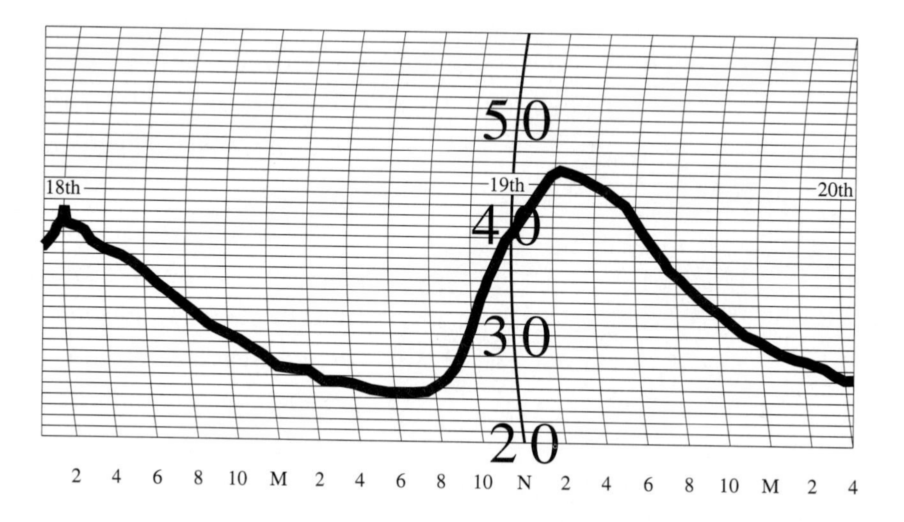

On January 20, 1993, the Carlisle *Sentinel* reported that the previous day's high was 46°F, the low was 24°F, and the average was 35°F. In fact, what the newspaper means by "average" is the average of the high and low temperatures for the day. Even without knowing anything about averaging continuous functions, you probably devised a process for computing a more meaningful average temperature.

For example, you might have estimated temperatures from the chart at two-hour intervals from midnight to midnight and come up with the following numbers:

$$27,\ 26,\ 25,\ 24,\ 25,\ 31,\ 40,\ 46,\ 43,\ 40,\ 35,\ 31,\ 29.$$

The average of these 13 numbers is 32. But wait—in a 24-hour day, there are only 12 two-hour periods. Should we count the midnight reading at both ends? If we drop either the starting or ending midnight reading, the average is 33°F.

The Exploration Activity does not have a unique answer. You might have decided to estimate temperatures every hour and average 24 numbers instead of 12. (And your reading of the graph may have been different from ours—you have to decide what that fuzzy line is really telling you.) You can easily imagine how to modify this procedure to improve your calculation of the average: Use half-hour intervals, quarter-hours, perhaps even readings every minute or every second. This means many more numbers to average, but the procedure is the same: Add them up, and divide by the number of them. If the data are being recorded by computer, that should be easy, even at one-second intervals.

Average Speed and Distance Traveled

We turn our attention now to a more familiar situation: your car's speedometer and odometer. We can only imagine that you have a continuous pen recorder for your speed, but speed has a distinct advantage over temperature because we already know an easy way to compute average speed for a trip—just divide total distance by total time. Whatever "average" means for a continuous function, it should give the same answer.

Exploration Activity 2

Suppose we attach a graphical recorder to the speedometer of your automobile, and on a 13-minute trip between two red traffic lights, it records speed (as a function of time) as shown in Figure 8.2. Roughly how far apart are the two traffic lights?

Figure 8.2 Speed (mph) as a function of time (minutes)

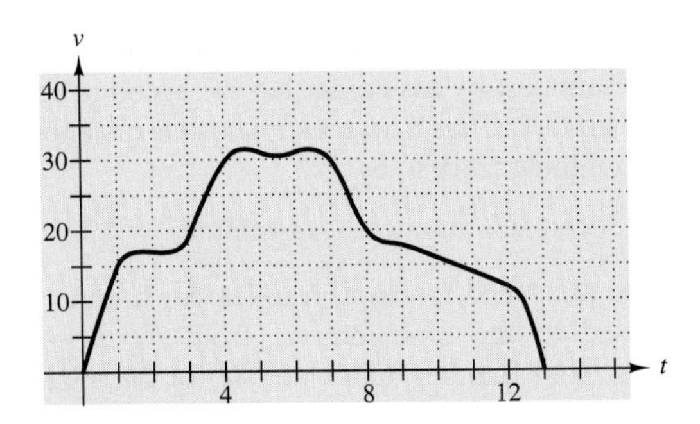

The obvious way to answer the question is to record the odometer readings at the start and end of the trip and subtract them. But we suppose you neglected to do that, so you have to find the answer from the data in Figure 8.2. If you could find the average speed, you would only have to multiply that number by the elapsed time—13 minutes—to get the distance. Conversely, if you can calculate the distance directly from the speed data, you can divide by the time to find the average speed. Thus finding distance and finding average speed are equivalent problems. If we can solve one, we can solve the other.

We will estimate the distance by dividing the time interval into 13 one-minute subintervals, estimating the distance traveled in each of these shorter intervals, and adding the estimates. The speeds at each minute mark are shown in Table 8.1. Since the minute is our unit of time, we have converted the speeds to miles per minute.

Table 8.1 Speed of a car between two traffic lights

Time	*Speed (mph)*	*Speed (miles/min)*
0	0	0
1	15	0.250
2	17	0.283
3	19	0.317
4	30	0.500
5	31	0.517
6	31	0.517
7	30	0.500
8	20	0.333
9	18	0.300
10	16	0.267
11	14	0.233
12	12	0.200
13	0	0

In a single one-minute interval—except for the first and last—there is not much change in speed. Thus we can estimate the distance traveled in that short time by assuming that the speed is constant. If the speed is constant, then that speed is the average speed, and the distance traveled is that speed times the time.

Therefore, we estimate the distance traveled in a particular time interval, say, from 2 minutes to 3 minutes, by assuming that the speed is constant at 0.283 miles per minute for the whole minute. That is, we estimate the distance traveled between the 2-minute mark and the 3-minute mark to be

$$(\text{Speed at 2 minutes}) \times (1 \text{ minute}) = 0.283 \times 1 = 0.283 \text{ miles.}$$

If we designate the speed function in miles per minute by $v(t)$, then our estimate of the distance traveled in the time interval from $t = k - 1$ minutes to $t = k$ minutes is $v(k-1) \cdot 1 = v(k-1)$ miles. Our estimate for the total distance is then the sum of these 13 estimates or

$$\text{Distance} \approx v(0) + v(1) + v(2) + v(3) + \cdots + v(11) + v(12).$$

This is just the sum of the numbers in the third column of Table 8.1 [because $v(13) = 0$], so our estimate of the distance is 4.22 miles.

Checkpoint 1

What is the approximate average speed for the whole trip—as best we can tell at this point? Record your answer both in miles per minute and miles per hour.

Lengthy sums like

$$v(0) + v(1) + v(2) + v(3) + \cdots + v(11) + v(12)$$

will turn up frequently in what follows. Even with only 13 terms, we didn't write all the terms explicitly. When we have sums with hundreds of terms, this wouldn't even be an option. Thus we will make a practice of writing sums in **sigma** notation. If you are unsure what sum is represented by a given sigma notation, you should write out explicitly at least the first few and the last few terms, as we did in our sum of table entries. Here is the same calculation expressed in sigma notation:

$$\text{Distance} \approx \sum_{k=1}^{13} v(k-1) \, .$$

We read the right-hand side of this equation as "the sum of terms of the form $v(k-1)$ where k runs from 1 to 13." Can you see why that means the same thing as the longhand sum at the start of this paragraph?

We would obtain a better estimate of the total distance if we divided the 13 minutes into 26 half-minute intervals. We could then estimate the distance traveled in each half-minute interval as we did before, taking the speed to be constant for that half-minute. The improvement in the estimate arises from the fact that our intervals are shorter than before, so the assumption of constant speed is closer to being true.

Checkpoint 2

In Table 8.2 we reproduce Table 8.1 with space to insert speeds at the half-minute times.

(a) Use Figure 8.2 to estimate the missing speeds in miles per hour.

(b) Convert these speeds to miles per minute.

(c) Estimate the distance traveled in each half-minute, and add the results to get a new estimate of the total distance.

Table 8.2 Speed of a car between two traffic lights

Time	*Speed (mph)*	*Speed (miles/min)*
0	0	0
0.5		
1	15	0.250
1.5		
2	17	0.283
2.5		
3	19	0.317
3.5		
4	30	0.500
4.5		
5	31	0.517
5.5		
6	31	0.517
6.5		
7	30	0.500
7.5		
8	20	0.333
8.5		
9	18	0.300
9.5		
10	16	0.267
10.5		
11	14	0.233
11.5		
12	12	0.200
12.5		
13	0	0

Checkpoint 3

Table 8.2 presents two different functions of time t: the speed of the car in miles per hour and the speed of the car in miles per minute. The former was graphed in Figure 8.2, but the latter is the function we have denoted $v(t)$. In Figure 8.3 we repeat Figure 8.2 without labels on the vertical scale. Fill in the labels so that Figure 8.3 becomes a graph of $v(t)$ as a function of t.

Figure 8.3 Speed $v(t)$ (miles per minute) as a function of time t (minutes)

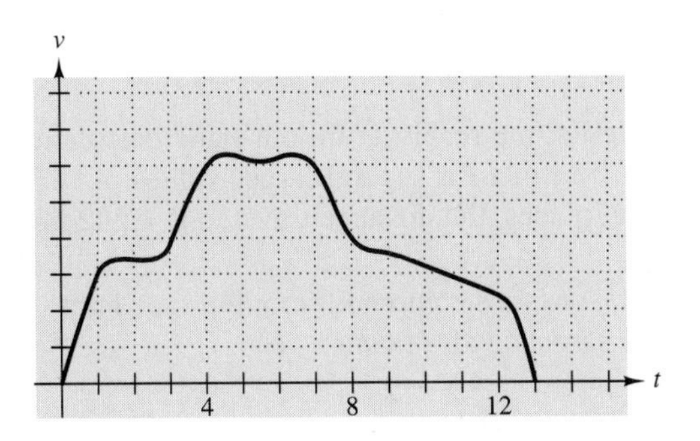

Exploration Activity 3

(a) Show that the sum you calculated in Checkpoint 2 can be written in sigma notation as

$$\sum_{k=1}^{26} v\left(\frac{k-1}{2}\right)\frac{1}{2}.$$

(b) Explain why this is the same as

$$\frac{1}{2}\sum_{k=1}^{26} v\left(\frac{k-1}{2}\right).$$

(c) Which of these expressions best describes what you actually calculated in Checkpoint 2?

(d) Which expression is easier to evaluate?

Because you were working with half-minute time intervals in Table 8.2, each distance estimate was

$$(\text{Starting speed for the interval}) \times (1/2 \text{ second}).$$

The kth interval starts at time $(k-1)/2$. For example, the seventh line in the table has time equal to 3 minutes, and $(7-1)/2 = 3$. Thus the kth starting speed is $v((k-1)/2)$, and the sum in part (a) is the appropriate sum of distance estimates. Each term of that sum has $1/2$ as a factor. The expression in part (b) is the result of factoring out the common factor. Only you can know the answer to part (c). You might have multiplied every speed by $1/2$ to get a distance estimate and added those, or you might have recognized that you could add all the starting speeds and then multiply the sum by $1/2$. The latter procedure is a little simpler, because the sum has the same number of terms, but there is only one multiplication instead of 26.

We can continue dividing the total time into more and more subintervals, in expectation that the resulting estimates would be better and better approximations of the total distance traveled. Here is the procedure: We divide the 13-minute time interval into n subintervals of length $\Delta t = 13/n$ minutes and name the endpoints of the subintervals t_0, t_1, t_2, and so on, up to t_n. For each k from 1 to n, we estimate the distance traveled in the kth subinterval, the one from t_{k-1} to t_k, by assuming the speed throughout that subinterval is constant. That is, we assume $v(t_{k-1})$ is the speed until time t_k. Then our estimate for the total distance traveled is

$$\sum_{k=1}^{n} v(t_{k-1})\,\Delta t.$$

In Table 8.3 we list the results from our hand calculations so far and from evaluation of this summation formula for several larger values of n. We conclude that, to an accuracy of three decimal places, the distance traveled is 4.262 miles.

Table 8.3 Estimates of total distance computed from the speed data in Figure 8.3 by sampling at n points

Value of n	*Estimated distance*	
13	4.22	(From Exploration Activity 2.)
26		(Fill in from Checkpoint 2.)
50	4.259	
100	4.261	
200	4.262	
400	4.262	

Exploration Activity 4

(a) What is the average speed for the trip?

(b) Draw a horizontal line on Figure 8.3 at the level of v equal to this average speed.

(c) What is the area of the rectangle under this line and above the t-axis?

(d) How do you think this area compares to the area under the speed curve and above the t-axis in Figure 8.3? Why?

Since we know the distance traveled in 13 minutes is 4.262 miles, we can calculate the average speed as $4.262/13 = 0.328$ miles per minute, or just under 20 mph. In Figure 8.4 we repeat Figure 8.3 with labels on the vertical axis and a horizontal line at $v = 0.328$. The area of the rectangle under this line is $0.328 \times 13 = 4.262$, or the same as the total distance traveled. A good guess is that this area is exactly the same as the area under the curve. You can see that the two areas are approximately equal by counting rectangles and parts of rectangles in the shaded parts of the figure. Later in this section we will see why the two areas are exactly the same.

Figure 8.4 Actual speed and average speed (miles per minute)

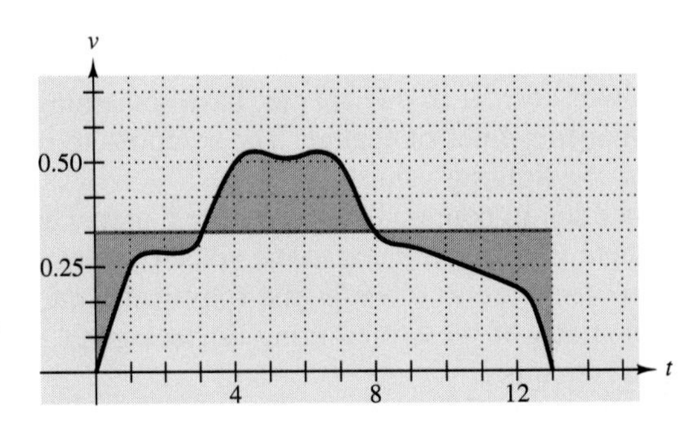

By the way, "area" has to be interpreted carefully here because our horizontal and vertical scales are so different. If we had made the scales the same, the picture would have the appearance of Figure 8.5. This figure obscures the features of the velocity curve that we wanted to see, but the true area you calculated in Exploration Activity 4 is the one shown in Figure 8.5.

Figure 8.5 Figure 8.3 redrawn with equal scales

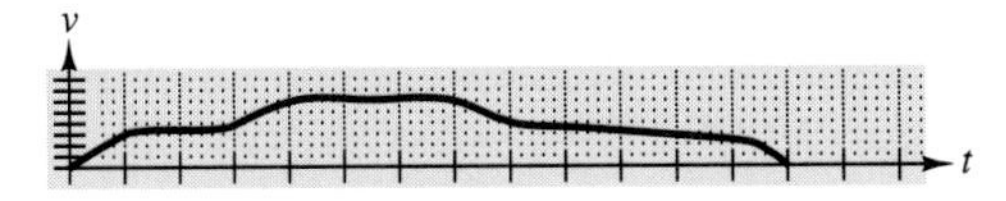

Subdivide and Conquer

Our technique for estimating distance using only speedometer readings can be applied to a wide range of problems. This class of problems can be described in the following way: *If all the measurements involved have constant values, then there is a simple product formula that gives the answer*. Here are some examples:

$$\begin{aligned} \text{distance} &= \text{rate} \times \text{time}. \\ \text{area (of a rectangle)} &= \text{length} \times \text{width}. \\ \text{volume (of a box)} &= \text{length} \times \text{width} \times \text{thickness}. \\ \text{mass} &= \text{density} \times \text{length} \times \text{width} \times \text{thickness}. \\ \text{force} &= \text{pressure} \times \text{area}. \\ \text{momentum} &= \text{mass} \times \text{velocity}. \end{aligned}$$

As was the case with the speed of our car, we often want to consider problems in which some measurement is *not* constant. The first step for calculating such product quantities when one or more of the factor measurements varies is what we call **subdivide and conquer**. If we can subdivide the problem into small pieces in such a way that the simple product formula applies on each piece, then we can add the results on all the pieces to get the answer we are looking for. More typically, we will find that the varying quantity is only *approximately* constant on each piece, so the simple product formula gives us only an *estimate* on each piece. We can still add the results on the individual pieces to get an estimate of the total. But to calculate the total exactly, we have to consider finer and finer subdivisions that produce better and better estimates, eventually arriving at a limiting value.

What takes this class of problems out of the reach of arithmetic or algebra and into the realm of calculus is the need to consider successively finer subdivisions. In this respect, the problems we consider in this chapter have something in common with the problem of finding instantaneous rates of change: The solutions to both types of problems require the determination of a limiting value.

The "something in common" is deeper than that: The distance-from-velocity problem we consider here is closely related to the velocity-from-distance problem in Chapter 2. In fact, these problems are as close as the (physical) connection between the speedometer and odometer. Each of these measuring devices gives enough information to determine the readings of the other.

When we find speeds from measured distances, our procedure is to subtract odometer readings (to find distances) and clock readings (to find elapsed times) and divide the differences. Then we sneak up on the instantaneous speed by taking smaller and smaller time intervals. To go in the other direction — to find distance from measured speeds — we use the opposite procedure: We multiply speeds by elapsed times and add the results. Then we sneak up on the exact distance by using smaller and smaller time intervals.

We already know one way to undo the differentiation process — by antidifferentiation. We have just discovered another way, by multiplying, adding, and finding a limiting value. It should come as no surprise that these two apparently different procedures, both of which have important applications, are closely related. We will find that relationship expressed in the central result of this chapter, the Fundamental Theorem of Calculus.

Area Under a Curve

We prepare the way for the Fundamental Theorem by giving our distance estimates a geometric interpretation. Consider again the estimate of total distance obtained from the 13 one-minute subintervals. The estimate for distance traveled in the particular subinterval $[2, 3]$ is $v(2) \times 1$. This is the area of a rectangle with base from $t = 2$ to $t = 3$ and height $v(2)$. In general, the estimate for distance traveled in the time interval $[k-1, k]$ is the area of the rectangle with base from $t = k - 1$ to $t = k$ and height $v(k-1)$. When we superimpose these rectangles on the graph of $v(t)$, we obtain part (i) of Fig-ure 8.6.

Our estimate using n equal subintervals of total distance traveled is the sum of the areas of n rectangles, the kth one with base from t_{k-1} to t_k and height $v(t_{k-1})$. We show the cases for $n = 26$, $n = 50$, and $n = 100$ in the remaining parts of Figure 8.6.

It is apparent from Figure 8.6 that the number 4.262 toward which the estimates are tending is the area of the region bounded above by the curve $v = v(t)$ and below by the t-axis for $0 \le t \le 13$. Think of approximating the region by placing a large number of rectangular strips side by side. As the number of strips increases, the rectangular polygon formed by them fits the region bounded by the curve better and better. Thus the sum of the areas of the rectangles comes closer and closer to the area of the region under the curve.

Figure 8.6 Geometric interpretation of successive estimates of distance

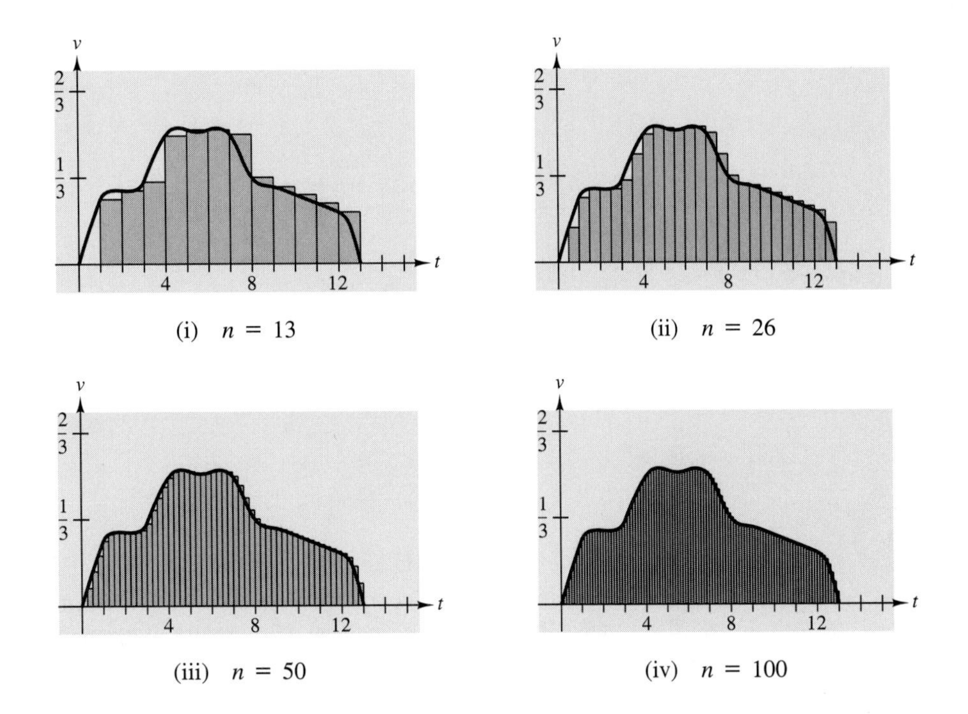

There is nothing special about the particular speed function $v(t)$ in Figure 8.3 or the particular time interval $[0, 13]$. For an arbitrary continuous speed function $v(t)$ and an arbitrary time interval $a \leq t \leq b$, we can describe the distance traveled and its relationship to area in the following way:

The **total distance** traveled by an object moving with speed $v(t)$ from time $t = a$ to time $t = b$ is the limiting value (as n becomes large) of the sums

$$\sum_{k=1}^{n} v(t_{k-1})\,\Delta t\,,$$

where $\Delta t = (b - a)/n$ and $t_k = a + k\Delta t$ for $k = 0, 1, \ldots, n$. Moreover, this distance is also the area of the region in the (t,v)-plane sketched in Figure 8.7.

Figure 8.7 Interpretation of distance as area under the velocity curve

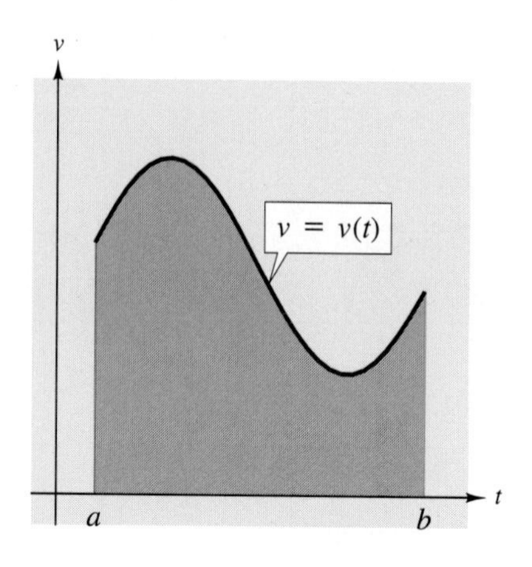

The Definite Integral

The subdivide-and-conquer calculation, which computes distance from velocity and area under the graph of a function, also occurs in many other applications. Thus it is useful to have a notation to summarize it so we don't have to keep repeating the boxed description just given.

Definition If $f(t)$ is a continuous function defined on the interval $a \le t \le b$, then **the definite integral of f from a to b** is the number to which the sums

$$\sum_{k=1}^{n} f(t_{k-1})\,\Delta t$$

converge as n becomes large. Here $\Delta t = (b-a)/n$ and $t_k = a + k\,\Delta t$ for $k = 0, 1, \ldots, n$.

Notation We use the symbol

$$\int_a^b f(t)\,dt$$

to stand for the definite integral of f from a to b.

Our definition of the definite integral uses the phrase "continuous function," which we have used informally on many previous occasions. This concept is an abstraction of the idea of "continuously varying quantity," such as temperature. The technical definition of "continuous" is a little complicated and would interfere with our development of integral calculus without adding much to our understanding at this point. In the next subsection we illustrate the meaning of this word by giving examples of familiar continuous functions.

A dictionary definition of the verb "integrate" is "to make into a whole by bringing all parts together." That's a pretty good description of what we are doing when we chop a computational problem into tiny parts, evaluate each part, and then add the results. The adjective "definite" refers to the appearance of definite endpoints of the interval being subdivided. These endpoints appear in both the computational description and the notation. Often the adjective is dropped, and we speak of **the integral** of a function over an interval. However, we will soon introduce the phrase "indefinite integral" for an entirely different concept, and then we will need the adjectives to distinguish the two concepts.

How do we make sense of the notation? Compare the symbols

$$\int_a^b f(t)\,dt \qquad \text{and} \qquad \sum_{k=1}^{n} f(t_{k-1})\,\Delta t.$$

There is a definite correspondence: The integral sign $\int$ —an elongated S standing for *sum*—corresponds to the sigma notation for the sum on the right. The endpoints a and b correspond to the first and last values of the summation index k, because $a = t_0$ and $b = t_n$. The function values $f(t)$ at every value of t between a and b correspond to the values $f(t_k)$ at equally spaced values of t in the sum. And the differential dt —so far, just part of the notation—corresponds to the difference Δt. The effect of this notation is to suggest the idea of summing infinitely many products of the form $f(t)\,dt$, each of which is infinitely small because of the infinitesimal factor dt. If you remember that this really means the result of a limiting process in which the number n of terms in the sum becomes arbitrarily large, then you will have a sensible meaning for the apparently nonsensical phrase "summing infinitely many infinitely small products." And the integral notation will serve as a constant reminder of the complicated process by which we defined the definite integral—the process by which we bring all the parts together.

Finally, our transformation of the distance problem into an area problem provides us with a visual image of the definite integral (Figure 8.8): If $f(t)$ is nonnegative for all values of t, then $\int_a^b f(t)\,dt$ represents the area of the region in the (t, y)-plane bounded below by the t-axis, bounded on the sides by the lines $t = a$ and $t = b$, and bounded above by the curve $y = f(t)$.

Figure 8.8 If $f(t) \geq 0$, then $\int_a^b f(t)\,dt$ is the area under the curve

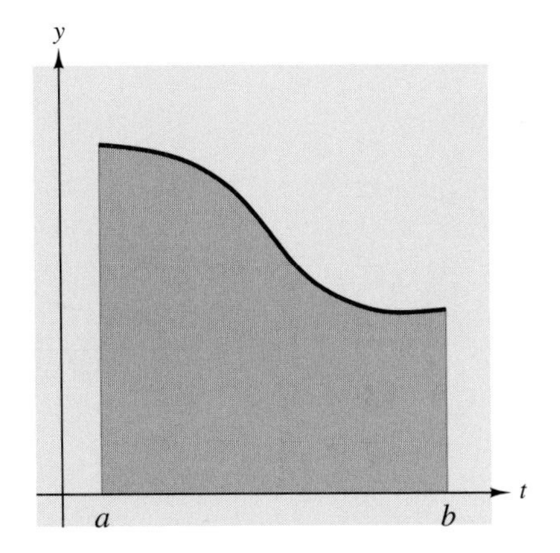

Continuous Functions

As promised in the preceding subsection, we now take up the concept of continuous function. Instead of giving a formal definition, we give some examples of continuous functions. These examples will reassure you that any function you are likely to see as a candidate for integration is, in fact, continuous. They will also emphasize — again — the importance of paying attention to the *domain* of a function, not just to its rule.

First, all *locally linear* functions are continuous. That is, if a function has a derivative at every point of its domain, it is a continuous function. Since most of our formula-based functions are locally linear, you should be reassured already.

Second, there are some familiar functions that are not locally linear (at certain points) but that are nevertheless continuous. For example, the absolute value function, $f(x) = |x|$, is one of these (see Figure 8.9). There is a sharp point on its graph at $(0, 0)$, but there is no break or jump in the graph at that point. This is one way to think of what it means to be continuous: *no breaks or jumps in the graph.*

Figure 8.9 Graph of $f(x) = |x|$

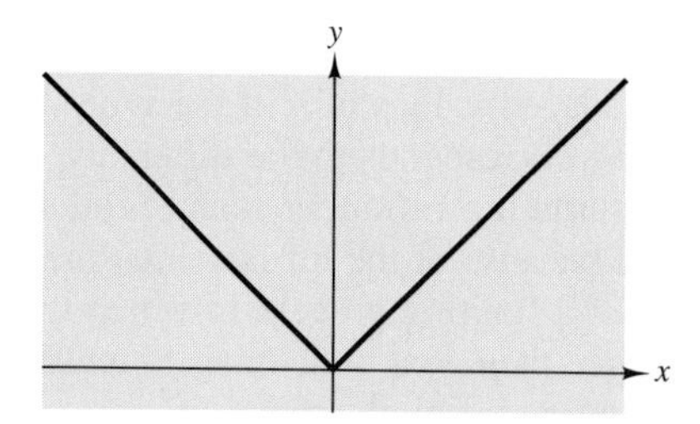

Third, there are some familiar functions whose graphs don't have corners or sharp points but that fail to have derivatives at certain points. For example, consider $g(x) = x^{1/3}$ (Figure 8.10). The derivative of this function (Figure 8.11) is $g'(x) = \frac{1}{3}x^{-2/3}$, which fails to have a value at $x = 0$. In fact, the graph of g is vertical at $x = 0$, so its tangent line at $(0, 0)$ can't have a slope. But again, there is no break or gap, and g is continuous.

Figure 8.10 $g(x) = x^{1/3}$

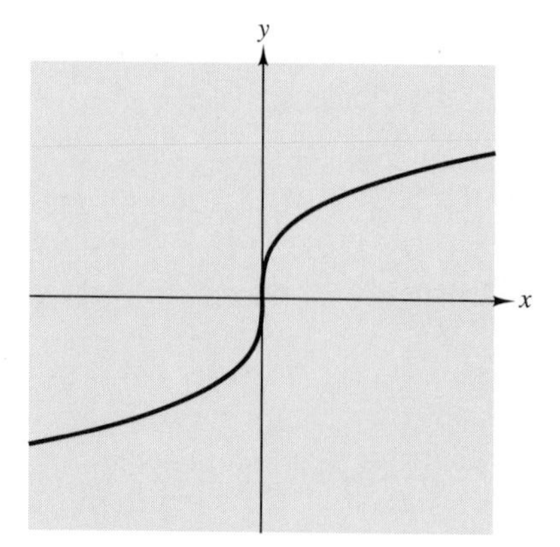

Figure 8.11 $g'(x) = \frac{1}{3}x^{-2/3}$

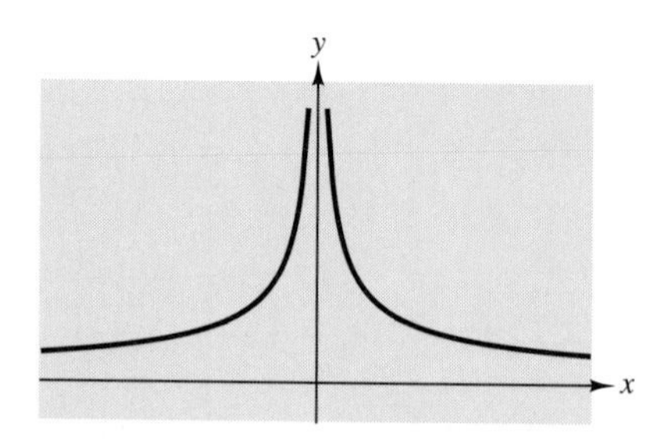

Similarly, the function $h(x) = \sqrt{1 - x^2}$ has a graph that is a semicircle of radius 1 on the domain $[-1, 1]$ (see Figure 8.12). At the endpoints of the domain, the tangent lines are vertical and hence have no slope — but the function is nevertheless continuous on its entire domain.

Figure 8.12 Graph of $h(x) = \sqrt{1 - x^2}$

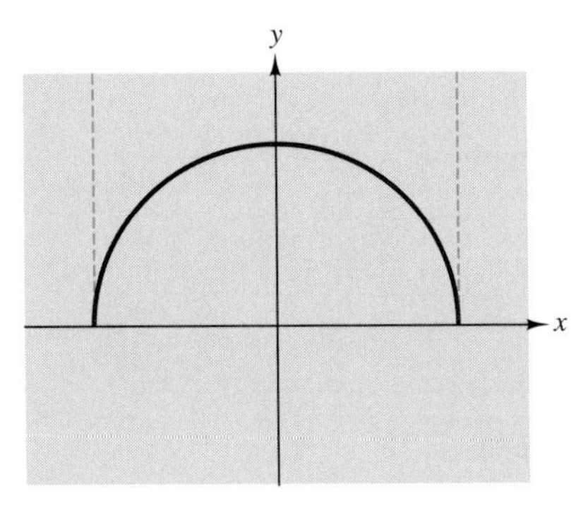

Fourth, and less intuitive, the function $q(x) = 1/x$ is continuous on its domain, even though its graph has a rather large gap in the middle (see Figure 8.13). The point is that 0 is not in the domain of this function. We can integrate this function between numbers a and b, but only if the interval $[a, b]$ does not contain 0. That is, a and b must both be positive or both be negative. On such intervals, the reciprocal function has no breaks or jumps. Similarly, any rational function (i.e., quotient of polynomial functions) is continuous, but only on intervals entirely contained in its domain.

Figure 8.13 Graph of the reciprocal function

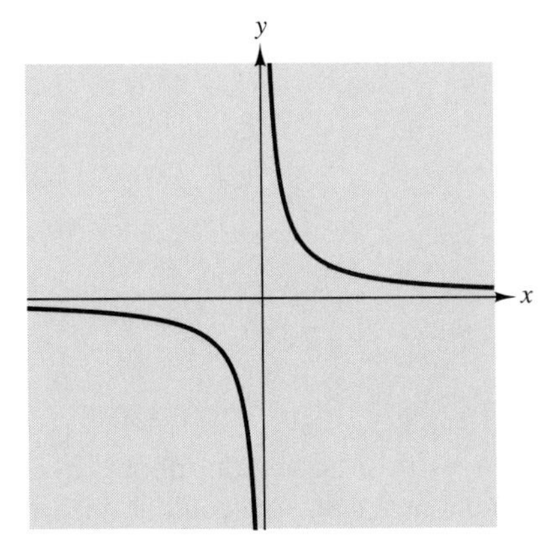

What functions are *not* continuous? A simple example is the first-class postage function (Figure 8.14), which has a value for every positive weight (up to the maximum weight) but which also has obvious jumps at whole numbers of ounces. Similarly, the approximate speed function represented by Table 8.1 — for which we assumed constant speeds for one-minute time intervals—also has jumps at whole numbers of minutes. It would be easy enough to define a definite integral for these *step functions*—just add up rectangular areas. We chose not to include that idea in our definition of the integral because it would complicate the definition without adding any benefit.

Figure 8.14 Graph of the first-class postage function

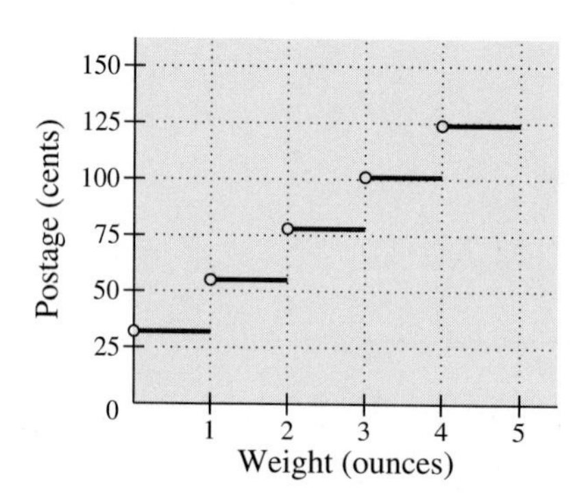

The bottom line is that you are unlikely to encounter a formula-based function f for which $\int_a^b f(t)\,dt$ is not already defined—or could be—as long as you make sure the interval $[a, b]$ is contained in the domain of f.

Checkpoint 4

Graph each of the following functions, and decide whether each function is continuous or not. Give a reason for each of your answers.

(a) $f_1(x) = \sin x$

(b) $f_2(x) = \sin \dfrac{1}{x}, \quad x > 0$

(c) $f_3(x) = \dfrac{1}{3} x^{-2/3}$

(d) $f_4(x) = \dfrac{x^2 + 1}{x^2 - 2}$

Average Value of a Continuous Function

We began this section with a question about averaging a continuously varying temperature function. You saw that you could find a good estimate by adding a large number of closely spaced temperature values and dividing by the number of them. But how do you get the answer when there are literally infinitely many of them?

We actually answered that question in the distance-from-speed calculation, for which we already had a meaning for average speed. In fact, we know that

$$\text{total distance traveled} = (\text{average speed}) \times (\text{length of time interval}),$$

or

$$\text{average speed} = \frac{\text{total distance traveled}}{\text{length of time interval}}.$$

In our new language of integrals, the total distance is the definite integral of the speed function. Thus

$$\text{average speed} = \frac{1}{b-a}\int_a^b v(t)\,dt,$$

where time runs from $t = a$ to $t = b$.

Checkpoint 5

Rewrite the three formulas in the previous paragraph in the context of area.

Definition The **average value** of a continuous function f on an interval $[a, b]$ is

$$\frac{1}{b-a}\int_a^b f(t)\,dt.$$

Checkpoint 6

Estimate the average value of the function $f(t) = t^2$ on the interval $[0, 2]$.

In this section we have seen that we can estimate the average of a continuous function, such as temperature or speed, by adding a large number of closely spaced values and dividing by the number of them. The limiting value of these estimates as the number of terms in the sum becomes arbitrarily large is the area under the graph of the function divided by the length of the interval.

We approached the problem of area under the graph of a nonnegative function in a different way, by adding up areas of thin rectangular strips. In this case, the limiting value — summing infinitely many infinitely small areas — is the concept we named the definite integral of the function. Thus the average value of a continuously varying function can be described as the definite integral of the function divided by the length of the interval.

Answers to Checkpoints

1. 0.324 miles per minute, 19.5 mph

2. (a), (b)

Time	*Speed (mph)*	*Speed (miles/min)*
0	0	0
0.5	8	0.133
1	15	0.250
1.5	17	0.283
2	17	0.283
2.5	17	0.283
3	19	0.317
3.5	25	0.417
4	30	0.500
4.5	31.5	0.525
5	31	0.517
5.5	30.5	0.508
6	31	0.517
6.5	31.5	0.525
7	30	0.500
7.5	25	0.417
8	20	0.333
8.5	18	0.300
9	18	0.300
9.5	17	0.283
10	16	0.267
10.5	15	0.250
11	14	0.233
11.5	13	0.217
12	12	0.200
12.5	9	0.150
13	0	0

(c) 4.25 miles

3.

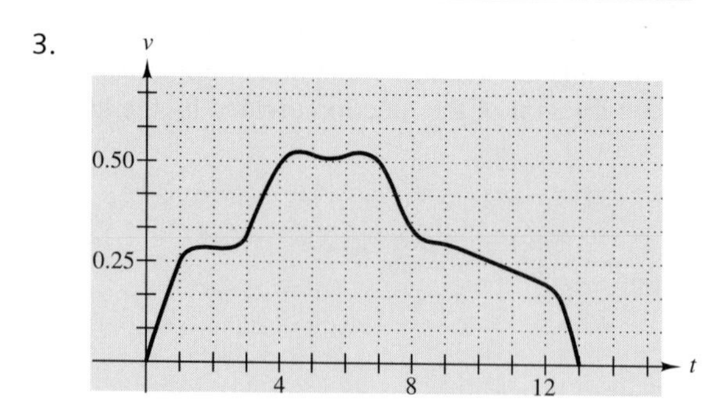

4. (a) Continuous on any interval of positive numbers because it has a derivative at every point of its domain.

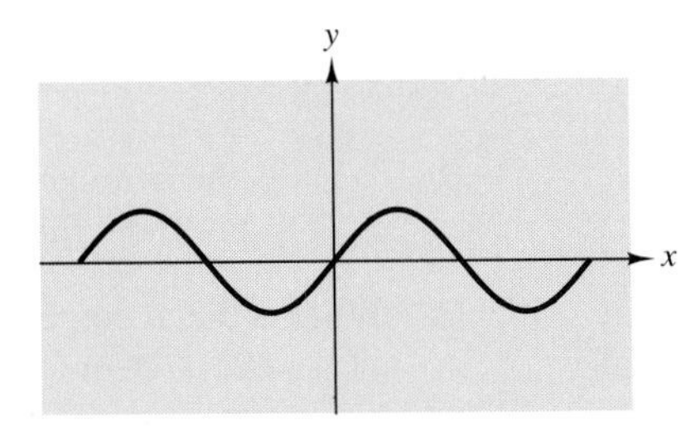

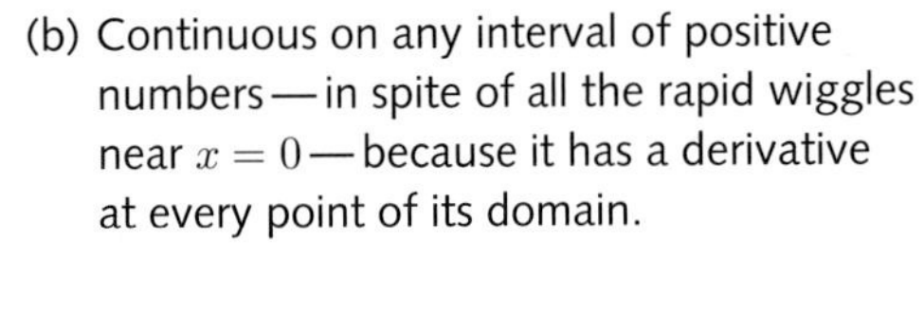
(b) Continuous on any interval of positive numbers—in spite of all the rapid wiggles near $x = 0$—because it has a derivative at every point of its domain.

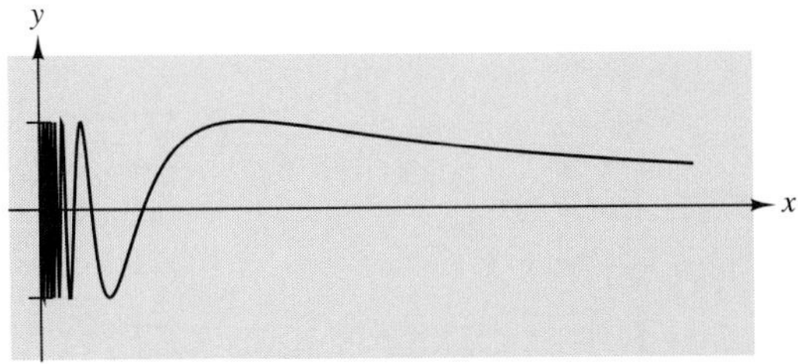

(c) Continuous on any interval that does not include $x = 0$ because it has a derivative at every point of its domain.

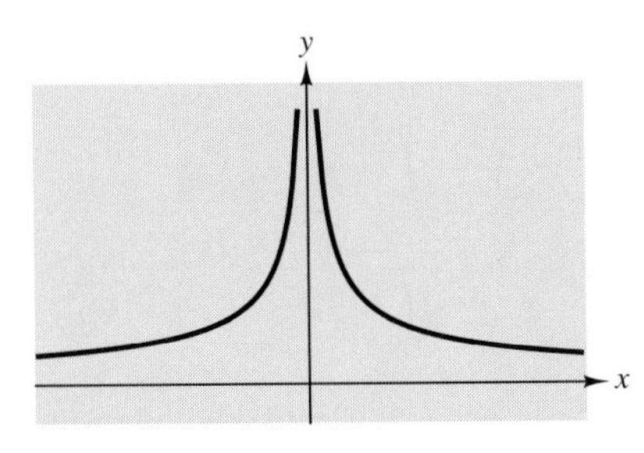

(d) Continuous on any interval that does not include either $x = \sqrt{2}$ or $x = -\sqrt{2}$ because it has a derivative at every point of its domain.

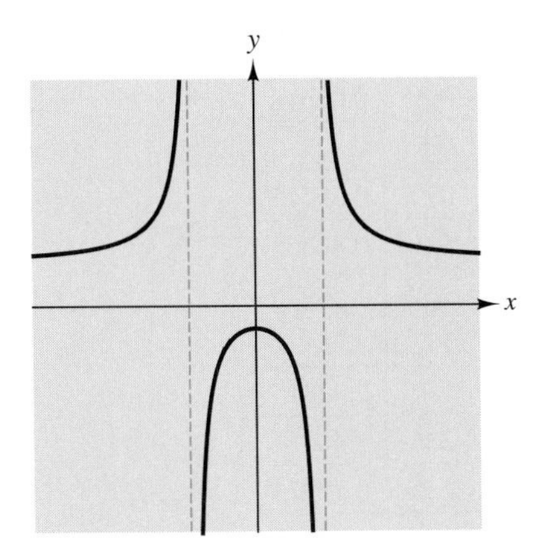

5. Area under curve = (average height) × (length of interval),

$$\text{average height} = \frac{\text{area under curve}}{\text{length of interval}},$$

$$\text{average height} = \frac{1}{b-a}\int_a^b f(t)\,dt.$$

6. If we subdivide the interval into 10 parts and average the values at the left endpoints, the estimate is 1.14. Your answer may be different if you used a different strategy.

Exercises 8.1

Calculate each of the following sums.

1. $\sum_{k=1}^{6} k$
2. $\sum_{k=1}^{6} 2k$
3. $\sum_{k=1}^{6} \frac{k}{2}$
4. $\sum_{k=1}^{6} \frac{k-1}{2}$
5. $\sum_{k=1}^{6} \frac{k+1}{2}$
6. $\sum_{k=1}^{6} \frac{k^2}{2}$

Exercises 7–11 concern a function f that is continuous on the interval $0 \le t \le 3$. Some values of f are given in Table 8.4.

Table 8.4 Values of an unknown function

t	$f(t)$
0	0
0.25	0.01618
0.5	0.125
0.75	0.39775
1.0	0.86603
1.25	1.50926
1.5	2.25
1.75	2.95815
2.0	3.4641
2.25	3.57973
2.5	3.125
2.75	1.95732
3	0

For each of the following values of n, calculate $\sum_{k=1}^{n} f(t_{k-1})\,\Delta t$, where $\Delta t = 3/n$, and $t_k = k\,\Delta t$ for $k = 0, 1, \ldots n$.

7. $n = 3$ 8. $n = 4$ 9. $n = 6$ 10. $n = 12$

11. For the function f whose values are given in Table 8.4, estimate
    (a) $\int_0^3 f(t)\,dt$,
    (b) the average value of f on the interval $[0, 3]$.

12. (a) What is the average value of the function $f(t) = 7$ for $-2 \le t \le 3$? How do you know?
    (b) What is the average value of a constant function $f(t) = c$ on the interval $[a, b]$? Explain.

13. (a) What is the average value of the function $f(t) = 2t$ for $0 \le t \le 3$? How do you know?
    (b) What is the average value of a linear function $f(t) = ct$ on the interval $[0, b]$? Explain.

14. (a) What is the average value of the function $f(t) = 2t + 7$ for $-2 \le t \le 3$? How do you know?
    (b) What is the average value of the function $f(t) = ct + p$ on the interval $[a, b]$? Explain.

Calculate the exact value of each of the following integrals.

15. $\int_{-2}^{3} 7\,dt$
16. $\int_0^3 2t\,dt$
17. $\int_{-2}^{3} (2t+7)\,dt$
18. $\int_a^b c\,dt$
19. $\int_0^b ct\,dt$
20. $\int_a^b (ct+p)\,dt$

21. (a) Divide the interval $[0, 1]$ into four parts, and estimate $\int_0^1 t^3\,dt$.
    (b) Divide the interval $[0, 1]$ into 10 parts, and estimate $\int_0^1 t^3\,dt$.
    (c) Make a guess at the exact value of $\int_0^1 t^3\,dt$.

Estimate each of the following definite integrals.

22. $\int_0^2 \sqrt{x}\,dx$
23. $\int_0^1 x^{4/3}\,dx$
24. $\int_0^2 e^{-2x}\,dx$
25. $\int_0^{0.8} x^4\,dx$
26. $\int_0^{\pi} \sin x\,dx$
27. $\int_{-1}^{1} (1-x^2)\,dx$

Estimate the average value of each of the following functions on the given interval.

28. $\sqrt{x},\ 0 \le x \le 2$
29. $x^{4/3},\ 0 \le x \le 1$
30. $e^{-2x},\ 0 \le x \le 2$
31. $x^4,\ 0 \le x \le 0.8$
32. $\sin x,\ 0 \le x \le \pi$
33. $1 - x^2,\ -1 \le x \le 1$

Learn how to use the integral key on your calculator, and use it to evaluate each of the following definite integrals. Use the results to check your estimates in Exercises 22–27.

34. $\int_0^2 \sqrt{x}\, dx$

35. $\int_0^1 x^{4/3}\, dx$

36. $\int_0^2 e^{-2x}\, dx$

37. $\int_0^{0.8} x^4\, dx$

38. $\int_0^{\pi} \sin x\, dx$

39. $\int_{-1}^1 (1 - x^2)\, dx$

40. Find $\int_a^b |x|\, dx$. Distinguish carefully the following cases.
   (a) $0 \le a \le b$
   (b) $a \le b \le 0$
   (c) $a \le 0 \le b$

41. Let $f(x) = |x|^{1/3}$.
   (a) Graph $f(x)$ on the interval $[-2, 2]$.
   (b) Explain why f is a continuous function.
   (c) Estimate the integral of f from -2 to 2.
   (d) Use the integral key on your calculator to evaluate the integral in part (c).

42. In the discussion of Exploration Activity 1, we observed that the newspaper average temperature—the average of high and low—could be different from an average based on 12 or 13 sample points or from a continuous average.
   (a) Sketch the graph of a continuous function whose average value must be *quite* different from the average of its high and low values on a given interval.
   (b) Explain how you know that the two averages are different.

43. In all our calculations of distance traveled by the car, we estimated the distance for a short time interval by multiplying the speed at the **left** endpoint (i.e., at the start of the interval) by the length of the interval. We could have just as well used the speed at the **right** endpoint of the interval.
   (a) Show that for the speed function in Figure 8.3, the **total** distance estimates would be the same using either choice of constant speed for the subintervals. That is, show that

$$\sum_{k=1}^{n} v(t_{k-1})\, \Delta t = \sum_{k=1}^{n} v(t_k)\, \Delta t.$$

   (b) What feature of this particular trip makes the preceding equation a true statement?
   (c) Would you expect this equation to be true for **every** velocity function on every time interval? Why or why not?

44. Suppose the values of a function $f(t)$ are positive for some values of t and negative for others, as in Figure 8.15. Think about how the negative values of f enter into the calculation of the definite integral and how they relate to areas of rectangles. What geometric interpretation can you give to the value $\int_a^b f(t)\, dt$?

Figure 8.15 How is $\int_a^b f(t)\, dt$ related to the shaded areas?

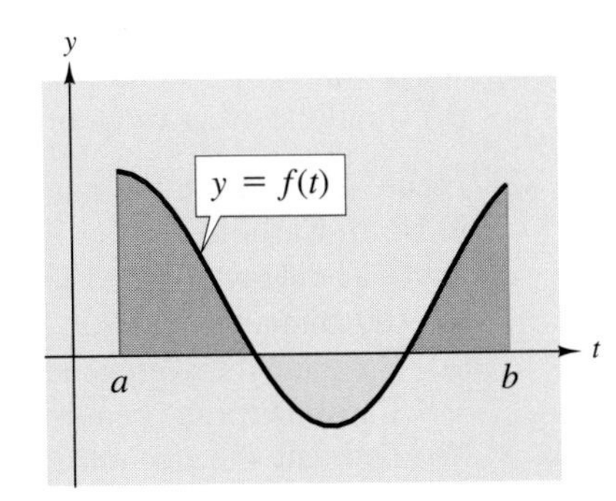

45. Suppose the function $f(t)$ in the previous exercise represents the velocity of a moving object at time t. (Velocity is positive if the object is moving forward, negative if it is moving backward.)
   (a) What physical interpretation can you give to the value $\int_a^b f(t)\, dt$? Is your answer to this question consistent with your answer to Exercise 44?
   (b) What would it mean if the value of the definite integral turned out to be zero? What would that mean in the geometric context of Exercise 44? What would that say about the average value of the function f between a and b?

46. To determine the average temperature for a day, a meteorologist decides to record the temperature at eight different times during the day. She further decides that these recordings do not have to be at equal time intervals, because she does not need several readings during times when the temperature is not changing very rapidly—and she doesn't want to get up in the middle of the night. Her data,

shown in Table 8.5, include one reading in each of the time intervals shown. Use her data to estimate the average temperature for the day.[1]

Table 8.5 Record of sample temperatures

Time	*Temperature*
12–5 A.M.	42°F
5–7 A.M.	57°F
7–9 A.M.	72°F
9 A.M.–1 P.M.	84°F
1–4 P.M.	89°F
4–7 P.M.	75°F
7–9 P.M.	66°F
9–12 P.M.	52°F

47. (a) Find a region of the xy-plane whose area is given by the integral $\int_0^1 \sqrt{1-x^2}\,dx$.
(b) Find the exact value of this integral.

48. Figure 8.16 shows the annual average temperature in North Carolina for the 100-year period ending in 1990. The heavy dashed line shows the average of the 100 annual averages.
(a) Estimate the average temperature in North Carolina for the century 1891–1990.
(b) Estimate the area under the data curve and above the time axis.

Figure 8.16 Average North Carolina temperature (°F), 1891–1990[2]

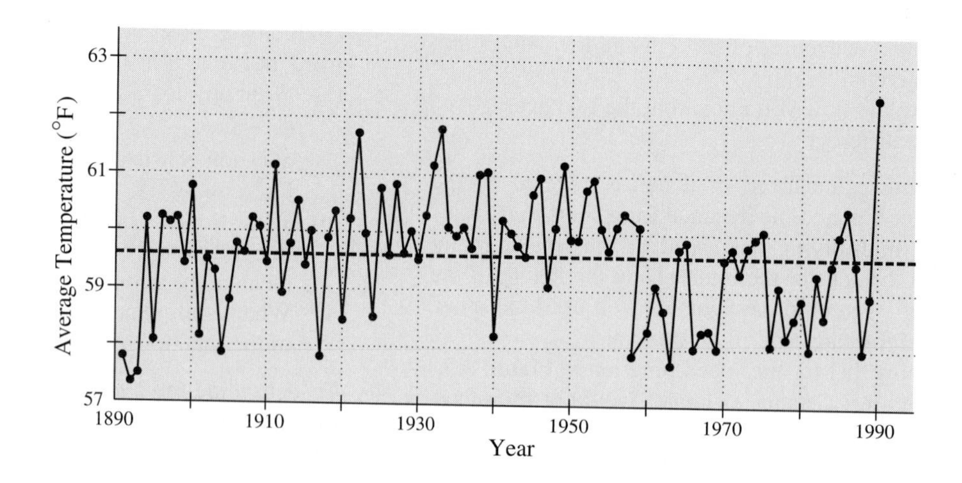

49. Our description of $\int_a^b f(t)\,dt$ for continuous functions works equally well for functions that can be described by a finite family of different continuous functions, one for each of a corresponding family of subintervals constituting $[a,b]$. Such functions are called **piecewise continuous**.

(a) Estimate the integral from 0 to 3 of the function f described by

$$f(t) = \begin{cases} t^2 & 0 \le t \le 1 \\ t-1 & 1 < t \le 2 \\ \sin t & 2 < t \le 3 \end{cases}$$

(b) Use your graphing tool to graph this function. Check that your estimate in part (a) is consistent with the area under the graph of the function. [*Hint*: For some graphing tools, this function can be defined by a formula of the form

$$\begin{aligned} f(t) = t^2 \cdot (0 \le t) \cdot (t \le 1) + \\ (t-1) \cdot (1 < t) \cdot (t \le 2) + \\ (\sin t) \cdot (2 < t) \cdot (t \le 3). \end{aligned}$$

Each of the inequality expressions in parentheses has the value 1 if the inequality is satisfied and 0 if it is not.]

50. Among the piecewise continuous functions, in the sense of the preceding exercise, are the *step functions*, that is, functions that are constant on particular intervals but not necessarily over the whole domain. An example of such a function is the first-class postage function, $p = f(w)$, where w is the weight of a letter in ounces, and p is the postage in cents. As of 1995, this function has the following values: For $0 < w \leq 1$, $p(w) = 32$. For each additional ounce, add 23 cents. Find

$$\int_0^5 f(w)\, dw.$$

1. Adapted from *Calculus Problems for a New Century*, edited by R. Fraga, MAA Notes no. 28, 1993.
2. Source: Southeast Regional Climate Center, as reported in *The News & Observer* (Raleigh, NC), December 30, 1990.

8.2 The Fundamental Theorem of Calculus: Evaluation of Integrals

Speedometer and Odometer

In Section 8.1 we worked very hard to solve the problem of finding distance from velocity and then to abstract the procedure into the concept of definite integral, which we can use to find average values of continuous functions. But we already knew how to get a distance function from a velocity function by a somewhat simpler process, that of antidifferentiation. What is the connection between the two ideas, integration and antidifferentiation?

We can answer this question by looking again at the distance-from-velocity problem, for which our two answers appear to be very different. The fact that they are the same will show us what the connection must be—not just in that specific context, but in a setting that applies to any continuous function.

Recall that, if $s(t)$ is the function giving the distance traveled at any time t after a starting time $t = a$, then $s(t)$ is the solution of the initial value problem

$$\frac{ds}{dt} = v(t)\,, \qquad s(a) = 0.$$

The total distance traveled from time $t = a$ to time $t = b$ is just $s(b)$. If we know the function that solves this initial value problem—a function that is a particular antiderivative of v—then we know the distance traveled at any time between a and b.

We can set the trip meter in our car to zero at the start of the trip, and the meter will record $s(t)$ as we travel. In particular, its reading at the end of the trip will be $s(b)$. On the other hand, we don't actually need to know the distance function to find distance traveled. That is, instead of using a trip meter, we can use the odometer. Suppose we let $O(t)$ stand for the odometer reading at time t —which is rarely zero in a real car. This function is also the solution of an initial value problem similar to the one that defines $s(t)$:

$$\frac{dO}{dt} = v(t)\,, \qquad O(a) = (\text{odometer reading at time } t = a).$$

If we record the odometer readings at the beginning and end of a trip (at times $t = a$ and $t = b$, respectively), then we compute the distance traveled as $O(b) - O(a)$.

The differential equations $s'(t) = v(t)$ and $O'(t) = v(t)$ are the same—each says that the derivative of the unknown function is the known function $v(t)$. [Think of $v(t)$ as the continuously recorded speedometer reading on the pen recorder.] Thus each initial value problem is asking for an antiderivative of $v(t)$. Since the odometer reading at time a can be anything, we are really asking for *any* antiderivative of $v(t)$. When we antidifferentiate, we find a family of functions of the form $F(t) + C$, where F is any function whose derivative is v, and C is any constant. In principle, every such function could be an odometer function, so the total distance traveled must be $[F(b) + C] - [F(a) + C]$, which is the same as $F(b) - F(a)$.

Notice that the answer obtained from this calculation does not depend on C. We know that from experience—no matter what the odometer says at the start of a trip, we can always use that number and the reading at the end of the trip to figure out how far we went.

The other answer we know to the distance-from-velocity question is the definite integral. That brings us to our first statement of an important part of the Fundamental Theorem:

$$\int_a^b v(t)\,dt = F(b) - F(a),$$

where F is any function whose derivative is v.

The First Half of the Fundamental Theorem

Our next task is to show that the relationship described in the equation above is not special to velocity-distance calculations but holds for integrals in general. To do this, we consider an approximate solution to the initial value problem

$$\frac{ds}{dt} = v(t)\,, \qquad s(a) = 0.$$

by means of Euler's Method.

Recall that Euler's Method generates a sequence of numbers s_1, s_2, s_3, and so on that approximates values of $s(t)$ at times t_1, t_2, t_3, and so on. Specifically, we can divide the interval $[a, b]$ into n subintervals by taking $\Delta t = (b - a)/n$ and choosing the points of subdivision to be

$$\begin{aligned} t_0 &= a, \\ t_1 &= a + \Delta t, \\ t_2 &= a + 2\Delta t, \\ t_3 &= a + 3\Delta t, \end{aligned}$$

and so on, up to

$$t_n = b.$$

Then, for $k = 1, 2, \ldots, n$, [starting with $s_0 = s(a) = 0$], each s_k is generated from the previous one by the "rise = slope × run" formula:

$$s_k = s_{k-1} + v(t_{k-1})\,\Delta t.$$

To make the connection between the Euler's Method approximate solution to the initial value problem and the definite integral in

$$\int_a^b v(t)\,dt = F(b) - F(a),$$

we write out explicitly the values of the s_k's:

$$\begin{aligned} s_1 &= s_0 + v(t_0)\,\Delta t = v(t_0)\,\Delta t, \\ s_2 &= s_1 + v(t_1)\,\Delta t = v(t_0)\,\Delta t + v(t_1)\,\Delta t, \\ s_3 &= s_2 + v(t_2)\,\Delta t = v(t_0)\,\Delta t + v(t_1)\,\Delta t + v(t_2)\,\Delta t \end{aligned}$$

and so on, up to

$$s_n = s_{n-1} + v(t_{n-1})\,\Delta t = v(t_0)\,\Delta t + v(t_1)\,\Delta t + \cdots + v(t_{n-1})\,\Delta t.$$

In sigma notation, this last equation is the same as

$$s_n = \sum_{k=1}^{n} v(t_{k-1})\,\Delta t\,.$$

That is, s_n, the Euler approximation to $s(t_n)$ [which is the same as $s(b)$], is *exactly* equal to the left-hand sum with n terms that approximates the definite integral of $v(t)$ from a to b.

Now, as n becomes large — and the step size Δt becomes small — two things happen to the equation

$$s_n = \sum_{k=1}^{n} v(t_{k-1})\,\Delta t\,.$$

First, the Euler approximation to $s(t)$ gets closer and closer to the true solution of the initial value problem

$$\frac{ds}{dt} = v(t), \qquad s(a) = 0.$$

In particular, s_n approaches $s(b)$. Second, the sum on the right-hand side approaches the definite integral of $v(t)$ from a to b. When we consider limiting values, we find

$$s(b) = \int_a^b v(t)\,dt.$$

This observation is not limited to the relationship between distance and velocity. In particular, nowhere in this discussion did we use any particular physical interpretation of the functions $v(t)$ and $s(t)$.

Checkpoint 1

Suppose f is any continuous function on the interval $[a,b]$. Explain why $\int_a^b f(t)\,dt = F(b)$, where $F(t)$ is the solution of the initial value problem

$$\frac{dF}{dt} = f(t), \qquad F(a) = 0.$$

We are now ready to examine the relationship between antidifferentiation and the definite integral. Suppose we want to calculate $\int_a^b f(t)\,dt$ and we know *one* antiderivative for $f(t)$ is $G(t)$. Then the solution $F(t)$ of the initial value problem

$$\frac{dF}{dt} = f(t), \qquad F(a) = 0,$$

is a function that differs from $G(t)$ by a constant. That constant might be zero, but all we know for sure is that $F(t) = G(t) + C$ for some constant C. Since $F(a) = 0$, we know that $0 = G(a) + C$, or $C = -G(a)$.

Checkpoint 2

Review Checkpoint 1 and the paragraph you just read. Then write a reason for each of the following statements.

$$\int_a^b f(t)\,dt = F(b)$$

$$F(b) = G(b) + C$$

$$F(b) = G(b) - G(a)$$

$$\int_a^b f(t)\,dt = G(b) - G(a)$$

This last equation is part of the **Fundamental Theorem of Calculus**. It says that one way to find the value of a definite integral $\int_a^b f(t)\,dt$ is to find any antiderivative $G(t)$ of $f(t)$ and then calculate $G(b) - G(a)$.

Example 1 Calculate $\displaystyle\int_0^1 t^3\,dt$.

Solution We need a function that has t^3 as its derivative. You already know that one such antiderivative is $t^4/4$. It follows that

$$\int_0^1 t^3\,dt = \frac{1}{4} - \frac{0}{4} = \frac{1}{4}.$$

■

Note that we could have used $t^4/4 + 7$ as an antiderivative, in which case the computation would have been

$$\int_0^1 t^3 dt = \left(\frac{1}{4} + 7\right) - \left(\frac{0}{4} + 7\right) = \frac{1}{4}.$$

Of course, if we had used any other antiderivative of t^3, the result would have been the same.

In this section we have begun to make the connection between the processes of antidifferentiation and integration. In so doing, we have established part of the

Fundamental Theorem of Calculus. In particular, we have found that one way to evaluate a definite integral $\int_a^b f(t)\,dt$ is to find an antiderivative $F(t)$ of $f(t)$ and then calculate $F(b) - F(a)$.

Answers to Checkpoints

1. The discussion preceding the Checkpoint works exactly the same way with f replacing v and F replacing s.
2. This is the result of Checkpoint 1.
   $F(t) = G(t) + C$ for all t in $[a, b]$—in particular, for $t = b$.
   We know that $C = -G(a)$.
   This is the result of substituting from the third line into the first.

Exercises 8.2

Use the Fundamental Theorem of Calculus to evaluate each of the following definite integrals.

1. $\int_1^3 t^3\,dt$
2. $\int_{-3}^0 t^3\,dt$
3. $\int_{-1}^3 t^3\,dt$
4. $\int_1^3 t^4\,dt$
5. $\int_{-3}^0 t^4\,dt$
6. $\int_{-1}^3 t^4\,dt$
7. $\int_0^{\pi/2} \cos t\,dt$
8. $\int_0^{\pi/2} 5\cos t\,dt$
9. $\int_0^{\pi/2} \sin t\,dt$
10. $\int_0^{\pi/2} 5\sin t\,dt$
11. $\int_1^3 \frac{1}{t}\,dt$
12. $\int_2^4 \frac{1}{t}\,dt$
13. $\int_2^4 \frac{1}{t+1}\,dt$
14. $\int_3^5 \frac{1}{t+1}\,dt$
15. $\int_0^1 e^t\,dt$
16. $\int_{-1}^1 e^t\,dt$
17. $\int_0^1 e^{-t}\,dt$
18. $\int_{-1}^1 e^{-t}\,dt$

For each of the following definite integrals,
(a) use the integral key on your calculator to evaluate the integral,
(b) graph the function being integrated on the indicated interval, and
(c) interpret the value of the integral in terms of area.

19. $\int_1^3 t^3\,dt$
20. $\int_{-3}^0 t^3\,dt$
21. $\int_{-1}^3 t^3\,dt$
22. $\int_1^3 t^4\,dt$
23. $\int_{-3}^0 t^4\,dt$
24. $\int_{-1}^3 t^4\,dt$
25. $\int_0^{\pi/2} \cos t\,dt$
26. $\int_0^{\pi/2} 5\cos t\,dt$
27. $\int_0^{\pi/2} \sin t\,dt$
28. $\int_0^{\pi/2} 5\sin t\,dt$
29. $\int_1^3 \frac{1}{t}\,dt$
30. $\int_2^4 \frac{1}{t}\,dt$
31. $\int_2^4 \frac{1}{t+1}\,dt$
32. $\int_3^5 \frac{1}{t+1}\,dt$
33. $\int_0^1 e^t\,dt$
34. $\int_{-1}^1 e^t\,dt$
35. $\int_0^1 e^{-t}\,dt$
36. $\int_{-1}^1 e^{-t}\,dt$

Find the average value of each of the following functions on the indicated interval.

37. t^3 on $[1, 3]$
38. t^3 on $[-3, 0]$
39. t^3 on $[-1, 3]$
40. t^4 on $[1, 3]$
41. t^4 on $[-3, 0]$
42. t^4 on $[-1, 3]$
43. $\cos t$ on $[0, \pi/2]$
44. $5\cos t$ on $[0, \pi/2]$

45. $\sin t$ on $[0, \pi/2]$

46. $5\sin t$ on $[0, \pi/2]$

47. $\dfrac{1}{t}$ on $[1, 3]$

48. $\dfrac{1}{t}$ on $[2, 4]$

49. $\dfrac{1}{t+1}$ on $[2, 4]$

50. $\dfrac{1}{t+1}$ on $[3, 5]$

51. e^t on $[0, 1]$

52. e^t on $[-1, 1]$

53. e^{-t} on $[0, 1]$

54. e^{-t} on $[-1, 1]$

55. (a) Find the average value of each of the following functions on the interval $[-1, 1]$.
    (i) $f(x) = x$
    (ii) $g(x) = x^3 - x$
    (iii) $h(x) = \sin x$
    (b) Why are all these so easy?

8.3 Powerful Notation: The Indefinite Integral

The Fundamental Theorem of Calculus gives added importance to the process of finding antiderivatives. You are probably tired of phrases such as "any antiderivative of e^{2t}." The process of antidifferentiation, like that of differentiation, has a notation to replace all those words, and it arises from the Fundamental Theorem, in which antidifferentiation is the most important step in the calculation of a definite integral. We refer to this main step in integration (via the Fundamental Theorem) as **indefinite integration**, and we give it a notation that suggests "do everything except the calculation with the endpoints":

$$\int f(t)\, dt.$$

There is a subtlety in this notation about which we need to be clear: It does not distinguish between some *particular* antiderivative of $f(t)$ and *all* the antiderivatives of $f(t)$. By convention, it usually stands for the latter.

Definition The **indefinite integral** of a given function means the family of all functions whose derivatives are the given function.

If $F(t)$ is any one antiderivative of a given function, then the entire family can be represented by $F(t) + C$, where C is an arbitrary constant. We illustrate the use of the indefinite integral notation in the following example.

Example 1 Find the indefinite integral of $f(t) = t^3$.

Solution

$$\int t^3 dt = \tfrac{1}{4}t^4 + C$$

■

This equation can be read "the antiderivatives of t^3 are all the functions of the form $t^4/4 + C$, where C is any constant." But we usually read it "the integral of t^3 is $t^4/4 + C$." The missing adjective here is "indefinite." Indeed, the statement and the notation have nothing directly to do with the *definite* integral introduced in Section 8.1. Contrast this with the statement "the integral of t^3 from 1 to 3 is 20," which we would write symbolically as

$$\int_1^3 t^3\,dt = 20.$$

In this case, the missing adjective is "definite," and it is signaled by the endpoints, "from 1 to 3."

The definite integral of a function over a particular interval is a *number*—think of area under a curve—whereas the indefinite integral is a whole family of *functions*—think of parallel curves passing through a slope field. The connecting link between the two is the Fundamental Theorem. Thus, to find $\int_1^3 t^3\,dt$, we pick any convenient antiderivative of t^3, say, $G(t) = t^4/4$, and we calculate

$$G(3) - G(1) = \frac{81}{4} - \frac{1}{4} = 20.$$

Finally, one more notational convenience will make the preceding calculation even simpler. The expression $G(b) - G(a)$, which turns up frequently in integral calculations, is often written

$$G(t)\Big|_a^b$$

which is read, "$G(t)$ evaluated from a to b." This notation has the advantage of allowing us to display explicitly the formula for G, not just a difference of numerical values. And, as we see in the next example, we don't even have to name our antiderivative.

Example 2 Use the evaluation notation with an antiderivative formula to calculate

$$\int_1^3 t^4\,dt.$$

Solution

$$\int_1^3 t^4\,dt = \frac{1}{5}t^5\Big|_1^3 = \frac{243}{5} - \frac{1}{5} = 48.4$$

Recall that the notation for the definite integral makes sense as a *suggestion* of summing infinitely many infinitely small terms of the form $f(t_k)\,\Delta t$. The notation for the indefinite integral also makes sense as a description of a *procedure* by which we can evaluate definite integrals without doing all the hard work of evaluating lengthy sums and finding a limiting value:

$$\int_a^b f(t)\,dt = \int f(t)\,dt\,\Big|_a^b.$$

The right side of this equation has to be interpreted to mean the same thing as the right side of

$$\int_a^b f(t)\,dt = G(b) - G(a),$$

that is, find an antiderivative G of f, and calculate $G(b) - G(a)$.

The indefinite integral notation is powerful in the sense that it enables us to carry out intricate, multistep calculations without writing down any words, and often without writing many of the intermediate steps. We will see later that the notation is even more powerful than it may now appear, especially when we interpret it in its differential form—using the dt—which we have so far ignored.

On the other hand, powerful notation also can be confusing, especially if the concepts it embodies are not yet familiar. The most confusing feature of the indefinite integral and its notation is that it obscures a very important fact: **The definite integral**—a limiting value of sums of products of a certain very special type—**is a completely different concept from the indefinite integral**—a new name for antidifferentiation. It is only because of a very remarkable fact—the Fundamental Theorem of Calculus—that tradition assigns similar names and notations to these very dissimilar ideas. Now that you have the powerful notations at your disposal, it is your responsibility to associate the appropriate concept with each notation.

In this section we have introduced the standard name and notation for the long-familiar process of antidifferentiation: the indefinite integral. This name and notation are suggested by the Fundamental Theorem of Calculus, which we have seen provides a way to evaluate definite integrals by first finding an antiderivative. We also introduced the evaluation bar notation, which, together with the indefinite integral, allows us to write this aspect of the Fundamental Theorem in a single compact formula:

$$\int_a^b f(t)\,dt = \int f(t)\,dt \Big|_a^b .$$

Exercises 8.3

Evaluate each of the following indefinite integrals.

1. $\int t^4\,dt$
2. $\int t^5\,dt$
3. $\int (t+1)^5\,dt$
4. $\int \sin t\,dt$
5. $\int \sin 2t\,dt$
6. $\int \cos t\,dt$
7. $\int \cos 2t\,dt$
8. $\int e^t\,dt$
9. $\int 3e^t\,dt$
10. $\int e^{3t}\,dt$
11. $\int \frac{1}{t}\,dt$
12. $\int \frac{1}{t+1}\,dt$
13. $\int \sqrt{t}\,dt$
14. $\int \frac{1}{\sqrt{t}}\,dt$
15. $\int t^{1/3}\,dt$

Evaluate each of the following definite integrals. Practice using the notation introduced in this section. Use the integral key on your calculator to check your evaluations.

16. $\int_{-1}^{1} t^4\,dt$
17. $\int_{-1}^{1} t^5\,dt$
18. $\int_{-1}^{1} (t+1)^5\,dt$
19. $\int_0^{\pi/2} \sin t\,dt$
20. $\int_0^{\pi/2} \sin 2t\,dt$
21. $\int_0^{\pi/2} \cos t\,dt$
22. $\int_0^{\pi/2} \cos 2t\,dt$
23. $\int_{-1}^{1} e^t\,dt$
24. $\int_{-1}^{1} 3e^t\,dt$
25. $\int_{-1}^{1} e^{3t}\,dt$
26. $\int_1^3 \frac{1}{t}\,dt$
27. $\int_1^3 \frac{1}{t+1}\,dt$

28. $\int_1^3 \sqrt{t}\, dt$

29. $\int_1^3 \frac{1}{\sqrt{t}}\, dt$

30. $\int_1^3 t^{1/3}\, dt$

Find the average value of each of the following functions on the indicated interval.

31. t^4 on $[-1, 1]$

32. t^5 on $[-1, 1]$

33. $(t+1)^5$ on $[-1, 1]$

34. $\sin t$ on $[0, \pi/2]$

35. $\sin 2t$ on $[0, \pi/2]$

36. $\cos t$ on $[0, \pi/2]$

37. $\cos 2t$ on $[0, \pi/2]$

38. e^t on $[-1, 1]$

39. $3e^t$ on $[-1, 1]$

40. e^{3t} on $[-1, 1]$

41. $\frac{1}{t}$ on $[1, 3]$

42. $\frac{1}{t+1}$ on $[1, 3]$

43. $\sqrt{t}$ on $[1, 3]$

44. $\frac{1}{\sqrt{t}}$ on $[1, 3]$

45. $t^{1/3}$ on $[1, 3]$

8.4 The Fundamental Theorem of Calculus: Representation of Functions

The More Important Half of the Fundamental Theorem

Sometimes the procedure

$$\int_a^b f(t)\, dt = \int f(t)\, dt \,\Big|_a^b$$

provided by the Fundamental Theorem is an enormous help in evaluating definite integrals. But for this to be the case, an antiderivative must be relatively accessible. Often we will find that this is not the case. Finding an antiderivative may be extremely difficult or even impossible if what we mean is finding a simple formula into which we can substitute a and b.

Often our real problem *is* to find an antiderivative—for example, if we are solving a differential equation. In this section we will see that the Fundamental Theorem can be an enormous help in that direction as well, showing us how to find antiderivatives when we have some other way to evaluate definite integrals.

As you have seen already, there *are* other ways to evaluate definite integrals. For example, almost every calculator or mathematical computer system has an integral button or menu choice. Thus, with the aid of technology, we can consider the problem of evaluating a definite integral to be solved—without any help from the Fundamental Theorem.

In a later chapter and in a laboratory exercise associated with this chapter we explore ways to get quick and accurate approximations of integrals—possibly including ways that may be programmed into your calculator or computer. For now we consider only the simple approximation on which we based the definition of the definite integral. We can estimate the value of

$$\int_a^b f(t)\, dt$$

by calculating the **left-hand sum**

$$\sum_{k=1}^{n} f(t_{k-1})\,\Delta t$$

for a large value of n. In Section 8.2 we observed that the process of forming left-hand sums coincides with the process of solving an initial value problem by Euler's Method, and we used that connection to relate evaluation of a definite integral to antidifferentiation. Now we look at the same process and connection, but with a slightly different interpretation—one that arises naturally if the definite integral problem is solved and the antidifferentiation problem is unsolved.

We think first of accumulating the sum $\sum_{k=1}^{n} f(t_{k-1})\,\Delta t$ one term at a time—just as you would if you were adding up the terms with a calculator. We begin with a starting sum of zero, which we call s_0. Then we add the first term $f(t_0)\,\Delta t$ to s_0 to get a first sum s_1:

$$s_1 = s_0 + f(t_0)\,\Delta t.$$

Now we add the second term, $f(t_1)\,\Delta t$, to get the second sum:

$$s_2 = s_1 + f(t_1)\,\Delta t.$$

We continue in this manner, adding a term at a time. At the kth step, we have the kth sum:

$$s_k = s_{k-1} + f(t_{k-1})\,\Delta t.$$

As we have observed already, this equation is Euler's Method for solving an initial value problem: The rise from s_{k-1} to s_k is obtained by multiplying the slope $f(t_{k-1})$ by the run Δt. Specifically, the problem whose solution is being approximated is

$$\frac{ds}{dt} = f(t), \qquad s(a) = 0.$$

In the abstract, we know the solution of this initial value problem: $s(t)$ is the antiderivative of $f(t)$ whose value is 0 at the left endpoint of the interval over which we are integrating $f(t)$.

Now, instead of using this connection between integral and antiderivative only at the right endpoint b, let's think about the function s—a particular antiderivative of f—over the entire interval from a to b. In particular, we want to find a way (other than solving the initial value problem explicitly) to calculate the values of s. We let x stand for a number in the interval $[a, b]$—a number that is temporarily fixed, but arbitrary. If the number n of steps in the summation $\sum_{k=1}^{n} f(t_{k-1})\,\Delta t$ is very large, we can find a t_k as close to x as we like. Thus after k steps the sum will approximate

$$\int_a^x f(t)\,dt.$$

But the sum also approximates $s(x)$, so these two expressions must be equal:

$$s(x) = \int_a^x f(t)\,dt.$$

That's it! To find a value of s at a particular number x, we integrate f—with help from a computer or calculator, if necessary—from a to x. And this gives us a way to find a particular antiderivative of f. If we need some other antiderivative of f, it must be of the form $s(x) + C$ for some constant C.

Example 1 Solve the initial value problem

$$\frac{ds}{dt} = \sin t, \qquad \text{with } s(0) = 0,$$

both by indefinite and definite integration.

Solution First we use indefinite integration,

$$s(t) = \int \sin t \, dt = -\cos t + C,$$

for some constant C. To match the initial condition, we must have

$$0 = s(0) = -\cos 0 + C,$$

so $C = 1$. Thus the unique solution is $s(t) = 1 - \cos t$.

To find the solution by definite integration, we calculate

$$s(x) = \int_0^x \sin t \, dt = -\cos t \Big|_0^x = -\cos x + 1,$$

which is clearly the same solution. ■

Checkpoint 1

Solve the initial value problem

$$\frac{ds}{dt} = e^{2t}, \qquad \text{with } s(0) = 0,$$

both by indefinite and definite integration.

The Fundamental Theorem of Calculus

Part I. If f is a continuous function from a to b, then f is the derivative of the function F defined by

$$F(x) = \int_a^x f(t) \, dt.$$

Part II. If F is any function whose derivative throughout the interval $[a, b]$ is f, then

$$\int_a^b f(t) \, dt = F(b) - F(a).$$

Part I (the more important part, which happened to come second) says that definite integrals can be used to find antiderivatives. Part II (the less important part, which happened to come first) says that antiderivatives can be used to find definite integrals. Each of these is a very useful thing to know—and a little bit surprising, since the processes leading to the two concepts are so different.

On the other hand, these statements are a little less surprising when we think of the discrete processes that approximate the continuous ones for small values of Δt, namely, taking quotients of differences and taking sums of products. In this discrete setting, Part I says that addition can be used to undo subtraction, and Part II says that subtraction can be used to undo addition. No surprise there!

This leads us to another way to interpret the Fundamental Theorem. We know that addition and subtraction are inverse processes, i.e., that each undoes the other. In much the same way, differentiation and *definite* integration are inverse processes—almost. It is no surprise that differentiation and *indefinite* integration (antidifferentiation) are inverse processes—that's in the way we defined the words. And since the Fundamental Theorem says that each kind of integration can be used to calculate the other, it is clear that there should be an inverse relationship between differentiation and definite integration.

But we need to be careful about the "almost." Specifically, suppose we define an antiderivative of f by definite integration:

$$s(x) = \int_a^x f(t)\,dt.$$

By proclaiming that $s(x)$ *is* an antiderivative of f, Part I is saying that differentiation undoes integration with a variable upper limit, or

$$\frac{d}{dx}\int_a^x f(t)\,dt = f(x).$$

That is, if you integrate first and then differentiate, you get back where you started.

To interpret Part II in the same way, we think of starting with a function F whose derivative is f. If, in Part II, we replace b with x —that is, we think of integrating up to each x, one at a time—then Part II says

$$\int_a^x \frac{dF}{dt}\,dt = F(x) - F(a).$$

Thus, if you differentiate first and then integrate—to a variable upper limit—you will not get exactly the same function back unless $F(a) = 0$. But the function you get back differs from the one you started with by an additive constant, namely, $-F(a)$. That's the sense in which differentiation and integration are *almost* inverse processes.

There is a familiar interpretation of that constant $-F(a)$. Think of F as distance traveled on a trip in your car. Its derivative is the velocity function measured by the speedometer. In order for your trip meter to record the right distance, you have to reset it to 0 at the start of the trip. Otherwise, if you are going to measure distance by your odometer, you have to subtract the initial reading from the final one. Thus $F(a)$ is the odometer reading at the start of your trip.

Example 2 Show by direct calculation that

$$\frac{d}{dx}\int_0^x \sin t\,dt = \sin x.$$

Solution We saw in Example 1 that

$$\int_0^x \sin t\,dt = 1 - \cos x.$$

The derivative of this function is clearly $\sin x$. ■

Checkpoint 2

Show by direct calculation that

$$\frac{d}{dx}\int_0^x e^{2t}\,dt = e^{2x}.$$

Example 3 If $F(t) = \cos t$, show by direct calculation that

$$\int_0^x \frac{dF}{dt}\,dt = F(x) - F(0).$$

Solution The derivative is $F'(t) = -\sin t$. Thus, from Example 1, the integral is $-(1 - \cos x) = \cos x - 1 = F(x) - F(0)$. ■

Checkpoint 3

If $F(t) = e^{2t}$, show by direct calculation that

$$\int_0^x \frac{dF}{dt}\,dt = F(x) - F(0).$$

In this section we have completed our development of the Fundamental Theorem of Calculus. We find that the theorem establishes a very close connection between the apparently rather different processes of definite integration and antidifferentiation. In particular, each can be used to compute the other. It follows that integration (continu-ous accumulation) and differentiation (instantaneous rate of change) are near-inverse processes.

Answers to Checkpoints

1. $s(t) = (e^{2t} - 1)/2$ or, equivalently, $s(x) = (e^{2x} - 1)/2$.
2. $\displaystyle \frac{d}{dx}\int_0^x e^{2t}\,dt = \frac{d}{dx}\,\frac{1}{2}\,e^{2t}\Big|_0^x = \frac{d}{dx}\left(\frac{1}{2}\,e^{2x} - \frac{1}{2}\right) = e^{2x}.$
3. $\displaystyle \int_0^x \frac{dF}{dt}\,dt = \int_0^x 2\,e^{2t}\,dt = e^{2t}\Big|_0^x = e^{2x} - 1 = F(x) - F(0).$

Exercises 8.4

Each of the following integrals describes a function f of x. Use your graphing tool to graph each of these functions.

1. $\int_0^x e^{t^2}\,dt$
2. $\int_1^x e^{t^2}\,dt$
3. $\int_{-1}^x e^{t^2}\,dt$
4. $\int_0^x t\sin(t^3)\,dt$
5. $\int_1^x t\cos(t^3)\,dt$
6. $\int_{-1}^x \sqrt{1+t^2}\,dt$

Each of the following integrals describes a function f of x. In each case, find $\frac{d}{dx}f(x)$.

7. $\int_0^x e^{t^2}\,dt$
8. $\int_1^x e^{t^2}\,dt$
9. $\int_{-1}^x e^{t^2}\,dt$
10. $\int_0^x t\sin(t^3)\,dt$
11. $\int_1^x t\cos(t^3)\,dt$
12. $\int_{-1}^x \sqrt{1+t^2}\,dt$

Evaluate each of the following indefinite integrals.

13. $\int (x^{2/3}-x^{1/3})\,dx$
14. $\int (x^{4/3}+x^{5/3})\,dx$
15. $\int (x^{1/2}-x^{3/2})\,dx$
16. $\int x^7\,dx$
17. $\int \sin 2x\,dx$
18. $\int (1+3x-x^2)\,dx$
19. $\int (1+x^2)^2\,dx$
20. $\int e^{-5x}\,dx$

Evaluate each of the following definite integrals. Use the integral key on your calculator to check your results.

21. $\int_0^2 (x^{2/3}-x^{1/3})\,dx$
22. $\int_1^2 (x^{4/3}+x^{5/3})\,dx$
23. $\int_0^1 (x^{1/2}-x^{3/2})\,dx$
24. $\int_5^{11} x^7\,dx$
25. $\int_0^{\pi} \sin 2x\,dx$
26. $\int_{-1}^3 (1+3x-x^2)\,dx$
27. $\int_{-2}^2 (1+x^2)^2\,dx$
28. $\int_{-1}^1 e^{-5x}\,dx$

Estimate each of the following integrals.

29. $\int_0^{\pi/2} \sin^2 x\,dx$
30. $\int_0^{\pi/2} \cos^2 x\,dx$
31. $\int_0^1 \frac{1}{1+x^2}\,dx$

32. (a) Use your graphing tool to sketch the graph of
$$f(t)=\sqrt{1+t^2}.$$
(b) Use your graphing tool and an exact formula (not Euler's Method) to sketch the solution of the initial value problem
$$\frac{dy}{dt}=\sqrt{1+t^2}, \qquad \text{with } y(0)=0.$$
(c) Make a sketch on paper of the solution of the initial value problem
$$\frac{dy}{dt}=\sqrt{1+t^2}, \qquad \text{with } y(0)=2.$$

33. (a) Use your graphing tool to sketch the graph of
$$f(t)=\sin(t^2).$$
(b) Use your graphing tool and an exact formula to sketch the solution of the initial value problem
$$\frac{dy}{dt}=\sin(t^2), \qquad \text{with } y(0)=0.$$
(c) Make a sketch on paper of the solution of the initial value problem
$$\frac{dy}{dt}=\sin(t^2), \qquad \text{with } y(0)=2.$$

34. Figure 8.17 shows the graphs of two functions we will call f and g. The function g is an antiderivative of f.
(a) Which graph is that of f and which is that of g?
(b) Write down a formula that defines g in terms of f.

Figure 8.17 Which is the antiderivative?

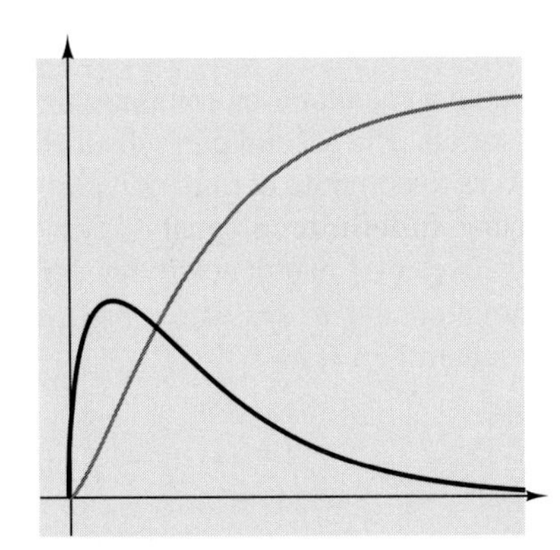

35. Figure 8.18 shows the graph of a function f on the interval $[0, 2]$. We define two more functions, $F(x)$ and $g(x)$, on the same interval:

$$F(x) = \int_0^x f(t)\,dt \qquad \text{for } 0 \le x \le 2,$$

and

$$g(x) = \int_0^x f'(t)\,dt \qquad \text{for } 0 \le x \le 2.$$

Figure 8.18 Graph of $f(x)$

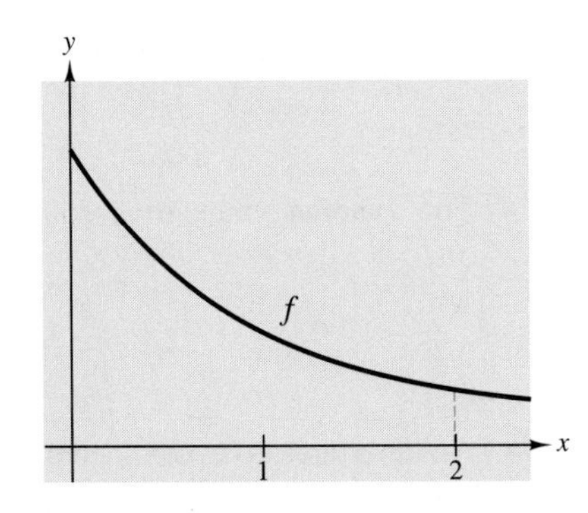

With the information given, respond as precisely as you can to each of the following:

(a) Complete the sentence: The value of F at 1 is ________________.

(b) Complete the sentence: The value of F' at 1 is ________________.

(c) Complete the sentence: The value of g at 1 is ________________.

(d) Complete the sentence: The value of g' at 1 is ________________.

(e) Sketch the graph of F.

(f) Sketch the graph of g.

36. Figure 8.19 shows the graph of a locally linear function $f(x)$ on the interval $[0, b]$ with the property that $f(0) = f(b)$. What is the average value of the derivative $f'(x)$ on the interval $[0, b]$? Explain.

Figure 8.19 $f(0) = f(b)$—What is the average slope?

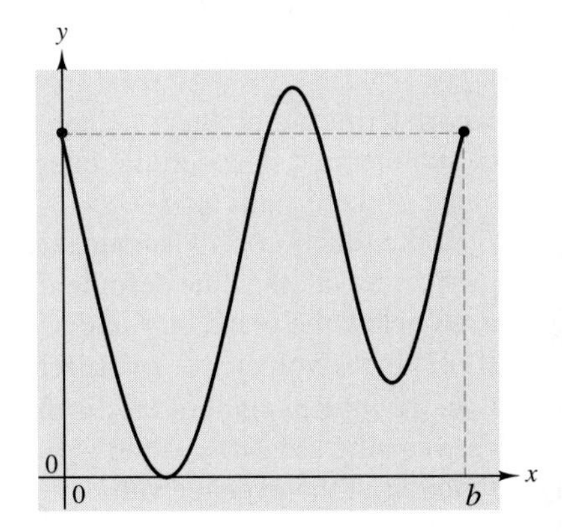

37. Let s be the function defined by

$$s(x) = \int_0^x e^{-t^2/2}\,dt.$$

If w is the function defined by $w(x) = s(x^2)$, show that $w'(x) = 2xe^{-x^4/2}$. (*Hint*: Use the Chain Rule.)

38. Find a function $f(x)$ that satisfies the functional equation

$$f(x)^2 = \int_0^x f(t)\,dt + 1.$$

Is there more than one such function?[3]

3. Problem contributed by Michael Reed.

Chapter 8 Summary

We have introduced the second key concept of calculus, that of "continuous accumulation," as expressed in the **definite integral**,

$$\int_a^b f(t)\,dt.$$

Roughly speaking, this concept embodies the idea of adding up infinitely many infinitely small quantities. More precisely, we form left-hand sums

$$\sum_{k=0}^{n} f(t_{k-1})\,\Delta t$$

whose terms are products of function values and widths of subintervals. The definite integral is then the limiting value of these sums as $n \to \infty$.

If the function $f(t)$ is nonnegative for all values of t in the interval, then the definite integral represents the **area** below the graph of f and above the interval $[a, b]$. If $v(t)$ is the velocity of an object moving in a straight line, then the integral of $v(t)$ is the **distance** traveled between the times $t = a$ and $t = b$. For any continuous function f, the **average value** of the function over the interval $[a, b]$ is

$$\frac{1}{b-a}\int_a^b f(t)\,dt.$$

Integration and differentiation are almost inverse operations. The exact relation between the two operations is contained in the **Fundamental Theorem of Calculus**:

Part I If f is a continuous function from a to b, then f is the derivative of the function F defined by

$$F(x) = \int_a^x f(t)\,dt.$$

Part II If F is any function whose derivative throughout the interval $[a, b]$ is f, then

$$\int_a^b f(t)\,dt = F(b) - F(a).$$

The first part tells us how to construct antiderivatives by definite integration—a powerful idea when using a calculator or computer that can automate integration. The second part tells us how to use antiderivatives to compute definite integrals. We introduced the name **indefinite integral** of f for the family of antiderivatives of f and denoted this $\int f(t)\,dt$. With this notation and an evaluation bar, we may write the formula in Part II as

$$\int_a^b f(t)\,dt = \int f(t)\,dt\,\Big|_a^b.$$

Concepts

- Continuous functions
- Average value of a continuous function
- The definite integral
- The Fundamental Theorem of Calculus
- The indefinite integral

Applications

- Distance traveled
- Area under a curve
- Average temperature

Formulas

- The average value of a continuous function f on $[a, b]$ is

$$\frac{1}{b-a}\int_a^b f(t)\,dt.$$

- An antiderivative of a continuous function f on $[a, b]$ is

$$F(x) = \int_a^x f(t)\,dt \quad \text{for } a \le x \le b.$$

- If f is a continuous function on $[a, b]$, then

$$\int_a^b f(t)\,dt = \int f(t)\,dt\,\Big|_a^b.$$

Chapter 8 Exercises

Evaluate each of the following indefinite integrals.

1. $\int x^{1/3}\,dx$
2. $\int x^{4/3}\,dx$
3. $\int x^{3/2}\,dx$
4. $\int x^4\,dx$
5. $\int \sin x\,dx$
6. $\int (1-x^2)\,dx$
7. $\int (1-x^2)^2\,dx$
8. $\int e^{-2x}\,dx$

Evaluate each of the following definite integrals.

9. $\int_0^1 x^{1/3}\,dx$
10. $\int_0^1 x^{4/3}\,dx$
11. $\int_0^1 x^{3/2}\,dx$
12. $\int_0^1 x^4\,dx$
13. $\int_0^\pi \sin x\,dx$
14. $\int_{-1}^1 (1-x^2)\,dx$
15. $\int_{-1}^1 (1-x^2)^2\,dx$
16. $\int_0^1 e^{-2x}\,dx$

Estimate each of the following integrals.

17. $\int_0^\pi \sin^2 x\,dx$
18. $\int_0^\pi \cos^2 x\,dx$
19. $\int_{-1}^1 \frac{1}{1+x^2}\,dx$

In Exercises 20–23,

(a) sketch the region of the plane described verbally,

(b) represent the area of the region using one or more definite integrals, and

(c) calculate the area of the region.[4]

20. The region bounded by $x=-1,\ x=1.5,\ y=x^3+8,$ and the x-axis.

21. The region above the t-axis and under the graph of $f(t)=\sqrt{t},$ where $1\le t\le 9.$

22. The region above the x-axis and under the graph of $g,$ where

$$g(x)=\begin{cases} -x & \text{if } -2\le x\le -1 \\ x^2 & \text{if } -1\le x\le 1 \end{cases}$$

23. The region bounded by the graph of $y=e^t$ and the lines $y=1$ and $t=1.5$.

4. Adapted from Unit 7 of *Workshop Calculus*, by Nancy Baxter, Dickinson College, 1994.

In Exercises 24–26,

(a) sketch a region whose area equals the given expression, and

(b) calculate the area of the region.[5]

24. $\int_0^{2\pi} (\sin x + 2)\, dx$ 25. $\int_0^2 x^2\, dx + \int_2^6 (6 - x)\, dx$ 26. $\int_{-1}^2 4\, dt - \int_{-1}^2 t^2\, dt$

Exercises 27–30 all refer to Figure 8.20, which shows the graph of a function, $y = f(t)$. In each case,

(a) determine whether the given statement is *true* or *false*, and

(b) carefully explain your answer to part (a).[6]

Figure 8.20 Graph of a function $y = f(t)$

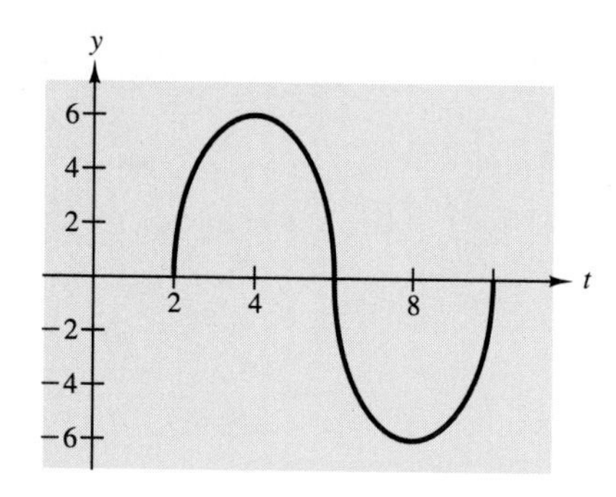

27. $\int_2^6 f(t)\, dt \approx -\int_6^{10} f(t)\, dt$ 28. $\int_2^6 f(t)\, dt + \int_6^{10} f(t)\, dt = \int_2^{10} f(t)\, dt$

29. $\int_2^6 f(t)\, dt < 24$ 30. $\int_4^6 f(t)\, dt > 0.5[f(4.5) + f(5.0) + f(5.5) + f(6.0)]$

31. (a) Find the average value of x over the interval $[0, 1]$; of x^2; of x^3.

(b) Find the average value of x^n over the interval $[0, 1]$, where n is any integer ≥ 1.

(c) What happens to the average value of x^n over the interval $[0, 1]$ as n gets larger and larger? Can you explain this?

32. (a) Find the average value of x over the interval $[0, 1]$; of $x^{1/2}$; of $x^{1/3}$.

(b) Find the average value of $x^{1/n}$ over the interval $[0, 1]$, where n is any integer ≥ 1.

(c) What happens to the average value of $x^{1/n}$ over the interval $[0, 1]$ as n gets larger and larger? Can you explain this?

5. Adapted from Unit 7 of *Workshop Calculus*, by Nancy Baxter, Dickinson College, 1994.
6. Adapted from Unit 7 of *Workshop Calculus*, by Nancy Baxter, Dickinson College, 1994.

33. Four calculus students disagree as to the value of the integral $\int_0^\pi \sin^8 x\,dx$. Jack says it is equal to π. Joan says it is equal to $35\pi/128$. Ed claims it is equal to $3\pi/90 - 1$, while Lesley says it is equal to $\pi/2$. One of them is right — which one? (You have no way yet to evaluate the integral exactly — try to eliminate the three wrong answers. If you get stuck, try using a numerical estimate of the integral.)[7]

34. The cost of digging a tunnel is the product of the length of the tunnel and the cost per unit length. If the cost per unit length is a constant, it is easy to calculate the cost of the tunnel. However, the cost per unit length increases as the tunnel gets longer because of the expense of carrying workers and tools in and carrying dirt and rock out. Assume that the dollar cost per foot varies as in Table 8.6.

Table 8.6 Cost per foot for digging a tunnel

Length (feet)	*Cost ($ per foot)*	*Length (feet)*	*Cost ($ per foot)*
0–100	500	500–600	2500
100–200	820	600–700	3020
200–300	1180	700–800	3580
300–400	1580	800–900	4180
400–500	2020	900–1000	4820

(a) Find the total cost for digging a tunnel 1000 feet long.

(b) How much could be saved by starting the 1000-foot tunnel from both ends and making the two halves meet in the middle?

(c) If the half-tunnels of part (b) don't meet in the middle, what can you do, and how much will it cost?

Explain the following equalities geometrically.

35. $\displaystyle\int_0^1 x^n\,dx + \int_0^1 x^{1/n}\,dx = 1$ for any positive integer n.

36. $\displaystyle\int_1^e \ln x\,dx + \int_0^1 e^x\,dx = e.$

7. This exercise and the three following are adapted from *Calculus Problems for a New Century*, edited by R. Fraga, MAA Notes no. 28, 1993.

Chapter 8 Projects

1. **Area of Your Body** Consider the problem of finding the front cross-sectional area of your body when you are standing erect with your hands at your sides.

   (a) In 60 seconds or less, make an estimate of this area. How did you arrive at your estimate?

   (b) In the next 20 minutes, possibly with the help of others, get a better estimate of this area. What method did you use, and how did it improve upon your previous estimate?

   (c) As you reflect on what you have done, are there any concepts from calculus that you have used in solving this problem?[8]

2. **Area of a Lake** This project requires a map with a lake of sufficient size that you can trace it accurately.

   (a) Trace the outline of the lake on a piece of paper. Using the scale of the map, draw a grid over the outline. By counting squares, give lower and upper bounds for the area of the lake.

   (b) What technique might you use to get a better estimate of the area of the lake? Do it!

3. **Flashbulb Illumination** The value of a flashbulb may be measured by the total amount of light produced by the bulb during the period when the lens of the camera is open. The rate at which light is produced by the bulb is measured in lumens, and this rate varies with time. A typical focal plane shutter opens 20 milliseconds after the button is pushed (and the bulb starts to burn) and closes 70 milliseconds after the button is pushed.

   Suppose we wish to compare two different flashbulbs, bulb A and bulb B, for which we have the data shown in Tables 8.7 and 8.8. The time is measured in milliseconds and the light output in millions of lumens. These data are also shown graphically in Figures 8.21 and 8.22.

   (a) If $L(t)$ is the rate at which light from a given bulb is produced as a function of time, define an integral that describes the total amount of light produced during the period the shutter is open.

   (b) Estimate the total light output of bulb A during the time the shutter is open.

   (c) Estimate the total light output of bulb B during the time the shutter is open.

   (d) Decide which is the better bulb.[9]

---

8. This project and the one following are adapted from *Calculus Problems for a New Century*, edited by R. Fraga, MAA Notes no. 28, 1993.

9. This project is based on an exercise in *Calculus and Analytic Geometry*, 7th ed., by G. B. Thomas, Jr., and R. L. Finney, Addison-Wesley, 1988. That exercise was itself adapted from *Integration*, by W. U. Walton, et al., Project CALC (not related to Duke University's Project CALC), Education Development Center, Inc., 1975. The source of the data is *Photographic Lamp and Equipment Guide*, P4-15P, General Electric Co. Bulb A is a no. 22 class M bulb, and bulb B is a no. 31 class FP bulb.

Table 8.7 Bulb A output

Time (ms)	*Light output (lumens)*
0	0
5	0.2
10	0.5
15	2.6
20	4.2
25	3.0
30	1.7
35	0.7
40	0.35
45	0.2
50	0

Table 8.8 Bulb B output

Time (ms)	*Light output (lumens)*	*Time (ms)*	*Light output (lumens)*
0	0	50	1.3
5	0.1	55	1.4
10	0.3	60	1.3
15	0.7	65	1.0
20	1.0	70	0.8
25	1.2	75	0.6
30	1.0	80	0.3
35	0.9	85	0.2
40	1.0	90	0
45	1.1		

Figure 8.21 Bulb A output

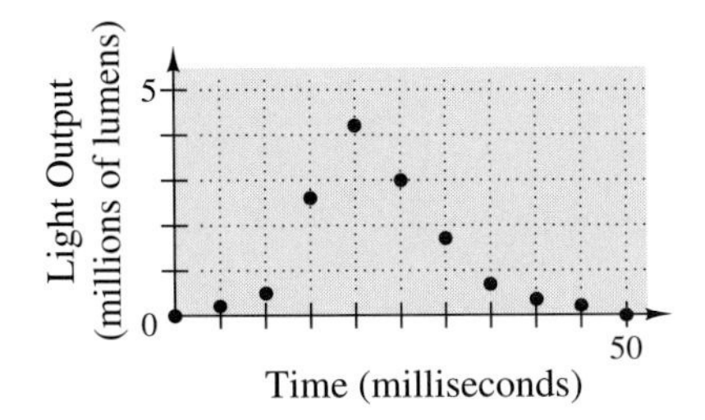

Figure 8.22 Bulb B output

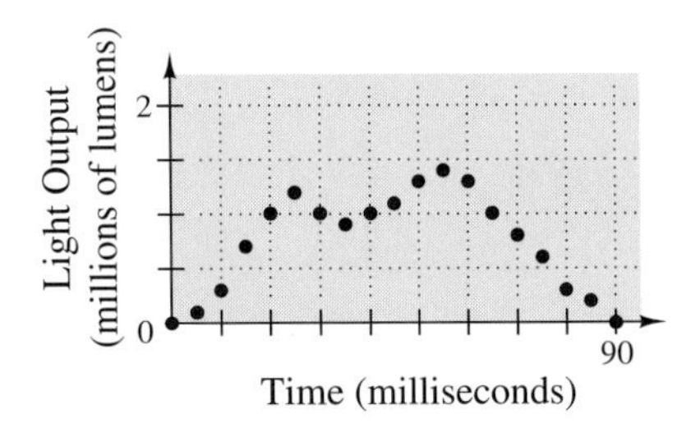

4. **Mass of a Bar** Figure 8.23 shows a rectangular bar that is four meters long and has a two-centimeter square cross section. (The figure is not drawn to scale.) In this project we will calculate the mass of this bar under varying assumptions about its density. If the density is constant, then

$$\text{mass} = \text{density} \times \text{volume}.$$

This is an immediate consequence of the meaning of the word "density": mass per unit volume.

Figure 8.23 A rectangular bar

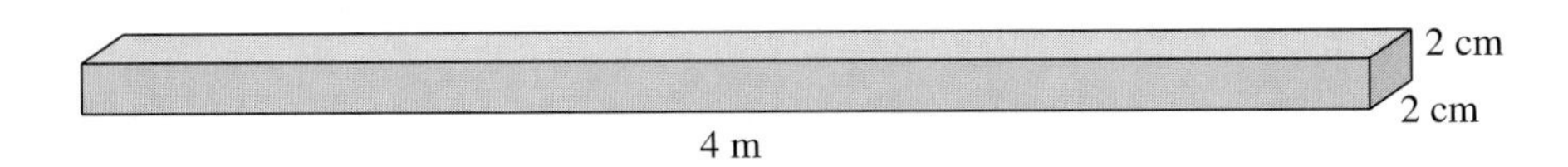

(a) Suppose the density throughout the bar is 0.5 grams per cubic centimeter. Find the mass of the bar.

(b) Now suppose the density varies continuously from one end of the bar to the other, and we have the information in Table 8.9 about the density at every half-meter. In the table x

measures distance from the left end of the bar, and $D(x)$ is the density at that distance from the end. Estimate the mass of the bar.

Table 8.9 Varying density of the bar

x *(centimeters)*	0	50	100	150	200	250	300	350	40
$D(x)$ *(gm/cm^3)*	2.5	3.3	3.4	3.1	2.8	2.5	2.1	1.6	1.0

(c) Plot the data in Table 8.9, and connect the data points with straight line segments. (Your graphing tool should have a mode that will do this.) We suppose that the resulting graph is exactly that of the function $D(x)$. That is, we assume the D varies linearly between the given data points. Find the exact density of the bar, and explain carefully how you are doing that.

(d) Express your answer for the exact density in terms of a definite integral.[10]

10. This project is adapted from Unit 7 of *Workshop Calculus*, by Nancy Baxter, Dickinson College, 1994.

CHAPTER 9

Integral Calculus and Its Uses

In this chapter we describe important applications of integration, and we develop procedures for evaluating both definite and indefinite integrals. We begin with the physical problem of finding the center of mass of an irregular object. This problem illustrates the importance of the divide-and-conquer process that leads to the definite integral.

Next, we discuss the process of evaluating definite integrals—first numerically and then symbolically. Our numerical methods start with the approximating-sum definition and lead to the type of method likely to be built into your calculator or computer. A fundamental technique for symbolic integration is substitution for the variable of integration, which is closely related to the Chain Rule for differentiation. Substitution is often the key to using a powerful low-tech tool for symbolic integration: a table of integrals. Many of the table entries point us to another technique, integration by parts, which is the inverse of the Product Rule for differentiation.

We finish the chapter by applying integration to the problem of representing a complicated periodic function as a sum of sine and cosine functions. This process, called "harmonic analysis," is basic in such diverse areas as synthesis of music and analysis of heat flow.

9.1 Moments and Centers of Mass

The **center of mass** of an object is the point at which the object balances in a gravitational field, no matter how the object is oriented. Another description, one that relates more directly to the integral, is that the center of mass is the average location of the mass distributed throughout the object. In general, whenever the problem is to find an average of something distributed in a continuous manner, the solution is likely to involve an integral.

We'll start by considering the center of mass of a simple object, a pool cue—basically, a tapered wooden rod. A typical pool cue is 58 inches long with circular cross sections that taper from a diameter of $1\frac{3}{8}$ inches at the butt to $\frac{9}{16}$ inches at the tip (see Figure 9.1).

Figure 9.1 A typical pool cue

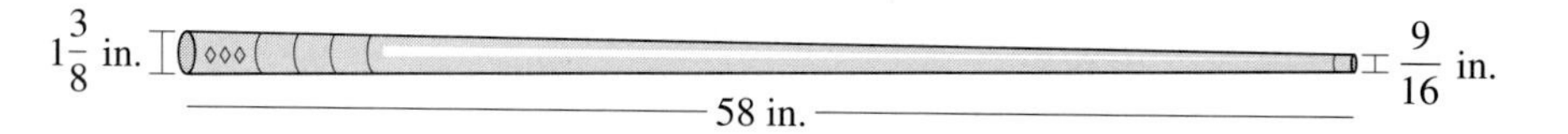

Now the balance point of a pool cue has a considerable effect on the "feel" of the cue. Our particular cue is weighted so that the center of mass is 17 inches from the larger end. How much does the weighting matter? Where would the center of mass be if the cue were shaped from a uniform piece of wood without the added weighting?

The fact that mass is distributed continuously along the length of the cue is what leads to the need for integration to solve this problem. Here is our strategy:

- We will work out how to find a center of mass in a discrete problem, such as hanging masses on a balance beam.
- We will approximate our continuous distribution of mass by a collection of discrete masses.
- We will improve the discrete approximation by using more, but smaller, masses.

If we choose our approximating terms appropriately, then as the discrete distribution of masses approaches the continuous one, the discrete average (a sum) will approach the appropriate continuous average (an integral).

Moments and Moment Arms

We formulate the discrete center of mass problem in terms of a **balance beam**—a rigid beam or rod that has negligible mass in comparison with the objects suspended from it. If you experiment with a balance beam, you will discover that different masses on opposite sides of the beam will balance if the products of mass and distance from the balance point are the same, as in Figure 9.2. For a single mass, this product of mass and distance is called a **moment** with respect to the balance point, and the distance factor alone is called the **moment arm**. On the left in Figure 9.2 we have a mass of three units and a moment arm of four units; on the right, a mass of four units and a moment arm of three units.

Figure 9.2 Unequal masses and moment arms but equal moments

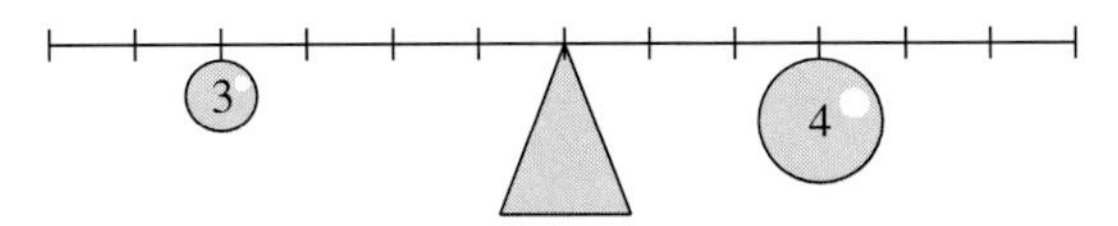

You also may learn from balance beam experiments that moments are additive. Thus, in Figure 9.3, we can calculate the total moment on the left as $2 \times 6 + 3 \times 2$ and on the right as $4 \times 2 + 2 \times 5$. Since each of these moments is 18, the beam is balanced.

Figure 9.3 Equal moments

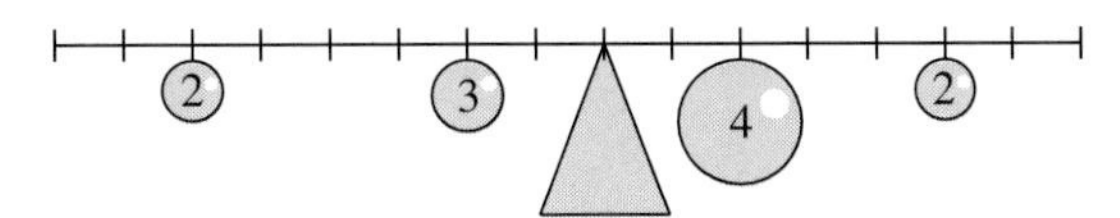

To make effective use of our information about balancing, in a context involving continuous objects, we need a coordinate system. It is convenient to let the signs of coordinates do the work of keeping track of left and right. Thus we revise our concept of the moment of a point mass to be the product of the mass and its *directed* distance from the balance axis. Then the condition for balancing is that *the sum of the moments is zero.*

Checkpoint 1

Copy Figure 9.3 and draw an x-axis along the balance beam with the origin at the balance point. Label each of the marked points on the x-axis with the appropriate coordinate. Write out explicitly the sum of the moments of the point masses, and show that the sum is indeed zero.

Center of Mass

In Checkpoint 1 we were able to make the balance point the center of the coordinate system because we already knew that point. Usually we must introduce the coordinate system before we find the balance point—because we will use the coordinates to find the balance point. We can use the balancing condition to locate an unknown center of mass by a simple trick: We create an artificial system consisting of the object of interest on one side of the balance point and a point mass of the same magnitude on the other. Then we are free to place the balance point for this enlarged system where we want it—for example, at the origin of the coordinate system.

We illustrate this idea in Figure 9.4 with our system of four objects on the right and a single object on the left with mass m equal to the sum of the masses on the right—in this case $m = 11$ units of mass. We denote the coordinate of the center of mass of the system on the right by $\overline{x}$. To achieve balance of the enlarged system at $x = 0$, we place the object of mass m at the symmetrical point on the left, i.e., at $-\overline{x}$.

Figure 9.4 Balancing an object with an equivalent point mass

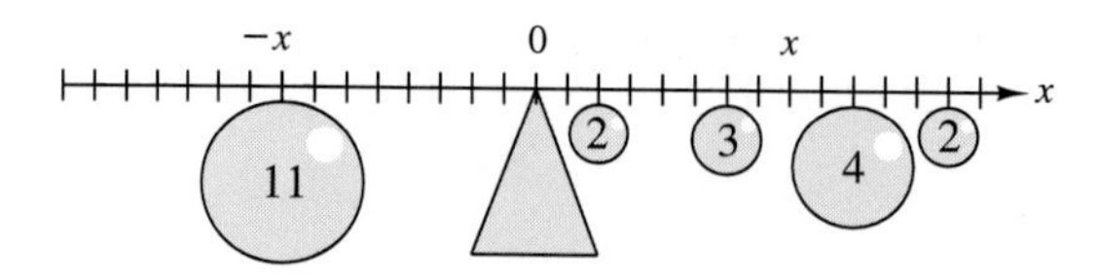

The moment of the system on the right in Figure 9.4 with respect to the balance point (or, equivalently, with respect to the origin) is unknown, so we just give it a name: M. The moment of the mass on the left is $-m\overline{x}$. The balancing condition is $M - m\overline{x} = 0$. Thus we can express the unknown coordinate $\overline{x}$ in terms of two other unknown quantities, a moment and a mass:

$$\overline{x} = \frac{M}{m}.$$

Checkpoint 2

Calculate M and $\overline{x}$ for the system on the right in Figure 9.4. Verify that this places the center of mass in the same place as in Checkpoint 1.

In general, if we have n objects with masses $m_1, m_2, \ldots, m_n$ distributed along a balance beam with corresponding x-coordinates $x_1, x_2, \ldots, x_n$, the total mass is given by

$$m = \sum_{k=1}^{n} m_k,$$

and the total moment is given by

$$M = \sum_{k=1}^{n} x_k\, m_k.$$

Then the center of mass is

$$\overline{x} = \frac{M}{m}.$$

Checkpoint 3

Find the center of mass of a system of five point masses distributed along the x-axis as in Table 9.1.

Table 9.1 A system of five point masses

Mass (grams)	*x-coordinate (centimeters)*
50	−10
20	25
10	30
30	45
20	60

Now we extend this calculation of center of mass to situations in which the mass is distributed continuously—e.g., to the pool cue. For now we assume that the cue is made of solid wood of uniform density—no added weights.

Calculation of Mass

To find $\overline{x}$, we need to calculate the moment M and the mass m. We start with the easier of the two calculations, finding the mass of the pool cue. If we know the volume of the cue and the density of the wood (say, in ounces per cubic inch), then the mass is just the product of the volume and the density. Unfortunately, we do not know the density. However, as we will see, the exact value for the density does not matter in the calculation of the center of mass. For the moment we simply designate this unknown density by δ ounces per cubic inch. (The symbol δ is the Greek "d," lower-case delta.)

We concentrate now on the calculation of volume. We place the cue along an x-axis with the butt at the origin and tip at 58 (inches), as in Figure 9.5.

Figure 9.5 Pool cue with a coordinate axis

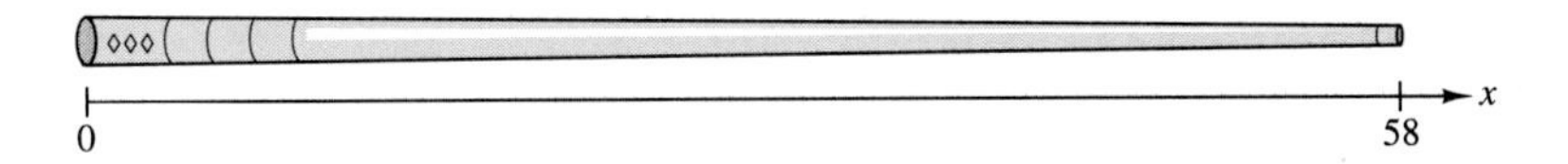

Now that we have a coordinate system, we may determine a function $r(x)$ that gives the radius of the circular cross section as a function of the x-coordinate. We know the radius varies linearly, and our diameter measurements of $\frac{9}{16}$ and $\frac{11}{8}$ inches give

$$r(0) = \frac{11}{16}$$

and

$$r(58) = \frac{9}{32}.$$

Checkpoint 4

Show that the function r is given by

$$r(x) = c - sx,$$

where $c \approx 0.688$ and $s \approx 0.00700$.

We illustrate the divide-and-conquer process in Figure 9.6. First, we mentally slice the cue into approximately cylindrical segments and approximate the volumes of these segments by volumes of actual cylinders. Next, we add the volumes of these cylinders to approximate the volume of the cue. Finally, we consider what happens to these sums as the number of terms becomes infinite and the size of each term becomes infinitesimal.

Figure 9.6 Subdivision of the pool cue into cylindrical slices

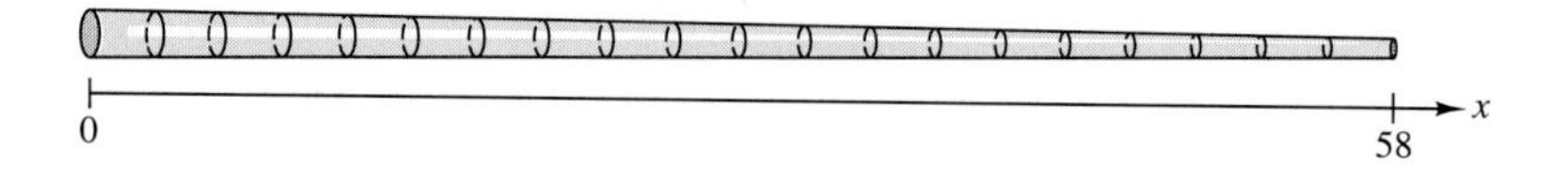

In Figure 9.6 we have drawn 20 cross-sectional cuts at equal intervals. Thus each cross-sectional piece extends along the x-axis for a distance of $\Delta x = 58/20 = 2.9$ inches. The x-coordinate of the kth cut point is

$$x_k = k\,\Delta x$$

for $k = 0, 1, 2, \ldots, 20$. We approximate the volume of the kth section (the one between x_{k-1} and x_k) by calculating the volume of the cylinder that has the constant cross-sectional radius $r(x_{k-1})$. In Figure 9.7 we show an exaggerated picture of this section of the cue and its cylindrical approximation.

Figure 9.7 The kth section and its approximation

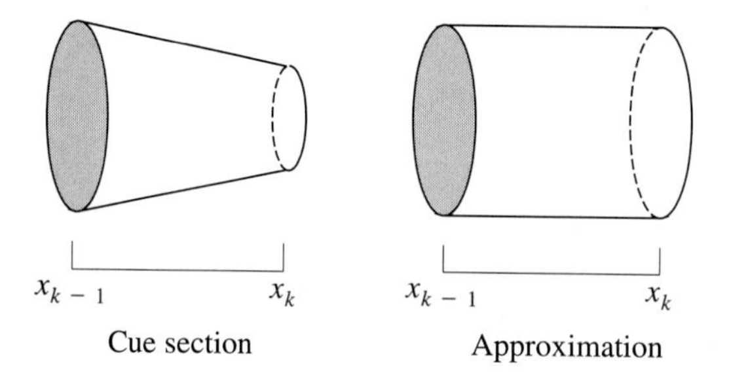

We obtain the volume of the approximation by the formula for the volume of a cylinder:

$$\text{Volume} = \text{area of the base} \times \text{height}.$$

Thus the volume of the cylinder approximating the kth cue section is

$$\pi\left[r(x_{k-1})\right]^2 \Delta x.$$

When we add up these cylindrical approximations, we obtain an approximation to the volume of the whole cue:

$$\text{Volume of the cue} \approx \sum_{k=1}^{20} \pi\left[r(x_{k-1})\right]^2 \Delta x.$$

Now there is nothing special about the number 20. Suppose we cut the cue into n of sections of equal length. Then the length of each section is

$$\Delta x = \frac{58}{n} \text{ inches.}$$

The corresponding cut points are

$$x_k = k\,\Delta x$$

for $k = 0, 1, \ldots, n$, and the approximate volume of each section is still

$$\pi\left[r(x_{k-1})\right]^2 \Delta x.$$

Thus the approximate volume of the cue using n cuts is

$$\sum_{k=1}^{n} \pi\left[r(x_{k-1})\right]^2 \Delta x.$$

As n increases, these approximations approach the true volume. But we already know that these sums approach the integral

$$\int_0^{58} \pi\left[r(x)\right]^2 dx = \pi \int_0^{58} (0.688 - 0.007x)^2\,dx.$$

Therefore, we know that the volume of the cue is the value of this definite integral.

Checkpoint 5

Show that the value of the integral

$$\int_0^{58} \pi\left[r(x)\right]^2 dx$$

is approximately 45.4. Use the integral key on your calculator to check your work.

Since the volume of the cue is approximately 45.4 cubic inches, the mass of the cue is approximately 45.4δ ounces.

Calculation of Moment

Now let's calculate the *moment* of the pool cue. We approach this calculation in the same way we approached the calculation of the volume. Again, we imagine the cue divided into n sections of width $\Delta x = 58/n$. Again, we approximate each section by a cylinder of constant cross-sectional radius. Then we calculate the moment of each of these approximating cylinders and add these moments to approximate the moment of the whole cue.

Recall that the volume of the kth approximating cylinder is

$$\pi\left[r(x_{k-1})\right]^2 \Delta x.$$

Thus the mass of the kth approximating cylinder is δ times this mass or

$$\delta\,\pi\left[r(x_{k-1})\right]^2 \Delta x.$$

Just as we assumed that the cross-sectional radius has the constant value $r(x_{k-1})$, we'll assume that all this mass is concentrated at the left endpoint x_{k-1} of the interval $[x_{k-1}, x_k]$. Then the moment of the kth section is approximately

$$x_{k-1} \times \text{the approximate mass of the } k\text{th section} = x_{k-1}\delta\,\pi\left[r(x_{k-1})\right]^2 \Delta x.$$

Now our approximation to the total moment is the sum of the point moments

$$\sum_{k=1}^{n} x_{k-1}\delta\,\pi\,[r(x_{k-1})]^2\Delta x.$$

As n becomes large, this sum approaches both the total moment of the cue and the integral

$$\int_0^{58} x\,\delta\,\pi\,[r(x)]^2 dx.$$

Checkpoint 6

Show that the integral

$$\int_0^{58} x\,\delta\,\pi\,[r(x)]^2 dx$$

is approximately equal to 969δ. Thus the moment is approximately 969δ inch-ounces. Use the integral key on your calculator to check your work.

Example 1 Calculate the center of mass coordinate $\overline{x}$ for the unweighted pool cue.

Solution Now that we have integral representations for both the mass and the moment of the cue, we find

$$\begin{aligned}\overline{x} &= \frac{\text{total moment}}{\text{total mass}}\\ &= \frac{\int_0^{58} \delta\,x\,\pi\,[r(x)]^2 dx}{\int_0^{58} \delta\,\pi\,[r(x)]^2 dx}\\ &\approx \frac{969\delta}{45.4\delta}\\ &\approx 21.3.\end{aligned}$$

Notice that the δs in the numerator and denominator canceled — we did not need to know a specific value for δ. Thus, for a uniform wooden cue, the center of mass is located 21.3 inches from the butt.

Recall that the real pool cue has a center of mass 17 inches from the butt. Apparently this is a more comfortable position for the cue to balance, and weights were placed in the butt end to make this shift in the balance point.

In this section we worked out the formula for the center of mass $\overline{x}$ of a system of n point masses $m_1, m_2, \ldots, m_n$ located at points $x_1, x_2, \ldots, x_n$ along an x-axis:

$$\overline{x} = \frac{\text{total moment}}{\text{total mass}} = \frac{\sum_{k=1}^{n} x_k\, m_k}{\sum_{k=1}^{n} m_k}.$$

We then used this to find the center of mass of a uniform tapered rod. If the rod lies along the x-axis between a and b, the cross-sectional radius $r(x)$ varies continuously with x, and the material composing the rod has density δ, then the center of mass coordinate $\overline{x}$ is

$$\overline{x} = \frac{\text{total moment}}{\text{total mass}} = \frac{\int_a^b \delta\, x\, \pi\, [r(x)]^2 dx}{\int_a^b \delta\, \pi\, [r(x)]^2 dx}.$$

Here we may think of $\pi\, [r(x)]^2 dx$ as an infinitesimal volume located at the point with coordinate x. When we multiply by δ, we obtain the infinitesimal mass dm given by $\delta\, \pi\, [r(x)]^2 dx$. Under this identification, our formula for $\overline{x}$ becomes

$$\overline{x} = \frac{\int_a^b x\, dm}{\int_a^b dm}.$$

Note, however, that m is not our variable of integration. To use this formula for computation, we also need a formula for the infinitesimal mass.

Answers to Checkpoints

1. 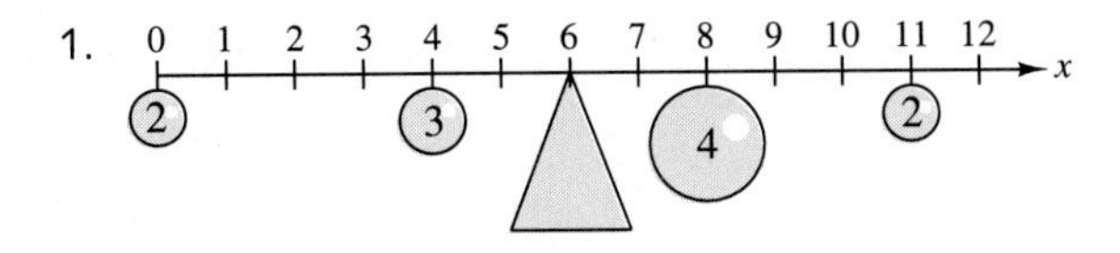

The moment calculation:

$$2 \times (-6) + 3 \times (-2) + 4 \times 2 + 2 \times 5 = -18 + 18 = 0.$$

2. $m = 2 + 3 + 4 + 2 = 11$, $M = 2 \times 2 + 6 \times 3 + 10 \times 4 + 13 \times 2 = 88$, and $\overline{x} = 88/11 = 8$, midway between masses 3 and 4 as before.

3. $m = 130$, $M = 2850$, $\overline{x} = 285/13 \approx 21.9$

4. We know that r is a linear function. Since $r(0) = 11/16 = 0.6875$ and the slope is

$$\frac{9/32 - 11/16}{58} \approx -0.00700,$$

we know that $r(x) \approx 0.688 - 0.00700\,x$.

5. $\displaystyle\int_0^{58} \pi\,[r(x)]^2 dx \approx \int_0^{58} \pi\,(c - s\,x)^2 dx = \pi\left(c^2 x - c\,s\,x^2 + s^2\frac{x^3}{3}\right)\Big|_0^{58} \approx 45.4$

6. $\displaystyle\int_0^{58} x\,\delta\,\pi\,[r(x)]^2 dx \approx \delta\,\pi \int_0^{58} x\,(c - s\,x)^2 dx = \pi\left(\frac{1}{2}c^2x^2 - \frac{2}{3}c\,s\,x^3 + \frac{1}{4}s^2x^3\right)\Big|_0^{58} \approx 968$

Exercises 9.1

Find the center of mass of each of the following systems of point masses. The masses are in grams and the distances in centimeters.

1.

Mass	x*-coordinate*
15	20
10	35
20	50

2.

Mass	x*-coordinate*
20	−25
10	−12
25	13

3.

Mass	x*-coordinate*
23	−27
18	−12
45	10
70	28
33	35

4.

Mass	x*-coordinate*
100	10
150	30
50	40
75	70

5. A stiff horizontal rod of negligible mass has 1-ounce fishing weights suspended at 6, 14, and 20 inches from the left end. In addition, 3-ounce fishing weights are suspended at 2 and 18 inches from the left end. Where is the balance point of this system?

6. A stiff horizontal rod of negligible mass has 1-ounce fishing weights suspended at 5, 15, and 25 inches from the right end. In addition, 2-ounce fishing weights are suspended at 10 and 20 inches from the right end. Where is the balance point of this system?

7. The following weights are suspended from a meter-long rod of negligible mass:

   50 grams at 10 centimeters from the left end
   60 grams at 40 centimeters from the left end
   20 grams at 70 centimeters from the left end

   Where should a 60-gram weight be suspended to place the balance point in the middle of the rod?

8. The following weights are suspended from a meter-long rod of negligible mass:

   80 grams at 20 centimeters from the left end
   60 grams at 50 centimeters from the left end
   40 grams at 80 centimeters from the left end

   Where should a 50-gram weight be suspended to place the balance point at 60 centimeters from the left end of the rod?

9. Find the center of mass of a 65-inch-long tapered rod that has a diameter of 0.75 inch at the larger end and a diameter of 0.25 inch at the smaller end.

10. Find the center of mass of a 60-inch long tapered rod that has a diameter of 0.70 inch at the larger end and a diameter of 0.15 inch at the smaller end.

11. Use an integral to explain the formula

$$V = \frac{1}{3}\pi R^2 h$$

   for the volume of a circular cone of radius R at the base and height h.

12. Find the y-coordinate of the center of mass of the cone in Figure 9.8.

Figure 9.8 A circular cone of radius R and height h

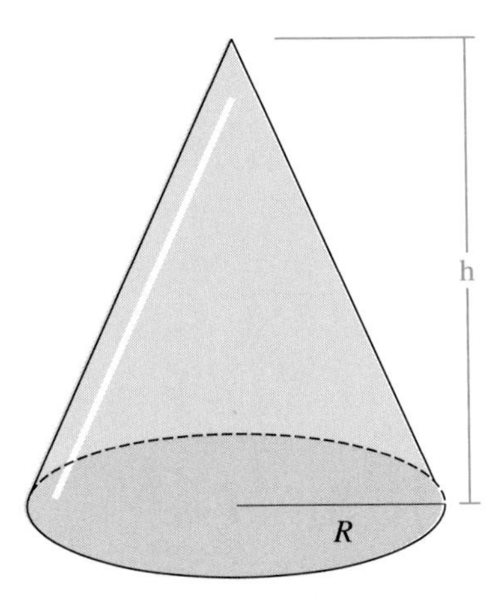

13. The horn-shaped solid in Figure 9.9 has circular cross sections. The radius r varies with x according to the formula

$$r(x) = \frac{1}{2}x^2 \text{ for } 0 \leq x \leq 1.$$

Find the volume of the solid.

Figure 9.9 A parabolic horn

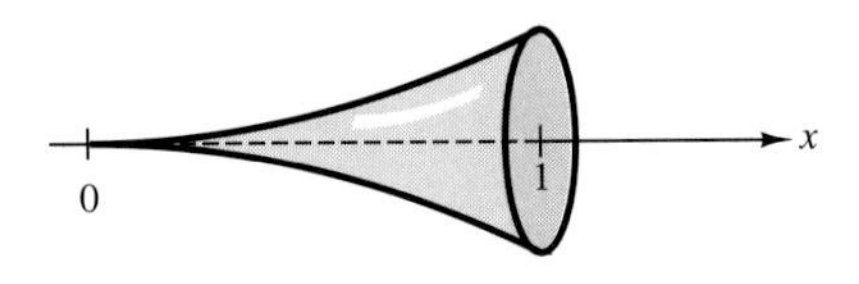

14. Find the x-coordinate of the center of mass of the solid in Figure 9.9.

15. The horn-shaped solid in Figure 9.10 has circular cross sections. The radius r varies with x according to the formula

$$r(x) = \frac{1}{2} + x^2 \text{ for } 0 \leq x \leq 1.$$

Find the volume of the solid.

Figure 9.10 Another parabolic horn

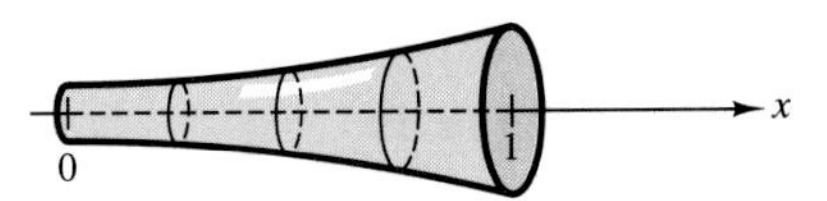

16. The horn-shaped solid in Figure 9.11 has circular cross sections. The radius r varies with x according to the formula

$$r(x) = \frac{1}{2} + \frac{1}{2}e^x \text{ for } 0 \leq x \leq 1.$$

Find the volume of the solid.

Figure 9.11 An exponential horn

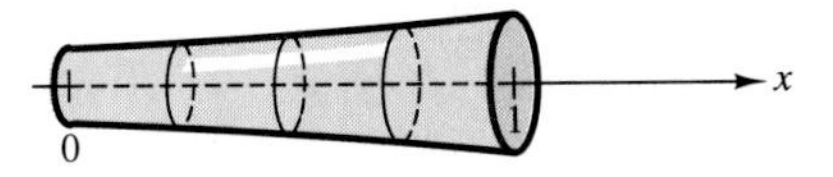

17. A thin cylindrical rod with diameter 10 centimeters and length 75 centimeters has a variable density. If we place the rod in a coordinate system as indicated in Figure 9.12, the density δ has the form

$$\delta(x) = 0.01 + 0.0001x$$

grams per cubic centimeter. Find the mass of the rod.

18. Find the center of mass of the rod in Exercise 17.

19. A thin cylindrical rod with diameter 5 centimeters and length 90 centimeters has a variable density. If we place the rod in a coordinate system as indicated in Figure 9.12, the density δ has the form

$$\delta(x) = 0.005 + 0.0002x$$

grams per cubic centimeter. Find the mass of the rod.

Figure 9.12 A cylindrical rod

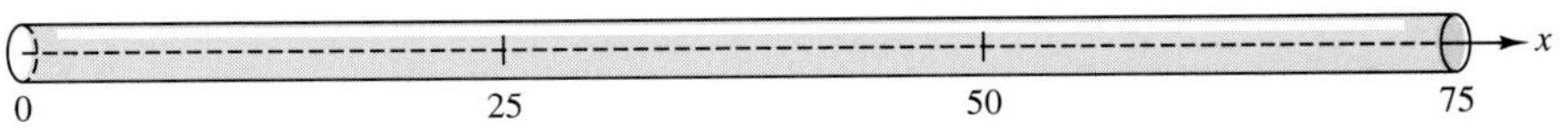

20. Find the center of mass of the rod in Exercise 19.

Evaluate each of the following indefinite integrals.

21. $\int \sin 3x\, dx$

22. $\int e^{3t+1}\, dt$

23. $\int \frac{du}{1+3u}$

24. $\int (1+x)^3\, dx$

25. $\int \frac{dt}{t(1+3t)}$

26. $\int \sqrt{u}\, du$

27. $\int \sqrt{1+x}\, dx$

28. $\int \cos \frac{t}{3}\, dt$

29. $\int \sin(u+\pi)\, du$

Evaluate each of the following definite integrals. Use the integral key on your calculator to check your work.

30. $\int_0^{\pi/3} \sin 3x\, dx$

31. $\int_0^1 e^{3t+1}\, dt$

32. $\int_0^1 \frac{du}{1+3u}$

33. $\int_1^2 (1+x)^3\, dx$

34. $\int_1^2 \frac{dt}{t(1+3t)}$

35. $\int_0^4 \sqrt{u}\, du$

36. $\int_{-1}^3 \sqrt{1+x}\, dx$

37. $\int_0^{\pi} \cos \frac{t}{3}\, dt$

38. $\int_0^{\pi/2} \sin(u+\pi)\, du$

9.2 Two-Dimensional Centers of Mass

In the preceding section we considered the problem of finding the center of mass of objects that, because of their symmetry, were essentially one-dimensional. For the pool cue, the only significant question was where the center of mass was located along its length.

The most important center-of-mass problems are three-dimensional. For example, the designer of an aircraft has to know where its center of mass will be—under all allowed loading conditions—in order to make sure that it is over the wings. Otherwise the plane won't fly. Furthermore, the center of mass has to be a little forward of the main landing gear; too far back, and the tail will drag on the runway. In Figure 9.13 we show an aircraft with wing-mounted engines. Aircraft with tail-mounted engines have the wings and landing gear much farther back—engines are heavy.

Figure 9.13 A typical object for which balance is an important problem

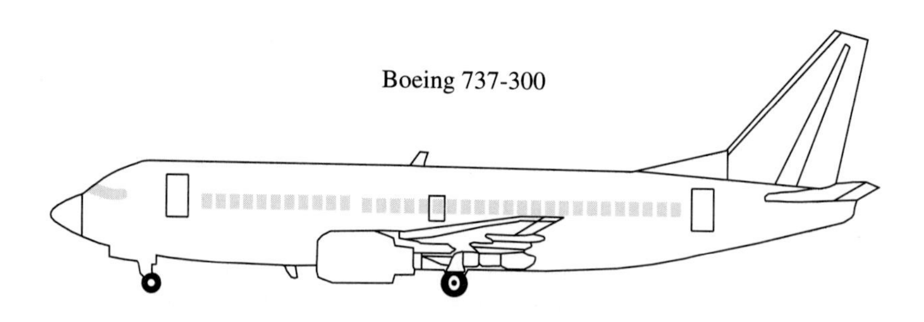

The problem of finding the center of mass in an object as complicated as a real aircraft is still beyond our capabilities, so we will simplify the problem to one of two dimensions. Suppose you draw Figure 9.13 on a cardboard cutout of the same shape and

hang it from a thread—for example, as part of a mobile displaying the aircraft being used by a given airline. Where would you place the thread?

Our problem is now simplified to one of finding the center of mass of a piece of cardboard in the shape shown in Figure 9.14.

Figure 9.14 A cardboard cutout for an aircraft mobile

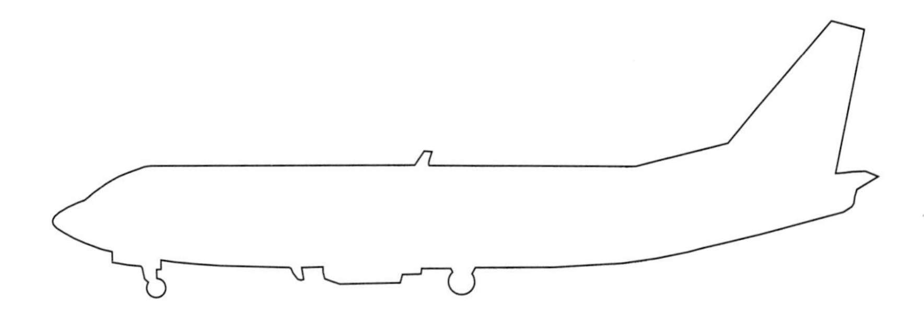

We'll assume that the cardboard has thickness θ (the Greek "th," lower-case theta) inches and a density δ ounces per cubic inch. We will see that we do not need to know the actual values of these two constants. The center of mass of a two-dimensional figure has two coordinates, $\overline{x}$ and $\overline{y}$. To find these, we need to find the mass and the moments in both horizontal and vertical directions. We will take this a step at a time.

Calculation of Mass

First, we need to calculate the mass m of the cutout. As before, this is just density $\times$ volume, and since the thickness is constant, the volume is area $\times$ thickness. Thus,

$$\begin{aligned} m &= \delta \times \text{volume} \\ &= \delta \times \theta \times \text{area}. \end{aligned}$$

So the calculation of the mass m is reduced to the calculation of the area.

We can determine the area by our divide-and-conquer strategy. First, we slice the shape into approximately rectangular segments. Then we approximate the areas of these segments by areas of actual rectangles. Finally, we consider what happens to the sum of the rectangular areas as the number of terms becomes infinite and the size of each term becomes infinitesimal.

Checkpoint 1

(a) Draw horizontal lines near the top and bottom of each strip in Figure 9.15 to create rectangles whose areas are approximately equal to those of the corresponding strips.

(b) Use a ruler to measure the height of each of your rectangles and their common width. Estimate the total area of the cardboard cutout.

Figure 9.15 Subdivision of the cardboard shape into vertical strips

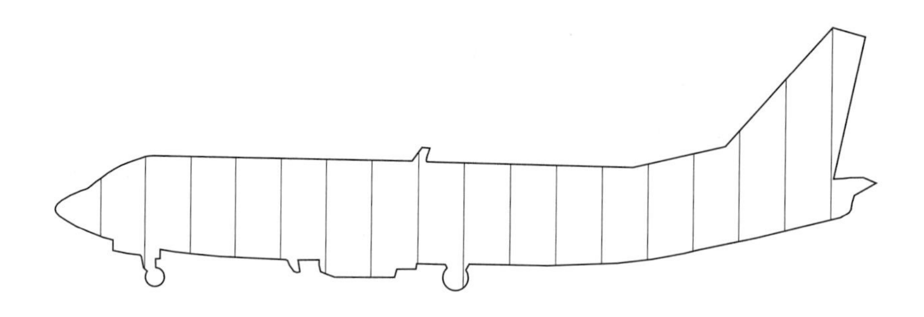

Notice that part (a) of the checkpoint task is much easier with 18 strips (the number shown) than it would be if we had subdivided into only 2 or 3 strips. As the number of strips increases, each strip looks more like a rectangle, regardless of the shape of the original object. If we drew the figure with 180 strips, it would be tedious to draw in all those rectangles, but it would be easy to decide where to draw the top and bottom of each strip.

After you have drawn your own rectangular approximation to the airplane shape, look at our solution in Figure 9.16, in which we have also added coordinate axes. It is not important that your rectangles match ours exactly or that your estimate of area based on 18 strips agree exactly with ours. We can only imagine the process of letting the number of strips become infinite, but as it does, our differences in estimation will disappear.

Let's call the points of subdivision on the x-axis x_0, x_1, x_2, and so on, up to x_{18}. (In this case, $x_0 = 0$ because of our placement of the origin.) As before, we write Δx for the constant spacing between the points; thus $\Delta x = x_k - x_{k-1}$ for every k from 1 to 18. The spacing between points on the x-axis is also the width of each of our rectangles, one of the things we need to know in order to calculate area. Now suppose that we know a function h whose value at x is the vertical distance (height) across our cutout at that particular x. Then the height of the kth rectangle is *approximately* $h(x_k)$. Thus the total area is approximately

$$h(x_1)\Delta x + h(x_2)\Delta x + h(x_3)\Delta x + \cdots + h(x_{17})\Delta x + h(x_{18})\Delta x.$$

This time we have written the approximation as a right-hand sum, i.e., we approximated by assuming that h has the constant value $h(x_k)$ on the interval $[x_{k-1}, x_k]$. In our previous applications of the divide-and-conquer process we used only left-hand sums. In a laboratory exercise associated with this chapter, you may see that it doesn't make much difference—if the number n of subdivisions is sufficiently large, the left- and right-hand sums would be very close to each other and to the actual area. Furthermore, as n becomes large, not only do all the area-approximating sums get closer to the actual area, but also they get closer to the definite integral of h from one end of the airplane cutout to the other. Thus the area of the cutout must be that integral.

Figure 9.16 Approximation of the cardboard mass by rectangular masses

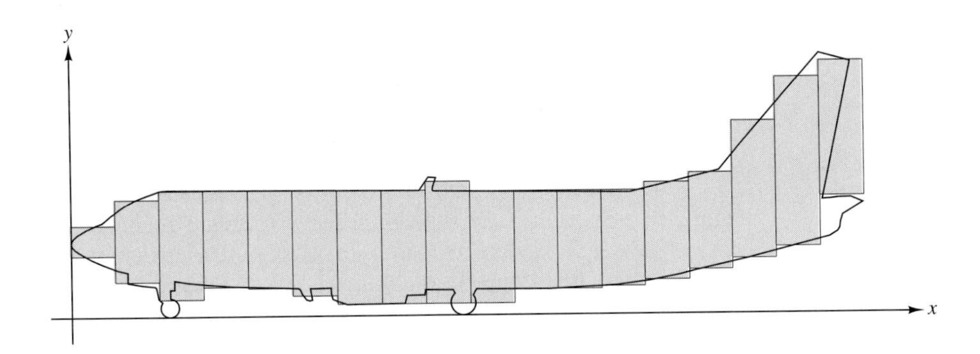

That's as far as we can take the explicit calculation right now, since we don't have a formula for the function h, and therefore, we have no way to actually evaluate the integral. However, we have made the following progress on a procedure for calculating a center of mass. To find the mass of a planar shape of uniform density and thickness:

- Find the area of the shape as

$$A = \int_a^b h(x)\,dx,$$

  where h is the function giving vertical height at an x, and a and b are the endpoints of the interval over which the shape extends.

- Calculate the mass as $m = \delta\theta A$, where δ is the density, and θ is the thickness.

Checkpoint 2

Consider the region in the xy-plane bounded above by $y = \sqrt{x}$, below by the x-axis, and on the right by the line $x = 1$.

(a) Make a sketch of the region.

(b) Find its area.

Calculation of the Horizontal Moment

The x-coordinate $\overline{x}$ is the location along the x-axis where, if you stuck pins in the edges of the cutout, it would balance (see Figure 9.17). As in the one-dimensional case, the horizontal moment of the figure is the product of $\overline{x}$ and the mass. Because $\overline{x}$ is measured from the y-axis, this moment is called **the moment with respect to the y-axis** and is denoted M_y. Thus, for a point mass m with x-coordinate $\overline{x}$, $M_y = \overline{x}\,m$. Our balance condition for $\overline{x}$ is that the moment with respect to the y-axis of a point mass m with x-coordinate $\overline{x}$ equals the moment M_y of the whole cutout.

Figure 9.17 Balancing the cutout along the line $x = \overline{x}$

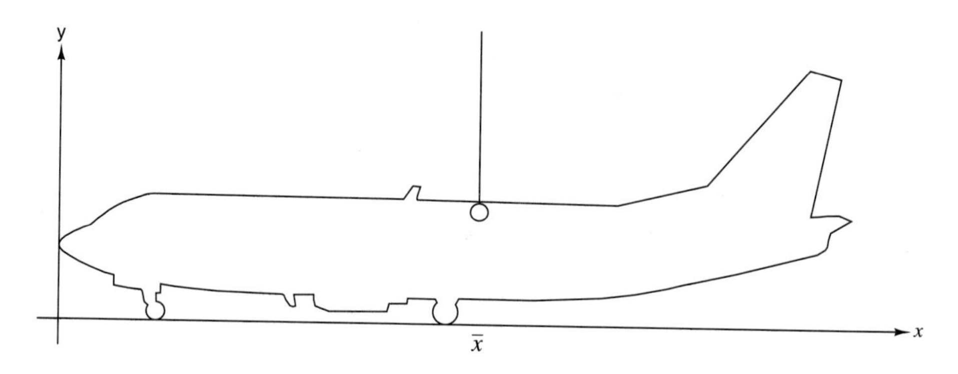

Now we consider the problem of calculating M_y. In Figure 9.18 we have selected and shaded a typical approximating rectangle and called it the kth rectangle. (It happens that k is 15 for this discussion.) It is easy to locate the center of mass of this rectangle: It must be at the geometric center. And it is easy to determine the coordinates $(\overline{x}_k, \overline{y}_k)$ of that center: $\overline{x}_k$ is the average of x_{k-1} and x_k, and $\overline{y}_k$ is the average of the y-coordinates of the top and bottom of the rectangle. We also know the mass of this rectangle:

$$\text{height} \times \text{width} \times \text{thickness} \times \text{density},$$

or $h(\overline{x}_k)(\Delta x)\theta\delta$, at least approximately. Thus we can approximate the moment for this rectangle by multiplying this mass by the moment arm, the distance of the center of mass from the y-axis, which is just $\overline{x}_k$.

Figure 9.18 The center of mass of the typical rectangle

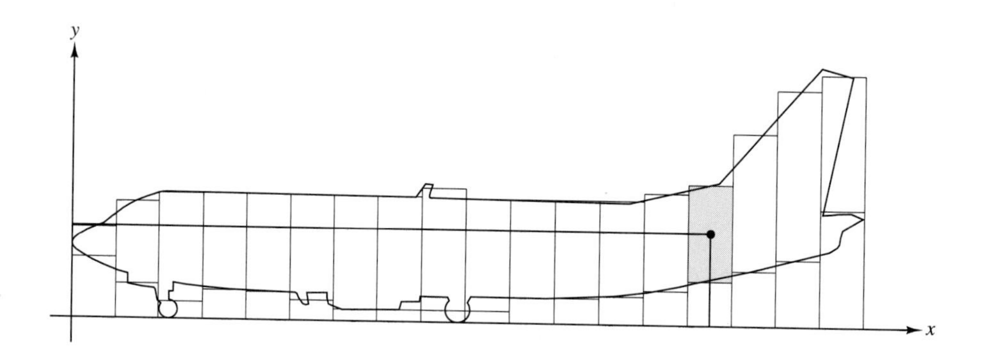

Now we can sum the moment contributions from all n rectangles to find that

$$M_y \approx \sum_{k=1}^{n} \overline{x}_k \, h(\overline{x}_k) \, (\Delta x) \, \theta \, \delta.$$

This too is an approximating sum for an integral, but not an integral of h. Rather, after we factor out the constant factors, θ and δ, we have a sum of the form

$$\theta \, \delta \sum_{k=1}^{n} w(\overline{x}_k) \, \Delta x,$$

where the new function w is defined by $w(x) = x\,h(x)$. As the number n of subdivisions increases, the approximating sum (including the constant factors of θ and δ) gets closer to the moment we seek and also to the integral it approximates, so we conclude that M_y *is* that integral:

$$M_y = \theta\delta \int_a^b x\,h(x)\,dx\,.$$

Calculation of $\overline{x}$

Now that we know how to find both the mass m and the horizontal moment M_y, the final step in finding the x-coordinate $\overline{x}$ is to divide the two:

$$\begin{aligned}\overline{x} &= \frac{M_y}{m}\\ &= \frac{\theta\delta \displaystyle\int_a^b x\,h(x)\,dx}{\theta\delta \displaystyle\int_a^b h(x)\,dx}\\ &= \frac{\displaystyle\int_a^b x\,h(x)\,dx}{\displaystyle\int_a^b h(x)\,dx}.\end{aligned}$$

We hesitate to dignify this equation with the name "formula," because it is not something you should think of as a universal solution to center-of-mass problems, and you certainly don't want to just plug something into it and start simplifying. Rather, you should think of it as a guide to the process of finding $\overline{x}$, more or less as we outlined the steps. Notice that θ and δ cancel out in this formula—because we took them to be constant, which meant that they factored out of the sums and the integrals.

The equation

$$\overline{x} = \frac{\displaystyle\int_a^b x\,h(x)\,dx}{\displaystyle\int_a^b h(x)\,dx}$$

has a compelling logic that should not be overlooked—and that will serve as a guide to finding other center of mass coordinates, e.g., $\overline{y}$.

First, locating the center of mass of a cardboard cutout should be a purely geometric (as opposed to physical) problem, in the sense that it should depend only on the geometry of the shape. Hence the disappearance of θ and δ. Alternatively, we can declare both θ and δ to be 1; that is, we can set the unit of density to be whatever the density of our shape is and the unit of distance (in the third dimension) to be whatever the thickness is. Clearly, these declarations should have nothing to do with the location of the two-dimensional center of mass. With either of these interpretations, area and mass become the same number, and the moment M_y is identified with the integral

$$\int_a^b x\,h(x)\,dx.$$

Second, we can read

$$\overline{x} = \frac{\displaystyle\int_a^b x\,h(x)\,dx}{\displaystyle\int_a^b h(x)\,dx}$$

in a less elegant but more meaningful way:

$$\text{Average } x = \frac{\displaystyle\int_a^b \text{(typical moment arm)(typical height)(typical width)}}{\displaystyle\int_a^b \text{(typical height)(typical width)}}.$$

Thus we accumulate mass (area) for the denominator by adding up height $\times$ width for infinitely many rectangles, each with infinitesimal area. We accumulate the total moment for the numerator by multiplying each small rectangle by its distance from the y-axis and adding up the results.

There is still another way to read this formula: The average x has to be a *weighted* average, because the distribution of mass (area) is not uniform. In the numerator, we can think of the height and width factors $[h(x)\,\Delta x]$ as weighting the typical x factor. Then all the weighted x's are added up—and to find an average, the sum of weighted x's is divided by the sum of the weights.

Checkpoint 3

(a) Find the moment with respect to the y-axis of a planar region of thickness θ and density δ that is bounded above by $y = \sqrt{x}$, below by the x-axis, and on the right by the line $x = 1$. (You found the area of this planar region in Checkpoint 2. Your calculation for M_y will contain θ and δ.)

(b) Find $\overline{x}$ for this region.

Calculation of the Vertical Moment and of $\overline{y}$

We return to our aircraft mobile scenario to discuss the y-coordinate of the center of mass, $\overline{y}$. We do not need to know $\overline{y}$ if we want the cutout to hang vertically—all we have to do is attach the string near the top edge along the line $x = \overline{x}$. On the other hand, we would need to know $\overline{y}$ if we want to balance the cutout in a horizontal position.

We have already done most—not all—of the hard work of figuring out how to calculate $\overline{y}$. Instead of thinking of the y-axis as a balance line, we need to think now of the x-axis in this way. We already know a formula for the mass m of our cardboard shape. Whatever $\overline{x}$ and $\overline{y}$ turn out to be, if we placed a point mass m at the point $(\overline{x}, -\overline{y})$ and connected it to the cutout with a very light rod, the combined figure should balance at the point $(\overline{x}, 0)$, i.e., along the x-axis. Thus, just as was the case with $\overline{x}$, we have

$$\overline{y} = \frac{M_x}{m},$$

where M_x is the moment with respect to the x-axis. The question is, how do we calculate M_x?

In Figure 9.19 we repeat the picture highlighting the kth rectangle. Observe that $\overline{y}_k$, the y-coordinate of the center of mass of the typical rectangle, has already been identified. In general, if we know y-coordinates for the top and bottom of the typical rectangle, the average of those coordinates will be $\overline{y}_k$. Since we may consider the mass $\theta\,\delta\,h(\overline{x}_k)\,\Delta x$ of the kth rectangle to be concentrated at $(\overline{x}_k, \overline{y}_k)$, the contribution of this rectangle to M_x is approximately $\overline{y}_k\,\theta\,\delta\,h(\overline{x}_k)\,\Delta x$. Thus

$$M_x \approx \theta\,\delta \sum_{k=1}^{n} \overline{y}_k\,h(\overline{x}_k)\,\Delta x.$$

Notice the similarity between this approximate equation and the corresponding formula for the horizontal moment,

$$M_y \approx \theta\,\delta \sum_{k=1}^{n} \overline{x}_k\,h(\overline{x}_k)\,\Delta x.$$

As the number n of subintervals becomes large, the right-hand side of

$$M_x \approx \theta\,\delta \sum_{k=1}^{n} \overline{y}_k\,h(\overline{x}_k)\,\Delta x.$$

approaches M_x. That right-hand side also approaches an integral, but to say what integral, we need some more notation.

Figure 9.19 The center of mass of the typical rectangle

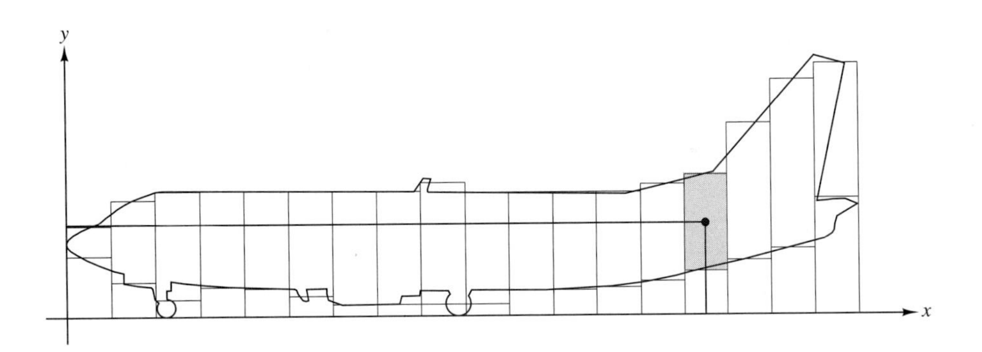

Let us suppose that the top edge of the airplane shape is the graph of a function $y = f(x)$, and the bottom edge is the graph of another function $y = g(x)$. (This is not strictly possible with our airplane shape, because there is one place where the top edge doubles back on itself, and there are several places where the bottom edge does so. We will ignore this complication; we could easily smooth off these rough edges without changing the center of mass much.) Then our height function $h(x)$ is $f(x) - g(x)$. Furthermore, we can express the y-coordinate of the center of mass of each rectangle in our subdivision in terms of f and g:

$$\overline{y}_k = \tfrac{1}{2}[f(\overline{x}_k) + g(\overline{x}_k)].$$

In this notation,

$$M_x \approx \theta\,\delta \sum_{k=1}^{n} \overline{y}_k\, h(\overline{x}_k)\, \Delta x$$

becomes

$$M_x \approx \frac{\theta\,\delta}{2} \sum_{k=1}^{n} [f(\overline{x}_k) + g(\overline{x}_k)]\,[f(\overline{x}_k) - g(\overline{x}_k)]\, \Delta x.$$

In terms of the functions f and g, it is clear how each component of the moment approximation depends on x, so it is now clear what integral is being approached by the sums — and therefore what integral represents M_x:

$$M_x = \frac{\theta\,\delta}{2} \int_a^b [f(x) + g(x)]\,[f(x) - g(x)]\, dx.$$

This equation contains a lot of symbols, but it says something familiar. First, there are the factors of θ and δ that will cancel with the corresponding factors in the mass calculation. Second, the integral has factors for the moment arm of each thin rectangle,

$$\frac{1}{2}[f(x) + g(x)],$$

for the height of each rectangle,

$$f(x) - g(x),$$

and for the width, dx, of each rectangle. The calculation of $\overline{y}$ therefore fits into the same pattern we saw for $\overline{x}$:

$$\text{Average } y = \frac{\displaystyle\int_a^b \text{(typical moment arm)(typical height)(typical width)}}{\displaystyle\int_a^b \text{(typical height)(typical width)}}.$$

Checkpoint 4

The equation

$$M_x = \frac{\theta\,\delta}{2} \int_a^b [f(x) + g(x)]\,[f(x) - g(x)]\, dx$$

may be written in an algebraically simpler form by multiplying out the factors involving f and g. Do it.

We can no more complete the calculation of $\overline{y}$ for our airplane shape than we could of $\overline{x}$, and for the same reason: We don't know formulas for the functions f and g. Thus, to illustrate the use of the formula from Checkpoint 4 for actually computing a $\overline{y}$, we return to the shape you considered in Checkpoints 2 and 3.

Example 1 Find $\overline{y}$ for the region the xy-plane bounded above by $y = \sqrt{x}$, below by the x-axis, and on the right by the line $x = 1$.

Solution The upper boundary curve is $f(x) = \sqrt{x}$, and the lower boundary curve is $g(x) = 0$, both defined for x running from 0 to 1. When $g(x) = 0$, our formula takes the simpler form

$$M_x = \frac{\theta\,\delta}{2} \int_a^b [f(x)]^2\,dx.$$

Thus

$$M_x = \frac{\theta\,\delta}{2} \int_0^1 [\sqrt{x}]^2\,dx = \frac{\theta\,\delta}{2} \int_0^1 x\,dx = \frac{\theta\,\delta}{4}.$$

You found in Checkpoint 2 that the area of the region is $2/3$ —equivalently, the mass is $2\,\theta\,\delta/3$— so $\overline{y} = 3/8$. ■

In this section we extended the subdivide-and-conquer approach from the preceding section to the two-dimensional problem of finding the center of mass of a thin planar shape. Assume the region is placed in an xy-coordinate system so that it may be described as the set of all (x, y) such that

$$a \le x \le b \qquad \text{and} \qquad g(x) \le y \le f(x).$$

Then the cross-sectional height $h(x) = f(x) - g(x)$. If δ is the density and θ the thickness, then the total mass is

$$\begin{aligned} m &= \delta\,\theta \times \text{area of the region} \\ &= \delta\,\theta \int_a^b [f(x) - g(x)]\,dx. \end{aligned}$$

The moment with respect to the y-axis is

$$\begin{aligned} M_y &= \int_a^b \text{(typical moment arm)(typical height)(typical width)} \\ &= \theta\,\delta \int_a^b x\,[f(x) - g(x)]\,dx, \end{aligned}$$

and the x-coordinate of the center of mass is

$$\begin{aligned} \overline{x} &= \frac{M_y}{m} \\ &= \frac{\displaystyle\int_a^b x\,[f(x) - g(x)]\,dx}{\displaystyle\int_a^b [f(x) - g(x)]\,dx}. \end{aligned}$$

The corresponding moment with respect to the x-axis is

$$M_x = \int_a^b \text{(typical moment arm)(typical height)(typical width)}$$
$$= \frac{\theta\,\delta}{2}\int_a^b \left[f(x)^2 - g(x)^2\right] dx,$$

and the y-coordinate of the center of mass is

$$\bar{y} = \frac{M_x}{m}$$
$$= \frac{\dfrac{1}{2}\displaystyle\int_a^b \left[f(x)^2 - g(x)^2\right] dx}{\displaystyle\int_a^b \left[f(x) - g(x)\right] dx}.$$

Answers to Checkpoints

1. (a) See Figure 9.16.

   (b) The estimate depends on where you drew your horizontal lines. Our estimate is about 14 square centimeters or about 2.1 square inches.

2. (a)

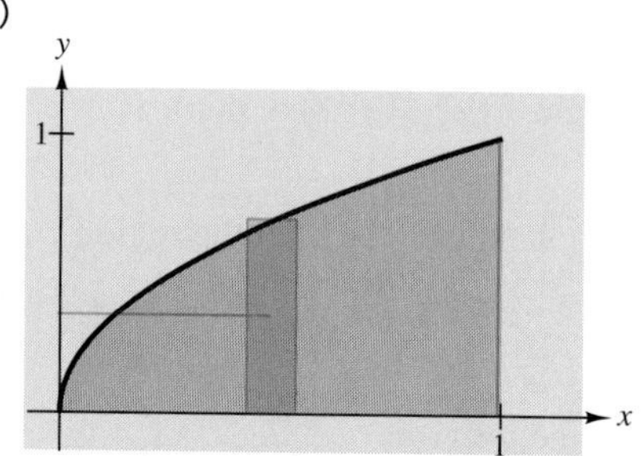

   This figure also shows a typical rectangle and a typical moment arm for use in the next checkpoint.

   (b) $\frac{2}{3}$.

3. (a) $\frac{2}{5}\,\delta\,\theta$ (b) $\frac{3}{5}$

4. $M_x = \dfrac{\theta\,\delta}{2}\displaystyle\int_a^b \left[f(x)^2 - g(x)^2\right] dx$

Exercises 9.2

Find the center of mass of each of the following regions. Calculate each integral by hand, and check your work with your calculator.

1. The region bounded above by the curve $y = 2 - x^2$ and below by the x-axis.
2. The region bounded above by the line $y = x$ and below by the curve $y = x^2$.
3. The region in the first quadrant bounded above by the curve $y = 2x^4$, below by the x-axis, and on the right by $x = 2$.
4. The region in the first quadrant bounded above by the curve $y = 1 + x^2$, below by the x-axis, on the left by $x = 0$, and on the right by $x = 1$.
5. The region in the first quadrant bounded above by the curve $y = \sqrt{x}$ and below by $y = x$.
6. The region in the first quadrant bounded above by the curve $y = x^3$, below by $y = x$, on the left by $x = 1$, and on the right by $x = 2$.
7. Find the center of mass of the half-disk under the graph of $x^2 + y^2 = 1$ and above the x-axis, following these steps:
   (a) First, without any calculation, what is $\overline{x}$? What is the area?
   (b) Now set up an integral for M_x and evaluate it.
   (c) What is $\overline{y}$?
8. Find the center of mass of the quarter-circular region under the graph of $x^2 + y^2 = 1$ and in the first quadrant. (*Hint*: $\overline{x}$ must be the same as $\overline{y}$ — why? Thus you need at most one moment calculation. Use as much of the previous exercise as you can—you may be able to solve this problem without any calculation at all.)

9.3 Numerical Approximation of Integrals

We have seen a variety of reasons for wanting to calculate definite integrals—to find distances, areas, volumes, and centers of mass. In future sections we will see many more. Thus we need to be able to calculate these integrals with confidence. One way of performing these calculations is close at hand: your calculator's integral key. You enter the function and the limits, and it does the rest, returning a numerical approximation to the integral. How does it do that?

In this section we explore numerical procedures for calculating definite integrals, procedures that are not difficult, just tedious. Your calculator or a computer uses procedures like these to calculate the numerical approximations. These procedures are effective tools for rapid enough calculation that we can generate numerical or graphical antiderivatives, even when we can't find formulas for them.

Left-Hand and Right-Hand Sums

Our starting point is the definition of the definite integral in terms of sums of products. Specifically, we formed the sum

$$\sum_{k=1}^{n} f(t_{k-1})\,\Delta t$$

to calculate the sum of areas of rectangles with height $f(t_{k-1})$ and width Δt, where Δt is the width of each subinterval when the interval $[a, b]$ is divided into n equal pieces. Then we defined the integral of f from a to b to be the limiting value of such sums as $n \to \infty$

and $\Delta t \to 0$. It follows that for large values of n, the sum should itself be a good approximation to the limiting value, i.e., to the integral.

The function values in this sum are calculated at the left-hand endpoints of the subintervals—i.e., at t_{k-1} for the kth subinterval $[t_{k-1}, t_k]$. For this reason, we call this sum a left-hand sum. Similarly, we can calculate a right-hand sum by using function values at the right-hand end of each subinterval.

Definitions If f is continuous on the interval $[a, b]$, n is a positive integer, and $\Delta t = (b - a)/n$, then

$$\sum_{k=1}^{n} f(t_{k-1})\,\Delta t$$

is the corresponding **left-hand sum** approximation to the integral

$$\int_a^b f(t)\,dt.$$

The sum

$$\sum_{k=1}^{n} f(t_k)\,\Delta t$$

is the corresponding **right-hand sum** approximation to the same integral.

We illustrate right- and left-hand sums by recalling a calculation from Chapter 8. There we estimated the distance traveled during a short automobile trip by integrating the velocity function over the time interval from 0 to 13 minutes. Thus we wanted the value of the definite integral

$$\int_0^{13} v(t)\,dt,$$

where $v(t)$ is the function graphed in Figure 9.20 and tabulated in Table 9.2, both repeated from Chapter 8.

Figure 9.20 Example of a left-hand sum

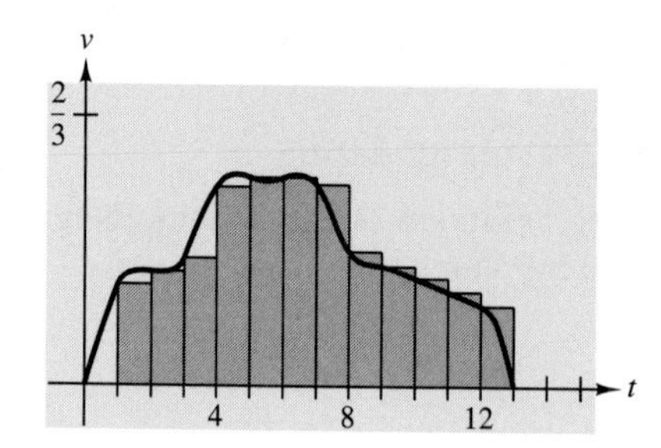

Table 9.2 Speed of a car between two traffic lights

Time (minutes)	*Speed (miles per minute)*	*Time (minutes)*	*Speed (miles per minute)*
0	0	7	0.500
1	0.250	8	0.333
2	0.283	9	0.300
3	0.317	10	0.267
4	0.500	11	0.233
5	0.517	12	0.200
6	0.517	13	0

Example 1 Find the left-hand sum approximation for $n = 13$ and $\Delta t = 1$ to

$$\int_0^{13} v(t)\, dt.$$

Solution Since $\Delta t = 1$, we need only add up the first 13 function values. The left-hand sum is 4.217.

Checkpoint 1

(a) Draw the companion to Figure 9.20 that illustrates the right-hand sum for $n = 13$; i.e., sketch in and shade the corresponding rectangles whose areas add up to the right-hand sum with 13 terms.

(b) Use the values in Table 9.2 to calculate the right-hand sum.

We saw in Table 8.3 that we could refine our estimate of the area under the curve in Figure 9.20 (i.e., the integral) by calculating left-hand sums with large numbers of terms. We turn now to an example for which we can compare sums to a known exact value for an integral—unlike the velocity example, for which we had no formula. Our purpose in estimating an answer we already know is to get some sense of how many terms are needed to achieve a given accuracy.

In Figure 9.21 we show the graph of $y = 1/x$, under which we have shaded the area from $x = 1$ to $x = 3$. According to the Fundamental Theorem, that area is

$$\int_0^3 \frac{1}{x}\, dx = \ln x\Big|_1^3 = \ln 3 - \ln 1 = \ln 3 \approx 1.098612288668110.$$

We have given the answer to 15 decimal places because we need a highly accurate answer to compare to our approximations.

Figure 9.21 Area under the graph of $y = \dfrac{1}{x}$ from $x = 1$ to $x = 3$

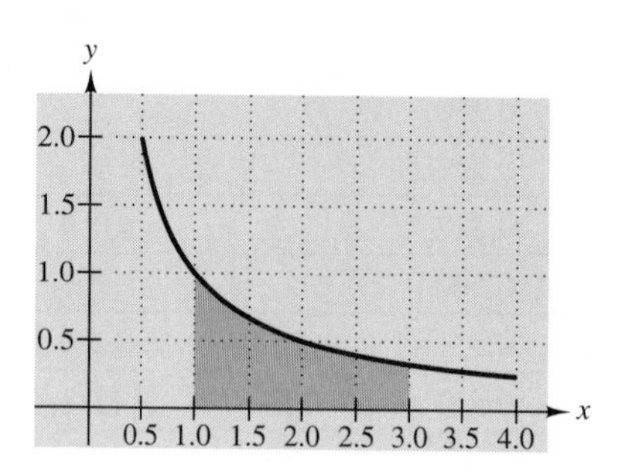

Exploration Activity 1

(a) The 10-term left-hand sum approximating the area in Figure 9.21 starts out

$$1 \cdot 0.2 + \frac{1}{1.2} \cdot 0.2 + \frac{1}{1.4} \cdot 0.2 + \cdots .$$

Fill in the other seven terms, and calculate the sum.

(b) Calculate the right-hand sum with ten terms.

(c) Subtract the known area ($\ln 3$) from the left-hand sum in part (a) and the right-hand sum in part (b) to find the errors in these two approximations.

(d) Explain the signs of the errors by referring to Figure 9.21.

In Table 9.3 we show the results of calculating 10-term, 100-term, and 1000-term left-hand sums and right-hand sums approximating the shaded area in Figure 9.21. In the last two columns of the table we show the errors in these approximations, obtained by subtracting the 15-place value of $\ln 3$. The $n = 10$ row in the table should confirm what you just did in Exploration Activity 1.

The region formed by the approximating rectangles for the left-hand sum contains the region under the curve. Thus the approximation is greater than the integral, and the error —for our chosen order of subtraction—is positive. For the right-hand sum, the region formed by the approximating rectangles is contained inside the region under the curve. Accordingly, this approximation is less than the integral and the error is negative.

Table 9.3 Approximations to $\int_1^3 \frac{1}{x}\,dx = \ln 3 \approx 1.098612$

n	*Left-hand sum (LHS)*	*Right-hand sum (RHS)*	*Error in LHS*	*Error in RHS*
10	1.168229	1.034896	0.0696	−0.0637
100	1.105309	1.091975	0.00669	−0.00664
1000	1.099279	1.097946	0.000667	−0.000666

Exploration Activity 2

(a) How do the errors in left-hand sum and right-hand sums compare in size?

(b) What happens to the error in LHS when the number n of subdivisions is made 10 times larger? In RHS?

For each value of n, the errors in left-hand and right-hand sums are approximately equal in size but of opposite signs. (This is in marked contrast to the situation with the velocity function in Figure 9.20, for which you showed in Checkpoint 1 that the LHS and RHS were equal. We saw in Table 8.3 that for this particular function both appeared to underestimate the integral.) Each time we increase n by a factor of 10 we appear to decrease both errors by a factor of $1/10$.

Exploration Activity 3

(a) Explain why, if f is any *decreasing* function, then any LHS will *over*estimate the integral and any RHS will *under*estimate the integral.

(b) What can you conclude if f is an *increasing* function? Explain your answer.

If f is a decreasing function, the value of f at the left endpoint t_{k-1} of any subinterval $[t_{k-1}, t_k]$ is always greater than the value of f at every other point in that subinterval. Accordingly, the LHS will always be larger than the value of the integral. Similarly, the value of f at the right endpoint t_k is always less than the value at every other point in that subinterval, so the RHS will always be less than the value of the integral. If f is increasing, the situation is reversed: The RHS is greater than the integral and the LHS is smaller.

For either increasing or decreasing functions, the situation in Table 9.3 is typical: One of the two sums will overestimate and the other underestimate the integral, both by about the same amount. That is, the value of the integral will lie *roughly* halfway between the two sums. This observation, together with a little algebra, leads to a simple explanation of the decrease in errors by a factor of 1/10 each time n is increased by a factor of 10.

Exploration Activity 4

(a) Explain why LHS and RHS can be written in the following way:

$$\text{RHS} = [\qquad f(x_1) + f(x_2) + \cdots + f(x_{n-1}) + f(x_n)]\,\Delta x;$$

$$\text{LHS} = [f(x_0) + f(x_1) + \cdots + f(x_{n-2}) + f(x_{n-1})\qquad]\,\Delta x.$$

(b) Subtract the LHS equation from the RHS equation, and make appropriate cancellations and substitutions to show that

$$\text{RHS} - \text{LHS} = \frac{[f(b) - f(a)]\,(b - a)}{n}.$$

(c) Explain why the calculation in part b implies that errors should decrease by a factor of $1/10$ each time n is increased by a factor of 10.

By definition, RHS $= \sum_{k=1}^{n} f(x_k)\,\Delta x$. When we factor out Δx, we obtain

$$\text{RHS} = \left[\sum_{k=1}^{n} f(x_k)\right]\Delta x.$$

When we express this without the summation notation, we have

$$\text{RHS} = [\quad f(x_1) + f(x_2) + \cdots + f(x_{n-1}) + f(x_n)]\Delta x.$$

A similar argument yields

$$\text{LHS} = [f(x_0) + f(x_1) + \cdots + f(x_{n-2}) + f(x_{n-1}) \quad]\Delta x.$$

Now we cancel terms in the difference to obtain

$$\text{RHS} - \text{LHS} = [f(x_n) - f(x_0)]\Delta x.$$

Since $x_n = b$, $x_0 = a$, and $\Delta x = (b-a)/n$, we may rewrite this as

$$\text{RHS} - \text{LHS} = [f(b) - f(a)]\frac{(b-a)}{n} = \frac{[f(b) - f(a)]\,(b-a)}{n}.$$

Thus, for a given n, the right-hand and left-hand sums differ by a *constant* times $1/n$. In particular, this says that increasing n by a factor of 10 must decrease the difference between these two sums by (exactly!) a factor of $1/10$. And, for increasing or decreasing functions, for which the integral lies roughly halfway between left and right sums, the distance to either sum (i.e., the error) also must go down by a factor of $1/10$.

Checkpoint 2

(a) For the shaded region in Figure 9.21, write down the value for each of the following:

$$a,\ b,\ f(b),\ b-a,\ f(b)-f(a),\ [f(b)-f(a)](b-a).$$

(b) For each line in Table 9.3, verify that the right-hand and left-hand sums differ by exactly $1/n$ times the last number in part (a)—as exactly as you can tell from the decimal approximations shown in the table.

(c) How big would n have to be for either LHS or RHS to approximate

$$\int_1^3 \frac{1}{x}\,dx$$

to six decimal place accuracy?

We've just run up against a reality check. It looked like we were making good progress on the problem of finding accurate approximations to integrals—all we have to do to decrease the error by a factor of $1/10$ (i.e., gain accuracy to one more place) is multiply the number of steps by 10. But to achieve even moderately good accuracy, this may require sums with *millions* of terms! That's simply not practical. We need a better idea.

The Midpoint and Trapezoidal Rules

In the discussion following Exploration Activities 1 and 2 we ignored a better idea staring us in the face. We observed that the errors for RHS and LHS are approximately equal with opposite signs; i.e., the value of the integral lies roughly halfway between these two approximations. Let's *average* the two sums and use that average as a new approximation to the integral.

As in Exploration Activity 4, we write the left-hand sums and right-hand sums in the form

$$\begin{aligned} \text{RHS} &= [\qquad\qquad f(x_1) + f(x_2) + \cdots + f(x_{n-1}) + f(x_n)]\,\Delta x; \\ \text{LHS} &= [f(x_0) + f(x_1) + \cdots + f(x_{n-2}) + f(x_{n-1}) \qquad\qquad]\,\Delta x. \end{aligned}$$

Now we add column by column and divide by 2 to find that the average of LHS and RHS is

$$\frac{\text{LHS} + \text{RHS}}{2} = \left[\frac{f(x_0) + f(x_1)}{2} + \frac{f(x_1) + f(x_2)}{2} + \cdots + \frac{f(x_{n-1}) + f(x_n)}{2}\right] \Delta x.$$

If you have already calculated LHS and RHS, you don't need a new formula to average them—just add and divide by 2. But if you are starting from scratch, the right-hand side of this equation reveals both a way to calculate the average and an interesting fact about it: It is also a sum of n terms, and the terms themselves (after factoring out Δx) are *averages* of the y-coordinates that were summed in LHS and RHS. In the notation of moments and centers of mass, this formula can be rewritten

$$\frac{\text{LHS} + \text{RHS}}{2} = (\overline{y}_1 + \overline{y}_2 + \cdots + \overline{y}_{n-1} + \overline{y}_n)\,\Delta x.$$

This suggests *another* good idea—which you also may have discovered in the laboratory: Instead of averaging y-coordinates, why not average x-coordinates? That is, instead of calculating function values at the endpoints (x_0, x_1, x_2, and so on), why not calculate them at the *midpoints* ($\overline{x}_0$, $\overline{x}_1$, $\overline{x}_2$, and so on)? This leads to an approximating sum for the definite integral that has the obvious name **Midpoint Rule** and the equally obvious abbreviation MR:

$$\text{MR} = [f(\overline{x}_1) + f(\overline{x}_2) + \cdots + f(\overline{x}_{n-1}) + f(\overline{x}_n)]\,\Delta x,$$

where, for each k from 1 to n, $\overline{x}_k = (x_{k-1} + x_k)/2$. The average of left and right sums has a less obvious name and abbreviation: For reasons we will explore in Exercise 23, it is called the **Trapezoidal Rule**, abbreviated TR. Thus

$$\text{TR} = (\overline{y}_1 + \overline{y}_2 + \cdots + \overline{y}_{n-1} + \overline{y}_n)\,\Delta x,$$

where, for each k from 1 to n, $\overline{y}_k = [f(x_{k-1}) + f(x_k)]/2$.

Now that we have two new ideas for approximating integrals, how good are they? In Table 9.4 we show the values for MR and TR that correspond to the entries of Table 9.3—same function, same interval, same values of n.

Table 9.4 Approximations to $\int_1^3 \frac{1}{x}\,dx = \ln 3 \approx 1.098612288668$

n	*Midpoint Rule (MR)*	*Trapezoid Rule (TR)*	*Error in MR*	*Error in TR*
10	1.097142	1.101562	−0.00147	0.00295
100	1.09859748	1.09864192	−0.0000148	0.0000296
1000	1.0986121405	1.0986125850	−0.000000148	0.000000296

Exploration Activity 5

(a) What patterns do you observe in the error columns of Table 9.4? Consider, in particular, how the errors decrease as n increases, which method underestimates and which overestimates, and how the MR errors are related to the TR errors.

(b) For this particular integral, how big does n have to be to achieve six-decimal-place accuracy with MR? with TR?

From Table 9.4 we see that the errors for the two methods have different signs, TR overestimates and MR underestimates, and that the magnitude of the error with MR is about half the magnitude of the error with TR. As we increase n by a factor of 10, both errors seem to decrease by a factor of $1/100$. We also see that $n = 1000$ is sufficient to obtain errors less than 0.5×10^{-6}, i.e., six-decimal-place accuracy.

Now we're getting somewhere! Sums with 1000 terms are easily evaluated by a personal computer—that's where the entries of the last two tables came from. On the other hand, you might have a little trouble getting such sums out of your calculator. You can probably tell from the speed of operation of your calculator's integral key that it isn't adding that many terms. Maybe we need another good idea.

Simpson's Rule

The patterns in Table 9.4, like those in the preceding table, suggest another good idea. As we observed, it appears that MR and TR tend to estimate on opposite sides of the exact answer and that the midpoint rule is roughly twice as good as the trapezoidal rule for a given n. That is, the error in MR is approximately half that of TR. These observations suggest that we can expect to find the value of the integral *between* MR and TR, about twice as far from TR as from MR—in other words, one-third of the way from MR to TR. What we want is an average that weights MR twice as much as TR:

$$\frac{2}{3}\,\text{MR} + \frac{1}{3}\,\text{TR}\,.$$

This good idea also has a name: **Simpson's Rule**, abbreviated SR. How good is Simpson's Rule? In Table 9.5 we display the results that correspond to those in the two preceding tables—same function, same interval, and same values of n.

Table 9.5 Approximations to $\int_1^3 \frac{1}{x}\,dx = \ln 3 \approx 1.098612288668110$

n	*Simpson's Rule (SR)*	*Error in SR*
10	1.09861550486	3.2×10^{-6}
100	1.098612288997	3.3×10^{-10}
1000	1.098612288668142	3.2×10^{-14}

Evidently Simpson's Rule is such a good idea that even a relatively short sum can approximate a "reasonable" integral to six-place accuracy, and somewhat longer sums—well within the range of computers or calculators—can do much better. Indeed, each increase in the number of steps by a factor of 10 decreases the error by a factor of $1/10,000$! It is very likely that your calculator carries out Simpson's Rule or something very much like it. And the same is likely to be true of commercial computer software that evaluates integrals.

Checkpoint 3

(a) Find the exact value of $\int_0^1 x^4\,dx$.

(b) With $n = 10$, calculate LHS, RHS, TR, MR, and SR.

(c) Calculate the errors in each of the five approximations.

(d) Do the errors fit the patterns you saw in the $n = 10$ lines in Tables 9.3, 9.4, and 9.5? If not, how do they differ?

In this section we have studied what your calculator or computer might be doing when it calculates definite integrals numerically. If f is a continuous function on the interval $[a, b]$, n is a positive integer, $\Delta x = (b - a)/n$, and $x_k = a + k\,\Delta x$ for $k = 0, 1, \ldots, n$, then the definition of the definite integral implies that the left-hand sum (LHS) and the right-hand sum (RHS) approximate the value of the integral:

$$\text{LHS} = \sum_{k=1}^{n} f(t_{k-1})\,\Delta t$$

and

$$\text{RHS} = \sum_{k=1}^{n} f(t_k)\,\Delta t.$$

However, these sums require too many calculations for reasonable accuracy in reasonable time. In fact, to get one more decimal place of accuracy with either LHS or RHS, n has to be increased by a factor of 10.

Better methods average the function values at the endpoints of the subintervals — the Trapezoidal Rule (TR) — or evaluate f at the midpoint of each subinterval — the Midpoint Rule (MR):

$$\text{TR} = \sum_{k=1}^{n} \frac{f(t_{k-1}) + f(t_k)}{2}\,\Delta t$$

and

$$\text{MR} = \sum_{k=1}^{n} f\left(\frac{t_{k-1} + t_k}{2}\right)\Delta t.$$

For each of these methods, increasing n by a factor of 10 tends to decrease error by a factor of $1/10^2$.

Simpson's Rule (SR) is a weighted average of the Midpoint and Trapezoidal Rules:

$$\text{SR} = \tfrac{2}{3}\text{MR} + \tfrac{1}{3}\text{TR}.$$

For SR, increasing n by a factor of 10 tends to decrease error by a factor of $1/10^4$. This makes Simpson's Rule an effective tool for rapid calculation of highly accurate approximations to definite integrals. The program in your calculator or computer is likely to be an adaptation of Simpson's Rule or something very much like it.

Answers to Checkpoints

1. 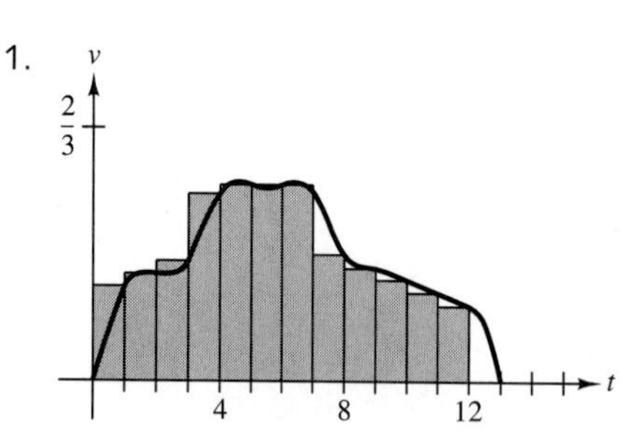

For this example, the right-hand sum is also 4.217.

2. (a) $a = 1,\ f(a) = 1, b = 3, f(b) = \frac{1}{3}, b - a = 2, f(b) - f(a) = \frac{2}{3}, [f(b) - f(a)](b - a) = \frac{4}{3}$

(b)

n	*RHS − LHS*
10	0.1333
100	0.01333
1000	0.001333

(c) It is sufficient to have $\frac{2}{3} \cdot \frac{2}{n} < \frac{1}{2} \times 10^{-6}$ or $n > \frac{8}{3} \times 10^6$, i.e., $n > 2,666,667$

3. (a) 0.2

(b) LHS $= 0.15333$, RHS $= 0.25333$, TR $= 0.20333$, MR $= 0.19834$,
SR $= 0.2000008333$ (to 10 decimal places)

(c)

Method	*Error*
LHS	−0.04667
RHS	0.05333
TR	0.00333
MR	−0.00166
SR	8.333×10^{-7}

(d) In general, the errors fit the patterns we expect. LHS and RHS have errors comparable in size and opposite in sign. TR and MR have errors another factor of $1/10$ smaller and MR has error magnitude half that of TR. The error for SR is smaller than we might have expected by at least a factor of $1/10$.

Exercises 9.3

For each of the following definite integrals use the Fundamental Theorem to calculate the exact value. (You may have already done this in Exercises 9.1.) Then calculate the Trapezoidal Rule, the Midpoint Rule, and the Simpson's Rule approximations for $n = 10$.

1. $\displaystyle\int_0^{\pi/3} \sin 3x\, dx$

2. $\displaystyle\int_0^1 e^{3t+1}\, dt$

3. $\displaystyle\int_0^1 \frac{du}{1+3u}$

4. $\displaystyle\int_1^2 (1+x)^3\, dx$

5. $\displaystyle\int_1^2 \frac{dt}{t(1+3t)}$

6. $\displaystyle\int_0^4 \sqrt{u}\, du$

7. $\displaystyle\int_{-1}^3 \sqrt{1+x}\, dx$

8. $\displaystyle\int_0^{\pi} \cos\frac{t}{3}\, dt$

9. $\displaystyle\int_0^{\pi/2} \sin(u+\pi)\, du$

10. (a) Use the trigonometric identity $\cos^2 x = \frac{1}{2}(1 + \cos 2x)$ to calculate

$$\int_0^{\pi/2} \cos^2 x\, dx$$

exactly.

(b) Estimate the integral in part (a) by left-hand sum and right-hand sums with four terms each. How close are your estimates to the exact answer?

(c) Estimate the integral in part (a) by midpoint and trapezoidal sums with four terms each. How close are your estimates to the exact answer?

(d) Estimate the integral in part (a) by Simpson's Rule, using the values of MR and TR from part (c). How close is your estimate to the exact answer?

For each of the following definite integrals, use a numerical method to approximate the value of the integral to 2 decimal places. Make sure your approximation has this accuracy by comparing it with the value determined by a calculator or computer program.

11. $\int_0^1 \sqrt{1+x^3}\, dx$

12. $\int_0^1 t\, e^{-t^2/2}\, dt$

13. $\int_0^{\pi} \theta \sin^2 \theta\, d\theta$

14. $\int_0^1 x^2 \sqrt{x^2+2}\, dx$

15. $\int_0^{\pi} \cos^2 x \sin^2 x\, dx$

16. $\int_0^1 \sqrt{9-x^2}\, dx$

17. $\int_0^1 \frac{x^2}{1+x^2}\, dx$

18. $\int_0^1 \frac{1}{\sqrt{1+x^2}}\, dx$

19. $\int_{-1}^1 \sqrt{1-\frac{x^2}{4}}\, dx$

20. $\int_0^1 e^{x^2}\, dx$

21. $\int_0^{\pi/2} \sqrt{9-4\sin^2\theta}\, d\theta$

22. $\int_{-\pi}^{\pi} \cos 2\theta \cos\theta\, d\theta$

23. Figure 9.22 shows four-step left-hand and right-hand approximations to the area under the graph of a function f on an interval $[a, b]$, where $a = x_0$ and $b = x_4$. Also shown are diagonal lines connecting the points $(x_{i-1}, f(x_{i-1}))$ and $(x_i, f(x_i))$ on each subinterval.

(a) Explain why the area under each diagonal line is the average of the areas of the corresponding left-hand and right-hand rectangles.

(b) Explain why the Trapezoidal Rule has the name it has.

Figure 9.22 Averaging areas of left-hand and right-hand rectangles

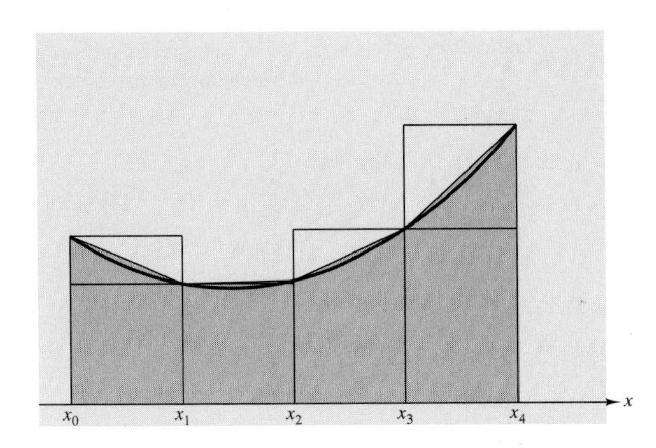

24. Figure 9.23 shows a four-step midpoint approximation to the area under the graph of a function f on an interval $[a, b]$, where $a = x_0$ and $b = x_4$. Also shown are the tangent lines at the points $(\overline{x}_i, f(\overline{x}_i))$ on each subinterval. Explain why the area under each tangent line is the same as the area of the corresponding midpoint rectangle.

Figure 9.23 Interpreting midpoint areas as areas under tangent lines

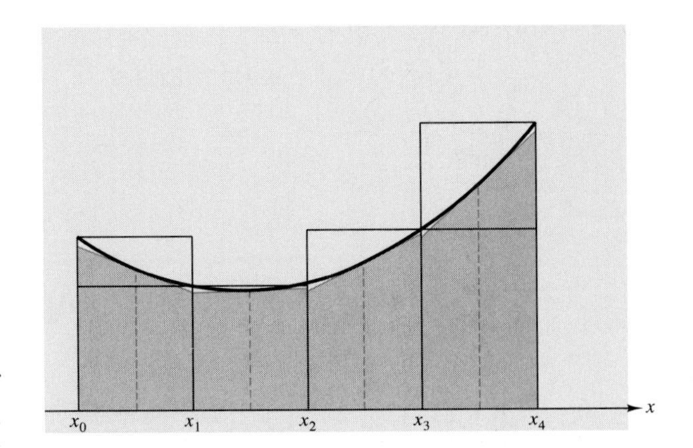

25. Figure 9.24 shows the graph of the quadratic polynomial function $f(x) = x^2 - 3x + 4$ on the interval $[-1, 2]$, along with one-step trapezoidal and midpoint approximations of area under the graph.

(a) Calculate the exact area under the graph and the one-step trapezoidal and midpoint approximations.
(b) Show that the error in the trapezoidal approximation is exactly twice the error in the midpoint approximation.

Figure 9.24 Errors in midpoint and trapezoidal approximations to the area under a parabola

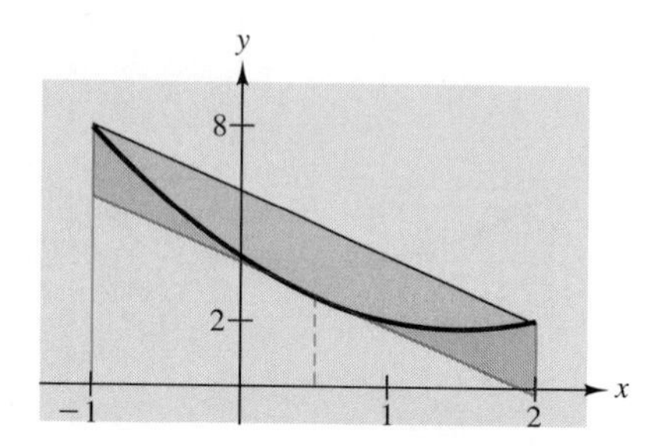

26. Suppose $f(x) = Ax^2 + Bx + C$ is an arbitrary quadratic polynomial function and $[a, b]$ is an arbitrary interval.
   (a) Calculate $\int_a^b f(x)\,dx$.
   (b) Calculate one-step trapezoidal and midpoint approximations to this integral.
   (c) Show that the error in the trapezoidal approximation is exactly twice the error in the midpoint approximation.
   (d) Explain why Simpson's Rule (for any number of steps) must give the exact value for the integral of any quadratic function.

27. Figure 9.25 shows the graph of

$$f(x) = 4x^3 - 4x^2 - 5x + 7$$

on the interval $[-1, 2]$, along with the graph of a quadratic function g that agrees with f at -1, at 2, and at the midpoint. That is, $f(-1) = g(-1)$, $f(2) = g(2)$, and $f(0.5) = g(0.5)$.
   (a) Calculate $\int_{-1}^2 f(x)\,dx$.
   (b) Calculate the one-step trapezoidal, midpoint, and Simpson approximations to this integral.
   (c) According to the preceding problem, the Simpson approximation is exactly the area under the graph of g. How does it compare to the area under the graph of f?
   (d) Choose another interval $[a, b]$, and calculate both $\int_a^b f(x)\,dx$ and a one-step Simpson approximation. How does the Simpson approximation compare with the exact value of the integral? (This result is no accident—it works the same way for all Simpson approximations to integrals of cubic functions.)

Figure 9.25 Error in Simpson's rule for area under a cubic polynomial

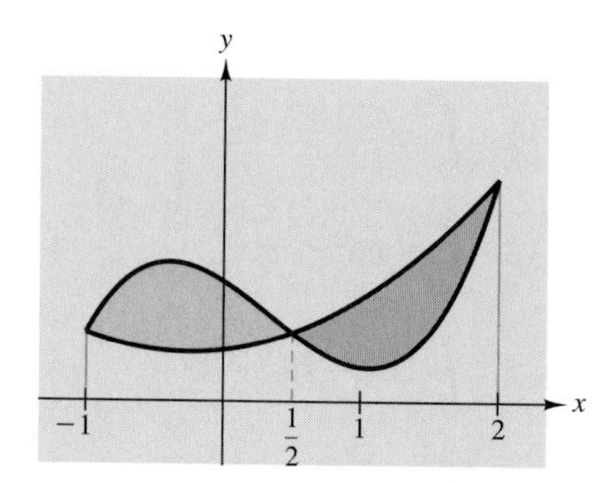

9.4 Substitution: Applying the Chain Rule to Integrals

We have just seen that given an appropriate calculating device, evaluating definite integrals is not difficult—and doesn't require the Fundamental Theorem. Our success notwithstanding, there are times and places at which we must have some ability to calculate integrals (definite or indefinite) without numerical or symbolic help from a calculator or computer. First, we have not yet reached that happy state in which every individual will have access to a calculator or computer whenever needed. Second, even with a computer at hand, there are some integration problems that are at least as easy to do with pencil and paper—including some that look hard. The way we learn to tell which ones we would just as soon do ourselves and which we really want to turn over to the computer is by *practice* — enough practice to discover the patterns that signal "hard" or "easy" integrals. Third, for some problems — differential equations, for example — the preferred form of a solution involves explicit antidifferentiation. That is, indefinite integration may be the key step in solving a problem that doesn't even involve definite integrals. True, the important part of the Fundamental Theorem says that we can always use definite integration to a variable limit as a way to find an antiderivative. But when a more explicit form for an answer is readily found, it can often save a lot of work in subsequent computations. In this section we will see that applying the Chain Rule to integrals — first indefinite, then definite — greatly expands our repertoire of "easy" integrals.

Algebraic Substitutions

We begin by asking you to calculate an improper integral, i.e., to guess an antiderivative.

Exploration Activity 1

Calculate $\int x\sqrt{1+x^2}\,dx.$

Your solution process for Exploration Activity 1 might be something like this:

"I need to find a function $F(x)$ with derivative equal to $x\sqrt{1+x^2}$. How could such a derivative have been produced? The result is a product, so possibly the Product Rule was used. But the Product Rule produces two terms, and I have only one. What other differentiation rule produces a product? The Chain Rule! So maybe the factor of x was produced by the second factor in the Chain Rule:

$$\frac{d}{dx}\,G(u(x)) = \frac{dG}{du}\,\frac{du}{dx}\,.$$

If so, then $u(x)$ should be something like x^2 or, better yet, $1+x^2$. So what should I use as G? Well, the derivative, dG/du, is the square root function, so G should be something like the $\frac{3}{2}$ power with maybe a $\frac{2}{3}$ out in front. Let's try that:

$$\frac{d}{dx}\,\frac{2}{3}(1+x^2)^{3/2} = \frac{2}{3}\,\frac{3}{2}(1+x^2)^{1/2}2x = 2x\sqrt{1+x^2}\,.$$

That was close — if I just divide my initial guess by 2, I'll have an antiderivative:

$$F(x) = \frac{1}{3}(1+x^2)^{3/2}\,.$$

Thus $$\int x\sqrt{1+x^2}\,dx = \frac{1}{3}(1+x^2)^{3/2} + C."$$

You probably noticed where we should have picked up the extra factor of $\frac{1}{2}$ as we went along.

This is an important technique, one that we should formalize for use with other problems as well. Let's start again. Let $u(x) = 1+x^2$. If we calculate the differential on each side, we have $du = 2x\,dx$. Thus

$$x\sqrt{1+x^2}\,dx = \sqrt{1+x^2}\,x\,dx = \frac{1}{2}\sqrt{u}\,du.$$

Now think of u as the *independent* variable, and try to evaluate $\int \frac{1}{2}\sqrt{u}\,du$. That's easy:

$$\int \frac{1}{2}\sqrt{u}\,du = \frac{1}{2}\,\frac{2}{3}\,u^{3/2} + C = \frac{1}{3}\,u^{3/2} + C.$$

Think of u as the *dependent* variable again— substitute $u = 1 + x^2$ to obtain

$$\int x\sqrt{1+x^2}\,dx = \int \frac{1}{2}\sqrt{u}\,du = \frac{1}{3}(1+x^2)^{3/2} + C.$$

The Chain Rule justifies our alternating between thinking of u first as a dependent variable, then as an independent variable, and then as a dependent variable again. If all this seems familiar, you are right. We did the same thing in Chapter 7 when we were finding solutions of differential equations by separating the variables.

Notice that we have finally justified our notational scheme for integration: The differential dx in the notation

$$\int x\sqrt{1+x^2}\,dx$$

helps us carry out substitution arguments. The process of antidifferentiation can be thought of either as finding anti*derivatives* or as finding anti*differentials*—it matters little in the final result. However, it matters a lot in terms of designing and using powerful notation to do much of the work for us. As we noted in Section 4.9, the particular powerful notation dx was one of Leibniz's many contributions to the calculus. The notation we use for integrals—both definite and indefinite—is also due to Leibniz.

Checkpoint 1

Use the substitution technique to evaluate $\displaystyle\int \frac{x}{1+x^2}\,dx$.

Next, we study application of substitution to definite integrals.

Exploration Activity 2

(a) Verify that

$$\int_1^3 x\sqrt{1+x^2}\,dx$$

and

$$\int_2^{10} \frac{1}{2}\sqrt{u}\,du$$

have the same value.

(b) How are the limits on the second integral related to those on the first?

(c) Fill in the limits on the second definite integral to make the equality a true statement:

$$\int_a^b x\sqrt{1+x^2}\,dx = \int_{\_\_}^{\_\_} \frac{1}{2}\sqrt{u}\,du.$$

(d) Use the substitution technique to evaluate $\displaystyle\int_0^{\pi/2} \sin^2\theta\cos\theta\,d\theta$.

Since $\frac{1}{3}(1+x^2)^{3/2}$ is an antiderivative for $x\sqrt{1+x^2}$, we know by the Fundamental Theorem that

$$\int_1^3 x\sqrt{1+x^2}\,dx = \frac{1}{3}(1+x^2)^{3/2}\Big|_1^3 = \frac{10\sqrt{10}-2\sqrt{2}}{3}.$$

Again using the Fundamental Theorem, we find

$$\int_2^{10} \frac{1}{2}\sqrt{u}\,du = \frac{1}{3}u^{3/2}\Big|_2^{10} = \frac{10\sqrt{10}-2\sqrt{2}}{3}.$$

If $u = 1+x^2$, then $2 = u(1)$ and $10 = u(3)$. In general, we want

$$\int_a^b x\sqrt{1+x^2}\,dx = \int_{u(a)}^{u(b)} \frac{1}{2}\sqrt{u}\,du.$$

For the definite integral

$$\int_0^{\pi/2} \sin^2\theta\cos\theta\,d\theta,$$

we may substitute $u = \sin\theta$. Then $du = \cos\theta\,d\theta$, $0 = u(0)$, and $1 = u(\pi/2)$. This gives us the transformed integral

$$\int_0^1 u^2\,du = \frac{1}{3}u^3\Big|_0^1 = \frac{1}{3}.$$

Trigonometric Substitutions

Now we consider a slightly different way to use substitution as an aid in evaluating integrals.

Example 1 Evaluate $\displaystyle\int \frac{1}{\sqrt{25-x^2}}\,dx$.

Solution We would like to make a substitution that would turn $25 - x^2$ into the square of something. An identity that looks useful is

$$1 - \sin^2\theta = \cos^2\theta.$$

For our purposes, it would be more useful in the form

$$25 - 25\sin^2\theta = 25\cos^2\theta$$

or

$$25 - (5\sin\theta)^2 = (5\cos\theta)^2.$$

Thus we could substitute $x = 5\sin\theta$. The new variable θ we are introducing into the problem is defined by the inverse relationship, $\theta = \arcsin x/5$. From the description $x = 5\sin\theta$, we obtain $dx = 5\cos\theta\, d\theta$. When we make the substitutions for x and dx, we obtain

$$\int \frac{1}{\sqrt{25 - (5\sin\theta)^2}}\, 5\cos\theta\, d\theta = \int \frac{1}{\sqrt{1 - \sin^2\theta}}\cos\theta\, d\theta$$
$$= \int \frac{1}{\sqrt{\cos^2\theta}}\cos\theta\, d\theta.$$

Now we would like to say that $\sqrt{\cos^2\theta} = \cos\theta$. However, that is not always true—it holds only if we know that θ is restricted to a range where $\cos\theta$ is nonnegative, such as the interval between $-\pi/2$ and $\pi/2$. But the θ's we are considering are already restricted to this interval because they are values of the arcsine function. Thus, in this setting, it is true that $\sqrt{\cos^2\theta} = \cos\theta$.

We are left with the simple indefinite integral

$$\int \frac{\cos\theta}{\cos\theta}\, d\theta = \int d\theta = \theta + C.$$

Replacing θ with $\arcsin x/5$, we find

$$\int \frac{1}{\sqrt{25 - x^2}}\, dx = \int d\theta = \theta + C = \arcsin\frac{x}{5} + C.$$ ■

Checkpoint 2

Recall that

$$\frac{d}{dx}\arcsin x = \frac{1}{\sqrt{1 - x^2}}.$$

Use this formula to check the result of Example 1.

Checkpoint 3

Evaluate $\displaystyle\int_2^4 \frac{1}{\sqrt{25 - x^2}}\, dx.$

Example 2 Evaluate $\displaystyle\int_0^{5/\sqrt{2}} \sqrt{25 - x^2}\, dx.$

Solution When we make the substitution $x = 5\sin\theta$, we obtain

$$\sqrt{25 - x^2}\, dx = 25\cos^2\theta\, d\theta.$$

Moreover, when $x = 0$, $\theta = \arcsin 0/5 = 0$, and when $x = 5/\sqrt{2}$, $\theta = \arcsin 1/\sqrt{2} = \pi/4$. Thus we may replace the original integral by

$$\int_0^{\pi/4} 25\cos^2\theta\, d\theta.$$

To evaluate this replacement integral, we need an antiderivative for $\cos^2\theta$. The following identities are helpful in integrating squares of sines and squares of cosines:

$$\sin^2\theta = \frac{1-\cos 2\theta}{2}$$

and

$$\cos^2\theta = \frac{1+\cos 2\theta}{2}.$$

Using the latter identity, we have

$$\int \cos^2\theta\, d\theta = \int \frac{1+\cos 2\theta}{2}\, d\theta = \frac{\theta}{2} + \frac{\sin 2\theta}{4} + C.$$

Thus the integral we want may be evaluated by using the Fundamental Theorem of Calculus:

$$\begin{aligned}\int_0^{\pi/4} 25\cos^2\theta\, d\theta &= 25\left(\frac{\theta}{2} + \frac{\sin 2\theta}{4}\right)\Bigg|_0^{\pi/4}\\ &= 25\left[\frac{\pi}{8} + \frac{\sin(\pi/2)}{4}\right]\\ &= 25\left(\frac{\pi}{8} + \frac{1}{4}\right) \approx 16.0675\end{aligned}$$

This is also the value of our original integral. Notice that, in this calculation, we never wrote down an antiderivative for $\sqrt{25 - x^2}$. ■

Checkpoint 4

Evaluate

$$\int_0^1 t^2\sqrt{1-t^2}\, dt.$$

Here you may find a double-angle formula helpful:

$$\sin 2\theta = 2\sin\theta\cos\theta.$$

Evaluating indefinite integrals, i.e., finding antiderivatives, can be a process of guessing and correcting—trial and error. Often it is not evident how to start the process, how to make the first guess. However, substitution can change a difficult problem into a more tractable one. This method is simply the Chain Rule in reverse. If you can rewrite the integral in question in the form $\int f(u(x))u'(x)\, dx$, i.e., as $\int f(u)\, du$, then you only need to find an antiderivative F for f. In that case,

$$\int f(u(x))u'(x)\,dx = \int f(u)\,du = F(u) + C = F(u(x)) + C.$$

There is a corresponding result for definite integrals:

$$\int_a^b f(u(x))\,u'(x)\,dx = \int_{u(a)}^{u(b)} F(u)\,du.$$

Answers to Checkpoints

1. Let $u = 1 + x^2$, so $du = 2x\,dx$. Then

$$\int \frac{x}{1+x^2}\,dx = \frac{1}{2}\int \frac{du}{u} = \frac{1}{2}\ln u + C = \frac{1}{2}\ln(1+x^2) + C.$$

2. $\dfrac{d}{dx}\left(\arcsin\dfrac{x}{5}\right) = \dfrac{1}{\sqrt{1-(x/5)^2}}\dfrac{1}{5} = \dfrac{1}{(1/5)\sqrt{25-x^2}}\dfrac{1}{5} = \dfrac{1}{\sqrt{25-x^2}}$

3. $\arcsin\dfrac{x}{5}\Big|_2^4 = \arcsin\dfrac{4}{5} - \arcsin\dfrac{2}{5} \approx 0.5158$

4. Set $t = \sin u$ —i.e., $u = \arcsin t$. Then the integral becomes

$$\begin{aligned}\int_0^{\pi/2} \sin^2\theta\cos^2\theta\,d\theta &= \int_0^{\pi/2}\left(\frac{\sin 2\theta}{2}\right)^2 d\theta \\ &= \int_0^{\pi/2} \frac{1}{4}\,\frac{1-\cos 4\theta}{2}\,d\theta \\ &= \left(\frac{\theta}{8} - \frac{\sin 4\theta}{32}\right)\Big|_0^{\pi/2} = \frac{\pi}{16}.\end{aligned}$$

Exercises 9.4

Evaluate each of the following indefinite integrals. Check your answers by differentiation.

1. $\displaystyle\int x^2\sqrt{1+x^3}\,dx$
2. $\displaystyle\int t\,e^{-t^2/2}\,dt$
3. $\displaystyle\int \sin^2\theta\,d\theta$
4. $\displaystyle\int x\sqrt{x^2+2}\,dx$
5. $\displaystyle\int \cos^2 x\sin x\,dx$
6. $\displaystyle\int x\sqrt{9-x^2}\,dx$
7. $\displaystyle\int \frac{x}{1+x^2}\,dx$
8. $\displaystyle\int \frac{x}{\sqrt{1+x^2}}\,dx$
9. $\displaystyle\int \sqrt{1-\frac{x}{4}}\,dx$
10. $\displaystyle\int \frac{x}{\sqrt{9-x^2}}\,dx$
11. $\displaystyle\int \sqrt{9-x^2}\,dx$
12. $\displaystyle\int \frac{x^2}{\sqrt{9-x^2}}\,dx$

Evaluate exactly each of the following integrals. If you already have an antiderivative from Exercises 1–12, use it. If not, use the substitution technique for definite integrals. Use your calculator to check your work.

13. $\displaystyle\int_0^1 x^2\sqrt{1+x^3}\,dx$
14. $\displaystyle\int_0^1 t\,e^{-t^2/2}\,dt$
15. $\displaystyle\int_0^{\pi} \sin^2\theta\,d\theta$
16. $\displaystyle\int_0^1 x\sqrt{x^2+2}\,dx$
17. $\displaystyle\int_0^{\pi} \cos^2 x\sin x\,dx$
18. $\displaystyle\int_0^1 x\sqrt{9-x^2}\,dx$
19. $\displaystyle\int_0^1 \frac{x}{1+x^2}\,dx$
20. $\displaystyle\int_0^1 \frac{x}{\sqrt{1+x^2}}\,dx$
21. $\displaystyle\int_{-1}^1 \sqrt{1-\frac{x}{4}}\,dx$
22. $\displaystyle\int_0^1 \frac{x}{\sqrt{9-x^2}}\,dx$

23. $\int_0^1 \sqrt{9-x^2}\,dx$

24. $\int_0^1 \frac{x^2}{\sqrt{9-x^2}}\,dx$

For each of the following integrals, graph the integrand on the interval of integration. Use the graph to make a visual estimate of the integral, preferably one of the form "It has to be at least this big, and it can't be any bigger than that." If you have an exact answer from Exercises 13–24, check that it fits with your estimate. Otherwise, use your calculator to check your estimate. If your estimate is not consistent with the actual value, check your calculation for possible errors.

25. $\int_0^1 x^2\sqrt{1+x^3}\,dx$

26. $\int_0^1 t\,e^{-t^2/2}\,dt$

27. $\int_0^\pi \sin^2\theta\,d\theta$

28. $\int_0^1 x\sqrt{x^2+2}\,dx$

29. $\int_0^\pi \cos^2 x\sin x\,dx$

30. $\int_0^1 x\sqrt{9-x^2}\,dx$

31. $\int_0^1 \frac{x}{1+x^2}\,dx$

32. $\int_0^1 \frac{x}{\sqrt{1+x^2}}\,dx$

33. $\int_{-1}^1 \sqrt{1-\frac{x}{4}}\,dx$

34. $\int_0^1 \frac{x}{\sqrt{9-x^2}}\,dx$

35. $\int_0^1 \sqrt{9-x^2}\,dx$

36. $\int_0^1 \frac{x^2}{\sqrt{9-x^2}}\,dx$

37. In the text we observed that a substitution of the form $x = a\sin\theta$ would turn an expression of the form $a^2 - x^2$ into a perfect square. (The value of a was 5 in our example.)

(a) Show that a substitution of the form $x = a\tan\theta$ will turn an expression of the form $a^2 + x^2$ into a perfect square.

(b) Use the idea in part (a) to show that

$$\int \frac{dx}{(25+x^2)^{3/2}}$$

can be transformed into

$$\frac{1}{25}\int \cos\theta\,d\theta = \frac{1}{25}\sin\theta + C.$$

(c) The antidifferentiation in part (b) isn't finished until we express the answer in terms of x. Draw a right triangle with sides x and 5, and label the angle θ in this triangle for which $x = 5\tan\theta$. What is the length of the hypotenuse of your triangle? What is $\sin\theta$ in terms of x? Complete your calculation of the indefinite integral.

(d) Check your answer to part (c) by differentiation.

38. (a) Apply the substitution in the preceding exercise to the integral

$$\int \frac{dx}{\sqrt{25+x^2}}.$$

What transformed integral do you get? Are you any closer to solution of the original problem?

(b) Calculate the derivative of the function $\ln|\sec\theta + \tan\theta|$. Now can you evaluate the transformed integral in part (a)? Do it.

(c) Complete the original problem.

39. In the text we argued that $\sqrt{\cos^2\theta} = \cos\theta$ (with no absolute value needed) whenever we make a trigonometric substitution of the form $x = a\sin\theta$, because each value of θ is a value of the arcsine function, and therefore lies between $-\pi/2$ and $\pi/2$. (We also implicitly assumed that a is the *positive* square root of a^2—an assumption we are always free to make.) In Exercise 37 we introduced the trigonometric substitution $x = a\tan\theta$ and blithely assumed that $\sqrt{\sec^2\theta} = \sec\theta$ (again with no absolute value). Explain why $\sec\theta$ is always nonnegative for the relevant values of θ.

Evaluate each of the following integrals.

40. $\int \frac{1}{9+x^2}\,dx$

41. $\int_0^1 \frac{1}{9+x^2}\,dx$

42. $\int_0^1 \frac{1}{7+x^2}\,dx$

9.5 Use of the Integral Table

Why "Old" Does Not Imply "Obsolete"

In contrast to antidifferentiation, differentiation is easy! Just follow the rules—to the letter—and the right answer is automatic. The procedure is so straightforward that machines can and do carry it out. Many calculator and computer programs can differentiate virtually any function that can be defined by a formula, and they almost always get the right answer. Often humans don't get the right answer, but that's because it is sometimes difficult to keep track of and follow all the algebraic rules correctly—not because there is anything tricky or mysterious about the procedure.

Integration is another matter altogether. Undoing differentiation often requires great ingenuity to find just the right combination of algebraic maneuvers, and for many problems there may be a lot of trial and error before that combination is found. Indeed, given a formula for a function that *is* the derivative of something, it is often not clear whether the something can be defined by a formula in the usual sense.

Some computer programs can find indefinite integrals as well as derivatives. However, you may have already discovered that the program you are using for this purpose will occasionally fail to find an integral that you know—or strongly suspect—does have a formula. You may need to use other tools besides the computer.

Long before there were computers, there were symbolic integration devices. They still exist, and they have advantages over most computers of being both totally portable and very cheap. They also have the disadvantage of being much less automated. Their collective name is *Integral Table.* A complete integral table—a small book—contains the collected wisdom of the ages on the subject of evaluating integrals. We have included a short integral table in the Appendix.

Before we begin our exploration of the use of an integral table, recall that two rather different types of problems have led us to a need for evaluating indefinite integrals. First, almost from the beginning of this course we have been concerned with the problem of solving differential equations, and we saw in Chapter 7 that this problem often can be turned into one of indefinite integration. Second, our study in Chapter 8 of the divide-and-conquer technique and the Fundamental Theorem has shown us that indefinite integrals can be used as a tool for evaluating definite integrals.

Matching Problems with Table Entries

Let's start with a problem we have encountered before, namely, the differential equation that models a falling body with air resistance proportional to the square of the speed:

$$\frac{dv}{dt} = g - cv^2.$$

When we separate the variables and integrate both sides, we turn this equation into

$$\int \frac{dv}{g - cv^2} = \int dt.$$

The integral on the right is easy: $\int dt = t + C$. The integral on the left is more of a challenge. What we need from our table is a formula that tells us how to integrate the reciprocal of a quadratic function of the variable. Our quadratic $g - cv^2$ has one special feature: There is no v (or first-power) term.

We find three entries in the Basic Forms section of the Integral Table that might be of use:

$$17 \quad \int \frac{du}{a^2 + u^2} = \frac{1}{a} \arctan \frac{u}{a} + C$$

$$19 \quad \int \frac{du}{a^2 - u^2} = \frac{1}{2a} \ln \left| \frac{u + a}{u - a} \right| + C$$

$$20 \quad \int \frac{du}{u^2 - a^2} = \frac{1}{2a} \ln \left| \frac{u - a}{u + a} \right| + C$$

Immediately we run into two sources of confusion! First, there are three entries where we only needed one—which one do we want? Second, the symbols for variables and constants are different from the ones in our problem.

Don't panic. Neither of these problems is particularly difficult. We'll come back to the problem of selecting the right formula after we deal with the symbols. Hardly ever will you find that the symbols in the table match the ones in the problem you are trying to solve. (How could the author of the table possibly know what variables and constants would appear in your problem?) If you have properly separated the variables in your differential equation, there can be only one variable in your integral. Our variable happens to be v; our table happens to use u. Thus we need to change all u's (in whichever formula we decide to use) to v's.

Next, none of the formulas has constants in front of the variable squared. We need to rewrite our integral so that it has the same form—which we can do by factoring the constant c out of the denominator:

$$\int \frac{dv}{g - cv^2} = \frac{1}{c} \int \frac{dv}{g/c - v^2}.$$

We know from the physical source of our problem that both g and c are positive constants. Thus we may identify $\sqrt{g/c}$ with the table constant a.

Example 1 Identify an appropriate table entry, and use it to evaluate

$$\int \frac{dv}{g - cv^2}.$$

Solution With the identifications $u = v$ and $a = \sqrt{g/c}$, we see that entry 19 is the one we want. When we rewrite this entry using these identifications, we obtain

$$\frac{1}{c} \int \frac{dv}{g/c - v^2} = \frac{1}{c} \left(\frac{1}{2\sqrt{g/c}} \ln \left| \frac{v + \sqrt{g/c}}{v - \sqrt{g/c}} \right| + C \right)$$

$$= \frac{1}{2c\sqrt{g/c}} \ln \left| \frac{v + \sqrt{g/c}}{v - \sqrt{g/c}} \right| + \frac{C}{c}.$$

Here C represents an arbitrary constant, and c has a definite nonzero value. Thus the quotient C/c can again be any constant, and we may denote it again by C. We also know that v is positive and less than $\sqrt{g/c}$, so

$$|v - \sqrt{g/c}| = \sqrt{g/c} - v.$$

When we put all the pieces together, we find

$$\int \frac{dv}{g - cv^2} = \frac{1}{2c\sqrt{g/c}} \ln\left(\frac{v + \sqrt{g/c}}{\sqrt{g/c} - v}\right) + C.$$

We may choose to use algebra to simplify further. For example, we could write $c\sqrt{g/c}$ as $\sqrt{cg}$. We also could resolve the compound fractions in the argument of the logarithm by multiplying numerator and denominator by $\sqrt{c}$. The result of these changes is

$$\int \frac{dv}{g - cv^2} = \frac{1}{2\sqrt{cg}} \ln\left(\frac{\sqrt{g} + \sqrt{c}\,v}{\sqrt{g} - \sqrt{c}\,v}\right) + C.$$

■

We have now evaluated the left-hand side of

$$\int \frac{dv}{g - cv^2} = \int dt\,,$$

and we already knew how to evaluate the right-hand side. If we set the two results equal, we now have an equation that solves the differential equation in the sense that no derivatives, differentials, or integrals remain. However, the algebraic problem of solving for v as an explicit function of t remains. We defer that calculation—which has nothing to do with calculus—to Project 2 at the end of the chapter.

Exploration Activity 1

(a) Use the Integral Table to evaluate

$$\int t^2\sqrt{1 - t^2}\,dt\,.$$

(b) Check your work by differentiating your antiderivative.

(c) Use your antiderivative from part (a) and the Fundamental Theorem to evaluate the definite integral

$$\int_0^1 t^2\sqrt{1 - t^2}\,dt\,.$$

(d) Graph the function being integrated in part (a). The integral in part (a) is an area—does the number you found in part (c) look about right?

(e) Use your calculator to evaluate the integral in part (c). Does the answer agree with your previous calculation?

Formula 32 in the Integral Table,

$$\int u^2 \sqrt{a^2 - u^2}\, du = -\frac{a^2}{8} u \sqrt{a^2 - u^2} + \frac{1}{4} u^3 \sqrt{a^2 - u^2} + \frac{a^4}{8} \arcsin \frac{u}{a} + C,$$

gives us what we need if we identify a with 1 and u with t:

$$\int t^2 \sqrt{1 - t^2}\, dt = -\frac{1}{8} t \sqrt{1 - t^2} + \frac{1}{4} t^3 \sqrt{1 - t^2} + \frac{1}{8} \arcsin t + C.$$

When we use this formula to evaluate

$$\int_0^1 t^2 \sqrt{1 - t^2}\, dt,$$

we obtain

$$\left[-\frac{1}{8} t \sqrt{1 - t^2} + \frac{1}{4} t^3 \sqrt{1 - t^2} + \frac{1}{8} \arcsin t \right] \Bigg|_0^1 = \frac{1}{8} \arcsin 1 = \frac{\pi}{16}.$$

Figure 9.26 shows a graph of the function being integrated. We can see that the area of the region under the graph is about half the area of the rectangle. Thus the integral should have a value approximately $\frac{1}{2} \times 0.4 \times 1 = 0.2$. Our calculated value of $\pi/16 \approx 0.196$ fits well with this rough estimate. A typical calculator integration routine confirms the answer $\pi/16$, but only to 6 significant digits (6SD). (The exact answer is *not* confirmed to machine accuracy because the vertical tangent at $t = 1$ causes problems for most integration routines.)

Figure 9.26 Graph of $y = t^2 \sqrt{1 - t^2}$

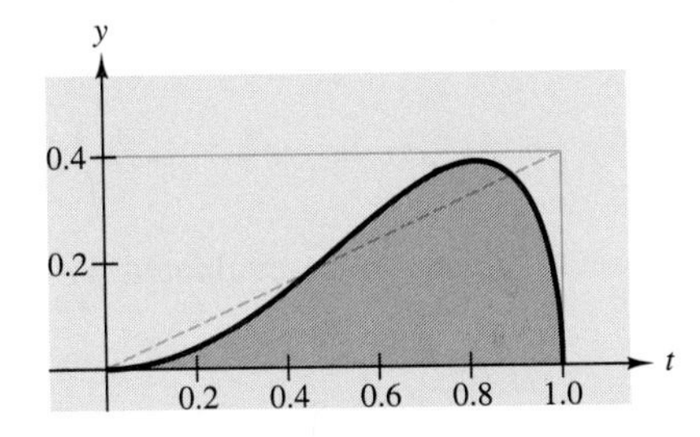

Parts (c) through (e) of Exploration Activity 1 suggest another way to check antiderivatives—in addition to calculating and simplifying a derivative. A wrong antiderivative formula is not likely to give correct answers for definite integrals, and you can check those either graphically or numerically with your calculator.

Reduction Formulas

Often you will need more than one formula from the table to solve a single problem. This is illustrated in the next exploration activity.

Exploration Activity 2

(a) Use the Integral Table to evaluate

$$\int x^2 e^{-x}\,dx.$$

(b) Check your calculation by differentiation.

(c) Graph $y = x^2 e^{-x}$ over the interval $[0, 3]$, and use this graph to estimate the value of

$$\int_0^3 x^2 e^{-x}\,dx.$$

(d) Use the antiderivative from part (a) and the Fundamental Theorem to evaluate

$$\int_0^3 x^2 e^{-x}\,dx.$$

Compare the result with your estimate in part (c).

(e) Check your answer in part (d) with your calculator.

There is no formula in the Integral Table for $x^2 e^{-x}$. However, these two formulas appear to be related:

$$98 \quad \int u e^{au}\,du = \frac{1}{a^2}(au - 1)e^{au} + \mathrm{C}$$

$$102 \quad \int u^n e^{au}\,du = \frac{1}{a}u^n e^{au} - \frac{n}{a}\int u^{n-1} e^{au}\,du$$

Formula 102 is a **reduction** formula—for each choice of the power n, it reduces a complicated integral to a slightly less complicated one. In particular, if we replace u by x and set $n = 2$ and $a = -1$, this formula becomes

$$\int x^2 e^{-x}\,dx = -x^2 e^{-x} + 2\int x e^{-x}\,dx.$$

The integral remaining on the right can be evaluated with Formula 98:

$$\int x e^{-x}\,dx = (-x - 1)e^{-x} + \mathrm{C}.$$

So

$$\begin{aligned}\int x^2\, e^{-x}\,dx &= -x^2 e^{-x} + 2[(-x - 1)e^{-x}] + C \\ &= -x^2 e^{-x} - 2(x + 1)e^{-x} + C.\end{aligned}$$

The differentiation check is a straightforward application of the Product Rule. We show the graph of $y = x^2 e^{-x}$ in Figure 9.27. The maximum value of the function is a little over $\frac{1}{2}$. Since the graph passes through $(0, 0)$, we can estimate that the value of the integral is somewhat less than half the area of the window, i.e., somewhat less than $\frac{3}{2}$.

Figure 9.27 Graph of $y = x^2e^{-x}$

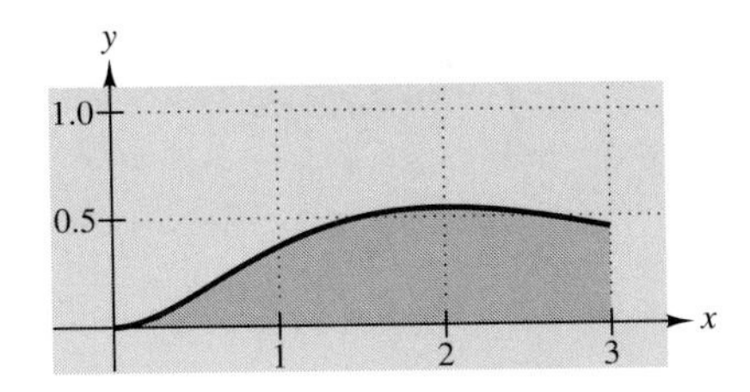

By the Fundamental Theorem,

$$\begin{aligned}\int_0^3 x^2e^{-x}\,dx &= \left[-x^2e^{-x} - 2(x+1)e^{-x}\right]\Big|_0^3 \\ &= -9 \times e^{-3} - 2 \times e^{-3} \times 4 - (-2) \\ &= 2 - 17e^{-3} \approx 1.154.\end{aligned}$$

This is in line with our rough estimate. Your calculator should confirm this value.

Often you will have to make a formal substitution to bring your integral into a form that can be identified with a formula in the Integral Table.

Exploration Activity 3

Make an appropriate substitution, and then use the Integral Table to evaluate

$$\int \theta^2 \sin 2\theta\, d\theta.$$

The substitution $u = 2\theta$ leads to the integral

$$\int \frac{u^2}{4}(\sin u)\,\frac{1}{2}\,du = \frac{1}{8}\int u^2 \sin u\, du.$$

Now we can use the reduction Formula 87 with $n = 2$:

$$\int u^2 \sin u\; du = -u^2 \cos u + 2\int u \cos u\; du.$$

The integral on the right may be evaluated with Formula 86:

$$\int u \cos u\; du = \cos u + u \sin\, u + \mathrm{C}.$$

Pulling this together, we find

$$\begin{aligned}\int \theta^2 \sin 2\theta\, d\theta &= \frac{1}{8}\left[-u^2 \cos u + 2(\cos u + u \sin\, u)\right] + C \\ &= -\frac{1}{8}u^2 \cos u + \frac{1}{4}\cos u + \frac{1}{4}u \sin\, u + C \\ &= -\frac{1}{2}\theta^2 \cos 2\theta + \frac{1}{4}\cos 2\theta + \frac{1}{2}\theta \sin 2\theta + C.\end{aligned}$$

Checkpoint 1

Use your answer to Exploration Activity 3 and the Fundamental Theorem of Calculus to evaluate

$$\int_0^{\pi/2} \theta^2 \sin 2\theta \, d\theta.$$

Check the result with your calculator.

It should be clear by now that no matter how much help you get from a table of integrals, both care and skill with the algebraic manipulations are required in order to get right answers. Sometimes you need more. You may need some ingenuity to think of appropriate algebraic manipulations that will transform your problem into a form that appears in your table.

For example, exponentials and logarithms with bases other than e often appear in scientific and technical work. In particular, the bases 2 and 10 have definite advantages for certain purposes, such as discussions of doubling times and orders of magnitude. However, integral tables often use only base e for *all* of their exponential and logarithmic forms. How can we use such a table to integrate, say, $x^2\, 2^{-x}$? Well, recall from Chapter 2 that *all* exponentials are related to natural exponentials in a very simple way: $b^t = e^{kt}$, where $k = \ln b$. In particular, $2^{-x} = e^{-(\ln 2)x}$. In the following checkpoint we ask you to use the existing table entries to create a new one.

Checkpoint 2

Evaluate the indefinite integral $\int x^2\, 2^{-x}\, dx$.

Different Answers to the Same Problem

Just as there is more than one way to skin a cat, there is often more than one way to evaluate an integral. Moreover, different ways of evaluating the same integral can lead to apparently different answers.

Example 2 Find an antiderivative of $\dfrac{\sin^3 5t}{\cos^3 5t}$.

Solution 1 Our table does not contain an entry for

$$\int \frac{\sin^n u}{\cos^m u}\, du.$$

However, when the power of sine in the numerator is *odd* and *positive*, there is an easy transformation of our problem to a form in which a simple substitution will suffice. Here are the steps — we ask you to supply the reasons in Checkpoint 3.

$$\begin{aligned}
\int \frac{\sin^3 5t}{\cos^3 5t}\, dt &= \int \frac{\sin 5t\ \sin^2 5t}{\cos^3 5t}\, dt \\
&= \int \frac{\sin 5t\,(1 - \cos^2 5t)}{\cos^3 5t}\, dt \\
&= \int \sin 5t \left(\frac{1}{\cos^3 5t} - \frac{1}{\cos 5t} \right) dt \\
&= -\frac{1}{5} \int \left(u^{-3} - u^{-1}\right) du \\
&= \frac{1}{10}\, u^{-2} + \frac{1}{5} \ln |u| + C \\
&= \frac{1}{10 \cos^2 5t} + \frac{1}{5} \ln |\cos 5t| + C
\end{aligned}$$

Solution 2 It probably occurred to you before we started this calculation that our integral also could be written in the form $\int \tan^3 5t\, dt$. And sure enough, our table has a relevant entry, Formula 74:

$$\int \tan^3 u\, du = \frac{1}{2} \tan^2 u + \ln |\cos u| + \mathrm{C}.$$

In Exploration Activity 4 we ask you to complete this solution and reconcile it with Solution 1. ■

Checkpoint 3

Provide a reason for each step in the Solution 1.

Exploration Activity 4

(a) Make the appropriate substitutions in Formula 74 to find (another) antiderivative for $\tan^3 5t$.

(b) Explain how the apparently different answers in Solutions 1 and 2 can both be right.

We start with the obvious substitution: $u = 5t$. Then

$$\int \frac{\sin^3 5t}{\cos^3 5t}\, dt = \frac{1}{5} \int \tan^3 u\, du.$$

By Formula 74, the integral on the right is

$$\frac{1}{5} \left(\frac{1}{2} \tan^2 u + \ln |\cos u| \right) + C.$$

When we substitute back $5t$ for u, we obtain

$$\frac{1}{10} \tan^2 5t + \frac{1}{5} \ln |\cos 5t| + C.$$

Although the two formulas for the indefinite integral look quite different, they describe the same family of functions. To see this, we subtract the two answers to see if they differ by a constant:

$$\begin{aligned}
&\frac{1}{10\cos^2 5t} + \frac{1}{5}\ln|\cos 5t| - \left(\frac{1}{10}\tan^2 5t + \frac{1}{5}\ln|\cos 5t|\right) \\
&\qquad = \frac{1}{10\cos^2 5t} - \frac{1}{10}\tan^2 5t \\
&\qquad = \frac{1}{10}\left(\frac{1}{\cos^2 5t} - \frac{\sin^2 5t}{\cos^2 5t}\right) \\
&\qquad = \frac{1}{10}\left(\frac{1-\sin^2 5t}{\cos^2 5t}\right) \\
&\qquad = \frac{1}{10}.
\end{aligned}$$

Thus the two antiderivative expressions from Solutions 1 and 2 differ by the constant $1/10$.

In this section we have seen that the Table of Integrals is a handy symbolic integration device. However, often we must use algebra together with substitution for the variable of integration to match our problem with a particular integration formula. Sometimes we need more than one integration formula to complete an integration. Sometimes there are several different formulas that may be used to integrate a given function, and it may not be immediately apparent that different-looking antiderivatives differ by a constant. As with most manipulative activities, one develops skill with this tool by practice.

Answers to Checkpoints

1. $\dfrac{\pi^2}{8} - \dfrac{1}{2}$

2. $-\dfrac{1}{\ln 2}x^2 2^{-x} - \dfrac{2^{1-x}}{(\ln 2)^3} - \dfrac{x}{(\ln 2)^2}2^{1-x} + C$

3. We factored $\sin^3 5t$.

   We used the identity $\sin^2 x = 1 - \cos^2 x$.

   We expanded $\dfrac{1-\cos^2 5t}{\cos^3 5t}$ and canceled $\cos^2 5t$ in the second term.

   We substituted $u = \cos 5t$, so $du = -5\sin 5t\, dt$.

   We used the Power Rule for integration.

   We substituted back $\cos 5t$ for u.

Exercises 9.5

Here again are Exercises 1–12 from Section 9.4. For each integral, determine whether there is a relevant entry in your integral table. If not, explain why the problem is too easy to need a table entry. If so, use the table entry to evaluate the integral. If you find more than one relevant entry, check to see if they all lead to the same answer—up to an additive constant.

1. $\int x^2\sqrt{1+x^3}\,dx$
2. $\int t\,e^{-t^2/2}\,dt$
3. $\int \sin^2\theta\,d\theta$
4. $\int x\sqrt{x^2+2}\,dx$
5. $\int \cos^2 x\sin x\,dx$
6. $\int x\sqrt{9-x^2}\,dx$
7. $\int \frac{x}{1+x^2}\,dx$
8. $\int \frac{x}{\sqrt{1+x^2}}\,dx$
9. $\int \sqrt{1-\frac{x}{4}}\,dx$
10. $\int \frac{x}{\sqrt{9-x^2}}\,dx$
11. $\int \sqrt{9-x^2}\,dx$
12. $\int \frac{x^2}{\sqrt{9-x^2}}\,dx$

Use your table to evaluate each of the following integrals. If necessary, make a substitution first to get the integrand into a form that can be found in the table.

13. $\int x\sqrt{1+3x}\,dx$
14. $\int t^2\,e^{-t/2}\,dt$
15. $\int \tan^2\theta\,d\theta$
16. $\int \frac{dx}{x\sqrt{x^2+2}}$
17. $\int \cos^6 2x\,dx$
18. $\int \frac{\sqrt{9-x^2}}{x}\,dx$
19. $\int \frac{x^2}{8+x^3}\,dx$
20. $\int \frac{x^2}{\sqrt{8+x^2}}\,dx$
21. $\int \sqrt{1-\frac{x^2}{4}}\,dx$
22. $\int x^2\sqrt{9-x^2}\,dx$
23. $\int \frac{x^2}{\sqrt{9+x^2}}\,dx$
24. $\int \frac{1}{(7+x^2)^2}\,dx$
25. $\int \frac{dx}{\sqrt{6x-x^2}}$
26. $\int \frac{t^2}{\sqrt{16-t^2}}\,dt$
27. $\int \sec^3\theta\,d\theta$
28. Find three entries in your integral table that look like they were calculated by an algebraic substitution, in the sense of Section 9.4. If the substitution variable were u, say, what function did u represent? How were the necessary factors for du provided?
29. Find three entries in your integral table that look like they were calculated by a trigonometric substitution, in the sense of Section 9.4. What trigonometric function was substituted for the original variable of integration?

9.6 Integration by Parts

One important integration technique is behind many of the formulas in the Integral Table—particularly the reduction formulas. Let's look at Formula 87:

$$\int t^n\sin t\,dt = -t^n\cos t + n\int t^{n-1}\cos t\,dt$$

(For a notational reason that soon will become evident, we have replaced u by t.) The function being integrated on the left is a product. Not surprisingly, the technique used to derive this formula is a method for finding an antiderivative for a product. To start with a simple example, we'll look at the case of Formula 87 with $n = 1$:

$$\int t \sin t \, dt = -t \cos t + \int \cos t \, dt.$$

Exploration Activity 1 will connect this formula with the Product Rule for differentiation,

$$\frac{d}{dt}[F(t)\,G(t)] = F(t)\frac{dG}{dt} + G(t)\frac{dF}{dt}.$$

Exploration Activity 1

(a) Find a function $G(t)$ such that

$$\frac{dG}{dt} = \sin t.$$

(b) Let $F(t) = t$, and write out the Product Rule for this choice of F and G.

(c) Solve the resulting equation for $t \sin t$.

(d) Use the result to fill in the blank

$$t \sin t = \frac{d}{dt}\Big(\underline{\hspace{6cm}}\Big)$$

(e) Write your result in part (c) in terms of an indefinite integral.

If we take $G(t) = -\cos t$ and $F(t) = t$, then the Product Rule states

$$\begin{aligned} \frac{d}{dt}[t(-\cos t)] &= t \sin t + (-\cos t) \times 1 \\ &= t \sin t - \cos t. \end{aligned}$$

When we solve for $t \sin t$, we obtain

$$t \sin t = \frac{d}{dt}(-t \cos t) + \cos t.$$

Now $\cos t$ is an antiderivative for $\sin t$, so

$$t \sin t = \frac{d}{dt}(-t \cos t + \sin t).$$

When we integrate both sides, we find

$$\int t \sin t \, dt = -t \cos t + \sin t + C.$$

If we had not replaced $\cos t$ by the derivative of $\sin t$, integration of both sides of

$$t \sin t = \frac{d}{dt}(-t \cos t) + \cos t$$

would have given us

$$\int t \sin t \, dt = -t \cos t + \int \cos t \, dt,$$

precisely the instance of Formula 87 we set out to explain. The Product Rule enabled us to replace the problem of finding an antiderivative for $t \sin t$ with the problem of finding an antiderivative for $\cos t$—which we already knew how to do. That's what this inverse use of the Product Rule enables us to do—trade in one antidifferentiation problem for another.

Now we need a better notation. For this we turn again to differentials. We rewrite the Product Rule in differential notation with f representing the derivative of F and g representing the derivative of G:

$$d\,[F(t)\,G(t)] = f(t)\,G(t)\,dt + F(t)\,g(t)\,dt.$$

We can rewrite this equation in the form

$$f(t)\,G(t)\,dt = d\,[F(t)\,G(t)] - F(t)\,g(t)\,dt.$$

When we integrate both sides, we obtain

$$\int f(t)\,G(t)\,dt = F(t)\,G(t) - \int F(t)\,g(t)\,dt.$$

Now, if we write $u = G(t)$ and $v = F(t)$, then $du = g(t)\,dt$ and $dv = f(t)\,dt$. In this notation, our equation takes the simpler form

$$\int u\,dv = u\,v - \int v\,du.$$

This inverse of the Product Rule is called the **integration-by-parts formula**. When we apply this formula, we first identify the parts u and dv—the second of which must be a differential, i.e., must include the factor dt. Then we calculate du (by differentiating u) and v (by antidifferentiating dv) and set up the right-hand side.

Example 1 Find $\int t\,e^{-3t}\,dt$.

Solution There are two obvious choices for u and dv:

(1) $u = t$ and $dv = e^{-3t}\,dt$ and

(2) $u = e^{-3t}$ and $dv = t\,dt$.

We will examine both.

If $u = t$ and $dv = e^{-3t}\,dt$, then $du = dt$, and we may set $v = -\frac{1}{3}e^{-3t}\,dt$. Then

$$\begin{aligned}\int t\,e^{-3t}\,dt &= t\left(-\frac{1}{3}e^{-3t}\right) - \int\left(-\frac{1}{3}e^{-3t}\right)dt \\ &= -\frac{1}{3}t\,e^{-3t} + \frac{1}{3}\int e^{-3t}\,dt.\end{aligned}$$

Since we just found an antiderivative for e^{-3t}, we can use it to evaluate the integral on the right:

$$\int t\,e^{-3t}\,dt = -\frac{1}{3}t\,e^{-3t} - \frac{1}{9}e^{-3t} + C.$$

Now suppose we set $u = e^{-3t}$ and $dv = t\,dt$. Then $du = -3e^{-3t}\,dt$, and we may set $v = t^2/2$. For this identification of u and dv, the integration-by-parts formula becomes

$$\begin{aligned}\int t\,e^{-3t}\,dt &= e^{-3t}\frac{t^2}{2} - \int \frac{t^2}{2}\left(-3e^{-3t}\right)dt \\ &= e^{-3t}\frac{t^2}{2} + \frac{3}{2}\int t^2 e^{-3t}\,dt.\end{aligned}$$

This formula is valid but not helpful. The new integral on the right is more complicated than the original one. ■

The integration-by-parts formula allows you to trade in one integration problem for another. There is no guarantee that the new problem is easier than the old. You may need to try several identifications of u and dv before you find one that is useful. Ideally, you want u to be a function whose differential will be simpler than itself and dv to be a differential whose integral will be simpler than itself. In Example 1, choice (1) has du simpler than u and v of about the same complexity as dv. In choice (2), du was of the same complexity as u, and v was more complicated than dv. This last fact is what caused choice (2) to fail.

Checkpoint 1

Use the integration-by-parts formula to evaluate

$$\int t\cos 2t\,dt.$$

Example 2 Use the integration-by-parts formula to establish the reduction formula

$$\int t^n \sin t\,dt = -t^n\cos t + n\int t^{n-1}\cos t\,dt.$$

Solution Since we want to reduce the exponent in t^n, we'll set $u = t^n$. Then $du = nt^{n-1}$, reducing the exponent as desired. Thus we must set $dv = \sin t\,dt$, and we may choose $v = -\cos t$. With these identifications, the integration-by-parts formula gives

$$\begin{aligned}\int t^n \sin t\,dt &= t^n(-\cos t) - \int(-\cos t)nt^{n-1}\,dt \\ &= -t^n\cos t + n\int t^{n-1}\cos t\,dt.\end{aligned}$$

■

Checkpoint 2

Derive Formula 95 in the Integral Table:

$$\int x^n \arcsin x\,dx = \frac{1}{n+1}x^{n+1}\arcsin x - \frac{1}{n+1}\int\frac{x^{n+1}}{\sqrt{1-x^2}}\,dx, \text{ for } n \neq -1.$$

In this section we have seen that the Product Rule for differentiation,

$$d(uv) = u\,dv + v\,du$$

(in differential form), leads to the integration-by-parts formula,

$$\int u\,dv = u\,v - \int v\,du.$$

This technique allows you to replace one integration problem with another. It is useful if the new problem is simpler than the original one. Thus, to apply this formula effectively, you need to consider what happens to each factor when you differentiate one and integrate the other. If it is not clear whether $\int v\,du$ is simpler than $\int u\,dv$, then you may have to try more than factorization to find one that works.

Answers to Checkpoints

1. $\int t\cos 2t\;dt = \frac{1}{2}t\sin 2t - \int \frac{1}{2}\sin 2t\;dt = \frac{1}{2}t\sin 2t + \frac{1}{4}\cos 2t + C$

2. Set $u = \arcsin x$, $dv = x^n\,dx$, and $v = \frac{1}{n+1}x^{n+1}$. Then

$$\begin{aligned}\int x^n \arcsin x\;dx &= \frac{1}{n+1}\,x^{n+1}\arcsin x - \int \frac{1}{n+1}x^{n+1}\frac{1}{\sqrt{1-x^2}}\,dx \\ &= \frac{1}{n+1}\,x^{n+1}\arcsin x - \frac{1}{n+1}\int \frac{x^{n+1}}{\sqrt{1-x^2}}\,dx.\end{aligned}$$

Exercises 9.6

Use integration by parts to evaluate each of the following indefinite integrals. In each case, think about what gets simpler when you differentiate one of the parts.

1. $\int t\,\sin 2t\;dt$

2. $\int t^2\,\sin 2t\;dt$

3. $\int t^3\,\sin 2t\;dt$

4. $\int \ln t\;dt$ (*Hint*: Consider $\ln t$ as a product of the functions 1 and $\ln t$.)

5. $\int t\ln t\;dt$

6. $\int \arcsin t\;dt$

7. $\int t\arcsin t\;dt$

Find an antiderivative for each of the following functions.

8. $\frac{\theta}{3}\sin\theta$

9. $(t-2)e^{-t}$

10. $\arcsin 3x$

11. $\theta\sin\frac{\theta}{3}$

12. $(t^2-2)e^{-t}$

13. $x\arcsin 3x$

Evaluate each of the following integrals. Don't assume that the best technique is necessarily integration by parts. Check your result by one of the checking techniques from Section 9.5:

- Differentiation
- Evaluation of a definite integral using the antiderivative and checking with your calculator's integral key

14. $\int \frac{\cos t}{2 + \sin t}\, dt$

15. $\int \theta^2 \cos 2\theta \, d\theta$

16. $\int \arctan 2x \, dx$

17. $\int x^2 \arctan 2x \, dx$

18. $\int x\sqrt{x+7}\, dx$

19. $\int x\sqrt{x^2+7}\, dx$

20. $\int \frac{\ln x}{x}\, dx$

21. $\int u\, e^u \, du$

22. $\int u\, e^{u^2}\, du$

Evaluate each of the following definite integrals. If you did the corresponding Exercises 14–22, use the result here. Check your answer by calculator.

23. $\int_0^{2\pi} \frac{\cos t}{2 + \sin t}\, dt$

24. $\int_0^{\pi} \theta^2 \cos 2\theta \, d\theta$

25. $\int_0^{1/2} \arctan 2x \, dx$

26. $\int_0^{1/2} x^2 \arctan 2x \, dx$

27. $\int_0^1 x\sqrt{x+7}\, dx$

28. $\int_0^1 x\sqrt{x^2+7}\, dx$

29. $\int_1^2 \frac{\ln x}{x}\, dx$

30. $\int_0^1 u\, e^u du$

31. $\int_0^1 u\, e^{u^2} du$

32. Use integration by parts, with $u = \arctan x$ and $dv = 2x\, dx$, to evaluate $\int 2x \arctan x \, dx$ two ways:

(a) with $v = x^2$, and

(b) with $v = x^2 + 1$.

How much difference does it make to choose your antidifferential of dv carefully?

Use integration by parts to establish each of the following formulas in the Table of Integrals.

33. 76 $\int \sin^n u \, du =$

$$-\frac{1}{n}\sin^{n-1} u \cos u + \frac{n-1}{n}\int \sin^{n-2} u \, du$$

34. 77 $\int \cos^n u \, du =$

$$\frac{1}{n}\cos^{n-1} u \sin u + \frac{n-1}{n}\int \cos^{n-2} u \, du$$

35. 93 $\int u \arccos u \, du =$

$$\frac{2u^2 - 1}{4} \arccos u - \frac{u\sqrt{1-u^2}}{4} + C$$

9.7 Representations of Periodic Functions

In Figure 9.28 we show an important periodic (or repeating) function: a trace from an electrocardiogram (EKG) of one of the authors. An EKG measures various aspects of the functioning of the heart. Each of us finds it important that the heart do essentially the same thing over and over, roughly once a second, for a large number of years. When the heart is not doing that, the EKG is a first-line diagnostic tool for discovering and interpreting deviations from regular behavior.

Figure 9.28 The rhythm strip trace from an electrocardiogram

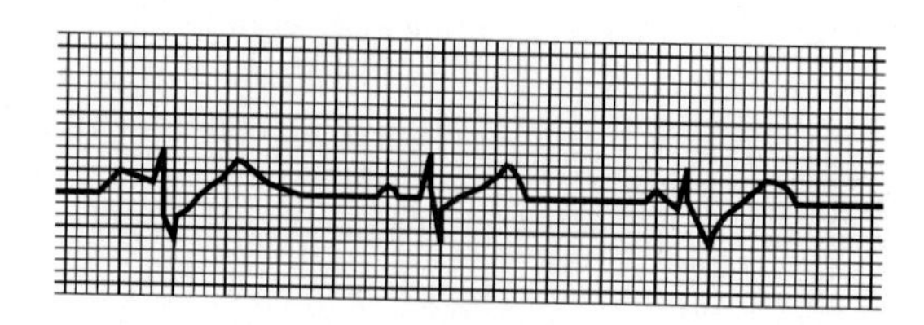

We first brought up the subject of repeating phenomena in Chapter 6, where we mentioned some other examples, such as the motion of planets and pendulums. To these we might add musical tones and other sounds, daily and yearly temperature fluctuations, light rays, and water waves. The central point of Chapter 6 was the observation of the role of trigonometric functions in representing these phenomena. That's the point we return to now.

Building Complicated Waveforms from Sines and Cosines

One way to analyze periodic functions is to try to break them down into simpler components. The reason this might work is that quite complicated periodic functions can be built up out of the **elementary** periodic functions: $\sin t$, $\sin 2t$, $\sin 3t$, $\dots$, and so on, together with 1, $\cos t$, $\cos 2t$, $\cos 3t$, $\dots$, and so on. (The constant function 1 can be thought of as $\cos 0t$.) For example, consider the sequence of functions whose graphs are shown in Figure 9.29. When we add terms that oscillate twice as fast and three times as fast as the earlier terms, the resulting functions became considerably more interesting. If we continued to add terms that oscillate even faster, such as $\cos 4t$ or $\sin 7t$, we could produce even more complicated graphs.

Figure 9.29 Building a waveform from elementary trigonometric functions

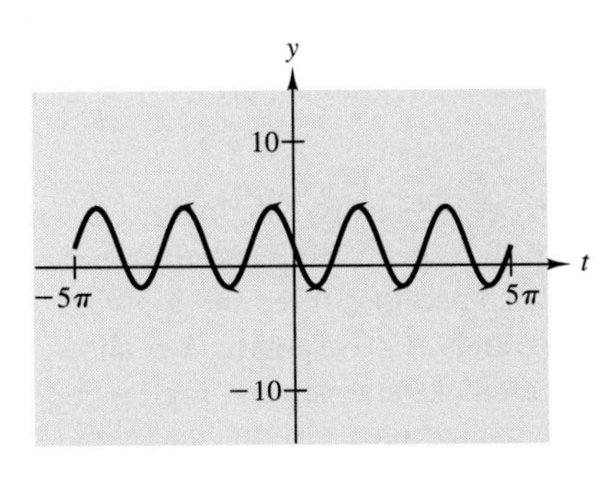

(a) $f(t) = 1.53 - 3.19 \sin t$

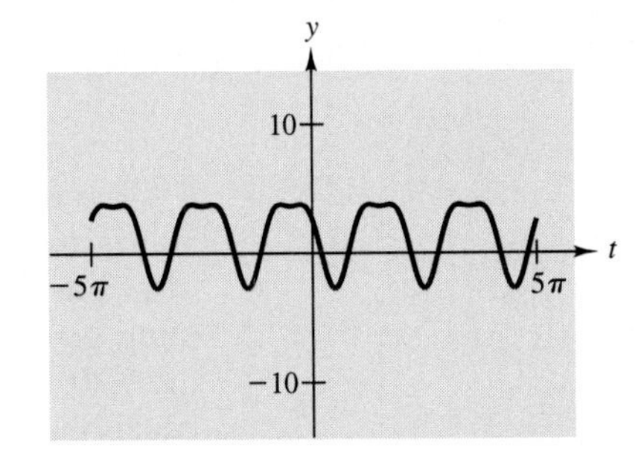

(b) $f(t) = 1.53 - 3.19 \sin t + 1.07 \cos 2t$

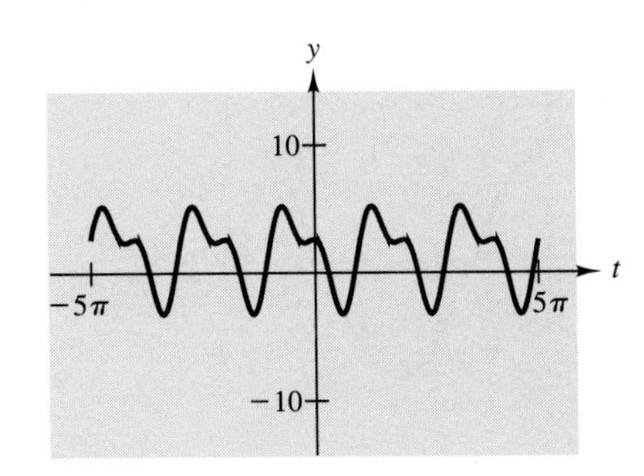

(c) $f(t) = 1.53 - 3.19 \sin t + 1.07 \cos 2t + 1.45 \sin 2t$

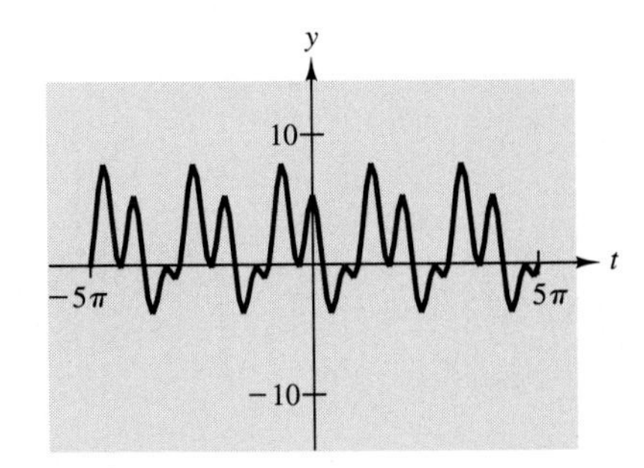

(d) $f(t) = 1.53 - 3.19 \sin t + 1.07 \cos 2t + 1.45 \sin 2t + 2.73 \cos 3t$

Using Integrals to Analyze Waveforms

We consider now the harder inverse problem: If you have a complicated periodic function, how can you break it down into simple trigonometric terms? We begin our attack on this problem with investigation of certain definite integrals of trigonometric functions over complete periods. Then we will see how this information about integrals can be used to solve a simplified version of the inverse problem. In a laboratory associated with this chapter you may solve a more substantial problem.

Exploration Activity 1

(a) Calculate each of the following integrals:

$$\int_{-\pi}^{\pi} 1\,dt, \qquad \int_{-\pi}^{\pi} \sin t\,dt, \qquad \int_{-\pi}^{\pi} \cos t\,dt, \qquad \int_{-\pi}^{\pi} \cos 2t\,dt.$$

(b) Use a graph of each of the integrands to explain why each of the integrals has the value you just calculated.

(c) Make a conjecture about the values of the integrals

$$\int_{-\pi}^{\pi} \sin nt\,dt \qquad \text{and} \qquad \int_{-\pi}^{\pi} \cos nt\,dt,$$

where n is any positive integer.

(d) Verify your conjecture by calculating the integrals in part (c).

The integrals in part (a) are easily calculated by the Fundamental Theorem—their values are 2π, 0, 0, and 0, respectively. The first integral is just the area of a rectangle of height 1 and length 2π. The other three integrals are easily seen to be zero by examining graphs of the integrands. For example, we show the graph of $y = \cos 2t$ in Figure 9.30. Notice that the area of the unshaded region under the curve and above the t-axis equals the area of the shaded region above the curve and below the t-axis.

Figure 9.30 Graph of $y = \cos 2t$

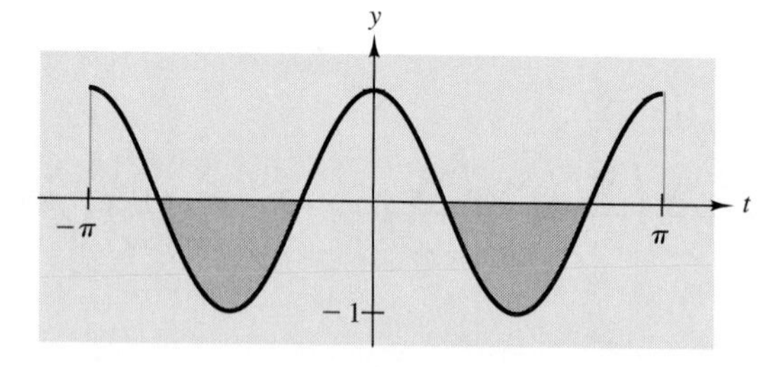

It appears likely that all integrals of $\sin nt$ and $\cos nt$ over the interval $[-\pi, \pi]$ will turn out to be zero for the same reason. We can verify this by direct calculation. For example,

$$\int_{-\pi}^{\pi} \cos nt\,dt = \left.\frac{\sin nt}{n}\right|_{-\pi}^{\pi} = \frac{\sin n\pi - \sin(-n\pi)}{n} = \frac{0-0}{n} = 0.$$

Now we consider integrals of products of these basic functions.

Exploration Activity 2

(a) Figure 9.31 shows the graph of

$$y = \cos t \cos 2t$$

on the interval from $-\pi$ to π. Guess the value of

$$\int_{-\pi}^{\pi} \cos t \cos 2t \, dt$$

from the graph. Explain your guess.

Figure 9.31 Graph of $y = \cos t \cos 2t$

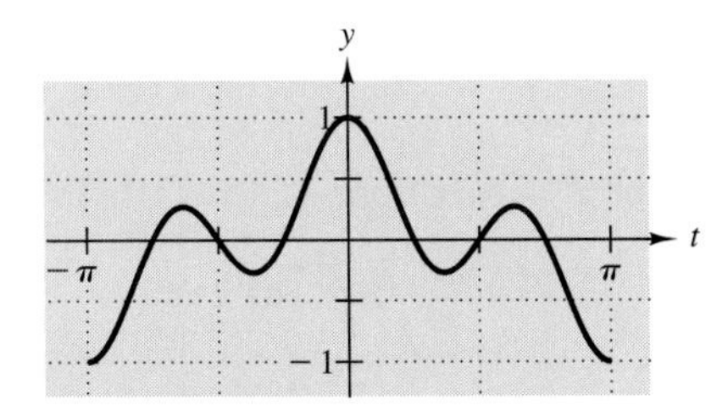

(b) Use the Integral Table to evaluate the integral in part (a).

(c) Assume that n and m are nonzero integers. Use the Integral Table to evaluate each of the following integrals.

$$\int_{-\pi}^{\pi} \cos nt \cos mt \, dt \text{ (for } n \neq m)$$

$$\int_{-\pi}^{\pi} \sin nt \sin mt \, dt \text{ (for } n \neq m)$$

$$\int_{-\pi}^{\pi} \cos nt \sin mt \, dt$$

Notice that there is no restriction that $n \neq m$ for the last integral. You may have to deal with the $n = m$ case separately.

Since the area below the curve and above the t-axis is matched by the area above the curve and below the axis, it appears that

$$\int_{-\pi}^{\pi} \cos t \cos 2t \, dt = 0.$$

We can verify this by Formula 83:

$$\int_{-\pi}^{\pi} \cos t \cos 2t \, dt = \left[\frac{\sin[(1-2)t]}{2(1-2)} + \frac{\sin[(1+2)t]}{2(1+2)} \right] \Bigg|_{-\pi}^{\pi}$$

$$= \left[\frac{\sin(-\pi)}{-2} + \frac{\sin 3\pi}{6} \right] - \left[\frac{\sin \pi}{-2} + \frac{\sin(-3\pi)}{6} \right] = 0.$$

For $n \neq m$, the three integrals in part (c) may be shown to be zero using Formulas 83, 82, and 84, respectively. For $n = m$ in the last integral, we can use the trigonometric identity $\sin 2x = 2 \sin x \cos x$:

$$\int_{-\pi}^{\pi} \cos nt \sin nt \, dt = \int_{-\pi}^{\pi} \frac{\sin 2nt}{2} \, dt = \frac{-\cos 2nt}{4n} \Bigg|_{-\pi}^{\pi} = 0.$$

The only integrals of products of basic trigonometric functions we have not examined yet are integrals of squares, $\cos^2 nt$ and $\sin^2 nt$. We take these up in the next exploration activity.

Exploration Activity 3

(a) Figure 9.32 shows the graph of $y = \cos^2 t$ from $-\pi$ to π. Use the graph to explain why

$$\int_{-\pi}^{\pi} \cos^2 t \, dt$$

cannot be zero.

Figure 9.32 Graph of $y = \cos^2 t$

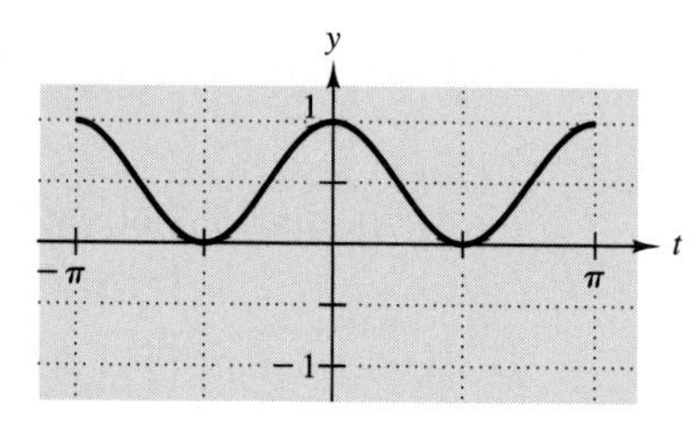

(b) Use the trigonometric identity

$$\cos^2 t = \tfrac{1}{2}(1 + \cos 2t)$$

to calculate

$$\int_{-\pi}^{\pi} \cos^2 t \, dt.$$

(c) Assume that n is a nonzero integer, and calculate

$$\int_{-\pi}^{\pi} \cos^2 nt \, dt.$$

(d) Assume that n is a nonzero integer, and calculate

$$\int_{-\pi}^{\pi} \sin^2 nt \, dt.$$

[Recall that $\sin^2 t = \frac{1}{2}(1 - \cos 2t)$.]

Since the graph of $\cos^2 t$ is always above the t-axis for $-\pi \le t \le \pi$ (except for the two tangent points), the integral must be positive. Using the identity, we may write

$$\int_{-\pi}^{\pi} \cos^2 t\, dt = \int_{-\pi}^{\pi} \frac{1}{2}(1 + \cos 2t)\, dt$$
$$= \frac{t}{2} + \frac{\sin 2t}{4}\bigg|_{-\pi}^{\pi} = \pi.$$

A similar calculation shows that

$$\int_{-\pi}^{\pi} \cos^2 nt\, dt = \pi$$

for all positive integers n. Using the identity for $\sin^2 t$, we may write

$$\int_{-\pi}^{\pi} \sin^2 nt\, dt = \int_{-\pi}^{\pi} \frac{1}{2}(1 - \cos 2nt)\, dt$$
$$= \frac{t}{2} - \frac{\sin 2nt}{4n}\bigg|_{-\pi}^{\pi} = \pi.$$

Here is the central puzzle of this section: In Figure 9.33 we show the graph of a function $f(t)$ that has the form $f(t) = b_0 + b_1 \cos t + b_2 \cos 2t$ for some choice of the constants b_0, b_1, and b_2. Suppose we can calculate numerical values of the function f but we do not know the formula, i.e., the values of the constants b_0, b_1, and b_2. How can we determine the values of the constants?

In a more realistic setting, such as the EKG in Figure 9.28, there *is* a way to get numerical values of the periodic function—read them from the graph—even though there is no formula in sight. Since the EKG appears to combine several different repeating phenomena, we might suspect that a good fit by a formula would require both sine and cosine terms of many different frequencies. The observations we shall make here about finding the coefficients would apply to that case as well. That is, the simple case illustrated in Figure 9.33 contains all the complexities required for dealing with more realistic problems—such as the one you may encounter in the laboratory.

Figure 9.33 Graph of $f(t) = b_0 + b_1 \cos t + b_2 \cos 2t$

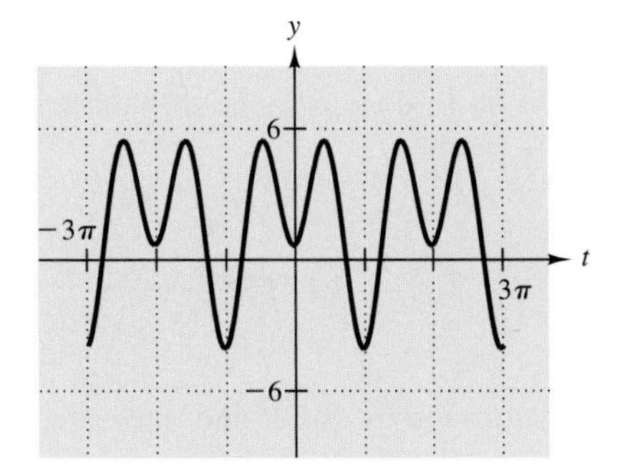

Exploration Activity 4

(a) Use the form of the unknown function $f(t)$ to write an expression for

$$\int_{-\pi}^{\pi} f(t)\, dt$$

in terms of integrals of the elementary functions 1, cos t, and cos $2t$. Then use this expression and the results of Exploration Activity 1 to show that

$$b_0 = \frac{1}{2\pi}\int_{-\pi}^{\pi} f(t)\, dt.$$

(b) Multiply both sides of the equation for f by an appropriate trigonometric function to find an expression for

$$\int_{-\pi}^{\pi} f(t) \cos t\, dt.$$

Use this expression and the results of Exploration Activities 1, 2, and 3 to find a formula for b_1.

(c) Now multiply both sides of the equation for f by an appropriate trigonometric function to find an expression for

$$\int_{-\pi}^{\pi} f(t) \cos 2t\, dt.$$

Use this expression and the results of Exploration Activities 1, 2, and 3 to find a formula for b_2.

When we integrate the formula for f, we obtain

$$\begin{aligned}\int_{-\pi}^{\pi} f(t)\, dt &= b_0\int_{-\pi}^{\pi} 1\, dt + b_1\int_{-\pi}^{\pi} \cos t\, dt + b_2\int_{-\pi}^{\pi} \cos 2t\, dt.\\ &= b_0 2\pi.\end{aligned}$$

Thus
$$b_0 = \frac{1}{2\pi}\int_{-\pi}^{\pi} f(t)\, dt.$$

When we multiply the formula for f by $\cos t$ and integrate, we obtain

$$\begin{aligned}\int_{-\pi}^{\pi} f(t) \cos t\, dt &= b_0\int_{-\pi}^{\pi} \cos t\, dt + b_1\int_{-\pi}^{\pi} \cos^2 t\, dt + b_2\int_{-\pi}^{\pi} \cos t \cos 2t\, dt.\\ &= b_1\pi.\end{aligned}$$

Thus
$$b_1 = \frac{1}{\pi}\int_{-\pi}^{\pi} f(t) \cos t\, dt.$$

Similarly, if we multiply through by $\cos 2t$ and integrate, we find

$$b_2 = \frac{1}{\pi}\int_{-\pi}^{\pi} f(t) \cos 2t\, dt.$$

Checkpoint 1

Assume that we can calculate the values of the three integrals

$$\int_{-\pi}^{\pi} f(t)\,dt, \qquad \int_{-\pi}^{\pi} f(t)\cos t\,dt, \qquad \int_{-\pi}^{\pi} f(t)\cos 2t\,dt$$

in some way, e.g., by numerical integration. If these values are, respectively, 10.87, 7.383, and -10.901, find b_0, b_1, and b_2. Graph the resulting function f, and check that your graph looks like Figure 9.33.

Functions f of the form

$$f(t) = b_0 + a_1 \sin t + b_1 \cos t + a_2 \sin 2t + b_2 \cos 2t + \cdots + a_n \sin nt + b_n \cos nt$$

are called **trigonometric polynomials**. Given such a function, we can obtain the coefficients by integration:

$$b_0 = \frac{1}{2\pi}\int_{-\pi}^{\pi} f(t)\,dt$$

$$b_k = \frac{1}{\pi}\int_{-\pi}^{\pi} f(t)\cos kt\ dt \qquad \text{for } k = 1, 2, \ldots, n$$

$$a_k = \frac{1}{\pi}\int_{-\pi}^{\pi} f(t)\sin kt\ dt \qquad \text{for } k = 1, 2, \ldots, n.$$

These observations represent the first steps in the general theory of representing complicated periodic functions as combinations of simpler basic periodic functions. This representation theory is generally known as the theory of Fourier series. We proceed a little farther into this theory in the exercises at the end of this section.

Answers to Checkpoint

1. $b_0 = 1.73$, $b_1 = 2.35$, and $b_2 = 3.470$

Exercises 9.7

Write down the value of each of the following integrals. You should be able to do this without any calculation.

1. $\int_{-\pi}^{\pi} \cos 5t\ dt$
2. $\int_{-\pi}^{\pi} \sin^2 3t\ dt$
3. $\int_{-\pi}^{\pi} \cos 4t \cos 6t\ dt$
4. $\int_{-\pi}^{\pi} \cos^2 5t\ dt$
5. $\int_{-\pi}^{\pi} \sin 2t \cos 7t\ dt$
6. $\int_{-\pi}^{\pi} \sin 3t \sin 5t\ dt$

In Exercises 7–9, assume f is a function of the form

$$f(t) = b_0 + a_1 \sin t + b_1 \cos t + a_2 \sin 2t.$$

7. Find a formula for f if

$$\int_{-\pi}^{\pi} f(t)\,dt = 2, \quad \int_{-\pi}^{\pi} f(t)\cos t\ dt = -1,$$

$$\int_{-\pi}^{\pi} f(t)\sin t\ dt = 3, \quad \int_{-\pi}^{\pi} f(t)\sin 2t\ dt = 4.$$

8. Find $\int_{-\pi}^{\pi} f(t)\sin 3t\,dt$.

9. Find $\int_{0}^{\pi} f(t)\ dt$.

10. (a) Show that the following functions are periodic with period 6:

$$\cos\frac{\pi}{3}t, \qquad \sin\frac{\pi}{3}t, \qquad \cos\frac{2\pi}{3}t.$$

(b) Generalize part (a): Show that for any positive integer n the functions

$$\cos\frac{n\pi}{3}t \text{ and } \sin\frac{n\pi}{3}t$$

are periodic with period 6.

Calculate each of the following integrals.

11. $\int_{-3}^{3} 1\,dt$

12. $\int_{-3}^{3} \sin\frac{\pi}{3}t\,dt$

13. $\int_{-\pi}^{\pi} \cos\frac{\pi}{3}t\,dt$

14. $\int_{-\pi}^{\pi} \cos\frac{2\pi}{3}t\,dt$

15. $\int_{-\pi}^{\pi} \sin^2\frac{\pi}{3}t\,dt$

16. $\int_{-\pi}^{\pi} \cos^2\frac{\pi}{3}t\,dt$

17. Assume the function f has the form

$$f(t) = b_0 + a_1\sin\frac{\pi}{3}t + b_1\cos\frac{\pi}{3}t + b_2\cos\frac{2\pi}{3}t.$$

Find a formula for f if

$$\int_{-3}^{3} f(t)\,dt = 6, \int_{-3}^{3} f(t)\cos\frac{\pi}{3}t\,dt = 2,$$

$$\int_{-3}^{3} f(t)\sin\frac{\pi}{3}t\,dt = -3, \int_{-3}^{3} f(t)\cos\frac{2\pi}{3}t\,dt = 9.$$

18. Assume the function f has the form

$$f(t) = b_0 + a_1\sin\frac{\pi}{5}t + b_1\cos\frac{\pi}{5}t + b_2\cos\frac{2\pi}{5}t.$$

Find a formula for f if

$$\int_{-5}^{5} f(t)\,dt = 5, \int_{-5}^{5} f(t)\cos\frac{\pi}{5}t\,dt = 4,$$

$$\int_{-5}^{5} f(t)\sin\frac{\pi}{5}t\,dt = -5, \int_{-5}^{5} f(t)\cos\frac{2\pi}{5}t\,dt = 20.$$

19. Look again at the trigonometric identity $\cos^2 t = \frac{1}{2}(1+\cos 2t)$. This is itself an example of representation of a periodic function, $\cos^2 t$, as a combination of elementary periodic functions, 1 and $\cos 2t$. Use Figure 9.34 to explain the significance of the 2 in $\cos 2t$, of the addition of a constant to $\cos 2t$, and of the coefficient $\frac{1}{2}$.

Figure 9.34 Graph of $y = \cos^2 t$

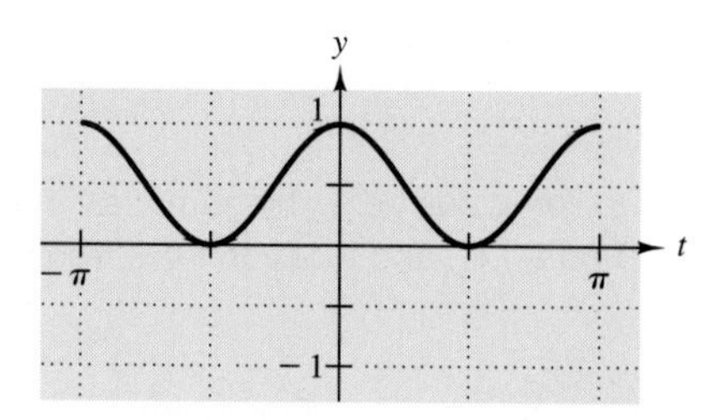

20. (a) There are infinitely many functions $y = f(x)$ defined implicitly by the equation

$$\sin^2\left(\frac{x}{2}\right) + \cos^2\left(\frac{y}{2}\right) = 1$$

Solve this equation for y to find one such function. Use your computer or graphing calculator to graph the solution you selected.

(b) If you haven't already done so, show that there is a function $y = f(x)$ defined by this equation that has domain $(-\infty, \infty)$ and that has the properties shown in Figure 9.35: It is **even** and **periodic** with period 2π.

Unlikely as it may seem, we are going to represent the triangle wave function $f(x)$ approximately by a trigonometric polynomial of the form

$$b_0 + b_1\cos x + b_2\cos 2x.$$

That is, we want to find coefficients b_0, b_1, and b_2 so that

$$b_0 + b_1\cos x + b_2\cos 2x$$

will approximate $f(x)$.

Figure 9.35 Triangular wave function

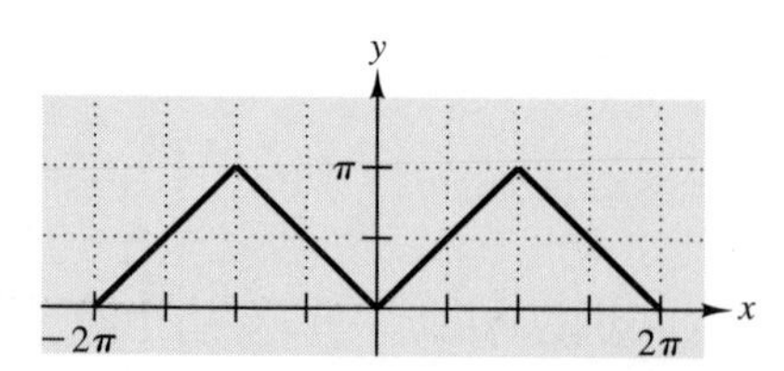

(c) Find b_0 by integrating both sides of

$$f(x) \approx b_0 + b_1\cos x + b_2\cos 2x$$

from $-\pi$ to π. You should be able to write down the integral of $f(x)$ by inspection from

Figure 9.35. In fact, you may be able to guess b_0 from the picture.

(d) To find additional coefficients, we need to know how to evaluate integrals such as

$$\int_{-\pi}^{\pi} f(x) \cos x \, dx.$$

Because both $f(x)$ and $\cos x$ have even symmetry, this integral is the same as

$$2\int_{0}^{\pi} f(x) \cos x \, dx.$$

And, on the interval $[0, \pi]$, $f(x)$ is just x (see Figure 9.35). Thus

$$\int_{-\pi}^{\pi} f(x) \cos x \, dx = 2\int_{0}^{\pi} x \cos x \, dx.$$

Use your integral table to evaluate the integral on the right.

(e) Solve for b_1 and b_2 as you did in Exploration Activity 4.

(f) Reality check: You already have the function $f(x)$ entered in your calculator or computer. Enter the function $b_0 + b_1 \cos x + b_2 \cos 2x$, with your calculated values for b_0, b_1, and b_2, and graph the two functions together. The second graph should snake along the first.

Chapter 9 Summary

This chapter, "Integral Calculus and Its Uses," has much the same place in our study of integration as Chapter 4, "Differential Calculus and Its Uses," had in our study of differentiation. In both chapters we study significant applications and gather together the important techniques of calculation.

The major applications in this chapter are at the beginning and the end. In the beginning we studied the problem of determining the center of mass of a complicated object. The subdivide-and-conquer strategy enabled us to extend summation formulas for discrete calculations to integral formulas for the corresponding continuous problem. At the end of the chapter we used definite integral calculations to represent complicated periodic functions as sums of basic sine and cosine functions.

Almost everyone has access to a quick and powerful computational device for evaluating routine definite integrals — a calculator or a computer that uses numerical integration programs. We investigated various numerical integration procedures, from the crude right- and left-hand sums to the more efficient Midpoint and Trapezoidal Rules, and then to the weighted average of these last two, Simpson's Rule. Some version of Simpson's Rule is probably what your calculator uses for its integration routine.

Sometimes symbolic evaluation of a definite integral, using the Fundamental Theorem of Calculus, is desirable or necessary. Although many powerful computer programs will generate antiderivatives for most common functions — and determine a symbolic evaluation of a definite integral in one step — we focused in this chapter on a valuable low-tech tool: an Integral Table. To use the Integral Table effectively, we often must rewrite the indefinite integral in question to match a particular entry. An important technique for doing this is substitution for the variable of integration. Many of the integral formulas themselves are obtained by using integration-by-parts. Both substitution and integration-by-parts are inverses of differentiation rules: Substitution is the inverse of the Chain Rule, and integration-by-parts is the inverse of the Product Rule.

Concepts

- Moments and centers of mass
- Numerical approximation to a definite integral
- Algebraic substitution in integration
- Trigonometric substitution in integration
- Reduction formulas (in a table of integrals)
- Integration by parts
- Representation of periodic functions by trigonometric polynomials

Applications

- Calculation of volume
- Calculation of mass
- Calculation of center of mass
- Fourier representation of periodic functions

Formulas

Centers of mass

For a uniform tapered rod lying between a and b on the x-axis with cross-sectional radius $r(x)$:

$$\overline{x} = \frac{\int_a^b x\,[r(x)]^2 dx}{\int_a^b [r(x)]^2 dx}$$

For a uniform thin plate described by $a \leq x \leq b$ and $g(x) \leq x \leq f(x)$:

$$\overline{x} = \frac{\int_a^b x\,[f(x) - g(x)]\,dx}{\int_a^b [f(x) - g(x)]\,dx}$$

$$\overline{y} = \frac{1}{2}\frac{\int_a^b [f(x)^2 - g(x)^2]\,dx}{\int_a^b [f(x) - g(x)]\,dx}$$

Numerical approximations to the definite integral $\int_a^b f(x)\,dx$

In each of the following formulas the function f is continuous on the interval $[a, b]$, n is a positive integer, $\Delta t = (b - a)/n$, and $t_k = a + k\,\Delta t$ for $k = 0, 1, \ldots, n$.

Left-Hand Sum

$$\text{LHS} = \sum_{k=1}^{n} f(t_{k-1})\,\Delta t$$

Right-Hand Sum

$$\text{RHS} = \sum_{k=1}^{n} f(t_k)\,\Delta t$$

Trapezoidal Rule

$$\text{TR} = \sum_{k=1}^{n} \frac{f(t_{k-1}) + f(t_k)}{2}\,\Delta t$$

Midpoint Rule

$$\text{MR} = \sum_{k=1}^{n} f\left(\frac{t_{k-1} + t_k}{2}\right)\Delta t$$

Simpson's Rule

$$\text{SR} = \frac{2}{3}\,\text{MR} + \frac{1}{3}\,\text{TR}$$

Integration techniques

Substitution in an indefinite integral

$$\int f(u(x))u'(x)\,dx = \int f(u)\,du$$

Substitution in a definite integral

$$\int_a^b f(u(x))\,u'(x)\,dx = \int_{u(a)}^{u(b)} F(u)\,du$$

Integration-by-parts

$$\int u\,dv = uv - \int v\,du$$

Fourier coefficients

If $f(t) = b_0 + a_1 \sin t + b_1 \cos t + a_2 \sin 2t + b_2 \cos 2t + \cdots + a_n \sin nt + b_n \cos nt$ is a trigonometric polynomial, then the coefficients can be found by the following formulas:

$$b_0 = \frac{1}{2\pi}\int_{-\pi}^{\pi} f(t)\,dt$$

$$b_k = \frac{1}{\pi}\int_{-\pi}^{\pi} f(t)\cos kt\;dt \quad \text{for } k = 1, 2, \ldots, n$$

$$a_k = \frac{1}{\pi}\int_{-\pi}^{\pi} f(t)\sin kt\;dt \quad \text{for } k = 1, 2, \ldots, n.$$

Chapter 9 Exercises

Use the Trapezoidal Rule with four subdivisions to approximate each of the following integrals. Then calculate the exact symbolic value and compare the two evaluations.

1. $\int_0^{\pi} \sin x \, dx$
2. $\int_{-1}^{1} (1 - x^2) \, dx$
3. $\int_{-1}^{1} (1 - x^2)^2 \, dx$
4. $\int_0^{1} e^{-2x} \, dx$

Use the Midpoint Rule with four subdivisions to approximate each of the following integrals. Compare your results to the corresponding approximation using the Trapezoidal Rule (Exer cises 9–12) and to the exact value.

5. $\int_0^{\pi} \sin x \, dx$
6. $\int_{-1}^{1} (1 - x^2) \, dx$
7. $\int_{-1}^{1} (1 - x^2)^2 \, dx$
8. $\int_0^{1} e^{-2x} \, dx$

Use your results from Exercises 9–0 and Simpson's Rule to approximate each of the following integrals.

9. $\int_0^{\pi} \sin x \, dx$
10. $\int_{-1}^{1} (1 - x^2) \, dx$
11. $\int_{-1}^{1} (1 - x^2)^2 \, dx$
12. $\int_0^{1} e^{-2x} \, dx$

Use the Trapezoidal Rule with four subdivisions to approximate each of the following integrals.

13. $\int_0^{1} \sqrt{1 + x^2} \, dx$
14. $\int_0^{1} e^{-t^2/2} \, dt$
15. $\int_0^{\pi} \theta^2 \sin^2\theta \, d\theta$

Use the Midpoint Rule with four subdivisions to approximate each of the following integrals.

16. $\int_0^{1} \sqrt{1 + x^2} \, dx$
17. $\int_0^{1} e^{-t^2/2} \, dt$
18. $\int_0^{\pi} \theta^2 \sin^2\theta \, d\theta$

Use your results from Exercises 13–18 and Simpson's Rule to approximate each of the following integrals.

19. $\int_0^{1} \sqrt{1 + x^2} \, dx$
20. $\int_0^{1} e^{-t^2/2} \, dt$
21. $\int_0^{\pi} \theta^2 \sin^2\theta \, d\theta$

Evaluate each of the following indefinite integrals.

22. $\int \cos^4 \theta \sin \theta \, d\theta$
23. $\int x^2(x^3 + 4) \, dx$
24. $\int \frac{2}{4 + u^2} \, du$
25. $\int y\sqrt{3y + 1} \, dy$
26. $\int \frac{x^2}{\sqrt{9 - x^2}} \, dx$

Evaluate each of the following definite integrals.

27. $\int_0^{1} x^2\sqrt{4x^3 + 2} \, dx$
28. $\int_0^{\pi} \cos^3 x \sin x \, dx$
29. $\int_0^{3} \sqrt{9 - x^2} \, dx$
30. $\int_0^{1} \frac{x}{(1 + x^2)^2} \, dx$
31. $\int_0^{\sqrt{3}} \frac{1}{\sqrt{1 + x^2}} \, dx$
32. $\int_{-2}^{2} \sqrt{1 - \frac{x^2}{4}} \, dx$

Evaluate each of the following integrals.

33. $\int (t^2 + \sin 3t)\, dt$

34. $\int \frac{7}{\sqrt{s+1}}\, ds$

35. $\int \frac{1}{x^2+4}\, dx$

36. $\int_3^7 \frac{4}{2y+1}\, dy$

37. $\int_{-\pi/4}^{\pi/4} \cos 2x\, dx$

38. $\int_0^2 e^{-0.3t}\, dt$

39. $\int 2x\,(1+x^2)^{3/4}\, dx$

40. $\int \sqrt{1-4w^2}\, dw$

41. $\int_0^{\pi/4} \sin^2 6\theta\, d\theta$

42. $\int_{-\pi}^{\pi} \cos^2 3t\, dt$

43. $\int t\, e^{5.3t}\, dt$

44. $\int_0^1 t\, e^{5.3t}\, dt$

45. $\int (t+3)\, e^{5.3t}\, dt$

46. $\int t \cos 2t\, dt$

47. $\int t^2 \sin 2t\, dt$

48. $\int_0^{\pi/2} t^2 \sin 2t\, dt$

49. $\int_0^1 \frac{x}{(1+x^2)^2}\, dx$

Evaluate each of the following integrals with aid of the Integral Table. Confirm the correctness of each answer by differentiating it.

50. $\int \sin^2 4x \cos^3 4x\, dx$

51. $\int x^3\, 3^x\, dx$

52. $\int x^2 \ln x\, dx$

53. $\int \frac{dx}{x^2+6x-20}$

54. $\int \frac{x+1}{x^2+6x-20}\, dx$

55. $\int \frac{x^2+x+1}{x^2+6x-20}\, dx$

56. In Figure 9.36 (repeated here from Figure 9.3), consider the "left system" consisting of the two masses on the left and the "right system" consisting of the two masses on the right. Assume that the origin of the coordinate system is at the balance point.

    (a) Show that $\overline{x}$ for the right system is 3.

    (b) Would you expect the left system to have $\overline{x} = -3$? Why or why not? Calculate $\overline{x}$ for the left system.

Figure 9.36 Equal moments

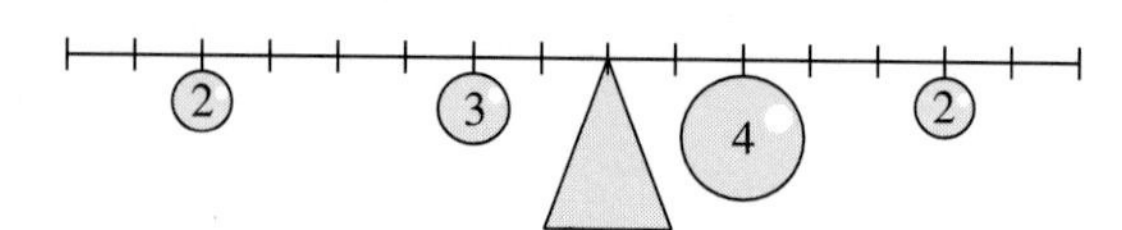

57. (a) Find the center of mass of the plane region bounded by the line $y = x$, the x–axis, and the line $x = 1$. (Can you do this without evaluating any integrals?)

    (b) Find the center of mass of the region bounded by the curve $y = x^2$, the x–axis, and the line $x = 1$.

    (c) Find the center of mass of the region bounded by the curve $y = x^3$, the x–axis, and the line $x = 1$.

(d) Find the center of mass of the region bounded by the curve $y = x^n$, the x–axis, and the line $x = 1$, where n is a positive integer.

(e) If we denote the center of mass in part (d) by $(\overline{x}_n, \overline{y}_n)$, what do you think the limiting values of $\overline{x}_n$ and $\overline{y}_n$ are as n becomes infinitely large? Calculate the limiting values from your answer to part (d). Do you find anything surprising in your results?

58. You are making a mobile, and one of the suspended parts is to have the shape shown in Figure 9.37. If the origin of the coordinate system is placed at the nose of the figure, the upper half of the figure is bounded by the graphs of $y = \sqrt{x}$, $y = x - 2$, and $x = 7$. You are going to cut the shape from a sheet of a thin material. At what point will you attach the thread so that the figure will hang in a horizontal plane?

Figure 9.37 Cutout for a mobile

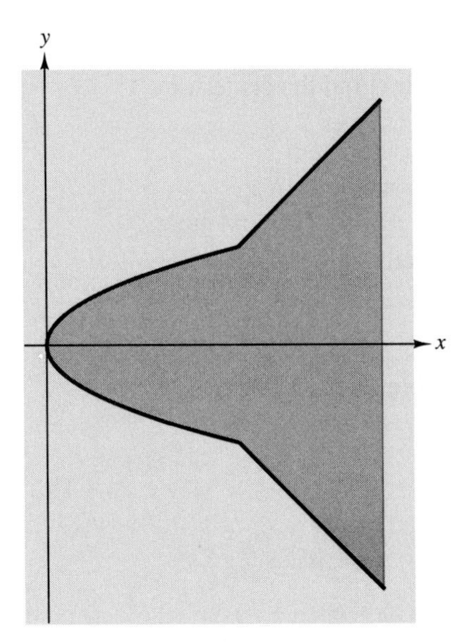

59. We saw in a laboratory associated with Chapter 7 that the historical data on world population has a growth rate that predicts an infinite population by about A.D. 2027. Specifically, if the population P is measured in billions and time t is measured in years, the data suggest that the growth rate of P can be modeled by $dP/dt = 0.005\ P^2$. We may also take as given that the population reached 5 billion about the start of 1987.

We consider now a fanciful scenario in which the world population problem is complicated by a small but steady immigration from outer space: Alien beings that are indistinguishable from humans (in every way) start arriving in 1987 at the rate of 5 million ($= 0.005$ billion) per year. Since the immigrants are traveling from a distant solar system (in a state of suspended animation, so they arrive with normal ages in Earth years), there is no way to turn off this steady flow — the travelers for the next several hundred years are already en route. Thus, from 1987 on, the growth rate for world population has an added component to account for immigration. That should mean that P is predicted to become infinite even earlier. How much earlier?

60. Figure 9.38 shows an idealized coat hanger made by bending a single piece of uniform wire into a triangle plus a hook. Ignoring the hook, find the center of mass of the wire triangle if the length of its base is b and the altitude to that base is a. (You should be able to find the center of mass without evaluating any integrals—but evaluate integrals if you must.)

Figure 9.38 Idealized coat hanger

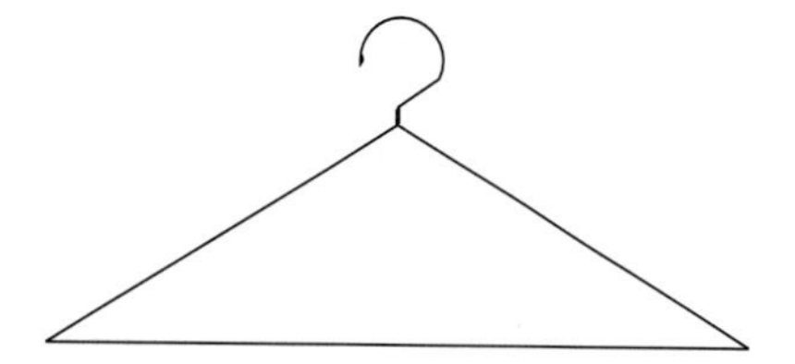

61. A wire one meter long and of uniform density is bent into a semicircle. Find the center of mass of the semicircular wire.

62. In Section 9.4 we observed that a substitution of the form $x = a \sin\theta$ would turn an expression of the form $a^2 - x^2$ into a perfect square. (The value of a was 5 in our example.) That helped us transform integrands that involve $\sqrt{a^2 - x^2}$, but it won't help much with integrands that involve $\sqrt{x^2 - a^2}$, because this expression is defined only when $|x| \geq |a|$, and $|a \sin\theta|$ is always $\leq |a|$.

    (a) Show that a substitution of the form $x = a \sec\theta$ will turn an expression of the form $x^2 - a^2$ into a perfect square.

    (b) Use the idea in part (a) to show that

    $$\int \frac{dx}{(x^2 - 25)^{3/2}}$$

    can be transformed into

    $$\frac{1}{25} \int (\sin\theta)^{-2} \cos\theta \, d\theta.$$

    Evaluate the transformed integral.

    (c) The antidifferentiation in part (b) isn't finished until we express the answer in terms of x. Draw a right triangle with hypotenuse x and one side 5, and label the angle θ in this triangle for which $x = 5 \sec\theta$. What is the length of the remaining side of your triangle? What is $\sin\theta$ in terms of x? Complete your calculation of the indefinite integral.

    (d) Check your answer to part (c) by differentiation.

63. In Exercise 62 we introduced the trigonometric substitution $x = a \sec\theta$ and blithely assumed that $\sqrt{\tan^2\theta} = \tan\theta$—which requires that $\tan\theta$ should be nonnegative for all relevant values of θ. Unlike the situation with the sine and tangent substitutions, our assumption is not always correct.

    (a) Resolution of the sign situation requires separate consideration of values of x greater than a and values of x less than $-a$. Note that $\sec\theta = x/a$ is equivalent to $\cos\theta = a/x$. Think separately about the positive and negative parts of the domain of the arccosine function and the corresponding ranges for θ. What is the sign of $\tan\theta$ for θ in each of these ranges?

(b) In Exercise 62 you evaluated

$$\int \frac{dx}{(x^2-25)^{3/2}}$$

and got an answer that you showed [in part (d) of the exercise] *must* be correct, no matter whether $x > 5$ or $x < -5$. But the substitution by which you got the answer is wrong in one of these two cases, because you left out a negative sign. Explain where in the calculation the missing sign gets compensated for.

64. In Exercise 20 of Section 9.7 you found the first three coefficients in a representation of the triangle wave function $f(x)$ (see Figure 9.39) by a trigonometric polynomial of the form

$$f(x) = b_0 + b_1 \cos x + b_2 \cos 2x + b_3 \cos 3x + \cdots + b_k \cos kx.$$

Here we explore the problem of finding additional coefficients $b_3, b_4, \ldots, b_k$ for as large a k as we wish.

Figure 9.39 Triangular wave function

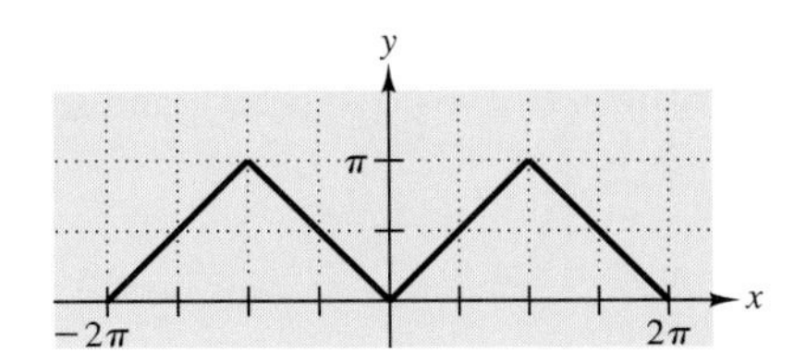

(a) Explain why, for every positive integer m,

$$\int_{-\pi}^{\pi} f(x) \cos mx \, dx = 2\int_0^{\pi} x \cos mx \, dx.$$

(b) Evaluate the integral on the right. (Use the Integral Table to check your antiderivative. You will need to make a distinction between odd and even values of m.)

(c) Find a formula for every b_m. (You will need to make a distinction between odd and even values of m.) Check that your calculation agrees with previous results for $m = 1$ and $m = 2$.

(d) Add enough terms to the trigonometric polynomial so that you cannot see any difference at all when you graph it along with $f(x)$ on the interval $[-\pi, \pi]$.

65. (a) *Guess* which of the following integrals will be larger:

$$\int_0^4 x\sqrt{16-x^2}\, dx \qquad \text{or} \qquad \int_0^4 \sqrt{16-x^2}\, dx.$$

Explain your reasoning.

(b) *Compute* which of the two integrals is actually larger.[1]

---

1. This and the remaining problems in this set are adapted from *Calculus Problems for a New Century*, edited by Robert Fraga, MAA Notes Number 28, 1993.

66. For each of the following functions f and intervals $[a, b]$, find the value of c for which the line $x = c$ divides the area of the region under the graph of f and over $[a, b]$ so that there is twice as much area to the left of $x = c$ as to the right.

    (a) $f(x) = (x-2)^2, \quad a = -1, \quad b = 1;$

    (b) $f(x) = 1 - 1/x^2, \quad a = 1, \quad b = 2.$

67. Consider the family of parabolas that are the graphs of $y = f(x)$, where $f(x)$ is any function of the form

$$f(x) = \frac{2}{a^2}\,x - \frac{1}{a^3}\,x^2$$

for some positive number a.

    (a) Show that the area of the region bounded by the graph of f and the x-axis does not depend on a. How large is this area?

    (b) What curve is determined by the vertices of all these parabolas?

68. Calculate $\int e^{-x}x^5\,dx$. [*Hint*: Save yourself some work by using a method that you will see again and again in your mathematics courses, the *Method of Undetermined Coefficients*. Make an educated guess that an antiderivative has the form $e^{-x}(a_0 + a_1x + \cdots + a_5x^5)$, and set the derivative of such a function equal to $e^{-x}x^5$.]

69. (a) Sketch the graph of $f(x) = 1/(x^2 - 4)$.

    (b) Is the integral $\displaystyle\int_{-1}^{1} \frac{dx}{x^2-4}$ positive or negative? Justify your answer.

    (c) Compute the integral in part (b).

    (d) A friend suggests that the integral in part (b) can be most easily evaluated by making the substitution $x = 2\sec\theta$ (see Exercise 62). What do you think?

70. Suppose $\displaystyle\int f(x)\sin x\,dx = -f(x)\cos x + \int 3x^2\cos x\,dx.$

    (a) Find a function that could be $f(x)$.

    (b) Find a *second* function that could be $f(x)$.

71. (a) Using Figure 9.40, give a geometric argument that the following identity is correct for every b between 0 and 1:

$$\int_0^b \arcsin x\,dx = b\arcsin b - \int_0^{\arcsin b} \sin y\,dy.$$

    (b) Use integration by parts to compute $\int_0^b \arcsin x\,dx$.

Figure 9.40 Area under the graph of the arcsine function

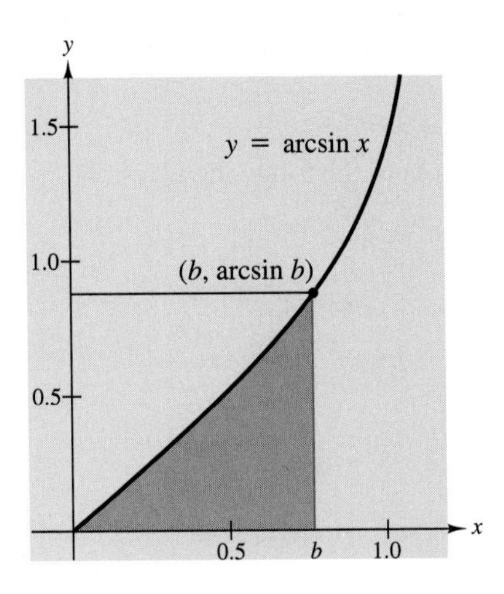

72. Four calculus students disagree as to the value of the integral

$$\int_0^{\pi} \sin^8 x \, dx.$$

Jack says it is equal to π. Joan says it is equal to $35\pi/128$. Ed claims it is equal to $3\pi/90 - 1$, while Lesley says it is equal to $\pi/2$. One of them is right; which one? (Without evaluating the integral, try to eliminate the three wrong answers.)

73. In Section 9.4 we suggested evaluating

$$\int_0^{\pi} \cos^2\theta \, d\theta$$

by using a double-angle formula to transform the integrand. Suppose instead that we make a substitution: $u = \sin\theta$. Then $\cos\theta = \sqrt{1-u^2}$, $du = \cos\theta \, d\theta$, and $u = 0$ at both $\theta = 0$ and $\theta = \pi$. That would seem to transform the integral into

$$\int_0^{0} \sqrt{1-u^2} \, du,$$

which is clearly 0. But the original function being integrated has only non-negative values, so the correct value of the integral must be positive. What went wrong? What is the correct value of the integral?

Chapter 9 Projects

1. **An alternate calculation of $\overline{y}$**

(a) Explain why the region in Figure 9.41 can be described in both of the following ways:

- All (x, y) such that $0 \leq x \leq 2$ and $0 \leq y \leq x^2$
- All (x, y) such that $0 \leq y \leq 4$ and $\sqrt{y} \leq x \leq 2$.

Figure 9.41 Region under the graph of $y = x^2$

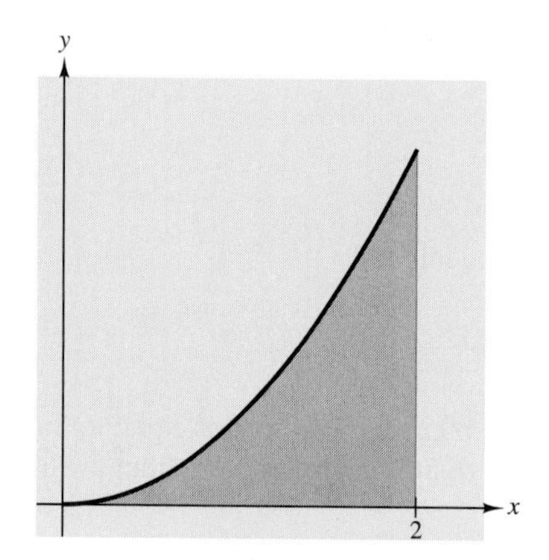

(b) Fill in the blanks in the definite integral so that its value equals the area of the region in Figure 9.41.

$$\int_0^{\_\_} \left(\_\_\_\_\_ - \_\_\_\_\_ \right) dy$$

(c) Check your answer in (b): Set up an integral with x as the independent variable that has value equal to the area of the region in Figure 9.41, and evaluate this integral.

(d) Fill in the blanks in the definite integral so that its value equals the moment M_x of the region in Figure 9.41.

$$\theta\delta \int_0^{\_\_} \_\_\_ \left(\_\_\_\_\_ - \_\_\_\_\_ \right) dy$$

(e) Check your answer in (d): Set up an integral with x as the independent variable that has value equal to the moment M_x of the region in Figure 9.41, and evaluate this integral.

(f) Find the y-coordinate $\overline{y}$ of the center of mass of the region shown in Figure 9.41.

(g) Find the center of mass of the region shown in Figure 9.42. The curve on the right side of the figure is $x = (\pi - y) \sin y$ and the curve on the left is $x = -(\pi - y) \sin y$.

Figure 9.42 Region between $x = (\pi - y)\sin y$ and $x = -(\pi - y)\sin y$

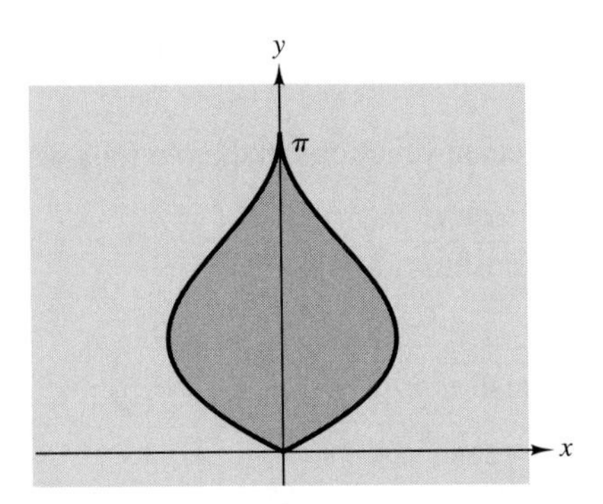

2. **Air resistance proportional to the velocity squared** In Example 1 of Section 9.5 we evaluated the left-hand side of

$$\int \frac{dv}{g - cv^2} = \int dt.$$

When you combine your result with the obvious evaluation of the right-hand side, you should get the equation

$$\frac{1}{2\sqrt{cg}} \ln\left(\frac{\sqrt{g} + v\sqrt{c}}{\sqrt{g} - v\sqrt{c}}\right) = t + C.$$

(a) Use an initial condition for the velocity of a dropped falling body to determine the constant C.

(b) Solve the equation for v as an explicit function of t. (You may have already done this in Project 2 at the end of Chapter 7.)

(c) Show that your answer can be written in the form

$$v(t) = \sqrt{\frac{g}{c}} \left(\frac{e^{\sqrt{cg}\,t} - e^{-\sqrt{cg}\,t}}{e^{\sqrt{cg}\,t} + e^{-\sqrt{cg}\,t}}\right).$$

(d) This equation gives us the rate of change of the position function for a falling body (such as a thunderstorm drop) subject to air resistance. If we assume $s(0) = 0$, then we have an initial value problem for the position function. Solve this initial value problem. (*Hint*: Differentiate the denominator of the fraction, and compare the result with the numerator.)

(e) Use the definitions of hyperbolic sine and hyperbolic cosine,

$$\sinh x = \frac{e^x - e^{-x}}{2}$$

and

$$\cosh x = \frac{e^x + e^{-x}}{2},$$

to rewrite the function $v(t)$ from part (c) in the form

$$v(t) = \sqrt{\frac{g}{c}} \frac{\sinh(\sqrt{gc}\ t)}{\cosh(\sqrt{gc}\ t)}.$$

(f) Define the hyperbolic tangent and hyperbolic secant functions in the obvious way:

$$\tanh u = \frac{\sinh u}{\cosh u}$$

and

$$\operatorname{sech} u = \frac{1}{\cosh u}.$$

Use the identity

$$1 + \sinh^2 u = \cosh^2 u$$

to show that

$$1 - \tanh^2 u = \operatorname{sech}^2 u.$$

(g) Show that

$$\frac{d}{du} \tanh u = \operatorname{sech}^2 u.$$

(h) Combine parts (e), (f), and (g) to show that

$$v(t) = \sqrt{\frac{g}{c}} \tanh\left(\sqrt{gc}\ t\right)$$

is a solution of

$$\int \frac{dv}{g - cv^2} = \int dt.$$

[Note that, once the right concepts and formulas are in place, the computation is much easier than the one you did in part (c).]

(i) What do you think an antiderivative of the hyperbolic tangent function should be? Check the Integral Table to be sure.

(j) Use the result of part (h) to solve for s again. (Notice that you can solve this problem more easily if you use hyperbolic functions.)

3. **Lengths of curves** In this project we will examine the use of integration to calculate the length of a curve. To have a particular curve in mind, consider the parabolic arc whose equation is $y = 3x^2$ for x ranging from 0 to 2.

(a) Use a ruler to approximate the length of the curve $y = 3x^2$ as shown in Figure 9.43.

(b) Let f be the function $f(x) = 3x^2$. Assume that we subdivide the interval $[0, 2]$ into n equal parts. As usual let $\Delta x = 2/n$ and $x_0, x_1, \ldots, x_n$ be the corresponding partition points. Assume that n is large enough that the portion of the curve in any one subinterval is essentially a straight line. Explain why the length of the portion of the curve between x_{k-1} and x_k can be approximated by

$$\sqrt{(\Delta x)^2 + [f'(x_{k-1})\Delta x]^2}.$$

(See Figure 9.44.)

Figure 9.43 Graph of $y = 3x^2$

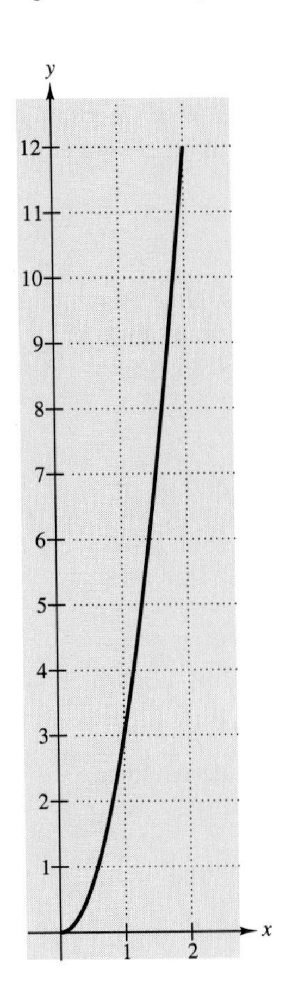

Figure 9.44 Line segment approximating a curve between two points

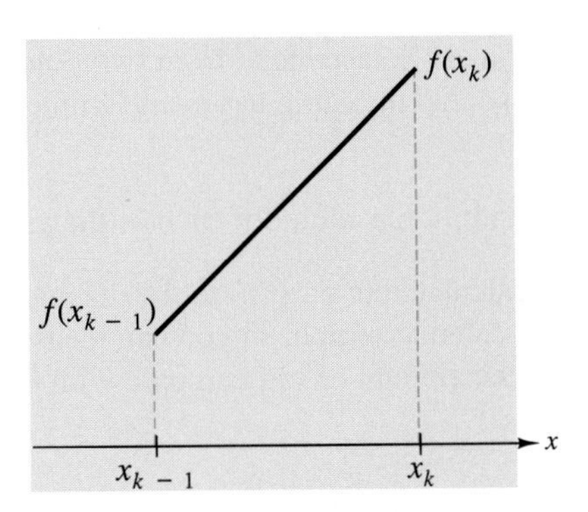

(c) Show that the approximation described in (b) leads to this integral formula for the length of the curve:

$$\text{Length of curve} = \int_0^2 \sqrt{1 + [f'(x)]^2}\, dx$$
$$= \int_0^2 \sqrt{1 + 36x^2}\, dx.$$

Calculate this integral using the integral key on your calculator, and compare to your estimate in part (a).

(d) The corresponding general formula for the length of the curve $y = f(x)$ for $a \leq x \leq b$ is

$$\text{Length of curve} = \int_a^b \sqrt{1 + [f'(x)]^2}\, dx.$$

Test this formula by showing that it leads to the same formula for the circumference of a circle as the one you learned in school.

(e) A cable suspended from its two ends hangs in the shape of a **catenary**, which is the graph of an equation of the form $y = a\cosh(x/a)$. Use your calculator or computer to graph a catenary for the following values of a: 50, 100, 200.

(f) Calculate the length of a cable suspended from two poles 100 meters apart. (*Hint*: Choose the coordinate system so that the cable has even symmetry. Your answer will be in terms of a.)

(g) Evaluate the actual length of the cable for each of the values of a in part (e).

(h) Compare the lengths calculated in part (h) to the distance between the poles. Note that the point $(0, a)$ lies on the catenary graph, so you can measure the maximum amount that the cable sags. How does the amount of sag compare with the length of the cable? Does this strike you as paradoxical?

4. **Calculating an entry in the Integral Table** Formula 21 in the Integral Table is

$$\int \sqrt{a^2 + u^2}\, du = \frac{u}{2}\sqrt{a^2 + u^2} + \frac{a^2}{2} \ln\left|u + \sqrt{a^2 + u^2}\right| + C.$$

In other tables you may see this formula:

$$\int \sqrt{a^2 + u^2}\, du = \frac{1}{2} u \sqrt{a^2 + u^2} + \frac{a^2}{2} \sinh^{-1} \frac{u}{a} + C.$$

In this project we will see why these two formulas are correct and show how to derive them both.

(a) We start with evaluation of $\int \sqrt{1 + x^2}\, dx$. Show that the obvious trigonometric substitution, $x = \tan\theta$, leads to another difficult integral.

(b) Another possibility is to substitute a hyperbolic function. The identity

$$\cosh^2\theta - \sinh^2\theta = 1$$

suggests the substitution $x = \sinh\theta$. Make the substitution, and show that this leads to the need to evaluate $\int \cosh^2\theta\, d\theta$.

(c) Use the definition of $\cosh\theta$ in terms of exponential functions to expand $\cosh^2\theta$. You should get an expression that you can easily integrate; do it.

(d) Show that your answer in part (c) can be written in the form $\frac{1}{4}\sinh 2\theta + \frac{1}{2}\theta + C$.

(e) Now we consider the problems of reverse substitution to get the answer in terms of x. From part (b), $\theta = \sinh^{-1}x$. However, substitution of this θ into the $\sinh 2\theta$ term is going to give us a mess unless we have a double angle formula of some sort. Show that $\sinh 2\theta = 2\sinh\theta\cosh\theta$. (*Hint*: Express both sides of the desired identity in terms of exponentials.) We know from part (b) that $\sinh\theta = x$ and $\cosh\theta = \sqrt{1 + x^2}$. Complete the substitution to show that the answer in part (d) can be written in the form $\frac{1}{2} x \sqrt{1 + x^2} + \frac{1}{2}\sinh^{-1} x + C$.

(f) Use the result of part (e) to show that

$$\int \sqrt{a^2+u^2}\,du = \frac{1}{2}u\sqrt{a^2+u^2} + \frac{a^2}{2}\sinh^{-1}\frac{u}{a} + C.$$

(g) Next, we consider how to express the inverse hyperbolic sine in terms of more familiar functions. We have already seen that $\theta = \sinh^{-1} x$ is equivalent to $x = \sinh\theta$. Rewrite this last equation using the definition of $\sinh\theta$ in terms of exponentials. Then multiply through both sides of the equation by e^{θ} and collect all the terms on one side of the equation. If you substitute $y = e^{\theta}$, you should get a quadratic equation in y (with x appearing in one of the coefficients). Solve for y. (There are two solutions because of the $\pm$ in the quadratic formula; keep the double sign for now.)

(h) The result of part (g) is a formula that gives e^{θ} as a function (or possibly two functions) of x. Take natural logs to solve for θ in terms of x. Explain why the algebraic solution with the minus sign is not a real solution and can therefore be dropped.

(i) Explain why the formula in part (f) also can be written in the form

$$\int \sqrt{a^2+u^2}\,du = \frac{1}{2}u\sqrt{a^2+u^2} + \frac{a^2}{2}\ln\left(u+\sqrt{a^2+u^2}\right) + C.$$

CHAPTER 10

Probability and Integration

When you pick up a package of light bulbs in a store, you might notice a statement about the average life of a typical bulb—say, 1000 hours. In this chapter we investigate what this means and how we could check the average lifetime by accumulating data on the actual lifetimes of a number of bulbs. This leads us to another application of integration: We find a formula for the average lifetime of a bulb in terms of an integral over all the possible lifetimes.

But **any** *length of time is a possible lifetime for a bulb. This means that we have to integrate from 0 to ∞! That involves finding limiting values of a process in which the upper limit of integration is a variable approaching ∞. Limiting values play an increasingly important role in this chapter and the next, so we introduce the standard notation for them and study them for their own sake.*

Having introduced a continuous probability distribution to study failure data (bulb burnouts), we move on to study several important distributions for modeling various other kinds of data. The most important of these is the normal distribution—the famous bell-shaped curve. Here the issue is not so much what the average value is—that's usually obvious from symmetry of the data—but rather how spread out the data are. We are led to study measures of spread, including standard deviation. We will find that we can study all normal distributions on a standardized scale by a two-step process of centering the data at the average value and then scaling to the standard deviation as the unit of distance.

The standard normal distribution function is defined by an integral that cannot be evaluated by simple formulas. This leads into the next chapter, in which we will study approximation of functions by polynomials—for which integration is always easy.

10.1 Reliability Theory: How Long Do Things Last?

In the information age we are absolutely dependent on electronic devices all around us — computers, calculators, voice mail, fuel injectors in our cars, automatic teller machines, product scanners in checkout lines, and so on, not to mention CD players and video games. These things have a way of wearing out or burning out at the most awkward times — for example, when a floppy disk drive goes haywire and trashes the only copy of your term paper. We're going to study the question of component failure, but we will start with a more prosaic — and more visible — component: the incandescent light bulb.

If you go to the store and buy a package of 100-watt light bulbs, you will see on the package a statement about the average life of the bulbs. Typical values are 750 to 1000 hours. What does this mean? What is the likelihood that such a bulb will burn out in, say, 7 weeks or 10 weeks?

Out of curiosity (and because we own stock in the power company) we bought 100 bulbs of the same kind and started them burning at the same time. We recorded the number that had burned out after the first day, the second day, and so on through day 200. We reproduce some of this data, mostly at one-week intervals through 16 weeks, in Table 10.1. By day 200, 99 of the 100 bulbs had burned out.

Table 10.1 Light bulb failure data

Day	*Bulbs burned out*	*Day*	*Bulbs burned out*
1	3	49	79
2	6	56	83
3	9	63	86
5	15	70	89
7	20	77	91
14	36	84	93
21	49	91	95
28	60	98	96
35	67	105	97
42	75	112	97

Modeling Failure Data

We'll begin our attempt to answer lifetime questions by trying to model the function $L(t)$ that represents the number of light bulbs burned out by the end of day t. We plot the data from Table 10.1 in Figure 10.1.

Figure 10.1 Light bulb data

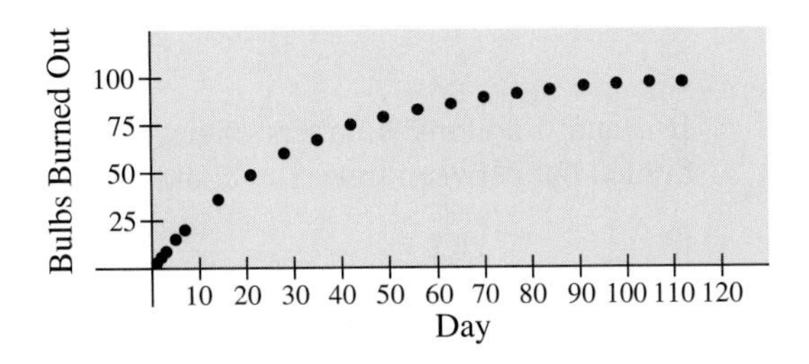

Exploration Activity 1

(a) What sort of function can we use to approximate L? Recall that we have seen curves like this before—in Newton's Law of Cooling, in free fall with air resistance, and in simple electrical circuits.

(b) Your proposed formula should have either one or two constants to be determined. Describe a procedure you could apply, if you had all the data on your computer or graphing calculator, to determine each of the constants.

In each of our previous encounters with curves like that in Figure 10.1, the function contained a term of the form e^{-rt}, where r is positive; the minus sign ensures that the term decays to zero as t increases. In the present case, it is reasonable to guess that a function of the form $M(1 - e^{-rt})$ should approximate $L(t)$. As t becomes large and the exponential term dies out, the function has to approach M. It is clear that from the fact that we had a total of 100 bulbs that M must be 100. That is, all the bulbs eventually burn out.

There are at least two ways we can find a value for r:

- Subtract each data value from 100, construct a semilog plot, and find the slope of the resulting line.
- Experiment with the function $100(1 - e^{-rt})$, varying r until the graph of the function matches the graph of the data.

You may have an opportunity in class to carry out either or both of these determinations. In the process, you will find that $r = 0.032$ gives a good fit.

We conclude that $L(t)$, the unknown function whose value at time t (in days) is the number of burned out bulbs at that time, is closely approximated by a function we call LA:

$$LA(t) = 100(1 - e^{-0.032\,t}).$$

Thus, for example, $LA(44)$ (about 76) approximates the number of bulbs burned out by the end of day 44, and $LA(5)$ (about 15) approximates the number of bulbs burned out by the end of day 5. It follows that $LA(44) - LA(5) \approx 61$ represents the number of bulbs burned out *between* day 5 and day 44. If we divide this number by 100, we obtain the *fraction* of the bulbs burned out between these two days: 0.61.

We also can obtain a fractions-burned-out function by *scaling* LA. If we define $F(t)$ as $LA(t)/100$, then we have

$$F(t) = 1 - e^{-0.032\,t}.$$

If a and b are any numbers of days, then $F(b) - F(a)$ represents the fraction of bulbs burned out between time $t = a$ and time $t = b$.

Checkpoint 1

Find the approximate fraction of bulbs that burned out between day 10 and day 85.

Another way of looking at this is to consider what might happen if we took another bulb of the same type out of its package, screwed it into a socket, turned it on, and left it on. Our experience with the first 100 bulbs suggests it is reasonable to assert that the *probability* the bulb would burn out between time $t = a$ and time $t = b$ is $F(b) - F(a)$.

We are now in a position to answer two of the questions raised at the beginning of the section.

Checkpoint 2

Find the probability that a typical bulb will burn out

(a) during the first seven weeks;

(b) during the first ten weeks.

Expected Lifetimes: Discrete Approximation

Now we return to the question of the expected life of a bulb. Intuitively, this is the same question as asking what the average lifetime would be if we burned all 100 bulbs until they failed—or, for that matter, what the average lifetime *was* for the 100 bulbs. Later we will move on to related questions, such as: Now that we know the average lifetime, if we pick a similar bulb "at random," how close to average is it likely to be?

Finding the average lifetime or expected performance is a challenging problem for two reasons. First, *lifetime* is not a discrete quantity: A bulb can burn out at any instant over a long continuous time interval. Second, the time interval is unbounded. In principle, an exceptional light bulb could burn "forever"—or at least for a very long time relative to the usual lifetimes of light bulbs. We will pay more attention in the next section to the problem of averaging over an "infinitely long" time interval.

As we have done with a number of other problems, we start by considering a discrete approximation to our continuous problem.

Exploration Activity 2

(a) Suppose we have a collection of (idealized) bulbs, all mixed up in one box, of exactly two types:

- Type A burns out in exactly 35 days, and we have 50 of these.
- Type B burns out in exactly 63 days, and we have 100 of these.

If we pick one bulb at random out of the box, we have a one-third probability of getting a bulb that will burn out in exactly 35 days and a two-thirds probability of getting a bulb that burns out in 63 days. What is the expected (or average) life of a bulb drawn at random from this collection?

(b) Suppose now that we have a collection of three types of bulbs, each with an exact lifetime and a probability of occurrence given in Table 10.2. What is the expected life of a bulb drawn at random from this collection?

Table 10.2 Bulb lifetimes and probabilities

Type	*Probability*	*Exact lifetime*
1	$p_1 = 1/4$	$t_1 = 23$ days
2	$p_2 = 1/2$	$t_2 = 47$ days
3	$p_3 = 1/4$	$t_3 = 65$ days

(c) If we have n types of bulbs, the kth occurring with probability p_k and having a lifetime of exactly t_k, for $k = 1, 2, \ldots, n$, what is the expected life of a bulb taken at random from this collection?

In part (a), if we average all the lifetimes, we have

$$\begin{aligned} \frac{50 \times 35 + 100 \times 63}{150} &= \frac{50}{150}35 + \frac{100}{150}63 \\ &= \frac{1}{3}35 + \frac{2}{3}63 \\ &\approx 53.7 \text{ days.} \end{aligned}$$

In part (b), we find an average lifetime of

$$p_1t_1 + p_2t_2 + p_3t_3 = 45.5 \text{ days.}$$

In the general case, the corresponding formula for average lifetime is

$$p_1t_1 + p_2t_2 + \cdots + p_nt_n = \sum_{k=1}^{n} p_kt_k \,.$$

Expected Lifetimes: Continuous Case

Now we return to our original problem—that of deciding what should be the expected lifetime of a random bulb—for which we are ready to do a divide-and-conquer argument. If we experiment with $F(t)$, we find that all the bulbs (except possibly one) should be burned out by day 200. Specifically, $F(200) = 0.9983$. We choose a large integer n, and we imagine the time interval from 0 to 200 divided into n subintervals, each of length $\Delta t = 200/n$. Our real bulbs don't have exact lifetimes, of course, but we will approximate the expected lifetime by pretending that they do: If a bulb burns out between the time $t_{k-1} = (k-1)\,\Delta t$ and the time $t_k = k\,\Delta t$, then we will count it as having burned for exactly t_k days. We are now in the situation described by the expression

$$p_1t_1 + p_2t_2 + \cdots + p_nt_n = \sum_{k=1}^{n} p_kt_k \,,$$

where $p_k = F(t_k) - F(t_{k-1})$ for $k = 1, 2, \ldots, n$. We can thus approximate the expected lifetime of a random bulb by the sum

$$\sum_{k=1}^{n} [F(t_k) - F(t_{k-1})]\, t_k.$$

This sum does not look quite like an approximating sum for an integral, because it is lacking a factor of Δt in each term—a situation you also may have encountered in measuring the length of a curve (see Project 3 in Chapter 9). Our solution to this problem is the same as it was for the length problem—and similar to solutions of other problems in which we failed to find a hoped-for algebraic form: We multiply and divide each term of the sum by Δt, which does not disturb the value of the sum. Then we factor Δt out of each term, which permits us to write the sum in the form

$$\sum_{k=1}^{n} \left[\frac{F(t_k) - F(t_{k-1})}{\Delta t}\, t_k \right] \Delta t.$$

For large values of n, Δt is small, and

$$\frac{F(t_k) - F(t_{k-1})}{\Delta t} \approx F'(t_k).$$

[The difference quotient is also approximately $F'(t_{k-1})$; when t_k and t_{k-1} are very close together, the values of F' also will be very close together.] If we abbreviate the derivative of F by f, then our approximate expression for the expected lifetime becomes

$$\sum_{k=1}^{n} t_k\, f(t_k)\, \Delta t.$$

As n gets large, this sum approaches the integral

$$\int_0^{200} t\, f(t)\, dt.$$

Exploration Activity 3

(a) Given $F(t) = 1 - e^{-0.032\,t}$, find $f(t)$.

(b) Find a symbolic evaluation of the integral

$$\int_0^{200} t\, f(t)\, dt.$$

(c) Check your calculation with the integral key on your calculator.

The function f is the derivative of the function F, so

$$f(t) = 0.032\, e^{-0.032\,t},$$

and the integral is

$$0.032 \int_0^{200} t\, e^{-0.032\,t}\, dt.$$

We may evaluate this by integration by parts. Let $u = t$ and $dv = e^{-0.032\,t}\,dt$. Then

$$0.032\int_0^{200} t\,e^{-0.032\,t}\,dt = 0.032\left[t\left(\frac{-1}{0.032}\,e^{-0.032\,t}\right)\Big|_0^{200} + \int_0^{200}\frac{1}{0.032}\,e^{-0.032\,t}\,dt\right]$$

$$= \int_0^{200} e^{-0.032\,t}\,dt - 200\,e^{-6.4}$$

$$= \frac{-1}{0.032}\,e^{-0.032\,t}\Big|_0^{200} - 200\,e^{-6.4} \approx 30.87 \text{ days.}$$

Measured in hours, the expected lifetime of a typical bulb is approximately $30.87 \times 24 = 741$ hours. At first glance, the 750 on the package seems to be a little exaggerated. However, $[0, 200]$ may not be a large enough interval.

Nevertheless, that interval is very large. More generally, to find expected lifetimes for things that fail, with a "cumulative failure fraction function" of the form

$$F(t) = 1 - e^{-rt},$$

we need to know how to calculate

$$r\int_0^T t\,e^{-rt}\,dt,$$

possibly for very large values of T.

Checkpoint 3

For another type of light bulb, the fraction-burned-out function F is

$$F(t) = 1 - e^{-0.024\,t},$$

where t is measured in days. Find the expected lifetime of a typical bulb of this type.

In this section we examined the expected time to failure of a particular type of light bulb. The data we saw are representative of failure data for electrical components. Such data are often modeled by describing the probability $F(a)$ that a random component will fail between times $t = 0$ and $t = a$. Typically, F has the form

$$F(t) = 1 - e^{-rt}.$$

The probability that a component will fail between $t = a$ and $t = b$ is $F(b) - F(a)$.

We modeled the expected lifetime of such a component by the familiar process of looking at a related discrete problem and then applying a subdivide-and-conquer analysis for the continuous problem. This led to the following formula for the expected lifetime:

$$\int_0^T t\,f(t)\,dt,$$

where f is the derivative of F and T is a large number. In the next section we will examine this sort of integral in more detail.

Answers to Checkpoints

1. 0.66
2. (a) 0.792 (b) 0.894
3. The expected life is approximated by an integral of the form

$$0.024 \int_0^T t\,e^{-0.024t}\,dt,$$

where T is large enough that the value of the integral doesn't change as we increase T still more. For $T = 400$, we obtain an expected life of approximately 41.6 days or 1000 hours.

Exercises 10.1

1. We have a collection of four types of bulbs, each with an exact lifetime and a probability of occurrence as given in Table 10.3. What is the expected lifetime of a bulb chosen at random from this collection?

Table 10.3 Lifetimes for idealized light bulbs

Type k	*Probability* p_k	*Exact lifetime* t_k *(days)*
1	1/4	22
2	1/3	46
3	1/4	64
4	1/6	82

2. We have a collection of five types of bulbs, each with an exact lifetime and a probability of occurrence given in Table 10.4. What is the expected lifetime of a bulb chosen at random from this collection?

Table 10.4 Lifetimes for idealized light bulbs

Type k	*Probability* p_k	*Exact lifetime* t_k *(days)*
1	1/10	15
2	1/10	22
3	1/5	41
4	1/5	52
5	2/5	60

3. Each of the numbers 1, 3, 6, 10, 15, and 21 appears on one face of a fair six-sided die. What is the average value that will appear on the top face in a random roll?

4. Suppose the die in Exercise 3 is "loaded" so that 1 and 3 each appear with probability $1/8$, 6 appears with probability $1/4$, and the remaining three values each appear with probability $1/6$. If the die is rolled many times, what is the average value that will appear?

5. A certain gambling game has a payoff of \$10 with probability $1/100$, a payoff of 10 cents with probability $1/3$, and otherwise no payoff. What is the expected per-play winning if this game is played many times? Is it reasonable to pay 25 cents to play?

6. A random bulb from a batch of light bulbs has a probability $F(t) = 1 - e^{-0.063t}$ of burning out before t days.
   (a) What is the probability that a random bulb will burn out after day 3 and before day 10?
   (b) Find the expected lifetime of a random bulb.

7. A random bulb from a batch of light bulbs has a probability $F(t) = 1 - e^{-0.081t}$ of burning out before t days.
   (a) What is the probability that a random bulb will burn out after day 2 and before day 15?
   (b) Find the expected lifetime of a random bulb.

8. A random bulb from a batch of light bulbs has a probability $F(t) = 1 - e^{-0.013t}$ of burning out before t days.
   (a) What is the probability that a random bulb will burn out after day 20 and before day 35?
   (b) Find the expected lifetime of a random bulb.

Evaluate each of the following indefinite integrals.

9. $\int t\,e^{-3t}\,dt$

10. $\int t^2 e^{-rt}\,dt$

11. $\int \frac{1}{1+4t^2}\,dt$

Evaluate each of the following definite integrals.

12. $\int_0^5 t\,e^{-3t}\,dt$

13. $\int_0^{10} t\,e^{-3t}\,dt$

14. $\int_0^{20} t\,e^{-3t}\,dt$

15. $\int_0^5 t^2 e^{-3t}\,dt$

16. $\int_0^{10} t^2 e^{-rt}\,dt$

17. $\int_0^{20} t^2 e^{-rt}\,dt$

18. $\int_0^{100} \frac{1}{1+4t^2}\,dt$

19. $\int_0^{200} \frac{1}{1+4t^2}\,dt$

20. $\int_0^{300} \frac{1}{1+4t^2}\,dt$

21. When $F(b) - F(a)$ turned up in our discussion of probability, you might have guessed there must be an integral lurking nearby—but the integral

$$\int_0^T t\,e^{-rt}\,dt$$

is not $F(b) - F(a)$, and it is also not a probability. Use the function f to express as an integral the probability that a particular bulb will fail between time $t = a$ and time $t = b$.

22. (a) Now that you have completed the development of the concept of expected lifetime, starting with a discrete approximation to the continuous problem, where have you seen this problem before? Where did integrals like

$$\int_0^T t\,f(t)\,dt$$

turn up in our previous work?

(b) Use the analogy with a physical problem to interpret expected lifetime as a "continuous weighted average of times." What function provides the weights, and how are the weights related to probabilities?

10.2 Improper Integrals

Integrating to Infinity

In our study of expected lifetime, we considered integrals of the form

$$\int_0^T g(t)\,dt,$$

where $g(t)$ was the specific function te^{-rt} and T was "large." What we really wanted to describe is the integral "from zero to infinity," i.e., the limiting case as T increases to ∞. For convenience, we use the abbreviated notation

$$\int_0^\infty g(t)\,dt$$

to describe this two-step process: (1) integrate from 0 to a large (but unspecified) upper limit T, and (2) find the limiting value as T grows infinitely large. Such an integral is called *improper* to distinguish it from a "proper" definite integral.

Definition If $g(t)$ is a continuous function for $t \geq a$, then the **improper integral**

$$\int_a^{\infty} g(t)\,dt$$

is the limiting value as $T \to \infty$ of integrals of the form

$$\int_a^{T} g(t)\,dt.$$

If the limiting value exists as a finite number, we say the improper integral **converges**. Otherwise, it **diverges**.

Example 1 (a) For light bulbs with cumulative failure fraction function

$$F(t) = 1 - e^{-rt},$$

show that the expected time to failure is exactly $1/r$.

(b) Find the expected lifetime of a light bulb for which $r = 0.032$ when time is measured in days.

Solution (a) As we saw in Section 10.1, the expected lifetime is the integral from 0 to ∞ of $tf(t)$, where $f(t) = F'(t) = r\,e^{-rt}$. As in Exploration Activity 3 in Section 10.1, we may integrate by parts — or refer to the Table of Integrals — to find that

$$\int t\,e^{-rt}\,dt = -\frac{1}{r}\,t\,e^{-rt} - \frac{1}{r^2}\,e^{-rt} = -\frac{1}{r^2}\,(rt+1)\,e^{-rt},$$

so

$$r\int_0^T t\,e^{-rt}\,dt = -\frac{1}{r}\,(rt+1)\,e^{-rt}\Big|_0^T = -\frac{1}{r}\,(rT+1)\,e^{-rT} + \frac{1}{r}.$$

Note that T appears only in the first of the two terms on the right — and the second term is our announced answer. To complete the argument that

$$r\int_0^{\infty} t\,e^{-rt}\,dt = \frac{1}{r},$$

we need to show that, as $T \to \infty$,

$$(rT+1)\,e^{-rT} = \frac{rT+1}{e^{rT}} \to 0.$$

In the next chapter we will establish firmly that exponential functions grow much more rapidly than linear functions — indeed, than any polynomial functions. For now, observe that $rT+1$ grows at a constant rate (slope $= r$) and e^{rT} grows at an exponentially increasing rate. Since the denominator grows much faster than the numerator, the quotient approaches 0.

(b) If $r = 0.032$, then the expected lifetime is $1/r = 31.25$ days or 750 hours. Note that this agrees with the manufacturer's claim on the package, and not with our preliminary calculation of 741 hours. ■

Checkpoint 1

Show that $\int_2^\infty \frac{1}{x^2}\,dx$ converges, and find its value.

Notation for Limiting Values

Improper integrals bring us — again — to the study of limiting behavior of a function as its independent variable becomes large. In particular, the first step in evaluating

$$\int_a^\infty g(t)\,dt$$

is to find the antiderivative

$$G(T) = \int_a^T g(t)\,dt.$$

The improper integral is then the limiting value of $G(T)$ as $T \to \infty$. The standard notation for this limiting value is

$$\lim_{T\to\infty} G(T).$$

With this notation we may condense the two-step definition of the improper integral into a single formula:

$$\int_a^\infty g(t)\,dt = \lim_{T\to\infty} \int_a^T g(t)\,dt.$$

Notation The limiting value of a function $f(x)$ as the independent variable x becomes large is denoted

$$\lim_{x\to\infty} f(x).$$

Example 2 Express in limiting notation the eventual behavior, as the independent variable becomes large, of the function

$$f(x) = \frac{2 - x^3}{x^3}.$$

Solution We have a numerator approaching $-\infty$ and a denominator approaching ∞, so we need to transform the fraction algebraically:

$$\begin{aligned}\lim_{x\to\infty} f(x) &= \lim_{x\to\infty} \frac{2 - x^3}{x^3}\\ &= \lim_{x\to\infty}\left(\frac{2}{x^3} - 1\right)\\ &= 0 - 1 = -1.\end{aligned}$$

Check this by using your graphing calculator to graph f for large values of x. ■

Example 3 Express in limiting notation the eventual behavior, as the independent variable becomes large, of the function

$$g(t) = \frac{2 - t^2}{t^3}.$$

Solution This example is similar, but the notation is slightly different, and the outcome is different:

$$\begin{aligned} \lim_{t \to \infty} g(t) &= \lim_{t \to \infty} \frac{2 - t^2}{t^3} \\ &= \lim_{t \to \infty} \left(\frac{2}{t^3} - \frac{1}{t} \right) \\ &= 0 - 0 = 0. \end{aligned}$$

Again, check with your calculator. ■

Exploration Activity 1

Make a list of previous occurrences in this text of limiting behavior as an independent variable becomes large. Express each item on your list in the new notation.

Here are some of the things your list might include. This list is not complete—you may have thought of many others.

- Decaying exponential functions (Section 3.2): If $k > 0$, then

$$\lim_{t \to \infty} e^{-kt} = 0.$$

- Temperature of a cooling body approaching ambient temperature (also Section 3.2): If $k > 0$, then

$$\lim_{t \to \infty} (21 + 9\,e^{-kt}) = 21.$$

- Raindrops approaching terminal velocity (Chapter 3 Lab Reading):

$$\lim_{t \to \infty} v(t) = v_{\text{term}}.$$

- Newton's Method for finding roots (Section 4.3):

$$\lim_{n \to \infty} t_n = r.$$

  Note that the independent variable does not have to vary continuously. In this case, the independent variable is n, which takes only integer values.

- The S-I-R model for spread of epidemics (Sections 5.2 and 5.3):

$$\lim_{t \to \infty} s(t) = s_\infty, \quad \lim_{t \to \infty} i(t) = 0, \quad \lim_{t \to \infty} r(t) = r_\infty.$$

  In this case, r_∞ is the fraction of the people who got sick, and s_∞ is the fraction remaining susceptible after the epidemic is over.

Checkpoint 2

Find each of the following limiting values. You may use your calculator—but be careful. A calculator cannot literally tell you about a function all the way to infinity.

(a) $\lim_{x \to \infty} x^{-1/2}$ (b) $\lim_{x \to \infty} \frac{2x^2 - 1}{x^2 + 7}$ (c) $\lim_{x \to \infty} \frac{\cos x}{x^2}$

A Geometric Look at Failure Times

Two functions played important roles in Section 10.1: The cumulative failure fraction function

$$F(t) = 1 - e^{-rt}$$

and its derivative

$$f(t) = r\,e^{-rt}.$$

For reasons we will explore in the next section, F is called an **exponential distribution function** and f is called an **exponential density function**. As we will see, it is no accident that the word "density" is turning up again. Recall that density played a role in our moment and center or mass calculations in Chapter 9—calculations that were about continuous distribution of mass. In this chapter the corresponding concept will be continuous distribution of probability.

Observe that $F(0) = 0$. (That is, at the instant we turn all the bulbs on, none can actually have burned out yet, even though there is a small probability that a given bulb will burn out almost immediately.) Thus $F(t)$ is the solution of the initial value problem

$$\frac{dy}{dt} = f(t), \qquad y(0) = 0.$$

According to the Fundamental Theorem, this means that

$$F(T) = \int_0^T f(t)\,dt.$$

For the light bulb failure functions, we could check this last formula by direct calculation. However, this is a correct conclusion about the solution F of *any* initial value problem of the form

$$\frac{dy}{dt} = f(t), \qquad y(0) = 0,$$

where f is a continuous function. We use this more general assertion in the next section.

We observed in Section 10.1 that the probability of a failure between time $t = a$ and time $t = b$ is $F(b) - F(a)$. But that is also the value of the integral of f from a to b. Thus *the probability of a failure between time $t = a$ and time $t = b$ is*

$$\int_a^b f(t)\,dt,$$

where f is the exponential density function. This in turn gives us an interpretation of probability as area (see Figure 10.2): The probability of failure in a given time interval is the area under the graph of the density function over that interval.

Figure 10.2 Probability of failure between $t = a$ and $t = b$

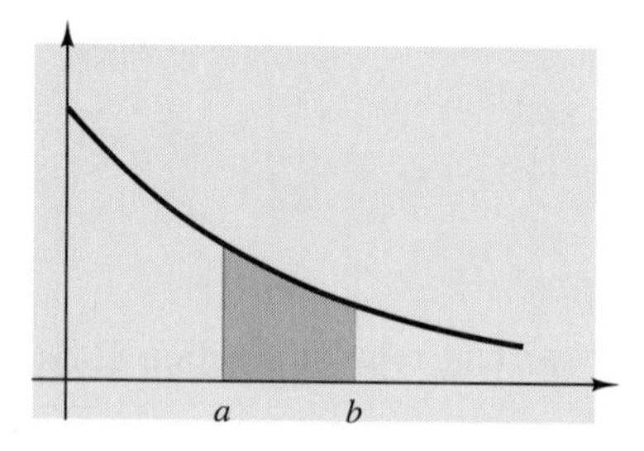

Checkpoint 3

If the area interpretation of probability is correct, then the area under the entire graph from $t = 0$ to ∞ — the probability of failure over the entire time interval after the bulb is turned on — should be 1. Show that this is the case.

We have just shown that it is possible for an infinitely long region of the plane to have a finite area. We saw in Chapter 5 that it is possible for an infinitely long geometric sum to add up to a finite number. Since integration is the continuous analogue of discrete addition, it should come as no surprise that these ideas are related. You may have an opportunity to explore that relationship in a laboratory exercise associated with the next chapter.

Example 4 Interpret expected time to failure as (a) a moment and (b) the center-of-mass coordinate $\overline{t}$ for the region shown in Figure 10.3.

Figure 10.3 Region under the graph of the exponential density function

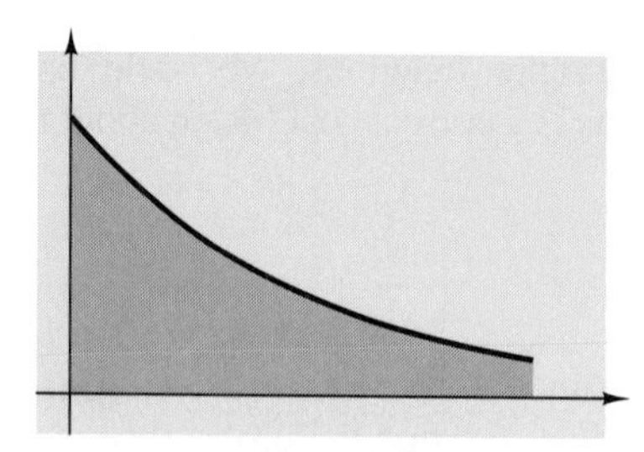

Solution The horizontal moment for the region shown in Figure 10.3 is

$$M_y = \int_0^\infty t\, f(t)\, dt.$$

But this is exactly the integral expression we derived for expected failure time in the preceding section. To find $\overline{t}$ from this expression, we need to divide by the area of the region. You just showed in Checkpoint 3 that the area of the region is 1, so in this case $\overline{t}$ and M_y are the same number:

$$\overline{t} = \int_0^\infty t\, f(t)\, dt.$$

One way to read this last formula is that the expected (or average) failure time is the *weighted average* of all possible failure times, where the weight for each t is given by the density function $f(t)$. If the sum of all the weights is 1, one may skip the step of dividing by the sum of the weights.

Checkpoint 4

For the region under the graph of $y = 2^{-x}$ from 0 to ∞, find

(a) the area;

(b) M_y;

(c) $\overline{x}$.

More Improper Integrals

Just as we can integrate to ∞, we can integrate to $-\infty$ and, for some functions, even from $-\infty$ to ∞.

Definitions If $g(t)$ is a continuous function for $t \leq a$, then the **improper integral**

$$\int_{-\infty}^a g(t)\, dt = \lim_{T \to -\infty} \int_T^a g(t)\, dt.$$

If the limiting value exists as a finite number, we say the improper integral **converges**. Otherwise, it **diverges**.
If $g(t)$ is a continuous function for all t, then the **improper integral**

$$\int_{-\infty}^\infty g(t)\, dt = \int_{-\infty}^0 g(t)\, dt + \int_0^\infty g(t)\, dt.$$

If both integrals on the right converge, we say the improper integral on the left **converges**. Otherwise, it **diverges**.

Example 5 Show that the area under the graph of $y = 1/(1 + t^2)$ is exactly π (see Figure 10.4).

Figure 10.4 Area under the graph of $y = \dfrac{1}{1+t^2}$

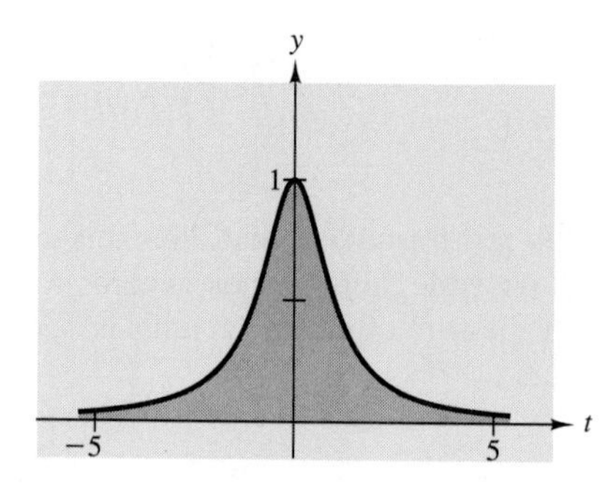

Solution The shaded area is

$$\begin{aligned}
\int_{-\infty}^{\infty} \frac{1}{1+t^2}\,dt &= \int_{-\infty}^{0} \frac{1}{1+t^2}\,dt + \int_{0}^{\infty} \frac{1}{1+t^2}\,dt \\
&= \lim_{T\to-\infty} \int_{T}^{0} \frac{1}{1+t^2}\,dt + \lim_{U\to\infty} \int_{0}^{U} \frac{1}{1+t^2}\,dt \\
&= \lim_{T\to-\infty} \arctan t\,\Big|_{T}^{0} + \lim_{U\to\infty} \arctan t\,\Big|_{0}^{U} \\
&= \lim_{T\to-\infty} (0 - \arctan T) + \lim_{U\to\infty} (\arctan U - 0) \\
&= \frac{\pi}{2} + \frac{\pi}{2} = \pi.
\end{aligned}$$

The choice of $t = 0$ as a dividing point between the left-hand and right-hand improper integrals is merely a matter of convenience, not of necessity—see Exercise 25. However, actually splitting the two-sided improper integral into two one-sided improper integrals is, in general, necessary for getting right answers—see Exercise 26 for an example in which failing to split gives a wrong answer. For the function in Example 5, we could have used symmetry to evaluate the integral as twice the integral from 0 to ∞ —see Exercise 27.

Checkpoint 5

Evaluate $\displaystyle\int_{-\infty}^{\infty} \frac{1}{7+x^2}\,dx$.

In this section we have expanded on and formalized our study of expected failure times in the preceding section. In particular, we have seen that the probability of failure of a component between times $t = a$ and $t = b$ can be expressed as

$$\int_{a}^{b} f(t)\,dt = F(b) - F(a),$$

where $f(t) = r\,e^{-rt}$ is the exponential density function and $F(t) = 1 - e^{-rt}$ is the exponential distribution function. Since every component fails eventually, we are led to the concept of improper integral and, in particular, the observation that

$$\int_0^\infty f(t)\,dt = 1.$$

We saw that the expected or average time for failure of a randomly selected component of a given type could be expressed as

$$\int_0^\infty t\,f(t)\,dt = \frac{1}{r},$$

and we interpreted this result geometrically as both a moment and a center-of-mass coordinate of an infinitely long region in the plane.

In preparation for the next section, we introduced left-hand and two-sided improper integrals to complement the right-hand integrals involved in the study of failure times. Because of the frequent appearance here of limiting values as an independent variable becomes large, we introduced the standard notation for such limiting values. At the end of this chapter we will take up variations of this notation for other limiting values, some of which will play an important role in the next chapter.

Answers to Checkpoints

1. $$\begin{aligned}\int_2^\infty \frac{1}{x^2}\,dx &= \lim_{T\to\infty} \int_2^T \frac{1}{x^2}\,dx \\ &= \lim_{T\to\infty} (-x^{-1})\Big|_2^T \\ &= \lim_{T\to\infty} \left(-\frac{1}{T} + \frac{1}{2}\right) = \frac{1}{2}\end{aligned}$$

2. (a) 0 (b) 2 (c) 0

3. $$\begin{aligned}\int_0^\infty r\,e^{-rt}\,dt &= \lim_{T\to\infty} (-e^{-rt})\Big|_0^T \\ &= \lim_{T\to\infty} \left(-e^{-rT} + 1\right) = 1\end{aligned}$$

4. (a) $A = 1/(\ln 2) \approx 1.4427$
   (b) $M_y = 1/(\ln 2)^2 \approx 2.0814$
   (c) $\overline{x} = 1/(\ln 2) \approx 1.4427$

5. $\pi/\sqrt{7} \approx 1.1874.$

Exercises 10.2

1. Why does it take two steps to evaluate an improper integral? In particular, what is it about the process of definite integration that requires an interval of finite length?

Find the cumulative failure fraction function $F(t)$, where time t is measured in days, for light bulbs with each of the following expected lifetimes.

2. 1000 hours
3. 750 hours
4. 350 hours

Decide whether each of the following integrals converges or diverges. If it converges, find its value.

5. $\displaystyle\int_1^\infty \frac{1}{x^4}\,dx$

6. $\displaystyle\int_2^\infty x^{-1/7}\,dx$

7. $\displaystyle\int_2^\infty x^{-7}\,dx$

8. (a) How big does x have to be in order that $\ln x > 100$?
   (b) How big does x have to be in order that $\ln x > 1{,}000{,}000$? ("My calculator won't do that" is not a useful answer. Outthink your calculator!)
   (c) How do you know that $\ln x \to \infty$ as $x \to \infty$?
   (d) How do you know that

$$\int_2^\infty \frac{1}{x}\,dx$$

   diverges?

9. (a) How big does x have to be in order that $\ln(\ln x) > 5$?
   (b) How big does x have to be in order that $\ln(\ln x) > 100$?
   (c) How do you know that $\ln(\ln x) \to \infty$ as $x \to \infty$?
   (d) How do you know that

$$\int_2^\infty \frac{1}{x \ln x}\,dx$$

   diverges?

10. (a) Find all numbers p for which

$$\int_1^\infty \frac{1}{x^p}\,dx$$

   converges.
   (b) For each such p, find the value of the integral in terms of p.

Evaluate each of the following integrals. Exact answers are preferable to approximations.

11. $\displaystyle\int_3^\infty x\,e^{-2x}\,dx$

12. $\displaystyle\int_1^\infty \frac{1}{x(x+1)}\,dx$

13. $\displaystyle\int_1^\infty \frac{1}{x^2+1}\,dx$

Guess each of the following limiting values. You may use a calculator.

14. $\displaystyle\lim_{x\to\infty} x^{1/x}$

15. $\displaystyle\lim_{x\to\infty} x \sin\frac{1}{x}$

16. $\displaystyle\lim_{x\to\infty} \left(1+\frac{1}{x}\right)^x$

17. $\displaystyle\lim_{x\to\infty} \frac{\ln x}{x}$

18. $\displaystyle\lim_{x\to\infty} \frac{x^2}{2^x}$

19. $\displaystyle\lim_{x\to\infty} \frac{x^{10}}{10^x}$

20. $\displaystyle\lim_{x\to\infty} \frac{\sqrt{1+x^2}-1}{x}$

21. $\displaystyle\lim_{x\to\infty} \ln\left(\frac{\sqrt{1+x^2}-1}{x}\right)$

22. $\displaystyle\lim_{x\to\infty} \left(1-\frac{1}{x}\right)^x$

23. (a) What is the probability that a light bulb with failure density function $f(t) = r\,e^{-rt}$ burns out before its expected lifetime?
    (b) What is the probability that it lasts longer than its expected lifetime?

24. The **median** lifetime of a light bulb with failure density function $f(t) = r\,e^{-rt}$ is the time T such that the probability of failure before $t = T$ is $\frac{1}{2}$. Find T. Is the median lifetime shorter, longer, or the same as the expected lifetime?

25. For a function $g(x)$ that is continuous for all x, we defined

$$\int_{-\infty}^\infty g(x)\,dx = \int_{-\infty}^0 g(x)\,dx + \int_0^\infty g(x)\,dx,$$

with the understanding that the two-sided improper integral has a value only if *both* integrals on the right converge. Explain why splitting at $x = 0$ is merely a matter of convenience—that is,

$$\int_{-\infty}^\infty g(x)\,dx = \int_{-\infty}^a g(x)\,dx + \int_a^\infty g(x)\,dx$$

for any number a, with exactly the same meaning. You have two things to show:
   (a) If the two-sided integral converges by either definition, it converges by the other as well.
   (b) If the two-sided integral converges, then both definitions give the same value.

26. In this exercise we explore why it is necessary to split a two-sided improper integral into two parts, with separate evaluations of limiting values, in order to get right answers.
   (a) Graph the function $f(x) = \dfrac{x}{1+x^2}$.
   (b) Show that

$$\int_0^\infty \frac{x}{1+x^2}\,dx$$

   does not converge. (*Hint*: See Exercise 8.)

(c) For any positive number T, what is the value of

$$\int_{-T}^{T} \frac{x}{1+x^2}\,dx\,?$$

Use part (a) to answer this graphically. Use your calculator (for selected values of T) to answer numerically. Use the antiderivative you calculated in part (b) to answer symbolically.

(d) Find

$$\lim_{T\to\infty} \int_{-T}^{T} \frac{x}{1+x^2}\,dx\,.$$

Explain why this limiting value is *not* the value of

$$\int_{-\infty}^{\infty} \frac{x}{1+x^2}\,dx\,.$$

27. (a) Suppose $f(x)$ is an even function that is continuous for all x and for which

$$\int_{0}^{\infty} f(x)\,dx$$

converges. Explain why

$$\int_{-\infty}^{\infty} f(x)\,dx = 2\int_{0}^{\infty} f(x)\,dx.$$

(b) Use part (a) to evaluate

$$\int_{-\infty}^{\infty} \frac{1}{1+x^2}\,dx.$$

Compare to Example 5.

28. (a) Suppose $f(x)$ is an odd function that is continuous for all x and for which

$$\int_{0}^{\infty} f(x)\,dx$$

converges. Explain why

$$\int_{-\infty}^{\infty} f(x)\,dx = 0.$$

(b) Use part (a) to evaluate

$$\int_{-\infty}^{\infty} x\,e^{-x^2}\,dx.$$

Be careful: You have to show that all the conditions on the function are satisfied.

When an improper integral cannot be evaluated directly, it may be possible to determine whether it converges or diverges by comparing it to another integral whose convergence or divergence is known. The next three exercises are about this idea.

29. (a) Explain why

$$\frac{1}{1+x^{3/2}} < \frac{1}{x^{3/2}}$$

for $x \geq 1$.

(b) Explain why

$$\int_{1}^{\infty} \frac{1}{1+x^{3/2}}\,dx$$

converges. (See Exercise 10.)

(c) Estimate the integral in part (b).

30. (a) Explain why

$$\frac{x}{1+x^{3/2}} > \frac{x}{1+x^2}$$

for $x \geq 1$.

(b) Explain why

$$\int_{1}^{\infty} \frac{x}{1+x^{3/2}}\,dx$$

diverges. (See Exercise 26.)

31. (a) Invent a test for convergence or divergence of integrals of the form $\int_a^\infty f(x)\,dx$ and $\int_a^\infty g(x)\,dx$, where f and g are assumed to be continuous and nonnegative for $x \geq a$.

(b) Give a geometric argument to justify your comparison test for convergence or divergence.[1]

Decide whether each of the following integrals converges or diverges (see Exercise 31).

32. $\displaystyle\int_{1}^{\infty} \frac{1}{1+x^4}\,dx$

33. $\displaystyle\int_{2}^{\infty} \frac{x}{\sqrt{1+x^3}}\,dx$

34. $\displaystyle\int_{0}^{\infty} e^{-x^3}\,dx$

35. $\displaystyle\int_{1}^{\infty} \frac{1}{x\sqrt{1+x^2}}\,dx$

36. $\displaystyle\int_{2}^{\infty} \frac{1}{(x+1)^2}\,dx$

37. $\displaystyle\int_{0}^{\infty} \frac{1}{1+e^x}\,dx$

For each of the following integrals that converges (see Exercises 32–37), use your calculator to estimate the value.

38. $\displaystyle\int_1^\infty \frac{1}{1+x^4}\,dx$

39. $\displaystyle\int_2^\infty \frac{x}{\sqrt{1+x^3}}\,dx$

40. $\displaystyle\int_0^\infty e^{-x^3}\,dx$

41. $\displaystyle\int_1^\infty \frac{1}{x\sqrt{1+x^2}}\,dx$

42. $\displaystyle\int_2^\infty \frac{1}{(x+1)^2}\,dx$

43. $\displaystyle\int_0^\infty \frac{1}{1+e^x}\,dx$

44. (a) Evaluate $\displaystyle\lim_{x\to\infty} \frac{\sin x}{x^2}$.
(b) The function $f(x) = (\sin x)/x^2$ has both positive and negative values. Expand on the idea in Exercise 31 to show that $\int_1^\infty f(x)\,dx$ converges. (Your calculator probably cannot help you evaluate this integral.)

45. (a) Explain carefully why

$$\int_1^\infty \frac{1}{x\sqrt{1+x^2}}\,dx$$

converges.
(b) Find the exact value of the integral in part (a).

1. Adapted from *Calculus Problems for a New Century*, edited by Robert Fraga, MAA Notes No. 28, 1993.

10.3 Continuous Probability: Distribution and Density Functions

In Section 10.1 we studied data on the numbers of failures in a large collection of similar light bulbs. The key functions in our analysis of that data were

$$F(t) = 1 - e^{-rt}$$

and its derivative,

$$f(t) = r\,e^{-rt}.$$

The value of F at time t is approximately the fraction of bulbs that burned out by that time, so the fraction that burned out between $t = a$ and $t = b$ is $F(b) - F(a)$. Since $f(t)$ is the derivative of $F(t)$, that fraction is also $\int_a^b f(t)\,dt$. We adopted the fraction of failures in a large sample as our meaning for the probability that a randomly selected bulb would fail in a given time interval. Thus the definite integral gives us a way to describe probability for events that occur on a continuous time scale.

The question we set out to answer in Section 10.1 was one of expected or average lifetime. We saw that average lifetime could be calculated much like a moment, as an integral giving weighted averages of times t, where the weights came from the function $f(t)$. We had to calculate these integrals over large intervals $[0, T]$, and this led to our discussion of improper integrals in Section 10.2. Now we resume our discussion of probability, and we describe the functions f and F in terms that we can use for other types of problems.

In general terms, the **distribution function** $F(t)$ describes the probability that a randomly selected data item will have a value less than t. (For light bulbs, "value" means "lifetime.") The **probability density function** $f(t)$ is the derivative of the distribution function. Thus the probability that a random data value will fall between $t = a$ and $t = b$ is $F(b) - F(a)$, which is the same thing as $\int_a^b f(t)\,dt$. If $F(t)$ is known, we find $f(t)$ by differentiation. On the other hand, if $f(t)$ is known, we find $F(t)$ as a particular antiderivative of $f(t)$, the one that has value 0 at the left end of the domain. Thus, a probability distribution can be specified by either its distribution function or its density function.

The **expected value** (also called **average value**) for data items from a large population with density f is the moment integral $\int_a^b t\, f(t)\, dt$, where the interval is selected to include *all* the possible outcomes. For example, to find the expected lifetime of light bulbs, the appropriate interval is $[0, \infty)$.

The distribution function $F(t) = 1 - e^{-rt}$ is an **exponential distribution**, and its derivative $f(t) = re^{-rt}$ is an **exponential density**. This model is the starting point for the study of reliability theory, which is useful for describing, among other things, failure times for electrical and electronic components such as chips in computers, batteries in toy rabbits, and bug zappers in backyards. In Example 1 in Section 10.2 we showed that the expected lifetime is exactly $1/r$. That's a measure of reliability.

Other kinds of data are distributed in other ways. In this section we study two more types of distributions and their density functions. The first of these is the Cauchy distribution (pronounced ko-SHEE), which may be defined by the **Cauchy probability density function**:

$$f(t) = \frac{1}{\pi}\frac{1}{1+t^2} \qquad \text{for } -\infty < t < \infty.$$

Exploration Activity 1

(a) Graph the Cauchy density function $f(t)$.

(b) Describe what the graph of the Cauchy density function says about the distribution of data values in a set modeled by this density function.

(c) Show that the integral of f over its entire domain is 1.

(d) Explain why, for any probability density function f, the integral of f over its entire domain must be 1.

(e) Find a formula for the Cauchy distribution function F. [Recall that $F(t)$ represents the probability that a random data value is less than t.]

(f) Evaluate $\lim_{t\to-\infty} F(t)$ and $\lim_{t\to\infty} F(t)$.

(g) Explain why any distribution function F must be an increasing function that approaches 0 at the left end of its domain and 1 at the right end.

We show the graph of the Cauchy density function in Figure 10.5. This graph suggests that data values selected at random would be tightly clustered around 0. For example, for two different intervals of length 1 we have

$$\int_{-1/2}^{1/2} f(t)\, dt = 0.295$$

and

$$\int_5^6 f(t)\, dt = 0.010.$$

Thus it is almost 30 times more likely to find a data value in the interval $[-1/2, 1/2]$ as to find one in $[5, 6]$.

Figure 10.5 Graph of the Cauchy density function

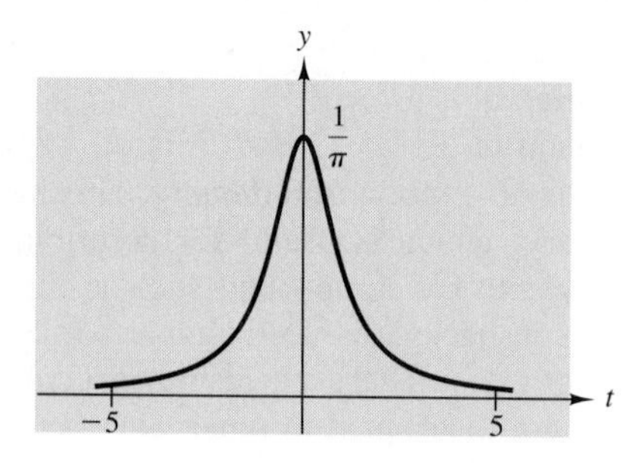

We showed in Example 5 in the preceding section that

$$\int_{-\infty}^{\infty} \frac{1}{1+t^2}\, dt = \pi,$$

so

$$\int_{-\infty}^{\infty} f(t)\, dt = \frac{1}{\pi} \int_{-\infty}^{\infty} \frac{1}{1+t^2}\, dt = 1.$$

For any density function f, the integral over the entire domain represents the probability that a random data value lies somewhere in the entire set of data values. This probability must be 1.

We find the distribution function F by integrating f from the left end of the domain up to an arbitrary t:

$$\begin{aligned} F(t) &= \frac{1}{\pi} \int_{-\infty}^{t} \frac{1}{1+s^2}\, ds \\ &= \lim_{T \to -\infty} \frac{1}{\pi} \int_{T}^{t} \frac{1}{1+s^2}\, ds \\ &= \frac{1}{\pi} \lim_{T \to -\infty} (\arctan t - \arctan T) \\ &= \frac{1}{\pi} \left[\arctan t - \left(-\tfrac{\pi}{2}\right)\right] \\ &= \frac{1}{2} + \frac{1}{\pi} \arctan t. \end{aligned}$$

We see from this formula that

$$\lim_{t \to -\infty} F(t) = \frac{1}{2} + \frac{1}{\pi}\left(-\frac{\pi}{2}\right) = 0$$

and

$$\lim_{t \to \infty} F(t) = \frac{1}{2} + \frac{1}{\pi}\left(\frac{\pi}{2}\right) = 1.$$

Also, F is an increasing function because its derivative f is always positive. Every distribution function must have these properties: The fraction of data values found to the left of the domain must be 0, and the fraction found to the left of the right endpoint of the domain must be 1. Furthermore, F describes a cumulative distribution, so it can never decrease. [It could stay level over some interval—if $f(t) = 0$ there—but that doesn't happen with the Cauchy distribution.] Now that you have a formula for the Cauchy distribution function, you can graph it. You should see an S shape similar to the graph of the arctangent function, but starting at 0 on the left and rising to 1 on the right.

We would like to discuss the expected value for the Cauchy distribution, but there is a small technical problem. From Figure 10.5 it certainly looks as though the average value must be 0. In fact, that is true: Whenever the graph of a density function is symmetric about a vertical line, the average value must be the horizontal coordinate for points on that line. In particular, an even density function must have expected value 0. The technical problem is that, in this case, the moment integral that defines expected value,

$$\int_{-\infty}^{\infty} t\, f(t)\, dt,$$

does not converge! (See Exercise 26 in the preceding section.)

Checkpoint 1

For the Cauchy distribution, find

(a) the probability that a random data value lies between $1/2$ and $3/2$.

(b) the probability that a random data value lies between $-3/2$ and $-1/2$. [Don't do any more calculation—use part (a).]

We turn now to an example of a probability distribution that is much simpler than either the exponential or the Cauchy distribution. In fact, this example is so simple it is child's play.

Suppose you have a spinner from a board game that randomly picks a value between 0 and 4, and you further subdivide the spinner into tenths of units. Thus you can decide whether the result of a spin is, say, greater than 2.3 and less than 3.6. We assume that this is a "fair" spinner—for example, the result of a spin is as likely to give a value greater than 2 as a value less than 2. More generally, given two intervals of equal length, the probability of landing in one is the same as that for landing in the other. We suppose that we have a data set that consists of the numerical results from a large number of spins.

Exploration Activity 2

(a) Explain why, for the spinner data just described, a reasonable density function is

$$f(t) = \frac{1}{4} \qquad \text{for } 0 \le t \le 4.$$

(b) What is the distribution function?

(c) Find the expected value for a random result from the spinner. Does your result agree with the value you would have assigned intuitively?

Since all the values are equally likely, the probability density weighting should be the same for all values in the interval $[0,4]$. Thus the density function should be constant. Since the integral of this constant c over the interval $[0,4]$ must be 1, we must set the constant c to be $1/4$. Strictly speaking, our description of the spinner is not continuous—there are lots of numbers between 2.3 and 2.4—so we are really modeling the distribution of continuous position around the circumference of the spinner.

The distribution function F for this model is

$$\begin{aligned} F(t) &= \text{probability of a value less than } t \\ &= \int_0^t \frac{1}{4}\, ds = \frac{t}{4} \qquad \text{for } 0 \le t \le 4. \end{aligned}$$

The expected value for this model is

$$\int_0^4 \frac{t}{4}\, dt = \frac{t^2}{8}\bigg|_0^4 = 2.$$

This is reasonable, since if all the values are weighted equally, we anticipate that the average value will be in the middle of the interval of possible values.

The type of distribution represented by the spinner data is called a **uniform distribution**, and the corresponding density function is called a **uniform density function**. A uniform density function must be constant on some interval of finite length, with the product of the length and the constant height equal to 1. The distribution function is linear with slope determined by the constant height of the density.

Checkpoint 2

A data set is generated by a computer random number generator with values uniformly distributed over the interval $[0,10]$.

(a) Find the density and distribution functions to model this data with a uniform distribution.

(b) Find the expected value for a random data value from this set.

(c) Find the probability that a random data value lies between 1.3 and 7.7.

In this section we have considered models for data distributed continuously over an interval, finite or infinite. Such a model consists of

- a distribution function F with the property that $F(t)$ is the probability that a random data value is less than t, and
- a probability density function f that is the derivative of F.

For such a model the probability that a random data value lies between a and b is

$$F(b) - F(a) = \int_a^b f(t)\, dt.$$

In particular the integral of f over the entire domain must be 1. The expected value for a random data value from a set with this distribution is the integral of $t\,f(t)$ over the entire domain.

Answers to Checkpoints

1. (a) 0.165 (b) 0.165
2. (a) $f(t) = 1/10,\ F(t) = t/10$ (b) 5 (c) 0.64

Exercises 10.3

1. If $F(t) = 1 - e^{-rt}$ is the failure distribution function for light bulbs with an expected lifetime of 800 hours and time is measured in days, what is r?

2. In the text we gave no application or physical interpretation of the Cauchy distribution. We ask you to explore that now with the game spinner we modeled by the uniform distribution (see Figure 10.6). However, instead of reading numbers from the spinner, we read *slopes*. Specifically, wherever the arrow ends up, look only at the right-hand half of it—whether that is the arrow end or not. Think of it as a line in the right half of the plane. For any run Δx there is a corresponding rise Δy (which might be negative). Calculate the slope $\Delta y/\Delta x$. That's the "value" of a spin.

Figure 10.6 A game spinner

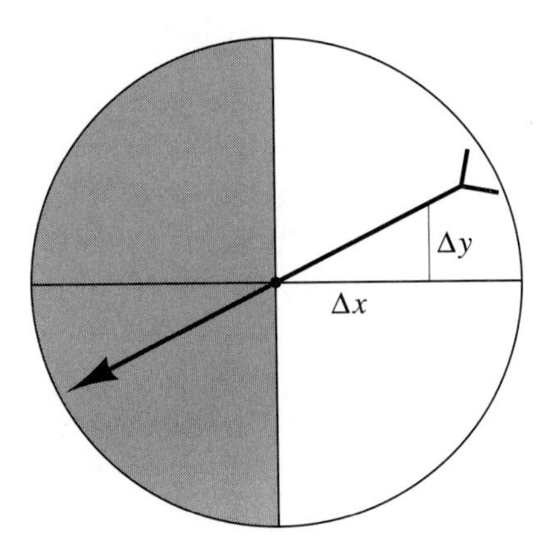

(a) Explain why data values can range from $-\infty$ to ∞. Where would you expect to find most of the data values clustered? What symmetry do you observe?

(b) Show that the slope data values are distributed according to the Cauchy distribution. That is, show that the probability of a slope between a and b is

$$\frac{1}{\pi}(\arctan b - \arctan a).$$

3. Let f be a probability density function given by

$$f(x) = cx \qquad \text{for } 1 \le x \le 5.$$

(a) Find c.
(b) Find a formula for the distribution function.
(c) What is the probability of the event occurring between 2 and 3.5?
(d) Find the expected value.

4. If $f(t) = c\,(6t - t^2)$ is a probability density function over the interval $0 \le t \le 6$, find c.

5. Figure 10.7 shows two graphs, one of which is a probability density function and the other of which is a probability distribution function. Which is which? Identify as many features as you can of each graph that help you make the proper identification.

Figure 10.7 Probability density and distribution functions

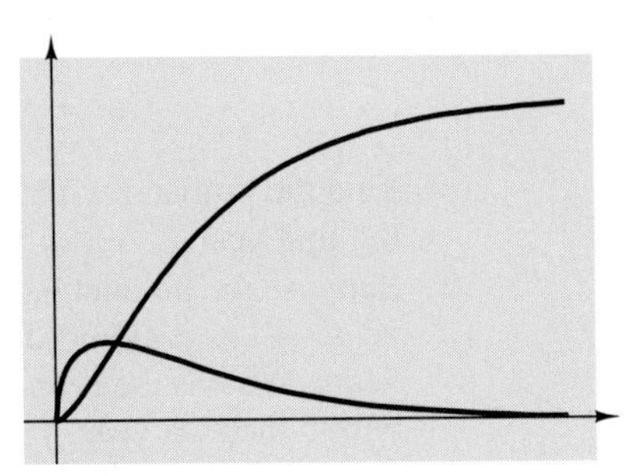

6. Suppose there are 3 red marbles, 5 green marbles, and 20 blue marbles mixed together in a bag. If you reach into the bag and draw out a marble at random, what is the probability that it's not green?

7. Suppose a bag contains 7 black balls, 6 yellow balls and 4 green balls.
   (a) If you reach in and remove one ball at random, what is the probability that it is black?
   (b) If you remove two balls, what is the probability that they are both black?
   (c) If you remove two balls, what is the probability that one is green and the other is black?
   (d) If you remove two balls, what is the probability that the second is black, given that the first is black?
   (e) If you remove two balls, what is the probability that the second is black, given that the first is green?

8. In this exercise we introduce the following notation: $P(A)$ stands for the probability that the event A happens. $P(A \text{ and } B)$ stands for the probability that both events A and B happen. $P(A \mid B)$, which we read "probability of A given B," stands for the probability that event A happens *given* that B has already happened.
   (a) Show that if $P(B) = \frac{1}{3}$ and $P(A \text{ and } B) = \frac{1}{6}$, then $P(A \mid B) = \frac{1}{2}$.
   (b) Explain why, if A and B are random events, then in general

$$P(A \mid B) = \frac{P(A \text{ and } B)}{P(B)}.$$

9. Suppose a product has a life span that is distributed exponentially with t measured in years, i.e.,

$$f(t) = r\,e^{-rt}.$$

   (a) Find the probability that the product will fail in the first year.
   (b) Suppose the product has already lasted five years—i.e., it did not fail during the first five years. What is the probability that it will fail within the next year? (See the preceding exercise.)
   (c) Show that the probability of failing in the first k years is the same as the probability of failing in k additional years, given it has already lasted n years.
   (d) Discuss the implications of the result in part (c).

10. Suppose f is a probability density function that models the following situation: x can take on any value between 0 and 10 and only those values. The probability of a value occurring in any two intervals of the same length is the same. Write down a formula for f.

11. Suppose that your company expects to sell about 300,000 computer chips over the next year. Your statisticians have determined that the following function represents the fraction of the total number of computer chips that will have failed as of time t (in months):

$$F(t) = 1 - e^{-0.008\,t}.$$

   (a) How many chips would you expect to last more than a year?
   (b) On average, how long would you expect a computer chip to last?
   (c) Does it make sense for your company to offer a one-year guarantee on these chips? Explain.

12. There is an old saying that "A watched pot never boils." In a recent study it was discovered that the probability that a watched pot boiled in the first minute is $\frac{1}{2}$. That is, if T represents the number of minutes until the pot boils, then

$$P(0 \le T < 1) = \frac{1}{2}.$$

Furthermore, scientists established that the probability density function is of the form

$$f(t) = \frac{c}{1 + t^2}.$$

(Note the we haven't said yet what the domain is.)
   (a) Find c by using the fact that $P(0 \le \mathrm{T} < 1) = \frac{1}{2}$.
   (b) Find the distribution function.
   (c) According to this research, what is the expected time for a pot to boil? What does your answer imply about the old saying?

13. In Table 10.5 we give data on cars failing emissions testing in 1992 in six populous counties in North Carolina. Unlike the light bulb failures, these are cars whose exhaust systems started operating in many different years. Nevertheless, if the percentage of cars of a given age failing the emissions test in a given year is relatively constant,

these data might represent a snapshot in time of an exponential failure situation. Let's explore that possibility.

Table 10.5 Emission inspection failures by auto model year[2]

Model year	*Number inspected*	*Emission failures*
1975	6,399	1,559
1976	13,452	3,182
1977	20,585	4,931
1978	25,807	7,217
1979	29,438	6,252
1980	25,641	5,329
1981	28,511	5,371
1982	32,943	5,749
1983	47,689	6,082
1984	73,929	7,504
1985	85,425	6,671
1986	99,250	5,556
1987	106,065	3,524
1988	112,358	1,869
1989	109,309	874
1990	94,654	386
1991	88,195	209
1992	11,875	16

(a) Plot the percentage of failures as a function of age. What characteristics does this plot share with an exponential distribution? What characteristics do not look like an exponential distribution?

(b) How can you test whether the percentage of failures fits an exponential distribution? Carry out your test.

(c) If the fit is good, what is the failure parameter r? If the fit is poor for some ages of cars, is there a substantial range of ages over which the fit is better? If so, what is r for that range?

2. Source: *The News and Observer*, Nov. 29, 1992.

10.4 Normal Distributions

A particularly important class of distributions for describing a wide range of phenomena is the class of normal distributions. The following are examples of data that tend to be distributed normally:

- Repeated weighings of an object using the same device
- Heights of individuals from a more or less homogeneous population
- Test scores on ability or aptitude tests (e.g., IQ, SAT, GRE)

Data of this sort tend to cluster around the mean or average value, with similar numbers of values at equal distances above and below the average. For a given length, as you move away from the average, you find fewer and fewer values in an interval of that length. These properties don't completely *characterize* normal distributions—e.g., the

Cauchy distribution has all these properties. However, we will quantify "clustering" and "spreading" in a way that distinguishes normal distributions from all others.

Histograms and Bell-Shaped Curves

In a lab associated with this chapter you may study a data set that consists of 100 heights of women. Figure 10.8 shows a **histogram** plot of this data. A histogram is constructed in two steps:

- First, subdivide an interval containing all the data values into a number of subintervals of equal length.
- Then, on each subinterval, plot a bar whose height represents the number of data points that fall in that subinterval.

If the data are normally distributed, and if the number of subintervals is not too large or too small, the histogram will have the characteristic bell shape that you see in Figure 10.8.

Figure 10.8 Histogram plot of the heights of 100 women

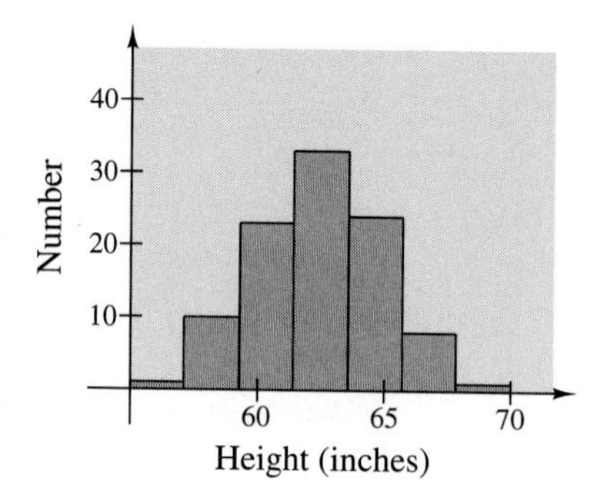

If we have a large sample from a normally distributed population, the histogram should have roughly the same shape as the *density* function; the height of each bar is proportional to the fraction of data points that fall in the corresponding interval, and that fraction should be approximately the probability of a random data value lying in that interval.

Exploration Activity 1

(a) Copy Figure 10.8, and sketch on it a smooth, bell-shaped curve that should have the shape of the probability density function $f(t)$. (The scale won't be right because the areas of the bars are too large to be probabilities.)

(b) Now make a rough sketch of the distribution function $F(t)$. All you need to know is that $F(t)$ is an antiderivative of $f(t)$, together with the values that $F(t)$ should have for very large and very small values of t.

Figure 10.9 is a sketch of a curve with the same shape as the density function superimposed on the histogram. The density function f is the derivative of the distribution function, and $F(t)$ represents the probability that a random data value is less than t. From this we can infer the following properties of the graph of F.

- The probability that a random data value is less than t drops to zero as $t \to -\infty$, so the graph of F must be tangent to the t-axis on the left.
- Since $f(t)$ is always positive, the graph of F must increase steadily.
- The slope of the graph at $(t, F(t))$ is $f(t)$, so the slope is steepest where $f(t)$ is largest, i.e., at the mean.
- The probability that a random data value is less than t rises to 1 as $t \to \infty$, so the graph must be tangent to the line $y = 1$ on the right.

Figure 10.10 is a sketch of the graph of a function F with all of these properties.

Figure 10.9 General shape of the density curve

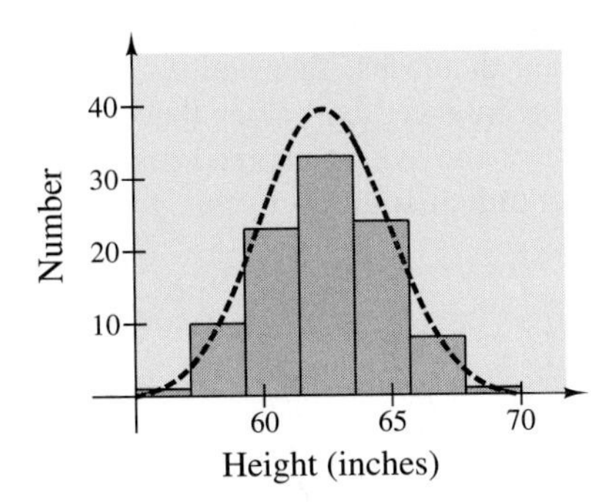

Figure 10.10 Sketch of the normal distribution curve

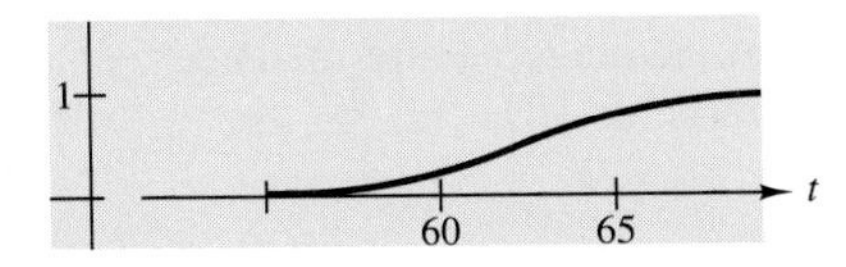

Checkpoint 1

Which of the four properties just stated for the graph of the normal distribution F are shared by the graph of the Cauchy distribution function

$$F(t) = \frac{1}{2} + \frac{1}{\pi} \arctan t?$$

Mean, Variance, and Standard Deviation

If you have a list of data values—say $v_1, v_2, \ldots, v_n$ —that you want to model with a normal distribution, you need to find both the average of the data and a measure of the average spread of the data away from the average. In statistical parlance the average is called the "mean."

Definition The **mean** of the data $v_1, v_2, \ldots, v_n$ is

$$\frac{1}{n}\sum_{k=1}^{n} v_k.$$

We denote the mean of a set of data by m.

The measure of average spread away from the mean requires more explanation. The directed distance of a data value v_k from the mean m is $v_k - m$. This is a *signed* distance: It is positive if v_k is larger than the mean and negative if v_k is smaller than the mean. If we average these signed distances, then values that are a large positive distance away from the mean would be balanced by values that are a large negative distance away from the mean. Our measure of the average spread might be small even though the values are widely spread out. A common solution to this problem is to average instead the *squares* of these distances.

Definition The **variance** of the data $v_1, v_2, \ldots, v_n$ is

$$\frac{1}{n}\sum_{k=1}^{n} (v_k - m)^2,$$

where m is the mean.

Since variance is an "average square distance," an appropriate "average distance" is the *square root* of the variance.

Definition The **standard deviation** of the data $v_1, v_2, \ldots, v_n$ is

$$\sqrt{\frac{1}{n}\sum_{k=1}^{n} (v_k - m)^2},$$

where m is the mean. We denote the standard deviation of a data set by sd.

Checkpoint 2

Find the mean, variance, and standard deviation of the data in Table 10.6.

Table 10.6 Heights in inches of the male students in a calculus class

62	67.5	68	68	70	70.5	71	71	71.5	72	74	79

Standardized Data and the Standard Normal Distribution

If a data set is distributed normally with mean m and standard deviation sd, then we can scale the data to obtain a new data set that is normally distributed with mean 0 and standard deviation 1.

Exploration Activity 2

Suppose m is the mean and sd is the standard deviation of data set with values $v_1, v_2, \ldots, v_n$.

(a) Show that the numbers $v_1 - m, v_2 - m, \ldots, v_n - m$ form a data set with mean 0.

(b) Show that the numbers

$$\frac{v_1 - m}{sd}, \frac{v_2 - m}{sd}, \ldots, \frac{v_n - m}{sd}$$

form a set with mean 0 and standard deviation 1.

The mean of the data set in part (a) is

$$\frac{1}{n}\sum_{k=1}^{n}(v_k - m) = \frac{1}{n}\sum_{k=1}^{n} v_k - \frac{nm}{n}$$
$$= m - m = 0.$$

Similarly, the mean of the data set in part (b) is

$$\frac{1}{n}\sum_{k=1}^{n}\frac{v_k - m}{sd} = \frac{1}{n\,sd}\sum_{k=1}^{n} v_k - \frac{nm}{n\,sd}$$
$$= \frac{m}{sd} - \frac{m}{sd} = 0.$$

The variance of this data set is

$$\frac{1}{n}\sum_{k=1}^{n}\left(\frac{v_k - m}{sd} - 0\right)^2 = \frac{1}{n\,sd^2}\sum_{k=1}^{n}(v_k - m)^2$$
$$= \frac{\text{variance of the original data set}}{sd^2}$$
$$= \frac{sd^2}{sd^2} = 1.$$

Thus the standard deviation of the data set in part (b) is $\sqrt{1} = 1$.

This process of scaling the data to obtain a new data set with mean 0 and standard deviation 1 is called "standardizing."

Definition For a data set $v_1, v_2, \dots, v_n$ with mean m and standard deviation sd, the set of **standardized** data values is

$$\frac{v_1 - m}{sd}, \frac{v_2 - m}{sd}, \dots, \frac{v_n - m}{sd}.$$

Any data set may be standardized. If the original data set was normally distributed with mean m and standard deviation sd, then the standardized data set also will be normally distributed—this time with mean 0 and standard deviation 1.

Checkpoint 3

(a) Find the standardized data values for the data in Checkpoint 2.

(b) Verify that the mean of the standardized data is 0 and the standard deviation is 1.

Because we can standardize any set of normally distributed data, we'll concentrate for the moment on describing a model for data that is normally distributed with mean 0 and standard deviation 1. The distribution for such a data set has a special name.

Definition The normal distribution with mean 0 and standard deviation 1 is called the **standard normal distribution.**

We need both a distribution function and the corresponding probability density function for the standard normal distribution. The bell shape of the histogram in Figure 10.8 (as well as your work in the laboratory, if you have done it already) suggests that the probability density function should have a graph that looks like Figure 10.11. The scale is yet to be determined, but it must be such that the total area under the curve turns out to be 1. Our standardization places the mean at 0, and we expect normally distributed data to peak at the mean and be symmetric with respect to the mean. The density should also be near zero for events far from the mean on either side.

Figure 10.11 Standard normal probability density function

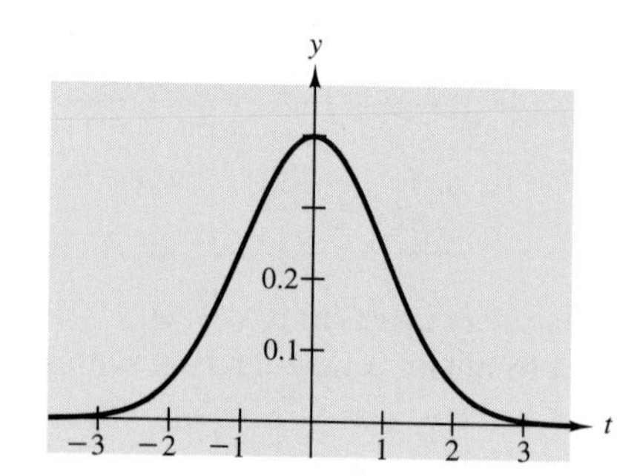

Figure 10.12 Standard normal distribution function

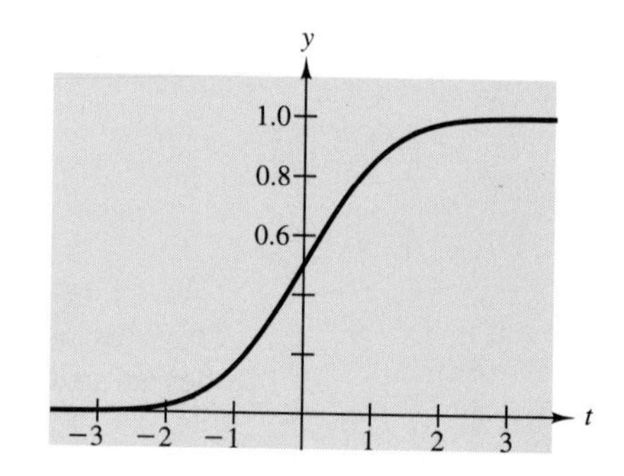

Except for scale, your response to Exploration Activity 1(b) should look like the curve in Figure 10.12, and the standard normal should have a distribution with the same shape. It should start at a zero level on the left, continuously increase, and level off at 1 (100%) on the right. The symmetry of the density function dictates a y-intercept of 0.5, since half the area under the density function must be on the left and half on the right. Furthermore, the steepest part of the distribution graph must be at the mean ($t = 0$), since the density peaks there.

Among functions that have graphs with the bell shape shown in Figure 10.11 are those with formulas of the form $f(t) = c\,e^{-t^2/2}$. You may have already found in the laboratory that, in order for the total area under the graph of f to be exactly 1, the scale factor c must be approximately 0.3989. In fact, the graph in Figure 10.11 *is* that of $f(t) = 0.3989\,e^{-t^2/2}$, which makes a very satisfactory probability density function for the standard normal distribution.

Exploration Activity 3

For the standard normal probability density function $f(t) = ce^{-t^2/2}$, explain why the distribution function must be

$$F(t) = \frac{1}{2} + \int_0^t ce^{-s^2/2}\,ds.$$

For the standard normal distribution function F, the value at t is the integral of the density function over all values less than t:

$$\begin{aligned} F(t) &= \int_{-\infty}^{t} ce^{-s^2/2}\,ds \\ &= \int_{-\infty}^{0} ce^{-s^2/2}\,ds + \int_0^t ce^{-s^2/2}\,ds. \end{aligned}$$

But

$$\int_{-\infty}^{0} ce^{-s^2/2}ds = F(0) = \frac{1}{2},$$

so

$$F(t) = \frac{1}{2} + \int_0^t ce^{-s^2/2}\,ds\,.$$

To illustrate the fit of the standard normal density function to a set of normally distributed data, we show in Figure 10.13 the result of

- standardizing the height data plotted in Figure 10.8,
- plotting a histogram of the standardized data divided by the number of data points, and
- superimposing a graph of the standard normal density function.

The division step replaces the *numbers* of standardized data points in each interval by *fractions* in each interval. Note the similarity of the standardized and scaled histogram to the original one in Figure 10.8.

Figure 10.13 Histogram plot of the standardized heights of 100 women, plus the standard normal probability density function

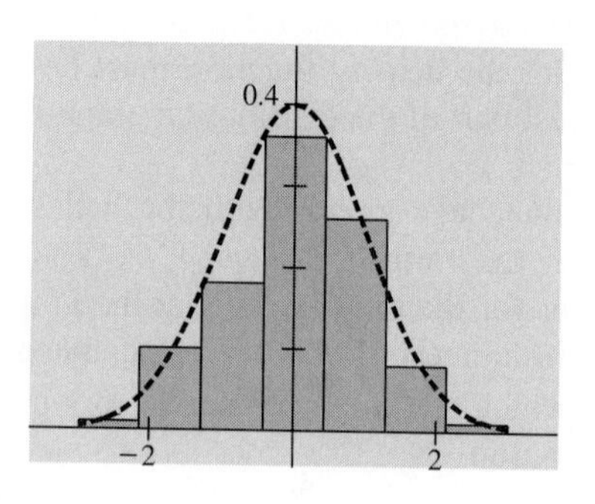

Normal Probabilities

It follows from Exploration Activity 3 that if we could evaluate the integral

$$\int_0^t e^{-s^2/2}\,ds$$

exactly, we could write down a formula for $F(t)$. However, $f(s) = e^{-s^2/2}$ is one of those functions that does not have an **elementary** antiderivative, that is, it does not have an antiderivative that can be described using algebraic combinations of the functions that we know. For the time being, if we want to know a value of F, we have to approximate an integral.

Example 1 Assume you have a data set with the standard normal distribution. Find the probability that a random data value from this distribution lies between 0 and 1.

Solution This probability is

$$F(1) - F(0) = \int_0^1 ce^{-s^2/2}\,ds.$$

Using the integral key on a calculator (with $c = 0.3989$), we find the value to be approximately 0.3413. ■

Example 2 Assume you have a data set with the standard normal distribution. Find the probability that a random data value lies between -1 and 0.

Solution This probability is

$$F(0) - F(-1) = \int_{-1}^0 ce^{-s^2/2}\,ds.$$

Since the integrand is an even function, by symmetry this is the same as the integral we estimated in Example 1: 0.3413. ■

Probability calculations involving the standard normal distribution are often done using a table. The values tabulated are $F(z) - F(0)$ for a range of z, i.e., the tables give the probability that a random data point will have a value between 0 and z. We provide such a table as Table 10.7.

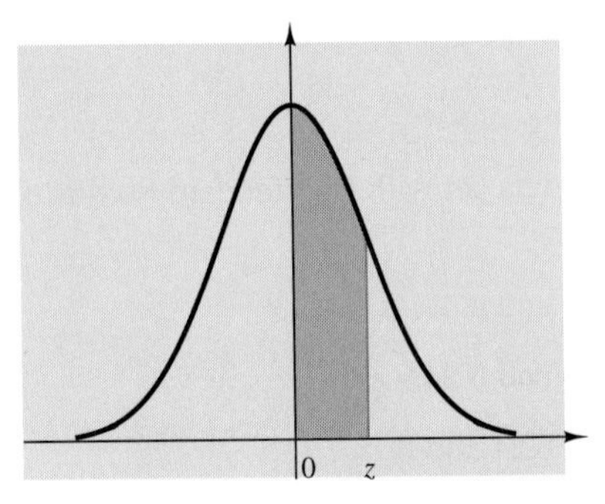

Table 10.7 Standard normal probabilities

z	.00	.01	.02	.03	.04	.05	.06	.07	.08	.09
0.0	.0000	.0040	.0080	.0120	.0160	.0199	.0239	.0279	.0319	.0359
0.1	.0398	.0438	.0478	.0517	.0557	.0596	.0636	.0675	.0714	.0753
0.2	.0793	.0832	.0871	.0910	.0948	.0987	.1026	.1064	.1103	.1141
0.3	.1179	.1217	.1255	.1293	.1331	.1368	.1406	.1443	.1480	.1517
0.4	.1554	.1591	.1628	.1664	.1700	.1736	.1772	.1808	.1844	.1879
0.5	.1915	.195	.1985	.2019	.2054	.2088	.2123	.2157	.2190	.2224
0.6	.2257	.2291	.2324	.2357	.2389	.2422	.2454	.2486	.2517	.2549
0.7	.2580	.2611	.2642	.2673	.2704	.2734	.2764	.2794	.2823	.2852
0.8	.2881	.2910	.2939	.2967	.2995	.3023	.3051	.3078	.3106	.3133
0.9	.3159	.3186	.3212	.3238	.3264	.3289	.3315	.3340	.3365	.3389
1.0	.3413	.3438	.3461	.3485	.3508	.3531	.3554	.3577	.3599	.3621
1.1	.3643	.3665	.3686	.3708	.3729	.3749	.3770	.3790	.3810	.3830
1.2	.3849	.3869	.3888	.3907	.3925	.3944	.3962	.3980	.3997	.4015
1.3	.4032	.4049	.4066	.4082	.4099	.4115	.4131	.4147	.4162	.4177
1.4	.4192	.4207	.4222	.4236	.4251	.4265	.4279	.4292	.4306	.4319
1.5	.4332	.4345	.4357	.4370	.4382	.4394	.4406	.4418	.4429	.4441
1.6	.4452	.4463	.4474	.4484	.4495	.4505	.4515	.4525	.4535	.4545
1.7	.4554	.4564	.4573	.4582	.4591	.4599	.4608	.4616	.4625	.4633
1.8	.4641	.4649	.4656	.4664	.4671	.4678	.4686	.4693	.4699	.4706
1.9	.4713	.4719	.4726	.4732	.4738	.4744	.4750	.4756	.4761	.4767
2.0	.4772	.4778	.4783	.4788	.4793	.4798	.4803	.4808	.4812	.4817
2.1	.4821	.4826	.4830	.4834	.4838	.4842	.4846	.4850	.4854	.4857
2.2	.4861	.4864	.4868	.4871	.4875	.4878	.4881	.4884	.4887	.4890
2.3	.4893	.4896	.4898	.4901	.4904	.4906	.4909	.4911	.4913	.4916
2.4	.4918	.4920	.4922	.4925	.4927	.4929	.4931	.4932	.4934	.4936
2.5	.4938	.4940	.4941	.4943	.4945	.4946	.4948	.4949	.4951	.4952
2.6	.4953	.4955	.4956	.4957	.4959	.4960	.4961	.4962	.4963	.4964
2.7	.4965	.4966	.4967	.4968	.4969	.4970	.4971	.4972	.4973	.4974
2.8	.4974	.4975	.4976	.4977	.4977	.4978	.4979	.4979	.4980	.4981
2.9	.4981	.4982	.4982	.4983	.4984	.4984	.4985	.4985	.4986	.4986
3.0	.4987	.4987	.4987	.4988	.4988	.4989	.4989	.4989	.4990	.4990

Example 3 Find the probability that a random data value from a set with the standard normal distribution is greater than 1.47.

Solution We first find the probability that a random value is *between* 0 and 1.47: Look in the row of Table 10.7 labeled 1.4, and scan over to the column headed .07. The entry there is 0.4292, and that is $F(1.47) - F(0)$. The probability of a value *less than* 1.47 is obtained by adding $F(0) = 0.5$ to the table entry: 0.9292. The probability of a value *greater than* 1.47 is obtained by subtracting the "less than" probability from $1 : 0.0708$. Equivalently, we could have found this answer by subtracting from 0.5 the table entry 0.4292 for 1.47. ■

Checkpoint 4

For a data set with the standard normal distribution, find the probability that a random data value is between

(a) 0 and 0.53,

(b) 0 and 0.24,

(c) −0.24 and 0,

(d) −0.24 and 0.53.

Example 4 Suppose we have a normally distributed data set with mean $m = 2.3$ and standard deviation $sd = 1.63$. Find the probability that a random data value from this data set lies between 1 and 3.

Solution We use the fact that the standardized data set obtained by subtracting the mean and dividing by the standard deviation has the standard normal distribution. Suppose v is a random value from the original data set. Then

$$1 < v < 3$$

exactly when

$$1 - m < v - m < 3 - m,$$

which is true exactly when

$$\frac{1-m}{sd} < \frac{v-m}{sd} < \frac{3-m}{sd},$$

which is true exactly when

$$\frac{-1.3}{1.63} < \frac{v-2.3}{1.63} < \frac{0.7}{1.63},$$

and that is true approximately when

$$-0.80 < \frac{v-2.3}{1.63} < 0.43.$$

Now the quantity $w = (v - 2.3)/1.63$ is a random value from a data set that has the standard normal distribution. Thus our original probability question has the same answer as the question of the probability that a random value from a standard normal data set lies between −0.80 and 0.43. Using Table 10.7, we see that this is $0.1664 + 0.2881 = 0.4545$. ■

Checkpoint 5

Give a reason for each line of inequalities in the solution of Example 4.

Checkpoint 6

Use Table 10.7 to find the probability that a random value from a normally distributed data set with mean 25 and standard deviation 4 lies between 20 and 30.

The Error Function

We have shown how to answer questions about normal distributions *without* obtaining a formula for the distribution function. Still, it would be nice to have such a formula. If we ask a typical computer algebra system for an antiderivative for $e^{-t^2/2}$, we get the answer

$$\frac{\sqrt{2\pi}}{2}\,\mathrm{erf}\left(\frac{\sqrt{2}\,t}{2}\right).$$

Here "erf" is a function that the computer algebra system knows about and is prepared to evaluate, in the same sense that it is prepared to evaluate sine, tangent, exponential, and logarithm. When we ask the system to differentiate $\mathrm{erf}(t)$, we obtain the derivative

$$\frac{2\,e^{-t^2}}{\sqrt{\pi}}.$$

Thus, whatever the function "erf" might be, it is an antiderivative of this function.

Checkpoint 7

A computer algebra system tells us that $\mathrm{erf}(0) = 0$. Explain why this means that

$$\mathrm{erf}(t) = \int_0^t \frac{2\,e^{-s^2}}{\sqrt{\pi}}\,ds.$$

The function erf is called the **error function**. Most computer algebra systems have some routine for calculating this function. It could be that such a system just approximates the integral in our last equation. We tested this conjecture by asking a typical computer algebra system to evaluate $\mathrm{erf}(0.753)$ and then to evaluate the integral

$$\frac{2}{\sqrt{\pi}}\int_0^{0.753} e^{-s^2}\,ds.$$

In addition to the answers, 0.713080 each way, this system gives the time of computation. The computation time for the integral was more than 8 times longer than the evaluation of $\mathrm{erf}(0.753)$. Clearly, the system is not evaluating the error function by approximating integrals.

What is the computer system doing? For that matter, how does it find values of the sine, tangent, exponential, and logarithm functions? For the last four functions, you can ask the same question about your calculator. To answer these questions, we need to look into how we can use polynomials to approximate functions—a subject we will take up in the next chapter.

In this section we have investigated the most important class of data distributions—normal distributions. We saw that the standard deviation of a data set is a measure of the spread of the data around the mean. Any data set can be standardized by subtracting the mean from each data value and then dividing each of the resulting values by the standard deviation. If the original data set was normally distributed, then the standardized data set is normally distributed with mean 0 and standard deviation 1—the standard normal distribution. Thus questions about the original data set may be recast as questions about the standard normal distribution.

The density function for the standard normal distribution is given by the formula

$$f(t) = c\,e^{-t^2/2},$$

where $c = 0.3989$. The corresponding distribution function is

$$F(t) = \frac{1}{2} + \int_0^t c\,e^{-s^2/2}\,ds.$$

Although we do not have a formula for this distribution function in terms of elementary functions, we can calculate its values by using a table, by numerical approximation to the integral, or by using a computer algebra system. Such a system will describe the distribution function in terms of a new (to us) function, the error function. In the next chapter we investigate how such a system might evaluate this function.

Answers to Checkpoints

1. All of them. In fact, all but one of these properties are shared by any distribution with positive density on the domain $(-\infty, \infty)$. The exception is having steepest slope at the mean, which is true of these two distributions, but not all.

2. mean $= 70.35$, variance ≈ 15.17, standard deviation ≈ 3.90

3. (a) The standardized data values are approximately

-2.15	-0.738	-0.610	-0.610	-0.0963	0.0321	0.161	0.161	0.289	0.417	0.931	2.21

   (b) The mean of this data set is -2.67×10^{-4} and the standard deviation is 0.999 to 3 SD.

4. (a) 0.2019, (b) 0.0948, (c) 0.0948, (d) 0.2967

5. We subtracted a constant from each term.

   We divided each term by the positive quantity sd.

   We substituted the value of the m.

   We calculated the quotients.

6. $2 \times 0.3944 = 0.7888$

7. Since the derivative of erf(t) is $2\,e^{-t^2}/\sqrt{\pi}$, the Fundamental Theorem of Calculus tells us that

$$\text{erf}(t) = \text{erf}(t) - \text{erf}(0) = \int_0^t \frac{2\,e^{-s^2}}{\sqrt{\pi}}\,ds.$$

Exercises 10.4

Find the mean and the standard deviation for each of the following data sets.

1. **Table 10.8** Scores on a mathematics examination with 200 points maximum

200	200	200	190	186	170
167	165	164	159	159	147
146	144	131	128	124	109

2. **Table 10.9** Maximum daily temperatures during a week in July

94	93	95	85	87	90	95

3. **Table 10.10** Annual rainfall in inches recorded at Raleigh-Durham International Airport from 1982–1992[3]

Year	*Rainfall (inches)*
1982	44.35
1983	47.23
1984	38.17
1985	36.95
1986	42.06
1987	37.66
1988	54.15
1989	37.55
1990	54.15
1991	35.46
1992	43.18

4. **Table 10.11** Numbers of Democratic governors in office from 1980–1993[4]

Year	*No. of governors*	*Year*	*No. of governors*
1980	31	1987	26
1981	27	1988	26
1982	27	1989	28
1983	34	1990	29
1984	35	1991	27
1985	34	1992	28
1986	34	1993	30

In Exercises 5–8 suppose you have a normally distributed data set with mean 10 and standard deviation 2. What is the probability that a random data value will be

5. Between 9 and 12?

6. Less than 9.7?

7. Greater than 9.7?

8. Between 9.3 and 11.5?

9. What is the probability that a random data value from a normally distributed data set will lie within one standard deviation of the mean? That is, what is the probability that $|\text{value} - m| < sd$?

10. What is the probability that a random data value from a normally distributed data set will lie within 1.5 standard deviations of the mean? within 2 standard deviations of the mean? within 3 standard deviations of the mean?

11. Suppose a data set has the standard normal distribution. Find a number w such that the probability that a random data value is greater than w is $\frac{1}{3}$.

12. Suppose a data set has the standard normal distribution. Find a number w such that the probability that a random data value is less than w is $\frac{1}{10}$.

13. Suppose a data set has a normal distribution with mean 3 and standard deviation 0.5. Find a num-ber w such that the probability that a random data value is greater than w is $\frac{1}{5}$.

14. Suppose a data set has a normal distribution with mean 3 and standard deviation 0.5. Find a num-ber w such that the probability that a random data value is less than w is $\frac{1}{10}$.

15. Write the standard normal distribution function,

$$F(t) = \frac{1}{2} + \int_0^t c\, e^{-s^2/2}\, ds,$$

in terms of the error function.

16. Toward the end of the 19th century, H. P. Bowditch measured the heights of 1253 eleven-year-old boys. We show this data in Table 10.12.[5]

Table 10.12 Heights of 11-year-old boys

Height (inches)	*Number*
46	1
47	3
48	7
49	13
50	31
51	73
52	111
53	176
54	189
55	198
56	162
57	106
58	77
59	49
60	31
61	10
62	9
63	4
64	1
65	1
66	1

(a) Find the mean and standard deviation of this data set.

(b) If the data are normally distributed, how many boys would you expect to have heights of 57 inches or more? How many boys *had* heights of 57 inches or more?

(c) Do you think these data are normally distributed? Carry out two more experiments similar to the one in (b) to see if your opinion is supported. (See Project 3 for another approach to this problem.)

17. Which of the data sets in Tables 10.8–10.11 do you think might be normally distributed? Explain your conclusions.

18. In 1983, Camilla Benbow and Julian Stanley published in *Science* a study of almost 40,000 seventh-grade students who had taken the mathematics section of the Scholastic Aptitude Test (SAT-M) in the three preceding years. Their results, separated by gender, are shown in Figure 10.14: Each group had normally distributed scores with a standard deviation of 60. The boys had a mean of 412 and the girls a mean of 385. A standard statistical test shows that these means are significantly different. Benbow and Stanley's controversial conclusion was that this showed "endogenous" male superiority in mathematical ability. For purposes of this exercise, we concede the point that there is a gender difference here, and we ignore questions of what the test measures, whether the boys and girls had the same educational experiences or levels of encouragement, and whether the sample of students was unbiased. The question we address is the *predictiveness* of the measurement.

Figure 10.14 Boys' and girls' SAT-M scores

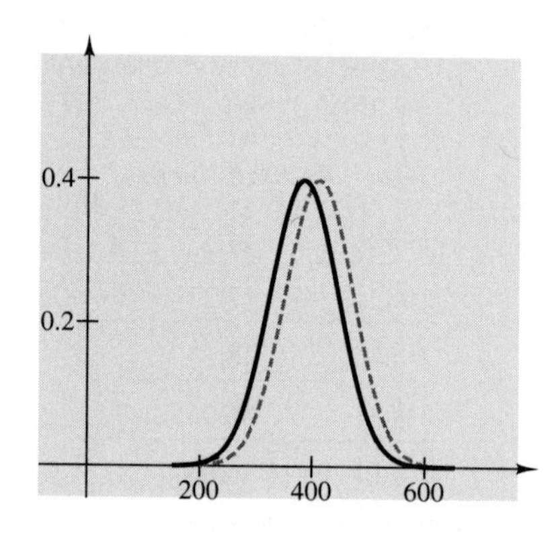

(a) Suppose you had no knowledge of the Benbow-Stanley study, and, for random boy-girl pairs of seventh-graders, you *guess* which will have the higher SAT-M score. How often would you expect to be right?

(b) Suppose instead, for your random boy-girl pairs, you always assume that the boy will have the higher score. How often would you expect to be right? Describe in your own words the gain in predictiveness that results from knowing the Benbow-Stanley result.

(c) How would you compare the difference between genders with the variability within genders? Think of appropriate numerical measures, and describe your comparison in words.[6]

19. Explain why any probability density function of the form $f(t) = c\,e^{-t^2/b}$ must have zero as its expected value (mean). You should be able to explain this without calculating an integral, but calculate an integral if you must.

20. Suppose m is the mean and sd is the standard deviation for a normal distribution. Define f by

$$f(t) = \frac{c}{sd}\exp\left[-\frac{(t-m)^2}{2\,sd^2}\right],$$

where c is the constant that appears in the definition of the standard normal density function.

(a) Show that for any $a < b$

$$\int_a^b f(t)\,dt = \int_c^d g(u)\,du,$$

where g is the probability density function for the standard normal distribution, $c = (a-m)/sd$, and $d = (b-m)/sd$.

(b) Explain why the calculation in (a) verifies that f is the probability density function for the normal distribution with mean m and standard deviation sd.

21. Recall that the mean or expected value of a continuous probability distribution with density function $f(t)$ defined from $-\infty$ to ∞ is

$$\mu = \int_{-\infty}^{\infty} t\,f(t)\,dt,$$

a weighted average of the values of t, with $f(t)$ serving as the weight function. [This mean is denoted by the Greek lower-case mu, μ, to distinguish it from the mean m of a sample of data from the population with density $f(t)$.] The **variance** of the distribution is the weighted average (in the same sense) of squared distances from the mean—that is, the variance is

$$\int_{-\infty}^{\infty} (t-\mu)^2 f(t)\,dt.$$

Not surprisingly, the square root of the variance is called the **standard deviation** of the distribution. Show that the standard deviation of the standard normal distribution with probability density function $f(t) = 0.3989\,e^{-t^2/2}$ is 1.

22. Consider the following process: We select an individual at random from among the population of healthy adults, and we have that individual run or jog for as long as he or she can and then stop. We don't know the distribution for this process, but we can imagine that the probability density function has the form shown in Figure 10.15, namely, a function of the form

$$f(t) = k\,e^{-t^2/2},$$

where k is a constant and t represents time in hours.

(a) What is the value of k? (Think about how this density function is related to the normal probability density function.)

(b) Make a rough sketch of the distribution function whose density function is $f(t)$.

(c) Find the mean (or expected) time that a random individual can run.

(d) Determine the probability that a randomly selected individual can run no longer than the mean time for the whole population.

Figure 10.15 $f(t) = k\,e^{-t^2/2}$ for $0 \le t < \infty$

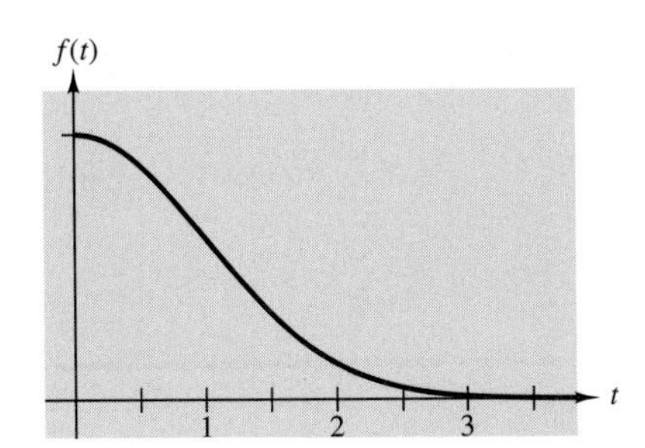

3. Source: National Weather Service, cited in *The News and Observer*, January 1, 1993.

4. Source: National Governors' Association, cited in *USA Today*, November 16, 1992.

5. Adapted from *Statistical Reasoning* by Gary Smith, Allyn and Bacon, Inc., 1985. The data appeared in H. P. Bowditch, "The Growth of Children," Report of the Board of Health of Massachusetts, VIII, 1877.

6. This exercise is based on "The Case of the Falling Nightwatchmen" by R. M. Sapolsky, *Discover*, July, 1987, pp. 42–45.

10.5 Gamma Distributions

In this section we study another family of important probability distributions, the gamma distributions. Various gamma distributions model times between malfunctions of a particular type of aircraft engine, times between arrivals in service lines—for example, a supermarket checkout, a fast-food counter, or a highway toll booth—and times to complete service for an individual in any of those lines. One type of gamma distribution coincides with the exponential distribution we studied earlier in connection with failure times of electrical or electronic components. Still another is the χ^2 (chi-square) distribution, which is used to test the "goodness of fit" of sample data to a theoretical distribution (e.g., normal) for the population from which the sample is drawn.

Gamma Density Functions

We start with a definition of the family of gamma density functions.

Definition Suppose a and b are constants such that $0 \leq a$ and $0 < b$. Then the **gamma probability density function** defined by a and b is the function f of the form

$$f(t) = c\,t^a e^{-t/b} \qquad \text{for } 0 \leq t < \infty,$$

where c is a positive constant such that

$$\int_0^\infty f(t)\,dt = 1.$$

Checkpoint 1

(a) For the case of $a = 0$, the gamma density functions are

$$f(t) = c\,e^{-t/b} \qquad \text{for } 0 \leq t < \infty,$$

which are the exponential density functions we studied in connection with light bulb failure. Sketch a typical graph of such a function.

(b) How is the constant c related to b?

Now we'll explore properties of the gamma density functions

$$f(t) = c\,t^a e^{-t/b}$$

for which $a > 0$. In this case, when we substitute $t = 0$, the factor t^a becomes 0. Thus

$$(1) \qquad f(0) = 0.$$

For $t > 0$, all three factors in the definition of $f(t)$ are positive, so

$$(2) \qquad f(t) \text{ is positive for all positive } t.$$

The function f also has a third property:

$$(3) \qquad \lim_{t \to \infty} f(t) = 0.$$

A full explanation of this property requires some knowledge of the polynomial approximations we will take up in the next chapter. Recall that we encountered a similar problem in Example 1 of Section 10.2. We make a start on the explanation here—in an exercise at the end of Section 11.2 we ask you to complete the explanation.

Here's the underlying idea for determining the behavior of $f(t)$ as $t \to \infty$: $f(t)$ has one factor that is growing big—the positive power of t —and another that is growing small as t becomes large—the decaying exponential. It is helpful to rewrite the formula for f as

$$f(t) = \frac{c\,t^a}{(e^t)^{1/b}} = c\left(\frac{t^{ab}}{e^t}\right)^{1/b},$$

that is, as a quotient of quantities that are both growing large. (This is still indeterminate in form.) Now, to show that

$$\lim_{t \to \infty} f(t) = 0,$$

all we have to do is show that e^t grows *faster* than any power of t. That's what we leave for you to show in Section 11.2, after we have developed polynomial approximations to e^t in Section 11.1. Specifically, we will ask you to show that

$$\lim_{t \to \infty} \frac{t^a}{e^{bt}} = 0$$

for all positive numbers a and b.

In Figure 10.16 we sketch the graph of a typical gamma density function for $a > 0$. Verify that the function whose graph is shown satisfies Properties (1)–(3).

Figure 10.16 $y = f(t)$ for a typical gamma density function with $a > 0$

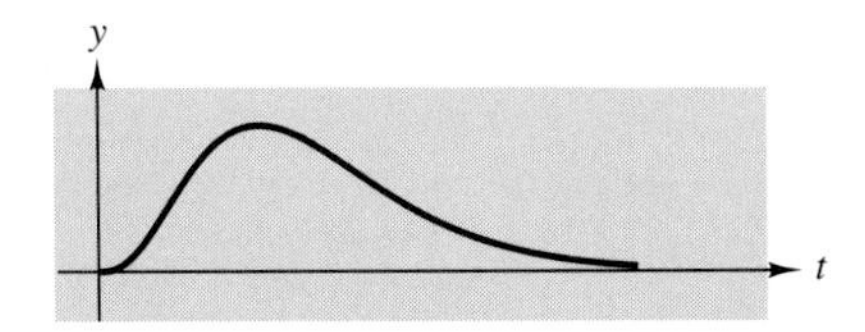

Checkpoint 2

Which of the properties (1), (2), and (3) of the gamma density are shared by the exponential density, and which are not? That is, if $a = 0$, which of these properties are true? Which are false? (You should be able to answer this immediately from the graph you drew for Checkpoint 1.)

In order to make

$$f(t) = c\,t^a e^{-t/b}$$

a probability density function, we need to choose c so that

$$\int_0^\infty f(t)\,dt = 1.$$

For the case of $a = 1$ and $b = 2$, we illustrate the computation to find c in the next example.

Example 1 Find the number c so that

$$f(t) = cte^{-t/2}$$

is a probability density function.

Solution We need an antiderivative for $f(t)$, for which we can use an integral table or integration by parts. Then

$$\int_0^T cte^{-t/2}dt = -2ce^{-t/2}(t+2)\Big|_0^T = -2ce^{-T/2}(T+2) + 4c.$$

Now we have to determine what happens to the first term in the answer as $T \to \infty$. But this is just like the problem of showing that property (3) is true: We need to know that exponential functions grow faster than power functions. Once we know that, we can assert that the first term decreases to zero as T grows large, and therefore the integral of $f(t)$ from 0 to ∞ is $4c$. In order for this integral—the probability of all possible events—to be 1, we must have $c = \frac{1}{4}$. ■

Expected Values of Gamma Distributions

Recall that the mean or expected value of a probability distribution is computed by integrating $tf(t)$ over the entire domain, where $f(t)$ is the probability density function.

Checkpoint 3

Use integration by parts or the Integral Table to carry out the following computations. Use your calculator or computer to check your results.

(a) Find the mean for the gamma distribution with $a = 1$ and $b = 2$.

(b) Find the constant c and the mean for the gamma distribution with $a = 3$ and $b = 2$.

Checkpoint 4

Use your calculator or computer to find the constant c and the mean for the gamma distribution with $a = \frac{1}{2}$ and $b = 2$.

Integration by parts or the Integral Table will work for evaluating gamma integrals when a is an integer—as in Checkpoint 3—but neither will work otherwise (why?). However, many of the important density functions—as in Checkpoint 4—have non-integer values of a. Your electronic tools may suffice to evaluate these integrals, but they may also fail. See Project 1 at the end of the chapter for a strategy for evaluating improper integrals in these cases. In Figure 10.17 we show a composite graph of the gamma density functions for $b = 2$ and $a = \frac{1}{2}$, 1, and 3. Check these graphs against your computations of total means in Checkpoints 3 and 4.

Figure 10.17 Gamma probability density functions for $a = \frac{1}{2}$, 1, and 3 and $b = 2$

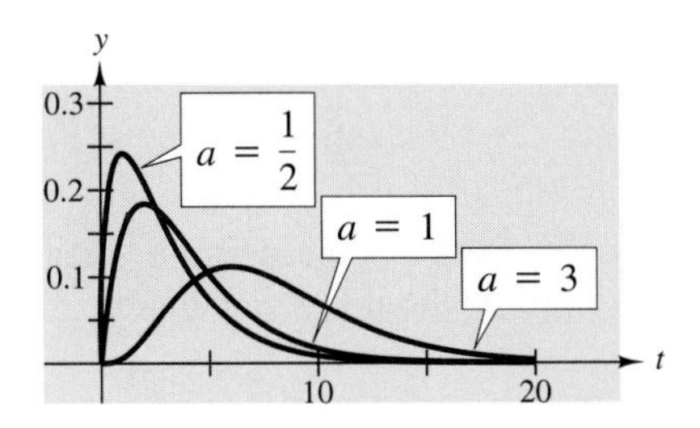

The Gamma Function (Optional)

The gamma distribution is closely related to an important function whose name is (surprise!) the **gamma function**. How do we know the function is important? First, many people worked very hard in the precomputer era to construct extensive tables of its values. Second (and not unrelated to the first reason), it turns up frequently in applications of mathematics in many different disciplines.

Definition The **gamma function** Γ is defined by the formula

$$\Gamma(x) = \int_0^{\infty} t^{x-1} e^{-t}\, dt.$$

The domain of Γ consists of all number x for which this expression makes sense.

Thus evaluating $\Gamma(x)$ for any particular x is just like finding c for a gamma distribution with $a = x - 1$ and $b = 1$. Of course, this can be difficult if x is not an integer.

Exploration Activity 1

(a) Show by direct calculation that $\Gamma(1) = 1$.

(b) Use integration by parts to show that $\Gamma(x) = (x - 1)\,\Gamma(x - 1)$.

(c) Combine parts (a) and (b) to show that if n is a positive integer, $\Gamma(n) = (n - 1)!$.

For $x = 1$ we find

$$\begin{aligned}\Gamma(1) &= \int_0^\infty e^{-t}\,dt \\ &= \lim_{T\to\infty} \int_0^T e^{-t}\,dt \\ &= \lim_{T\to\infty} (1 - e^{-T}) = 1.\end{aligned}$$

The definition of $\Gamma(x)$ is

$$\begin{aligned}\Gamma(x) &= \int_0^\infty t^{x-1}e^{-t}\,dt \\ &= \lim_{T\to\infty} \int_0^T t^{x-1}e^{-t}\,dt.\end{aligned}$$

Now we use integration by parts on the definite integral. Keep in mind that t is the variable of integration, i.e., throughout this calculation x is a constant:

$$\begin{aligned}\int_0^T t^{x-1}e^{-t}\,dt &= (-e^{-t})t^{x-1}\Big|_0^T - \int_0^T (x-1)t^{x-1}(-e^{-t})\,dt \\ &= -\frac{T^{x-1}}{e^T} + (x-1)\int_0^T t^{x-2}e^{-t}\,dt.\end{aligned}$$

Now, by the definition of Γ,

$$\Gamma(x) = \lim_{T\to\infty}\left[-\frac{T^{x-1}}{e^T} + (x-1)\int_0^T t^{x-2}e^{-t}\,dt\right].$$

Since we know that the exponential function grows faster than any power function, we know $\lim_{T\to\infty} T^{x-1}/e^T = 0$. Thus

$$\Gamma(x) = (x-1)\int_0^\infty t^{x-2}e^{-t}\,dt = (x-1)\Gamma(x-1).$$

With the results of (a) and (b) we can bootstrap our way forward:

$$\begin{aligned}\Gamma(2) &= 1\Gamma(1) = 1 \\ \Gamma(3) &= 2\Gamma(2) = 2 \\ \Gamma(4) &= 3\Gamma(3) = 3!\end{aligned}$$

and, in general, $\Gamma(n) = (n-1)!$.

What a remarkable discovery! Γ is a continuous function whose values at positive integers are factorials—*all* the factorials!

Checkpoint 5

(a) Use your calculator to compute factorials up to $10!$.

(b) Sketch what you think the graph of $\Gamma(x)$ should look like for $x \geq 0$.

Now that you know about the gamma function, you can find an expression for the constant c for *every* gamma density function.

Exploration Activity 2

(a) Here again is the definition of the gamma probability density functions:

$$f(t) = ct^a e^{-t/b} \qquad \text{for } 0 \le t < \infty.$$

Show that no matter what a and b are, c is $1/[b^{a+1}\Gamma(a+1)]$.

(b) Check that the expression for c in part (a) agrees with our calculations for $a = 1$ and 3 and $b = 2$ (Example 1 and Checkpoint 3).

(c) Show that the mean of the gamma distribution is $(a+1)b$, and check this against your answers to Checkpoint 3 for $a = 1$ and 3 and $b = 2$.

For a general gamma distribution we know that c is the reciprocal of

$$\int_0^\infty t^a e^{-t/b}\, dt = \lim_{T\to\infty} \int_0^T t^a e^{-t/b}\, dt.$$

We make the substitution $u = t/b$ in the definite integral to find

$$\begin{aligned}\int_0^T t^a e^{-t/b}\, dt &= \int_0^{T/b} b^a t^a e^{-u}\, b\, du \\ &= b^{a+1} \int_0^{T/b} t^a e^{-u}\, du.\end{aligned}$$

Since b is positive, $T/b \to \infty$ as $T \to \infty$, so

$$\begin{aligned}\int_0^\infty t^a e^{-t/b}\, dt &= b^{a+1} \int_0^\infty t^a e^{-u}\, du \\ &= b^{a+1}\Gamma(a+1).\end{aligned}$$

Thus $c = 1/[b^{a+1}\Gamma(a+1)]$.

For $a = 1$ and $b = 2$, this formula gives $c = \frac{1}{4}$, as in Example 1. For $a = 3$ and $b = 2$, we have $c = 1/[2^4\,\Gamma(4)] = 1/96$, which should agree with your calculation in Checkpoint 3.

For a general gamma density function the mean is

$$\frac{1}{b^{a+1}\Gamma(a+1)} \int_0^\infty t^{a+1} e^{-t/b}\, dt.$$

Following the steps just described for the calculation of c, we find

$$\int_0^\infty t^{a+1} e^{-t/b}\, dt = b^{a+2}\,\Gamma(a+2).$$

So the mean is

$$\frac{b^{a+1}\,\Gamma(a+1)}{b^{a+2}\Gamma(a+2)}.$$

Since $\Gamma(a+2) = (a+1)\Gamma(a+1)$, this simplifies to $1/[(a+1)b]$. We leave it to you to verify the earlier calculations.

Our study of gamma distributions extends our investigation of the exponential distribution and shows again the role of improper integrals. The gamma density functions are functions of the form

$$f(t) = c\,t^a e^{-t/b} \qquad \text{for } 0 \le t < \infty,$$

where $a \ge 0$ and $b > 0$. Here c is determined by the condition

$$\int_0^\infty c\,t^a e^{-t/b}\,dt = 1.$$

The gamma function is closely related to the gamma distributions. This function, which may be thought of as a continuous extension of the factorial function, is defined by

$$\Gamma(x) = \int_0^\infty t^{x-1} e^{-t}\,dt$$

for all x at which the improper integral converges. Using the gamma function, we obtained the following formulas for the constant c in the gamma density function and the mean of the gamma distribution:

$$c = \frac{1}{b^{a+1}\Gamma(a+1)}$$

and

$$\text{mean} = b(a+1).$$

Answers to Checkpoints

1. (a)

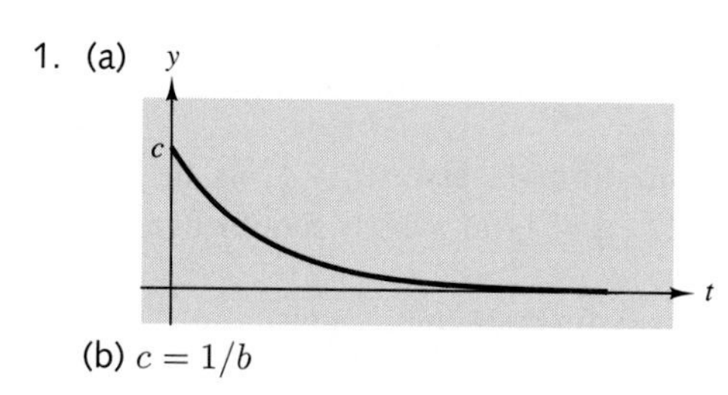

(b) $c = 1/b$

2. (1) is not true, but (2) and (3) are.

3. (a) mean $= 4$ (b) $c = 1/96$, mean $= 8$

4. $c = 0.3989$, mean $= 3$

5. (a) $10! = 3,628,800$

(b)

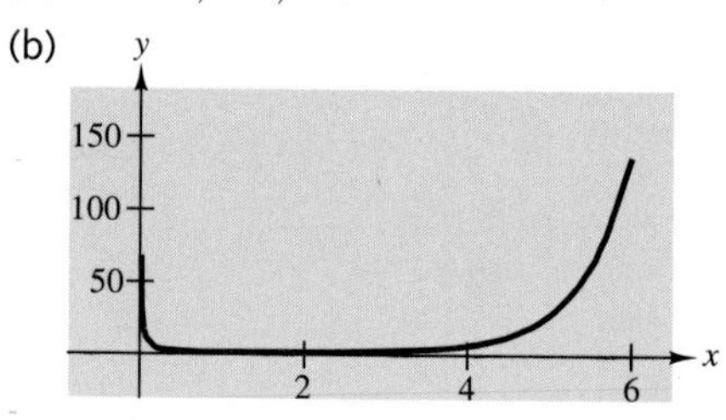

Exercises 10.5

For each of the following gamma density functions, calculate the constant c.

1. $a = 1$ and $b = 4$
2. $a = 2$ and $b = 1/2$

For each of the following gamma density functions, calculate the mean.

3. $a = 1$ and $b = 4$
4. $a = 2$ and $b = 1/2$

For the next two exercises, assume a data set is distributed according to the gamma distribution with $a = 1$ and $b = 2$. Find the probability that a random data value lies between

5. 0 and 3.
6. 1 and 3.

For the next two exercises, assume a data set is distributed according to the gamma distribution with $a = 3$ and $b = 2$. Find the probability that a random data value lies between

7. 0 and 7.
8. 4 and 7.

For the next two exercises, assume a data set is distributed according to the gamma distribution with $a = 1$ and $b = 3$. Find the probability that a random data value

9. is greater than 10.
10. lies between 0 and 10.

Find the following values of the gamma function.

11. $\Gamma(3)$
12. $\Gamma(5)$
13. $\Gamma(8)$
14. Use numerical integration to verify that $\Gamma(3/2) = \sqrt{\pi}/2$.
15. Use the exact value $\Gamma(3/2) = \sqrt{\pi}/2$ to calculate $\Gamma(7/2)$.
16. Let $f(t)$ be the gamma probability density function with $a = 1$ and $b = \frac{1}{2}$:

$$f(t) = cte^{-2t} \qquad \text{for } 0 \leq t < \infty.$$

(a) Find c.
(b) Find the distribution function $F(t)$.
(c) Use your calculator or computer to graph $f(t)$ and $F(t)$.
(d) This density function might model the lifetime of a system that contains an exponentially distributed component for which there is a backup component in the system. Find the expected time-to-failure for the system.
(e) Find the probability that the system fails in the time interval $[1, 5]$.[7]

17. In Figure 10.17 we sketched the graphs of three gamma density functions. The graphs suggest that a gamma density function with $a > 0$ has exactly one maximum point. Differentiate the formula defining the gamma density functions,

$$f(t) = ct^a e^{-t/b} \qquad \text{for } 0 \leq t < \infty,$$

and use the derivative to show that this is true.

---

7. Adapted from "Reliability and the Cost of Guarantees," by K. J. Hastings, in *Applications of Calculus* (P. Straffin, ed.), MAA Notes No. 29, 1993. This article is an excellent source for moving beyond our superficial discussion of reliability earlier in this chapter.

10.6 More Notation for Limiting Values

In Section 10.2 we introduced the notation

$$\lim_{x \to \infty} f(x)$$

for the limiting value of a function f as the independent variable goes to infinity. We noted in passing that the same notation could be used for discrete sequences of numbers, such as those generated by Newton's Method for finding roots:

$$\lim_{n \to \infty} t_n = r,$$

where g is a function, r is a root of $g(t) = 0$, and

$$t_{n+1} = t_n - \frac{g(t_n)}{g'(t_n)}.$$

In this case, the infinite sequence of numbers

$$t_0,\ t_1,\ t_2,\ \ldots$$

is itself a function of the index variable n. That is, for each nonnegative integer n there is a unique value t_n. If Newton's Method converges, then the limiting value must in fact be a zero of g. We haven't added anything new here except notation for such limiting values.

Checkpoint 1

Find each of the following limiting values.

(a) $\lim_{x \to \infty} \frac{\sin x}{x}$ (b) $\lim_{n \to \infty} \left(2 - \frac{1}{n^2}\right)$ (c) $\lim_{T \to \infty} \int_0^T \frac{1}{1+3t^2}\,dt$

Example 1 Find the limiting value of the geometric sum

$$1 - \frac{1}{3} + \frac{1}{9} - \frac{1}{27} + \cdots.$$

Solution The ellipsis notation $\cdots$ indicates that this sum goes on forever. This is a sum of the form

$$1 + \beta + \beta^2 + \beta^3 + \cdots,$$

where $\beta = -1/3$. As we saw in Chapter 5, the sum of the first n terms is

$$s_n = 1 + \beta + \beta^2 + \cdots + \beta^{n-1} = \frac{1-\beta^n}{1-\beta}.$$

Since $|\beta| < 1$,

$$\lim_{n \to \infty} \beta^n = 0.$$

Thus the total sum is

$$\lim_{n\to\infty} s_n = \lim_{n\to\infty} \frac{1-\beta^n}{1-\beta} = \frac{1}{1-\beta} = \frac{1}{4/3} = \frac{3}{4}.$$

Limiting values have appeared in a number of other contexts, notably in the definition of the derivative. In this case the independent variable Δt approaches 0 rather than ∞, but the notation works the same way:

$$f'(t) = \lim_{\Delta t\to 0} \frac{f(t+\Delta t) - f(t)}{\Delta t}.$$

This is a concise notation for a complicated process: At each fixed value of t, for each increment Δt in t, form the difference quotient

$$\frac{\text{change in values of } f}{\text{change in } t},$$

and take the limiting value as the change in t shrinks to 0.

Example 2 Express the process of differentiating the exponential function in the notation for limiting values.

Solution

$$\begin{aligned}\frac{d}{dt} e^t &= \lim_{\Delta t\to 0} \frac{e^{t+\Delta t} - e^t}{\Delta t} \\ &= e^t \lim_{\Delta t\to 0} \frac{e^{\Delta t} - 1}{\Delta t}.\end{aligned}$$

Because we know the answer is e^t, we also know the limiting value of the remaining quotient:

$$\lim_{\Delta t\to 0} \frac{e^{\Delta t} - 1}{\Delta t} = 1.$$

Note that this last equation is actually an evaluation of $f'(0)$, where $f(t) = e^t$.

Checkpoint 2

Find each of the following limiting values. In each case, make a connection with a derivative calculation for which you already know the answer.

(a) $\lim_{\Delta t\to 0} \frac{\sin \Delta t}{\Delta t}$

(b) $\lim_{\Delta t\to 0} \frac{b^{\Delta t} - 1}{\Delta t}$, where b is a positive number.

There is nothing special about the variable Δt. The limiting value notation works the same way no matter what the independent variable is — and no matter what number is being approached by that variable.

Exploration Activity 1

Find each of the following limiting values. Start by exploring each of the functions with your graphing calculator or computer. Once you have a value, try to explain why it must be the right value.

(a) $\lim_{x\to 2} \dfrac{x^3-8}{x-2}$ (b) $\lim_{x\to 0} \dfrac{\ln\left(\frac{2+x}{2}\right)}{x}$

(c) $\lim_{x\to 0} \dfrac{1-\cos x}{x}$ (d) $\lim_{x\to 0} \dfrac{1-\cos x}{x^2}$

We show the graph of the function $(x^3-8)/(x-2)$ in Figure 10.18. It looks remarkably like the graph of a quadratic polynomial, and you may have noticed that $x-2$ divides evenly into x^3-8. In fact, except at $x=2$,

$$\frac{x^3-8}{x-2} = x^2+2x+4.$$

The graph is missing the point at $(2, 12)$, but there can be little doubt that the function values are close to 12 when x is close to 2. Thus

$$\lim_{x\to 2} \frac{x^3-8}{x-2} = \lim_{x\to 2} (x^2+2x+4) = 12.$$

You also might have looked at this as a derivative calculation for a well-known function, $f(x) = x^3$:

$$f'(2) = \lim_{x\to 2} \frac{f(x)-f(2)}{x-2} = \lim_{x\to 2} \frac{x^3-8}{x-2}.$$

But $f'(2) = 3\cdot 2^2 = 12$, which is another way to get the same answer.

Our second function is also a difference quotient for a well-known function, but that may not be evident until you write the quotient in another form:

$$\begin{aligned}\frac{\ln\left(\frac{2+x}{2}\right)}{x} &= \frac{\ln(2+x)-\ln 2}{x}\\ &= \frac{g(2+\Delta t)-g(2)}{\Delta t},\end{aligned}$$

where $g(t) = \ln t$ and $x = \Delta t$. Thus the limiting value is $g'(2) = \frac{1}{2}$. We show a graph of the given function in Figure 10.19.

Similarly, if we take $h(x) = \cos x$, then we know that $h'(x) = -\sin x$, so $h'(0) = 0$. On the other hand, if we calculate the derivative at 0 by difference quotients, we find

$$\begin{aligned}0 &= h'(0)\\ &= \lim_{x\to 0} \frac{\cos x - \cos 0}{x}\\ &= \lim_{x\to 0} \frac{\cos x - 1}{x}\\ &= -\lim_{x\to 0} \frac{1-\cos x}{x}.\end{aligned}$$

If the negative of our answer is 0, so is the answer itself. We show the graph of the function $(1 - \cos x)/x$ in Figure 10.20.

We don't want to give the impression that all interesting limiting value calculations are just derivatives in disguise. Part (d) of the Exploration Activity is a case in point. It is easy to determine — for example, by tracing the graph — that

$$\lim_{x \to 0} \frac{1 - \cos x}{x^2} = \frac{1}{2}.$$

It's not at all easy to explain why — not yet. Early in the next chapter this will be a relatively easy calculation. There we will see that we can represent $\cos x$ by a formula of the form

$$\cos x = 1 - \frac{1}{2}\,x^2 + \text{other terms with higher powers of } x.$$

When we subtract this expression from 1 and divide by x^2, we are left with

$$\frac{1 - \cos x}{x^2} = \frac{1}{2} + \text{other terms with factors of } x.$$

When we let $x \to 0$, all that is left is the constant term, $\frac{1}{2}$. We show a graph of the relevant function in Figure 10.21.

Figure 10.18 Graph of $y = \dfrac{x^3 - 8}{x - 2}$

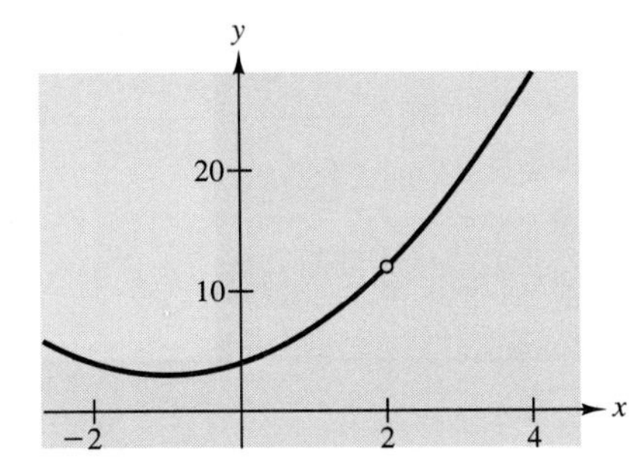

Figure 10.19 Graph of $y = \dfrac{\ln[(2 + x)/2]}{x}$

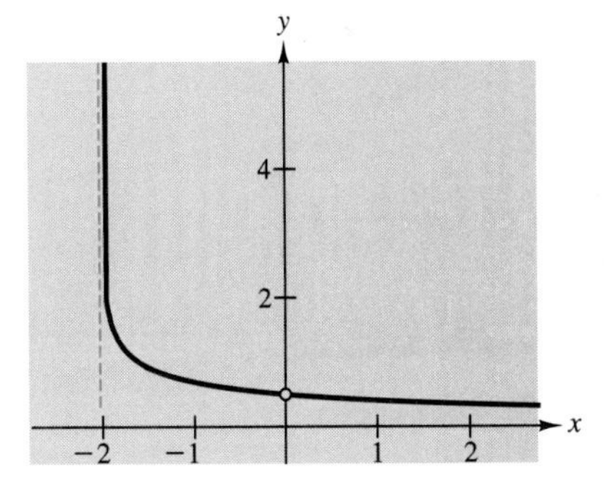

Figure 10.20 Graph of $y = \dfrac{1 - \cos x}{x}$

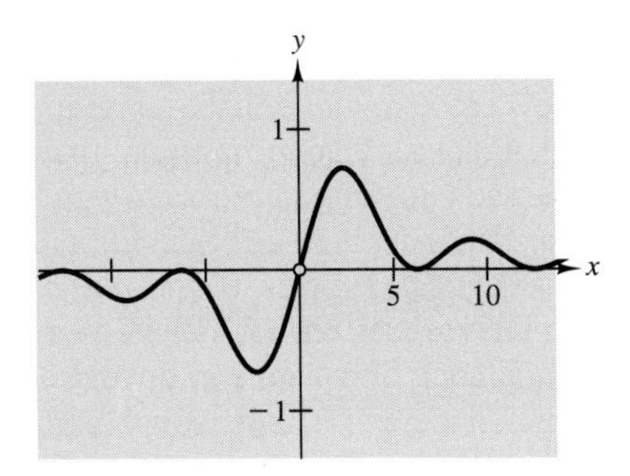

Figure 10.21 Graph of $y = \dfrac{1 - \cos x}{x^2}$

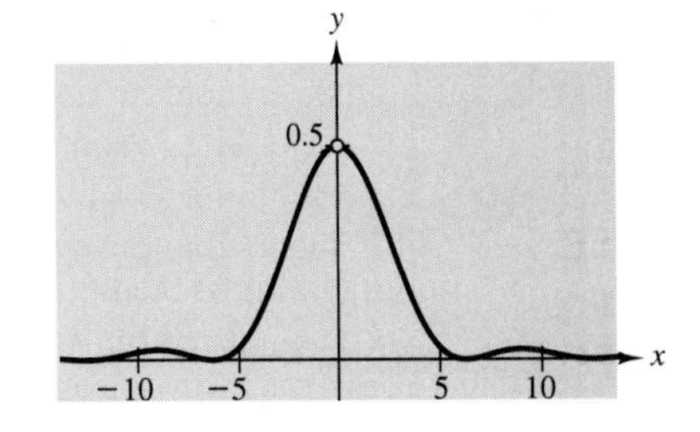

All of our examples of the use of the notation

$$\lim_{x \to a} f(x)$$

have involved numbers a at which the function f is undefined. Functions have—or may have—limiting values at numbers in their domain as well. Usually it's not much of a challenge to find those limiting values: Just evaluate the function. That's because practically all our functions are continuous in the sense described in Chapter 8. Indeed, the definition we avoided giving there may be expressed in terms of limiting values: A function f is **continuous at a number** a if

$$\lim_{x \to a} f(x) = f(a).$$

This simple-looking equation actually has three parts: First, it asserts that a is in the domain of f. That is, $f(a)$ means something. Second, the values of $f(x)$ approach something as x gets close to a—the same something on both sides, if a is not an endpoint of the domain. Third, the number being approached as limiting value and the number $f(a)$ are the same. Finally, a function is **continuous** if it is continuous at each number in its domain. As we noted in Chapter 8, almost every formula-based function we see is continuous.

Example 3 Evaluate $\lim_{x \to 1} \dfrac{x^3 - 8}{x - 2}$.

Solution The function $(x^3 - 8)/(x - 2)$ is continuous at 1. (In fact, it is continuous at every number except 2.) Thus

$$\lim_{x \to 1} \frac{x^3 - 8}{x - 2} = \frac{1 - 8}{1 - 2} = 7.$$

Checkpoint 3

Evaluate $\lim_{x \to \pi/2} \dfrac{\sin x}{x}$.

In this section we have extended the notation for limiting values first introduced in Section 10.2. From that section and this one we now have notation for

- limiting behavior of a function f as the independent variable goes to infinity:

$$\lim_{x \to \infty} f(x),$$

- limiting behavior of a function f as the independent variable goes to negative infinity:

$$\lim_{x \to -\infty} f(x),$$

- limiting behavior of a discrete sequence with terms t_n:

$$\lim_{n \to \infty} t_n,$$

- limiting behavior of a function f as the independent variable approaches a finite number a — from both sides, if a is not an endpoint of the domain:

$$\lim_{x \to a} f(x).$$

In the exercises we will introduce similar notation for

- limiting behavior of a function as the independent variable approaches a finite number from one side or the other, and
- behavior of a function going to ∞ or $-\infty$ as the independent variable approaches a number or goes to ∞ or $-\infty$.

Answers to Checkpoints

1. (a) 0 (b) 2 (c) $\pi/(2\sqrt{3}) \approx 0.9069$
2. (a) 1 (b) $\ln b$
3. $2/\pi \approx 0.6366$

Exercises 10.6

Find the each of the following limiting values. Start by exploring each of the functions with your graphing calculator or computer. Once you have a value, try to explain why it must be the right value. In some cases you may not have adequate tools — beyond the visual evidence — to establish that your answer is correct. We will consider some of these exercises again at the end of the next chapter.

1. $\lim_{x \to 0} \dfrac{\tan 2x}{x}$
2. $\lim_{x \to 0} \dfrac{\sin 2x}{\tan 3x}$
3. $\lim_{x \to 0} \dfrac{\arctan 5x}{\arcsin 3x}$
4. $\lim_{x \to \infty} \dfrac{\ln x}{x}$
5. $\lim_{x \to 0} \dfrac{\sin^2 x}{x}$
6. $\lim_{x \to 0} \dfrac{\cos x - 1}{x^2 e^x}$
7. $\lim_{x \to 0} \dfrac{e^x - e^{-x}}{1 - e^x}$
8. $\lim_{k \to \infty} \dfrac{5^k}{k!}$
9. $\lim_{x \to 0} \left(\dfrac{1}{x} - \dfrac{1}{e^x - 1} \right)$
10. $\lim_{x \to 0} \dfrac{(\sin x - x)^3}{x(1 - \cos x)^4}$
11. $\lim_{x \to \infty} \dfrac{\arctan 5x}{\arcsin 3x}$
12. $\lim_{x \to 0} \dfrac{2^x}{3^x}$
13. $\lim_{x \to 0} \dfrac{2^x - 1}{3^x - 1}$
14. $\lim_{x \to \infty} \dfrac{2^x}{3^x}$
15. $\lim_{x \to 0} x \ln |x|$
16. $\lim_{x \to \pi/2} (2x - \pi) \tan x$
17. $\lim_{k \to \infty} \dfrac{k^5}{5^k}$
18. $\lim_{x \to 0} \left(\dfrac{1}{x} - \dfrac{1}{\sin x} \right)$
19. When we defined the definite integral in Chapter 8, we could have written the defining formula in the form

$$\int_a^b f(t)\, dt = \lim_{n \to \infty} \sum_{k=1}^{n} f(t_{k-1}) \Delta t.$$

Make sense of this terse notation by explaining carefully all the places that the index variable n is hidden in the expression on the right. Write out as explicitly as you can the first several terms of the sequence to which the "lim" notation is being applied.

Sometimes we are interested in the behavior of a function as the independent variable approaches a number just from the right or just from the left. In particular, if the limiting values from left and right are different, then there is no two-sided limiting value. We use the notation

$$x \to a^+$$

to indicate values of x approaching a from the right

and the notation

$$x \to a^-$$

to indicate values of x approaching a from the left. Determine each of the following limiting values. If you can, give a reason for your conclusion.

20. $\lim_{x \to 0^+} \frac{|x|}{x}$

21. $\lim_{x \to 0^-} \frac{|x|}{x}$

22. $\lim_{x \to 0^+} x \ln x$

23. $\lim_{x \to 1^-} (x - 1) \ln(1 - x)$

24. $\lim_{x \to 0^+} (x - 1) \ln(1 - x)$

25. $\lim_{x \to -1^+} (x - 1) \ln(1 - x)$

26. $\lim_{x \to 1^-} \frac{\ln x}{1 - x}$

27. $\lim_{x \to 1^+} \frac{\ln x}{1 - x}$

28. $\lim_{x \to 0^+} \left(\frac{1}{x} - \frac{1}{\sqrt{x}} \right)$

We can use the notations ∞ and $-\infty$ to describe behavior of function values as well as values of the independent variable. For example, we write

$$\lim_{x \to 0^+} \frac{1}{x} = \infty$$

and

$$\lim_{x \to 0^-} \frac{1}{x} = -\infty$$

to describe the limiting behavior of the reciprocal function as x approaches 0 from the right and from the left, respectively. Describe the limiting behavior represented by each of the following notations. If there is a limiting value, state the value. Whether there is a limiting value or not, give a reason for your conclusion.

29. $\lim_{x \to 0} \frac{3}{|x|}$

30. $\lim_{x \to 0} \frac{2}{x^2}$

31. $\lim_{x \to \pi/2^+} \tan x$

32. $\lim_{x \to \infty} \frac{e^x}{x}$

33. $\lim_{x \to -\infty} \frac{e^x}{x}$

34. $\lim_{x \to 0^-} \frac{\sin x}{x^2}$

35. $\lim_{x \to 0^+} \frac{\ln x}{x}$

36. $\lim_{x \to 1^-} \frac{\ln(1 - x)}{1 - x}$

37. $\lim_{x \to \infty} \left(\frac{1}{x} - \frac{1}{\sqrt{x}} \right)$

38. $\lim_{x \to \pi/2^-} (x - \pi) \tan x$

39. $\lim_{k \to \infty} \frac{5^k}{k^5}$

40. $\lim_{x \to 0^-} \left(\frac{1}{x} - \frac{1}{e^x} \right)$

Chapter 10 Summary

Our investigation of reliability theory opens up a new area of application of the definite integral. This study introduced us to the concept of exponential distribution of data and to more general continuous probability distributions. Each of these distributions may be characterized by the probability distribution function F, where $F(t)$ is the probability that a random data value from a set with this distribution is less than t. Equivalently, the distribution may be characterized by the probability density function f, the derivative of F.

Our attempt to calculate the expected value or mean of the exponential distribution led us to investigate integrals from 0 to ∞. These "improper" integrals are defined as limiting values of definite integrals $\int_0^T g(t)\,dt$ as T approaches infinity. We used this opportunity to introduce formal notation for this and similar limiting processes:

$$\int_0^\infty g(t)\,dt = \lim_{T \to \infty} \int_0^T g(t)\,dt.$$

The most important continuous probability distributions are the normal distributions. In order to reduce the study of this class of distributions to the study of a single distribution, we considered standardization of a finite data set: Each data value v is replaced by the value

$$(v - \text{mean})/(\text{standard deviation}).$$

This transforms a general normally distributed data set into one with the standard normal distribution.

The probability density function for the standard normal distribution is $f(t) = c\,e^{-t^2/2}$ for $-\infty < t < \infty$, where $c \approx 0.3989$. Our attempt to describe the corresponding standard normal distribution function led us to a new function, the error function. The question of how a computer might evaluate this function sets the stage for our study in Chapter 11 of approximation of functions by polynomials.

The gamma probability distributions are a general class of continuous distributions that includes the exponential distributions as a special case. Description of these distributions brought us to another new function, the gamma function. The gamma function Γ is defined by an improper integral:

$$\Gamma(x) = \int_0^\infty t^{x-1} e^{-t}\, dt.$$

Since $\Gamma(n) = (n-1)!$ for all positive integers n, we may think of Γ as the natural continuous extension of the factorial function from the positive integers to all positive real numbers.

We concluded the chapter by extending our notation for limiting values to include those for which the independent variable approaches a finite value, $\lim_{x \to a} f(x)$. We also explored the problem of calculating various limiting values, a problem we will return to at the end of Chapter 11.

Concepts

- Histogram of a data set
- Mean of a data set
- Standard deviation of a data set
- Standardized data values
- Probability distribution
- Distribution function
- Density function
- Mean or expected value of a distribution
- Limiting value of a function as the independent variable approaches infinity
- Improper integral

Applications

- Reliability theory
- Representation of data

Notation

$\lim_{x \to \infty} f(x)$	limiting value of $f(x)$ as x goes to ∞
$\lim_{x \to -\infty} f(x)$	limiting value of $f(x)$ as x goes to $-\infty$
$\lim_{x \to a} f(x)$	limiting value of $f(x)$ as x approaches a

Formulas

Statistics for a set of data values $\{v_1, v_2, \ldots, v_n\}$

Mean: $$m = \frac{1}{n} \sum_{k=1}^{n} v_k$$

Variance: $$var = \frac{1}{n} \sum_{k=1}^{n} (v_k - m)^2$$

Standard deviation: $$sd = \sqrt{var} = \sqrt{\frac{1}{n} \sum_{k=1}^{n} (v_k - m)^2}$$

Continuous Probability

- **Exponential Distributions** $(0 \le t < \infty, r > 0)$

  Distribution function: $F(t) = 1 - e^{-rt}$

  Density function: $f(t) = re^{-rt}$

  Mean: $m = \dfrac{1}{r}$

- **Cauchy Distribution** $(-\infty < t < \infty)$

  Distribution function: $F(t) = \dfrac{\pi}{2} + \arctan t$

  Density function: $f(t) = \dfrac{1}{\pi} \dfrac{1}{1+t^2}$

  Mean: $m = 0$

- **Uniform Distributions** $(a \le t \le b)$

  Distribution function: $F(t) = \dfrac{t-a}{b-a}$

  Density function: $f(t) = \dfrac{1}{b-a}$

  Mean: $m = \dfrac{a+b}{2}$

- **Standard Normal Distribution**
  $(-\infty < t < \infty,\ c \approx 0.3989)$

Distribution function: $F(t) = \dfrac{1}{2} + \displaystyle\int_0^t c\,e^{-s^2/2}\,ds$

Density function: $f(t) = c\,e^{-t^2/2}$

Mean: $m = 0$

- **Gamma Distributions** $(0 \le t < \infty,\ a \ge 0,\ b > 0)$

Distribution function: $F(t) = \dfrac{1}{b^{a+1}\Gamma(a+1)} \displaystyle\int_0^t s^a e^{-s/b}\,ds$

Density function: $f(t) = \dfrac{t^a\,e^{-t/b}}{b^{a+1}\Gamma(a+1)}$

Mean: $m = (a+1)b$

Error Function

$$\operatorname{erf}(t) = \int_0^t \frac{2\,e^{-s^2}}{\sqrt{\pi}}\,ds$$

Gamma Function

$$\Gamma(x) = \int_0^\infty t^{x-1}e^{-t}\,dt,$$ for all x such that the integral makes sense

Improper Integrals

$$\int_a^\infty g(t)\,dt = \lim_{T\to\infty} \int_a^T g(t)\,dt$$

$$\int_{-\infty}^a g(t)\,dt = \lim_{T\to-\infty} \int_T^a g(t)\,dt$$

$$\int_{-\infty}^\infty g(t)\,dt = \int_{-\infty}^0 g(t)\,dt + \int_0^\infty g(t)\,dt$$

Chapter 10 Exercises

Evaluate each of the following integrals. Exact answers are preferable to approximations.

1. $\displaystyle\int_0^\infty \frac{1}{1+t^2}\,dt$
2. $\displaystyle\int_0^\infty \frac{1}{(1+t^2)^2}\,dt$
3. $\displaystyle\int_{\sqrt{3}}^\infty \frac{1}{x^2+1}\,dx$
4. $\displaystyle\int_3^\infty x\,e^{-5x}\,dx$
5. $\displaystyle\int_2^\infty \frac{1}{x(x-1)}\,dx$
6. $\displaystyle\int_3^\infty \frac{1}{x^2+1}\,dx$

Decide whether each of the following integrals converges or diverges. If it converges, find its value.

7. $\displaystyle\int_1^\infty \frac{x^2}{1+x^4}\,dx$
8. $\displaystyle\int_{100}^\infty x^{-1/3}\,dx$
9. $\displaystyle\int_3^\infty x^{-100}\,dx$

10. Each of the numbers 1, 3, 7, 11, 13, and 19 appears on one face of a fair six-sided die. What is the average value that will appear on the top face in a random roll?

11. Suppose the die in Exercise 10 is "loaded" so that 1, 3 and 7 each appear with probability $\frac{1}{6}$, 11 appears with probability $\frac{1}{4}$, and the other two values each appear with probability $\frac{1}{8}$. If the die is rolled many times, what is the average value that will appear?

Find each of the following limiting values. You may use a calculator.

12. $\lim_{x\to\infty} x^{-1/x}$

13. $\lim_{x\to\infty} \frac{\sin x}{x}$

14. $\lim_{x\to 0} (1+x)^{1/x}$

15. $\lim_{x\to\infty} \frac{\ln x}{x^2}$

16. $\lim_{x\to\infty} \frac{x^2}{x^x}$

17. $\lim_{x\to 0^+} \frac{x^2}{x^x}$

18. $\lim_{x\to\infty} \frac{\sqrt{1+x^2}+1}{x}$

19. $\lim_{x\to 0^-} \frac{\sqrt{1+x^2}+1}{x}$

20. $\lim_{x\to 0} (1-x)^{1/x}$

21. (a) Write the standard normal distribution function,

$$F(t) = \frac{1}{2} + \int_0^t c\, e^{-s^2/2}\, ds,$$

in terms of the error function,

$$\operatorname{erf}(t) = \int_0^t \frac{2\, e^{-s^2}}{\sqrt{\pi}}\, ds.$$

(b) Our computer algebra system told us that an antiderivative for $e^{-t^2/2}$ is

$$\frac{\sqrt{2\pi}}{2} \operatorname{erf}\left(\frac{\sqrt{2}\, t}{2}\right).$$

This fact, together with your answer to part (a), gives a symbolic formula for the constant c. What is it? Check that your symbolic formula agrees with $c \approx 0.3989$.

Find the mean and the standard deviation for each of the following data sets.

22. **Table 10.13** Scores on a physics test with 100 points maximum

100	100	97	96	91	88
88	85	84	79	79	79
76	64	31	28	24	20

23. **Table 10.14** Maximum daily temperatures during a week in February

20	15	18	30	27	32	38

24. The age distribution of North Carolina drivers involved in accidents in 1993 is shown in Table 10.15. Note that the age groups are not of equal size. For purposes of this exercise, take the last age group to mean 75–84.

(a) What was the median age of North Carolina drivers involved in accidents in 1993?

(b) How would you estimate from the table the probability that a randomly selected driver involved in an accident was 23 years old? 37 years old?

(c) What was the average age of North Carolina drivers involved in accidents in 1993?

(d) Graph the probability density function for drivers involved in accidents as a function of age.

(e) Graph the distribution function for drivers involved in accidents as a function of age.

(f) What is the probability that a randomly selected driver involved in an accident was younger than the average age for all such drivers?

(g) How would you interpret this data in terms of public policy implications? Explain.

Table 10.15 Percentage of drivers by age group involved in accidents in 1993[8]

Age group	*Percent*
16	24.7
17	17.6
18-19	15.1
20-24	11.5
25-34	7.8
35-44	5.7
45-54	4.7
55-64	4.2
65-74	4.1
75+	4.6

25. Explain the following rules of thumb for normal distributions: About 68% of the data values are within one standard deviation of the mean, about 95% within two standard deviations, and about 99% within three standard deviations.

26. IQ scores are supposed to be normally distributed with mean 100 and standard deviation 15. What percentage of the population has an IQ

    (a) between 100 and 110?

    (b) less than 90?

    (c) between 115 and 125?

    (d) more than 140?

27. Observations of traffic on limited access highways have shown that the time gaps between successive cars tend to be exponentially distributed. Suppose the observed average time gap on a certain highway is 10 seconds.

    (a) What is the probability density function for time gaps?

    (b) What is the distribution function?

    (c) What is the median time gap?

    (d) What is the probability that a gap between cars will exceed 30 seconds?

    (e) What is the probability that a gap between cars will be 5 seconds or less?

28. Suppose the probability of failure of an electronic component in the first year is 8%.

    (a) What is the probability density function?

8. Source: N.C. Division of Motor Vehicles, reported in *The News and Observer*, March 4, 1995.

(b) What is the probability that the component will fail in the second year?

(c) What is the expected lifetime of the component?

29. Find the points of inflection of the graph of the standard normal density function.

30. A manufacturer is considering making a new product. The only information available about the production process is that the minimal cost per unit is \$50, the maximal cost per unit is \$97, and the most likely cost per unit is \$78. We model the uncertainty in production cost with a continuous probability density function $f(c)$ for which $f(c) = 0$ if $c \leq 50$ or $c \geq 97$, $f(78) = h$ for some number h, and otherwise the graph of f consists of straight lines. Such a density function is called **triangular**.

    (a) Find h.

    (b) Find equations for the straight lines that make up the graph of f.

    (c) Find the expected cost according to this model.

    (d) Find the median cost.

    (e) Find the probability that the cost per unit is no more than \$70.

    (f) Find a formula for the distribution function in two parts, first for $50 \leq c \leq 78$, then for $78 \leq c \leq 97$. Sketch the graph of the distribution function.

Chapter 10 Projects

1. **Integrating to Infinity** Your calculator or computer may do an adequate job of calculating improper integrals by just integrating to a "big number." However, if your upper limit of integration is too small, you may miss part of the integral; if it is too large, you may get a ridiculous answer or no answer at all. In this project, we explore one way to be sure of the quality of your answer.

    The gamma distribution with $a = \frac{1}{2}$ and $b = 2$ has a density function of the form $f(t) = c\sqrt{t}\, e^{-t/2}$. To find c, we need to integrate $\sqrt{t}\, e^{-t/2}$ from 0 to ∞, but our integral tables and other symbolic tools are no help for finding an antiderivative. Numerical integration techniques don't work very well on very long intervals. However, they can work adequately on long intervals for functions that are nearly constant, as $\sqrt{t}\, e^{-t/2}$ is when the exponential factor makes all of its values very close to zero.

    A possible strategy: Integrate from 0 to, say, 10 to find most of the area under the curve. Then integrate from 10 to 100, from 100 to 1000, and so on, until you are sure you have accounted for all the area under the curve, up to the desired accuracy. Add the results to get the total integral. (Each integral may require several evaluation steps with larger numbers of subdivisions until you are sure you have the answer for that piece of the area to the desired accuracy.) The reason for using increasing powers of 10 for the endpoints is that you can often find patterns in the results that will make it unnecessary to evaluate more integrals. However, you may have to vary the strategy if you find that each integral requires a very large number of steps.

(a) Evaluate $\int_0^\infty \sqrt{t}\, e^{-t/2}\, dt$.

(b) Find c. (The answer should look familiar from your computations with the normal density.)

(c) Use a similar strategy to evaluate $\int_0^\infty \frac{1}{1+x^{3/2}}\, dx$.

(d) Compare your answers in parts (a) and (c) with answers from your calculator or computer. How good are these electronic tools?

2. **Projecting Social Security Revenues**[9] The Social Security Administration regularly estimates the flow of money into its trust funds in order to project present and future benefits and to compare those benefits to present and future revenues. The amount of money flowing in depends on the taxable maximum T, which was \$42,000 in 1986, \$51,300 in 1990, and \$60,600 in 1994. The larger T is, the more money is exposed to taxation, and thus the more that flows into the trust funds.

For simplicity, we will consider only the wages and salaries for employees. We could apply a similar approach for self-employed workers, but some adjustments would be necessary for workers who have both wages and self-employment income. We have to consider separately the Medicare part of the Social Security tax, which is not subject to the same taxable maximum T. In 1991, the taxable maximum for Medicare was \$125,000. In 1994, the maximum was removed—all wages are now subject to Medicare tax. The combined tax rate for Social Security and Medicare for 1991–1994 was 15.3%, half paid by the employee and half by the employer. Of this, 12.4% went to Social Security and the rest to Medicare. Until we get to part (g) of the project we will focus on Social Security only, so there is just one T.

We denote by F the distribution function for wages and salaries and by f the corresponding density function. Our goal is a formula for the proportion of wages that are taxable, expressed in terms of F but not f, because data is available for the distribution function but not for the density function. We simplify the calculations by assuming that the mean for f is 1. That is, we take as our unit of money the average wage for a given year.

(a) Explain carefully why the amount of wages subject to taxation is

$$N \int_0^T x\, f(x)\, dx + NT[1 - F(T)],$$

where N is the total number of workers.

(b) Use integration by parts to rewrite the expression in part (a), and simplify to show that the amount of wages subject to taxation is

$$N \int_0^T [1 - F(x)]\, dx.$$

Since the average wage is the unit of money, N is also the total of all wages. It follows

9. Based on "A Surprising Application of the Integral $\int_0^T (1 - F(x))\, dx$ to Revenue Projection" by A. Kroopnik, *Mathematics Magazine* 66 (1993), 254–256.

that the *proportion* of wages subject to taxation is

$$\int_0^T [1 - F(x)]\, dx.$$

(c) If this model makes sense, the taxable proportion of wages should approach 1 as $T \to \infty$. Show that this is the case. [You may want to go back to the formula in part (a).]

(d) In 1986 the average wage was $16,361. Calculate T in the appropriate units. Suppose F is the exponential distribution $F(x) = 1 - e^{-x}$, and calculate the proportion of wages subject to taxation. Compare with the published figure of 0.913. (There is no reason to think that F can be represented by a single exponential. In practice, one fits a model with several exponential terms with different decay rates.)

(e) The average wage in 1991 was $21,150. Making the same assumption about F as in part (d), find the proportion subject to taxation.

(f) Write a formula for the total amount paid into Social Security in 1991. Your formula will involve N.

(g) Write a formula for the combined total paid into Social Security and Medicare in 1991.

(h) Write a formula for the combined total paid into Social Security and Medicare in 1994. This formula will have unknowns for both the number of workers and the average wage.

3. **Human Measurements** In this project you will gather data on a particular human measurement — for example, height, length of forearm, circumference of wrist — and then test the data to determine if it is normally distributed. You may want to consider data from males and females separately, or perhaps gather data only from males or only from females.

(a) Decide the measurements you want to make, and have your choice approved by your instructor.

(b) You should gather data from approximately 100 subjects. (This is a group effort.) Describe the method you use to ensure that your sample is reasonably random.

(c) Sort the data in ascending order. Compute the mean m and standard deviation sd of the data.

(d) Find the fraction of data values that fall within one standard deviation of the mean — i.e., the fraction that are greater than $m - sd$ and less than $m + sd$. Display and explain your calculations.

(e) If the data were normally distributed, what fraction of the data would you expect to find in the interval $[m - sd, m + sd]$? Explain your answer.

(f) What conclusion do your results so far suggest with regard to the question of whether this data is normal? Explain your answer.

(g) Standardize your data, and then compute an approximation to the distribution function. Make a plot of the approximate distribution function, and superimpose a graph of the standard normal distribution function. Describe your work, and attach a copy of your graph.

(h) Compute an approximation of the corresponding density function. Make a plot of the approximate density function, and superimpose a graph of the standard normal density function. Describe your work, and attach a copy of your graph.

(i) What evidence do the graphs in parts (g) and (h) provide regarding the distribution of your data?

CHAPTER 11

Polynomial and Series Representations of Functions

In Chapter 9 we studied a technique for representing periodic functions as combinations of trigonometric functions; in this chapter we take up the equally important problem of representing functions—periodic or not—by polynomials. These polynomial representations will enable us to address—if not completely answer—such questions as

- *How is it possible for your calculator or computer to calculate numeric values of logarithmic, exponential, trigonometric, and inverse trigonometric functions?*

If you have come to take those buttons for granted, you may prefer to focus instead on these questions:

- *How accurate are those computations?*
- *Can you believe all the digits such a machine produces?*

We left unanswered in Chapter 10 the question of how to calculate values of the error function in a fraction of the time that would be required for numerical integration. [The question was suggested by observation that our computer could provide values of erf(t) much faster than it could evaluate the corresponding integral.] What we learn about polynomial representation of simpler functions will show us how to find approximating polynomials for the error function as well, and that will turn out to be the tool we need for rapid calculation.

Throughout this chapter you will see the fundamental concepts of derivative and integral being put to use in a variety of ways, and we will also make frequent use of their connecting link, the Fundamental Theorem of Calculus.

11.1 Taylor Polynomials

Polynomials are nice functions. We easily see how to calculate their values, and we know what sorts of graphs these functions should have. Polynomials are easy to integrate and differentiate. Life would be much simpler if every important function were a polynomial. However, there are many important functions that are definitely not polynomials: exponential, sine, cosine, logarithm, inverse tangent, inverse sine, and the error function, to name a few. Although the functions just listed are not polynomials, they can be approximated by polynomials. In this section we study one procedure for finding approximating polynomials.

Approximation of Functions by Polynomials

Let's begin by considering polynomials themselves. Suppose we have a polynomial of degree 4. Here "have" means that we can obtain the values of the function and any of its derivatives at any particular number in the domain, say, at 0. How can we obtain a formula for the polynomial? We know that it has a formula of the form

$$p(x) = c_0 + c_1 x + c_2 x^2 + c_3 x^3 + c_4 x^4.$$

Our problem is to determine the coefficients c_0, c_1, c_2, c_3, and c_4.

It is easy to see that, if we substitute zero for x, we find $c_0 = p(0)$. To obtain c_1, we consider the derivative of $p(x)$. The derivative has the form

$$p'(x) = c_1 + 2c_2 x + 3c_3 x^2 + 4c_4 x^3.$$

Again, it is easy to see that $c_1 = p'(0)$. You should notice a pattern beginning to de-velop —but don't jump to a conclusion too quickly.

Exploration Activity 1

(a) Calculate the second derivative of $p(x)$, substitute $x = 0$, and find a formula for c_2. Does this formula fit with the pattern you just conjectured? What do you think the pattern is now?

(b) Find a formula for c_3.

(c) Find a formula for c_4.

(d) Make a conjecture about a general formula that fits all the coefficients.

The second derivative of $p(x)$ has the form

$$p''(x) = 2c_2 + 6c_3 x + 12c_4 x^2.$$

Thus $2c_2 = p''(0)$, so $c_2 = p''(0)/2$. Differentiating again, we obtain

$$p^{(3)}(x) = 6c_3 + 24c_4 x,$$

which implies that $c_3 = p^{(3)}(0)/6$. (The prime notation for derivatives quickly gets out of hand for higher derivatives. Instead of adding another prime for each new derivative, we put the order of the derivative in parentheses.) Differentiating one more time, we find that

$c_4 = p^{(4)}(0)/24$. The general pattern here is that, for each k, the coefficient c_k satisfies the formula

$$c_k = \frac{p^{(k)}(0)}{k!},$$

where $k! = 1 \cdot 2 \cdot 3 \cdots (k-1) \cdot k$.

The symbol $k!$ is read "k factorial." As a matter of convenience, for $k = 0$, we define $0! = 1$. We also define the 0th derivative of a function to be the function itself, in this case $p(x)$. Then the formula given here for c_k is correct for $k = 0$ as well: $c_0 = p(0)$.

More generally, suppose

$$p(x) = c_0 + c_1 x + c_2 x^2 + \cdots + c_{n-1}x^{n-1} + c_n x^n = \sum_{k=0}^{n} c_k\, x^k$$

is any polynomial of degree n. Then for each k, the coefficient c_k satisfies

$$c_k = \frac{p^{(k)}(0)}{k!}.$$

Example 1 Find a formula for the fourth-degree polynomial function $p(x)$ that has the following values for the function and its derivatives at 0:

$$p(0) = 3, \quad p'(0) = -4, \quad p''(0) = 8, \quad p^{(3)}(0) = 4, \quad \text{and} \quad p^{(4)}(0) = -3.$$

Solution The polynomial has a formula

$$p(x) = 3 - 4x + \frac{8}{2}\,x^2 + \frac{4}{6}\,x^3 - \frac{3}{24}\,x^4,$$

or, in simplified form,

$$p(x) = 3 - 4\,x + 4x^2 + \frac{2}{3}\,x^3 - \frac{1}{8}\,x^4.$$

■

Checkpoint 1

For the polynomial $p(x)$ just found in Example 1, confirm that

$$p(0) = 3, \quad p'(0) = -4, \quad p''(0) = 8, \quad p^{(3)}(0) = 4, \quad \text{and} \quad p^{(4)}(0) = -3.$$

This is all very good, but we do not anticipate often having to discover a formula for a polynomial by asking a tight-lipped guru questions about the value of the polynomial and its derivatives at zero. The real value of this calculation comes from the following act of faith. We reason that if $f(x)$ is a function that is not a polynomial, and if we calculate the numbers $c_k = f^{(k)}(0)/k!$ for $k = 0, 1, 2, \ldots$, then the polynomials

$$
\begin{aligned}
P_0 &= c_0, \\
P_1(x) &= c_0 + c_1 x, \\
P_2(x) &= c_0 + c_1 x + c_2 x^2, \\
P_3(x) &= c_0 + c_1 x + c_2 x^2 + c_3 x^3, \\
&\vdots \\
P_n(x) &= c_0 + c_1 x + c_2 x^2 + c_3 x^3 + \cdots + c_n x^n
\end{aligned}
$$

should approximate $f(x)$ near 0, with each polynomial in the list providing a better approximation than the one before.

In general, this hope turns out to be well-justified, and the reason is not difficult to understand. P_0 is the constant function whose value is $f(0)$. P_1 is the linear function that has the same value at zero as f and the same first derivative at zero as well—i.e., the graph of P_1 is the tangent line to the graph of f at zero. P_2 is a quadratic polynomial that has the same value, the same first derivative, and the same second derivative as f—all at $x = 0$—so its graph should fit the graph of f better than the tangent line. With each addition of a term of higher degree, we get a polynomial with one more derivative that exactly matches the corresponding derivative of f at the special point of close fit. Thus, as the degree increases, we expect that the shapes of the polynomial graphs more and more resemble the shape of the graph of f.

Example 2 For $f(x) = e^x$, calculate the first three polynomial approximations P_0, P_1, and P_2.

Solution For this function, $f'(x) = e^x$ and $f''(x) = e^x$, so

$$f(0) = f'(0) = f''(0) = 1.$$

Thus

$$P_0 = 1,$$

$$P_1(x) = 1 + x,$$

and

$$P_2(x) = 1 + x + \frac{1}{2}x^2.$$

■

Checkpoint 2

(a) For $f(x) = e^x$, calculate $P_3(x)$ and $P_4(x)$.

(b) In a single plotting window, graph $f(x)$ and all five of the polynomials, P_0 through P_4.

Taylor Polynomials for e^x

We now give a name to the approximating polynomials we constructed in the preceding subsection.

Definition The **Taylor polynomials** for a function $f(x)$ **at the reference point** $x = 0$ are the polynomials of the form

$$P_n(x) = f(0) + f'(0)\,x + \frac{f''(0)}{2}\,x^2 + \cdots + \frac{f^{(n)}(0)}{n!}\,x^n.$$

In sigma notation, we may write the defining formula as

$$P_n(x) = \sum_{k=0}^{n} \frac{f^{(k)}(0)}{k!}\,x^k.$$

A note about the terminology: The Taylor polynomials at 0 are also called **Maclaurin polynomials** for $f(x)$. We will not use this terminology because one name is enough— because we will have occasional need for Taylor polynomials at other reference points— and because computers and calculators that know about these polynomials generally use "Taylor."

In Figure 11.1 we plot the graphs of the exponential function and its first five approximating Taylor polynomials. [This is an answer to Checkpoint 2(b).] In Figure 11.2 we plot the *errors* in the polynomial approximations, i.e., the functions $e^x - P_n(x)$, up to $n = 4$. Notice the change of scale from Figure 11.1: In the error plot we consider only the interval from $x = -1$ to $x = 1$, and we have exaggerated the vertical scale to make the errors show up. Thus we can see clearly, for example, that $P_1(x)$ moves quickly away from the exponential curve for values of x not very far from zero, but $P_4(x)$ stays within 0.01 of e^x over the entire interval from -1 to 1.

Figure 11.1 Exponential function and Taylor polynomials up to degree 4

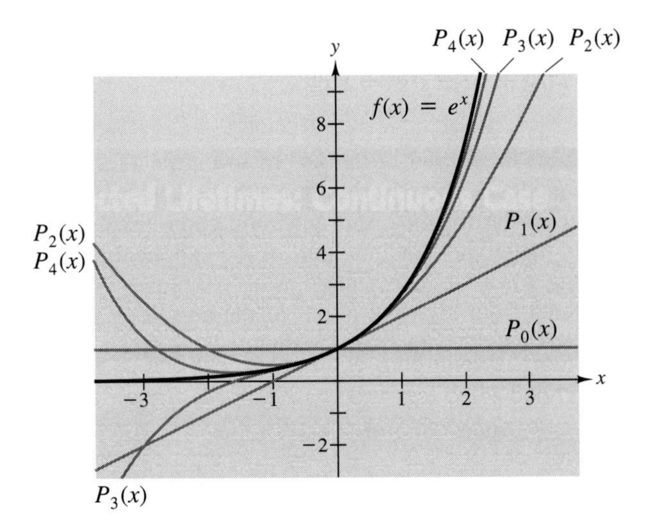

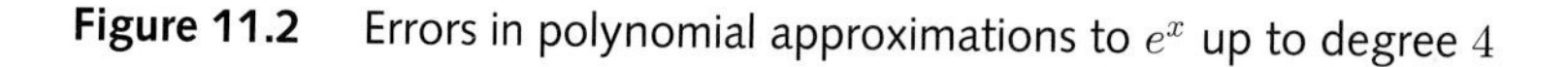

Figure 11.2 Errors in polynomial approximations to e^x up to degree 4

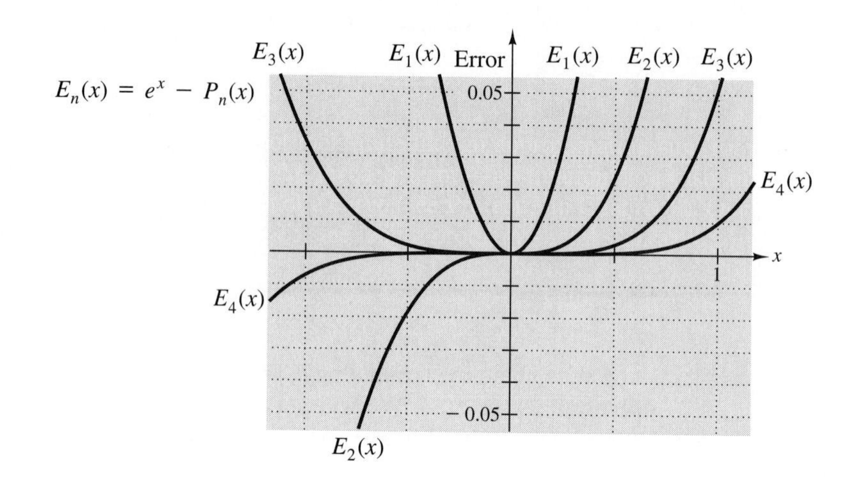

We can do only so much with two-color graphics. In the next checkpoint we ask you to enhance our graphics by adding your own colors. You can do that either in the book or on a photocopy of this page. You will need pencils or pens in five different colors.

Checkpoint 3

(a) Color each of the five polynomial curves in Figure 11.1. Use our legend to identify and keep track of the curves, and color each legend entry to match the corresponding curve.

(b) For each of your "colorized" curves in part (a), draw a box of matching color around the corresponding formula in Example 2 or Checkpoint 2.

(c) Color the curves and legend entries in Figure 11.2 using colors that match the corresponding polynomials in Figure 11.1.

Notice the *shapes* of the graphs in Figure 11.2: The error in the linear approximation looks quadratic, the error in the quadratic approximation looks cubic, and so on. There is a good reason for this. The quadratic approximation, for example, has been selected to exactly match the quadratic part of the exponential function—which will have a precise meaning when we represent e^x as a "very long polynomial." When that quadratic part is subtracted from e^x, what is left of the very long polynomial is the part of degree 3 and higher—and the cubic part dominates all the rest.

Taylor Polynomials for sin x

We turn now to the sine function, for which calculation of the Taylor polynomials is only slightly more difficult than for the exponential function. The corresponding calculations for the cosine function are very similar and will be left to the exercises at the end of this section.

Exploration Activity 2

(a) Calculate the first six Taylor polynomials for $f(x) = \sin x$. How many different functions are there among these six polynomials?

(b) Graph $f(x)$ and these Taylor approximations in a single plotting window.

(c) Graph the error functions $\sin x - P_n(x)$ in a plotting window that is scaled so you can see the distinct error curves.

As you calculate successive derivatives of $f(x)$, you should find a repeating pattern: $\cos x,\ -\sin x,\ -\cos x,\ \sin x$. Since the fourth derivative is $f(x)$ itself, you now know all the derivatives. When you evaluate these functions at $x = 0$ (starting with the 0th derivative, $\sin x$), you get a repeating pattern of numbers: $0,\ 1,\ 0,\ -1,\ 0,\ 1,$ and so on. Thus the first six coefficients are $0,\ 1,\ 0,\ -1/3!,\ 0,$ and $1/5!$. This gives us the following six polynomials:

$$P_0 = 0,$$

$$P_1(x) = x,$$

$$P_2(x) = x,$$

$$P_3(x) = x - \frac{1}{6}x^3,$$

$$P_4(x) = x - \frac{1}{6}x^3,$$

$$P_5(x) = x - \frac{1}{6}x^3 + \frac{1}{120}x^5.$$

Notice that, because the even-degree coefficients are all zero, P_2 is the same as P_1, P_4 is the same as P_3, and so on. Counting the constant function P_0, there are four distinct functions among the first six Taylor approximations. We plot the graphs of $\sin x$ and these Taylor polynomials in Figure 11.3. In the following figure we plot the errors in the approximations.

Checkpoint 4

(a) As you did in Checkpoint 3, color each of the curves in Figure 11.3. Use our legend to identify and keep track of the curves, and color each legend entry to match the corresponding curve.

(b) For each of your "colorized" curves in part (a), draw a box of matching color around the corresponding formula in our discussion of Exploration Activity 2.

(c) Color the curves and legend entries in Figure 11.4 using colors that match the corresponding polynomials in Figure 11.3.

Figure 11.3 Sine function and Taylor polynomials up to degree 5

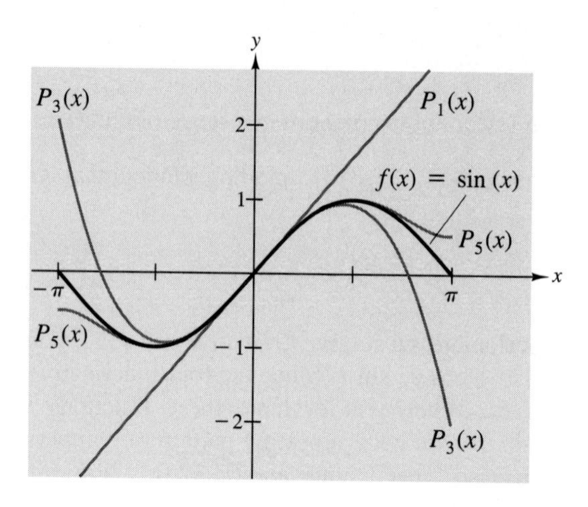

Figure 11.4 Errors in the polynomial approximations of $\sin x$ up to degree 5

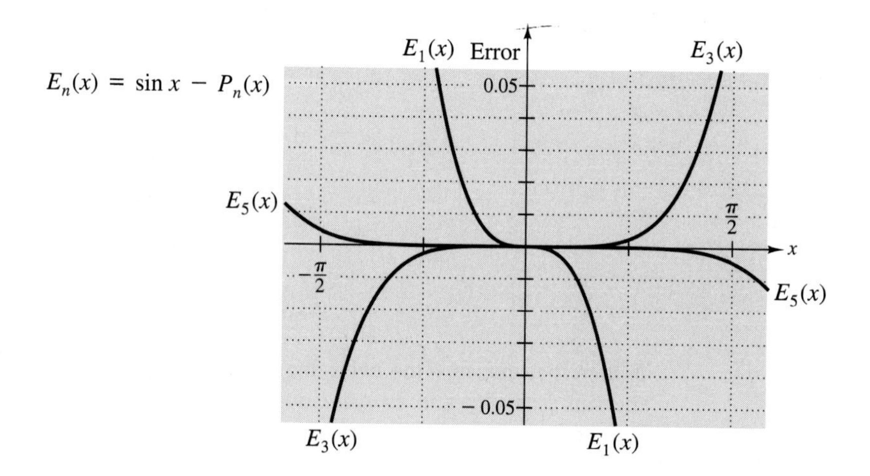

In this section we have seen that we can approximate a nonpolynomial function by polynomials whose coefficients can be calculated by matching derivative values with the original function at a particular reference point, $x = 0$. We have computed these approximating polynomials explicitly for the exponential and sine functions; in the exercises you will extend these ideas and computations to several other functions. We find that our approximations, even for low degree, are quite good near the reference point, but they seem to deteriorate as we move away from the reference point. On the other hand, it appears that we can extend the interval of close approximation by taking higher-degree approximations. That will be our point of departure in the next section.

Answers to Checkpoints

1. Here are $p(x)$ and its first four derivatives:

$$p(x) = 3 - 4x + 4x^2 + \frac{2}{3}x^3 - \frac{1}{8}x^4$$
$$p'(x) = -4 + 8x + 2x^2 - \frac{1}{2}x^3$$
$$p''(x) = 8 + 4x - \frac{3}{2}x^2$$
$$p^{(3)}(x) = 4 - 3x$$
$$p^{(4)}(x) = -3$$

The values of these functions at 0 are the constant terms, 3, -4, 8, 4, -3, as required.

2. (a) $P_3(x) = 1 + x + x^2/2 + x^3/6$, $P_4(x) = 1 + x + x^2/2 + x^3/6 + x^4/24$

(b) See Figure 11.1.

Exercises 11.1

1. If $f(x) = \sin x$, what is the coefficient of x^5 in the seventh-degree polynomial approximation $P_7(x)$ of $f(x)$?

2. For any function $f(x)$ with enough derivatives, what is the coefficient of x^5 in the seventh-degree polynomial approximation $P_7(x)$ of $f(x)$?

3. For any function $f(x)$ with enough derivatives and any positive integers k and n ($n \geq k$), what is the coefficient of x^k in the nth-degree polynomial approximation $P_n(x)$ of $f(x)$?

4. What is the seventh-degree Taylor polynomial approximation to $f(x) = \sin x$?

5. What is the fourth-degree Taylor polynomial approximation to $g(x) = \cos x$?

6. What is the sixth-degree Taylor polynomial approximation to $h(x) = e^x$?

7. Why are only three error curves plotted in Figure 11.4?

8. Interpret each of the curves in Figure 11.4 as we did for the exponential function. What can you say about their shapes?

9. For approximately what interval does the cubic approximation to the sine function match $\sin x$ to within 0.01?

10. (a) What is the largest error (in absolute value) in the fifth-degree approximation to the sine function over the entire interval $[-\pi/2, \pi/2]$?
    (b) For what x's does that largest error occur?
    (c) For each such x, is $\sin x$ larger than or smaller than $P_5(x)$?

11. Find the Taylor polynomials of degrees 0 through 6 for the function $f(x) = \cos x$.

12. In a single plotting window, graph $f(x) = \cos x$ and the Taylor polynomials you calculated in Exercise 11.

13. (a) In a single plotting window, graph the errors in the Taylor polynomials you calculated in Exercise 11.
    (b) Interpret each of the error curves as we did for the exponential function. What can you say about their shapes?

14. For approximately what interval does the sixth-degree approximation to the cosine function match $\cos x$ to within 0.01?

15. (a) What is the largest error (in absolute value) in the sixth-degree approximation to the cosine function over the entire interval $[-\pi/2, \pi/2]$?
    (b) For what values of x does that largest error occur?
    (c) For each such x, is $\cos x$ larger than or smaller than $P_6(x)$?

16. (a) What calculus operation, when applied to $\sin x$, gives $\cos x$?
(b) What happens if you perform the same operation on the Taylor polynomials for $\sin x$?
17. (a) What happens if you differentiate the Taylor polynomials for e^x?
(b) Is your answer to part (a) consistent with your answer to Exercise 16?
18. (a) Calculate e ($= e^1$) by substituting $x = 1$ into the appropriate polynomials $P_n(x)$ for e^x with increasing values of n until you see no change in the sixth decimal place. [*Hint*: $P_n(1) = P_{n-1}(1) + 1/n!$ for every value of n, so you can get each new approximation by adding a single term to the preceding one.]
(b) Check your result in part (a) by comparing your partial sums with your calculator's value for e.
19. (a) Calculate $\sin 1$ (radian) by substituting $x = 1$ into the appropriate polynomials $P_n(x)$ for increasing values of n until you see no change in the sixth decimal place.
(b) Check your result in part (a) by comparing your partial sums with your calculator's value for $\sin 1$.
20. (a) Calculate $\cos 1$ (radian) by substituting $x = 1$ into the appropriate polynomials $P_n(x)$ for increasing values of n until you see no change in the sixth decimal place.
(b) Check your result in part (a) by comparing your partial sums with your calculator's value for $\cos 1$.
21. (a) Here, courtesy of a computer algebra system, are a well-known function and its first five derivatives:

$$f(x) = \tan x$$

$$f'(x) = 1 + \tan^2 x$$

$$f''(x) = 2\tan x + 2\tan^3 x$$

$$f^{(3)}(x) = 2 + 8\tan^2 x + 6\tan^4 x$$

$$f^{(4)}(x) = 16\tan x + 40\tan^3 x + 24\tan^5 x$$

$$f^{(5)}(x) = 16 + 136\tan^2 x + 240\tan^4 x + 120\tan^6 x$$

Find the fifth-degree Taylor approximation $P_5(x)$ to $f(x)$.
(b) Use your graphing tool to find an interval over which the graph of $P_5(x)$ appears to fit the graph of $f(x)$ well.
(c) Using an appropriate y-scale, graph $f(x) - P_5(x)$ on the interval you determined in part (b), and find the maximum error in the approximation on that interval.

22. With your graphing calculator, graph $y_1 = e^x$ and $y_2 = 1 + x$ for $-2 \leq x \leq 2$ and $-5 \leq y \leq 5$. One term at a time, change y_2 to higher degree Taylor polynomial approximations for e^x, and regraph. That is, first add $x^2/2$, then $x^3/6$, and so on. As y_2 begins to look very close to e^x, enlarge the window so you can see clearly how and where the two functions differ. Continue up to at least degree 5. Describe in your own words the characteristics that distinguish all the Taylor polynomials from e^x.

23. As in the preceding exercise, start with $y_1 = e^x$ and $y_2 = 1 + x$. Deselect both these functions, and enter $y_3 = y_1 - y_2$, $y_4 = x^2/2$ (with both y_3 and y_4 selected for graphing). We have seen that y_3 is the error in the approximation of y_1 by y_2; y_4 is the first term left out of the approximation. Set the window at $-1 \leq x \leq 1$ and $-1 \leq y \leq 1$ with both x-scale and y-scale equal to 0.1. If your calculator has a grid option, turn it on. Watch carefully as the error curve y_3 is drawn and as the next term y_4 is drawn.
(a) What is the approximate shape of the error curve?
(b) For what values of x is the error smaller (in absolute value) than the next term?
(c) For what values of x is the error larger (in absolute value) than the next term?
(d) What is the largest error (in absolute value) on the interval $[-1, 1]$, and where does this largest error occur?
(e) In what interval on the x-axis is $|\text{error}|$ no larger than the value of y-scale?
(f) Change y_2 by adding the next term, $x^2/2$, to it. Change y_4 to the new first term left out, $x^3/6$. Change the y-range to $[-0.25, 0.25]$ with a y-scale of 0.025. Answer the same questions as in parts (a)–(e). Record your numerical results in Table 11.1.
(g) Continue adding terms to y_2 and changing the next term y_4, answering the questions of parts (a)–(e), and filling in each line of Table 11.1. Use the y-ranges and y-scales indicated in the table.

Table 11.1 Errors in polynomial approximations to e^x

Degree	*Max* y	y-*scale*	*Max* $\lvert error \rvert$ *on* $[-1,1]$	*Interval for* $\lvert error \rvert \leq y$-*scale*
1	1.0	0.1	0.72	$[-0.47, 0.41]$
2	0.25	0.025		
3	0.1	0.01		
4	0.01	0.001		
5	0.0025	0.00025		
6	0.00025	0.000025		

In Exercises 24–30, give a reason why the indicated function is not a polynomial function.

24. The exponential function
25. The sine function
26. The cosine function
27. The natural logarithm function
28. The inverse tangent function
29. The normal probability density function
30. The error function

31. (a) Find the Taylor polynomials of degrees 0, 1, 2, 3, and 4 that approximate the function $f(x) = (1+x)^4$.
    (b) However you answered part (a), there is at least one more way to arrive at the same answer. What's a second way to answer part (a)?

32. (a) Find a fourth-degree polynomial approximation for the function

$$f(x) = \frac{\sin x}{x}.$$

(*Hint*: Don't calculate any derivatives!)
    (b) Graph both $f(x)$ and your approximating polynomial.
    (c) Use your approximating polynomial to estimate

$$\int_0^1 \frac{\sin x}{x}\, dx.$$

    (d) Use the integral key on your calculator to evaluate the integral in part (c). How accurate was your estimate?

33. (a) Find the first six Taylor polynomial approximations to $f(x) = e^{-x}$.
    (b) In the first six Taylor polynomials for $g(x) = e^x$, substitute $-x$ for x. How are the resulting polynomials related to the ones you calculated in part (a)?
    (c) Graph $f(x)$ and all six of the approximating polynomials in a single graphing window.

In the exercises for Chapter 6 we introduced the hyperbolic sine and hyperbolic cosine functions:

$$\sinh x = \frac{1}{2}(e^x - e^{-x})$$

and

$$\cosh x = \frac{1}{2}(e^x + e^{-x}).$$

Recall that each of these functions is the derivative of the other. Exercises 34 and 35 are about these functions.

34. (a) Find the Taylor polynomial of degree 6 that approximates the function $f(x) = \sinh x$.
    (b) However you answered part (a), there is at least one more way to arrive at the same answer. What's a second way to find the approximating polynomial? (*Hint*: See Exercise 33.)
    (c) Graph both $f(x)$ and your approximating polynomial.
    (d) Find the maximum error in the approximation on the interval $[-1, 1]$.

35. (a) Find the Taylor polynomial of degree 6 that approximates the function $g(x) = \cosh x$.

(b) However you answered part (a), there is at least one more way to arrive at the same answer. What's a second way to find the approximating polynomial?

(c) Graph both $g(x)$ and your approximating polynomial.

(d) Find the maximum error in the approximation on the interval $[-1, 1]$.

11.2 Taylor Series

In the preceding section we saw that (1) some familiar important functions can be approximated by polynomial functions and (2) the coefficients of these polynomials can be calculated from derivatives of the original functions. We found that these approximations can be quite accurate close to the reference point $x = 0$, but the accuracy deteriorates as we move away from that point. On the other hand, accuracy appears to improve as the degree of the approximating polynomial increases. This suggests that we may want to consider what happens as the degree becomes quite large—perhaps even to consider limiting values as the degree goes to infinity. That will be the subject of this section.

We note in passing two features of Taylor polynomials that emerge from our study thus far:

- We don't waste anything by computing low-degree polynomials on the way to finding higher-degree polynomials. All the low-degree terms are also terms in the higher-degree approximations. Indeed, each P_n can be computed from P_{n-1} by adding one more term. And that next term can be calculated by an easily automated process: Take one more derivative, evaluate at $x = 0$, divide by $n!$, and multiply by x^n.

- On the other hand, this direct calculation of additional terms depends on symbolic calculation of higher-order derivatives. So far we have done that only for functions whose higher-order derivatives are easy to calculate—but most functions aren't like that. Thus, if we are going to find approximations of high degree, we need to be on the lookout for easier ways to find the coefficients.

What Happens as the Degree Becomes Large?

If $P_n(x)$ is the nth degree Taylor approximation to the function e^x, then for each real number a, the numbers $P_n(a)$ approach e^a as the degree n increases. Similarly, if $P_n(x)$ is the nth degree Taylor approximation to the function $\sin x$, then for each real number a, the numbers $P_n(a)$ approach $\sin a$ as the degree n increases. To get a sense of the meanings of these statements, we will carry out an exploration on Figures 11.1 and 11.3, which are repeated here for your convenience (Figures 11.5 and 11.6).

Figure 11.5 Exponential function and Taylor polynomials up to degree 4

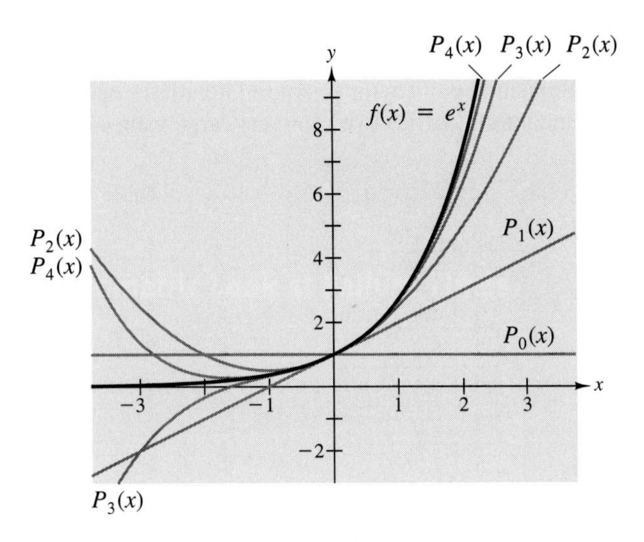

Figure 11.6 Sine function and Taylor polynomials up to degree 5

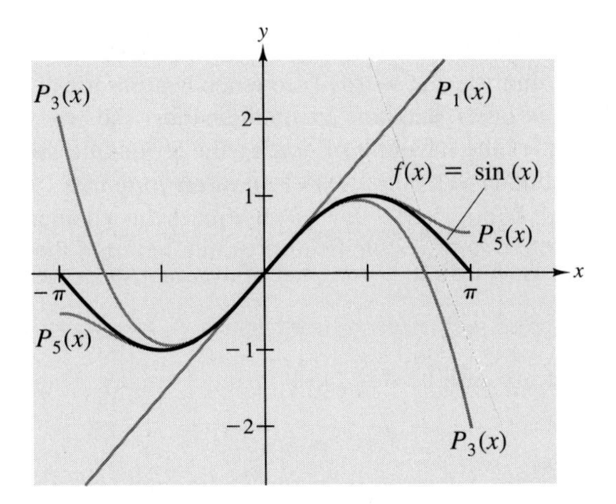

Exploration Activity 1

(a) In Figure 11.5 consider the successive values of the Taylor polynomials along the vertical line at $x = 2$ — that is, set a equal to 2. Mark the successive values P_0, $P_1(2)$, $P_2(2)$, $P_3(2)$, $P_4(2)$ along that line. Then locate the point at $(2, e^2)$ on the exponential curve.

(b) Make a table of the approximating y-coordinates as a function of the degree n, and extend the table to $n = 8$. Compare these values with the numerical value of e^2.

(c) Carry out the steps of parts (a) and (b) for the sine function in Figure 11.6 at $a = -\pi/2$.

The actual numbers being approximated graphically and numerically in Exploration Activity 1 are $e^2 = 7.389056$ and $\sin(-\pi/2) = -1$. The numerical values are shown in Tables 11.2 and 11.3. In these two examples, our values of a are not terribly far from zero, and the approach to $f(a)$ is fairly rapid in each case. For reasons we will see later, the statement about Taylor polynomial values approaching values of the exponential and sine functions is correct even for very large values of a.

Table 11.2 Approaching e^2

n	$P_n(2)$
0	1.0
1	3.0
2	5.0
3	6.3333
4	7.0
5	7.2667
6	7.3556
7	7.3810
8	7.3873

Table 11.3 Approaching $\sin(-\pi/2)$

n	$P_n(-\pi/2)$
0	0.0
1	−1.5708
2	−1.5708
3	−0.9248
4	−0.9248
5	−1.0045
6	−1.0045
7	−0.9998
8	−0.9998

Sequences and Series

In ordinary English, the words "sequence" and "series" are synonyms. Each means a list of things, one after another. In mathematics, the two words have different meanings: "Series" carries the meaning of *adding* the terms of a sequence. Furthermore, both words are used primarily in the context of *infinitely long* lists.

Definitions A **sequence** is an infinitely long list of numbers or expressions. If

$$b_0, b_1, b_2, \ldots, b_k, \ldots$$

is such a list, we call the sum (with infinitely many terms),

$$b_0 + b_1 + b_2 + \cdots + b_k + \cdots$$

an **infinite series**. Often the adjective is dropped, and we call an infinitely long sum a **series**. As with ordinary addition, the list items b_k are called **terms** of the series. The sum of the first n terms,

$$b_0 + b_1 + b_2 + \cdots + b_n,$$

is called the nth **partial sum** of the series.

Note carefully the use of the three-dot (ellipsis) notation in these definitions. When an ellipsis occurs *between* two similar things, it means the pattern on both ends is continued in the middle, and we are just not bothering to fill in all the intervening items. When it occurs at the *end* of an expression, it means the preceding pattern *continues forever*.

We observe that the partial sums of a series constitute another sequence:

$$b_0,\ b_0+b_1,\ b_0+b_1+b_2,\ldots,\ b_0+b_1+b_2+\cdots+b_n,\ldots.$$

The terms in this sequence are found by ordinary addition of finitely many terms. If this sequence has a limiting value, we take that value to be the sum of the entire series.

Definitions If the sequence of partial sums of a series has a finite limiting value S, we say the series **converges**, and we call S the **sum** of the series. If the sequence of partial sums does not have a finite limiting value, we say the series **diverges**.

Example 1 Find (a) the limit of the sequence of terms, (b) the sequence of partial sums, and (c) the sum of the series

$$1+\frac{1}{2}+\frac{1}{2^2}+\cdots+\frac{1}{2^k}+\cdots.$$

Solution This is a geometric series with term-to-term ratio $\frac{1}{2}$. The limiting value of the terms is

$$\lim_{k\to\infty}\frac{1}{2^k}=0.$$

In Chapter 5 we evaluated the nth partial sum of such a series and found that

$$1+\frac{1}{2}+\frac{1}{2^2}+\cdots+\frac{1}{2^n}=\frac{1-(1/2)^{n+1}}{1-(1/2)}=2-\frac{1}{2^n}.$$

From this we easily find the sum of the series:

$$\begin{aligned}1+\frac{1}{2}+\frac{1}{2^2}+\cdots+\frac{1}{2^k}+\cdots&=\lim_{n\to\infty}\left(1+\frac{1}{2}+\frac{1}{2^2}+\cdots+\frac{1}{2^n}\right)\\&=\lim_{n\to\infty}\left(2-\frac{1}{2^n}\right)=2.\end{aligned}$$

■

This example is not typical: Geometric series are the only series we have seen for which we know how to find all the partial sums by an explicit formula. For most series of interest, finding the sum — or even determining whether there is a finite sum — will not be so easy.

By contrast, finding a limit of the sequence of terms often is easy. In fact, the terms $1,\ 1/2,\ 1/4,\ldots,\ 1/2^k,\ldots$ of the series in Example 1 are just y-coordinates at integer values of x for the function $y=2^{-x}$. We illustrate this in Figure 11.7, which shows both the graph of the function and the points for nonnegative integer values of x up to $x=5$. Thus, if you are not sure of a limiting value for terms given by a formula, you can often determine the limiting value by graphing the function defined by the formula. This figure also illustrates that a sequence actually *is* a special kind of function, namely one whose domain consists of an infinite list of integers.

Figure 11.7 Graphs of the function $1/2^x$ and the sequence $1/2^k$

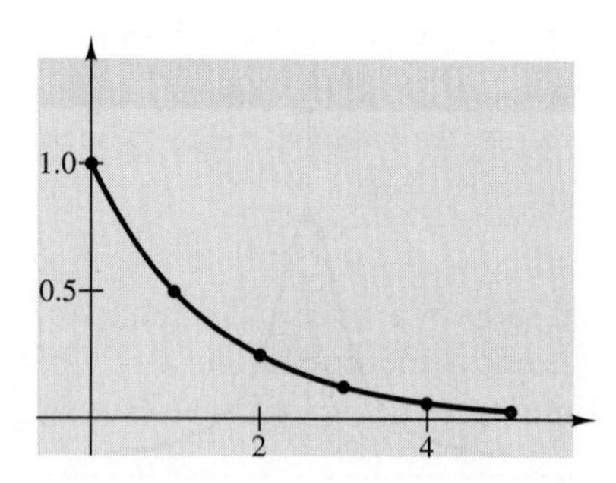

Checkpoint 1

Graph the partial sums of the geometric series

$$1+\frac{1}{2}+\frac{1}{2^2}+\cdots+\frac{1}{2^k}+\cdots.$$

We will often find it convenient, as we have in earlier chapters, to use sigma notation for the sums—both partial and total—that occur in our study of series. Thus, in Example 1, we could write

$$\sum_{k=0}^{n}\frac{1}{2^k}=2-\frac{1}{2^n}$$

for the nth partial sum and

$$\sum_{k=0}^{\infty}\frac{1}{2^k}=2$$

for the total sum. In general, a sum with upper limit ∞ is defined by

$$\sum_{k=0}^{\infty}b_k=\lim_{n\to\infty}\sum_{k=0}^{n}b_k.$$

As usual, whenever you find the sigma notation even a little confusing, we encourage you to write out enough terms explicitly to be sure you understand what the notation represents.

Taylor Series for e^x and $\sin x$

In Exploration Activity 1 you explored the sequence of numbers

$$P_0, P_1(2), P_2(2), P_3(2), P_4(2), \ldots$$

where $P_n(x)$ is the nth Taylor polynomial for e^x. It appeared that this sequence has e^2 as its limit. In fact, Taylor polynomials are partial sums of an infinite series, because each polynomial is the preceding polynomial plus one more term. Specifically, for the exponential function, the terms are the power functions

$$1,\ x,\ \frac{1}{2}x^2,\ \frac{1}{3!}x^3, \ldots,$$

and the partial sums are

$$P_n(x) = 1 + x + \frac{1}{2}x^2 + \cdots + \frac{1}{n!}x^n = \sum_{k=0}^{n} \frac{1}{k!}\,x^k.$$

We summarize the convergence of these partial sums to e^x by writing

$$e^x = 1 + x + \frac{1}{2}\,x^2 + \cdots + \frac{1}{k!}\,x^k + \cdots = \sum_{k=0}^{\infty} \frac{1}{k!}\,x^k.$$

We have certainly not shown yet that this assertion about limiting values is true—that will come later in the chapter. But we have some graphic and numerical evidence that it is true for $x = 2$. We further illustrate convergence of partial sums with the case of $x = 1$, which leads to the following interesting formula for e:

$$e = 1 + 1 + \frac{1}{2} + \cdots + \frac{1}{k!} + \cdots = \sum_{k=0}^{\infty} \frac{1}{k!}.$$

Example 2 Calculate the first ten partial sums of this series representation for e, and show that the resulting sequence appears to be approaching e.

Solution Here are the sums:

$$1 = 1$$

$$1 + 1 = 2$$

$$1 + 1 + \frac{1}{2} = 2.5$$

$$1 + 1 + \frac{1}{2} + \frac{1}{6} = 2.667$$

$$1 + 1 + \frac{1}{2} + \frac{1}{6} + \frac{1}{24} = 2.70833$$

$$1 + 1 + \frac{1}{2} + \frac{1}{6} + \frac{1}{24} + \frac{1}{120} = 2.71667$$

$$1 + 1 + \frac{1}{2} + \frac{1}{6} + \frac{1}{24} + \frac{1}{120} + \frac{1}{720} = 2.718056$$

$$1 + 1 + \frac{1}{2} + \frac{1}{6} + \frac{1}{24} + \frac{1}{120} + \frac{1}{720} + \frac{1}{5040} = 2.7182540$$

$$1 + 1 + \frac{1}{2} + \frac{1}{6} + \frac{1}{24} + \frac{1}{120} + \frac{1}{720} + \frac{1}{5040} + \frac{1}{40320} = 2.71827877$$

$$1 + 1 + \frac{1}{2} + \frac{1}{6} + \frac{1}{24} + \frac{1}{120} + \frac{1}{720} + \frac{1}{5040} + \frac{1}{40320} + \frac{1}{362880} = 2.71828153$$

The error after adding up 10 terms (i.e., the difference between our last partial sum and e) is about 3×10^{-7} — that is, the first six decimal places agree with e. (Check this with your calculator.)

Exploration Activity 2

(a) Interpret the Taylor polynomials for $\sin x$ in the preceding section as partial sums of a series.

(b) Interpret your calculations in Exploration Activity 1(c) as partial sums of a series.

(c) Calculate sin 1 (radian) by substituting $x = 1$ in the series for sin x and adding up partial sums until you see no change in the sixth decimal place. Check your result by comparing your partial sums with your calculator's value for sin 1.

There is a small complication in the Taylor polynomials for the sine function in that all the even-numbered terms are zero. Specifically, as they were originally derived, the terms are

$$0,\ x,\ 0,\ -\frac{1}{3!}\,x^3,\ 0,\ \frac{1}{5!}\,x^5, \ldots .$$

Our record keeping will be simpler if we drop all the zero terms — which contribute nothing to the sum — and renumber the remaining terms by $k = 0, 1, 2, \ldots$. This makes the degree of the kth term $2k + 1$. Thus the $(2n + 1)$th partial sum is

$$P_{2n+1}(x) = x - \frac{1}{3!}\,x^3 + \frac{1}{5!}\,x^5 - \cdots + \frac{(-1)^n}{(2n+1)!}\,x^{2n+1} = \sum_{k=0}^{n} \frac{(-1)^k}{(2k+1)!}\,x^{2k+1}.$$

The corresponding series [the limiting value of $P_{2n+1}(x)$ as $n \to \infty$] is

$$\sin x = x - \frac{1}{6}\,x^3 + \frac{1}{120}\,x^5 - \cdots + \frac{(-1)^k}{(2k+1)!}\,x^{2k+1} + \cdots.$$

In particular, for $x = -\pi/2$, the series interpretation is the improbable (but true) formula

$$-1 = -\frac{\pi}{2} + \frac{\pi^3}{6 \cdot 2^3} - \frac{\pi^5}{120 \cdot 2^5} + \cdots + \frac{(-1)^{k+1}\pi^{2k+1}}{(2k+1)! \cdot 2^{2k+1}} + \cdots.$$

The approximations in Exploration Activity 1(c) are partial sums of the series on the right. For $x = 1$, we find consecutive sums:

$$1 = 1$$

$$1 - \frac{1}{6} = 0.8333333$$

$$1 - \frac{1}{6} + \frac{1}{120} = 0.8416667$$

$$1 - \frac{1}{6} + \frac{1}{120} - \frac{1}{5040} = 0.8414683$$

$$1 - \frac{1}{6} + \frac{1}{120} - \frac{1}{5040} + \frac{1}{362880} = 0.8414710$$

$$1 - \frac{1}{6} + \frac{1}{120} - \frac{1}{5040} + \frac{1}{362880} - \frac{1}{39916800} = 0.8414710$$

The last value is the correct value of sin 1 to six decimal places.

It should come as no surprise that the series whose partial sums are Taylor polynomials are called Taylor series.

Definition The **Taylor series** for a function $f(x)$ **at the reference point** $x = 0$ is the series

$$f(0) + f'(0)\, x + \frac{f''(0)}{2}\, x^2 + \cdots + \frac{f^{(k)}(0)}{k!}\, x^k + \cdots.$$

In sigma notation, we may write the defining formula as

$$\sum_{k=0}^{\infty} \frac{f^{(k)}(0)}{k!}\, x^k.$$

Here are the formulas for the Taylor series we know so far. For completeness sake, we include a formula for the cosine function — see Exercise 1.

$$e^x = 1 + x + \frac{1}{2}\, x^2 + \cdots + \frac{1}{k!}\, x^k + \cdots = \sum_{k=0}^{\infty} \frac{1}{k!}\, x^k$$

$$\sin x = x - \frac{1}{6}\, x^3 + \frac{1}{120}\, x^5 - \cdots + \frac{(-1)^k}{(2k+1)!}\, x^{2k+1} + \cdots = \sum_{k=0}^{\infty} \frac{(-1)^k}{(2k+1)!}\, x^{2k+1}$$

$$\cos x = 1 - \frac{1}{2}\, x^2 + \frac{1}{24}\, x^4 - \cdots + \frac{(-1)^k}{(2k)!}\, x^{2k} + \cdots = \sum_{k=0}^{\infty} \frac{(-1)^k}{(2k)!}\, x^{2k}$$

Example 3 Use the Taylor series for e^x to explain why, for any positive integer m,

$$\lim_{x \to \infty} \frac{x^m}{e^x} = 0.$$

That is, show that the exponential function grows much faster than any power function. [This is the result we needed to complete the assertion in Chapter 10 that the values of a gamma density function $f(t)$ approach zero as t gets large.]

Solution Since the terms in the exponential series are all positive when x is positive, e^x is bigger than every term of the series. In particular,

$$e^x > \frac{1}{(m+1)!}\, x^{m+1}.$$

Thus

$$\frac{x^m}{e^x} < \frac{x^m}{x^{m+1}/(m+1)!} = \frac{(m+1)!}{x}.$$

No matter how big $(m+1)!$ is, it's constant, and x will eventually be bigger. Thus

$$0 \le \lim_{x \to \infty} \frac{x^m}{e^x} \le \lim_{x \to \infty} \frac{(m+1)!}{x} = 0.$$

Hence the limit must be 0. ■

In Figure 11.8 we illustrate the significance of Example 3 for the case of $m = 50$. Take careful note of the scales in the figure. The function x^{50}/e^x peaks at about $x = 50$ with a value of about 1.7×10^{63}. Pretty big! If you started to graph this function without knowing what to look for, you might never guess that this is a function that eventually goes to zero. If you try to produce a comparable picture for $m = 100$, your graphing tool will fail. (Don't take our word for it—try it!) Nevertheless, we can solve problems of such enormous magnitude with relatively simple calculations—and powerful concepts.

Figure 11.8 Graph of $y = \dfrac{x^{50}}{e^x}$

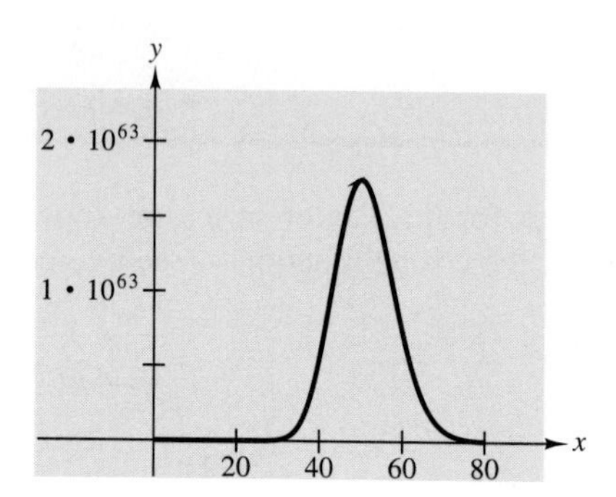

In this section we have formalized the observations of the preceding section about values of polynomial approximations approaching values of nonpolynomial functions as the degree of the approximation is increased. We have introduced the language of infinite series, previously encountered only in the very special case of geometric series (Chapter 5). In particular, the concept of Taylor polynomial leads in a natural way to Taylor series as a "polynomial of infinite degree." This is a theme that we will develop in subsequent sections.

We find that the question of convergence of a sum with infinitely many terms is intimately related to convergence of a sequence, namely, the sequence of partial sums. The idea of sequence convergence has now appeared many times—in Euler's Method, Newton's Method, evolution of prices (Chapter 5), discrete logistic growth (Chapter 7), approximating sums for definite integrals (Chapter 9), and so on.

As a side benefit of having introduced the Taylor series for the exponential function, we have now been able to show that this function grows faster than any power function.

Answer to Checkpoint

1. We take advantage of the fact that the partial sums have an explicit formula. We show the graphs of both $2 - 2^{-x}$ and $2 - 2^{-n}$ for integer values of n up to $n = 5$. We also show the limiting value at $y = 2$.

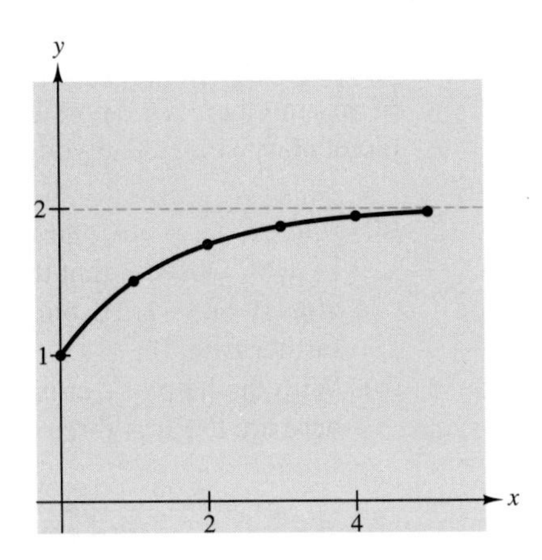

Exercises 11.2

1. Derive the Taylor series for $\cos x$.

2. (a) Calculate $\cos 1$ (radian) by substituting $x = 1$ in your answer to Exercise 1 and adding up partial sums until you see no change in the sixth decimal place.
   (b) Check your result in part (a) by comparing your partial sums with your calculator's value for $\cos 1$.

3. If $f(x) = \sin x$, what is the coefficient of x^{13} in the Taylor series for $f(x)$?

4. For any function $f(x)$ with infinitely many derivatives, what is the coefficient of x^{13} in the Taylor series for $f(x)$?

5. (a) Find the Taylor series for the function
$$f(x) = \frac{\sin x}{x}.$$
(*Hint*: Don't calculate any derivatives!)
   (b) Use your Taylor series to estimate
$$\int_0^1 \frac{\sin x}{x}\, dx.$$
   (c) Use the integral key on your calculator to evaluate the integral in part (b). How accurate was your estimate?

6. (a) Find the Taylor series for $f(x) = e^{-x}$.
   (b) In the Taylor series for $g(x) = e^x$, substitute $-x$ for x. How is the resulting series related to the one you calculated in part (a)?

In the exercises for Chapter 6 we introduced the hyperbolic sine and hyperbolic cosine functions:
$$\sinh x = \frac{1}{2}(e^x - e^{-x})$$
and
$$\cosh x = \frac{1}{2}(e^x + e^{-x}).$$

Recall that each of these functions is the derivative of the other. Exercises 7 and 8 are about these functions.

7. (a) Find the Taylor series for the function $f(x) = \sinh x$.
   (b) However you answered part (a), there is at least one more way to arrive at the same answer. What's a second way to find the Taylor series? (*Hint*: See Exercise 6.)

8. (a) Find the Taylor series for the function $g(x) = \cosh x$.
   (b) However you answered part (a), there is at least one more way to arrive at the same answer. What's a second way to find the Taylor series?

9. What does the graph of an exponential function look like on a log-log graph? What does the graph of a power function look like on a semilog graph? How are these observations related to Example 3?

10. Show that 2^x grows faster than any power function. (*Hint*: Write 2^x as an exponential with base e.)

11. Graph $f(x) = x^{30}/2^x$ on a scale that shows clearly the limiting behavior as $x \to \infty$.

12. In Chapter 10 we left it to you to finish the explanation of the following property of a gamma density function $f(t)$ with positive exponent a:

    $f(t)$ decreases to 0 as t increases to ∞.

    We took the explanation as far as writing $f(t)$ in the form

$$f(t) = \frac{c\, t^a}{(e^t)^{1/b}} = c\left(\frac{t^{ab}}{e^t}\right)^{1/b}.$$

    Use the result of Example 3 to show that $f(t)$ approaches zero as t gets large. The product ab is not necessarily an integer. Think about replacing t^{ab} by t^m, where m is any integer larger than ab.

13. (a) Use your calculator to find the value of $\sin 100$.
    (b) Find an angle between $-\pi/2$ and $\pi/2$ whose sine is the same as $\sin 100$.
    (c) Use your small angle from part (b) in the series for $\sin x$ to find $\sin 100$.
    (d) In general, given a number x, how would you find a number α between $-\pi/2$ and $\pi/2$ such that $\sin \alpha = \sin x$? Describe a procedure that could be automatic enough to be programmed into a calculator and that could evaluate the sine of any number.

14. The values of $\cos x$ are not all found at numbers x between $-\pi/2$ and $\pi/2$. Why not? Describe an effective automatic procedure for evaluating cosine of any number. (One possibility: Use sine of a complementary angle. Can you think of others?)

15. (a) Graph $f(x) = \sin x$ and $g(x) = x\sqrt[3]{\cos x}$ on $[-3, 3]$. Observe that these functions agree closely on $[-1, 1]$, but less so when x is farther from 0.
    (b) With the help of a computer algebra system, here are the first three derivatives of g:

$$g'(x) = \frac{3\cos x - x\sin x}{3(\cos x)^{2/3}}$$

$$g''(x) = -\frac{3x\cos^2 x + 6\sin x\cos x + 2x\sin^2 x}{9(\cos x)^{5/3}}$$

$$g'''(x) = -\frac{27\cos^3 x + 9x\sin x\cos^2 x}{27(\cos x)^{8/3}} + \frac{18\sin^2 x\cos x + 10x\sin^3 x}{27(\cos x)^{8/3}}$$

    Find the Taylor polynomials for $g(x)$ through degree 3.
    (c) Use Taylor polynomials for both functions to explain your graphical observations in part (a).[1]

1. Adapted from *Calculus Problems for a New Century*, edited by Robert Fraga, MAA Notes No. 28, 1993.

11.3 More Taylor Polynomials and Series

Geometric Sums and Series

You may not have noticed it at the time, but in Chapter 5 we investigated the approximation of one particular function by polynomials. We were not trying to solve an approximation problem at the time, so there is no particular reason why you should have noticed this connection. We begin this section by reviewing—in more detail than in the preceding section—what we know about geometric sums.

If β is any number other than 1, and n is any nonnegative integer, then

$$1 + \beta + \beta^2 + \cdots + \beta^n = \frac{1 - \beta^{n+1}}{1 - \beta}.$$

If $|\beta| < 1$, then, in the notation of this chapter,

$$\lim_{n\to\infty} \beta^{n+1} = 0,$$

so

$$\lim_{n\to\infty} (1 + \beta + \beta^2 + \cdots + \beta^n) = \frac{1}{1 - \beta}.$$

This says that the function

$$f(x) = \frac{1}{1 - x}$$

is approached by the geometric polynomials

$$1 + x + x^2 + \cdots + x^n$$

as n approaches infinity, at least for values of x with $|x| < 1$. In the notation of infinite series,

$$\frac{1}{1 - x} = 1 + x + x^2 + \cdots = \sum_{k=0}^{\infty} x^k \quad \text{for } |x| < 1.$$

Representation of this function by approximating polynomials is not especially important because its values can be computed by simple arithmetic from the formula $f(x) = 1/(1 - x)$ and without any artificial restriction on the size of x. Indeed, when the subject of geometric series first came up, our *problem* was the infinitely long sum, and its *solution* was to find the simple quotient expression—not the other way around. However, we will soon see that this function is intimately related to other important functions, especially the natural logarithm and inverse tangent functions. The relationships carry over to the approximating polynomials, and that enables us to find polynomial approximations to these other functions from the geometric polynomials, $1 + x + x^2 + \cdots + x^n$.

Exploration Activity 1

(a) Find the Taylor polynomials of degrees 0 through 5 for the function $f(x) = 1/(1-x)$. How do these polynomials compare to the geometric polynomials?

(b) In a single graphing window, graph $f(x)$ and the six Taylor polynomials from part (a).

(c) Evaluate the Taylor polynomials in part (a) at each of the following values of x: $0.9,\ -0.9,\ 1,\ -1,\ 1.1,\ -1.1$. For each choice of x, describe in words how the polynomial value at x compares with $f(x)$.

If we write $f(x) = (1-x)^{-1}$, we can quickly find derivatives from the Power Rule and Chain Rule:

$$\begin{aligned} f'(x) &= (-1)(1-x)^{-2}(-1) = (1-x)^{-2}, \\ f''(x) &= (-2)(1-x)^{-3}(-1) = 2(1-x)^{-3}, \\ f'''(x) &= (-3)(1-x)^{-4}(-1) = 3!\,(1-x)^{-4}, \end{aligned}$$

and so on. In general, we find that

$$f^{(k)}(x) = k!\,(1-x)^{-(k+1)}.$$

When we substitute $x = 0$, we find

$$f^{(k)}(0) = k!\,.$$

Since $c_k = f^{(k)}(0)/k!$, it follows that every coefficient in every Taylor polynomial is 1, i.e., the first six polynomials are

$$\begin{aligned} P_0 &= 1, \\ P_1(x) &= 1 + x, \\ P_2(x) &= 1 + x + x^2, \\ P_3(x) &= 1 + x + x^2 + x^3, \\ P_4(x) &= 1 + x + x^2 + x^3 + x^4, \\ P_5(x) &= 1 + x + x^2 + x^3 + x^4 + x^5. \end{aligned}$$

These are precisely the geometric polynomials. We show the graphs of these polynomials, along with the graph of $f(x)$, in Figure 11.9.

The graphs in Figure 11.9 suggest that the Taylor approximations are not very good near the endpoints of the interval $[-1, 1]$, and the polynomials are not approximations at all outside that interval. In Table 11.4 we show the requested values of the various functions at numbers x near these endpoints. The numbers in the 0.9 and -0.9 rows are consistent with the assertion that

$$\lim_{n\to\infty} P_n(\pm 0.9) = f(\pm 0.9),$$

but they are not very compelling evidence. The convergence is evidently very slow. The remaining rows show that, for $x \geq 1$ or $x \leq -1$, the polynomial values have nothing to do with values of $f(x)$.

Figure 11.9 Graphs of $f(x) = 1/(1-x)$ and Taylor polynomials to degree 5

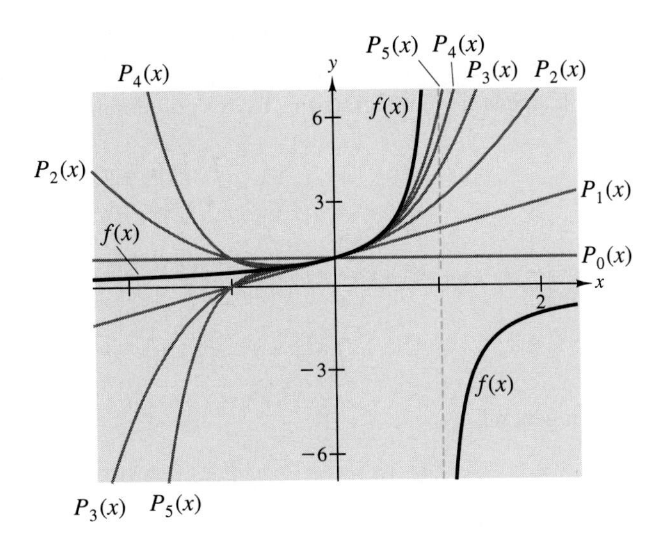

Table 11.4 Values of $f(x) = 1/(1-x)$ and its Taylor polynomials

x	P_0	$P_1(x)$	$P_2(x)$	$P_3(x)$	$P_4(x)$	$P_5(x)$	$f(x)$
0.9	1	1.9	2.71	3.44	4.10	4.69	10.0
−0.9	1	0.1	1.0	0.27	0.93	0.34	0.53
1	1	2	3	4	5	6	(no value)
−1	1	0	1	0	1	0	0.5
1.1	1	2.1	3.31	4.64	6.11	7.72	−10.0
−1.1	1	−0.1	1.11	−0.22	1.24	−0.37	0.48

For the function $f(x) = 1/(1-x)$, we find that the Taylor polynomials do not converge to the function for all x, but only for a limited range of x's: $-1 < x < 1$. In fact, for $x \geq 1$, we see that the values of $1 + x + x^2 + \cdots + x^n$ grow without bound as n becomes large. Similarly, for $x < -1$, the successive polynomial values alternate in sign and eventually become large in absolute value. And for the special value $x = -1$, the values alternate between 0 and 1, never settling down to a limiting value.

Taylor Polynomials and Series for ln$(1 + x)$

We can use our knowledge of the Taylor polynomials for $1/(1-x)$ to obtain the Taylor polynomials for other functions. For example, since

$$\frac{1}{1+t} = \frac{1}{1-(-t)},$$

we can substitute $x = -t$ in the geometric series to find

$$\frac{1}{1+t} = 1 - t + t^2 - t^3 + \cdots, \qquad \text{for } |t| < 1.$$

We know from the more important part of the Fundamental Theorem that

$$\ln(1+x) = \int_0^x \frac{1}{1+t}\,dt.$$

What happens if we integrate the Taylor polynomials for $f(t) = 1/(1+t)$? We find

$$\int_0^x 1\,dt = x,$$

$$\int_0^x (1-t)\,dt = x - \frac{x^2}{2},$$

$$\int_0^x (1-t+t^2\,)\,dt = x - \frac{x^2}{2} + \frac{x^3}{3},$$

and, in general,

$$\int_0^x \left[1-t+t^2-t^3+\cdots+(-1)^n\,t^n\right] dt = x - \frac{x^2}{2} + \frac{x^3}{3} - \frac{x^4}{4} + \cdots + (-1)^n\,\frac{x^{n+1}}{n+1}.$$

Checkpoint 1

Verify that the Taylor polynomials of degrees $1, 2, 3, \ldots, n+1$ for $\ln(1+x)$ are exactly the polynomials resulting from the integrations above.

You will probably not be surprised to learn that

$$\ln(1+x) = x - \frac{x^2}{2} + \frac{x^3}{3} - \frac{x^4}{4} + \cdots = \sum_{k=0}^{\infty} (-1)^k\,\frac{x^{k+1}}{k+1}, \qquad \text{if } |x| < 1.$$

On the other hand, if $|x| > 1$, the polynomials in Checkpoint 1 do not approximate $\ln(1+x)$. We can see this in Figure 11.10, which shows $f(x) = \ln(1+x)$ and its first seven approximating polynomials on the interval $[-2, 2]$. As x approaches -1 from the right, $f(x)$ runs away to $-\infty$. The successive polynomials $P_n(x)$ all continue across the line $x = -1$, where there are no values of f to approximate. As x approaches 1 from the left, we see the successive polynomials getting closer to f at each x less than 1. Then, as the curves cross $x = 1$, the higher degree polynomials run away from f faster than the lower degree ones.

Figure 11.10 $\ln(1+x)$ and polynomial approximations up to degree 7

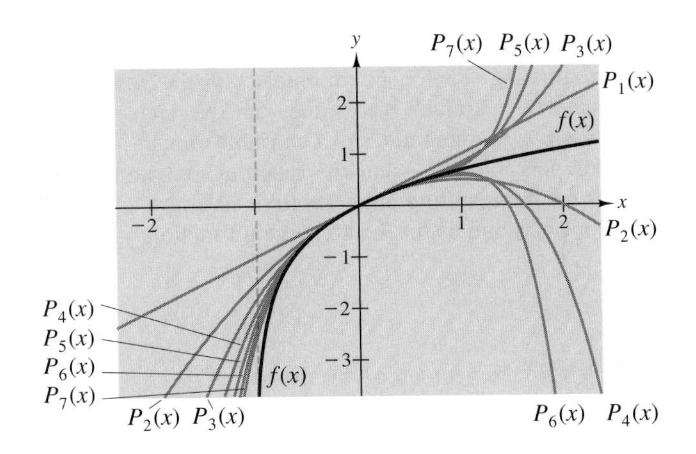

In Figure 11.11 we show the errors in the polynomial approximations plotted in Figure 11.10, again on an exaggerated scale. These low-degree polynomials have errors close to zero (i.e., fit well) only within about $\frac{1}{2}$ unit of the origin. To get a good fit closer to $x = 1$ or $x = -1$, much higher degree polynomials are required.

Checkpoint 2

(a) As you did in Checkpoint 1 in Section 11.1, use our legend to identify the individual curves in Figure 11.10. Then trace each curve in a different color.

(b) As you did in Checkpoint 4 in Section 11.1, use our legend to identify the individual curves in Figure 11.11. Then trace each error curve in a color that matches the corresponding curve in part (a).

Figure 11.11 Errors in polynomial approximations of $\ln(1+x)$

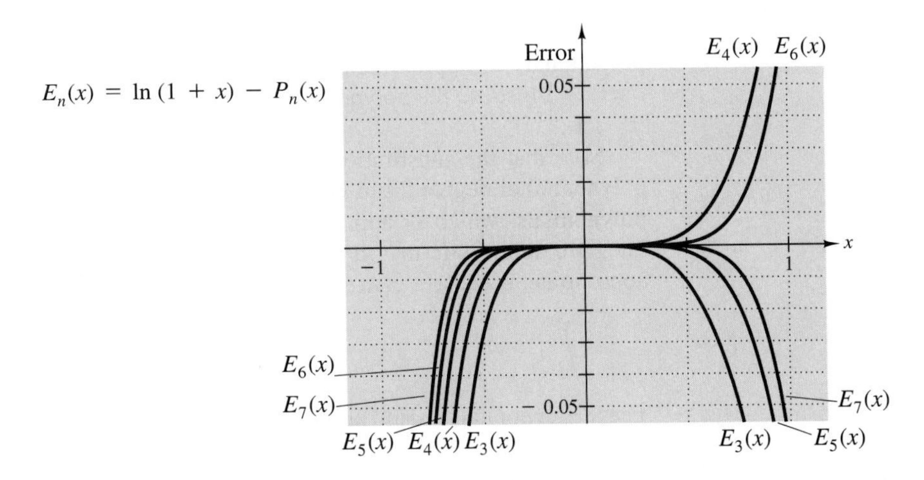

Taylor Polynomials and Series for $\arctan x$

We have seen that it is permissible to substitute for the independent variable in a function and in its approximating Taylor polynomials, and the resulting polynomials will be approximating Taylor polynomials for the resulting function—as long as the new independent variable has values in the right range. We also have seen that it is permissible to integrate (to a variable upper limit) a function and its approximating Taylor polynomials—and the resulting polynomials are approximating Taylor polynomials for the resulting antiderivative. We now use the same ideas to find approximating Taylor polynomials for the arctangent function—an antiderivative of $1/(1+t^2)$.

Exploration Activity 2

(a) Substitute $x = -t^2$ in the geometric polynomials

$$1 + x + x^2 + \cdots + x^n$$

that approximate $1/(1-x)$ to find approximating polynomials for $1/(1+t^2)$.

(b) For what values of t would you expect these polynomials to converge to $1/(1+t^2)$ as the degree $2n$ becomes large?

(c) Integrate the typical polynomial of degree $2n$ (from 0 to x) to find a polynomial of degree $2n+1$ that should approximate $\arctan x$.

First we observe that

$$\frac{1}{1-(-t^2)} = \frac{1}{1+t^2},$$

so the substitution in $1/(1-x)$ produces the right function. The same substitution in the geometric polynomial of degree n gives

$$1 - t^2 + t^4 - \cdots + (-1)^n t^{2n}.$$

For values of t between -1 and 1, $-t^2$ is also in this interval, so we would expect the polynomials to approximate $1/(1+t^2)$ if $|t| < 1$. On the other hand, if $|t| \geq 1$, then $|-t^2| \geq 1$ also, so the polynomials cannot be expected to approximate outside the interval $|t| < 1$.

Note that the substitution turns a polynomial of degree n (in the variable x) into a polynomial of degree $2n$ in the variable t. Indeed, only even-degree terms appear in these polynomials, which is appropriate, because $1/(1+t^2)$ is an even function. When we integrate term-by-term, we get polynomials with only odd-degree terms—which is appropriate, because $\arctan t$ is an odd function:

$$\int_0^x [1 - t^2 + t^4 - \cdots + (-1)^n t^{2n}] = x - \frac{1}{3}x^3 + \frac{1}{5}x^5 - \cdots + \frac{(-1)^n}{2n+1}x^{2n+1}.$$

These polynomials should approximate $\arctan x$.

Example 1 Verify that the Taylor coefficients of degrees 0, 1, and 2 match the coefficients of the polynomials computed in Exploration Activity 2.

Solution For $f(x) = \arctan x$, we have

$$f'(x) = \frac{1}{1+x^2}$$

and

$$f''(x) = \frac{-2x}{(1+x^2)^2},$$

so $f(0) = 0$, $f'(0) = 1$, and $f''(0) = 0$. Thus the coefficients of the first three terms in a Taylor polynomial for $\arctan x$ are $0, 1, 0$. That means the constant term and the quadratic term are both 0, and the linear term is x—exactly what we saw in Exploration Activity 2. ■

Checkpoint 3

Given that

$$\frac{d^3}{dx^3} \arctan x = \frac{2(3x^2 - 1)}{(x^2+1)^3},$$

verify that the Taylor coefficient of x^3 also matches the coefficient you calculated in Exploration Activity 2.

Checkpoint 3 suggests that it would be difficult to calculate more than the first few Taylor coefficients for $\arctan x$ by direct evaluation of $f^{(k)}(0)$, because the successive derivatives get more and more complicated—and only the odd-numbered ones contribute new information. However, Exploration Activity 2 shows that it is quite easy to find all of the Taylor polynomials for the arctangent function by substitution into and integration of a known set of Taylor polynomials. In a laboratory project associated with this chapter you may have an opportunity to generate, plot, and study these polynomial approximations to the arctangent. On the basis of the exercises above and the laboratory project, you should be prepared to believe that

$$\arctan x = x - \frac{1}{3}x^3 + \frac{1}{5}x^5 - \cdots = \sum_{k=0}^{\infty} (-1)^k \frac{x^{2k+1}}{2k+1}, \quad \text{if } |x| < 1.$$

In this section we have used known formulas for geometric sums to derive Taylor polynomials and Taylor series for the important functions $\ln(1+x)$ and $\arctan x$. Along the way we have made or confirmed the following discoveries:

- If $P_n(x)$ is a Taylor polynomial for $f(x)$, then we can substitute a power function $x = p(t)$ in both formulas, and $P_n(p(t))$ will be a Taylor polynomial for $f(p(t))$. [The degree of this new Taylor polynomial will not be n unless $p(t)$ is linear.]
- If $P_n(t)$ is a Taylor polynomial of degree n for $f(t)$, then

$$\int_0^x P_n(t)\,dt$$

is a Taylor polynomial of degree $n+1$ for

$$\int_0^x f(t)\,dt.$$

- Substitution and integration can be used to find Taylor polynomials that would be very difficult to find by using the derivative formula for Taylor coefficients.
- Some Taylor series have limited intervals of convergence. Outside these intervals, the Taylor polynomials do not approximate the function from which they were derived, and therefore the corresponding Taylor series does not converge to that function.
- Taylor series found by substitution and integration inherit their intervals of convergence in a natural way. (Note that we have not yet settled all the questions that might arise about convergence at endpoints of these intervals.)

Answers to Checkpoints

1. For $f(x) = \ln(1+x)$, we have

$$f'(x) = (1+x)^{-1}, \quad f''(x) = -(1+x)^{-2}, \quad f^{(3)}(x) = 2(1+x)^{-3}, \quad f^{(4)}(x) = -3!\,(1+x)^{-4},$$

and so on. When we substitute $x = 0$, we find

$$f(0) = 0, \quad f'(0) = 1, \quad f''(0) = -1, \quad f^{(3)}(0) = 2, \quad f^{(4)}(0) = -3!\,,$$

and so on. In general, after the 0th entry,

$$f^{(k)}(0) = (-1)^{k+1}(k-1)!\,.$$

When we divide by $k!$ we find that

$$c_k = \frac{(-1)^{k+1}}{k}.$$

These coefficients give the same terms in the polynomials as those we computed by integration.

3. $f^{(3)}(0) = -2$, so $c_3 = -2/3! = -1/3$, which matches the coefficient calculated in Exploration Activity 2.

Exercises 11.3

Evaluate each of the following expressions. Exact answers are preferable to decimal approximations.

1. $\displaystyle\sum_{k=0}^{\infty} \left(\frac{2}{3}\right)^k$

2. $\displaystyle\sum_{k=0}^{\infty} (0.9)^k$

3. $\displaystyle\int_0^{\infty} \frac{1}{1+t^2}\,dt$

4. $\displaystyle\int_0^{\infty} \frac{1}{(1+t^2)^2}\,dt$

5. $\displaystyle\int_3^{\infty} x\,e^{-2x}\,dx$

6. $1 + 3 + \dfrac{3^2}{2} + \dfrac{3^3}{3!} + \dfrac{3^4}{4!} + \cdots$

7. $\displaystyle\int_1^{\infty} \frac{1}{x\sqrt{1+x^2}}\,dx$

8. $1 - 3 + \dfrac{3^2}{2} - \dfrac{3^3}{3!} + \dfrac{3^4}{4!} - \cdots$

9. (a) Find the Taylor polynomials of degrees 0 through 5 for the function

$$f(t) = \frac{1}{(1+t)}.$$

How do these polynomials compare to the geometric polynomials

$$1 + x + \cdots + x^n$$

for $x = -t$?

(b) In a single graphing window, graph $f(t)$ and the six Taylor polynomials from part (a).

(c) How do the values of the Taylor polynomials compare with $f(t)$ at $t = 1$? at $t = -1$?

(d) For what range of t do these polynomials converge to $f(t)$?

10. If $f(x) = \ln(1 + x)$, what is the coefficient of x^5 in the seventh-degree polynomial approximation $P_7(x)$ of $f(x)$?

11. What is the seventh-degree Taylor polynomial approximation to $f(x) = \ln(1 + x)$?

12. If $g(x) = \arctan x$, what is the coefficient of x^5 in the seventh-degree polynomial approximation $P_7(x)$ of $g(x)$?

13. What is the fourth-degree Taylor polynomial approximation to $g(x) = \arctan x$?

14. What is the sixth-degree Taylor polynomial approximation to $h(x) = 1/(1 - x)$?

15. What is the sixth-degree Taylor polynomial approximation to $q(x) = 1/(1 + x^2)$?

16. Suppose that 80 cents of every dollar spent in the United States is spent again in the United States. (Economists call this the **multiplier effect**.) If the federal government pumps an extra billion dollars into the economy, how much total spending in the United States occurs as a result?

17. For what value or values of x is

$$x - \frac{1}{2}x^2 + \frac{1}{3}x^3 - \frac{1}{4}x^4 + \cdots = 0.5?$$

18. (a) Calculate ln 1.5 by substituting $x = \frac{1}{2}$ in the formula

$$\ln(1 + x) = \sum_{k=0}^{\infty} (-1)^k \frac{x^{k+1}}{k + 1}$$

and adding up partial sums until you see no change in the third decimal place.

(b) Check your result in part (a) by comparing your partial sums with your calculator's value for ln 1.5.

19. Find the Taylor series for the function $f(x) = (1 + x)^4$.

20. In the preceding exercise you found a Taylor series for a function of the form

$$f(x) = (1 + x)^m,$$

where m happened to be a positive integer. However, the calculation of Taylor coefficients can be carried out even if m is not a positive integer.

(a) Calculate the first three or four derivatives of $f(x)$. When you see the pattern, write down a formula for the kth derivative.

(b) Evaluate each of your derivatives at $x = 0$, and divide by the appropriate factorial to find the coefficients in a Taylor expansion for $f(x)$. The usual notation for the kth coefficient is

$$\binom{m}{k}.$$

The numbers $\binom{m}{k}$ found in part (b) are called **binomial coefficients**, and the Taylor series

$$\sum_{k=0}^{\infty} \binom{m}{k} x^k$$

is called a **binomial series**. [We will see later that, for any number m, the binomial series converges to $f(x) = (1 + x)^m$ if $|x| < 1$.]

(c) Suppose m is a positive integer. Show that the binomial coefficients $\binom{m}{k}$ are zero for $k > m$. (The nonzero coefficients are the binomial coefficients you learned about in high school.) Explain why, in this case, the series is an mth degree polynomial that equals $f(x)$ for all real numbers x.

21. (a) For $f(x) = \sqrt{1 + x}$ (i.e., for $m = \frac{1}{2}$), calculate the first five coefficients in the binomial series defined in Exercise 20. Notice that the first two coefficients are positive, but thereafter they alternate in sign.

(b) Use your five-term polynomial to estimate $\sqrt{1.2}$. Use your calculator to check your answer. How accurate is the polynomial estimate?

(c) Graph $f(x) = \sqrt{1 + x}$ and your polynomial from part (b) in the same window. Describe in your own words what you see.

22. (a) Use the result of Exercise 21 to find the first five terms of the Taylor series for

$$g(x) = \sqrt{1 + x^3}.$$

(b) Use the result of part (a) to estimate

$$\int_0^{1/2} \sqrt{1 + x^3}\, dx.$$

(c) Use the integral key on your calculator to evaluate the integral in part (b). How accurate is your polynomial estimate?

23. Here, courtesy of a computer algebra system, are a function $f(x)$ and its first six derivatives:

$$f(x) = \ln(1 + x^2)$$

$$f'(x) = \frac{2x}{1 + x^2}$$

$$f''(x) = \frac{2(1 - x^2)}{(1 + x^2)^2}$$

$$f^{(3)}(x) = -\frac{4x(3 - x^2)}{(1 + x^2)^3}$$

$$f^{(4)}(x) = -\frac{12(1 - 6x^2 + x^4)}{(1 + x^2)^4}$$

$$f^{(5)}(x) = \frac{48x(5 - 10x^2 + x^4)}{(1 + x^2)^5}$$

$$f^{(6)}(x) = \frac{240(1 - 15x^2 + 15x^4 - x^6)}{(1 + x^2)^6}$$

(a) Find the sixth-degree Taylor approximation $P_6(x)$ to $f(x)$.
(b) Use your computer or graphing calculator to find an interval over which the graph of $P_6(x)$ appears to fit the graph of $f(x)$ well.
(c) Using an appropriate y-scale, graph $f(x) - P_6(x)$ on the interval you determined in (b), and determine the maximum error in the approximation on that interval.

24. Confirm the result in part (a) of the preceding exercise—without any help from a computer algebra system—by completing the following steps.
(a) Recall the approximating polynomials for $1/(1 + t^2)$ calculated in Exploration Activity 2(a); multiply each by $2t$ to find approximating polynomials for $f'(t) = 2t/(1 + t^2)$.
(b) Integrate term by term from 0 to x to find approximating polynomials for $\ln(1 + x^2)$. In particular, verify that the sixth-degree approximating polynomial is the one you computed in the preceding exercise.

25. Graph $y_1 = \ln(1 + x)$ and $y_2 = x$ for $-2 \le x \le 2$ and $-5 \le y \le 5$. One term at a time, change y_2 to higher degree Taylor polynomial approximations for $\ln(1 + x)$, and regraph. That is, first subtract $x^2/2$, then add $x^3/3$, and so on. Continue up to at least degree 6. Describe in your own words the characteristics that distinguish all the Taylor polynomials from $\ln(1 + x)$.

26. As in the preceding exercise, start with $y_1 = \ln(1 + x)$ and $y_2 = x$. Deselect both these functions, and enter $y_3 = y_1 - y_2$, $y_4 = -x^2/2$ (with both y_3 and y_4 selected for graphing). As we saw in the text, y_3 is the error in the approximation of y_1 by y_2, and y_4 is the first term left out of the approximation. Set the window at $-0.5 \le x \le 0.5$ and $-0.2 \le y \le 0.2$ with x-scale equal to 0.05 and y-scale equal to 0.02. If your calculator or computer software has a grid option, turn it on. Watch carefully as the error curve y_3 is drawn and as the next term y_4 is drawn.
(a) What is the approximate shape of the error curve?
(b) For what values of x is the error smaller (in absolute value) than the next term?
(c) For what values of x is the error larger (in absolute value) than the next term?
(d) What is the largest error (in absolute value) on the interval $[-0.5, 0.5]$, and where does this largest error occur?
(e) In what interval on the x-axis is $|\text{error}|$ no larger than the value of y-scale?
(f) Change y_2 by adding the next term, $-x^2/2$, to it. Change y_4 to the new first term left out, $x^3/3$. Change the y-range to $[-0.1, 0.1]$ with a y-scale of 0.01. Answer the same questions as in parts (a)–(e). Record your numerical results in Table 11.5.
(g) Continue adding terms to y_2 and changing the next term y_4, answering the questions of parts (a)–(e), and filling in each line of Table 11.5. Use y-ranges and y-scales indicated in the table.

Table 11.5 Errors in polynomial approximations to $\ln(1+x)$

Degree	*Max* y	y*-scale*	*Max* $\lvert error \rvert$ *on* $\left[-\frac{1}{2}, \frac{1}{2}\right]$	*Interval for* $\lvert error \rvert \leq y$*-scale*
1	0.2	0.02	0.19	$[-0.182, 0.206]$
2	0.1	0.01		
3	0.025	0.0025		
4	0.001	0.0001		
5	0.0005	0.00005		
6	0.0002	0.00002		

27. What is the largest interval on which Taylor polynomials could possibly approximate $\arcsin x$?

28. Here, thanks to a computer algebra system, are the first seven derivatives of the function $f(x) = \arcsin x$:

$$f'(x) = \frac{1}{(1-x^2)^{1/2}}$$

$$f''(x) = \frac{x}{(1-x^2)^{3/2}}$$

$$f^{(3)}(x) = \frac{2x^2+1}{(1-x^2)^{5/2}}$$

$$f^{(4)}(x) = \frac{3x(2x^2+3)}{(1-x^2)^{7/2}}$$

$$f^{(5)}(x) = \frac{3(8x^4+24x^2+3)}{(1-x^2)^{9/2}}$$

$$f^{(6)}(x) = \frac{15x(8x^4+40x^2+15)}{(1-x^2)^{11/2}}$$

$$f^{(7)}(x) = \frac{45(16x^6+120x^4+90x^2+5)}{(1-x^2)^{13/2}}$$

(a) Find the Taylor polynomials of degrees 0 through 7 for $\arcsin x$.

(b) Let $E_7(x) = \arcsin x - P_7(x)$, where $P_7(x)$ is the seventh-degree polynomial you found in part (a). Use your calculator to make a table of x, $\arcsin x$, $P_7(x)$, and $E_7(x)$ for $x = 0.1, 0.3, 0.5, 0.7, 0.9$.

(c) Without any additional calculation, make a table of x, $\arcsin x$, $P_7(x)$, and $E_7(x)$ for $x = -0.1, -0.3, -0.5, -0.7, -0.9$.

(d) Complete the following formula: For $-1 < x < 1$,

$\arcsin x = \arctan$ ________________ .

(*Hint*: Draw a triangle in which one angle is $\arcsin x$.)

(e) Some computer algebra systems convert all arcsines to arctangents for evaluation. Can you think of a reason why this conversion might be built into the design of the system?

The **Taylor polynomials** for a function $f(x)$ **at the reference point** $x = a$ are the polynomials of the form

$$P_n(x) = f(a) + f'(a)(x-a) + \frac{f''(a)}{2}(x-a)^2 + \cdots + \frac{f^{(n)}(a)}{n!}(x-a)^n.$$

In sigma notation, we may write the defining formula as

$$P_n(x) = \sum_{k=0}^{n} \frac{f^{(k)}(a)}{k!}(x-a)^k.$$

The **Taylor series** for $f(x)$ **at the reference point** $x = a$ is the infinite series

$$f(a) + f'(a)(x-a) + \frac{f''(a)}{2}(x-a)^2 + \cdots + \frac{f^{(n)}(a)}{n!}(x-a)^n + \cdots = \sum_{k=0}^{\infty} \frac{f^{(k)}(a)}{k!}(x-a)^k.$$

Find the Taylor polynomials of degrees 0 through 6 for each of the following functions and reference points.

29. $\ln x$, $a = 1$

30. $\sin x$, $a = \frac{\pi}{2}$

31. $\cos x$, $a = \frac{\pi}{2}$

32. e^x, $a = 1$

33. e^{-x}, $a = 1$

34. $\arctan x$, $a = 2$

For each of the following functions and reference points, find the interval on which the Taylor polynomial of degree 6 approximates the function with an error no greater than 0.01.

35. $\ln x, \quad a = 1$

36. $\sin x, \quad a = \dfrac{\pi}{2}$

37. $\cos x, \quad a = \dfrac{\pi}{2}$

38. $e^x, \quad a = 1$

39. $e^{-x}, \quad a = 1$

40. $\arctan x, \quad a = 2$

Find the Taylor series for each of the following functions and reference points.

41. $\ln x, \quad a = 1$

42. $\sin x, \quad a = \dfrac{\pi}{2}$

43. $\cos x, \quad a = \dfrac{\pi}{2}$

44. $e^x, \quad a = 1$

45. $e^{-x}, \quad a = 1$

46. $\arctan x, \quad a = 2$

As best you can, estimate the interval of convergence of the Taylor series for each of the following functions and reference points.

47. $\ln x, \quad a = 1$

48. $\sin x, \quad a = \dfrac{\pi}{2}$

49. $\cos x, \quad a = \dfrac{\pi}{2}$

50. $e^x, \quad a = 1$

51. $e^{-x}, \quad a = 1$

52. $\arctan x, \quad a = 2$

11.4 Series of Constants

In this section we study some of the specific values of functions defined by infinite series, such as the Taylor series for $\ln(1+x)$ and for $\arctan x$. We also begin our study of whether a given series actually has a value—that is, whether its terms get small fast enough that infinitely many terms can be added to get a finite answer. Some aspects of this question are easy. For example, we already know that the geometric series, which represents $1/(1-x)$, has a finite sum if $-1 < x < 1$, but not otherwise. Here we will see the first instance in which the question becomes subtle: a series whose terms get small, but not fast enough to have a finite sum.

Convergence and Divergence

We start with a less subtle example, a series whose terms do not get small. For your convenience, we repeat the formula for the logarithmic series here:

$$\ln(1+x) = x - \frac{x^2}{2} + \frac{x^3}{3} - \frac{x^4}{4} + \cdots = \sum_{k=0}^{\infty} (-1)^k \frac{x^{k+1}}{k+1}, \qquad \text{if } |x| < 1.$$

The subject of the next activity is the result of substituting $x = 2$ in this series:

$$2 - \frac{2^2}{2} + \frac{2^3}{3} - \frac{2^4}{4} + \cdots = \sum_{k=0}^{\infty} (-1)^k \frac{2^{k+1}}{k+1}.$$

We will see that this series does not add up to $\ln 3$—or to anything else, for that matter.

Exploration Activity 1

(a) Sketch the graph of $f(x) = 2^x/x$ for $x > 0$. Describe the behavior of $f(x)$ as $x \to \infty$.

(b) Explain why $2^m/m \to \infty$ as $m \to \infty$. (Be careful: Both numerator and denominator are approaching ∞.)

(c) Why does this behavior of $2^m/m$ imply that the infinite series

$$\sum_{k=0}^{\infty}(-1)^k \frac{2^{k+1}}{k+1}$$

does not converge to anything? What are the partial sums doing as the number of terms increases? Generate the first 10 partial sums to confirm your conclusion.

(d) In general, if an infinite series

$$\sum_{k=0}^{\infty} b_k$$

converges, what can you say about

$$\lim_{k\to\infty} b_k?$$

We show the graph of $f(x) = 2^x/x$ in Figure 11.12. This graph suggests that the function has a local minimum between 1 and 2, and thereafter its values get steadily bigger as we go to the right. Furthermore, it appears that the graph is always concave upward, which means it can't level out as x gets large. Thus $f(x) \to \infty$ as $x \to \infty$. Now the values of $2^m/m$ are values of f at integer values of x, so we conclude that

$$\lim_{m\to\infty} \frac{2^m}{m} = \infty.$$

If we substitute $m = k + 1$ in this formula, it tells us that the terms of the series in part (c) grow without bound (in absolute value) as $k \to \infty$. Thus, after the first few terms, we are alternately adding and subtracting larger and larger numbers. The effect of this is to generate partial sums that oscillate wildly—in particular, sums that cannot possibly settle down to a limiting value. We show those sums in Table 11.6.

Figure 11.12 Graph of $f(x) = 2^x/x$

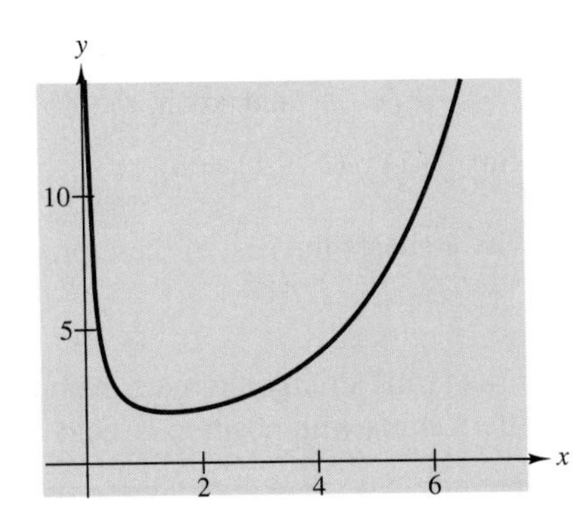

Table 11.6 Partial sums of $\sum (-1)^k 2^{k+1}/(k+1)$

k	kth term	kth sum
0	2	2
1	−2	0
2	2.667	2.667
3	−4	−1.333
4	6.4	5.067
5	−10.667	−5.6
6	18.286	12.686
7	−32	−19.314
8	56.889	37.575
9	−102.4	−64.825

What you should have observed by now in our examples of infinite series is that there is no hope of getting a finite sum from infinitely many terms unless those terms get small. That is, in order for $\sum b_k$ to converge, we must have $b_k \to 0$ as $k \to \infty$. Since the terms of the series in Exploration Activity 1 do not approach zero, the series cannot have a finite sum. We state this condition as a formal test for divergence.

Divergence Test If the sequence $b_0,\ b_1,\ b_2,\ \ldots$ does not converge to 0, then

$$\sum_{k=0}^{\infty} b_k$$

diverges.

Notice that we use the symbolism

$$\sum_{k=0}^{\infty} b_k$$

to mean two different things. On the one hand, it designates an infinite series — a sum with infinitely many terms — which may or may not converge to something. On the other hand, if the series converges, this notation represents the actual sum of the series — a number. This is not the first time you have seen one notation standing for two different concepts, nor will it be the last.

We know that the logarithmic series converges for $-1 < x < 1$, and the Divergence Test tells us that it diverges for $|x| > 1$. We turn our attention now to the two interesting series that result from substituting $x = 1$ and $x = -1$ into the logarithmic series. These numbers are the endpoints of the interval of convergence. Depending on what we find out about the series that result from these substitutions, an endpoint may or may not be in the interval of convergence. That is, the resulting series of constants may or may not converge.

Here is the series we get by substituting $x = 1$ into the Taylor series for $\ln(1 + x)$:

$$1 - \frac{1}{2} + \frac{1}{3} - \frac{1}{4} + \cdots + \frac{(-1)^k}{k+1} + \cdots = \sum_{k=0}^{\infty} \frac{(-1)^k}{k+1}.$$

Notice in particular the following features of this series:

- The terms alternate in sign: $+\quad -\quad +\quad -$ and so on.
- The terms approach zero: $\lim_{k\to\infty} (-1)^k/(k+1) = 0$.

The series in Exploration Activity 1 shares the first of these properties but not the second.

Definition A series is called **alternating** if its terms strictly alternate in sign. It does not matter whether the alternating sign pattern starts with positive or negative.

The alternating series we obtained by substituting $x = 1$ into the Taylor series for $\ln(1+x)$ has a special name.

Definition The **alternating harmonic series** is the infinite series

$$1 - \frac{1}{2} + \frac{1}{3} - \frac{1}{4} + \cdots .$$

When we substitute $x = -1$ in the expression $(-1)^k x^{k+1}$, we get $(-1)^{2k+1}$. Because the power is odd for every k, this expression is always -1. Thus the series that results from substituting -1 in the logarithmic series is

$$-1 - \frac{1}{2} - \frac{1}{3} - \frac{1}{4} - \cdots - \frac{1}{k+1} - \cdots = \sum_{k=0}^{\infty} \frac{-1}{k+1} .$$

Definition A series is called **monotonic** if its terms all have the same sign. It does not matter whether the constant sign is positive or negative.

The series

$$-1 - \frac{1}{2} - \frac{1}{3} - \frac{1}{4} - \cdots - \frac{1}{k+1} - \cdots$$

looks like a close relative of the alternating harmonic series. The corresponding series with all plus signs is one of the most important of all infinite series.

Definition The **harmonic series** is the infinite series

$$1 + \frac{1}{2} + \frac{1}{3} + \frac{1}{4} + \cdots .$$

Thus the second endpoint series we get from the logarithmic series is the negative of the harmonic series. Both the harmonic series and its negative are monotonic. Both have terms that approach zero. But so far all we know about either the harmonic series or the alternating harmonic series is a name. Do these series converge or diverge? If either converges, to what number does it converge? In spite of the similarity of the two series, these questions have to be answered in quite different ways for the two cases. We will study the harmonic series and the alternating harmonic series separately in the next two subsections.

The Harmonic Series

The question of convergence of the harmonic series brings us face to face with subtlety at a level we have not experienced previously in this course. We are adding up terms that get smaller and smaller. After a million such terms, every new term must be smaller than one-millionth. After a billion terms, every term is smaller than one-billionth. Moreover, the partial sum of the first billion terms is only about 20. (Don't try to check that on your calculator — trust us.) How could such a sum not converge? But the fact is, it does not! In the next activity you can convince yourself that the sum of all the terms is larger than every positive integer; i.e., the sum is infinite! In a laboratory exercise associated with this chapter you may see a completely different way to come to the same conclusion.

Exploration Activity 2

(a) Show that the harmonic series may be written in the form

$$\sum_{k=1}^{\infty} \frac{1}{k}.$$

(b) Show that

$$\frac{1}{3}+\frac{1}{4}>\frac{1}{2},$$

and use this to conclude that

$$1+\frac{1}{2}+\frac{1}{3}+\frac{1}{4}>\frac{4}{2}.$$

(*Hint*: You don't need a calculator for this.)

(c) Show that

$$\frac{1}{5}+\frac{1}{6}+\frac{1}{7}+\frac{1}{8}>\frac{1}{2},$$

and use this to show that

$$\sum_{k=1}^{8} \frac{1}{k}>\frac{5}{2}.$$

(d) Find an integer n so that

$$\sum_{k=1}^{n} \frac{1}{k}>\frac{6}{2}.$$

(e) Explain why, for any positive integer r, there is an integer n such that

$$\sum_{k=1}^{n} \frac{1}{k}>\frac{r}{2}.$$

(f) Explain why the harmonic series does not converge.

(g) Does this contradict the Divergence Test? Why or why not?

In part (a) we are making a change of notation simply as a matter of convenience for the rest of this activity. The first few terms of the harmonic series are

$$1+\frac{1}{2}+\frac{1}{3}+\frac{1}{4}+\cdots.$$

The only reason these terms were numbered from 0 was because we obtained the series originally by substitution in a Taylor series. It is clearly simpler to start the numbering with 1, in which case the sigma notation is

$$\sum_{k=1}^{\infty}\frac{1}{k}.$$

For part (b), we observe that $\frac{1}{3}>\frac{1}{4}$, so $\frac{1}{3}+\frac{1}{4}>\frac{1}{4}+\frac{1}{4}=\frac{1}{2}$ and

$$1+\frac{1}{2}+\frac{1}{3}+\frac{1}{4}>\frac{3}{2}+\frac{1}{2}=\frac{4}{2}.$$

Continuing in this manner, we see that

$$\frac{1}{5}+\frac{1}{6}+\frac{1}{7}+\frac{1}{8}>\frac{1}{8}+\frac{1}{8}+\frac{1}{8}+\frac{1}{8}=\frac{1}{2},$$

so the sum of the first eight terms is more than $\frac{5}{2}$. Eight more terms add more than $\frac{1}{2}$, bringing the total for 16 terms above $\frac{6}{2}$. At each step we double the number of terms from the previous step, and we add more than $\frac{1}{2}$. To get the sum above $r/2$, we need 2^{r-2} terms. No matter how large r is, we can always add up 2^{r-2} terms, so there is no finite bound for the sum of the harmonic series. That is, the series diverges.

The harmonic series is an example of a series whose terms decrease to zero but whose sum nevertheless diverges to infinity. This does not contradict the Divergence Test because it is a *one-way* test. It says that a series whose terms do not approach zero must diverge, but it does not say anything about a series whose terms approach zero.

The Alternating Harmonic Series

The presence of alternating signs in the alternating harmonic series produces a quite different result from what we have just seen with the harmonic series. We start by introducing notation for the partial sums of the series:

$$s_1=1,\quad s_2=1-\frac{1}{2},\quad s_3=1-\frac{1}{2}+\frac{1}{3},$$

and, in general,

$$s_n=1-\frac{1}{2}+\frac{1}{3}-\frac{1}{4}+\cdots+(-1)^{n+1}\,\frac{1}{n}.$$

The question of whether or not the infinite series

$$\sum_{k=1}^{\infty}(-1)^{k+1}\,\frac{1}{k}$$

converges to a limiting value S is the same question as whether

$$\lim_{n\to\infty} s_n = S.$$

We investigate this question in the next two activities.

Exploration Activity 3

(a) Draw a horizontal line the width of a piece of paper. Near the left end, mark a point 0, and near the right end mark a point 2. Carefully measure the midpoint, and label it 1. Now calculate and place on your number line the numbers $s_1, s_2, s_3, s_4, s_5, s_6, s_7,$ and s_8.

(b) Use your graphing tool to graph, in the same window, both the terms

$$b_n = (-1)^{n+1}\frac{1}{n}$$

and the partial sums

$$s_n = 1 - \frac{1}{2} + \frac{1}{3} - \frac{1}{4} + \cdots + (-1)^{n+1}\frac{1}{n}$$

for $n = 1, 2, \ldots, 20.$

(c) Show that

$$s_1 > s_3 > s_5 > \cdots$$

for odd-numbered partial sums, and

$$s_2 < s_4 < s_6 < \cdots$$

for even-numbered ones.

(d) Describe the general pattern of the sums $s_1, s_2, \ldots, s_n$ for an arbitrary positive integer n.

(e) Show that

$$\lim_{n\to\infty} |s_n - s_{n+1}| = 0.$$

(f) Use these observations to show that the alternating harmonic series converges, i.e., that there really is a limiting value S.

(g) Explain why S is smaller than every odd-numbered partial sum and larger than every even-numbered partial sum.

(h) The error after summing n terms is $|S - s_n|$. Explain why this error is smaller than the absolute value of the $(n+1)$th term, that is, $1/(n+1)$.

We show possible responses to parts (a) and (b) in Figures 11.13 and 11.14, respectively. These figures illustrate two different ways to represent a sequence graphically, by plotting successive points on a number line, and by plotting terms as a function of the subscript. We have used the latter representation for sequences many times. The former is introduced here as a way to visualize partial sums squeezing down on a limiting value.

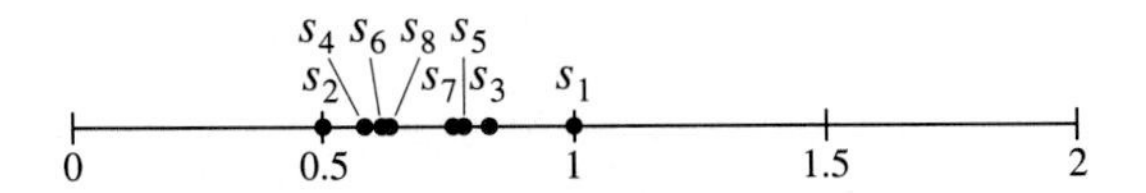

Figure 11.13 Number line plot of partial sums

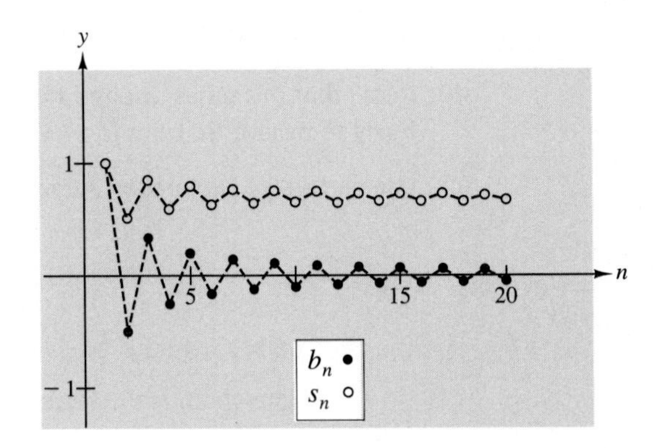

Figure 11.14 Terms and partial sums of the alternating harmonic series

Both figures suggest that odd-numbered partial sums decrease and even-numbered ones increase as n increases. We can see that algebraically in the following way. Each partial sum is obtained from the previous one by adding one more term:

$$s_{n+1} = s_n + b_{n+1}.$$

At the next step, the same formula becomes

$$s_{n+2} = s_{n+1} + b_{n+2}.$$

Thus, to get from s_n to s_{n+2}, we first add $b_{n+1} = (-1)^n/(n+1)$ and then add $b_{n+2} = (-1)^{n+1}/(n+2)$. If n is odd, then $1/(n+1)$ is being *subtracted* to get to s_{n+1}, and $1/(n+2)$ is being *added* to get to s_{n+2}. The amount being added is smaller than the amount being subtracted, so the next odd-numbered partial sum must be smaller. The argument for even-numbered sums is similar.

In general, the pattern for consecutive sums is that you jump down on even-numbered steps and back up on odd-numbered steps, with each jump being smaller than the one before it. Specifically, the size of the jump at step $n+1$ is

$$|s_{n+1} - s_n| = |b_{n+1}| = \frac{1}{n+1},$$

which approaches 0 as $n \to \infty$. This shows that the even-numbered and odd-numbered sums steadily approach each other—that there is no "room" between them—so the sequence of partial sums must actually squeeze down on a limiting value S. Furthermore, because the sequence of even-numbered sums increases, S must be bigger than every even-numbered sum. Similarly, S must be smaller than every odd-numbered sum. At step n, if n is odd, then

$$s_{n+1} < S < s_n,$$

so

$$|S - s_n| < |s_{n+1} - s_n| = \frac{1}{n+1}.$$

At an even-numbered step, $s_n < S < s_{n+1}$, but the conclusion is the same.

Checkpoint 1

(a) You have already calculated partial sums of the alternating harmonic series up to s_8. Extend your numerical calculation out to s_{20}.

(b) Recall that this series resulted from substituting $x = 1$ into the Taylor series for $\ln(1+x)$. Are you ready to believe that S is $\ln 2$? Why or why not?

(c) Use part (h) of Exploration Activity 3 to estimate how close s_{20} must be to S? Is it that close to $\ln 2$?

The Leibniz Series

Here again is the formula for the arctangent series:

$$\arctan x = x - \frac{1}{3}x^3 + \frac{1}{5}x^5 - \cdots = \sum_{k=0}^{\infty} (-1)^k \frac{x^{2k+1}}{2k+1}, \qquad \text{if } |x| < 1.$$

The value $x = 1$ is not in the presumed domain $|x| < 1$ for this series. However, if we substitute this value into the series, we get another interesting series, which has a name.

Definition The **Leibniz series** is the series of constants

$$1 - \frac{1}{3} + \frac{1}{5} - \frac{1}{7} + \cdots + (-1)^k \frac{1}{2k+1} + \cdots = \sum_{k=0}^{\infty} (-1)^k \frac{1}{2k+1}.$$

The question of immediate interest is whether the Leibniz series actually converges to $\arctan 1 = \pi/4$. If so, it produces an interesting formula for π:

$$\pi = 4\left[1 - \frac{1}{3} + \frac{1}{5} - \frac{1}{7} + \cdots + (-1)^k \frac{1}{2k+1} + \cdots\right].$$

Exploration Activity 4

(a) Explain why the Leibniz series converges to something. (*Hint*: See Exploration Activity 3.)

(b) Use your calculator to add up the first 10 partial sums. Are you ready to believe that the limiting value might be $\pi/4$? Why or why not? (If you have a computer available, change 10 to 100 or 1000, and answer the same questions.)

(c) Whatever the actual sum of the Leibniz series is, let's call it S. Explain why S is smaller than every even-numbered partial sum and larger than every odd-numbered partial sum.

(d) Explain why the error after summing n terms (numbered 0 through $n-1$) is smaller than the absolute value of the $(n+1)$th term, that is, $1/(2n+1)$.

The explanations here are exactly like those for Exploration Activity 3. The sums after 10, 100, and 1000 terms are shown in Table 11.7, along with the value of $\pi/4$, the difference $|s_{n-1} - \pi/4|$, and the size of the nth term. The table does not prove that

$S = \pi/4$, but it provides strong evidence for that conclusion. (If you add only the first 10 terms, the evidence is a little shaky.) Notice that the next-term bound on the error is rather conservative — in fact, the actual error is approximately half of the estimate. Also notice that each factor of 10 in the number of terms produces about one more decimal digit of the eventual answer. Thus summing up terms of the Leibniz series would not be a very efficient way to compute digits of π.

Table 11.7 Partial sums of the Leibniz series

n	s_{n-1}	$\pi/4$	$\lvert s_{n-1} - \pi/4\rvert$	$1/(2n+1)$
10	0.760460	0.785398	0.024938	0.04762
100	0.782898	0.785398	0.002500	0.00498
1000	0.785148	0.785398	0.000250	0.00050

In this section we have studied series of constants, i.e., series that might represent specific values of functions that can be represented by series. However, a series cannot represent a specific value unless it converges. In particular, we saw that a series can't converge unless its terms approach 0.

Our sample series were generated by substituting specific values of x into the Taylor series for $\ln(1+x)$ and for $\arctan x$. Interesting series arise from choosing values of x at the endpoints of intervals of convergence, because it is not clear a priori whether such series converge or diverge. In fact, we found that the alternating harmonic series and Leibniz series both converge, but the harmonic series diverges. In particular, the harmonic series is the prototypical example of a series whose terms approach 0 but that nevertheless diverges.

In the next section we will formalize into a test for convergence what we have learned here about alternating series. We also will study why Taylor series have an interval of convergence. The outcome of that study will be a convergence test we can use for series that are not necessarily alternating.

Answers to Checkpoint

1. (a) $s_{20} = 0.6688$

   (b) $\ln 2 = 0.6931$ — it's possible that this is the limiting value.

   (c) The error must be less than $1/21 = 0.0476$. The actual error is $|s_{20} - \ln 2| = 0.0244$.

Exercises 11.4

Give examples — other than those discussed in the text — of each of the following.

1. A series that does not converge because its terms do not approach zero.
2. A series that does not converge even though its terms do approach zero.
3. A series that does converge.

For each of the following series, decide whether the series converges or diverges, and state how you know.

4. $\displaystyle\sum_{k=1}^{\infty} (-1)^k \frac{1}{1+k}$

5. $\displaystyle\sum_{k=1}^{\infty} (-1)^k \frac{1}{k}$

6. $\displaystyle\sum_{k=1}^{\infty} \frac{1}{k+3}$

7. $\displaystyle\sum_{k=1}^{\infty} (-1)^k \frac{k}{k+1}$

8. $\sum_{k=1}^{\infty} \sin k$

9. $\sum_{k=1}^{\infty} (-1)^k \frac{1}{2k+1}$

For each of the following series, give its value if there is one and you know it. If the series converges, but you don't know its value, estimate the value with your calculator.

10. $\sum_{k=1}^{\infty} (-1)^k \frac{1}{1+k}$

11. $\sum_{k=1}^{\infty} (-1)^k \frac{1}{k}$

12. $\sum_{k=1}^{\infty} \frac{1}{k+3}$

13. $\sum_{k=1}^{\infty} (-1)^k \frac{k}{k+1}$

14. $\sum_{k=1}^{\infty} \sin k$

15. $\sum_{k=1}^{\infty} (-1)^k \frac{1}{2k+1}$

16. Ask a music major or musician in your class (yourself if you are a music major or musician) what the fractions $1/k$ have to do with harmonics.

17. (a) Substitute $x = -1$ into the arctangent series. How is the resulting series related to the Leibniz series?
 (b) Does the series in part (a) converge or diverge? How do you know? If it converges, what is its value?
 (c) Is the arctangent function even, odd, or neither? Are its Taylor polynomials even, odd, or neither?
 (d) What does part (c) have to do with parts (a) and (b)?

18. What is the sum of the series

$$1 + \frac{1}{\sqrt{2}} + \frac{1}{\sqrt{3}} + \frac{1}{\sqrt{4}} + \cdots ?$$

Explain carefully. (*Hint*: Compare the terms and partial sums with those of the harmonic series.)

19. Does the series

$$\frac{1}{1{,}000{,}000} + \frac{1}{1{,}000{,}001} + \frac{1}{1{,}000{,}002} + \cdots$$

converge or diverge? Explain carefully.

20. Does the series

$$\frac{1}{1{,}000{,}000} - \frac{1}{1{,}000{,}001} + \frac{1}{1{,}000{,}002} - \cdots$$

converge or diverge? Explain carefully.

11.5 Convergence of Series

The Divergence Test in the preceding section gives us an easy way to decide that some series do not converge. However, it tells us nothing about convergence or divergence of series whose terms approach zero. In this section we develop several ways to tell whether or not a given series converges. The first of these arises naturally from the examples in the preceding section. It is also the easiest convergence test to apply, but its scope is rather limited. From there we move on to tests that are more robust—and also more difficult to carry out. With each convergence test we also will acquire an error estimate, that is, a way to tell how close any given partial sum is to the total sum. By the end of this section we will have resolved most of the convergence issues for Taylor series representations of important functions.

The Alternating Series Test

Your study of the harmonic series in the preceding section shows that you have to be careful not to jump to conclusions about convergence of series. There is no hope of convergence unless the terms approach zero—the Divergence Test—but "terms approaching zero" is not enough to ensure convergence. In particular, the harmonic series has terms approaching zero, but the sum fails to converge. On the other hand, we can abstract from the examples of the alternating harmonic series and the Leibniz series (both also studied in the preceding section) a general statement about convergence that applies to any alternating series in which the absolute values of the terms steadily decrease to zero.

The Alternating Series Test Suppose a series has the form

$$a_1 - a_2 + a_3 - \cdots + (-1)^{k+1} a_k + \cdots,$$

where

- each a_k is positive,
- each a_k is larger than a_{k+1}, and
- $\lim_{k\to\infty} a_k = 0$.

Then

(i) the series converges to some number S, and

(ii) the error $|S - s_n|$ after summing n terms is less than a_{n+1}.

Exploration Activity 1

Use the arguments you have already developed in Exploration Activities 3 and 4 of the preceding section to explain why the statement of the Alternating Series Test is true.

First note how the notation works in the statement of the Alternating Series Test. The symbol a_k stands for the *absolute value* of the kth term. The term itself is $b_k = (-1)^{k+1} a_k$. The first bulleted condition ensures that the terms really alternate in sign; that is, there are no zero terms counted among the numbered terms. The second condition ensures that each "jump" in the sum—up or down—is smaller than the one before. And the third condition ensures that the sizes of the jumps decrease to zero; that is, the series doesn't fail to converge because of the Divergence Test. These are precisely the properties of the alternating harmonic series and the Leibniz series that we used in Section 11.4 to establish their convergence.

Specifically, we find that odd-numbered partial sums decrease and even-numbered ones increase as n increases: To get from s_n to s_{n+2}, we first add $b_{n+1} = (-1)^n a_{n+1}$ and then add $b_{n+2} = (-1)^{n+1} a_{n+2}$. If n is odd, then a_{n+1} is being *subtracted* to get to s_{n+1} and a_{n+2} is being *added* to get to s_{n+2}. The amount being added is smaller than the amount being subtracted, so the next odd-numbered partial sum must be smaller. The argument for even-numbered sums is similar.

In general, the pattern for consecutive sums is to jump down on even-numbered steps and back up on odd-numbered steps, with each jump being smaller than the one before it. In fact, the size of the jump at step $n + 1$ is

$$|s_{n+1} - s_n| = |b_{n+1}| = a_{n+1},$$

which we know approaches 0 as $n \to \infty$. This shows that the even-numbered and odd-numbered sums steadily approach each other, so the sequence of partial sums must actually squeeze down on a limiting value S. Furthermore, because the sequence of even-numbered sums increases, S must be bigger than every even-numbered sum. Similarly, S must be smaller than every odd-numbered sum. At step n, if n is odd, then

$$s_{n+1} < S < s_n,$$

so
$$|S - s_n| < |s_{n+1} - s_n| = a_{n+1}.$$

At an even-numbered step, $s_n < S < s_{n+1}$, but the conclusion is the same.

Be careful not to read too much into the Alternating Series Test. In one sense, it says that any alternating series that looks like it converges probably does—and it's pretty easy to tell how close any partial sum is to the total sum. But the test certainly doesn't say that all alternating series converge. In fact, it doesn't even say that all alternating series whose terms approach zero must converge. The requirement that absolute values of the terms actually decrease is important. Exceptional cases—in which the absolute values approach zero but not in a strictly decreasing way—are not very important. If you are curious about such things, see Exercise 29.

In order to decide that a given series converges, it's enough that its terms *eventually* satisfy the conditions of a convergence test such as the Alternating Series Test. That is, we only need to consider all the terms from some point on, a **tail** of the series. If the tail is a convergent series, then adding in a finite number of terms at the beginning won't affect convergence—although those terms do affect the sum. In a later section we will use the Alternating Series Test to demonstrate convergence of some important series whose tails have the form described here.

Convergence of the Arctangent Series

Once more, here is the formula for the arctangent series:

$$\arctan x = x - \frac{1}{3}x^3 + \frac{1}{5}x^5 - \cdots = \sum_{k=0}^{\infty} (-1)^k \frac{x^{2k+1}}{2k+1}, \qquad \text{if } |x| < 1.$$

The presumed domain in which this series converges, namely, $|x| < 1$, was inherited from the geometric series. But do we really know that the series converges for these values of x? Do we know that it diverges for other values of x? Those are questions we can now answer. You have already partially answered one of these questions: In Section 11.4 you showed that the series converges for $x = 1$, which is *not* in the presumed domain.

Note that the arctangent series is alternating for every value of x. That's clear for positive values of x; if x is negative, so is every odd power of x, which means the signs still alternate, but in the opposite pattern, $- + - + \cdots$. In fact, wherever the series converges, it defines an odd function, which is no surprise, since arctangent is also an odd function. Since the series for $x = -1$ is the negative of the series for $x = 1$, you already know that it converges for $x = -1$ —apparently to $-\pi/4$.

Example 1 Show that the arctangent series converges if $|x| \le 1$.

Solution We already know the series converges if $x = \pm 1$. If x is a number *between* -1 and 1, then the powers of x approach zero, so the terms of the series do too. Is each term larger (in absolute value) than the one after it? First we look at the numerators of consecutive terms:

$$|x|^{2k+3} = |x|^2 |x|^{2k+1} < |x|^{2k+1}$$

because $|x| < 1$. We already know that the denominators are increasing, so their reciprocals are decreasing. Thus the absolute value of the kth term is the product of two factors, $|x|^{2k+1}$ and $1/(2k+1)$, both of which are larger than the corresponding factors in the $(k+1)$th term. Hence the terms really do decrease in size. It follows from the Alternating Series Test that the arctangent series converges for all numbers x in the interval $[-1, 1]$. ■

Checkpoint 1

(a) Explain why the terms in the arctangent series do not approach zero if $|x| > 1$.

(b) Explain why the arctangent series diverges if $|x| > 1$.

Convergence of the Logarithmic Series

Here again is the Taylor series for $\ln(1 + x)$, with terms renumbered for convenience:

$$\ln(1+x) = x - \frac{x^2}{2} + \frac{x^3}{3} - \frac{x^4}{4} + \cdots = \sum_{k=1}^{\infty} (-1)^{k+1} \frac{x^k}{k}, \qquad \text{if } |x| < 1.$$

Notice that this series does not have odd symmetry—or any symmetry at all, since every positive power of x appears in it. No surprise there—the natural logarithm function doesn't have any symmetry either. This series is alternating for *some* values of x, but not for all.

Exploration Activity 2

(a) For what values of x will the Taylor series for $\ln(1 + x)$ satisfy all the bulleted conditions of the Alternating Series Test?

(b) For what values of x does the Alternating Series Test ensure that the Taylor series for $\ln(1 + x)$ converges? (You already determined one such value of x in the preceding section.)

(c) For each such x and each fixed value of n, how big can the error be in the nth degree Taylor polynomial approximation to $\ln(1 + x)$? In particular, for $x = \frac{1}{2}$ and $n = 10$, how close must $P_{10}(\frac{1}{2})$ be to $\ln \frac{3}{2}$? Calculate both $P_{10}(\frac{1}{2})$ and $\ln \frac{3}{2}$ to be sure.

(d) Explain why the Taylor series for $\ln(1 + x)$ must diverge if $|x| > 1$.

The sign of each term in the logarithmic series depends on the sign of x. If x is positive, then so is x^k for every k, and the sign of the term is the sign of $(-1)^{k+1}$, positive for odd k and negative for even k. In particular, the signs alternate in this case.

On the other hand, if x is negative, then $x = -|x|$, and

$$(-1)^{k+1}x^k = (-1)^{k+1}(-1)^k|x|^k = (-1)^{2k+1}|x|.$$

In this case, every term is negative, so the signs do not alternate. This means that the first condition of the Alternating Series Test is satisfied only for $x > 0$.

If $0 < x \leq 1$, then

$$\frac{x^{k+1}}{k+1} = \frac{x \cdot x^k}{k+1} < \frac{x^k}{k},$$

so the terms do decrease in size. Furthermore, $x^k \leq 1$, so $x^k/k \leq 1/k$, and

$$\lim_{k\to\infty} \frac{x^k}{k} = 0.$$

Thus all three conditions of the Alternating Series Test are satisfied if $0 < x \leq 1$. If x is any fixed number greater than 1, then x^k, *as a function of* k, is an exponential function with base $x > 1$, which must grow faster than any linear function of k, so the terms x^k/k do not approach zero—in fact, they become arbitrarily large. Conclusion: The Alternating Series Test applies to the logarithmic series *only* for $0 < x \leq 1$.

This leaves a gap in your knowledge of exactly where the logarithmic series converges—although what happens in that gap is at least suggested by Figures 11.15 and 11.16, repeated here from Figures 11.10 and 11.11. The fact that the series does not always have alternating signs (e.g., at $x = -\frac{1}{2}$) does not tell us that the series diverges there—it merely says that we can't apply the Alternating Series Test to find out. Exercise 19 at the end of the chapter will show you how to calculate natural logarithms for every positive number without worrying about filling the gap in Exploration Activity 2.

Figure 11.15 $\ln(1+x)$ and polynomial approximations up to degree 7

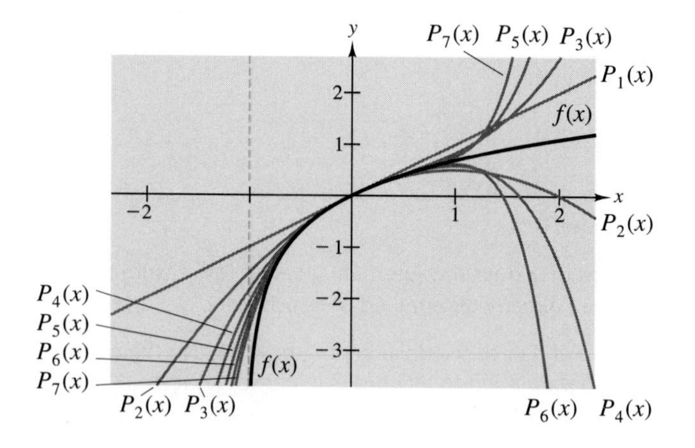

Figure 11.16 Errors in polynomial approximations of $\ln(1+x)$

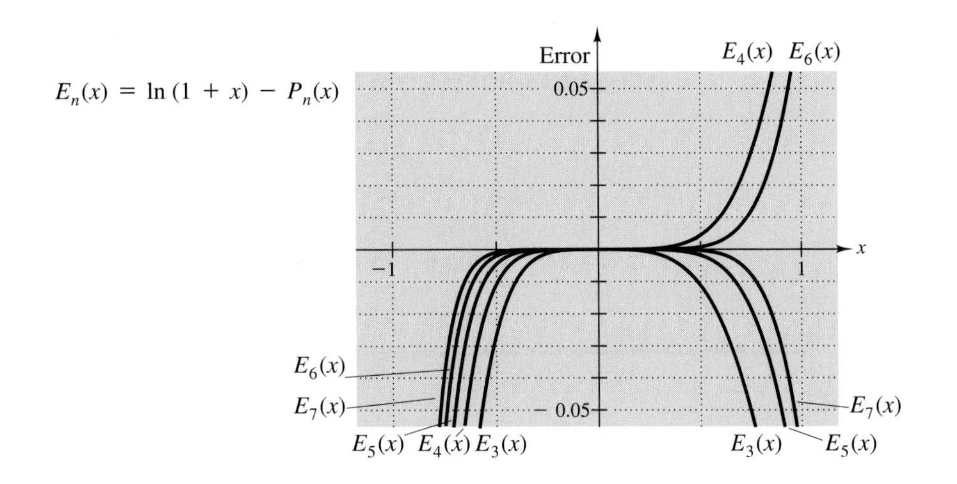

Using Geometric Series to Estimate Tails: Exponential Series

The Alternating Series Test makes it fairly easy to show that some important Taylor series do indeed converge, at least for some values of the input variable. However, several of our important series are not alternating, e.g., the series for $\ln(1+x)$ with $x<0$ and the series for e^x with $x>0$. Furthermore, there are important series whose terms always alternate in sign, but for which it is not yet clear that the terms decrease to zero, e.g., the series for $\sin x$ and for $\cos x$ when x is large. In this subsection we develop another way — based on geometric series — to detect whether such a Taylor series converges. This geometric series method is not particularly sensitive to sign patterns of the terms — in fact, it often works in the worst-case scenario of a series whose terms are all of the *same* sign.

To illustrate our problem, here again is the Taylor series for the exponential function:

$$e^x = 1 + x + \frac{1}{2}x^2 + \frac{1}{6}x^3 + \cdots + \frac{1}{k!}x^k + \frac{1}{(k+1)!}x^{k+1} + \cdots.$$

Suppose we knew nothing about the origin of this series, and we simply defined a new function $G(x)$ by

$$G(x) = 1 + x + \frac{1}{2}x^2 + \frac{1}{6}x^3 + \cdots + \frac{1}{k!}x^k + \frac{1}{(k+1)!}x^{k+1} + \cdots$$

for all numbers x for which the series converges — no matter what it converges to. G is a *new* function because we have not shown yet that it equals the exponential function. From our graphical studies of the Taylor polynomials for the exponential function, we have some reason to believe that every number x may be in the domain of G — that's one of the things we will establish in this section. Our immediate task, then, is to determine the domain of the function G.

Because we need to get away from Taylor series as the source of all of our examples, we will use the term "power series" to stand for any "infinite polynomial."

Definitions A **power series** in x is any expression of the form

$$c_0 + c_1 x + c_2 x^2 + c_3 x^3 + \cdots + c_k x^k + \cdots,$$

or, in sigma notation, of the form

$$\sum_{k=0}^{\infty} c_k x^k,$$

where the coefficients c_0, c_1, c_2, and so on, may be any real numbers. We will say that a function g is **defined by a power series** if

$$g(x) = c_0 + c_1 x + c_2 x^2 + c_3 x^3 + \cdots + c_k x^k + \cdots.$$

By convention, the domain of the function g consists of all the numbers x for which the series above converges. Thus our function

$$G(x) = 1 + x + \frac{1}{2}x^2 + \frac{1}{6}x^3 + \cdots + \frac{1}{k!}x^k + \frac{1}{(k+1)!}x^{k+1} + \cdots$$

is defined by a power series, namely, the Taylor series for the exponential function. We will show in this section that the domain of G contains all real numbers x and in the next section that G really is the exponential function.

Given a power series, the sum of the first n terms (the terms numbered from 0 to $n-1$) is

$$s_n = c_0 + c_1 x + c_2 x^2 + c_3 x^3 + \cdots + c_{n-1} x^{n-1}.$$

What's left over—not included in the nth partial sum—we call the nth **tail** of the series:

$$n\text{th tail} = c_n x^n + c_{n+1} x^{n+1} + c_{n+2} x^{n+2} + c_{n+3} x^{n+3} + \cdots.$$

Now the nth tail is itself a power series. For a fixed value of n, if the nth tail converges, then so does the original series, because what's left to be added on, the partial sum s_n, is an *ordinary* sum. To show that the original power series converges, we only have to find an n for which the nth tail converges.

We recall that, for any numbers A and B,

$$|A + B| \le |A| + |B|.$$

That is, the size of a sum cannot be any bigger than the sum obtained by first taking absolute value of each of the terms—a sum in which no cancellation can occur. The implication for nth tails is

$$\begin{aligned} |n\text{th tail}| &= |c_n x^n + c_{n+1} x^{n+1} + c_{n+2} x^{n+2} + c_{n+3} x^{n+3} + \cdots| \\ &\le |c_n||x|^n + |c_{n+1}||x|^{n+1} + |c_{n+2}||x^{n+2}| + |c_{n+3}||x|^{n+3} + \cdots. \end{aligned}$$

Thus, if the sum of absolute values is finite, so is the nth tail.

Example 2 For any number x, show that

$$G(x) = 1 + x + \frac{1}{2}x^2 + \frac{1}{6}x^3 + \cdots + \frac{1}{k!}x^k + \frac{1}{(k+1)!}x^{k+1} + \cdots$$

converges.

Solution For this power series,

$$n\text{th tail} = \frac{x^n}{n!} + \frac{x^{n+1}}{(n+1)!} + \frac{x^{n+2}}{(n+2)!} + \cdots.$$

If this infinite sum converges, it has the same value as the expression we get by factoring out all the common factors: x^n in the numerators and $n!$ in the denominators:

$$\begin{aligned} |n\text{th tail}| &= \left|\frac{x^n}{n!}\left[1 + \frac{x}{(n+1)} + \frac{x^2}{(n+1)(n+2)} + \cdots\right]\right| \\ &\le \frac{|x|^n}{n!}\left[1 + \frac{|x|}{(n+1)} + \frac{|x|^2}{(n+1)(n+2)} + \cdots\right] \end{aligned}$$

The value of this last expression is no bigger than the sum we get by replacing each term in the brackets with a term that is at least as large. Our plan will be to replace *all* the factors $n+2$, $n+3$, and so on, in the denominators of the third and later terms with $n+1$. This gives us the estimate

$$|n\text{th tail}| \le \frac{|x|^n}{n!}\left[1 + \frac{|x|}{n+1} + \left(\frac{|x|}{n+1}\right)^2 + \left(\frac{|x|}{n+1}\right)^3 + \cdots\right].$$

The series inside the brackets is a geometric series $1 + r + r^2 + r^3 + \cdots$ with term-to-term ratio

$$r = \frac{|x|}{n+1}.$$

We learned in Chapter 5—and recapitulated at the start of Section 11.3—all there is to know about convergence of these series. In particular, a geometric series with term-to-term ratio r converges if $|r| < 1$. Furthermore, when $|r| < 1$, the geometric series converges to $1/(1-r)$. Thus no matter what x is, as soon as n is large enough that

$$\frac{|x|}{n+1} < 1,$$

the tail converges. Now, for any x we can find such an n, namely, the first integer as big as $|x|$. For example, if $x = 6.7$, then we can take $n = 7$, and $6.8/(7+1)$ is certainly less than 1. This shows that the exponential series which defines $G(x)$ converges for every real number x. ■

We turn now to the question of estimating the error in a polynomial approximation, i.e., the error in a partial sum of a power series. For a given value of x, if the series happens to converge to the number S, then the error after summing n terms, $|S - s_n|$, is the absolute value of the nth tail. Our job then is to estimate the size of the nth tail—

essentially the same task we illustrated in Example 2 as a key step in showing that power series converges.

Example 3 Estimate the largest possible error on the interval $-2 \le x \le 2$ if

$$G(x) = 1 + x + \frac{1}{2}\,x^2 + \frac{1}{6}\,x^3 + \cdots + \frac{1}{k!}\,x^k + \frac{1}{(k+1)!}\,x^{k+1} + \cdots$$

is approximated by

$$P_9(x) \;=\; 1 + x + \frac{1}{2}\,x^2 + \frac{1}{6}\,x^3 + \cdots + \frac{1}{8!}\,x^8 + \frac{1}{9!}\,x^9.$$

Solution The ninth-degree approximation is the sum of the first 10 terms of $G(x)$, so the error is $|\text{10th tail}|$. In Example 2 we saw that

$$\begin{aligned}
|n\text{th tail}| &\le \frac{|x|^n}{n!}\left[1 + \frac{|x|}{n+1} + \left(\frac{|x|}{n+1}\right)^2 + \left(\frac{|x|}{n+1}\right)^3 + \cdots\right] \\
&= \frac{|x|^n}{n!}(1 + r + r^2 + r^3 + \cdots), \qquad \text{where } r = \frac{|x|}{n+1}, \\
&= \frac{|x|^n}{n!}\cdot\frac{1}{1-r} \\
&= \frac{|x|^n}{n!\,(1-r)}.
\end{aligned}$$

For this example, $n = 10,\ |x| \le 2,$ and $r \le 2/11$. Thus $1 - r \ge 9/11$, and

$$|\text{10th tail}| \le \frac{2^{10}}{10!\,(9/11)} = 0.000344895\ldots < 0.0004.$$

It follows that any value of $P_9(x)$ for $-2 \le x \le 2$ will approximate $G(x)$ to three-decimal-place accuracy. ■

We haven't established yet that G is actually the exponential function, but our calculator tells us that $e^2 = 7.389056$ and $P_9(2) = 7.3887125$, so the actual error is about 0.0003436. In this case, the geometric-series estimate is close to the actual error at the endpoint. On the other hand, $e^{-2} = 0.1353353$ and $P_9(-2) = 0.1350970$, so the actual error is about 0.0002383. In this case, the error estimate is somewhat conservative. For values of x closer to 0, the actual error will be much smaller than our estimate. We illustrate this point with Figures 11.17 and 11.18, which show, respectively, e^x and $P_9(x)$ plotted together on the interval $[-3, 3]$ and the error on the interval $[-2, 2]$. You see the graph of only one function in Figure 11.17 because the two graphs are visually indistinct at the scale shown. Figure 11.18 proves that the two functions are not identical, but the small size of the error also shows why the graphs in Figure 11.17 are visually indistinct.

Figure 11.17 Graphs of e^x and $P_9(x)$

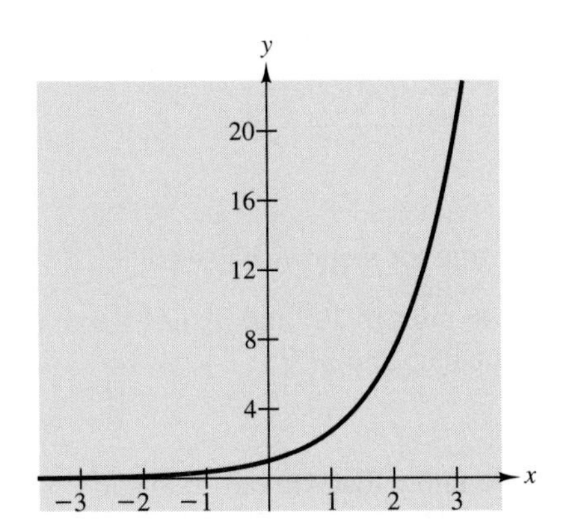

Figure 11.18 Graph of $e^x - P_9(x)$

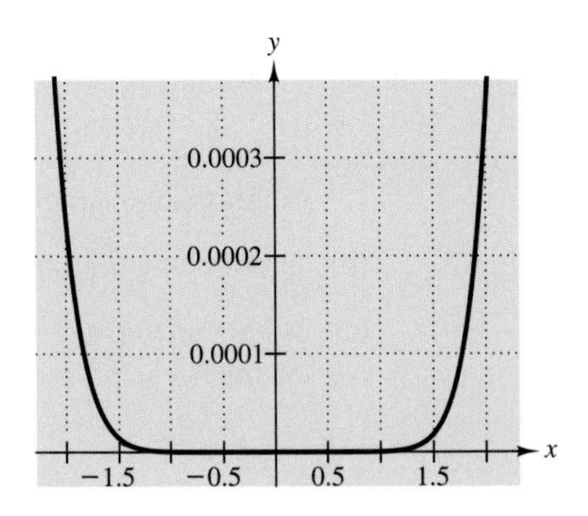

Checkpoint 2

(a) Graph e^x and $P_9(x)$ on a large enough interval to see the two graphs separate.

(b) Graph the error on the same interval as in part (a). Choose an appropriate vertical scale.

(c) Explain what you see in parts (a) and (b).

Figure 11.18 reminds us that we could have solved the problem in Example 3 with our graphing tool—no algebra needed—provided we knew that $G(x)$ really is e^x. But the algebraic procedure doesn't require that we know the sum of the series—we can estimate error in partial sums without that knowledge, which means we can estimate sums numerically and be sure that we are reasonably close to the right answer. The engineer who designed your calculator has to be able to do that in order for it to give you accurate values of the exponential (or any other) function. Furthermore, the algebraic procedure doesn't require that we choose n or the interval ahead of time. We explore both of these observations in Exercise 30.

Our estimation of tails of the exponential power series provides a bonus that is important in its own right. For positive values of x, the equation

$$n\text{th tail} = \frac{x^n}{n!}\left[1 + \frac{x}{(n+1)} + \frac{x^2}{(n+1)(n+2)} + \cdots\right]$$

shows that the nth tail is always *larger* than $x^n/n!$. We also know that the power series converges, that is,

$$\lim_{n\to\infty} s_n = S$$

or

$$\lim_{n\to\infty} S - s_n = 0.$$

But $S - s_n = n$th tail, which is bigger than $x^n/n!$. Now we know what happens to the ratio of nth powers to factorials of n as $n \to \infty$.

Exploration Activity 3

(a) Finish the preceding paragraph: For any positive value of x,

$$\lim_{n\to\infty} \frac{x^n}{n!} = \underline{\qquad\qquad}.$$

Finish the explanation as well.

(b) Explain why the same conclusion must be true for negative values of x.

(c) What can you say about the relative growth rates of 10^n and $n!$, i.e., about the relative sizes of these quantities as n becomes large? What about 100^n and $n!$?

For a positive value of x, we know that nth tail $\to 0$ and

$$0 < \frac{x^n}{n!} < n\text{th tail}.$$

It follows that $x^n/n! \to 0$ as well. For a negative value of x, we may look instead at $|x^n/n!| = |x|^n/n!$ and apply the result of part (a)—because $|x|$ is positive. If $|x^n/n!| \to 0$, then $x^n/n!$ also must approach 0 as $n \to \infty$.

This may seem a bit surprising if we look only at relatively small values of n. For example, for $n = 5$, $10^n = 100,000$ and $100^n = 10,000,000,000$—whereas $n!$ is only 120. But no matter how fast the exponentials x^n start growing, the factorials always catch up and overtake the exponentials, eventually growing so much faster that their ratio decreases to 0. We already knew that exponentials (constant base, growing exponent) grow very fast. Now we see that factorials grow much faster. The result of Exploration Activity 3 is demonstrated in an entirely different way in Exercise 31.

Using Geometric Series to Estimate Tails: Logarithmic Series

Here again is the Taylor series for $\ln(1 + x)$:

$$x - \frac{x^2}{2} + \frac{x^3}{3} - \frac{x^4}{4} + \cdots.$$

From Exploration Activity 2 we know that this power series converges if $0 < x \le 1$ and diverges if $x > 1$ or $x < -1$. For $x = -1$, this series is the negative of the harmonic series, which also diverges. But we don't know for sure whether it converges for $-1 < x < 0$. Our technique of comparing tails with geometric series will resolve this question also.

For x in the interval $-1 < x < 0$, let's write $t = -x = |x|$. Then $0 < t < 1$. When we substitute $x = -t$ in the power series for $\ln(1 + x)$, we get

$$-t - \frac{t^2}{2} - \frac{t^3}{3} - \frac{t^4}{4} - \cdots = -\left(t + \frac{t^2}{2} + \frac{t^3}{3} + \frac{t^4}{4} + \cdots\right).$$

We can now study the power series in parentheses without worrying about keeping track of minus signs. If this series converges for $0 < t < 1$, so does the logarithmic series for $-1 < x < 0$. The nth tail of the subject series is

$$
\begin{aligned}
\frac{t^{n+1}}{n+1}+\frac{t^{n+2}}{n+2}+\frac{t^{n+3}}{n+3}+\frac{t^{n+4}}{n+4}+\cdots &= t^{n+1}\left(\frac{1}{n+1}+\frac{t}{n+2}+\frac{t^2}{n+3}+\cdots\right)\\
&< t^{n+1}\left(\frac{1}{n+1}+\frac{t}{n+1}+\frac{t^2}{n+1}+\cdots\right)\\
&= \frac{t^{n+1}}{n+1}(1+t+t^2+t^3+\cdots)\\
&= \frac{t^{n+1}}{(n+1)(1-t)}.
\end{aligned}
$$

Thus, for every positive integer n and every fixed number t between 0 and 1, the nth tail converges. (In contrast to the error calculation for the exponential series, we did not need to introduce a new variable r for the term-to-term ratio in the geometric series; t itself is r.) Furthermore, the expression

$$\frac{t^{n+1}}{(n+1)(1-t)}$$

is a bound on the error after summing n terms.

Example 4 Determine how close

$$-0.75-\frac{0.75^2}{2}-\frac{0.75^3}{3}-\frac{0.75^4}{4}-\cdots-\frac{0.75^{100}}{100}$$

is to $\ln 0.25$.

Solution The sum here is the first 100 terms of the series for $\ln(1+x)$ with $x=-0.75$, so $t=0.75$ and $n=100$. The error after summing 100 terms is no bigger than $0.75^{101}/(101\times 0.25)$, which is approximately 10^{-14}. Thus adding 100 terms of the series for $\ln 0.25$ produces accuracy that exceeds that of most calculators. ■

Checkpoint 3

(a) Show that the size of the nth tail of the logarithmic power series with $x=-\frac{1}{2}$ is less than $1/[2^n(n+1)]$.

(b) Substitute $x=-\frac{1}{2}$ in the logarithmic series and simplify to find an explicit expression for the power series whose nth tail is estimated in part (a).

(c) Find an n for which the estimate in part (a) is less than 0.001.

(d) Add up n terms of the series in part (b), where n is the number you identified in part (c).

(e) What logarithm should be approximated to within 0.001 by your answer to part (d)? Use your calculator to check that the answer in (d) is within 0.001 of the appropriate logarithm.

(f) How big is the actual error in your answer to part (d)? How good was your estimate of the error in part (a)?

Convergence of the Taylor Series for Sine and Cosine

The sine and cosine Taylor series have terms of alternating sign for *every* value of the input variable, but we can not apply the Alternating Series Test to show that these series always converge unless we know that the terms of these series decrease to zero in size. Now we can show this — and, therefore, show that these series do indeed converge for every real input. As with the exponential and logarithmic series, we postpone until the next section the question of whether the series actually converge to "the right stuff," namely, values of the sine and cosine functions.

Here again is the series for the sine function:

$$\sin x = x - \frac{1}{6}\,x^3 + \frac{1}{120}\,x^5 - \cdots + \frac{(-1)^k}{(2k+1)!}\,x^{2k+1} + \cdots.$$

Note first that the series has the *form* of a power series: We may consider the even-numbered powers all to have 0 as coefficient. Having made that observation, we now ignore all those even-power terms because they don't fit our pattern for terms of the series. Thus a convenient numbering is the one indicated in the notation above: Start with $k = 0$, and only count odd-power terms.

Checkpoint 4

Explain why the power series for $\sin x$ is an alternating series for every nonzero value of x, positive or negative.

In order to apply the Alternating Series Test, we have to decide whether the terms of the series *eventually* get small. The question is not trivial for large values of x — check the sizes of the first few terms with $x = 10$. In fact, this was the same problem we had with the exponential series: The size of the kth term is $|x|^m/m!$, where $m = 2k + 1$. But we saw in Exploration Activity 3 that the limiting value of such terms, as m (or k) becomes large, is zero — no matter how big x is.

The Alternating Series Test requires more — not just that the terms approach zero, but that they do so in a disciplined way, with each term larger (in absolute value) than the one following it. Is that eventually true? Let's compare the term numbered k with the term numbered $k + 1$. The absolute values of these terms are, respectively,

$$a_k = \frac{1}{(2k+1)!}\,|x|^{2k+1}$$

and

$$a_{k+1} = \frac{1}{(2k+3)!}\,|x|^{2k+3}.$$

The ratio of a_{k+1} to a_k is

$$\frac{a_{k+1}}{a_k} = \frac{(2k+1)!}{(2k+3)!}\,\frac{|x|^{2k+3}}{|x|^{2k+1}} = \frac{|x|^2}{(2k+3)(2k+2)}.$$

Now, for any fixed x, we can make the expression on the right as small as we like by taking k sufficiently large. In particular, we can make it less than 1 by choosing k to be, say, bigger than $|x|$. (Half that big would be big enough because of the factors of 2 in the denominator.) Thus, eventually, we have a_{k+1} smaller than a_k for all the remaining terms of the series. Now we have satisfied all three conditions of the Alternating Series Test, so we can conclude that (1) the Taylor series for $\sin x$ converges for every real number x, and (2) the error after summing n terms (the ones numbered from 0 to $n-1$) is no bigger than

$$a_n = \frac{1}{(2n+1)!}\,|x|^{2n+1}.$$

Checkpoint 5

(a) Explain the simplification

$$\frac{a_{k+1}}{a_k} = \frac{(2k+1)!}{(2k+3)!}\,\frac{|x|^{2k+3}}{|x|^{2k+1}} = \frac{|x|^2}{(2k+3)(2k+2)}.$$

(b) Calculate the error estimate

$$a_n = \frac{1}{(2n+1)!}\,|x|^{2n+1}$$

for $x = 5$ and $n = 7$.

(c) Calculate the sum of the first 7 terms of the series for $\sin x$ with $x = 5$.

(d) Compare the sum in part (c) with your calculator's approximation to $\sin 5$. How close is the estimate in part (b) to the actual error?

The cosine series is treated in exactly the same way as the sine series. We leave the details to you in Exercise 9.

The Ratio Test

Our estimations of nth tails of series by comparison to geometric series leads to a consistent procedure that we can apply to any series with infinitely many terms. There are cases in which this procedure will not tell us anything about convergence or divergence, but in many other cases it completely resolves the problem. In particular, this procedure will show us why the interval of convergence of a power series must be symmetric with respect to the reference point, except possibly at the endpoints of the interval.

In all our comparison examples we started by taking absolute values. These comparison techniques work only if all the inequalities go in the same direction. We wouldn't be saying much about the size of an nth tail if we said the tail was smaller than something and "something" turned out to be -10^{10}. It's important that all terms of the series being tested are nonnegative.

The effect of always taking absolute values is that we are always dealing with the worst case—no cancellations among the terms being added, so the sum is always getting bigger as we add more terms. If any of the terms of the original series are negative, that can only produce a smaller sum—but not too small, because none of the partial sums can be smaller than the *negative* of the absolute value series. In symbols,

$$-\sum_{k=0}^{\infty} |b_k| \leq \sum_{k=0}^{\infty} b_k \leq \sum_{k=0}^{\infty} |b_k|.$$

If the third of these sums is finite, then so are the other two. If the one in the middle happens to have oscillations, they must die out, because convergence of the absolute value series tells us

$$\lim_{k\to\infty} |b_k| = 0.$$

A geometric series has the special property that the ratio of each term to the preceding term is *constant*. That is, the kth term in

$$1 + r + r^2 + r^3 + \cdots$$

is r^k, and $b_{k+1}/b_k = r^{k+1}/r^k = r$ for every k. The series converges when $|r| < 1$ and diverges otherwise. Thus we can expect a successful comparison to a convergent geometric series if the terms in the series under consideration satisfy

$$\left|\frac{b_{k+1}}{b_k}\right| \leq r$$

for some constant $r < 1$ and for all sufficiently large values of k. In particular, that will be the case if

$$\lim_{k\to\infty} \left|\frac{b_{k+1}}{b_k}\right|$$

is a number less than 1. We can then take r to be a slightly larger number—still less than 1—and our ratios will eventually be that small. We illustrate this idea with the series we have already compared to geometric series.

Exploration Activity 4

(a) For the exponential series,

$$1 + x + \frac{1}{2}\,x^2 + \frac{1}{6}\,x^3 + \cdots + \frac{1}{k!}\,x^k + \frac{1}{(k+1)!}\,x^{k+1} + \cdots,$$

calculate $|b_{k+1}/b_k|$ and simplify. How is this ratio related to the r we found in estimating the size of the nth tail in Example 2?

(b) Find

$$\lim_{k\to\infty} \left|\frac{b_{k+1}}{b_k}\right|$$

for this series. How do you know that $|b_{k+1}/b_k|$ must eventually be smaller than some constant less than 1?

(c) For the logarithmic series,

$$x - \frac{x^2}{2} + \frac{x^3}{3} - \frac{x^4}{4} + \cdots,$$

calculate $|b_{k+1}/b_k|$ and simplify.

(d) Find

$$\lim_{k\to\infty}\left|\frac{b_{k+1}}{b_k}\right|$$

for this series. For what values of x will $|b_{k+1}/b_k|$ eventually be smaller than some constant less than 1? How do you know?

For the exponential series, we find that

$$\left|\frac{b_{k+1}}{b_k}\right| = \frac{k!\,|x|^{k+1}}{(k+1)!\,|x|^k} = \frac{|x|}{k+1}.$$

For $k = n$, this is precisely the ratio of terms in the geometric series we use to estimate the nth tail. Since x is fixed—and independent of k—the limiting value of the ratios is 0, no matter how big x is. As we saw already, for any $k > |x|$, all the ratios will be less than 1 from then on, so the tail has a finite sum. Specifically, we can take r in the geometric series to be $|x|/(|x|+1)$.

For the logarithmic case,

$$\left|\frac{b_{k+1}}{b_k}\right| = \frac{k\,|x|^{k+1}}{(k+1)\,|x|^k} = \frac{k}{k+1}|x|.$$

Recall that in this case we took r to be $|x|$. What we see from our ratio calculation is that the ratio of consecutive terms is always smaller than $|x|$. Thus we can be sure of convergence if $|x| < 1$. We find that the limiting value of the ratios is also $|x|$, which means we can't expect these ratios to stay less than 1 unless $|x| < 1$.

While these calculations are superficially similar, the results are quite different. In the first case, we get a limit of ratios that is 0 no matter what $|x|$ is—so we get a convergent series for every value of x. In the second case, the limiting value depends on x—so we get convergence only if $|x|$ is small enough. In both cases, however, we can say that the interval in which the absolute value series converges has the form

$$-a < x < a,$$

provided we allow the possibility that $a = \infty$.

Exploration Activity 5

(a) What can you say about convergence or divergence of a series $\sum b_k$ if

$$\lim_{k\to\infty}\left|\frac{b_{k+1}}{b_k}\right| > 1?$$

(b) What is

$$\lim_{k\to\infty}\left|\frac{b_{k+1}}{b_k}\right|$$

for the harmonic series

$$1+\frac{1}{2}+\frac{1}{3}+\frac{1}{4}+\cdots+\frac{1}{k}+\cdots ?$$

(c) What is

$$\lim_{k\to\infty}\left|\frac{b_{k+1}}{b_k}\right|$$

for the alternating harmonic series

$$1-\frac{1}{2}+\frac{1}{3}-\frac{1}{4}+\cdots+\frac{(-1)^{k+1}}{k}+\cdots ?$$

(d) What can you say about convergence or divergence of a series $\sum b_k$ if

$$\lim_{k\to\infty}\left|\frac{b_{k+1}}{b_k}\right|=1?$$

If $|b_{k+1}/b_k|$ approaches a limit greater than 1 —or diverges to infinity—then from some point on, $|b_{k+1}|$ is always bigger than $|b_k|$. This means the terms b_k can't decrease in size to 0, so the series must diverge.

For both the harmonic series and the alternating harmonic series, we have

$$\lim_{k\to\infty}\left|\frac{b_{k+1}}{b_k}\right|=\lim_{k\to\infty}\frac{1/(k+1)}{1/k}=\lim_{k\to\infty}\frac{k}{k+1}=1.$$

But the harmonic series diverges, and the alternating harmonic series converges. Thus we cannot conclude anything about convergence of a series from the observation that

$$\lim_{k\to\infty}\left|\frac{b_{k+1}}{b_k}\right|=1.$$

We summarize our observations in the last two Exploration Activities in the Ratio Test.

Ratio Test Given an infinite series $\sum b_k$, let

$$L=\lim_{k\to\infty}\left|\frac{b_{k+1}}{b_k}\right|.$$

If $L<1$, then the series converges.
If $L>1$, then the series diverges.
If $L=1$, then this test gives no information about convergence or divergence.

Checkpoint 6

Determine what the Ratio Test says about convergence of the arctangent series

$$\arctan x = x - \frac{1}{3}x^3 + \frac{1}{5}x^5 - \cdots + (-1)^k \frac{x^{2k+1}}{2k+1} + \cdots.$$

Next we explore what the Ratio Test tells us about convergence of power series in terms of the coefficients of those series. Suppose we have a power series

$$\sum_{k=0}^{\infty} c_k\, x^k$$

for which all the coefficients c_k are different from 0. Then

$$\left|\frac{b_{k+1}}{b_k}\right| = \left|\frac{c_{k+1}\, x^{k+1}}{c_k\, x^k}\right| = \left|\frac{c_{k+1}}{c_k}\right| |x|,$$

and

$$L = \lim_{k\to\infty} \left|\frac{b_{k+1}}{b_k}\right| = |x| \lim_{k\to\infty} \left|\frac{c_{k+1}}{c_k}\right|.$$

Let's write R for the *reciprocal* of the limit on the right:

$$R = \frac{1}{\lim_{k\to\infty} \left|\frac{c_{k+1}}{c_k}\right|}.$$

If the limit of ratios of consecutive coefficients happens to be 0, we let $R = \infty$. Then

$$L = \frac{|x|}{R},$$

and $L < 1$ exactly when

$$|x| < R.$$

If R happens to be ∞, this means $L < 1$ for every x. In fact, $L = 0$ for every x, and 0 is certainly less than 1. Thus the Ratio Test tells us that the series converges for every x in the symmetrical interval

$$-R < x < R$$

and diverges for $x > R$ or $x < -R$. For $x = \pm R$, we get no information. Those are the values of x for which we have to resort to other tests of convergence. For example, if the power series happens to have alternating signs at an endpoint, we can try the Alternating Series Test.

Ratio Test for Power Series Suppose that

$$\sum_{k=0}^{\infty} c_k\, x^k$$

is a power series with all coefficients c_k different from 0. Let

$$R = \frac{1}{\lim\limits_{k\to\infty} \left|\frac{c_{k+1}}{c_k}\right|}.$$

Then the series converges for $-R < x < R$ and diverges for $|x| > R$. The test gives no information about convergence or divergence for $x = \pm R$.

Suppose that $R = 1$ for a particular power series. Then we know that the series converges for each x in the open interval $(-1, 1)$. Without more information, we cannot say whether the series converges at either or both of the endpoints, -1 and 1. The collection of x for which the series converges will be *one* of the following four intervals: $(-1, 1)$, $[-1, 1)$, $(-1, 1]$, or $[-1, 1]$.

Definitions If

$$\sum_{k=0}^{\infty} c_k\, x^k$$

is a power series with all coefficients c_k different from 0, and

$$R = \frac{1}{\lim\limits_{k\to\infty} \left|\frac{c_{k+1}}{c_k}\right|},$$

then R is called the **radius of convergence** of the series, and the interval

$$(-R, R)$$

is called the **open interval of convergence** of the series.

Example 5 Find the open interval of convergence of the series

$$\sum_{k=0}^{\infty} (-1)^k \frac{x^k}{(k+1)\,2^k}.$$

What can be said about convergence at the endpoints of the interval?

Solution We first calculate the limit of absolute values of ratios of consecutive coefficients:

$$\lim_{k\to\infty}\left|\frac{c_{k+1}}{c_k}\right| = \lim_{k\to\infty}\frac{1/[(k+2)\,2^{k+1}]}{1/[(k+1)\,2^k]} = \lim_{k\to\infty}\frac{k+1}{(k+2)\,2} = \frac{1}{2}.$$

The radius of convergence, therefore, is 2. Thus the series converges for

$$-2 < x < 2.$$

At $x = 2$, the series is

$$\sum_{k=0}^{\infty}\frac{(-1)^k}{k+1},$$

which is the alternating harmonic series. We know this series converges—in fact, we know its sum is $\ln 2$. At $x = -2$, the series is

$$\sum_{k=0}^{\infty}\frac{1}{k+1},$$

the harmonic series, which we know diverges.

Conclusion: The series converges for $-2 < x \le 2$ and diverges otherwise. ■

Checkpoint 7

(a) Find the open interval of convergence of the series $\sum_{k=0}^{\infty}\frac{3^k x^k}{k+1}$.

(b) What can you say about convergence of the series at each of the endpoints of the open interval of convergence?

So far we have considered the Ratio Test only for powers series in which every power of x has a nonzero coefficient. The general Ratio Test works for other power series as well, as we illustrate in the next example. We will pursue this further in the exercises.

Example 6 Find the open interval of convergence of the series

$$\sum_{k=0}^{\infty}(-1)^k\frac{x^{2k+1}}{(k+1)\,2^k}.$$

What can be said about convergence at the endpoints of the interval?

Solution This series has only odd-power terms. It starts out

$$x - \frac{x^3}{2\cdot 2} + \frac{x^5}{3\cdot 2^2} - \frac{x^7}{4\cdot 2^3} + \cdots.$$

Using the Ratio Test for series in general, we find

$$\lim_{k\to\infty}\left|\frac{b_{k+1}}{b_k}\right| = \lim_{k\to\infty}\frac{|x|^{2k+3}/[(k+2)\,2^{k+1}]}{|x|^{2k+1}/[(k+1)\,2^k]} = \lim_{k\to\infty}\frac{|x|^2(k+1)}{(k+2)\,2} = \frac{|x|^2}{2}.$$

This limit will be less than 1 when $|x|^2 < 2$, that is, for

$$-\sqrt{2} < x < \sqrt{2}\,.$$

At $x = \sqrt{2}$ we have $x^{2k+1} = (\sqrt{2})^{2k+1} = (\sqrt{2})^{2k}(\sqrt{2}) = 2^k\sqrt{2}$. Thus the series becomes

$$\sum_{k=0}^{\infty} (-1)^k \frac{\sqrt{2}}{k+1},$$

or $\sqrt{2}$ times the alternating harmonic series. We know this converges, and in fact the sum is $\sqrt{2}\ln 2 \approx 0.98$. At $x = -\sqrt{2}$ the signs no longer alternate, and we have a multiple of the harmonic series, which diverges.

Conclusion: The series converges for $-\sqrt{2} < x \leq \sqrt{2}\,$. ■

We see in Example 6 that the effect of having only odd-power terms in a power series is to introduce a factor of $|x|^2$ in the limit of ratios, which leads to a square root in the calculation of endpoints of the open interval of convergence. The same thing happens if there are only even-power terms. We could work out special ratio tests for these cases, but it is simpler to refer back to the general Ratio Test rather than have a plethora of special tests.

In this section we have developed three tests for convergence of series with infinitely many terms. If the series is a power series with variable x, each of these tests leads to an interval of values of x for which the series converges. If x has already been assigned a value (i.e., we are testing a series of constants), then each test may or may not yield an answer about convergence.

In the order considered, our tests are

- the Alternating Series Test (AST),
- Comparison to a Geometric Series (CGS), and
- the Ratio Test (RT).

In a laboratory activity associated with this chapter you may have an opportunity to explore another test,

- the Integral Test (IT).

The order of presentation was appropriate for development of new ideas from familiar ones, but it is not the order in which one would usually use these tests. Confronted with a power series whose radius of convergence is unknown, your best starting point is the Ratio Test, which is just a reorganization and simplification of CGS. It identifies an interval in which the power series definitely converges. If that interval is the entire number line, there is nothing else to do. If the interval has endpoints, the test also tells us that the series diverges outside those endpoints. That leaves only the endpoints to be tested. The Divergence Test is the next thing to think about—if the terms of an endpoint series don't go to zero, it can't converge. If an endpoint series has alternating signs and terms that go to zero in a disciplined way, AST tells you that it converges. If the terms do not alternate in sign, you can check to see if the series is a multiple of a series whose converge or divergence is known, such as the harmonic series. You also can attempt to apply IT.

Answers to Checkpoints

1. (a) For the arctangent series, $|b_k| = |x|^{2k+1}/(2k+1)$. If $|x| > 1$, then the numerator is an increasing exponential function of k, which grows faster than the linear function of k in the denominator. Thus $|b_k| \to \infty$.

   (b) The Divergence Test.

2. (a) We use two graphs, one for negative values of x and one for positive values, because the scales are so different. The left figure shows clear visual separation of $P_9(x)$ from e^x between -3 and -4. On the other side, the separation at this distance from 0 is not so obvious, although it would be if we zoomed in on the right-hand portion of the graph.

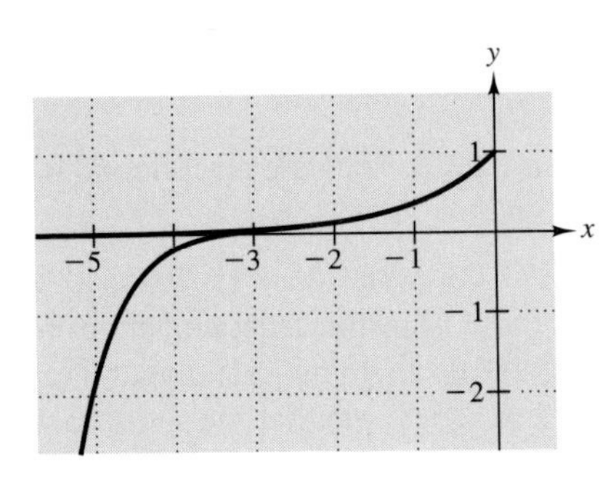

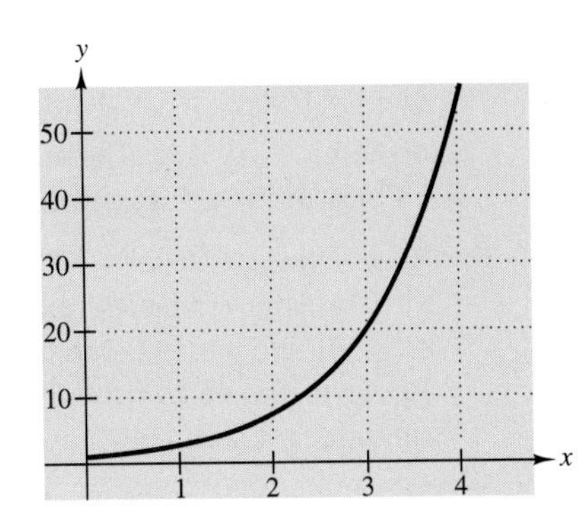

   (b) The graph of $e^x - P_9(x)$ shows that the error is actually larger on the right. However, the error at $x = 4$ is less than 1% of the function value, and at $x = -4$ it is many times the function value.

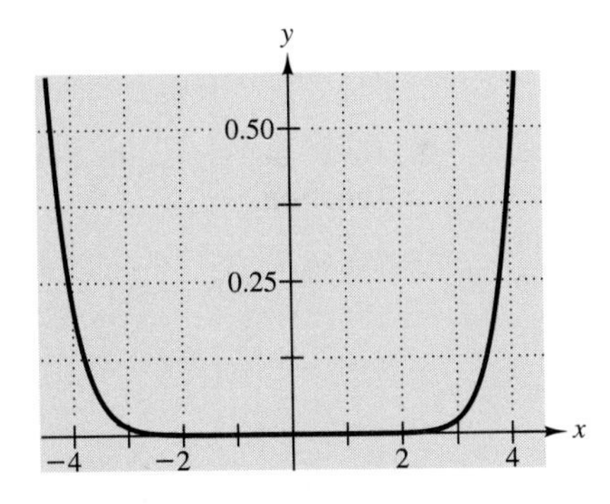

   (c) The error is always positive. That is, $e^x > P_9(x)$ at every x. For positive values of x, this is because the polynomial is the first 10 terms of a series for e^x in which all of the terms are positive. The tail also must be positive. For negative values of x, the explanation is harder, but eventually it is because the odd-degree polynomial must turn down and take negative values, whereas e^x never has negative values. This is why the separation is visually more evident on the left even though the error is larger on the right.

3. (a) For every negative x the error is bounded by $t^{n+1}/[(n+1)(1-t)]$, where $t = |x|$. For $x = -1/2$, this bound is $(1/2^{n+1})/[(n+1)(1/2)] = 1/[2^n(n+1)]$.

   (b) $-\left(\frac{1}{2} + \frac{1}{2 \cdot 2^2} + \frac{1}{3 \cdot 2^3} + \frac{1}{4 \cdot 2^4} + \cdots\right)$

(c) $n = 7$

(d) -0.69226

(e) $\ln 0.5 = -0.69314$

(f) The actual error is 0.000885 as opposed to the estimated error of 0.000977. The error bound is fairly close to the actual error—just over 10% greater than the actual error.

4. For positive values of x, it is clear that the signs alternate because of the factor $(-1)^k$. For negative values of x, $x^{2k+1} = (-|x|)^{2k+1} = (-1)^{2k+1}|x|^{2k+1}$, which is always negative. Thus the sign is always opposite to that of $(-1)^k$, and again the signs alternate.

5. (a) There are two parts of the simplification: First, $|x|^{2n+3} = |x|^2|x|^{2n+1}$. Second, $(2n+3)! = (2n+3)(2n+2)(2n+1)\cdots 3 \cdot 2 \cdot 1 = (2n+3)(2n+2)(2n+1)!$.

(b) 0.0233

(c) -0.93758

(d) $\sin 5 = -0.95892$. The actual error is 0.0213. Thus the error bound is fairly close—within about 10% of the actual error.

6. $|b_{k+1}/b_k| = [|x|^{2k+3}(2k+1)]/[|x|^{2k+1}(2k+3)] = |x|^2(2k+1)]/(2k+3)$. As $k \to \infty$, this approaches $|x|^2$. This limit will be less than 1 for $|x| < 1$. Thus the series converges for $-1 < x < 1$ and diverges for $x > 1$ and $x < -1$. The test doesn't tell us anything about what happens when $x = 1$ or $x = -1$. (However, we know from the Alternating Series Test that the series converges at both endpoints. The sums are $\pi/4$ and $-\pi/4$, respectively.)

7. (a) $-\frac{1}{3} < x < \frac{1}{3}$

(b) At $x = \frac{1}{3}$, the series is the harmonic series—divergent. At $x = -\frac{1}{3}$, the series is the alternating harmonic series—convergent with sum $\ln 2$.

Exercises 11.5

For each of the following series, (a) decide whether the series converges or diverges, and state how you know; (b) if the series converges, state whether you know its value, and give that value if you know one; and (c) if the series converges but you don't know its value, estimate the value with your calculator.

1. $\sum_{n=1}^{\infty} \frac{(-1)^{n+1}}{n^2}$

2. $\sum_{n=1}^{\infty} \frac{(-1)^n}{\sqrt{n}}$

3. $\sum_{n=1}^{\infty} \frac{(-1)^{n+1}}{2^n}$

4. $\sum_{n=1}^{\infty} \frac{(-1)^n}{(\sqrt{2})^n}$

5. $\sum_{n=1}^{\infty} \frac{1}{n+1}$

6. $\sum_{n=1}^{\infty} \frac{(-1)^n n}{n+10}$

7. $\sum_{n=1}^{\infty} \frac{(-1)^{n+1}}{\sin n}$

8. $\sum_{n=1}^{\infty} \frac{(-1)^n}{2n+3}$

9. Here again is the Taylor series for the cosine function:

$$\cos x = 1 - \frac{1}{2}x^2 + \frac{1}{24}x^4 - \cdots + \frac{(-1)^k}{(2k)!}x^{2k} + \cdots.$$

(a) Explain why, for any fixed x,

$$\lim_{k\to\infty} \frac{x^{2k}}{(2k)!} = 0.$$

(b) Explain why, for any fixed x and for sufficiently large values of k,

$$\frac{x^{2k}}{(2k)!} > \frac{x^{2k+2}}{(2k+2)!}.$$

(c) Explain why the cosine series converges for every real number x.

10. (a) Estimate the size of the nth tail of the cosine series, i.e., the error after summing the terms numbered 0 to $n-1$.
   (b) Calculate your error estimate for $x=5$ and $n=7$.
   (c) Calculate the sum of the first 7 terms of the series for $\cos x$ with $x=5$.
   (d) Compare the sum in part (c) with cos 5. How close is the estimate in part (b) to the actual error?

Use the Ratio Test to determine the interval of convergence of each of the following series—even if you already know the interval. If the interval has endpoints, use some other test(s) to determine convergence or divergence at the endpoints.

11. The exponential series

12. The sine series

13. The cosine series

14. The logarithmic series

15. The binomial series (See Exercise 20 in Section 11.3.)

Find the interval of convergence of each of the following series. If the interval has endpoints, determine convergence or divergence at the endpoints as best you can.

16. $\sum_{k=1}^{\infty} k^2 x^k$

17. $\sum_{k=0}^{\infty} \frac{x^k}{3^k}$

18. $\sum_{k=1}^{\infty} \frac{k\,x^k}{2^k}$

19. $\sum_{k=0}^{\infty} (-1)^k x^{2k+1}$

20. $\sum_{k=0}^{\infty} \frac{(-1)^k x^{2k}}{3^k(k+1)}$

21. $\sum_{k=1}^{\infty} \frac{(-1)^{k+1} k!\,x^k}{2^k}$

22. $\sum_{k=0}^{\infty} \frac{(-1)^k x^{2k}}{k!}$

23. $\sum_{k=0}^{\infty} \frac{(-1)^k x^{2k}}{3^k(k+1)!}$

24. $\sum_{k=1}^{\infty} \frac{(-1)^{k+1} 10^k\,x^k}{k!}$

25. Determine how close

$$-\left(0.75 + \frac{0.75^2}{2} + \frac{0.75^3}{3} + \cdots + \frac{0.75^{50}}{50}\right)$$

is to ln 0.25. If your calculator will easily add a sum of 50 terms, use it to check your answer.

26. Determine how close

$$0.75 + \frac{0.75^2}{2} + \frac{0.75^3}{3} + \cdots + \frac{0.75^{50}}{50}$$

is to ln 4. If your calculator will easily add a sum of 50 terms, use it to check your answer.

27. What logarithm is approximated by

$$0.65 + \frac{0.65^2}{2} + \frac{0.65^3}{3} + \cdots + \frac{0.65^{50}}{50}\ ?$$

Determine how close the sum is to that logarithm. If your calculator will easily add a sum of 50 terms, use it to check your answers.

28. (a) Estimate the size of the nth tail of the Taylor series for $\ln(1+x)$ with $x=-\frac{1}{3}$.
   (b) Substitute $x=-\frac{1}{3}$ in the logarithmic series and simplify to find an explicit expression for the series whose nth tail you estimated in part (a).
   (c) Find an n for which the estimate in part (a) is less than 0.0001.
   (d) Add up n terms of the series in part (b), where n is the number you identified in part (c).
   (e) What logarithm should be approximated to within 0.0001 by your answer to part (d)? Use your calculator to check that the answer in (d) is within 0.0001 of the appropriate logarithm.
   (f) How big is the actual error in your answer to part (d)? How good was your estimate of the error in part (a)?

29. Construct an alternating series in the following way: Take the positive terms to be the terms of the harmonic series,

$$1, \frac{1}{2}, \frac{1}{3}, \frac{1}{4}, \ldots .$$

Take the negative terms to be the negatives of terms from a geometric series with ratio $\frac{1}{2}$, that is,

$$1, \frac{1}{2}, \frac{1}{2^2}, \frac{1}{2^3}, \ldots .$$

Thus the alternating series starts out

$$1 - 1 + \tfrac{1}{2} - \tfrac{1}{2} + \tfrac{1}{3} - \tfrac{1}{4} + \tfrac{1}{4} - \tfrac{1}{8} + \tfrac{1}{5} - \tfrac{1}{16} + \cdots .$$

Notice that the terms strictly alternate in sign. Notice also that the limiting value for the terms is 0.

(a) What is the limiting behavior of the partial sums? Why?

(b) Why doesn't this example contradict the Alternating Series Test?

30. (a) Using Example 3 as a model, find a bound for the error when the exponential power series $G(x)$ is approximated by the nth partial sum $P_{n-1}(x)$ on the interval $-a \le x \le a$. Your answer should involve n and a. Check that your formula gives the right estimate if $n = 10$ and $a = 2$.

(b) Estimate the error in the 49th degree approximation on the interval $[-10, 10]$.

(c) Would $P_{49}(x)$ be an adequate way to program the exponential function for a calculator if we assume that only values for $-10 \le x \le 10$ are needed? Explain.

31. Suppose x is any real number, and let M be any fixed integer bigger than x. For any k larger than M, write

$$\frac{x^k}{k!} = \underbrace{\frac{x \cdot x \cdot x \cdots x}{1 \cdot 2 \cdot 3 \cdots M}}_{M \text{ factors}} \times \underbrace{\frac{x \cdot x \cdot x \cdots x}{(M+1) \cdot (M+2) \cdot (M+3) \cdots k}}_{k - M \text{ factors}}.$$

Use this factorization to explain the limiting behavior of $x^k/k!$ as $k \to \infty$.

32. In Exercise 21 of Section 11.3 you calculated five terms of a power series for $f(x) = \sqrt{1+x}$ and found that the terms alternate in sign after the first two. Now you know how to estimate the error when a partial sum of this series is used to approximate the total sum.

(a) Find a bound for the error if you use five terms of this series to approximate $\sqrt{1.2}$.

(b) How does the bound in part (a) compare with the actual error you found in Exercise 21 of Section 11.3?

(c) How many terms of this series would be required to calculate $\sqrt{1.2}$ to three-decimal-place accuracy? Use your calculator to check your answer.

33. (a) Find the first five terms of the Taylor series for $g(x) = \sqrt{1+x^3}$. (You may have done this already in Exercise 22 of Section 11.3.)

(b) Calculate $\int_0^{1/2} \sqrt{1+x^3}\,dx$, accurate to three decimal places. Explain how you know—without relying on the integral key of your calculator—that the answer has that accuracy. Use your calculator's integral key to check your answer.

34. (a) Why doesn't the cube-root function $x^{1/3}$ have a Taylor series at the reference point $x = 0$?

(b) Find a Taylor series for the cube-root function at the reference point $x = 1$.

(c) Find the interval of convergence of the series in part (b). If the interval has endpoints, determine what you can about convergence at the endpoints.

35. Here's an idea for getting better estimates from your calculator. In Table 11.8 we give terms and partial sums of the alternating harmonic series for moderate-sized values of n, along with errors in estimating the total sum, which happens to be ln 2. Notice that each added term overshoots the eventual answer, on the high side if the term is positive, on the low side if it is negative. Since the terms are roughly the same size, we could get much closer by adding only half the next term to a given partial sum. Fill in the last column of the table to see what we mean.

36. Use the idea of the preceding exercise to estimate the sum of the Leibniz series,

$$1 - \frac{1}{3} + \frac{1}{5} - \frac{1}{7} + \cdots + \frac{(-1)^k}{(2k+1)} + \cdots.$$

Record your results in Table 11.9. The exact sum of this series happens to be $\pi/4$. Fill in the last column to see how close your estimates in the next-to-last column really are.

Table 11.8 Estimating ln 2 by corrected partial sums

n	$\frac{(-1)^{n+1}}{n}$	*partial sum*	$\lvert$*ln 2 − partial sum*$\rvert$	*partial sum* + $\frac{1}{2}$ *(next term)*	$\lvert$*error*$\rvert$
5	0.2000	0.7833	0.0902	0.7000	0.00685
6	− 0.1667	0.6167	0.0764		
7	0.1429	0.7595	0.0664		
8	− 0.1250	0.6345	0.0586		
9	0.1111	0.7456	0.0525		
10	− 0.1000	0.6456	0.0475		

Table 11.9 Estimating $\frac{\pi}{4}$ by corrected partial sums

n	$\frac{(-1)^n}{(2n+1)}$	*partial sum*	$\left\lvert\frac{\pi}{4}-\right.$ *partial sum*$\rvert$	*partial sum* + $\frac{1}{2}$ *(next term)*	$\lvert$*error*$\rvert$
5	−0.0909	0.7440	0.0414		
6					
7					
8					
9					
10					

37. Use the idea of the two preceding exercises to estimate

$$1-\frac{1}{4}+\frac{1}{9}-\frac{1}{16}+\cdots+\frac{(-1)^{k+1}}{k^2}+\cdots.$$

Record your results in Table 11.10. How many digits in your last estimate do you think match the exact answer? Why?

Table 11.10 Estimating a total sum by corrected partial sums

n	$\frac{(-1)^{n+1}}{n^2}$	*partial sum*	*partial sum* + $\frac{1}{2}$ *(next term)*
5	0.0400	0.8386	
6			
7			
8			
9			
10			

38. Here's another idea for getting better estimates from your calculator. To establish convergence of the Taylor series for e^x, we used a geometric series to calculate a bound on the nth tail that works as soon as n is as large as $|x|$. We then showed that the bound on the tail approaches zero as n becomes large. But even if n is not very large, our calculated bound,

$$|n\text{th tail}| \le \frac{|x|^n}{n!(1-r)} \qquad \text{where } r = \frac{|x|}{n+1},$$

can be a good *estimate* of the actual size of the tail, especially if the terms are all positive. Thus, for positive values of x, we conclude that

$$n\text{th tail} \approx \frac{x^n}{n!(1-r)} \qquad \text{where } r = \frac{x}{n+1}.$$

We can get a good estimate of the total sum by adding a good estimate of the tail to the corresponding partial sum. Check this out (with $x = 2$) by filling in the blanks in Table 11.11 to see how corrected partial sums of

$$1 + 2 + \frac{2^2}{2!} + \frac{2^3}{3!} + \frac{2^4}{4!} + \cdots$$

can approximate e^2. Recall that the nth partial sum includes the terms numbered from 0 through $n - 1$, so its value is $P_{n-1}(2)$. The correction

$$n\text{th tail} \approx \frac{x^n}{n!(1-r)} \qquad \text{where } r = \frac{x}{n+1}.$$

amounts to $1/(1 - r)$ times what would be the next term, i.e., the highest-power term in $P_n(2)$.

Table 11.11 Estimating e^2 by corrected partial sums

	Partial sum		*Correction*	*Partial sum + correction*	*Error in corrected sum*
n	$P_{n-1}(2)$	$r = \frac{2}{n+1}$	$\frac{2^n}{n!(1-r)}$	$P_{n-1}(2) + \frac{2^n}{n!(1-r)}$	$e^2 - \left[P_{n-1}(2) + \frac{2^n}{n!(1-r)}\right]$
5	7.000	0.1818	0.3259	7.3259	0.0631
6	7.2667				
7					
8					
9					
10					

11.6 Convergence to the Right Function

Early in this chapter we began writing equals signs between function formulas such as e^x and the Taylor series generated by letting the degrees of Taylor polynomials run on to infinity. We did this on the strength of visual evidence that suggested the Taylor series really does converge to values of the function from which the series was generated. In this section, we examine two questions that look quite different but that turn out to have a strong connection:

1. Are the functions defined by Taylor series really *equal* to the functions from which the series were generated — and how can we tell?

2. Can we use functions defined directly by series to solve important problems — e.g., initial value problems?

The connecting link between these two questions is the basic assumption that has undergirded much of our course: *Every initial value problem has a unique solution.* If we can show that Taylor series satisfy the same initial value problems as the functions from which they were derived, then they will have to be the same functions.

In order to know whether a function defined by a power series satisfies a differential equation, we will have to know something about how such functions are differentiated. We take up that question first. Then, in the following subsection, we answer questions 1 and 2.

Derivatives of Functions Defined by Power Series

As we did in Section 11.5, we define a function G by the exponential power series:

$$G(x) = 1 + x + \frac{1}{2}\,x^2 + \frac{1}{6}\,x^3 + \cdots + \frac{1}{k!}\,x^k + \frac{1}{(k+1)!}\,x^{k+1} + \cdots.$$

Since we now know that this series converges for every real number x, the domain of G contains all real numbers. Consider the series we obtain by differentiating each term and adding the resulting derivatives:

$$0 + 1 + x + \frac{1}{2}\,x^2 + \cdots + \frac{1}{(k-1)!}\,x^{k-1} + \frac{1}{k!}\,x^k + \cdots.$$

It is the same series! This is not too surprising, as we know that the derivative of e^x is again e^x.

Our derivations of Taylor series for the natural logarithm and arctangent functions (Section 11.3) suggest that it is legitimate to integrate power series term by term, and we have just suggested that it is legitimate to differentiate power series term by term. Both these assertions are true, at least in the open interval of convergence, $-R < x < R$, where R is the radius of convergence. In fact, the power series that results from either integration or differentiation will have the same radius of convergence. We are not in a position to prove any of this — these facts would be established in a course in advanced calculus or real analysis. But the analogy between polynomials and power series ("polynomials of infinite degree") is compelling: *Power series can be differentiated and integrated term by term,* so the calculus of functions defined by power series is no more complicated than the calculus of polynomials.

Our term-by-term differentiation shows that $G'(x) = G(x)$. That is, G is a solution of the differential equation $y' = y$. Furthermore, it is clear that $G(0) = 1$, so G is a solution of the same initial value problem that led us to the natural exponential function back in Chapter 2. According to our fundamental assumption for this course, there can be only one solution to any initial value problem, so we must have $G(x) = e^x$ for every real number x —that is, G *is* the exponential function.

Checkpoint 1

Here again is the series for the sine function:

$$\sin x = x - \frac{1}{6}x^3 + \frac{1}{120}x^5 - \cdots + \frac{(-1)^{k-1}}{(2k-1)!}x^{2k-1} + \frac{(-1)^k}{(2k+1)!}x^{2k+1} + \cdots .$$

Differentiate the series term by term, and explain the results.

We summarize our assertion about derivatives of functions defined by power series:

Derivative formula for power series If f is a function defined by a power series, i.e.,

$$f(x) = c_0 + c_1x + c_2x^2 + c_3x^3 + \cdots,$$

then the derivative of f is given by

$$f'(x) = c_1 + 2c_2x + 3c_3x^2 + 4c_4x^3 + \cdots .$$

This series converges and the formula is valid for all numbers x such that $|x| < R$, where R is the radius of convergence for the series defining $f(x)$.

The exponential example illustrates how we will answer Questions 1 and 2, and it also shows how the two questions are related. We now apply these ideas to the sine and cosine functions — simultaneously.

Exploration Activity 1

(a) Show that sin x and cos x are both solutions of the differential equation $y'' + y = 0$.

(b) Show that the functions defined by the series

$$f(x) = x - \frac{1}{6}x^3 + \frac{1}{120}x^5 - \cdots + \frac{(-1)^k}{(2k+1)!}x^{2k+1} + \cdots$$

and

$$g(x) = 1 - \frac{1}{2}x^2 + \frac{1}{24}x^4 - \cdots + \frac{(-1)^k}{(2k)!}x^{2k} + \cdots$$

are both solutions of the differential equation $y'' + y = 0$. (See Checkpoint 1.)

(c) Show that $g(x)$ is a solution of the initial value problem

$$y'' + y = 0 \qquad \text{with } y(0) = 1 \text{ and } y'(0) = 0.$$

(d) What initial value problem is satisfied by $f(x)$?

(e) Explain why $f(x) = \sin x$ and $g(x) = \cos x$.

We learned back in Chapter 6 that both $\sin x$ and $\cos x$ are solutions of $y'' + y = 0$. As we saw in Checkpoint 1, $f'(x) = g(x)$, and a similar calculation shows that $g'(x) = -f(x)$. It follows that $f''(x) = -f(x)$ and $g''(x) = -g(x)$. Thus both f and g are also solutions of $y'' + y = 0$. By direct substitution, we see that $g(0) = 1$ and $g'(0) = -f(0) = 0$. Similarly, $f(0) = 0$ and $f'(0) = g(0) = 1$, so $f(x)$ is a solution of the initial value problem $y'' + y = 0$, with $y(0) = 0$ and $y'(0) = 1$. But this means that $f(x)$ and $\sin x$ are both solutions of the same initial value problem, so they must be the same function. Similarly, $g(x)$ and $\cos x$ are solutions of the same initial value problem, so they must be the same function.

Series Solutions of Initial Value Problems

We turn our attention now to an initial value problem similar to that in Exploration Activity 1:

$$y'' - y = 0 \qquad \text{with } y(0) = 1 \text{ and } y'(0) = 0.$$

This is a problem for which we can find a solution defined by a series, but we may not recognize the series as the Taylor series for some well-known function.

Checkpoint 2

Show that the exponential function e^x is a solution of the differential equation $y'' - y = 0$, but that it does not satisfy the initial conditions.

Comparison of our new initial value problem with the one in Exploration Activity 1(c),

$$y'' + y = 0 \qquad \text{with } y(0) = 1 \text{ and } y'(0) = 0,$$

suggests a way to find a solution of the new one: We might be able to describe a solution by means of a series similar to that for $\cos x$, but without the minus signs. The factorial coefficients should work out right to make $y'' = y$, the constant term will make $y(0) = 1$, and the absence of a first-power term will make $y'(0) = 0$.

Checkpoint 3

Show that the function $y = h(x)$ defined by

$$h(x) = 1 + \frac{x^2}{2} + \frac{x^4}{4!} + \cdots + \frac{x^{2k}}{(2k)!} + \cdots$$

is the unique solution of the initial value problem

$$y'' - y = 0 \qquad \text{with } y(0) = 1 \text{ and } y'(0) = 0.$$

Checkpoint 3 leaves us with two questions:

- Where is $h(x)$ defined, i.e., for what values of x does its series converge?
- How can we calculate values $h(x)$ for specific values of x? In other words, if we plan to use the partial sums of the series to approximate $h(x)$, how many terms do we need?

We can't answer these questions as we did for the sine or cosine series—using the Alternating Series Test—because the terms of the series defining h all have the same sign. We could answer the first question by using the Ratio Test—but not the second. However, we can use comparison with a geometric series to answer both questions at the same time. In this case the terms of the series are positive for every value of x, so we don't have to worry about signs or absolute values.

So how big is the nth tail of the $h(x)$ series, i.e., what's left after summing the terms numbered 0 to $n-1$? Here it is:

$$n\text{th tail} = \frac{x^{2n}}{(2n)!} + \frac{x^{2n+2}}{(2n+2)!} + \frac{x^{2n+4}}{(2n+4)!} + \cdots.$$

When we factor out the leading term, we get

$$n\text{th tail} = \frac{x^{2n}}{(2n)!}\left[1 + \frac{x^2}{(2n+2)(2n+1)} + \frac{x^4}{(2n+4)(2n+3)(2n+2)(2n+1)} + \cdots\right].$$

Checkpoint 4

Write a reason for each of the following assertions.

(a)
$$n\text{th tail} \le \frac{x^{2n}}{(2n)!}\left[1 + \left(\frac{x}{2n+1}\right)^2 + \left(\frac{x}{2n+1}\right)^4 + \left(\frac{x}{2n+1}\right)^6 + \cdots\right].$$

(b) For any $n > |x|/2$,

$$n\text{th tail} \le \frac{x^{2n}}{(2n)!\,(1-r)} \qquad \text{where } r = \left(\frac{x}{2n+1}\right)^2.$$

(c) The series defining $h(x)$ converges for all real numbers x.

We have answered the first question about $h(x)$: The series converges for all x. Now we consider approximating $h(x)$ by polynomials. Suppose we want to approximate $h(x)$ on the interval $-3 \le x \le 3$, and we want the error in the approximation to be less than 0.0001. If we approximate by the polynomial

$$1 + \frac{x^2}{2} + \frac{x^4}{4} + \cdots + \frac{x^{2n-2}}{(2n-2)!},$$

then the error in the approximation is the nth tail, for which we have the estimate

$$n\text{th tail} \le \frac{x^{2n}}{(2n)!\,(1-r)} \qquad \text{where } r = \left(\frac{x}{2n+1}\right)^2.$$

For any one value of n, this estimate is as large as possible in the interval $-3 \le x \le 3$ when $x = 3$ or -3, because that makes x^{2n} as large as possible and $1 - r$ as small as possible. Let ρ denote the value of r when $x = 3$: $\rho = 9/(2n+1)^2$. Then, for any n that is 2 or larger,

$$n\text{th tail} \le \frac{9^n}{(2n)!(1-\rho)} \qquad \text{where } \rho = \frac{9}{(2n+1)^2}.$$

Checkpoint 5

(a) Why did we say "any n that is 2 or larger"? (Think about the size of ρ for $n = 0$ and $n = 1$.)

(b) Use the inequality above to fill in the following table of estimates for tails of the $h(x)$ series in the interval $-3 \le x \le 3$.

n	ρ	***Error estimate***
2	0.36	5.27
3	0.1837	
4		0.183
5		
6		
7		
8		

(c) How many terms of the series are needed to calculate values of $h(x)$ on the interval $[-3, 3]$ with an error no worse than 0.0001? Calculate $h(-3)$ and $h(2)$ to within this accuracy.

(d) Graph your polynomial function on the interval $[-3, 3]$. Have you seen this function somewhere before?

We have now solved our initial value problem

$$y'' - y = 0 \qquad \text{with } y(0) = 1 \text{ and } y'(0) = 0.$$

We have found the function

$$h(x) = 1 + \frac{x^2}{2} + \frac{x^4}{4!} + \cdots + \frac{x^{2k}}{(2k)!} + \cdots \quad \text{for all } x$$

that is the unique solution of the problem. And we have demonstrated that we can calculate values of $h(x)$ without knowing a closed form formula for those values. As it happens, there is a closed form formula for this function, and we have seen it before—see Exercise 2 in this section.

Series Calculation of the Error Function

In Chapter 10 we introduced the error function:

$$\operatorname{erf}(t) = \frac{2}{\sqrt{\pi}} \int_0^t e^{-s^2}\, ds.$$

According to the Fundamental Theorem of Calculus, $\operatorname{erf}(t)$ is a constant multiple of a particular antiderivative of the function e^{-t^2}. We have seen that $\operatorname{erf}(t)$ can be used to calculate values of the standard normal distribution function. If you ask a computer algebra system to integrate $e^{-t^2/2}$, it will probably respond with an answer that involves the error function.

Also in Chapter 10 we cited computer evidence that the integral definition is not an efficient way to evaluate $\operatorname{erf}(t)$—nor is it the way the computer program itself evaluates this function. This suggests the possibility of a formula for $\operatorname{erf}(t)$ that requires much less arithmetic than would be required for numerical evaluation of an integral. We raised a similar issue with respect to the elementary functions $\exp x$, $\sin x$, $\cos x$, $\ln(1+x)$, and $\arctan x$ (and for all the other function buttons on your calculator): How can these functions be calculated quickly by a device that does nothing but arithmetic? For the five functions just mentioned, we have an answer of sorts in their Taylor series formulas, at least for values of x that are not too far from zero. Our answer is, "Add up enough terms of the series to get a partial sum that is acceptably close to the total sum."

The tools developed in Section 11.5—the Alternating Series Test and comparison with a geometric series—give us ways to determine a priori the number of terms needed to produce any desired accuracy with any of the important polynomial approximations to functions. For values of x that are too far from zero for use of these formulas, we can first relate the input x to another number close to zero (e.g., by using the cycles of the trigonometric functions) and then apply the series formula. Exactly how to do this depends on the function in question, and in some cases this transformation is quite intricate. We develop this idea in more detail in Exercises 18 through 20 at the end of the chapter. Also see Exercises 13 and 14 in Section 11.2

Returning to the question of evaluating $\operatorname{erf}(t)$, we find that we have hints scattered through the chapter. We know how to find a series for the exponential function:

$$e^x = 1 + x + \frac{1}{2}x^2 + \frac{1}{6}x^3 + \cdots + \frac{1}{k!}x^k + \cdots.$$

We know that we can substitute expressions for the independent variable in series expansions, as we did in Section 11.3 to find series for $1(1+t)$ and $1/(1+t^2)$. So we should be able to find a series for e^{-t^2} by substituting $-t^2$ for x in the exponential series. We know that we can integrate a series expansion term by term, as we did in Section 11.3, to find a Taylor series for the antiderivative. Let's follow the steps just outlined and see what happens.

When we substitute $-t^2$ for x in the series for e^x, we get

$$\begin{aligned} e^{-t^2} &= 1 + (-t^2) + \frac{1}{2}(-t^2)^2 + \cdots + \frac{1}{k!}(-t^2)^k + \cdots \\ &= 1 - t^2 + \frac{1}{2}t^4 + \cdots + \frac{(-1)^k}{k!}t^{2k} + \cdots. \end{aligned}$$

Since we know that the exponential series converges to e^x for every number x, we also know that the new series converges to e^{-t^2} for every number t. Now we integrate both sides from , say, 0 to x:

$$\begin{aligned}\int_0^x e^{-t^2}dt &= \int_0^x 1\,dt - \int_0^x t^2\,dt + \frac{1}{2}\int_0^x t^4\,dt + \cdots + \frac{(-1)^k}{k!}\int_0^x t^{2k}\,dt + \cdots \\ &= t\Big|_0^x - \frac{t^3}{3}\Big|_0^x + \frac{t^5}{2!\cdot 5}\Big|_0^x - \frac{t^7}{3!\cdot 7}\Big|_0^x + \cdots + \frac{(-1)^k\,t^{2k+1}}{k!\cdot(2k+1)}\Big|_0^x + \cdots \\ &= x - \frac{x^3}{3} + \frac{x^5}{2!\cdot 5} - \frac{x^7}{3!\cdot 7} + \cdots + \frac{(-1)^k\,x^{2k+1}}{k!\cdot(2k+1)} + \cdots.\end{aligned}$$

Thus $\operatorname{erf}(x)$ can be evaluated by substituting a value for x in this series, calculating enough terms to achieve the desired accuracy, and multiplying the result by $2/\sqrt{\pi}$.

Is this correct? A computer algebra system tells us that $\operatorname{erf}(1) = 0.84270079$, to the eight decimal places shown. When we divide this by $2/\sqrt{\pi}$, we get 0.74682413, which should be the result of substituting $x = 1$ in the series shown above. Let's make the substitution and start calculating partial sums:

$$\begin{aligned}&1 &&= 1\\ &1-\frac{1}{3} &&= 0.66666667\\ &1-\frac{1}{3}+\frac{1}{10} &&= 0.76666667\\ &1-\frac{1}{3}+\frac{1}{10}-\frac{1}{42} &&= 0.74285714\\ &1-\frac{1}{3}+\frac{1}{10}-\frac{1}{42}+\frac{1}{216} &&= 0.74748677\\ &1-\frac{1}{3}+\frac{1}{10}-\frac{1}{42}+\frac{1}{216}-\frac{1}{1320} &&= 0.74672920\\ &1-\frac{1}{3}+\frac{1}{10}-\frac{1}{42}+\frac{1}{216}-\frac{1}{1320}+\frac{1}{9360} &&= 0.74683603\\ &1-\frac{1}{3}+\frac{1}{10}-\frac{1}{42}+\frac{1}{216}-\frac{1}{1320}+\frac{1}{9360}-\frac{1}{75600} &&= 0.74682281\\ &1-\frac{1}{3}+\frac{1}{10}-\frac{1}{42}+\frac{1}{216}-\frac{1}{1320}+\frac{1}{9360}-\frac{1}{75600}+\frac{1}{685440} &&= 0.74682427\\ &1-\frac{1}{3}+\frac{1}{10}-\frac{1}{42}+\frac{1}{216}-\frac{1}{1320}+\frac{1}{9360}-\frac{1}{75600}+\frac{1}{685440}-\frac{1}{6894720} &&= 0.74682412\end{aligned}$$

Thus with only ten terms of the series we have seven-place accuracy in the number we expected to calculate. It should not be hard to see now how values of $\operatorname{erf}(t)$ — or of the normal distribution function — can be calculated by simple arithmetic. This is much less arithmetic than would be required for a numerical integration involving many values of an exponential function.

Checkpoint 6

(a) Check our calculation of the denominators of the fractions in the preceding table. Our formula says they should be the numbers $k!\,(2k+1)$ for $k = 0, 1, \ldots, 9$. Are they?

(b) Explain why the series for erf (x) will have alternating signs for every value of x, positive or negative.

(c) Use the Alternating Series Test to show that, for large enough values of n, the error after summing n terms must be smaller than the absolute value of the $(n+1)$th term. The error in our calculation above is about 10^{-8}. Is that smaller than the eleventh term?

In this section we have seen that we can use functions defined by power series to solve important problems—in particular, initial value problems. We have used power series solutions of differential equations in two ways. First, by solving an initial value problem satisfied by a known function, we have shown that the Taylor series for that function converges to the function it's supposed to converge to. Second, by estimating tails of power series representations for unknown functions, we have found polynomial approximations—with error bounds—for solutions to initial value problems. Since any arithmetic machine can evaluate a polynomial expression, this could be how your calculator or computer finds very accurate values of nonpolynomial functions. The actual routines programmed in these devices are somewhat more complicated—for greater efficiency—but the idea is the same. Finally, we combined ideas developed throughout the chapter to solve a problem left over from Chapter 10: how to calculate values of the error function efficiently.

Answers to Checkpoints

1. The term by term derivative is $1 - \frac{1}{2}x^2 + \frac{1}{24}x^4 - \cdots + \frac{(-1)^k}{(2k)!}x^{2k} + \cdots$. This is the Taylor series for the cosine function—no surprise, since the cosine function is the derivative of the sine function.

2. $y(x) = y'(x) = y''(x) = e^x$, so e^x satisfies $y'' - y = 0$, but $y'(0) = 1 \neq 0$, so one of the initial conditions is not satisfied.

3. If $y = h(x)$, then

$$h'(x) = x + \frac{x^3}{3!} + \frac{x^5}{5!} + \cdots + \frac{x^{2k-1}}{(2k-1)!} + \cdots.$$

and

$$h''(x) = 1 + \frac{x^2}{2} + \frac{x^4}{4!} + \cdots + \frac{x^{2k-2}}{(2k-2)!} + \cdots.$$

Thus $y'' = y$. Also, by direct substitution, $h(0) = 1$ and $h'(0) = 0$.

4. (a) In each denominator, we replaced $2n+2$, $2n+3$, etc., by $2n+1$. The smaller denominator makes each fraction bigger.

   (b) The series inside the brackets is the geometric series $1 + r + r^2 + \cdots$. If $n > |x|/2$, then $2n+1 > x$, and $r < 1$, so the series converges to $1/(1-r)$.

   (c) For any real number x there is an integer $n > |x|/2$. Part (b) shows that the tail after summing that many terms is finite, so the entire series is finite. That is, $h(x)$ is defined for every real number.

5. (a) For $n = 0$, $\rho = 9$, and for $n = 1$, $\rho = 1$. The bound on the nth tail is not correct until n is large enough that $\rho < 1$. That is the case for $n \geq 2$.

   (b) For $n = 8$, $\rho = 0.03114$, and the error bound is less than 3×10^{-6}.

(c) Seven terms are needed. Using seven terms, we find $h(-3) = 10.0676$ and $h(2) = 3.7621955$ to within 0.0001. In fact, all the digits shown here are correct except the last digit of each number. The actual errors are about 0.00006 at $x = -3$ and 2×10^{-7} at $x = 2$. (See Exercise 2 for a more accurate way to calculate values of h.)

(d)

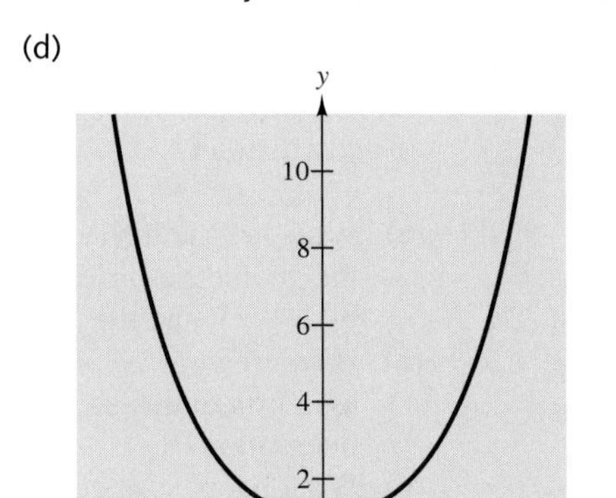

You may have seen this function before—see Exercise 6.

6. (a) For example, $9! \times 19 = 6,894,720$

(b) For $x > 0$, x^{2k+1} is always positive. For $x < 0$, x^{2k+1} is always negative. In either case, the factor $(-1)^k$ causes the product $(-1)^k x^{2k+1}$ to alternate in sign.

(c) Part (b) ensures that the series is alternating. The ratio of absolute values of consecutive terms is

$$\frac{|x|^{2k+3} k!\,(2k+1)}{(k+1)!\,(2k+3)|x|^{2k+1}} = \frac{|x|^2(2k+1)}{(k+1)(2k+3)} < \frac{|x|^2}{k+1},$$

and this approaches 0 for every x. This ensures that eventually each term is smaller in absolute value than the preceding one. It also ensures that the series converges, so the terms have to approach zero. Hence the Alternating Series Test applies, and the error after summing n terms must be no bigger than the next term. The size of the eleventh term is $1/(10! \times 21) = 1.3 \times 10^{-8}$. The actual error is about 1×10^{-8}, which is in fact smaller than the eleventh term.

Exercises 11.6

1. Here again is the series for the cosine function:

$$\cos x = 1 - \frac{1}{2}\,x^2 + \frac{1}{24}\,x^4 - \cdots + \frac{(-1)^{k-1}}{(2k-2)!}\,x^{2k-2} + \frac{(-1)^k}{(2k)!}\,x^{2k} + \cdots.$$

Differentiate the series term by term, and explain the results.

2. (a) Recall from Exercise 46 at the end of Chapter 6 that

$$\cosh x = \frac{1}{2}(e^x + e^{-x}).$$

Show that $y = \cosh x$ is a solution of the initial value problem

$$y'' - y = 0 \qquad \text{with } y(0) = 1 \text{ and } y'(0) = 0.$$

(b) Show that the function

$$h(x) = 1 + \frac{x^2}{2} + \frac{x^4}{4!} + \cdots + \frac{x^{2k}}{(2k)!} + \cdots$$

is $\cosh x$ for all x.

3. Show directly that if you average the Taylor series for e^x and for e^{-x}, you obtain the series that defines the function $h(x)$ in the preceding exercise.

4. Graph $y = \cosh x$ together with your polynomial from Checkpoint 5. Start with the interval $[-3, 3]$—do you see one graph or two? How big does the interval have to be before these functions become visually distinct?

5. Find a power series solution of the initial value problem

$$y'' - y = 0 \qquad \text{with } y(0) = 0 \text{ and } y'(0) = 1.$$

6. (a) Recall from Exercise 46 at the end of Chapter 6 that

$$\sinh x = \frac{1}{2}(e^x - e^{-x}).$$

Show that $y = \sinh x$ is a solution of the initial value problem

$$y'' - y = 0 \qquad \text{with } y(0) = 0 \text{ and } y'(0) = 1.$$

(b) Show that the function

$$s(x) = x + \frac{x^3}{3!} + \frac{x^5}{5!} + \cdots + \frac{x^{2k+1}}{(2k+1)!} + \cdots$$

is $\sinh x$ for all x.

7. Show directly that, if you average the Taylor series for e^x and for $-e^{-x}$, you obtain the series that defines the function $s(x)$ in the preceding exercise.

8. Estimate the nth tail of the series in Exercise 6(b). How big does n have to be, relative to $|x|$, to be sure that the nth tail is finite?

9. (a) How many terms of the series in Exercise 6(b) are needed to approximate $\sinh x$ on the interval $[-3, 3]$ to within 0.00005?
(b) Use a polynomial with the number of terms determined in part (a) to approximate $\sinh 3$ and $\sinh(-2)$. Check that your answers have the required accuracy by using the hyperbolic sine function on your calculator or computer.

10. Graph $y = \sinh x$ together with your polynomial from Exercise 9. Start with the interval $[-3, 3]$—do you see one graph or two? How big does the interval have to be before these functions become visually distinct?

11. (a) What are the intervals of convergence of the geometric series, the Taylor series for $\ln(1 + x)$, and the Taylor series for $\arctan x$?
(b) How are your answers to part (a) related to the assertion in this section about term-by-term integration?
(c) What happens at each endpoint of the interval of convergence for each of the three series in part (a)? What can you conclude about how term-by-term integration affects convergence or divergence at endpoints of the interval of convergence?

12. (a) Differentiate the Taylor series for $\ln(1 + x)$ term by term. The resulting power series is a Taylor series for some function. What function?
(b) How is your answer to part (a) related to the assertion in this section about term-by-term differentiation?
(c) What happens at each endpoint of the interval of convergence for each of the logarithmic series and its derivative? What can you conclude about how term-by-term differentiation affects convergence or divergence at endpoints of the interval of convergence?

13. (a) Differentiate the Taylor series for $\arctan x$ term by term. The resulting power series is a Taylor series for some function. What function?
(b) How is your answer to part (a) related to the assertion in this section about term-by-term differentiation?
(c) What happens at each endpoint of the interval of convergence for each of the arctangent series and its derivative? What can you conclude about how term-by-term differentiation affects convergence or divergence at endpoints of the interval of convergence? Is your answer here consistent with your answer to Exercise 12(c)?

14. (a) Show that the function $y = \ln(1 + x)$ satisfies the initial value problem

$$(x + 1)\frac{dy}{dx} = 1, \qquad y(0) = 0.$$

(b) Show that, on the interval $-1 < x < 1$, the Taylor series for $\ln(1 + x)$ satisfies the same initial value problem. It follows that, on its interval of convergence, the Taylor series converges to the function from which it was derived.

15. (a) Show that the function $y = \arctan x$ satisfies the initial value problem

$$(x^2 + 1)\frac{dy}{dx} = 1, \qquad y(0) = 0.$$

(b) Show that, on the interval $-1 < x < 1$, the Taylor series for $\arctan x$ satisfies the same initial value problem. It follows that, on its interval of convergence, the Taylor series converges to the function from which it was derived.

Find the sum of each of the following series. (*Hint*: Find a familiar Taylor series similar to the given one. Then convert one to the other by integrating, differentiating, making a substitution, or some combination of these.)[2]

16. $x + x^3 + \dfrac{x^5}{2!} + \dfrac{x^7}{3!} + \cdots$

17. $2 - 3 \cdot 2x + 4 \cdot 3x^2 - 5 \cdot 4x^3 + \cdots$

18. $\dfrac{x^2}{2} + \dfrac{x^4}{4} + \dfrac{x^6}{6} + \dfrac{x^8}{8} + \cdots$

19. $1 - \dfrac{3x^2}{2!} + \dfrac{5x^4}{4!} - \dfrac{7x^6}{6!} + \cdots$

Find the sum of each of the following series. (*Hint*: Find a power series in x for which some specific value of x yields the given series. Sum the power series as for Exercises 16–19.)[3]

20. $\dfrac{1}{2} + \dfrac{2}{2^2} + \dfrac{3}{2^3} + \dfrac{4}{2^4} + \cdots$

21. $\dfrac{1}{2} + \dfrac{1}{2 \cdot 2^2} + \dfrac{1}{3 \cdot 2^3} + \dfrac{1}{4 \cdot 2^4} + \cdots$

22. $1 + \dfrac{2}{2!} + \dfrac{2^2}{3!} + \dfrac{2^3}{4!} + \dfrac{2^4}{5!} + \cdots$

23. $(0.98)^2 + (0.98)^3 + (0.98)^4 + \cdots$

24. (a) Find a power series representation for $\sin(t^2)$.

(b) Find a power series representation for

$$\int_0^x \sin(t^2)\, dt.$$

(c) Use the result of part (b) to estimate

$$\int_0^1 \sin(t^2)\, dt$$

with six-decimal-place accuracy.

(d) Check your result in part (c) by using the integral key on your calculator.

25. (a) Find a power series representation for

$$\frac{1 - \cos t}{t}.$$

(b) Find a power series representation for

$$\int_0^x \frac{1 - \cos t}{t}\, dt.$$

(c) Use the result of part (b) to estimate

$$\int_0^2 \frac{1 - \cos t}{t}\, dt$$

with four-decimal-place accuracy.

(d) Check your result in part (c) by using the integral key on your calculator.

26. Suppose you are working on a final exam problem that asks for the probability that a random selection from a normally distributed population will lie within 0.8 standard deviations of the mean. You turn to your Standard Normal Probabilities Table, and you find to your horror that your dog has taken a bite out of the left-hand edge of the page. A two-decimal place answer will suffice for the problem you are trying to solve. Thus you need quickly to find a value, accurate to two decimal places, for the integral

$$\frac{1}{\sqrt{2\pi}} \int_0^{0.8} e^{-t^2/2}\, dt.$$

Decide on a strategy for evaluating this integral without using the Standard Normal Probabilities Table, and carry it out. How will you know that your answer is accurate enough to give a correct answer to the final exam problem?

Find the interval of convergence of the Taylor series for each of the following functions and reference points. (See Exercises 41–52 in Section 11.3.)

27. $\ln x, \quad a = 1$

28. $\sin x, \quad a = \dfrac{\pi}{2}$

29. $\cos x, \quad a = \dfrac{\pi}{2}$

30. $e^x, \quad a = 1$

31. $e^{-x}, \quad a = 1$

32. $\arctan x, \quad a = 2$

33. Suppose a function g is defined by a power series:

$$g(x) = c_0 + c_1 x + c_2 x^2 + c_3 x^3 + \cdots + c_k x^k + \cdots.$$

Show that the Taylor series for g is the same power series.

---

2. Adapted from *Calculus Problems for a New Century*, edited by Robert Fraga, MAA Notes No. 28, 1993.
3. Adapted from *Calculus Problems for a New Century*, edited by Robert Fraga, MAA Notes No. 28, 1993.

Chapter 11 Summary

Our study of normal distributions in Chapter 10 led us to the error function,

$$\operatorname{erf}(x) = \frac{2}{\sqrt{\pi}} \int_0^x e^{-t^2}\, dt,$$

which led to the question of how calculators and computers evaluate such functions. That in turn led us to ask how they evaluate more familiar functions, such as $\sin x$ and e^x. In this chapter we have seen one way to evaluate such functions by approximating polynomials.

In particular, we studied the Taylor polynomials (at $x = 0$) for general functions f. These polynomials have the form

$$P_n(x) = \sum_{k=0}^{n} c_k\, x^k \qquad \text{where } c_k = \frac{f^{(k)}(0)}{k!}$$

for $k = 0, 1, 2, \ldots, n$.

For many functions, including the exponential, sine, and cosine functions, as n gets large, $P_n(x)$ approaches $f(x)$ for all x. For other functions, including $\ln x$ and $\arctan x$, this convergence occurs only for x's in a particular interval.

Our investigation of approximating polynomials led to the concept of a "polynomial of infinite degree" or infinite series. For example, the formula

$$e^x = 1 + x + \frac{x^2}{2} + \cdots + \frac{x^k}{k!} + \cdots = \sum_{k=0}^{\infty} \frac{x^k}{k!}$$

contains the same information on approximation of e^x as does listing the individual polynomials $1 + x + x^2/2 + \cdots + x^n/n!$, but in a more succinct form.

We studied several special series that illustrate the subtleties of the notion of convergence. The harmonic series $\sum_{k=1}^{\infty} 1/k$ does not converge — the partial sums grow without bound, but very slowly. This series diverges because the terms $1/k$ do not become small quite fast enough as k becomes large. On the other hand, as you may have seen in a laboratory exercise, $\sum_{k=1}^{\infty} 1/k^2$ does converge (to $\pi^2/6$). The terms $1/k^2$ do become small fast enough to add up infinitely many of them.

A case of special interest is that of alternating series. The prototype is the alternating harmonic series,

$$1 - \frac{1}{2} + \frac{1}{3} - \frac{1}{4} + \frac{1}{5} - \cdots.$$

Here we observed that, in addition to alternating in sign, the terms steadily decrease to 0 in absolute value. The Alternating Series Test tells us that series with these properties always converge, and the error in any partial sum is smaller than the absolute value of the first term not used in the partial sum.

Some of our important series representations do not have alternating sign patterns — or do not have them for all numbers in their domains. To estimate error in these cases, we introduced a tool that is not sensitive to sign patterns: comparison of the tail of a series with a geometric series. This tool is harder to apply than the Alternating Series Test, but it resolved all the remaining questions about whether important series did in fact converge. We consolidated our knowledge of comparison to geometric series in the Ratio Test, a method that is easier to apply but that does not directly provide an error estimate for a partial sum. It does, however, show that all power series have an interval of convergence that is symmetric about the reference point. In fact, it resolves all convergence questions for power series except for what happens at endpoints of the interval of convergence.

Finally, we took up the question of whether a convergent series converges to the right thing, namely, the function its partial sums are supposed to approximate. Our resolution of this question rested on an assumption we have used many times throughout the course: uniqueness of solutions of initial value problems. By showing that both the function being approximated and its Taylor series are solutions of the same initial value problem, we can conclude that the function and its series representation are in fact the same. As a by-product of this investigation, we found that we could describe solutions of some initial value problems by series even if we didn't know (in any other form) what functions were represented by the series.

We concluded the chapter by answering the question that started it — how to calculate the error function without using numerical integration. The answer drew on results and ideas scattered through the chapter: the exponential series, substitution to construct new series, integration term by term, the Alternating Series Test, and approximation of the sum by partial sums.

Concepts

- Taylor polynomials
- Taylor series
- Convergence of sequences and series
- Power series
- Series of constants
- Harmonic and alternating harmonic series
- Leibniz series
- Alternating Series Test
- Comparison to a Geometric Series
- Ratio Test
- Substitution in power series
- Differentiation of power series
- Integration of power series
- Error estimates for approximations
- Series solutions of differential equations

Applications

- Approximation of functions by polynomials

Formulas

Coefficients for Taylor polynomials and series

$$c_k = \frac{f^{(k)}(0)}{k!} \quad \text{for } k = 0,\ 1,\ 2,\ \ldots.$$

Exponential series

$$e^x = 1 + x + \frac{1}{2}\,x^2 + \cdots + \frac{1}{k!}\,x^k + \cdots = \sum_{k=0}^{\infty} \frac{1}{k!}\,x^k$$

Sine series

$$\sin x = x - \frac{1}{6}\,x^3 + \frac{1}{120}\,x^5 - \cdots + \frac{(-1)^k}{(2k+1)!}\,x^{2k+1} + \cdots = \sum_{k=0}^{\infty} \frac{(-1)^k}{(2k+1)!}\,x^{2k+1}$$

Cosine series

$$\cos x = 1 - \frac{1}{2}\,x^2 + \frac{1}{24}\,x^4 - \cdots + \frac{(-1)^k}{(2k)!}\,x^{2k} + \cdots = \sum_{k=0}^{\infty} \frac{(-1)^k}{(2k)!}\,x^{2k}$$

Error function series

$$\operatorname{erf} x = \frac{2}{\sqrt{\pi}}\left[x - \frac{1}{3}x^3 + \cdots + \frac{(-1)^k}{k!\,(2k+1)}\,x^{2k+1} + \cdots\right] = \frac{2}{\sqrt{\pi}}\sum_{k=0}^{\infty}\frac{(-1)^k}{k!\,(2k+1)}\,x^{2k+1}$$

Natural logarithm series

$$\ln(1+x) = x - \frac{1}{2}x^2 + \frac{1}{3}x^3 - \cdots + \frac{(-1)^{k+1}}{k}\,x^k + \cdots = \sum_{k=1}^{\infty}\frac{(-1)^{k+1}}{k}\,x^k, \qquad \text{for } |x| < 1$$

Arctangent series

$$\arctan x = x - \frac{1}{3}x^3 + \frac{1}{5}x^5 - \cdots + \frac{(-1)^k}{2k+1}\,x^{2k+1} + \cdots = \sum_{k=0}^{\infty}\frac{(-1)^k}{2k+1}\,x^{2k+1}, \qquad \text{for } |x| \le 1$$

Geometric series

$$\frac{1}{1-x} = 1 + x + x^2 + x^3 + \cdots + x^k + \cdots = \sum_{k=0}^{\infty} x^k, \qquad \text{for } |x| < 1$$

Binomial series

$$(1+x)^m = 1 + mx + \frac{m(m-1)}{2!}\,x^2 + \frac{m(m-1)(m-2)}{3!}\,x^3 + \cdots + \frac{m(m-1)\cdots(m-k+1)}{k!}\,x^k + \cdots$$

$$= \sum_{k=0}^{\infty}\binom{m}{k}x^k, \text{ for } |x| < 1$$

Harmonic series (divergent)

$$1 + \frac{1}{2} + \frac{1}{3} + \cdots + \frac{1}{k} + \cdots$$

Alternating harmonic series

$$1 - \frac{1}{2} + \frac{1}{3} - \cdots + \frac{(-1)^k}{k} + \cdots = \ln 2$$

The Leibniz series

$$1 - \frac{1}{3} + \frac{1}{5} - \cdots + \frac{(-1)^k}{2k+1} + \cdots = \frac{\pi}{4}$$

Formulas you may have studied in a laboratory activity

p-Series, $p = 2$

$$1 + \frac{1}{4} + \frac{1}{9} + \cdots + \frac{1}{k^2} + \cdots = \frac{\pi^2}{6}$$

p-Series (divergent if $p \le 1$, convergent if $p > 1$)

$$1 + \frac{1}{2^p} + \frac{1}{3^p} + \cdots + \frac{1}{k^p} + \cdots$$

Chapter 11 Exercises

1. (a) Find the Taylor polynomials of degrees 0 through 6 for $\cos x$.

   (b) Use these polynomials to give a pencil-and-paper calculation of $\cos(1)$ that is accurate to within 0.001.

   (c) Use your calculator to check your calculation.

For which of the following series can you conclude that the series converges by using the Alternating Series Test?

2. $\sum_{k=1}^{\infty} (-1)^k \frac{1}{1+\ln k}$

3. $\sum_{k=1}^{\infty} (-1)^k \frac{1}{\sqrt{k}}$

4. $\sum_{k=1}^{\infty} \frac{1}{\sqrt{k}}$

5. $\sum_{k=1}^{\infty} (-1)^k \frac{k}{k+1}$

6. $\sum_{k=1}^{\infty} (-1)^k \, |\sin k|$

To answer Exercises 7 – 14, recall how you used your calculator or computer in Exercise 23 at the end of Section 11.1 and Exercise 26 at the end of Section 11.3.

7. What is the maximum error on the interval $[-1, 1]$ in the Taylor polynomial approximation to $\sin x$ of the indicated degree?

   (a) 1 (b) 3 (c) 5 (d) 7

8. What is the maximum error on the interval $[-1, 1]$ in the Taylor polynomial approximation to $\cos x$ of the indicated degree?

   (a) 2 (b) 4 (c) 6 (d) 8

9. What is the maximum error on the interval $[-\frac{1}{2}, \frac{1}{2}]$ in the Taylor polynomial approximation to $\ln(1+x)$ of the indicated degree?

   (a) 2 (b) 3 (c) 4 (d) 5

10. What is the maximum error on the interval $[-\frac{1}{2}, \frac{1}{2}]$ in the Taylor polynomial approximation to $\arctan x$ of the indicated degree?

    (a) 1 (b) 3 (c) 5 (d) 7

11. On what interval can you be sure that approximation of $\sin x$ by a Taylor polynomial of the indicated degree n has an error no worse than 0.001?

    (a) $n = 1$ (b) $n = 3$ (c) $n = 5$ (d) $n = 7$

12. On what interval can you be sure that approximation of $\cos x$ by a Taylor polynomial of the indicated degree n has an error no worse than 0.001?

    (a) $n = 2$ (b) $n = 4$ (c) $n = 6$ (d) $n = 8$

13. On what interval can you be sure that approximation of $\ln(1+x)$ by a Taylor polynomial of the indicated degree n has an error no worse than 0.001?

(a) $n = 2$ (b) $n = 3$ (c) $n = 4$ (d) $n = 5$

14. On what interval can you be sure that approximation of $\arctan x$ by a Taylor polynomial of the indicated degree n has an error no worse than 0.001?

(a) $n = 1$ (b) $n = 3$ (c) $n = 5$ (d) $n = 7$

15. Give a reason why the error function is not a polynomial function.

16. (a) Graph the function $f(x) = \dfrac{\arctan x}{x}$ on the interval $[-3, 3]$.

(b) Find the fourth-degree Taylor polynomial $P_4(x)$ that approximates $f(x)$ near 0.

(c) Graph $P_4(x)$ together with $f(x)$. Indicate which is which.

(d) Use $\int_0^x P_4(t)\,dt$ to approximate an antiderivative of f on the interval $[-0.5, 0.5]$. What is the maximum possible error in this approximation?

(e) Find $\int_0^{0.7} P_4(x)\,dx$, $\int_0^{0.7} f(x)\,dx$, and the actual error when the first of these integrals is used to approximate the second.

17. (a) Devise a strategy for finding the scale factor c for the gamma density function

$$f(t) = c\sqrt{t}\,e^{-t/2}.$$

There are possible strategies associated with the names "Simpson" and "Taylor." Choose one and carry it out.

(b) Find the mean for this gamma distribution.

18. The nth degree Taylor approximation to e^x is

$$P_n(x) = 1 + x + \frac{x^2}{2!} + \cdots + \frac{x^n}{n!}.$$

For every value of x, $P_n(x)$ converges to e^x as n becomes large, and $P_n(-x)$ converges to e^{-x} as n becomes large. However, for large values of x, the convergence of either family of polynomials will be slow, so there is good reason to do our calculations (in a calculator or computer) with numbers that are close to zero. Devise a method to calculate e^x when x is large (or e^{-x}, if you prefer) by using $P_n(z)$ for some z near zero. [One possibility: Use $z = x/m$, where m is the next integer larger than x, so $z < 1$. Then $e^x = (e^z)^m \approx P_n(z)^m$. But note that the act of raising the approximate value to a large power can magnify error, so you have to be careful about how big that error is to begin with.]

19. (a) Use the Taylor polynomials for $\ln(1+x)$ to derive Taylor polynomials for $\ln(1-x)$ and then for $\ln[(1+x)/(1-x)]$.

(b) Sketch the graph of $y = (1+x)/(1-x)$. Show that as x varies between 0 and $\frac{1}{2}$, y varies between 1 and 3. Thus the Taylor polynomials in part (a), which converge fairly rapidly for $0 \le x \le \frac{1}{2}$, can be used for effective computation of logarithms of numbers from 1 to 3.

(c) Show that the inverse of the function $y = (1+x)/(1-x)$ for x between 0 and $\frac{1}{2}$ is the function $x = (y-1)/(y+1)$ for y between 1 and 3.

(d) What property of the logarithm ensures that logarithms of numbers between 0 and 1 can be computed from logarithms of numbers bigger than 1?

(e) If x is a number bigger than 1, and n is a non-negative integer such that $e^n \le x \le e^{n+1}$, show that $\ln x = n + \ln y$, where y is a number between 1 and 3.

(f) Put all the pieces together to describe an effective way to calculate values of $\ln x$ for arbitrary positive numbers x.

20. (a) Use a standard half-angle formula for tangent to show that

$$\tan\frac{u}{2} = \frac{\tan u}{1+\sqrt{1+\tan^2 u}}.$$

(b) Use part (a) to show that

$$u = 2\arctan\frac{\tan u}{1+\sqrt{1+\tan^2 u}}.$$

(c) Substitute $u = \arctan x$ in the equation in part (b) to show that

$$\arctan x = 2\arctan\frac{x}{1+\sqrt{1+x^2}}.$$

(d) Show that $\dfrac{|x|}{1+\sqrt{1+x^2}} < 1$ for every real number x.

(e) Show that, if $|x| < 1$, then $\dfrac{|x|}{1+\sqrt{1+x^2}} < \dfrac{1}{2}$.

(f) Show that the Taylor polynomial for $\arctan x$ can be used for effective computation of $\arctan x$ for every x by computing instead $4\arctan z$, where

$$z = \frac{y}{1+\sqrt{1+y^2}}$$

and

$$y = \frac{x}{1+\sqrt{1+x^2}}.$$

21. Consider the problem of finding the limiting value as x approaches 0 of a quotient

$$\frac{f(x)}{g(x)},$$

where $f(x)$ and $g(x)$ both approach 0. (This is one of the limiting situations we have called indeterminate. We have encountered this problem often, starting with our earliest attempts to calculate derivatives.) Suppose $f(0)$ and $g(0)$ are both 0, suppose $g'(0)$ is not 0, and suppose $f(x)$ and $g(x)$ both have Taylor series expansions that are valid near $x = 0$. Show that

$$\lim_{x\to 0}\frac{f(x)}{g(x)} = \frac{f'(0)}{g'(0)}.$$

[This is a special case of a result known as **l'Hôpital's Rule**, after the Marquis G. F. A. de l'Hôpital (1661–1701), pronounced "lo-pee-TAHL." The Marquis was the author of the first calculus book (1696), but the discoverer of most of its contents, including his rule, was his tutor, Johann Bernoulli (1667–1748).]

Find each of the following limiting values.

22. $\lim_{x\to 0} \frac{\tan 2x}{x}$

23. $\lim_{x\to 0} \frac{\sin 2x}{\sinh 3x}$

24. $\lim_{x\to 0} \frac{\arctan 5x}{\arcsin 3x}$

L'Hôpital's Rule is not always the best tool for evaluating indeterminate limits. Other options include expanding functions in Taylor series, tracing with a graphing calculator, and using inequalities as we did in Example 3 in Section 11.2. Find each of the following limiting values.

25. $\lim_{x\to \infty} \frac{\ln x}{x}$

26. $\lim_{x\to 0} \frac{\sin x}{x}$

27. $\lim_{x\to 0} \frac{\cos x - 1}{x^2 e^x}$

28. $\lim_{x\to 0} \frac{e^x - e^{-x}}{1 - e^x}$

29. $\lim_{k\to \infty} \frac{99^k}{k!}$

30. $\lim_{x\to 0} \left(\frac{1}{x} - \frac{1}{e^x - 1}\right)$

31. $\lim_{x\to 0} \frac{(\sin x - x)^3}{x\,(1 - \cos x)^4}$

32. $\lim_{x\to \infty} \frac{\arctan 5x}{\arcsin 3x}$

33. $\lim_{x\to 0} \frac{2^x}{3^x}$

34. $\lim_{x\to 0} \frac{2^x - 1}{3^x - 1}$

35. $\lim_{x\to \infty} \frac{2^x}{3^x}$

Chapter 11 Project

Dice We will answer the question: How many rolls of a pair of dice do you expect to make before the first 7 turns up?

(a) Roll a pair of dice until the first 7 appears. Write down the number of rolls. Repeat until you have 10 observations of how long it takes to get the first 7. Write down your conjecture about the average number of rolls to get a 7.

(b) Explain why, each time you roll a pair of dice, the probabilities for the possible outcomes are given by Table 11.12.

Table 11.12 Probabilities for a pair of dice

Outcome n	2	3	4	5	6	7	8	9	10	11	12
Probability of n	$\frac{1}{36}$	$\frac{2}{36}$	$\frac{3}{36}$	$\frac{4}{36}$	$\frac{5}{36}$	$\frac{6}{36}$	$\frac{5}{36}$	$\frac{4}{36}$	$\frac{3}{36}$	$\frac{2}{36}$	$\frac{1}{36}$

(c) Explain why the probability of not rolling a 7 on a given roll of the dice is $5/6$.

(d) Successive rolls of the dice are independent events, so, if the probability of outcome A on one roll is p and the probability of outcome B is q, then the probability of outcome A followed by outcome B is pq. In particular, the probability of not rolling a 7 once is $5/6$, of not rolling a 7 twice in a row is $(5/6)^2$, and so on. What is the probability of not rolling a 7 on any of three consecutive rolls? Four? What is the probability of not rolling a 7 on any of n consecutive rolls?

(e) Now we study the distribution of outcomes t, where t is the number of the roll on which the first 7 turns up. This is similar to our study of failure times in Chapter 10, but this time t can take only integer values $1,\ 2,\ 3,\ \ldots$. What is the probability that $t = 1$? In order for t to be 2, we must fail to roll a 7 on the first roll and then roll a 7 on the second roll — what is the probability that this happens? What is the probability that $t = 3$? That $t = 4$? In general, what is the probability of a given outcome t?

(f) If you answered part (e) correctly, the sum of the probabilities for all possible outcomes should be 1 — is it? [If you get a sum other than 1, go back and straighten out part (e) before continuing.]

(g) Let's denote the answer to the last question in part (e) by $p(t)$. That is, for each possible number t, $p(t)$ is the probability that the first 7 appears on the tth roll of the dice, and, from part (e), you should now know a formula for $p(t)$. As in the continuous case (failure times), the expected value of t is the sum over all values of t of the products $t\,p(t)$, which in this case is really a sum, not an integral. Your formula from part (e) should show that this sum has the form

$$c \sum_{t=1}^{\infty} t\ x^{t-1},$$

where c and x are already known — our task is to evaluate this sum. The sum is the derivative of some other infinite series — what series?

(h) What function of x is the sum of the series in your answer to (g)? What is the derivative of that function? What is the sum of the series in part (g) as an explicit function of x? For what values of x is your answer valid?

(i) Substitute your known values of c and x into the function form you derived in part (h) — the answer also should answer our original question about the expected number of rolls to the first 7. Does the answer agree with your conjecture in part (a)? Does the answer agree with your intuition about the problem? Why or why not?

Appendix: A Short Table of Integrals

Basic Forms

1. $\displaystyle\int u^n\,d\,u = \frac{1}{n+1}u^{n+1} + C \quad (\mathrm{n} \neq -1)$

2. $\displaystyle\int \frac{du}{u} = \ln|u| + \mathrm{C}$

3. $\displaystyle\int e^u\,du = e^u + \mathrm{C}$

4. $\displaystyle\int a^u\,du = \frac{1}{\ln a}\,a^u + \mathrm{C}$

5. $\displaystyle\int \sin u\,du = -\cos u + \mathrm{C}$

6. $\displaystyle\int \cos u\,du = \sin u + \mathrm{C}$

7. $\displaystyle\int \sec^2 u\,du = \tan u + \mathrm{C}$

8. $\displaystyle\int \csc^2 u\,du = -\cot u + \mathrm{C}$

9. $\displaystyle\int \sec u \tan u\,du = \sec u + \mathrm{C}$

10. $\displaystyle\int \csc u \cot u\,du = -\csc u + \mathrm{C}$

11. $\displaystyle\int \tan u\,du = \ln|\sec u| + \mathrm{C}$

12. $\displaystyle\int \cot u\,du = \ln|\sin u| + \mathrm{C}$

13. $\displaystyle\int \sec u\,du = \ln|\sec u + \tan u| + \mathrm{C}$

14. $\displaystyle\int \csc u\,du = \ln|\csc u - \cot u| + \mathrm{C}$

15. $\displaystyle\int \ln u\,du = u\ln u - u + \mathrm{C}$

16. $\displaystyle\int \frac{du}{\sqrt{a^2-u^2}} = \arcsin\frac{u}{a} + \mathrm{C}$

17. $\displaystyle\int \frac{du}{a^2+u^2} = \frac{1}{a}\arctan\frac{u}{a} + \mathrm{C}$

18. $\displaystyle\int \frac{du}{u\sqrt{u^2-a^2}} = \frac{1}{a}\arccos\frac{a}{u} + C$

19. $\displaystyle\int \frac{du}{a^2-u^2} = \frac{1}{2a}\ln\left|\frac{u+a}{u-a}\right| + \mathrm{C}$

20. $\displaystyle\int \frac{du}{u^2-a^2} = \frac{1}{2a}\ln\left|\frac{u-a}{u+a}\right| + \mathrm{C}$

Forms Involving $\sqrt{a^2+u^2}$

21. $\displaystyle\int \sqrt{a^2+u^2}\,du = \frac{u}{2}\sqrt{a^2+u^2} + \frac{a^2}{2}\ln\left|u+\sqrt{a^2+u^2}\right| + \mathrm{C}$

22. $\displaystyle\int \frac{du}{\sqrt{a^2+u^2}} = \ln\left|u+\sqrt{a^2+u^2}\right|u + \mathrm{C}$

23. $\displaystyle\int \frac{\sqrt{a^2+u^2}}{u}\,du = \sqrt{a^2+u^2} - a\ln\left|\frac{a+\sqrt{a^2+u^2}}{u}\right| + \mathrm{C}$

24. $\displaystyle\int \frac{du}{u\sqrt{a^2+u^2}} = -\frac{1}{a}\ln\left|\frac{\sqrt{a^2+u^2}+a}{u}\right| + \mathrm{C}$

25. $\displaystyle\int u^2\sqrt{a^2+u^2}\,du = \frac{u}{4}\sqrt{(a^2+u^2)^3} - \frac{a^2u}{8}\sqrt{a^2+u^2} - \frac{a^4}{8}\ln\left|u+\sqrt{a^2+u^2}\right| + \mathrm{C}$

26. $\displaystyle\int \frac{du}{u^2\sqrt{a^2+u^2}} = -\frac{\sqrt{a^2+u^2}}{a^2u} + \mathrm{C}$

27. $\displaystyle\int \frac{u^2}{\sqrt{a^2+u^2}}\,du = \frac{u}{2}\sqrt{a^2+u^2} - \frac{a^2}{2}\ln\left|u+\sqrt{a^2+u^2}\right| + C$

28. $\displaystyle\int \frac{\sqrt{a^2+u^2}}{u^2}\,du = -\frac{\sqrt{a^2+u^2}}{u} + \ln\left|u+\sqrt{a^2+u^2}\right| + C$

Forms Involving $\sqrt{a^2-u^2}$

29. $\displaystyle\int \sqrt{a^2-u^2}\,du = \frac{u}{2}\sqrt{a^2-u^2} + \frac{a^2}{2}\arcsin\frac{u}{a} + C$

30. $\displaystyle\int \frac{\sqrt{a^2-u^2}}{u}\,du = \sqrt{a^2-u^2} - a\ln\left|\frac{a+\sqrt{a^2-u^2}}{u}\right| + C$

31. $\displaystyle\int \frac{du}{u\sqrt{a^2-u^2}} = -\frac{1}{a}\ln\left|\frac{a+\sqrt{a^2-u^2}}{u}\right| + C$

32. $\displaystyle\int u^2\sqrt{a^2-u^2}\,du = -\frac{a^2}{8}\,u\sqrt{a^2-u^2} + \frac{1}{4}u^3\sqrt{a^2-u^2} + \frac{a^4}{8}\arcsin\frac{u}{a} + C$

33. $\displaystyle\int \frac{\sqrt{a^2-u^2}}{u^2}\,du = -\frac{\sqrt{a^2-u^2}}{u} - \arcsin\frac{u}{a} + C$

34. $\displaystyle\int \frac{du}{u^2\sqrt{a^2-u^2}} = -\frac{\sqrt{a^2-u^2}}{a^2u} + C$

35. $\displaystyle\int \frac{u^2}{\sqrt{a^2-u^2}}\,du = -\frac{u}{2}\sqrt{a^2-u^2} + \frac{a^2}{2}\arcsin\frac{u}{a} + C$

Forms Involving $\sqrt{u^2-a^2}$

36. $\displaystyle\int \sqrt{u^2-a^2}\,du = \frac{u}{2}\sqrt{u^2-a^2} - \frac{a^2}{2}\ln\left|u+\sqrt{u^2-a^2}\right| + C$

37. $\displaystyle\int \frac{du}{\sqrt{u^2-a^2}} = \ln\left|u+\sqrt{u^2-a^2}\right| + C$

38. $\displaystyle\int u^2\sqrt{u^2-a^2}\,du = \frac{u}{4}\left(u^2-a^2\right)^{3/2} + \frac{a^2}{8}\,u\sqrt{u^2-a^2} - \frac{a^4}{8}\ln\left|u+\sqrt{u^2-a^2}\right| + C$

39. $\displaystyle\int \frac{du}{u^2\sqrt{u^2-a^2}} = \frac{\sqrt{u^2-a^2}}{a^2u} + C$

40. $\displaystyle\int \frac{\sqrt{u^2-a^2}}{u^2}\,du = -\frac{\sqrt{u^2-a^2}}{u} + \ln\left|u+\sqrt{u^2-a^2}\right| + C$

41. $\displaystyle\int \frac{u^2}{\sqrt{u^2-a^2}}\,du = \frac{u}{2}\sqrt{u^2-a^2} + \frac{a^2}{2}\ln\left|u+\sqrt{u^2-a^2}\right| + C$

42. $\displaystyle\int \frac{\sqrt{u^2-a^2}}{u}\,du = \sqrt{u^2-a^2} - a\arccos\frac{a}{u} + C$

Forms Involving $\sqrt{2au - u^2}$

43. $\int \sqrt{2au - u^2}\, du = \frac{u-a}{2}\sqrt{2au - u^2} + \frac{a^2}{2} \arcsin\left(\frac{u-a}{a}\right) + C$

44. $\int \frac{du}{\sqrt{2au - u^2}} = \arcsin\left(\frac{u-a}{a}\right) + C$

45. $\int u\sqrt{2au - u^2}\, du = \frac{2u^2 - au - 3a^2}{6}\sqrt{2au - u^2} + \frac{a^3}{2} \arcsin\left(\frac{u-a}{a}\right) + C$

46. $\int \frac{\sqrt{2au - u^2}}{u}\, du = \sqrt{2au - u^2} + a \arcsin\left(\frac{u-a}{a}\right) + C$

47. $\int \frac{du}{u\sqrt{2au - u^2}} = -\frac{\sqrt{2ua - u^2}}{au} + C$

48. $\int \frac{u}{\sqrt{2au - u^2}}\, du = -\sqrt{2au - u^2} + a \arcsin\left(\frac{u-a}{a}\right) + C$

49. $\int \frac{u^2}{\sqrt{2au - u^2}}\, du = -\frac{(u+3a)}{2}\sqrt{2au - u^2} + \frac{3a^2}{2} \arcsin\left(\frac{u-a}{a}\right) + C$

Forms Involving $a + bu$

50. $\int \frac{u}{a + bu}\, du = \frac{u}{b} - \frac{a}{b^2} \ln |a + bu|) + C$

51. $\int u\sqrt{a + bu}\, du = \frac{2}{15b^2}(3bu - 2a)(a + bu)^{3/2} + C$

52. $\int \frac{u^2}{a + bu}\, du = \frac{1}{b^3}\left[\frac{1}{2}(a + bu)^2 - 2a(a + bu) + a^2 \ln |a + bu|\right] + C$

53. $\int \frac{du}{u(a + bu)} = \frac{1}{a} \ln \left|\frac{u}{a + bu}\right| + C$

54. $\int \frac{du}{u^2(a + bu)} = -\frac{1}{au} + \frac{b}{a^2} \ln \left|\frac{a + bu}{u}\right| + C$

55. $\int \frac{u}{(a + bu)^2}\, du = \frac{a}{b^2(a + bu)} + \frac{1}{b^2} \ln |a + bu| + C$

56. $\int \frac{du}{u(a + bu)^2} = \frac{1}{a(a + bu)} - \frac{1}{a^2} \ln \left|\frac{a + bu}{u}\right| + C$

57. $\int \frac{u^2}{(a + bu)^2}\, du = \frac{1}{b^3}\left[a + bu - \frac{a^2}{a + bu} - 2a \ln |a + bu|\right] + C$

58. $\int \frac{u}{\sqrt{a + bu}}\, du = \frac{2}{3b^2}(bu - 2a)\sqrt{a + bu} + C$

59. $\int \frac{\sqrt{a + bu}}{u}\, du = 2\sqrt{a + bu} + \sqrt{a} \ln\left(\frac{\sqrt{a + bu} - \sqrt{a}}{\sqrt{a + bu} + \sqrt{a}}\right) + C \quad (a > 0)$

60. $\displaystyle\int \frac{\sqrt{a+bu}}{u}\,du = 2\sqrt{a+bu} - 2\sqrt{-a}\ \arctan\sqrt{\frac{a+bu}{-a}} + C \quad (a<0)$

61. $\displaystyle\int \frac{u^2}{\sqrt{a+bu}}\,du = \frac{2}{15b^3}\left(8a^2+3b^2u^2-4abu\right)\sqrt{a+bu}+C$

62. $\displaystyle\int \frac{\sqrt{a+bu}}{u^2}\,du = -\frac{\sqrt{a+bu}}{u} + \frac{b}{2}\int \frac{du}{u\sqrt{a+bu}} + C$

63. $\displaystyle\int u^n\sqrt{a+bu}\,du = \frac{2u^n(a+bu)^{3/2}}{b(2n+3)} - \frac{2na}{b(2n+3)}\int u^{n-1}\sqrt{a+bu}\,du$

64. $\displaystyle\int \frac{u^n}{\sqrt{a+bu}}\,du = \frac{2u^n\sqrt{a+bu}}{b(2n+1)} - \frac{2na}{b(2n+1)}\int \frac{u^{n-1}}{\sqrt{a+bu}}\,du$

65. $\displaystyle\int \frac{du}{u^n\sqrt{a+bu}} = -\frac{\sqrt{a+bu}}{a(n-1)u^{n-1}} - \frac{b(2n-3)}{2a(n-1)}\int \frac{du}{u^{n-1}\sqrt{a+bu}} \quad (n\neq 1)$

66. $\displaystyle\int \frac{du}{u\sqrt{a+bu}} = \frac{1}{\sqrt{a}}\ln\left|\frac{\sqrt{a+bu}-\sqrt{a}}{\sqrt{a+bu}+\sqrt{a}}\right| + C \quad (a>0)$

67. $\displaystyle\int \frac{du}{u\sqrt{a+bu}} = \frac{2}{\sqrt{-a}}\arctan\sqrt{\frac{a+bu}{-a}} + C \quad (a<0)$

Trigonometric Forms

68. $\displaystyle\int \sin^2 u\,du = \frac{1}{2}u - \frac{1}{4}\sin 2u + C$

69. $\displaystyle\int \cos^2 u\,du = \frac{1}{2}u + \frac{1}{4}\sin 2u + C$

70. $\displaystyle\int \tan^2 u\,du = \tan u - u + C$

71. $\displaystyle\int \cot^2 u\,du = -\cot u - u + C$

72. $\displaystyle\int \sin^3 u\,du = -\frac{1}{3}(2+\sin^2 u)\cos u + C$

73. $\displaystyle\int \cos^3 u\,du = \frac{1}{3}(2+\cos^2 u)\sin u + C$

74. $\displaystyle\int \tan^3 u\,du = \frac{1}{2}\tan^2 u + \ln|\cos u| + C$

75. $\displaystyle\int \cot^3 u\,du = -\frac{1}{2}\cot^2 u - \ln|\sin u| + C$

76. $\displaystyle\int \sin^n u\,du = -\frac{1}{n}\sin^{n-1}u\cos u + \frac{n-1}{n}\int \sin^{n-2}u\,du$

77. $\displaystyle\int \cos^n u\,du = \frac{1}{n}\cos^{n-1}u\sin u + \frac{n-1}{n}\int \cos^{n-2}u\,du$

78. $\displaystyle\int \tan^n u\,du = \frac{1}{n-1}\tan^{n-1}u - \int \tan^{n-2}u\,du$

79. $\displaystyle\int \cot^n u\,du = \frac{-1}{n-1}\cot^{n-1}u - \int \cot^{n-2}u\,du$

80. $\displaystyle\int \sec^n u\,du = \frac{1}{n-1}\tan u\sec^{n-2}u + \frac{n-2}{n-1}\int \sec^{n-2}u\,du \quad (n\neq 1)$

81. $\displaystyle\int \csc^n u\,du = \frac{-1}{n-1}\cot u\csc^{n-2}u + \frac{n-2}{n-1}\int \csc^{n-2}u\,du \quad (n\neq 1)$

82. $$\int \sin mu \sin nu \, du = \frac{\sin(n-m)u}{2(n-m)} - \frac{\sin(n+m)u}{2(n+m)} + C \quad (m^2 \neq n^2)$$

83. $$\int \cos mu \cos nu \, du = \frac{\sin(m-n)u}{2(m-n)} + \frac{\sin(m+n)u}{2(m+n)} + C \quad (m^2 \neq n^2)$$

84. $$\int \sin mu \cos nu \, du = -\frac{\cos(m-n)u}{2(m-n)} - \frac{\cos(m+n)u}{2(m+n)} + C \quad (m^2 \neq n^2)$$

85. $$\int u \sin u \, du = \sin u - u \cos u + C$$

86. $$\int u \cos u \, du = \cos u + u \sin u + C$$

87. $$\int u^n \sin u \, du = -u^n \cos u + n \int u^{n-1} \cos u \, du$$

88. $$\int u^n \cos u \, du = u^n \sin u - n \int u^{n-1} \sin u \, du$$

Inverse Trigonometric Forms

89. $$\int \arcsin u \, du = u \arcsin u + \sqrt{1-u^2} + C$$

90. $$\int \arccos u \, du = u \arccos u - \sqrt{1-u^2} + C$$

91. $$\int \arctan u \, du = u \arctan u - \frac{1}{2}\ln(1+u^2) + C$$

92. $$\int u \arcsin u \, du = \frac{2u^2-1}{4} \arcsin u + \frac{u\sqrt{1-u^2}}{4} + C$$

93. $$\int u \arccos u \, du = \frac{2u^2-1}{4} \arccos u - \frac{u\sqrt{1-u^2}}{4} + C$$

94. $$\int u \arctan u \, du = \frac{u^2+1}{2} \arctan u - \frac{u}{2} + C$$

95. $$\int u^n \arcsin u \, du = \frac{1}{n+1} u^{n+1} \arcsin u - \frac{1}{n+1} \int \frac{u^{n+1}}{\sqrt{1-u^2}} \, du \quad (n \neq -1)$$

96. $$\int u^n \arccos u \, du = \frac{1}{n+1} u^{n+1} \arccos u + \frac{1}{n+1} \int \frac{u^{n+1}}{\sqrt{1-u^2}} \, du \quad (n \neq -1)$$

97. $$\int u^n \arctan u \, du = \frac{1}{n+1} u^{n+1} \arctan u - \frac{1}{n+1} \int \frac{u^{n+1}}{1+u^2} \, du \quad (n \neq -1)$$

Exponential and Logarithmic Forms

98. $\int ue^{au}\,du = \frac{1}{a^2}(au-1)e^{au} + C$

99. $\int \frac{1}{u\ln u}\,du = \ln|\ln u| + C$

100. $\int e^{au}\sin bu\,du = \frac{e^{au}}{a^2+b^2}(a\sin bu - b\cos bu) + C$

101. $\int e^{au}\cos bu\,du = \frac{e^{au}}{a^2+b^2}(a\cos bu + b\sin bu) + C$

102. $\int u^n e^{au}\,du = \frac{1}{a}u^n e^{au} - \frac{n}{a}\int u^{n-1}e^{au}\,du$

103. $\int u^n \ln u\,du = \frac{u^{n+1}}{n+1}\ln u - \frac{u^{n+1}}{(n+1)^2} + C$

Hyperbolic Forms

104. $\int \sinh u\,du = \cosh u + C$

105. $\int \cosh u\,du = \sinh u + C$

106. $\int \tanh u\,du = \ln\cosh u + C$

107. $\int \coth u\,du = \ln|\sinh u| + C$

108. $\int \operatorname{sech} u\,du = \arctan|\sinh u| + C$

109. $\int \operatorname{csch} u\,du = \ln\left|\tanh\frac{u}{2}\right| + C$

110. $\int \operatorname{sech}^2 u\,du = \tanh u + C$

111. $\int \operatorname{csch}^2 u\,du = -\coth u + C$

112. $\int \operatorname{sech} u\,\tanh u\,du = -\operatorname{sech} u + C$

113. $\int \operatorname{csch} u\coth u\,du = -\operatorname{csch} u + C$

Answers

Chapter 1

Exercises 1.1

1.

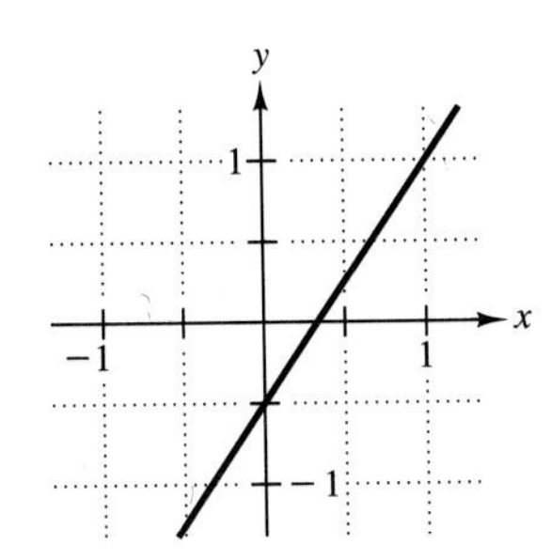

5. $y = \frac{1}{2}x$ **8.** $y = x + 0.5$

13. (a)

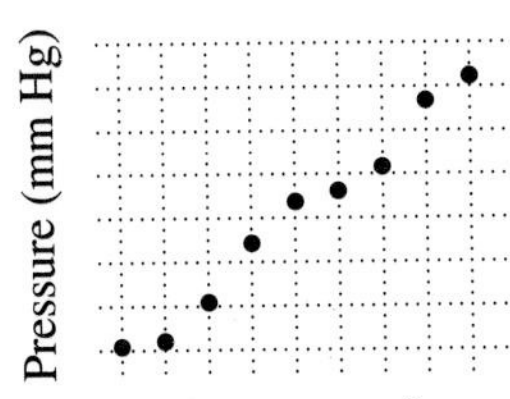

(b) essentially straight **(c)** linear relationship

18. (a)x

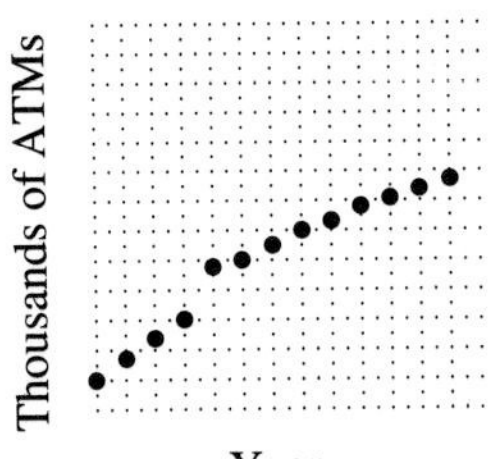

(b) increasing, but leveling off **(c)** increasing relationship

22. (a) 86, 87, 88, 89, 90, 91, 93 **(b)** 83, 84, 85 **(c)** 89

Exercises 1.2

1. (a) Time is independent; height is dependent.

(b)

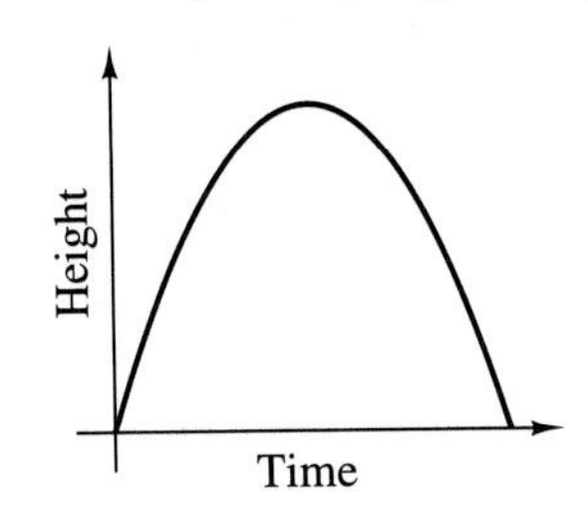

(c) The ball rises to a maximum height and then comes back to the ground. **5. (a)** Time is independent; amount of money is dependent.

(b)

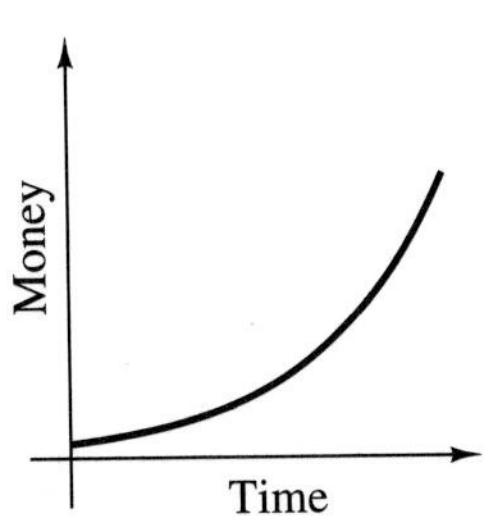

(c) The amount earned is increasing at an increasing rate.
9. (a) Day is independent; number absent is dependent.
(b) Answers will vary, possibly cyclical. **13. (a)** Speed is independent; distance is dependent. **(c)** This is the linear relationship of 5 car lengths for every 10 miles per hour in speed.

Exercises 1.3

1. (a) function **(c)** The domain consists of positive real numbers—possibly with some upper bound. The rule is to assign to each time the height of the baseball at that time.
5. (a) function **(c)** The domain is the period of time in question. The rule is to assign to each time the amount of money in the individual's account at that time.
9. (a) function **(c)** The domain is the days of the academic year. The rule is to assign to each day the number of students absent on that day. **12. (a)** not a function
(b) Requiring that all pizzas have the same toppings and type of crust and be ordered from the same restaurant would make this a function. **29. (b)** Each value of x is assigned one and only one value of y. **(c)** The domain is the set of x-values in the top row. The rule is to assign to each x-value the y-value directly beneath it.
31. (a) $F = 1.8C + 32$ **(b)** $C = (F - 32)/1.8$
(c) Both of these functions are linear. **(d)** Yes, solve $C = 1.8C + 32$ for C.

Exercises 1.4

1. function of time **5.** function of time
10. function of time **12.** not a function of time; unless there are other restrictions, not a function at all. With sufficient restrictions cost is a function of diameter.
17. The domain is the interval of time during which you were in the car. The rule is to assign to each time the speedometer reading at that time. **22.** The function is the distance traveled from the time the car was built until the time of reading. The measurement by the rolling, numbered

wheels is continuous, but the number displayed (to a tenth of a mile) is discrete; the function is continuous.

Exercises 1.7

1. difference of x^3 and $4x^2$or product of x^2 and $x - 4$.
5. cube of the square root of t or square root of the cube of t.
9. (first choice)

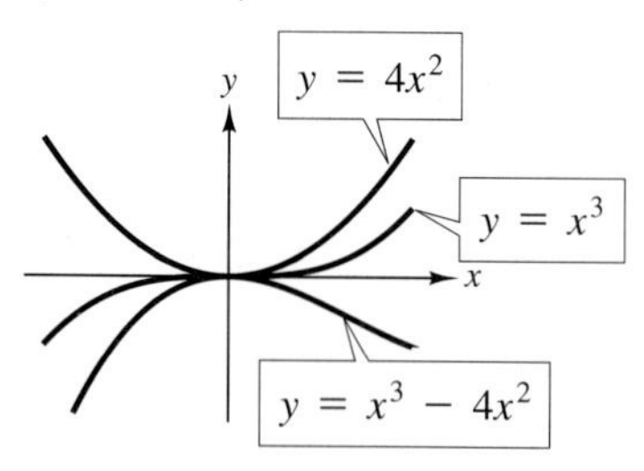

17. No. For example $f(7) = 56$, but $f(3) + f(4) = 16 + 23 = 39$. **21. (b)** The data indicate a linear relationship. (See Exercise 13 in Section 1.1.)
(c) One line that approximates this relationship is $y = 19.2 + 2.75T$. A one-degree change in temperature brings about a 2.75 mm change in pressure.
(d) $1/2.75 \approx 0.364$ degrees **28.** logarithms, e.g., $\log_{10}(ab) = \log_{10}(a) + \log_{10}(b)$

Exercises 1.8

1. (c) The inverse is $x^{1/3}$, which is a function.
5. (c) The inverse is not a function. **10. (c)** The inverse is a function. **15.** The domain is all x; the inverse is a function with the same domain. **20.** The domain is all $x \geq -1$; the inverse is a function with domain all $x \geq 0$.
22. The domain is all positive t; the inverse is the function 10^t with domain all t. **27.** $f^{-1}(x) = (x-3)^{3/2}$ for $x \geq 3$
32. (b) We do not have a formula for real values of x that satisfy $\frac{1}{8}x^3 + x - y = 0$ for a given y. **35.** no inverse; $f(x) = 1$ for at least 3 different values of x.

Exercises 1.9

2. (b) 5.655

Chapter 1 Exercises

3. $f(x) = (4x - 9)/3$ **6.** There is an inverse given by the table

Time (seconds)	Distance (meters)
1	5
2	19
3	44
4	78
5	123
6	176
7	240
8	313
9	396
10	489

9. (a) (v) **(b)** in year 4 to 5 **(c)** decreasing at a rate of approximately 70 deer per year **12. (a)** 18,400 gallons **(b)** 2710 kilowatt-hours **(c)** \$270 **(d)** 18,500,000 Btus **(e)** depends on the individual tuition

Chapter 2

Exercises 2.1

5. $-\frac{1}{3}$ **9.** $y = 1.5x + 3.8$ **13.** $y = 1.7x - 4.85$
17.

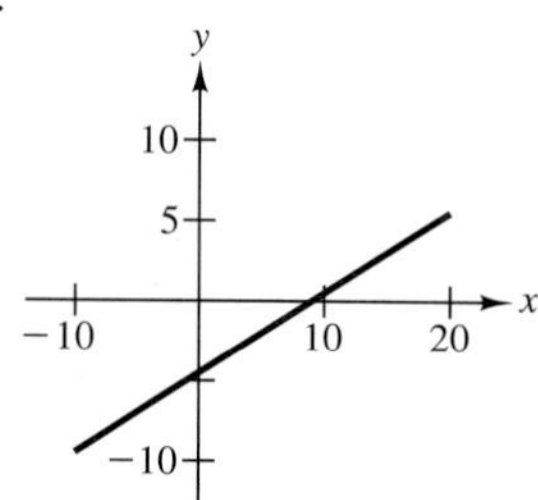

21. 1.5 **27.** 0.90032 **31.** 0.01571
35. 0.37292 **39.** 0.00651

Exercises 2.2

3. (b) $-\dfrac{1}{t(t+\Delta t)}$, approaches $-\dfrac{1}{t^2}$
(c) the negative of the reciprocal of the squaring function
6. (b) $3(t^2 + t\,\Delta t)$, approaches $3t^2$ **(c)** $3t^2$ **8.** 22.932
12. 42.00 **17. (a)** $9/5$ **(b)** $5/9$ **(c)** They are reciprocals. **19.** 0.434 and 0.217 **22.** 3.43 and 5.47

Exercises 2.3

1. $10t^4 - 21t^2 + 8t - 6$ **5.** $12t^3 - 3t^2 + 14t - 1$
9. $5t^4 - 28t^3 + 4t - 6$ **12.** $-27 + 10t + 3t^2$
17. (c) $d/dt\,(1/t^2) = -2/t^3$

Exercises 2.4

1. \$2063.75 **4.** $9^{1/2} = 3$ **8.** $10^2 = 100$ **11.** 1 **15.** -2 **17.** $t = \ln 10 \approx 2.303$ **19.** $-2e^{-2t}$ **22.** $2 - 5e^t$ **27.** $2 \cdot e^{(\ln 3)t}$ **31.** 0 **35.** $10t^9 + (\ln 10)10^t$ **39.** $4\ln 2 \approx 2.773$ **43.** $9\ln 3 \approx 9.888$ **51.** $\ln b$ **55. (b)** $e^t + e^{-t}$ **(c)** even

Exercises 2.5

7. (a) any three functions of the form $P = C\,e^{0.02t}$, for example $e^{0.02t}$, $5\,e^{0.02t}$, $-3\,e^{0.02t}$ **(b)** $P = C\,e^{0.02t}$ **(c)** $5\,P = C\,e^{0.02t}$ **(d)** $4.80\,e^{0.02t}$ **11.** $P = 500\,e^{0.12t}$ **13.** $P = 393e^{0.12t}$ **15.** $y = 100\,e^{0.35t}$ **19.** 18.70 **23. (a)** 0.0198 **(b)** $\ln(1 + r/100)$ **27. (a)** \$1080 **(b)** 1.08 **(c)** $1.08^2 \cdot 1000 = 1166.40$ **(d)** $1.08^{20} \cdot 1000 = 4660.96$ **(e)** $1.08^t \cdot 1000$ **28. (a)** \$1082.43, 1.08243, 1171.70, 4875.44, $(1 + 0.08/2)^{4t} \cdot 1000 = 1.02^{4t} \cdot 1000$ **32. (a)** $(\ln 2)/(1 + r/100)$

Exercises 2.6

1. $y = 7.2\,t^{0.45}$ **5. (a)** A straight line in a Cartesian plot implies a linear function, in a semilog plot implies an exponential function, in a log-log plot implies a power function. **(b)** If none of the plots is a straight line, then the function is some other type. **(c)** The points fit $y = 3x$, so both the Cartesian and log-log plots lie on a straight line—the function is both linear and a power function. The semilog plot does not lie on a straight line. **(d)** The points fit $y = 10^x$, so the semilog plot lies on a straight line. The other two plots are not straight. **9.** If $P = P_0 e^{kt}$, then $\log_b(P) = \log_b(P_0) + (k\log_b(e))t$, so $\log_b(P)$ is a linear function of t. **14.** See answer to Exercise 9.

Chapter 2 Exercises

1. 2.0959 **4.** $3/2$ **7.** 1.4542 **9.** 1.301 **13.** 0 **15.** -5.419 **19.** $2e^{2t}$ **23.** $(-4/3)e^{-4/3t}$ **27.** 0 **31.** $c = 17$, $b = e^{0.07}$ **35.** $5t^4 - 28e^{4t} + 4t + 6e^{-t}$ **41. (a)** 11:50 A.M.

Chapter 3

Exercises 3.1

1. $3t^2$ **5.** $1 - 3e^{3t}$ **9.** $3e^{3t} - 6t + 3$ **13.** (iv) **17.** any three functions of the form Ce^{8t} **21.** Graph three functions of the form Ce^{8t}. **25.** $2e^{8t}$ **29.** Graph $y = 2e^{8t}$. **40. (a)** $6e^{-3t}$ **(b)** $6e^{-3t} = 6 - 3(2 - 2e^{-3t})$

Exercises 3.2

1. $-2t$ **5.** $3e^{-3t}$ **9.** any three functions of the form Ce^{-8t} **13.** Graph any three functions of the form Ce^{-9t}. **17.** $20e^{-8t}$ **21.** $10 + 10e^{-8t}$ **25.** Sketch the graph of $y = 20e^{-8t}$. **29.** Sketch the graph of $y = 10 + 10e^{-8t}$. **33.** $\ln(1/2) = -\ln 2$ **36.** 0.409 **41 (a)** 3:37 A.M.

Exercises 3.3

1. (a) 80.3 feet per second **(b)** 2.49 seconds **4.** $1 - 2t$ **8.** $2t - 3e^{-3t}$ **11.** $-21e^{-3t} - 6t + 3t^2$ **20.** $y = t - t^2 + 3$ **23.** $w = -e^{1.7t} + 3$ **26.** $P = t^2 + e^{-3t}$ **29.** $x = 7e^{-3t} - 3t^2 + t^3 - 4$ **32.** 5.46 seconds; 186 feet per second **35. (a)** $a(t) = 1.4$, $v(t) = 1.4t$, $s(t) = 0.7t^2$ for $0 \leq t \leq 60$ **(b)** 57 miles per hour **(c)** 2520 feet

Chapter 3 Exercises

1. $3e^{3t}$ **5.** $-3e^{3t}$ **9.** $3\ln 2\ 2^{3t}$ **13.** $28t^3 + 15t^2 - 8t + 3$ **19.** $y = 3e^{(1/2)t}$ **23.** $y = 0.45t^2 0.1t + 3$ **27.** (ii) and (iv) **31.** (iv) **35.** three functions of the form $-0.85t^2 + C$ **39.** Graph three functions of the form $-0.85t^2 + C$. **43.** $y = -0.85t^2 + 2.3$ **48. (a)** 64.4 feet **(b)** 359 feet per second **(c)** 11.1 seconds **52.** $y = Ce^{2t} - 1$ **56.** $y = 4e^{2t} - 1$

Chapter 4

Exercises 4.1

1. $3e^{3t}$ **5.** $2t + 1 - 1/t^2$ **13.** 3.96°C **17. (a)** height $= 8.14$ in.; radius $= 1.57$ in. **(b)** 14.4¢ **21.** $x = 1$

Exercises 4.2

1. $9e^{3t}$ **5.** 2 **9.** concave up for $t < 0.204$ and concave down for $t > 0.204$ **15. (a)** 0, 6, -6 **(b)** $3t^2 - 36$; zeros are $\pm\sqrt{12} \approx \pm 3.46$ **(c)** $6t$ at 0 is 0. **(d)** $\sqrt{12}$ is a local minimum and $-\sqrt{12}$ is a local maximum. **(e)** yes

Exercises 4.3

1. (a) -3.43, -2.10, 1.27, 4.23 **5. (a)** 0.4839 **7.** 0.2541, 5.023 **11.** -0.1681 **12. (a)** -3.162, -0.2476, 0.5662, 1.352

Exercises 4.4

1. $2te^t + t^2e^t$ **5.** $-e^{2x} + 2x^2e^{2x}$ **9.** $\left(\frac{3}{5}\right)^t \ln\frac{3}{5}$ **10.** $2e^t + 4te^t + t^2e^t$ **14.** $-2e^{2x} + 4xe^{2x} + 4x^2e^{2x}$ **18.** $\left(\frac{3}{5}\right)^t \left(\ln\frac{3}{5}\right)^2$ **20.** $2e^x + xe^x$ **23.** $-(1 + 6t)e^{-3t}$

Exercises 4.5

1. $1/\sqrt{2t+3}$ **5.** $2te^{t^2}$ **8.** $(2t^2+1)e^{t^2}$
11. $2(t^3+t)e^{t^2}$ **14.** $\frac{1}{2\sqrt{t}}e^{t^2} + 2t\sqrt{t}\,e^{t^2}$
17. $(2x+1)e^{(x^2+x)}$ **21.** $9(3x+1)^2+3$
24. -6 **26.** 40 **31.** The maximum volume is $\frac{2\pi h^3}{9\sqrt{3}}$.

Exercises 4.6

2. The point P should be $1/\sqrt{3}$ of the way from B to C.

Exercises 4.7

2. $-1/9$ **4.** $1/(x+2)$ **8.** $2x\ln(x+2) + \frac{x^2}{x+2}$
12. $2x\,e^{x^2}$ **16.** $-1/(x+5)$ **20.** $-(x^2+5)^{-3/2}$
22. 1.296

Exercises 4.8

1. $\frac{5}{2}x^{3/2}$ **5.** $5x(x^2+3)^{3/2}$ **9.** $\frac{p^2}{(p^2+x^2)^{3/2}}$
12. $\frac{1}{3}(1+x^2)^{-2/3}$ **17.** The chip should be 10 mm by 20 mm, with the longer edges corresponding to the sides with the circuitry 1 mm from the edge. **21. (c)** At $x=1/e$ the function x^x has the minimum value ≈ 0.692.
23. (a) $\left(\frac{f(t)}{g(t)}\right)' = \frac{g(t)(f(t))' - f(t)\cdot(g(t))'}{(g(t))^2}$
24. $-\frac{2}{3}\frac{-1+2t^2}{t^{1/3}(1+t^2)^2}$

Exercises 4.9

3. $\frac{5}{2}\,x^{3/2}dx$ **7.** $5x(x^2+3)dx$ **11.** $\frac{p^2}{(p^2+x^2)^{3/2}}dx$
15. $(\ln x+1)dx$ **19.** $[\ln 2\,2^x(x^2+1) + 2x\,2^x]dx$
23. $5dx$

Chapter 4 Exercises

1. $30t^4-12t^2$ **5.** $\frac{3t^2}{2\sqrt{t^3+1}}$
9. $\frac{3}{2}\sqrt{x\ln x+1}(\ln x+1)$ **13.** $\frac{2x}{x^2+2}$
17. $\frac{2+x-\ln x}{2x^{3/2}}$ **21.** $1+3e^{-t}$ **23.** $-3e^{-t}$
32. any function of the form $\frac{1}{2}t^2 + \frac{1}{3}e^{-3t}$ + constant
36. The maximum area is $2e^{-2}$ generated by the tangent at $(1, e^{-1})$.

Chapter 5

Exercises 5.1

1. 1.1, 1.221, 1.370 **5.** 2.05, 2.1020, 2.1562
9. 1, 1.000000001, 1.000000005 **13.** The exact values to 14 SD are 1.0000000003333, 1.0000000002667, and 1.0000000090000. **17.** $P(t) = t^3/3+t+2$

Exercises 5.2

1. Since $S+I+R=$ constant, we know $dS/dt + dI/dt + dR/dt = 0$ or $dR/dt = -dS/dt - dI/dt$. Substituting from the differential equations for dS/dt and dI/dt, we obtain $dR/dt = \alpha I(t)S(t) - (\alpha I(t)S(t) - \lambda I(t)) = \lambda I(t)$.

Exercises 5.3

1. 95% **5.** 87% **7.** 100% **11.** 92% **15.** 99%

Exercises 5.4

1. 88573 **5.** $\frac{1-\frac{1}{4^9}}{\frac{3}{4}} = \frac{87381}{65536}$ **9.** $\frac{1-\frac{1}{2^{13}}}{\frac{1}{2}} = \frac{8191}{4096}$
13. $\frac{13}{6}\left[\left(\frac{13}{7}\right)^{25} - 1\right]$
19. (a)

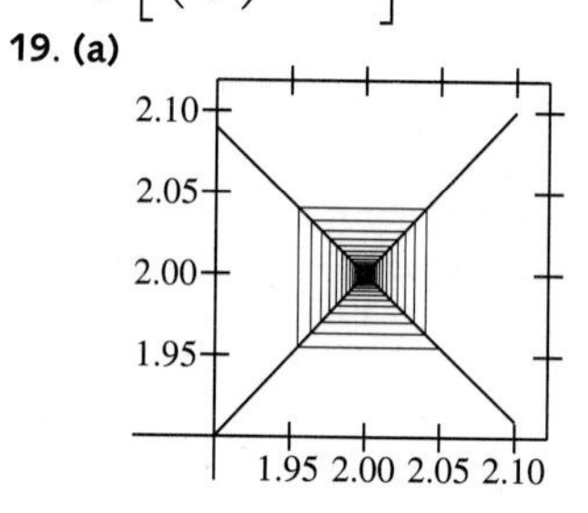

Chapter 5 Exercises

1. $2/t$ **5.** $\frac{2}{t}\ln t$ **9.** $-\frac{2}{t^2}$ **13.** $\left(9\ln t + \frac{6}{t} - \frac{1}{t^2}\right)e^{3t}$
18. (b) $35/99$ **21. (b)** 7 meters

Chapter 6

Exercises 6.1

1. $2\,e^{2t}\sin t + e^{2t}\cos t$ **5.** $-\cos t$
9. $3\,e^{2t}\sin t + 4\,e^{2t}\cos t$ **10.** $2\cos 2t + 3\sin t$
14. $-2/(t-1)^2$ **18.** $3x^2\sin 2x + 2x^3\cos 2x$
28. $\frac{1}{2}e^{2t}+C$ **31.** $\sin t + C$ **35.** $\frac{1}{7}\ln|1+7t| + C$
39. Maximum is $\sqrt{2}$; minimum is -1.
44. (a) $\frac{d}{dt}\cot t = -\frac{1}{\sin^2 t} = -\csc^2 t$

Exercises 6.2

1. $60t^5 - 12t^2$ **5.** $\frac{\cos\theta}{\cos 2\theta} + 2\frac{\sin 2\theta \sin\theta}{\cos^2 2\theta}$ **9.** $(t-2)e^{-t}$

10. $\frac{10x^9 - 3}{3(x^{10} - 3x)}$ **13.** $1/(\cos x + 1)$ **16.** $2x\cos x^2$

26. $y = -0.5e^{1.3t}$

29. $y = -\frac{0.5}{\sqrt{1.3}}\sin\left(\sqrt{1.3}\,t\right) \approx -0.439\sin(1.14t)$

32. 120π **36. (c)** Ignoring damping, the mass will be closest to the ceiling after $\pi/2$ seconds and every $4\pi/3$ seconds thereafter.

Exercises 6.3

1. $2(t^6 - 4t^3 + 1)(6t^5 - 12t^2)$

5. $2\frac{\cos 2\theta}{\cos\theta} + \frac{\sin 2\theta \sin\theta}{\cos^2\theta} = 2\cos\theta$

9. $(2 - 4t + t^2)e^{-t}$ **10.** $\frac{1}{3}\frac{\cos x}{\sin x}$ **13.** $\frac{\sin x}{(\cos x - 1)^2}$

16. $e^x\cos(e^x)$ **26.** $y = -0.5e^{-0.2t}$

29. $-\frac{0.5}{\sqrt{0.2}}\sin\left(\sqrt{0.2}\,t\right)$

34. (c) $A = \sqrt{\frac{L}{g}} \times$ (the first derivative at $t = 0$)

37. (a) The period of both is $\frac{2\pi}{\omega}$.

Exercises 6.4

1. $-e^{-t}\sin 5t + 5e^{-t}\cos 5t$ **5.** $4 \cdot 3^{2t}(\ln 3)^2$

9. $(-24\sin 5t - 10\cos 5t)e^{-t}$

13. $dy/dt = 3\cos t\cos 3t - 9\sin t\sin 3t$

17. $dz/dt = -4/(2t+1)^3$ **21.** $dz/dt = 14/(3+2t)$

24. $(2x + x^2\ln 3)3^x$

Chapter 6 Exercises

1. $-3e^{-3t}\sin 2t + 2e^{-3t}\cos 2t$ **5.** $6\sin t\cos^2 t - 3\sin^3 t$

9. $\cos^3 x - 2\sin^2 x\cos x$

13. $dy/dt = 2\cos 2t\cos 5t - 5\sin 2t\sin 5t$

17. $dy/dx = 5(\ln 2)\,2^{5x}$ **28.** $-\frac{1}{2}e^{-2t} + C$

32. $\frac{1}{3}\sin 3t + C$ **36.** $\frac{7}{2}\ln|3 + 2t| + C$

37. $y = -0.5e^{-1.07t}$

40. $y = -0.5\cos\left(\sqrt{1.07}\,t\right) + \frac{0.5}{\sqrt{1.07}}\sin\left(\sqrt{1.07}\,t\right) \approx -0.5\cos(1.03t) + 0.483\sin(1.03t)$

47. (a) $x^2 - y^2 = 1$

Chapter 7

Exercises 7.1

1. $P = e^{2t}$ **5.** $P = -1/((2t+1)$ **9.** P is the constant zero function. **10.** $\ln t$ on $(0, \infty)$ **14.** $y = 2\ln\left(\frac{t+1}{2}\right)$

17. $7 = -e^{-t} + 2$

Exercises 7.2

7. graph $P = \frac{4}{3}\frac{e^{8t}}{1 + \frac{1}{3}e^{8t}}$ **10.** $A = 1/5,\ B = 1/5$

13. $A = 1/12,\ B = -1/12$ **15.** $A = 1/6,\ B = 1/6$

17. $\frac{1}{5}\ln\left(\frac{x}{5-x}\right)$ **19.** $\frac{1}{12}\ln\left(\frac{x}{x+4}\right)$

21. $\frac{1}{12}\ln\left(\frac{3+2x}{3-2x}\right)$

24. $\frac{1}{5}\ln(-x) - \frac{1}{5}\ln(5+x) = \frac{1}{5}\ln\left(\frac{-x}{5+x}\right)$

28. $\frac{1}{4}\ln(x+2) - \frac{1}{4}\ln(x-2) = \frac{1}{4}\ln\left(\frac{x+2}{x-2}\right)$

32. $\frac{1}{12}\ln(-3-2x) - \frac{1}{12}\ln(3-2x) = \frac{1}{12}\ln\left(\frac{-3-2x}{3-2x}\right)$

Exercises 7.3

1. (a) limit point **4. (a)** 2-cycle **8. (a)** chaotic range

12. (a) limit point **14. (a)** 4-cycle **18. (a)** $r = 2$

Chapter 7 Exercises

1. any function of the form $y = \sqrt{2t + C}$ or $y = -\sqrt{2t + C}$ **3.** any function of the form $y = -e^{-t} + C$ **5.** any function of the form $y = \pm 1/\sqrt{C - 2t}$ **7.** $y = \sqrt{2t + 4}$

9. $y = -e^{-t} + 3$ **11.** $y = 1/\sqrt{3 - 2t}$

18. 2-cycle **20.** 8-cycle

Chapter 8

Exercises 8.1

1. 21 **4.** 7.5 **6.** 45.5 **7.** 4.33 **9.** 4.915065

15. 35 **17.** 40 **20.** $p(b-a) + \frac{c}{2}(b^2 - a^2)$

40. (a) $(b^2 - a^2)/2$ **(b)** $(a^2 - b^2)/2$ **(c)** $(a^2 + b^2)/2$

47. (b) $\pi/4$

Exercises 8.2

1. 20 **4.** $242/5$ **7.** 1 **11.** $\ln 3 \approx 1.099$

15. $e - 1 \approx 1.718$ **18.** $e - e^{-3} \approx 2.688$

21. (a) 2.6684948

(b)

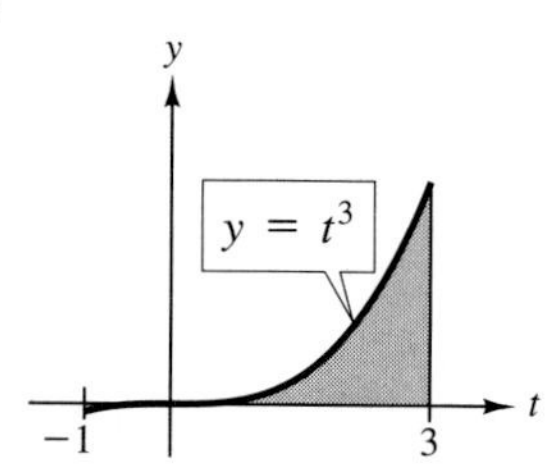

(c) The value of the integral is equal to the area of the lightly shaded region minus the area of the darkly shaded region.
24. 48.8 **30.** 0.6931472 **35.** 0.6321206 **37.** 10
41. 16.2 **45.** 0.6366 **49.** 0.2554
53. 0.6321

Exercises 8.3

1. $t^5/5 + C$ **5.** $-\frac{1}{2}\cos(2t) + C$ **9.** $3e^t + C$
13. $\frac{2}{3}t^{3/2} + C$ **16.** $2/5$ **20.** 1
24. $3(e - 1/e) \approx 7.051$ **28.** $2\sqrt{3} - 2/3 \approx 2.797$
31. $1/5$ **35.** $2/\pi \approx 0.6366$ **39.** $\frac{3}{2}(e - 1/e) \approx 3.526$
43. $\sqrt{3} - 1/3 \approx 1.399$

Exercises 8.4

1.

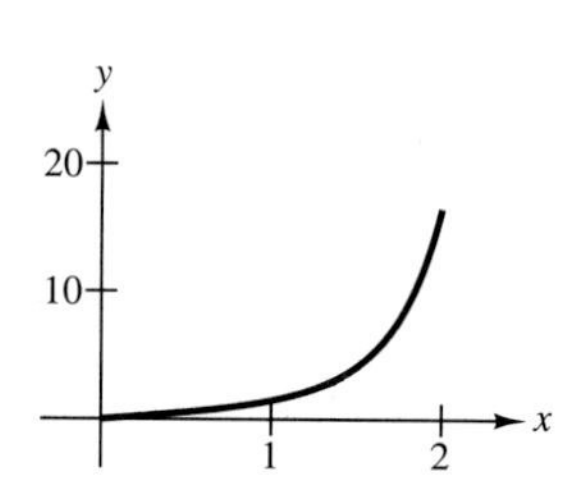

4.

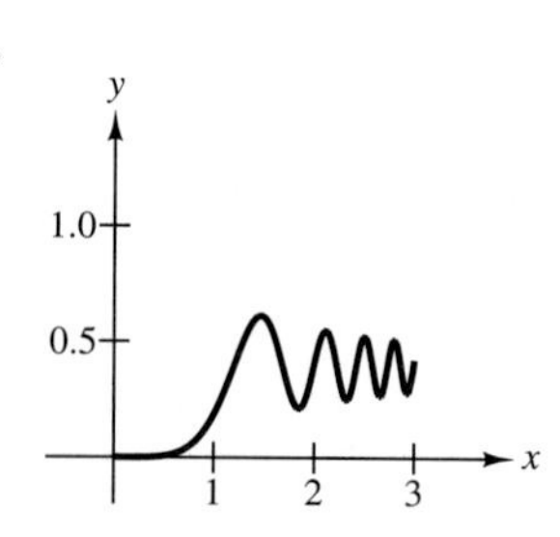

7. e^{x^2} **10.** $x \sin x^3$ **13.** $\frac{3}{5}x^{5/3} - \frac{3}{4}x^{4/3} + C$
17. $-\frac{1}{2}\cos(2x) + C$ **20.** $-\frac{1}{5}e^{-5x} + C$ **21.** 0.01500
25. 0 **28.** $\frac{1}{5}(e^5 - e^{-5}) \approx 29.68$

Chapter 8 Exercises

1. $\frac{4}{3}x^{1/3} + C$ **5.** $-\cos x + C$ **8.** $-e^{-2x} + C$
9. $3/4$ **13.** 2 **16.** $(1 - e^{-2})/2 \approx 0.4323$
17. ≈ 1.57

20. (a)

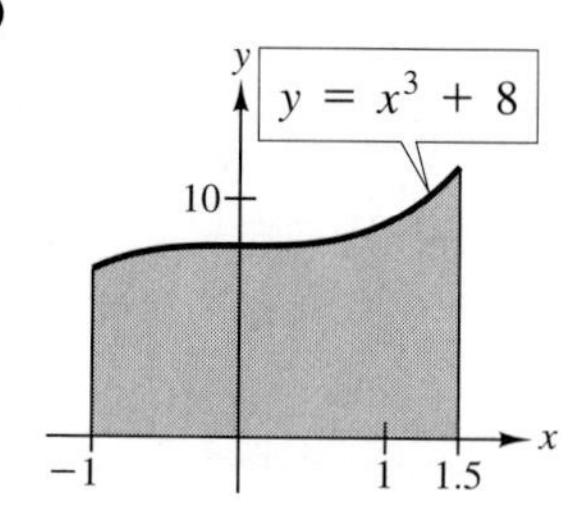

(b) $\int_{-1}^{1.5} (x^3 + 8)\, dx$ **(c)** ≈ 21.02

25. (a)

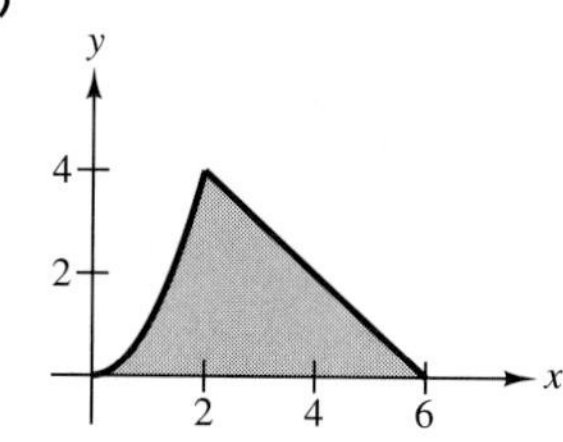

(b) $32/3$ **31. (b)** $1/n$

Chapter 9

Exercises 9.1

1. $36\,\frac{2}{3}$ **5.** $11\,\frac{1}{9}$ inches from the left end **9.** 22.5 inches from the larger end **13.** 0.157 **14.** 0.833 **17.** 3240 grams **21.** $-\frac{1}{3}\cos(3x) + C$ **25.** $\ln\left(\frac{|t|}{|1 + 3t|}\right) + C$
29. $-\cos(u + \pi) + C$ **30.** $2/3$ **34.** $3\ln 2 - \ln 7$
38. -1

Exercises 9.2

1. $\overline{x} = 0, \overline{y} = 4/5$ **5.** $\overline{x} = 2/5, \overline{y} = 1/2$

Exercises 9.3

1. exact $= 2/3$, TR $= 0.6611745125$, MR $= 0.6694161360$, SR $= 0.6666689281$
5. exact $= \ln(8/7) = 0.1335313926$, TR $= 0.1338401613$, MR $= 0.1333772114$, SR $= 0.1335315281$
9. exact $= -1$, TR $= -0.9979429864$, MR $= -1.0010288241$, SR $= -1.0000002115$ **11.** 1.11
15. 0.39 **19.** 1.91 **22.** 0

Exercises 9.4

1. $\frac{2}{9}(1 + x^3)^{3/2} + C$ **3.** $(\theta - \sin\theta\cos\theta)/2 + C$
6. $-\frac{1}{3}(9 - x^2)^{3/2} + C$ **8.** $\sqrt{1 + x^2} + C$
11. $\frac{x}{2}\sqrt{9 - x^2} + \frac{9}{2}\arcsin\left(\frac{x}{3}\right) + C$
13. $\frac{2}{9}(2^{3/2} - 1) \approx 0.4063$ **20.** $\frac{1}{2}\ln 2 \approx 0.3466$
23. $\sqrt{2} + \frac{9}{2}\arcsin\frac{1}{3} \approx 2.943$ **41.** $\frac{1}{3}\arctan\frac{1}{3} \approx 0.1073$

Exercises 9.5

1. $\frac{2}{9}(1+x^3)^{3/2}+C$ (Substitute $u=1+x^3$ and use the Power Rule.) **5.** $-\frac{1}{3}\cos^3 x+C$ (Substitute $u=\cos x$ and use the Power Rule.) **8.** $\sqrt{1+x^2}+C$ (Substitute $u=1+x^2$ and use the Power Rule.)
12. $-\frac{1}{2}x\sqrt{9-x^2}+\frac{9}{2}\arcsin\frac{x}{3}+C$ (Entry 35)
13. $\frac{2}{135}(9x-2)(1+3x)^{3/2}+C$ (Entry 51)
17. $\frac{1}{6}\cos^5 x\sin x+\frac{5}{24}\cos^3 x\sin x+\frac{5}{32}\sin(2x)+C$ (Entry 77 twice and Entry 69)
20. $\frac{x}{2}\sqrt{8+x^2}-4\ln\left(x+\sqrt{8+x^2}\right)+C$ (Entry 27)
24. $-\dfrac{1}{7x}+\dfrac{1}{49}\ln\left|\dfrac{7+x}{x}\right|+C$
27. $\frac{1}{2}\tan\theta\sec\theta+\frac{1}{2}\ln|\sec\theta+\tan\theta|+C$ (Entries 80 and 13)

Exercises 9.6

1. $-\dfrac{1}{2}t\cos(2t)+\dfrac{\sin(2t)}{4}+C$ **5.** $\dfrac{t^2}{2}\ln t-\dfrac{t^2}{4}+C$
10. $x\arcsin(3x)+\frac{1}{3}\sqrt{1-9x^2}+C$
13. $\left(\dfrac{x^2}{2}-\dfrac{1}{36}\right)\arcsin(3x)+\dfrac{x}{12}\sqrt{1-9x^2}+C$
14. $\ln(2+\sin t)+C$
18. $\dfrac{2}{3}x(x+7)^{3/2}-\dfrac{4}{15}(x+7)^{5/2}+C$
22. $\frac{1}{2}e^{u^2}+C$ **23.** 0 **27.** $\frac{196}{15}\sqrt{7}-\frac{32}{5}\sqrt{2}\approx 1.384$
31. $\dfrac{e-1}{2}\approx 0.8591$

Exercises 9.7

1. 0 **4.** π **6.** 0 **8.** 0 **11.** 6 **14.** 0 **16.** 3

Chapter 9 Exercises

1. 1.896119 **4.** 0.441302 **5.** 2.052344 **8.** 0.427862
9. 2.000269 **12.** 0.432342 **22.** $-\dfrac{1}{5}\cos^5\theta+C$
26. $-\frac{x}{2}\sqrt{9-x^2}+\frac{9}{2}\arcsin\frac{x}{3}+C$
27. $\dfrac{6\sqrt{6}-2\sqrt{2}}{18}\approx 0.659362$ **29.** $9\pi/4$
31. $\ln\left(\sqrt{3}+2\right)\approx 1.316958$
33. $\dfrac{t^3}{3}-\dfrac{\cos(3t)}{3}+C$ **37.** 1 **41.** $\pi/8$
45. $\dfrac{1}{5.3}(t+3)e^{5.3t}-\dfrac{1}{(5.3)^2}e^{5.3t}+C$ **49.** $1/4$
50. $-\frac{1}{20}\cos^4(4x)\sin(4x)+\frac{1}{60}\cos^2(4x)\sin(4x)+\frac{1}{30}\sin(4x)+C$ **53.** $\dfrac{1}{\sqrt{11}}\arctan\left(\dfrac{x+3}{\sqrt{11}}\right)+C$
66. (a) $2-(29/3)^{1/3}\approx -0.130226$

Chapter 10

Exercises 10.1

1. 50.5 days **4.** $25/6$ **6. (a)** 0.295 **(b)** 15.87 days
10. $-\dfrac{rt^2+2rt+2}{r^3}e^{-rt}+C$ **13.** $\dfrac{1}{9}\left(1-31e^{-30}\right)+C$
16. $\dfrac{2}{r^3}-2\dfrac{1+10r+50r^2}{r^3}+C$ **19.** $\dfrac{1}{2}\arctan 400\approx\pi/4$

Exercises 10.2

4. $1-e^{-0.002857t}$ **5.** converges **11.** $\frac{7}{4}e^{-6}$ **13.** $\pi/4$
14. 1 **16.** e **19.** 0 **21.** 0 **24.** The median lifetime is $\frac{1}{r}\ln 2$. **32.** converges **33.** diverges **36.** converges
38. 0.24375 **42.** $1/3$

Exercises 10.3

1. $24/800=0.03$ **4.** $1/36$ **7. (a)** $7/17\approx 0.412$
(b) $21/136\approx 0.154$ **(c)** $7/34\approx 0.206$
10. $f(x)=1/10$

Exercises 10.4

1. $m=160.5$, $sd=26.82$ **4.** $m\approx 50.44$, $sd\approx 39.09$
5. 0.5328 **8.** 0.4102 **9.** 0.6826 **13.** $w\approx 3.218$

Exercises 10.5

1. $1/16$ **3.** 8 **5.** 0.4422 **7.** 0.2716 **9.** 0.1546
11. 2 **15.** $15\pi/8$

Exercises 10.6

1. 2 **5.** 0 **9.** $1/2$ **13.** $\dfrac{\ln 2}{\ln 3}\approx 0.6309$ **17.** 0
20. 1 **24.** 0 **28.** no limit (or ∞; see the next set of exercises) **29.** ∞ **33.** ∞ **35.** $-\infty$ **37.** 0

Chapter 10 Exercises

1. $\pi/2$ **4.** $\frac{16}{25}e^{-15}$ **6.** $\pi/2-\arctan 3$ **7.** converges to approximately 0.8670 **10.** 9 **12.** 1 **16.** 0 **20.** $1/e$
23. $m=25.71$, $sd=7.722$

Chapter 11

Exercises 11.1

1. $1/5!$ **5.** $1-\frac{x^2}{2}+\frac{x^4}{24}$ **9.** $[-1.05,1.05]$
14. $[-0.7,0.7]$ **21. (a)** $x+\frac{1}{3}x^3+\frac{2}{15}x^5$
32. (a) $1-\frac{1}{2}x^2+\frac{1}{120}x^4$

Exercises 11.2

3. $1/(13)!$
5. (a) $1-\dfrac{1}{6}x^2+\dfrac{1}{5!}x^4+\ldots+\dfrac{(-1)^k}{(2k+1)!}x^{2k}+\ldots$

7. (a) $x + \frac{1}{6}x^3 + \frac{1}{5!}x^5 + \ldots + \frac{1}{(2k+1)!}x^{2k+1} + \ldots$
8. (a) $1 + \frac{1}{2}x^2 + \frac{1}{4!}x^4 + \ldots + \frac{1}{(2k)!}x^{2k} + \ldots$
15. (b) Taylor polynomial of degree 3 is $x - \frac{1}{6}x^3$.

Exercises 11.3

1. 3 **3.** $\pi/2$ **4.** $\pi/4$ **6.** e^3
9. $1 - t + t^2 - t^3 + t^4 - t^5$ **12.** 1/5
15. $1 - x^2 + x^4 - x^6$ **17.** $e^{\frac{1}{2}} - 1$
21. (a) The polynomial is $1 + \frac{x}{2} - \frac{x^2}{8} + \frac{x^3}{16} - \frac{5}{128}x^4$ **23. (a)** $x^2 - \frac{x^4}{2} + \frac{x^6}{3}$ **29.** The 6th degree polynomial is $(x-1) - \frac{1}{2}(x-1)^2 + \frac{1}{3}(x-1)^3 - \frac{1}{4}(x-1)^4 + \frac{1}{5}(x-1)^5 - \frac{1}{6}(x-1)^6$. **31.** The 6th degree polynomial is $-\left(x - \frac{\pi}{2}\right) + \frac{1}{6}\left(x - \frac{\pi}{2}\right)^3 - \frac{1}{120}\left(x - \frac{\pi}{2}\right)^5$.
34. The 6th degree polynomial is $\frac{\pi}{4} + \frac{1}{2}(x-1) - \frac{1}{4}(x-1)^2 + \frac{1}{12}(x-1)^3 - \frac{1}{40}(x-1)^5 + \frac{1}{48}(x-1)^6$. **35.** $[0.205, 2.04]$
40. $[-0.528, 2.47]$
41. $(x-1) - \frac{1}{2}(x-1)^2 + \frac{1}{3}(x-1)^3 - \frac{1}{4}(x-1)^4 + \ldots + (-1)^{k+1}\frac{1}{k}(x-1)^k + \ldots$
43. $-\left(x - \frac{\pi}{2}\right) + \frac{1}{6}\left(x - \frac{\pi}{2}\right)^3 - \frac{1}{120}\left(x - \frac{\pi}{2}\right)^5 + \ldots + \frac{(-1)^k}{(2k+1)!}\left(x - \frac{\pi}{2}\right)^{2k+1} + \ldots$ **46.** $(0, 2]$
48. $(-\infty, \infty)$

Exercises 11.4

4. converges; Alternating Harmonic Series less first term
7. diverges; Divergence Test **9.** converges; Leibniz Series less first term **10.** $\ln 2 - 1$
15. $\frac{\pi}{4} - 1 \approx -0.215$

Exercises 11.5

1. (a) converges by Alternating Series Test **(c)** 0.822
3. (a) converges; a geometric series with $|\text{ratio}| < 1$
(b) 1/3 **6. (a)** diverges; Divergence Test
8. (a) converges by Alternating Series Test
(b) The series is the negative of the Leibniz Series with first two terms missing. The sum is $-\left(\frac{\pi}{4} - 1 + \frac{1}{3}\right) \approx -0.119$.
16. $(-1, 1)$ **19.** $(-1, 1)$ **22.** $(-\infty, \infty)$

Exercises 11.6

5. See Exercise 6. **9.** 6 **12. (a)** $1/(1+x)$ **16.** xe^{x^2}
19. $\cos x - x \sin x$ **20.** 2 **22.** e^2 **27.** $(0, 2]$
29. $(-\infty, \infty)$

Chapter 11 Exercises

2. converges by Alternating Series Test **4.** Terms do not alternate. **5.** Terms do not approach 0. **7. (a)** 0.159
(b) 0.00814 **(c)** $1.96 \cdot 10^{-4}$ **(d)** $2.73 \cdot 10^{-6}$
10. (a) 0.0364 **(b)** 0.00531 **(c)** $9.36 \cdot 10^{-4}$
(d) $1.80 \cdot 10^{-4}$ **13. (a)** $[-0.138, 0.149]$
(b) $[-0.238, 0.263]$ **(c)** $[-0.325, 0.365]$
(d) $[-0.397, 0.449]$ **22.** 2 **24.** 5/3 **28.** -2
30. 1/2 **34.** 0

Index